Chemical Catalysts for Biomass Upgrading

Chemical Catalysts for Biomass Upgrading

Edited by
Mark Crocker
Eduardo Santillan-Jimenez

WILEY-VCH

Editors

Mark Crocker
University of Kentucky
Department of Chemistry and Center for Applied Energy Research
2540 Research Park Drive
Lexington
KY 40511
United States

Eduardo Santillan-Jimenez
University of Kentucky
Center for Applied Energy Research
2540 Research Park Drive
Lexington
KY 40511
United States

Library of Congress Card No.:
applied for

British Library Cataloguing-in-Publication Data
A catalogue record for this book is available from the British Library.

Bibliographic information published by the Deutsche Nationalbibliothek
The Deutsche Nationalbibliothek lists this publication in the Deutsche Nationalbibliografie; detailed bibliographic data are available on the Internet at <http://dnb.d-nb.de>.

Print ISBN: 978-3-527-34466-6
ePDF ISBN: 978-3-527-81481-7
ePub ISBN: 978-3-527-81480-0
oBook ISBN: 978-3-527-81479-4

Typesetting SPi Global, Chennai, India
Printing and Binding Markono Print Media Pte Ltd, Singapore

Printed on acid-free paper

10 9 8 7 6 5 4 3 2 1

Contents

Preface

For most of human history biomass has been the principal energy source powering human development. Indeed, biomass can be utilized for the production of heat, steam, motive power and electricity, and can be converted via thermal, biological, or chemical pathways into chemicals and fuels. Although the twentieth century can be regarded as the "petroleum" century, i.e. the time when petroleum came to the fore as the predominant source of fuels and chemicals, in recent years interest in biomass conversion to chemicals and fuels has grown steadily, driven by the issues of sustainability and environmental protection. Increasing concerns surrounding global warming and the contribution of fossil fuel use to atmospheric CO_2 levels mean that humanity is once again looking to biomass resources for the production of essential commodities.

Given the complexity of biomass, as reflected in the wide range of functional groups present, along with its recalcitrance – as exemplified by the lignin component of biomass – catalysis has a major role to play in the conversion of biomass to useful molecules. However, although catalysis is integral to modern life (it has been estimated that catalysis is involved at some stage in the production of ca. 80% of all manufactured goods) the use of catalysis is much less developed for the production of chemicals from biomass compared to petroleum.

Against this backdrop, this book aims to provide the reader with a detailed description of the catalysts and catalytic processes employed in the synthesis of chemicals and fuels from biomass, the information being organized in a way that covers the most abundant and important types of biomass feedstock. The issue of catalyst design is emphasized throughout, bearing in mind that catalysts used for biomass processing must often function in aqueous environments and in the presence of potential poisons such as mineral components.

Two general approaches can be discerned for the conversion of biomass to chemicals and fuels, involving either (i) conversion of the whole biomass into an intermediate product such as pyrolysis oil or syngas that can then be catalytically upgraded to useful products, or (ii) fractionation of the biomass into its main components, followed by the upgrading of these fractions using tailored conversion processes. Following this rationale, Chapters 1–4 of this book focus on the application of catalysts to the pyrolysis of whole biomass and to the upgrading of bio-oils. Subsequent chapters focus on the valorization of biomass constituents. Chapters 5–7 discuss catalytic approaches to the processing of biomass-derived oxygenates – as exemplified by sugars – via reactions such as reforming, hydrogenation, oxidation, and condensation reactions. Lignin is considered next, Chapters 8 and 9 providing an overview of catalysts for lignin valorization via oxidative and reductive methods. After that, Chapters 10–12 consider the conversion of fats and oils to fuels and terminal olefins by means of esterification/transesterification, hydrodeoxygenation, and decarboxylation/decarbonylation processes. Chapters 13 and 14 provide an overview of conversion processes based on terpenes and chitin, two emerging feedstocks with a rich chemistry. Finally, Chapter 15 summarizes some of the emerging trends in the field of catalysis for biomass valorization and looks ahead to future developments.

The editors of this book are deeply indebted to the authors of the chapters. Without their time, effort and expertise this book would not have been possible. We would also like to thank Leslie Hughes at the Center for Applied Energy Research for her unflagging help in the preparation of the manuscript.

July 2, 2019 — *Mark Crocker*
Lexington, Kentucky, U.S.A. — *Eduardo Santillan-Jimenez*

1

Upgrading of Biomass via Catalytic Fast Pyrolysis (CFP)

Charles A. Mullen

USDA-Agricultural Research Service, Eastern Regional Research Center, 600 E. Mermaid Lane, Wyndmoor, PA, USA

1.1 Introduction

Defined as heating of an organic material in a nonoxidative environment, pyrolysis has been recognized for decades as the most efficient process for converting lignocellulosic biomass into a dense liquid, commonly called *pyrolysis oil* or *bio-oil* [1–3]. The most commonly used conditions for conversion of biomass to liquid have been high heating rates to temperatures of around 500 °C, at atmospheric pressure, the so-called fast pyrolysis process [1–3]. The fast pyrolysis process offers many advantages that make it attractive for conversion of biomass to bio-fuel intermediates and production of renewable chemicals. These advantages include high liquid yields (>60% in some cases) and production of a potentially valuable coproduct in *bio-char*. This solid, consisting of fixed carbon and minerals, has been shown to be a good soil amender and a potential route to sequester carbon [4–6]. With the potential utilization of the combustible off gases, and if needed some of the bio-char, pyrolysis can be powered by its own energy, making it a nearly self-sufficient process requiring few other inputs [3].

Bio-oil contains hundreds of oxygenated compounds derived from the cellulose, hemicellulose, and lignin that comprise the biomass. In recent years, much has been made of bio-oil as a potential intermediate to the production of advanced hydrocarbon transportation fuels or as a feedstock from which to isolate renewable chemicals. However, commercial or even precommercial success for utilization of these bio-oils has been limited to lower value applications such as use as boiler-type fuels for heat and power [3, 7] or utilization as an asphalt-like material [8]. The technical reason for these limitations is that the composition of the bio-oil, comprising high concentrations of reactive oxygenated functional groups, plus the presence of catalytic microsolids, makes the mixture thermally unstable [9–11]. Therefore, processing technologies requiring even moderate heating of the bio-oil mixture, such as distillation, result in production of intractable materials [12]. While catalytic hydrodeoxygenation (HDO) has been the post-production upgrading choice for refining of bio-oil to hydrocarbons to be used as fuels, the unstable nature of the bio-oil also makes

Chemical Catalysts for Biomass Upgrading, First Edition. Edited by Mark Crocker and Eduardo Santillan-Jimenez.

this HDO process difficult. The most effective post-production deoxygenation processes developed require multiple catalytic, high pressure hydrotreating steps, at significant cost, making the production of a low margin fuel product of questionable economic viability [13–15].

Because of these limitations, researchers have sought to develop processes that alter the chemical pathways during pyrolysis to produce a more stable bio-oil product with more favorable compositions for various end-use applications, including HDO. Utilization of heterogeneous catalysts during the pyrolysis process, termed *catalytic fast pyrolysis* (CFP), has received the most attention. Because of the interest in advanced hydrocarbon bio-fuels, the most common goal of catalytic pyrolysis has been to produce a partially deoxygenated, thermally stable pyrolysis oil that is more amenable to final HDO-type upgrading to fuel-range hydrocarbons. However, alternative processes have aimed to converge chemical pathways toward production of various individual compounds or groups of compounds for petrochemical or fine chemical uses. In this chapter we will discuss CFP processes both aimed at general deoxygenation and those aimed at targeted classes of molecules. For the purposes of this chapter, we will consider catalytic pyrolysis processes that fall within the following definitions: (i) heterogeneous catalytic processing of biomass pyrolysis vapors either *in situ* (pyrolysis and catalysis occur in the same reactor zone) or *ex situ* (pyrolysis and vapor phase catalysis are decoupled) and (ii) reactions taking place in inert or reactive (but non-oxidative) atmospheres at near atmospheric pressure. Therefore, catalytic hydrothermal or solvent liquefaction [16] and other technologies such as pressurized hydropyrolysis [17] are outside the scope of this chapter.

As mentioned earlier, CFP encompasses processes where solid biomass contacts the catalyst and both pyrolysis and vapor upgrading occur in the same reactor, the so-called *in situ* methods, and processes where the pyrolysis and catalytic vapor upgrading occur in separate reactors, called *ex situ* or *vapor upgrading* methods. Simplified schematics of the two processes are presented in Figure 1.1. The main advantage of the *in situ* method is its simplicity – the "one-pot" reaction system saves capital cost as it does not require a second reactor. However, there are several advantages that decoupling the two steps allows for [18]. While it is well known that temperatures in the 500 °C range produce the maximum yield of condensable range species from pyrolysis of most biomass, the ideal range for various catalytic processes may be significantly different, depending on the catalyst and the desired end products. Furthermore, because the *in situ* process requires contact of solid biomass with solid catalyst, reactors for the unit operation tend to be limited to only fluidized beds. Decoupling the process allows the catalysis step to occur in either a fixed or fluidized bed and also allows for other various reactor types to accomplish the pyrolysis. Another very important factor is that the *ex situ* process allows for removal of bio-char and its associated metal content prior to introduction of the catalyst. This has important implications for catalyst lifetimes as inorganic materials contained in the biomass, particularly Group 1 and Group 2 metals, can poison catalysts, leading to more frequent catalyst replacement or replenishment and a higher catalyst demand. This is an especially important consideration for zeolite-catalyzed processes because they are highly susceptible to this type of deactivation, as described later.

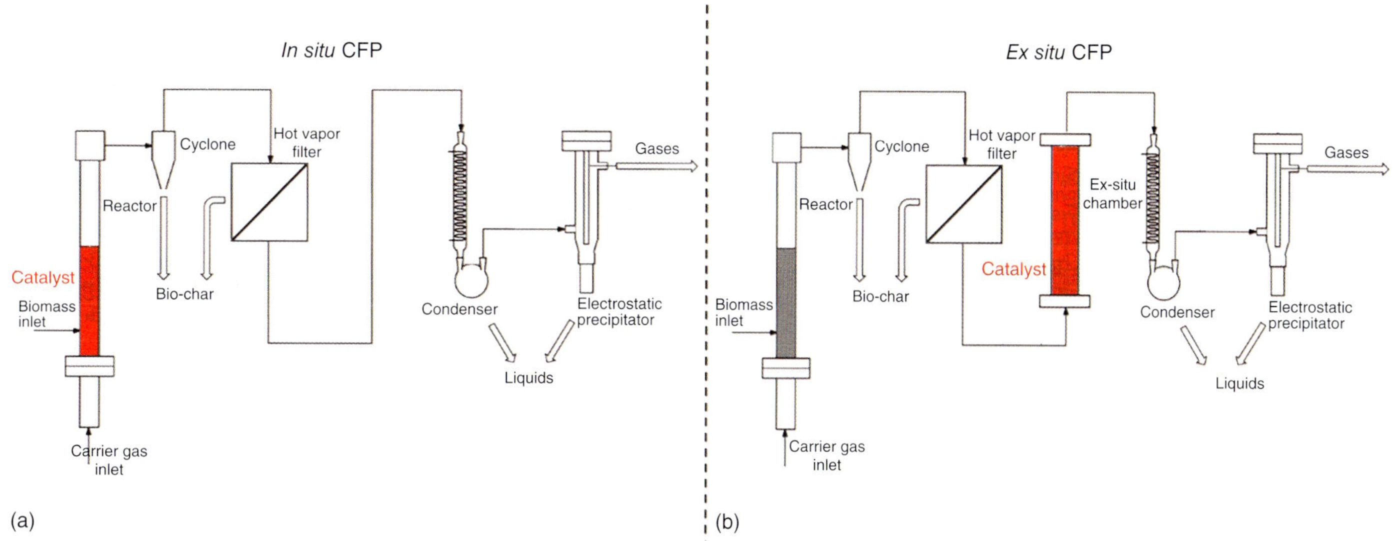

Figure 1.1 Simplified process schematics comparing (a) *in situ* and (b) *ex situ* catalytic pyrolysis.

1.1.1 Catalytic Pyrolysis Over Zeolites

Zeolites and similar materials have been by far the most common catalysts employed in CFP processes. Zeolites are crystalline substances with a structure characterized by a framework of linked tetrahedra, each consisting of four oxygen atoms surrounding a cation [19–21]. Zeolites occur naturally, but the advent of synthetic zeolites in the 1950s, free of the defects and impurities found in nature, is when their use in chemical catalysis took off [20]. Industrial scale catalytic use of zeolites started in 1962, with the use of zeolites X and Y for fluid catalytic cracking (FCC) of heavy petroleum fractions, a process that is still today one of the highest volume chemical processes used. Other petrochemical uses include hydrocracking, isomerization, disproportionation, and alkylation of aromatics. The framework of zeolites is usually silica and alumina based. This framework contains open cavities in the form of channels and cages [19]. These are usually occupied by H_2O molecules and extra-framework cations that are commonly exchangeable [19]. In silica–alumina-based materials, the need for the extra-framework cation is to balance the charge imbalance created by the four coordinate Al(III) sites, as opposed to the neutral Si(IV) sites (Figure 1.2) [19–21]. Often, in the active catalysts, some or all of the extra-framework cations are Brønsted acid (H^+) active sites, which along with the Lewis acidity of other sites are responsible for initiating the catalytic reactions. The channels allow the passage of guest species to these active sites [19–21]. In biomass CFP these guest species are molecules derived from the breakdown of the biopolymers caused by the initial pyrolysis reactions. The pore size and shape, Brønsted acid strength and site density, and the presence of other types of active sites are important factors in the activity and selectivity of the catalysts.

1.1.1.1 Catalytic Pyrolysis Over HZSM-5

Several different types of zeolites have been tested as catalysts for the deoxygenation of pyrolysis vapors including Zeolite Socony Mobil-5 (ZSM-5), Y, beta, mordenite, and ferrierite [22–26]. Among these types of zeolites, the ZSM-5 (also called MFI for mordenite framework inverted)-based materials have shown the most promise and received the most attention and will be the focus of this section. ZSM-5-type zeolites have been shown to selectively convert molecules from a wide variety of sources to aromatic hydrocarbons. Among these processes are methanol to olefins and gasoline and cracking of waste hydrocarbon plastics [27, 28]. Aromatic hydrocarbons are good target product molecules from biomass, because, like the biomass starting materials, they have low H/C ratios, meaning the biomass carbon can be more efficiently converted without the

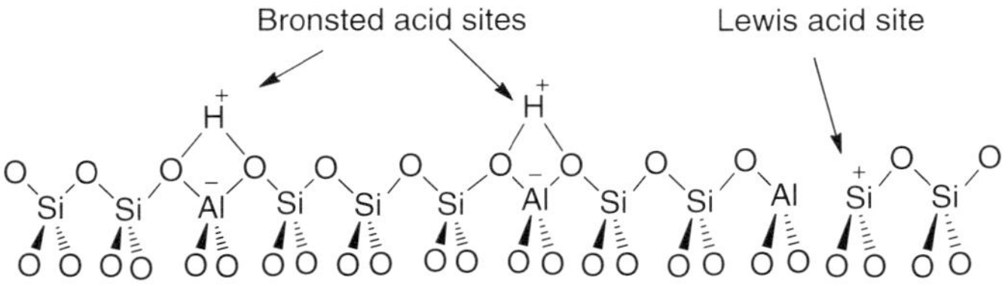

Figure 1.2 Brønsted and Lewis acid sites of zeolites.

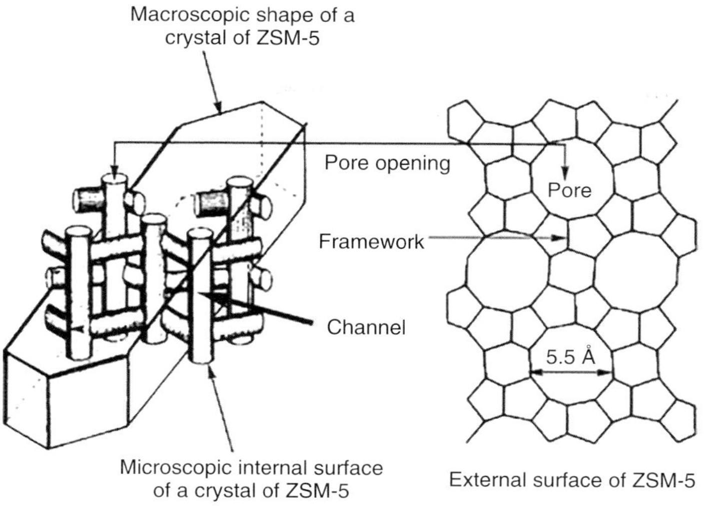

Figure 1.3 Structure of ZSM-5. Source: Lei et al. 2003 [30]. Reproduced with permission of Royal Society of Chemistry.

addition of an external source of hydrogen [29]. The shape selectivity of ZSM-5 is responsible for convergence toward aromatics. ZSM-5 has a three-dimensional pore system comprising straight 10-membered ring 5.2 × 5.7 Å channels connected by sinusoidal 5.3 × 5.6 Å channels (Figure 1.3) [25, 30]. This puts it in the category of a medium pore size zeolite. This pore structure creates a mass transfer effect, limiting the molecules that enter and interact with the chemical functionality of the acid site based on size. At the same time, the confined space limits the geometry that can occur in the reaction transition states, forcing the chemistry through certain pathways, creating the observed product selectivity [24].

Biomass catalytic pyrolysis, whether the physical process is performed in the *in situ* or *ex situ* configuration, encompasses two distinct chemical processes. First, pyrolytic decomposition of the biopolymers produces oxygenated vapors, aerosols, and bio-char. The catalyst is involved in the second step wherein some of the oxygenated gas phase products interact with the catalyst, initiating reactions that alter the composition of the gas stream and hence the condensed product that is collected. The various oxygenated pyrolysis products that are the substrates in the catalytic reactions have differing interactions with the catalyst, depending on several factors including whether their molecular size allows entry into the pores of the catalyst, their diffusion rate through the channels, and the interaction of their chemical functionality with the catalyst active site and other reactants present. During catalytic pyrolysis over HZSM-5, most of the aromatic hydrocarbons are ultimately derived from the carbohydrate portions (cellulose and hemicellulose) of the biomass, while lignin-derived vapors are considered primarily, but not entirely, as precursors to coke formation [30, 31]. The portions of the lignin that do contribute to the hydrocarbon pool are mostly derived from the carbon chain linkers or side chains rather than the phenolic units themselves [32]. The chemical pathways that occur to produce aromatic hydrocarbons are summarized in Figure 1.4. The key step after production of oxygenated vapors

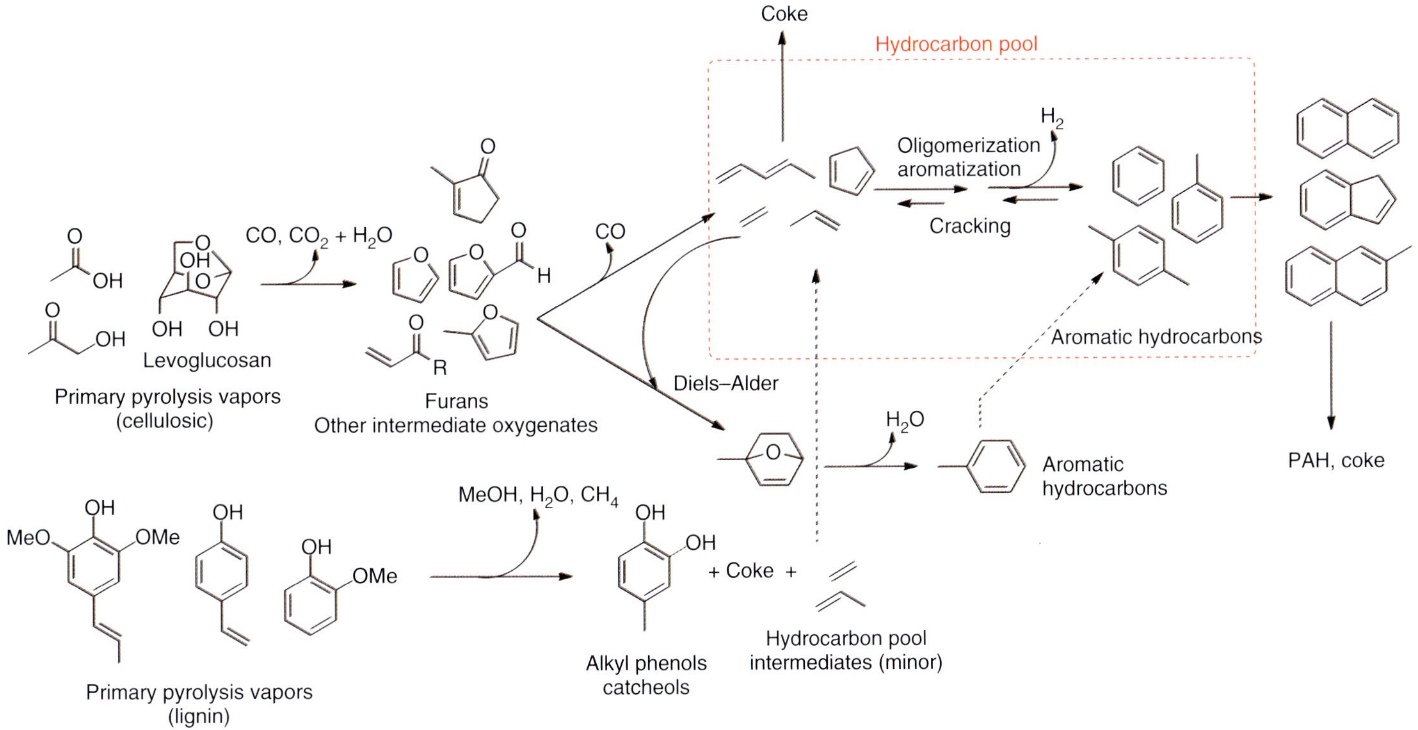

Figure 1.4 Major chemical pathways from primary pyrolysis vapors to aromatic hydrocarbons via CFP over HZSM-5.

via the initial pyrolytic depolymerization is the dehydration of compounds such as anhydrosugars to form furans [33]. This step can occur on the surface of the catalysts, producing smaller molecules that can diffuse into the micropores [25, 34, 35]. Here, two main pathways toward aromatics become operative. The first pathway involves decarbonylation to form olefins which can go on to oligomerize and aromatize. This mixture of olefins and product hydrocarbons is referred to as the *hydrocarbon pool* [33]. Alternatively, Diels–Alder-type cycloaddition reactions of furans with olefins can directly produce aromatics upon dehydration of the bicyclic cycloaddition adduct [36]. Related reactions of other primary pyrolysis products can also produce CO_2 along with olefins and alkanes.

Many studies have been conducted on the effects of process conditions for catalytic pyrolysis of biomass and their components over HZSM-5, focused either on the goal of producing partially deoxygenated bio-oil (for direct bulk use or further processing), or for the production of aromatic hydrocarbons and light olefins [22–26, 29, 37–49]. An estimated carbon balance for CFP of biomass over HZSM-5 is presented in Figure 1.5 [49], and Table 1.1 summarizes the details of selected studies where product samples were produced (i.e. larger than analytical or micropyrolyzer studies). While there are many factors that contribute to the variation in the reported results, generally the level of deoxygenation achieved follows an inverse trend with bio-oil yield [50]. Lower biomass to catalyst ratios, corresponding to higher catalyst activity, result in more deoxygenation at the expense of bio-oil yield; potential organic liquid yield is lost to gas phase products, water, and coke [18, 49]. While overall organic liquid yield decreases due to elimination of oxygenated species, yields of aromatic hydrocarbons generally increase with increasing biomass to catalyst ratio. Residence time and space velocity considerations can also affect the conversion to aromatic hydrocarbons. In a study on CFP of cellulose over HZSM-5 in an *in situ* fluidized bed process, at equal weight hourly space velocity (WHSV), a gas residence time of about 8.6 seconds (studied over a range of 5.6 to ~10 seconds by changing carrier gas flow rates) maximized aromatic yield [46]. Selectivity for benzene and toluene were also maximized at this residence time, while selectivity to xylenes and naphthalenes was minimized.

As mentioned earlier, the cellulosic portion of the biomass is most efficiently converted to aromatic hydrocarbons, while only a small portion of lignin can be converted to aromatic hydrocarbons over HZSM-5. A number of other factors related to the biomass composition can also affect the production of aromatics. This gives rise to variation in yields and quality of liquid products and aromatic hydrocarbons. Biomass composition, along with process conditions and catalyst properties are also important variables in catalyst deactivation rates, which are discussed in detail later. Studies of biomass within a small range of compositions have found, unsurprisingly, that increased ash content (particularly potassium) correlates with decreased conversion of carbon to aromatics during CFP, due to an increased production of gases, and increased lignin is also correlated with decreased aromatics and increased coke yield [51].

With regards to the properties of a standard HZSM-5 zeolite catalyst that influence biomass vapor upgrading, the number of Brønsted acid sites the biomass is exposed to, controlled by the acid site density (inversely proportional to the

Table 1.1 Selected results of biomass CFP over typical HZSM-5 catalysts.

Biomass	Scale (kg/h)	Configuration	SAR[a]	Temperature (°C)	C/B (mass)	WHSV[b] (h^{-1})	Bio-oil yield (C%)[c]/oxygen content (wt%)	AHC[d] yield (C%)[c]	References
Cellulose	~0.06	*In situ*	30	500		0.25	—	39.5	[46]
Lignin	mg	*In situ*[e]	23	650	15		—	8.2	[32]
Pine	~0.165	*In situ*	30	600	6	0.3		15.5	[43]
Pine	2	*In situ*	—	475	0.33	2	13.3/19.42		[45]
Hybrid poplar	2	*In situ*	—	475	0.33	2	12.2/20.25		[45]
Corn stover	2	*In situ*	—	475	0.3	2	8.9/13.99		[45]
Switchgrass	2	*In situ*	—	475	0.33	2	15.8/14.7		[45]
Juniper	2	*In situ*	—	475	0.33	2	15.8/12.3		[45]
Pine bark	2	*In situ*	—	475	0.33	2	8.9/18.9		[45]
Switchgrass	0.3	*In situ*	30	450–500	0.5	1.2	9.4/23.5		[41]
Corn cob	0.036	*In situ*	48	550	5		25.5/14.69		[38]
Pine wood	20	*In situ*	50	550	7		24/21.5		[42]
Beech wood	0.5	*In situ*	50	500	11		27/19.5		[47]
Beech wood	0.5	*In situ*	50	500	21		23/17.5		[47]
Pine	0.150	*Ex situ*	30	500	2		14.3/4.0		[48]
Pine	0.150	*Ex situ*	30	500	0.67		17.2/14.2		[48]
Pine	0.150	*Ex situ*	30	500	0.48		23.1/17.7		[48]

a) SAR, silica alumina ratio.
b) WHSV, weight hourly space velocity.
c) C%, carbon yield.
d) AHC, aromatic hydrocarbons.
e) Micropyrolyzer.

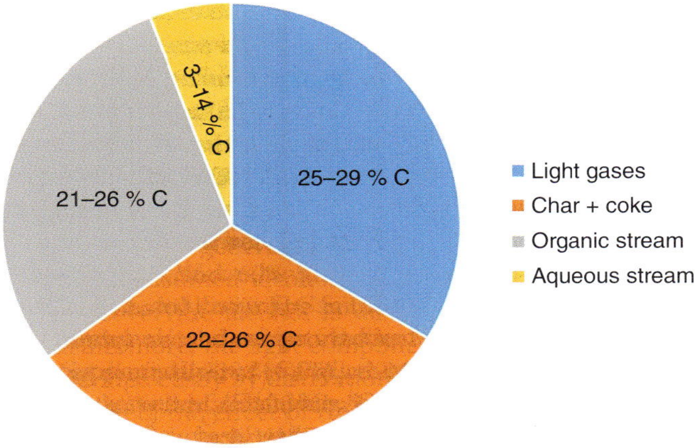

Figure 1.5 Typical carbon distribution from CFP of biomass over HZSM-5 and related catalysts. Source: Starace et al. 2017 [49]. Reproduced with permission of American Chemical Society.

silica/alumina ratio) is the most important variable controlling activity [40], along with the biomass to catalyst ratio. In the absence of other variations, the higher the acid site density, the more active the catalyst. This generally leads to higher yields of aromatics; however, at very high acid site densities, the small distance between active sites can promote the formation of coke, decreasing the initial yield of aromatic hydrocarbons and also leading to more rapid catalyst deactivation as discussed in more detail in the succeeding text [52]. One study looking at aromatic hydrocarbon yields when using HZSM-5 with $SiO_2/Al_2O_3 = 23$, 30, 50, and 80, found 30 as the optimum to maximize the yield of aromatic hydrocarbons [40]. Lower SiO_2/Al_2O_3 also correlated with higher production of CO_2 (but not CO), indicating Brønsted acid dependence for any decarboxylation reactions, although this is a relatively minor deoxygenation pathway in CFP over zeolites. High acidity also increased selectivity for benzene and toluene over C8+ aromatics including xylenes, ethyl benzene, and indanes [40].

1.1.1.2 Deactivation of HZSM-5 During CFP

Catalyst deactivation is a significant concern for biomass CFP. Zeolites are subjected to three major deactivation mechanisms during the process. The first is formation of coke that blocks access to the catalyst active site. This is the most rapid form of catalyst deactivation and can have noticeable effects on the product distribution almost immediately at cumulative biomass/catalyst mass ratios of <1. Fortunately, this type of deactivation is also reversible, via combustion of the coke. A second form of deactivation results from poisoning of the catalyst active sites with inorganic species from the biomass, particularly alkali metals. This type of deactivation is, for practical purposes, irreversible. A third type of deactivation is physical degradation of the catalyst, most commonly dealumination. This can occur due to exposure to high temperatures and steam and can also occur during combustion process used to regenerate catalysts from coke deposits. This type of deactivation is also irreversible.

There are two main mechanisms of coke formation active during biomass CFP. The first is similar to coke formation that occurs during other zeolite catalyzed reactions (including petroleum processing and methanol to olefins), i.e. molecular weight growth of already formed aromatics by reaction with other aromatics or olefins present in the hydrocarbon pool [50, 53]. Because biomass pyrolysis vapors are deficient in hydrogen, any deoxygenation occurring by dehydration reactions will further reduce the hydrogen content of the vapors favoring molecular weight growth reactions that form large aromatic molecules with low H/C ratios that comprise coke. This process can be further exacerbated by slow diffusion of products out of the zeolite [25]. Another source of coke is direct condensation or deposition of lignin-derived phenolics, which have poor reactivity over ZSM-5 and tend to be too bulky to navigate the pores of the catalyst [54]. If HZSM-5 catalysts are not regenerated by combustion of the coke, complete deactivation of the catalyst and production of product vapors with similar oxygen content and composition to noncatalytic pyrolysis are produced at cumulative biomass to catalyst ratios of about 3 or 4 to 1 for biomass such as wood and grasses [50, 55, 56]. Lignin deactivates catalysts much quicker than cellulose and use of feedstocks with high lignin concentration may result in more rapid catalyst deactivation via coking. The rapid coking effect can account for a significant portion of the variability in results and the effect of biomass to catalyst ratios as seen in Table 1.1, depending on run lengths and regeneration protocols.

The effect of coking on the product vapors can be seen almost immediately, at very low cumulative biomass to carbon ratios. During CFP over HZSM-5, the first oxygenates to appear are not primary pyrolysis vapors but partially deoxygenated species including furans, phenols, and alkyl phenols (mostly cresols) [50]. Furans are intermediates on the pathway from anhydrosugars (cellulose primary pyrolysis products) to aromatics, and their appearance is reflective of the loss of catalytic activity for the decarbonylation step that results in their conversion to olefins that make up part of the hydrocarbon pool (see Figure 1.4). A decrease in CO formation has been observed to occur at low cumulative/biomass ratios [56]. Phenols have also been found as a partially deoxygenated product from both cellulose and lignin [51, 55–57]. Unlike furans, phenols are not intermediates on the pathway from cellulosic primary pyrolysis vapors to aromatic hydrocarbons. There is evidence that they are formed via a reaction of deoxygenated species still bound to the catalyst with water, confirmed by observing the incorporation of ^{17}O in phenolic products from labeled water added in a CFP experiment [57]. These products of partially deactivated catalysts reach a maximum and then begin to become less concentrated as the activity for even the initial dehydration step becomes nullified by coke blocking both internal and surface acid sties [50]. Primary pyrolysis products are then observed at higher cumulative biomass/catalyst ratios. The product vapors consist of a mixture of the three groups until the complete deactivation point is reached. The results of one study monitoring the composition of the vapors over the course of CFP up to a cumulative biomass to catalyst ratio of 4.5/1 is presented in Figure 1.6 [50].

Coking rates can vary with catalyst properties, particularly with acid site density. As discussed earlier, the activity of fresh catalysts generally trends with Brønsted acid site density, but yields of aromatic hydrocarbons were

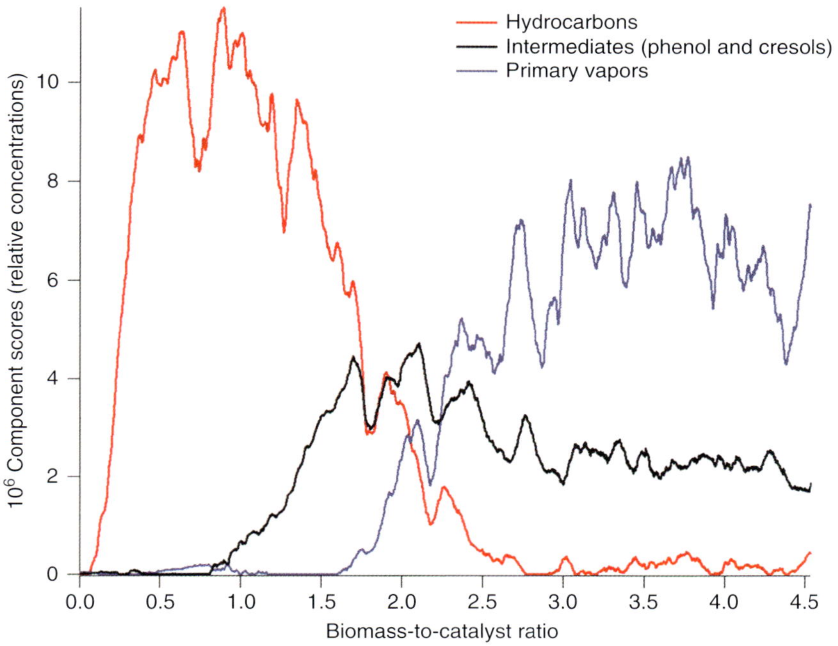

Figure 1.6 Trends in concentration of aromatic hydrocarbons, intermediate products from pine pyrolysis over HZSM-5 (microreactor-MBMS system) vs. cumulative biomass to catalyst ratio. Source: Mukarakate et al. 2014 [50]. Reproduced with permission of Royal Society of Chemistry.

optimized at $SiO_2/Al_2O_3 = 30$ over an even more acid-site-dense catalyst with $SiO_2/Al_2O_3 = 23$ [40]. The reason for this is likely a more rapid deactivation of the more acidic catalyst due to increased coking over the short course of the experiment. One study, which decoupled activity and deactivation from the total number of acid sites, found that HZSM-5 with a SiO_2/Al_2O_3 of 40 had comparable initial activity for the production of aromatic hydrocarbons, but the deactivation rate was significantly slower than catalysts with higher acid site density (Figure 1.7) [52]. Some studies on model compounds have shown that close proximity of acid sites can facilitate the condensation (C—C bond and H_2O forming) reactions between oxygenated intermediates, leading to coking [52, 58]. A change in the amount of extra-framework alumina, which can serve as nucleation sites for molecular weight growth, can also lead to coking. In the following Sections 1.1.3 and 1.1.4 that discuss modification of HZSM-5, the effects of various changes to the catalyst structure with respect to deactivation are considered.

Alkali metals present in the biomass are also a problematic source of catalyst deactivation. Alkali metals can accumulate on catalysts rapidly and can ion exchange with Brønsted acid sites, deactivating the sites. In two studies of *in situ* CFP over HZSM-5 (where catalysts were regenerated from coking, one by use of a circulating fluidized bed and one by intermittent regeneration) mineral

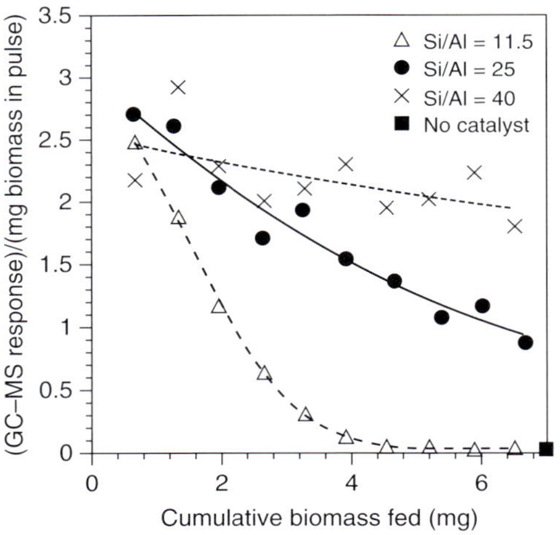

Figure 1.7 Comparison of alkylbenzenes production on HZSM-5 Si/Al = 11.5, 25, and 40 from oak pyrolysis vapors with fixed total sites at 500 °C. Source: Wan et al. 2015 [52]. Reproduced with permission of John Wiley and Sons.

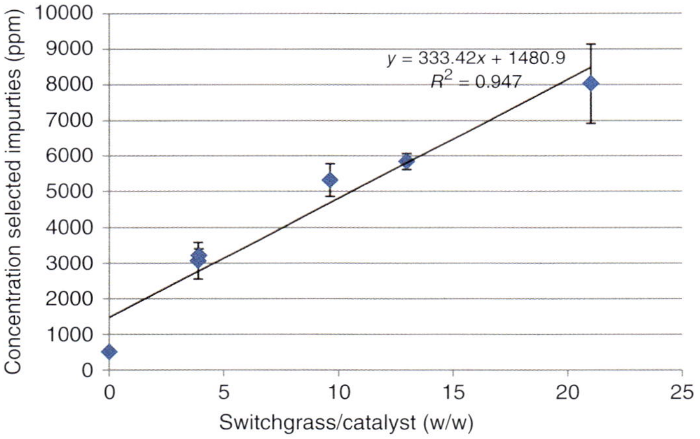

Figure 1.8 Concentration of total Ca, Cu, Fe, K, Mg, Na, and P on HZSM-5 at different levels of catalyst exposure to switchgrass in a fluidized bed at pyrolysis conditions (500 °C, N_2). Values are the average of three inductively coupled plasma spectroscopy (ICP) measurements. Error bars are one standard deviation. Source: Mullen and Boateng 2013 [41]. Reproduced with permission of American Chemical Society.

accumulation including Ca, K, Mg, Na, and P was found to be linear with time on stream (or cumulative biomass/catalyst ratio) [41, 42]. Each study, one with switchgrass (~2.6% ash) and one with pine wood (~0.4 wt% ash), found alkali concentration on the catalyst was around 1 wt% at cumulative biomass to catalyst ratios of around 20 by mass (Figure 1.8). A linear correlation was also found between the concentration of the inorganic species and the Brønsted acidity, suggesting that at least some ion exchange with acidic sites had occurred. This had the effect of decreasing the catalyst activity, both studies report observing increases in bio-oil oxygen content with increasing cumulative biomass to catalyst ratios. Because these catalysts were subjected to high temperatures during regeneration from coking, physical damage may be another reason for activity loss, as described in more detail hereafter. However, another study specifically looked at the effect of K by preparing ion exchanged catalysts (2.95 wt% loading) and found a drastic reduction in catalyst acidity and yield of aromatic hydrocarbons via CFP of cellulose and switchgrass [59]. The same study also found an increase in furans and phenolics, and the same products are found when catalysts begin to deactivate as a result of coking. Use of *ex situ* CFP methods, whereby the large majority of the inorganic species are removed from the vapor stream with the bio-char, can greatly reduce the impact of this type of deactivation.

In either *in situ* or *ex situ* CFP, however, zeolite catalysts need to be regenerated from coking, likely continuously, if maximized aromatic hydrocarbon production is the goal. Regeneration usually consists of combustion of the coke deposits in an air atmosphere. Because combustion is an exothermic process, it is possible that hot spots may occur during the regeneration processes, making it more likely for catalyst damage to occur. In one study where a commercial spray-dried ZSM-5 catalyst was used in 30 successive reaction/regeneration cycles where the regeneration was performed offline at 580 °C over a long period (15–20 hours), no evidence of dealumination or loss of crystallinity was found, only macroscopic physical damage – some breaking apart of the spray dried particles – was observed [43]. A small loss in performance was attributed to alkali metal poisoning. However, in the aforementioned work using a circulating fluidized bed where the regeneration conditions were much more rapid at measured temperatures of 650–670 °C, evidence of dealumination was evident and a decrease in micropore volumes was observed [42]. The negative effects on performance were limited at 100 hours time on stream, but may become more significant with additional exposure to the reaction/regeneration conditions. Only recently has work been done to optimize the conditions that control temperature during regeneration to prevent catalyst damage. A recent report found that the measured temperature just above a fixed bed regeneration reactor correlated well with the excess of oxygen over the combustion stoichiometry in the atmosphere [60]. Limiting the oxygen to 15 vol% during combustion significantly increased the catalyst lifetime for conversion of furan to aromatics. Further, controlling the temperature via addition of steam further improved the lifetime of the catalyst. This study also explored the idea of a controlled regeneration where a targeted amount of coke remained on the catalyst. Some other processes use this type of partial deactivation (equilibrium state) to control product selectivity, an idea that warrants further study for biomass CFP.

1.1.1.3 Modification of ZSM-5 with Metals

There have been several attempts to modify ZSM-5 to further improve yields, change the selectivity of the products, and/or decrease catalyst deactivation rates. One facile way to modify the zeolite is via ion exchange of the Brønsted acid sites with an active metal cation. There are a few methods to do this, including ion-exchange and incipient wetness impregnation. Some metals can be incorporated into the framework of the zeolite during synthesis. Metals can not only adjust the acidity of ZSM-5 by replacing the proton with a metal cation but also promote desirable reactions during pyrolysis, potentially increasing the yield of desirable compounds while reducing coke formation on the catalyst. Among the metals that have been studied for this purpose are Zn, Mo, Ga, Ni, Co, Cu, Mn, and Fe [61–75]. Some metal-modified ZSM-5 catalysts are currently being used commercially for the conversion of light hydrocarbons to aromatics. For example, a Ga-modified ZSM-5 catalyst has been used for the aromatization of light alkanes in the Cyclar process by UOP and BP [76–78], while Zn/HZSM-5 is used for the production of aromatics from olefin-rich hydrocarbons in the Alpha process from Asahi Kasei Chemicals Corp. [79]. Modification with gallium has been among the most studied approaches for biomass CFP [43, 56, 68, 70–73]. Gallium incorporation by ion exchange or impregnation has been shown to effect 21–50% increases in yields of aromatic hydrocarbons over parent ZSM-5 catalysts. Catalysts synthesized with Ga in the zeolite framework were less successful for production of aromatics from a model compound (furan) [68]. Table 1.2 summarizes results from various selected CFP studies using GaZSM-5.

The presence of gallium increases the number of Lewis acid sites present in the catalyst, increasing its dehydrogenation activity. Ion exchange methods for preparation of GaZSM-5 have been shown to only effect a minimal amount of actual ion exchange, and much of the gallium exists as Ga_2O_3 on the surface of the catalyst [76]. The small amount of gallium that is exchanged is likely responsible for the increased yields. In a similar process, conversion of ethanol to aromatics, a physical mixture of Ga_2O_3 and HZSM-5 showed no improvement, but the ion exchanged zeolite did [77]. Furthermore, reductive pretreatment appears to mobilize the gallium to ion exchange, further increasing activity. The exchanged Ga(III) sites are most important for the observed activity. The reduction that allows this mobility may not actually perform a reduction to Ga(I) but produce low-coordinate Ga(III) hydride species [77]. Preservation of these states may further enhance the aromatic yield for alkane aromatization reactions.

In CFP, gallium-modified ZSM-5 catalysts that have been prereduced have shown especially strong dehydrogenation activity (as measured by production of H_2), although this rapidly decreases upon exposure to pyrolysis vapors as a result of either oxidation of Ga(I) to Ga(III) or conversion of low coordinate active Ga(III) species to fully ligated species. In either case, the resultant gallium(III)-exchanged catalyst continued to show superior performance for production of aromatic hydrocarbons. The increased dehydrogenation and/or decarbonylation activity afforded increases in olefin formation and in their rate of aromatization. Compared with parent HZSM-5 catalysts, those modified with gallium showed more selectivity toward less alkylated aromatics (benzene and toluene) and away from xylenes [68, 71]. In terms of production of deoxygenated bio-oil, one study of the CFP of pinewood showed a slight increase in yield of organic phase bio-oil (from ~25 to ~26–28 C%) with GaZSM-5 over HZSM-5 with a decrease in oxygen content of the product (from 18.4 to 17.1–17.7 wt%) [73]. Furthermore, GaZSM-5 catalysts have shown longer lifetimes with respect to deactivation due to coking than standard HZSM-5 [56]. Figure 1.9 depicts how GaZSM-5 (SiO_2/Al_2O_3 = 30) continued to exhibit superior deoxygenation activity at higher cumulative biomass to catalyst ratios than did its parent HZSM-5 (SiO_2/Al_2O_3 = 30) or HZSM-5 (SiO_2/Al_2O_3 = 80) during CFP of eucalyptus wood [56].

Table 1.2 Comparison of CFP results for Ga-modified and standard ZSM-5 catalysts.

Catalyst	Biomass	Configuration	SAR[a)]	Temperature (°C)	WHSV[b)] (h^{-1})	Bio-oil yield (C%)[c)]/oxygen content (wt%)	AHC[d)] yield (C%)[c)]	References
HZSM-5	Pine	*In situ*	30	550	0.35	—	15.5	[68]
Ga/HZSM-5	Pine	*In situ*	30	550	0.35	—	23.2	
HZSM-5	Pine	*Ex situ*	30	500	7	25.1/18.4	—	[73]
Ga/HZSM-5	Pine	*Ex situ*	30	500	7	28.5/17.1	—	
HZSM-5	Eucalyptus	*In situ*	23	600	C/B = 10	—	14.4	[71]
Ga/ZSM-5	Eucalyptus	*In situ*	23	600	C/B = 10	—	19.7	

a) SAR, silica alumina ratio.
b) WHSV, weight hourly space velocity.
c) C%, carbon yield.
d) AHC, aromatic hydrocarbons.

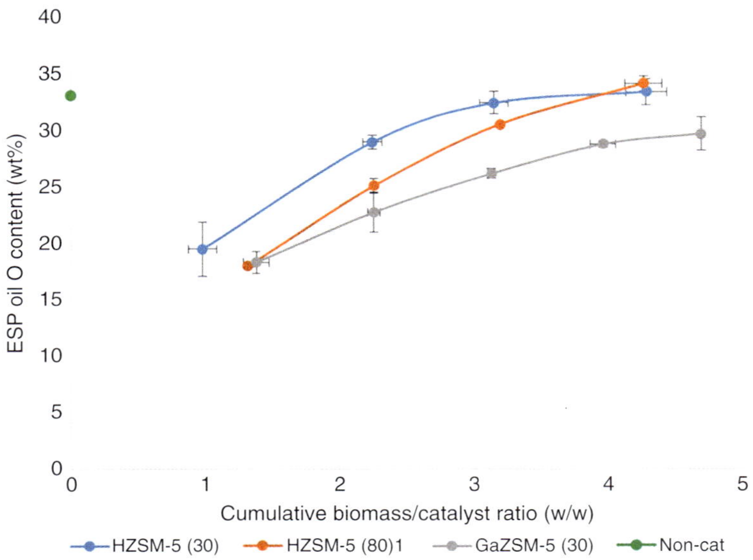

Figure 1.9 Comparison in trend in oxygen content of bio-oil produced via CFP of eucalyptus wood over HZSM-5 ($SiO_2/Al_2O_3 = 30$), HZSM-5 ($SiO_2/Al_2O_3 = 80$) and GaZSM-5 ($SiO_2/Al_2O_3 = 30$), with increasing cumulative exposure of catalysts to biomass. Source: Mullen et al. 2018 [56]. Reproduced with permission of American Chemical Society.

Other metals that have shown positive effects when used to modify ZSM-5 zeolites for CFP are nickel, cobalt, zinc, and iron (Table 1.3) [67, 73–75]. A CoZSM-5 catalyst produced via wet impregnation of ZSM-5 diluted with silica–alumina (for particle size and fluidization properties) was shown to have fewer Brønsted acid sites but greatly increased Lewis acidity. Use of this catalyst produced bio-oil that had lower oxygen content, but at a lower yield than HZSM-5 in a circulating fluidized bed system. The role of the Co was proposed to be similar of that of the Ga [74], i.e. enhancement of dehydrogenation activity to enhance aromatization; however, the reported results with Ga are superior. Nickel has been shown to increase yields of hydrocarbons, up to ~25% over the parent HZSM-5 for conversion of pine wood, a level similar to that of Ga, but the pyrolysis may have to occur in the presence of H_2 for the Ni to retain its activity [73]. The role of Ni seems to be to decrease coke formation, perhaps by hydrogenation. Pyrolysis over NiZSM-5 was more selective for naphthalenes than was HZSM-5 or GaZSM-5. In terms of bio-oil yield and oxygen content, NiZSM-5 produced bio-oil with decreased oxygen content, but with a decreased yield compared with HZSM-5 or GaZSM-5 [73]. Other metals have shown some, but lesser, effects. Iron-modified ZSM-5 has been shown to increase the yield of aromatics in CFP of cellulose and cellobiose (a glucose dimer) compared with HZSM-5, but only at very low Fe loadings where the Brønsted acidity level of the catalyst remained high [67]. Furthermore, the increases noted with isolated carbohydrates did not extend to switchgrass biomass. Fe was shown to change the selectivity toward benzene and naphthalenes and away from production of alkylated benzenes. Other metals have shown little effect on yields but have changed selectivity some; for example, the presence of zinc was shown to increase selectivity for toluene [71].

Table 1.3 Comparison of CFP results for metal-modified and standard ZSM-5 catalysts.

Catalyst	Biomass	Configuration	SAR[a)]	Temperature (°C)	Atmosphere	WHSV[b)](h^{-1})	Bio-oil yield (C%)[c)]/oxygen content (wt%)	AHC yield[d)] (C%)[c)]	References
HZSM-5	Pine	*Ex situ*	30	500	N_2	7	25.1/18.4		[73]
Ni/HZSM-5	Pine	*Ex situ*	30	500	30/70 H_2/N_2	7	24.0/14.4		
Co/HZSM-5	Beech wood	*In situ*		500	N_2	C/B = 12.2	19.9/11.3		[74]
Co/HZSM-5	Beech wood	*In situ*		500	N_2	C/B = 19	16.4/9.8		
HZSM-5	Eucalyptus	*In situ*	23	600	He	C/B = 10	—	14.4	[71]
Zn/HZMS-5	Eucalyptus	*In situ*	23	600	He	C/B = 10	—	12.6	
HZSM-5	Cellulose	*In situ*	23	500	He	C/B = 10	—	14.7	[67]
Fe/HZSM-5	Cellulose	*In situ*	23	500	He	C/B = 10	—	18.0	
HZSM-5	Switchgrass	*In situ*	23	500	He	C/B = 10	-	18.2	
Fe/HZSM-5	Switchgrass	*In situ*	23	500	He	C/B = 10	—	18.3	

a) SAR, silica alumina ratio.
b) WHSV, weight hourly space velocity.
c) C%, carbon yield.
d) AHC, aromatic hydrocarbons.

1.1.1.4 Modifications of ZSM-5 Pore Structure

The microporous structure of the HZSM-5 zeolite is what provides the strong selectivity toward aromatics as well as hinders the diffusion of larger molecules with kinetic diameters >5 Å [25]. These are present in significant quantities in biomass pyrolysis vapor streams. Levoglucosan (6.7 Å) readily undergoes dehydration reactions on the surface of the catalyst to produce molecules smaller than the pore size of HZSM-5; however, lignin-derived guaiacols (~8.1 Å and up) and syringols (~7.9 Å and up) do not have the same reactivity [78, 79]. The poor diffusion of these molecules, and product molecules, contributes to the propensity to form coke, meaning that they are not converted to useful products and also block acid sites, resulting in catalyst deactivation. With this in mind, mesoporous materials, i.e. those with pore sizes of >2 nm, such as SBA-15 or MCM-41, have been well studied for biomass pyrolysis in an effort to overcome the diffusional limitations of the microporous zeolites [26, 80, 81]. However, these materials tend to have weaker and fewer acid sites, can be hydrothermally unstable at biomass pyrolysis conditions, and they lack the selectivity toward aromatics that ZSM-5 exhibits. The result is that these materials tend to promote the dehydration step from carbohydrates resulting in furans, but do not further deoxygenate the latter to the hydrocarbon pool intermediates that are the precursors to aromatics; however, some mesoporous materials have shown increased production of phenols.

There have been efforts to introduce larger pores into zeolitic materials to make mesoporous/microporous materials (possessing what is sometimes referred to as a "hierarchical" pore structure) with the aim of increasing the diffusion rate for bulky materials but preserving the strong acidity and selectivity of the microporous active sites [70, 82–87]. Most of the activity testing on these materials for CFP has been limited to the analytical scale (i.e. micropyrolysis-GC or similar techniques), and selected results are summarized in Table 1.4. The simplest method to produce such materials is to desilicate presynthesized zeolites by treatment with NaOH solutions. Studies using this method have indicated that mild treatment (0.2–0.3 M NaOH at 65–70 °C) of HZSM-5 optimized the catalyst, increasing its performance for the production of aromatic hydrocarbons from wood [82, 83]. Compared with the parent catalysts, the mildly treated catalysts did not show significant variation in elemental composition, crystallographic structure, or microporosity [82]. They did, however, show an increased Brønsted acid site density and increased mesoporosity. More aggressive treatments (using more concentrated NaOH solutions) led to a decrease in micropore volume and decreased performance. Interestingly, two separate studies found that the optimized ZMS-5 catalyst resulted in no change in the aromatic yield and very little change in the selectivity for the CFP of cellulose, but significant increases in aromatic hydrocarbon yield for CFP of beech or red oak wood [82, 83]. This result may indicate improved conversion of bulky lignin-derived substrates with the introduction of mesopores, which is supported by the results of a test using lignin in one of the studies [82].

More recently, some researchers have taken a more systematic approach to synthesizing these hybrid porosity materials in an effort to design an improved catalyst for biomass CFP. Templating techniques to incorporate mesopores

Table 1.4 Comparison of CFP results for ZSM-5 catalysts containing mesopores with standard HZSM-5 catalysts.

Treatment[a)]	Micropore volume (cm^3/g)	Mesopore volume (cm^3/g)	SAR[b)]	Biomass	AHC[c)] yield (C%)	References
None	0.164	0.058	25.5	Cellulose	31.1[d)]	[82]
				Lignin	9.89[d)]	
				Beech wood	23.7[d)]	
0.3 M NaOH	0.133	0.127	24.0	Cellulose	32.1[d)]	[82]
				Lignin	13.2[d)]	
				Beech wood	30.1[d)]	
0.5 M NaOH	0.116	0.210	21.1	Beech wood	26.2[d)]	[82]
None	0.128	0.074	23.2	Cellulose	28.5[e)]	[83]
				Lignin	11.8[e)]	
				Red oak	23.9[e)]	
0.2 M NaOH	0.128	0.123	23.6	Cellulose	29.4[e)]	[83]
				Lignin	7.7[e)]	
				Red oak	27.9[e)]	
0.5 M NaOH	0.110	0.222	26.1	Cellulose	25.3[e)]	[83]
1 M NaOH	0.122	0.174	15.3	Cellulose	25.2[e)]	[83]
None	0.14	0.12	20.3	Cellulose	20.4[f)]	[84]
				Miscanthus	21.8[f)]	
Synthesized	0.12	0.24	9.6	Cellulose	26.1[f)]	[84]
				Miscanthus	24.8[f)]	
None	0.127	—	23	Cellulose	27.5[g)]	[85]
Synthesized	0.113	—	34.4	Cellulose	32.0[g)]	[85]

a) None, commercially sourced HZSM-5; Synthesized, mesoporosity was generated during zeolite synthesis.
b) Silica alumina ratio.
c) AHC = aromatic hydrocarbons (carbon yield).
d) Micropyrolyzer, 550 °C, C/B = 10.
e) Micropyrolyzer, 550 °C, C/B = 20.
f) Micropyrolyzer, 600 °C. C/B = 5.
g) Micropyrolyzer, 700 °C, C/B = 20.

during zeolite synthesis (a bottom-up approach) can offer more control to tailor the properties of the catalysts than washing zeolites with NaOH (a top-down approach) [84]. While some attempts to use bottom-up approaches have had less success than the desilication technique, a combination of techniques has allowed researchers to develop optimized catalysts for CFP. A study using 10 variations in the synthesis of hierarchical mesoporous ZSM-5 led to an optimized catalyst for CFP of cellulose [85]. In addition to the presence of mesopores, high crystallinity was found to be important, as a small concentration of amorphous silica–alumina surface defects was found to impact the diffusion of bulky substrates. The optimized, highly crystalline catalysts improved production of

Table 1.5 Properties of a ZSM-5 catalyst containing mesopores optimized for aromatic hydrocarbon yield via cellulose CFP.

	SAR[a)]	Surface area (m^2/g)			Volume (cm^3/g)		RC[b)]	^{27}Al FWHM[c)] (ppm)	NH_3-TPD BAS peak (°C/area)
		Total	Micro	Meso	Total	Micro			
ZSM-5[d)]	23	372	274	98	0.202	0.127	100	5.9	408/86
ZSM5-OPT	34.4	318	244	74	0.159	0.113	100.7	4.9	432/147

a) Silica-alumina ratio.
b) Relative crystallinity.
c) Framework ^{27}Al NMR signal.
d) Commercial from Zeolyst, CBV2314.
Source: Hoff et al. 2016 [85]. Adapted with permission of John Wiley and Sons.

aromatics from CFP of cellulose by 12%. The reported properties of this catalyst are summarized in Table 1.5 [85]. Another strategy to incorporate mesoporosity is to form nanosheets; the sheets wind up stacked in random orientations, creating a mesoporous/microporous structure [86]. In one study on cellulose CFP using ZSM-5 nanosheets, aromatic hydrocarbon yield was not improved over commercial HZSM-5 using fresh catalysts, but the catalyst lifetime was improved in the case of the nanosheet catalyst due to reduced production of coke [87]. The stability of hybrid materials under regeneration conditions has not yet been explored.

1.1.2 CFP with Metal Oxide Catalysts

Another class of materials receiving considerable attention as potential catalysts for biomass CFP are various forms of metal oxides [88–98]. Metal oxides used for this purpose can be divided into three main groups, namely, acidic metal oxides, basic metal oxides, and transition metal oxides. Some transition metal oxides are being considered for use in reactive catalytic fast pyrolysis (RCFP) processes having a non-inert atmosphere (usually some concentration of H_2). A summary of some typical results of CFP with metal oxide catalysts can be found in Table 1.6.

Acidic metal oxides are somewhat analogous to zeolites in that they promote changes in the pyrolysis pathways via their acid functionality. This class of catalysts includes Lewis acids such as amorphous Al_2O_3, SiO_2, TiO_2, or ZrO_2 [26, 88]. These catalysts are not as effective in the production of oxygen-free hydrocarbons as zeolites, particularly ZSM-5 zeolites; however, bulk reduction of oxygen content through conversion of highly oxygenated species such as anhydrosugars and methoxylated phenols to lesser oxygenated species such as furans, cyclopentenones, and simple phenols is observed. In a large pilot scale study (~1 ton/d) on CFP of loblolly pine, γ-Al_2O_3 was used in a fluidized bed with continuous regeneration of the catalyst from coking. Bio-oil was produced in a yield of 11.5 C% containing 23 wt% oxygen, comparable to some results using HZSM-5 but with lower concentrations of oxygen-free hydrocarbons in the bio-oil [88].

Base metal oxides, particularly MgO and CaO, as well as TiO_2, have long been used in thermal processes for various reasons and have also been studied as

Table 1.6 CFP results using metal oxide and related catalysts.

Catalyst	Biomass	Scale	Configuration	Temperature (°C)	WHSV (h^{-1})	Bio-oil yield (C%)[a]/ oxygen content (wt%)	References
γ-Al_2O_3	Pine	1 Mt/d	*In situ*[b]	520	0.33	11.5/23	[88]
Red mud[c]	Juniper	0.150 kg/h	*In situ*	500	1.5	26.4/25.0	[89]
CaO	Pine	0.060 kg/h	*In situ*[d]	500	C/B = 0.6	22/9.9	[90]
$Ca(COOH)_2$	Pine	0.060 kg/h	*In situ*[d]	500	C/B = 1.4	39/16.3	[90]
MgO	Beech wood	1.5 g batch	*Ex situ*	500	C/B = 0.5	31.6/28.4	[91]
TiO_2	Beech wood	1.5 g batch	*Ex situ*	500	C/B = 0.5	39/35.8	[92]
ZrO_2	Beech wood	1.5 g batch	*Ex situ*	500	C/B = 0.5	39/33.0	[92]
Al_2O_3	Beech wood	1.5 g batch	*Ex situ*	500	C/B = 0.5	23.0/24.0	[92]

a) C%, carbon yield.
b) Catalyst was continuously regenerated.
c) Red mud is a mixture of metal oxides (mainly Fe_2O_3 and Al_2O_3).
d) Catalyst/biomass premixed.

catalysts for biomass CFP. These materials are known catalysts for ketonization and aldol condensation reactions of carboxylic acids and carbonyls [91, 92]. They are therefore able to partially deoxygenate these types of compounds during biomass CFP via these C—C bond forming reactions, releasing CO_2 (as opposed to zeolite catalysis where CO is the main oxygen-carrying gas produced) and H_2O, along with production of new, longer chain ketones. Chain lengthening of small molecules prior to further refining of the bio-oil (e.g. via HDO) is important, so that when the skeletons are deoxygenated the carbon is not lost to gas and is in the correct chain length to operate as a hydrocarbon fuel [91, 92]. Elimination of acids is also a benefit of these reactions. An example of this type of reaction is the conversion of 2 mol of acetic acid (C2) generated via primary pyrolysis reactions to 1 mol of acetone (C3, which can continue to react to larger ketones, including cyclopentenones, under the CFP conditions). In a study of several MgO catalysts for CFP of beech wood, materials with increased porosity and surface area were more effective catalysts [91]. A slightly deoxygenated bio-oil of 28.4 wt% oxygen in about 31% carbon yield was achieved with the most active material screened [91]. Pine wood premixed with CaO was pyrolyzed to produce a highly deoxygenated bio-oil (9.9 wt% oxygen) with a carbon yield of 22 wt% [90]. Use of calcium formate rather than CaO increased the yield; it was suggested that the formate acted as a hydrogen donor, allowing the increase in yield. Elimination of acids and furans was observed, and like the MgO-catalyzed case, an increased production of cyclopentenones was noted. In both cases, increased non-methoxylated phenolic content, particularly *m*-substituted alkyl phenols, was also observed [90, 91]. Subsequent work revealed that both cellulose and lignin were converted to these phenols in the presence of Ca catalysts, but a mechanism for their formation is not known [93]. Similar CFP trends have been reported using TiO_2 and ZrO_2 as well, but with lower performance for deoxygenation [94].

Certain transition metal oxides bring another type of functionality to biomass CFP. Some inspiration for testing these types of catalysts has come from recent success in using molybdenum oxide and molybdenum carbide as effective formulations for the HDO of certain bio-oil model compounds [95, 99–101]. Supported transition metal oxides have been used as effective catalysts for CFP in inert atmospheres or under H_2 at atmospheric pressure (Table 1.7). Various versions of supported molybdenum oxide have been the most successful catalysts in this class found thus far, but some based on iron and tungsten oxides have also been tested. In one study using MoO_3/TiO_2 or MoO_3/ZrO_2 in an *ex situ* CFP method run in a micropyrolysis system under ~70 vol% H_2, a hydrocarbon yield of 27 C% from pine was achieved [96]. This catalyst was much more selective for alkanes (over alkenes and aromatics) than zeolite catalysts, producing these hydrocarbons in a 19 : 2 : 7 ratio, respectively. However, much of the product mixture consisted of small carbon chains with ethane and butane comprising 67–80% of the alkane product. Very high yields of hydrocarbons (C1–C10) via CFP of cellulose and corn stover over prereduced bulk MoO_3 at very high catalyst loadings have also been reported [97]. Another study performed *in situ* CFP using a laboratory scale continuous fluidized bed under H_2 with a supported molybdenum oxide catalyst, yielding 43.2% of highly deoxygenated liquid range bio-oil

Table 1.7 RCFP results using reducible metal oxide catalysts.

Catalyst	Biomass	Configuration	Temperature (°C)	Atmosphere	C/B	Bio-oil yield (C%)[a)]/ oxygen content (wt%)	HC yield[b)](C%)	References
MoO_3/TiO_2	Pine	*Ex situ*	500	71/29 H_2/He	26		28	[96]
MoO_3/ZrO_2	Pine	*Ex situ*	500	71/29 H_2/He	26		26	[96]
MoO_3	Cellulose	*Ex situ*	500	50/50 H_2/He	~166		72.6	[97]
MoO_3	Lignin	*Ex situ*	500	50/50 H_2/He	~166		35.7	[97]
MoO_3	Corn stover	*Ex situ*	500	50/50 H_2/He	~166		49.5	[97]
W-RMO[c)]	Pine	*In situ*	500	N_2		16.9/18.5		[98]
W-RMO[c)]	Pine	*In situ*	500	60/40 H_2/N_2		25.0/17.3		[98]
Fe-RMO[c)]	Pine	*In situ*	500	N_2		22.0/18.8		[98]
Fe-RMO[c)]	Pine	*In situ*	500	60/40 H_2/N_2		36.8/14.8		[98]
Mo-RMO[c)]	Pine	*In situ*	450	N_2		26.6/21.4		[98]
Mo-RMO[c)]	Pine	*In situ*	450	80/20 H_2/N_2		43.0/6.2		[98]

a) Includes alkanes, alkenes and aromatics including gas and liquid range molecules, carbon yield.
b) Includes products ≥C4, carbon yield.
c) RMO, reducible metal oxide (specific catalyst composition was not disclosed).

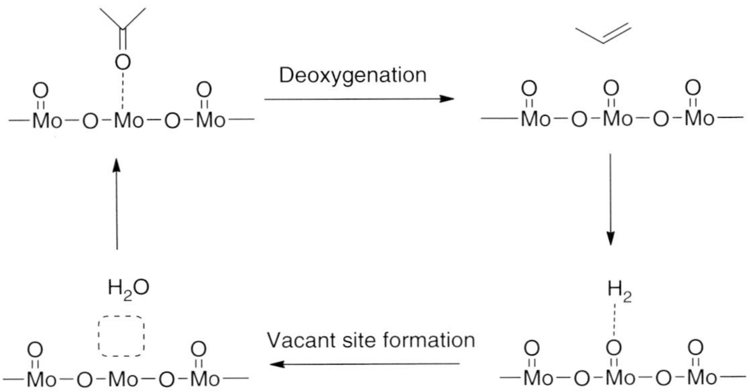

Figure 1.10 Reverse Mars–Van Krevelin mechanism for hydrodeoxygenation (e.g. acetone to propene) over MoO_3. Source: Prasomsri et al. 2013 [101]. Adapted with permission of Royal Society of Chemistry.

product ($\geq C_4$ hydrocarbons, 6.2 wt% oxygen) from pine [98]. The consumption of H_2 during the reaction led to the formation of alkanes and concurrently increased the H/C ratio from what is typically produced from zeolite-catalyzed CFP. The deoxygenation mechanism is thought to be based on oxygen vacancies produced by removal of surface oxygen from MoO_3 via reaction with H_2 to form water. The oxygen vacancies are then filled by removal of oxygen from biomass-derived substrates (a reverse Mars–van Krevelen mechanism, Figure 1.10) [101]. Some deactivation of the catalysts due to coking is observed [96]. Deactivation may also occur due to formation of inactive MoO_2 [96, 97]. Not enough information is yet available to assess how biomass inorganics or exposure to combustion conditions for regeneration would affect the structure of catalysts of this type.

1.1.3 CFP to Produce Fine Chemicals

While complete deoxygenation is desired for fuel purposes, preserving some functionality, and chemical structure of the biomass can lead to the production of more valuable chemical products. Supported Brønsted acids are used as catalysts to perform selective dehydration reactions during pyrolysis. Of particular interest is the direction of cellulose pyrolysis toward levoglucosenone (LGO) rather than levoglucosan (Figure 1.11). LGO is a highly dehydrated sugar monomer that retains one chiral center from cellulose and contains reactive moieties, namely, the alkene double bond and the carbonyl group [102, 103]. This makes it valuable in the preparation of chiral pharmaceutical precursors and for the synthesis of biologically active α,β-unsaturated ketones. The production of LGO from pyrolysis of acid-impregnated cellulose was first reported in 1973, but more recently advances in heterogeneous acid catalysts for these reactions have been made (Table 1.8). These reactions are typically performed at lower temperatures than other CFP processes. One catalyst reported was sulfated zirconia (prepared by impregnation of ZrO_2 with H_2SO_4); at the optimum temperature of 335 °C, the yield of LGO was about 8 wt% from catalytic pyrolysis

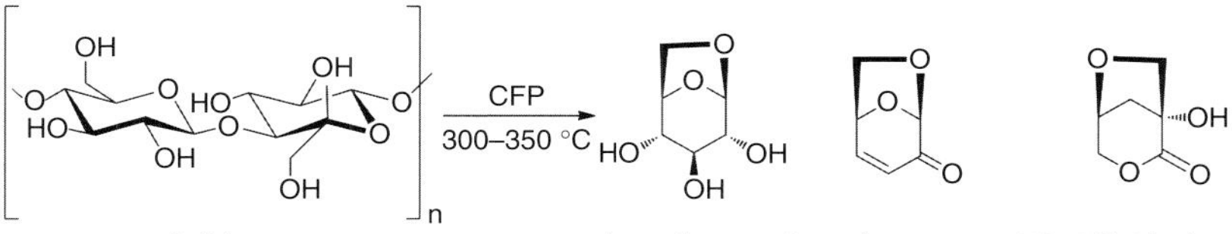

Cellulose Levoglucosan Levoglucosenone (LGO) (1R, 5S)-1-hydroxy-3,6-dioxabicyclo [3.2.1] octan-2-one (LAC)

Figure 1.11 Depolymerization and dehydration of cellulose to chiral chemical building blocks via low temperature CFP.

Table 1.8 Low temperature CFP for specific chemicals.

Catalyst[a)]	Biomass	C/B	Temperature (°C)	Targeted product[b)]/ yield (wt%)	References
H_3PO_4	Cellulose	0.03[c)]	335	LGO, 5.6%	[102]
SO_4^{2-}/ZrO_2	Cellulose	1	335	LGO, 8.14%	[102]
SO_4^{2-}/TiO_2-Fe_2O_3	Cellulose	0.33	300	LGO, 15.4	[103]
H_3PO_4/AC	Cellulose	0.33	300	LGO, 18.1	[103]
H_3PO_4/AC	Pine	0.33	300	LGO, 9.1	
H_3PO_4/AC	Poplar	0.33	300	LGO, 8.3	
H_3PO_4/AC	Bagasse	0.33	300	LGO, 6.2	
NP AlTi[d)]	Cellulose	0.3	350	LAC, 8.6	[104]
Pd/SBA-15	Bagasse	0.3	350	4-EG, 0.73	[105]
Pd/SBA-15	Poplar	0.3	350	4-EG, 0.11	
Pd/SBA-15	Pine	0.3	350	4-EG, 0.18	[106]
AC	Bagasse	1.5	300	4-EP, 2.49	

a) AC, activated carbon.
b) LGO, levoglucosenone; LAC, (1*R*, 5*S*)-1-hydroxy-3,6-dioxabicyclo[3.2.1]octan-2-one; 4-EG, 4-ethyl guaiacol; 4-EP, 4-ethyl phenol.
c) Premixed.
d) Nanoparticle Al_2O_3/TiO_2.

of cellulose [102]. Recycling of the catalyst was studied in a limited fashion. Catalysts could be regenerated by combustion of any accumulated char or coke, and the structure of the ZrO_2 support remained intact, but the sulfur content significantly decreased and the support had to be reimpregnated with H_2SO_4 to fully reactivate the catalyst [102]. Similarly, sulfated TiO_2 has also been shown to be effective for this transformation [103]. Another reported catalyst for this transformation was an activated carbon prepared by chemical activation with H_3PO_4 [103]. At 300 °C this catalyst was able to produce LGO in yields of 14.7–18 wt% from cellulose and about 7 wt% from wood. Upon reuse of the catalyst, the yield dropped. Analysis of the spent catalyst indicated a drop in surface area, perhaps from coking [103]. The use of a carbon support is a limitation in this case, as the coke cannot be removed by combustion without destruction of the support.

Another potential chiral synthon from cellulose produced from CFP is (1*R*,5*S*)-1-hydroxy-3,6-dioxabicyclo[3.2.1]octan-2-one (LAC). It was produced in about 8 wt% yield via low temperature (350 °C) pyrolysis of cellulose over nanopowder TiO_2/Al_2O_3 [104]. Further optimization of catalysts for the synthesis of this potentially valuable product has not been reported.

Other targets for biomass catalytic pyrolysis processes include phenols produced by improved selectivity during the decomposition of lignin. Increased yields of 4-ethyl phenol and 4-ethyl guaiacol have been reported for CFP of biomass over Pd on SBA-15, depending on the biomass source [105, 106]. 4-Ethyl phenol appears to form from the net hydrogenation of 4-vinylphenol, a common pyrolytic breakdown product of *p*-hydroxyphenyl (H-type) lignin units, and similarly, 4-ethyl guaiacol is derived from hydrogenation of 4-vinyl guaiacol, derived from G-type lignin units [105]. The hydrogen may be donated from the cellulosic portions of the biomass, as addition of external H_2 is not necessary. An activated carbon catalyst without metal functionality has also been reported for the transformation to ethyl phenol. The best conditions were CFP at a temperature of 300 °C and a catalyst/biomass ratio of 1.5/1, producing 4-ethyl phenol in 2.5 wt% yield from a lignin-enriched bagasse material [106].

1.1.4 Outlook and Conclusions

Zeolite-catalyzed processes for the production of deoxygenated bio-oil as an intermediate to the production of advanced biofuels have received considerable attention from industry. One company, Kior, advanced to the point of running a 500 ton/d plant based on a CFP technology using a proprietary blend of zeolites (including ZSM-5 types) to produce bio-gasoline from pine wood. Kior had planned further expansion and scale-up, but the company went bankrupt in 2014 due to lower than expected product yields [107, 108]. The most complete economic analyses from the US Department of Energy National Renewable Energy Laboratory (DOE NREL) of the zeolite-catalyzed biomass pyrolysis pathway to advanced bio-fuels (including hydrotreatment of the CFP bio-oil) suggests that a price around \$3.30 or \$3.50/gal of gasoline equivalent for the *in situ* or *ex situ* case, respectively (at biomass cost of \$80/ton), is possible at carbon yields of 44% and oxygen content of 6.0–6.4 wt% for the bio-oil intermediate [18]. These targets have not yet been reached in the open literature. The use of biofuel from such a process is estimated to reduce greenhouse gas (GHG) emissions 90% compared with the 2005 petroleum baseline. This reduction in GHG emissions is well above the reduction requirement to be considered as a cellulosic biofuel by the Renewable Fuel Standard (minimum 60% reduction) [18].

While more development is needed to bring CFP of biomass to true economic competiveness with petroleum for production of renewable transportation fuels, one of the appeals of using catalytic pyrolysis is the potential to tailor the pyrolysis process to targeted higher value products over lower margin fuel products. In the realm of zeolite catalysis, one company, Anellotech, is positioning itself to use biomass CFP over zeolites to produce renewable aromatics (particularly BTX: benzene, toluene, xylenes) as a chemical feedstock rather than a fuel feedstock [107] (www.anellotech.com).

As seen in this chapter, other efforts are ongoing to target other high value chemicals, but these are at a much earlier stage in their development than zeolite-based biomass CFP processes. These other bio-based products may become the economic driver that allow other biomass carbon to find its way to the fuel supply. There is plenty of opportunity for development of not only new and improved catalysts, but of process steps including separation and isolation of targeted products.

References

1 Mohan, D., Pittman, C.U. Jr., and Steel, P.H. (2006). Pyrolysis of wood/biomass for bio-oil: a critical review. *Energy Fuels* 20: 848–889.
2 Huber, G.W., Iborra, S., and Corma, A. (2006). Synthesis of transportation fuels from biomass: chemistry, catalysts, and engineering. *Chem. Rev.* 106: 4044–4098.
3 Meier, D., van del Beld, D., Bridgwater, A.V. et al. (2013). State-of-the-art of fast pyrolysis in IEA bioenergy member countries. *Renewable Sustainable Energy Rev.* 20: 619–641.
4 Weber, K. and Quicker, P. (2018). Properties of biochar. *Fuel* 217: 240–261.
5 Qambrani, N.A., Rahman, M.M., Won, S. et al. (2017). Biochar properties and eco-friendly applications for climate change mitigation, waste management, and wastewater treatment: a review. *Renewable Sustainable Energy Rev.* 79: 255–273.
6 Qian, K., Kumar, A., Zhang, H. et al. (2015). Recent advances in utilization of biochar. *Renewable Sustainable Energy Rev.* 42: 1055–1064.
7 Ensyn Fuels, Inc. (2015). Ensyn, Youngstown Thermal sign RFO biofuel supply agreement, *Biomass Magazine* (08 June), http://biomassmagazine.com/articles/12029/ensyn-youngstown-thermal-sign-rfo-biofuel-supply-agreement.
8 Williams, R.C., Satrio, J., Rover, M.R. et al. (2009). Utilization of fractionated bio oil in asphalt. In: *Transportation Research Board 88th Annual Meeting*. Washington, DC: Transportation Research Board https://trid.trb.org/View/882150.
9 Oasmaa, A., Fonts, I., Pelaaez-Samaniego, M.R. et al. (2016). Pyrolysis oil multiphase behavior and phase stability: a review. *Energy Fuels* 30: 6179–6200.
10 Diebold, J.P. (2000). A Review of the Chemical and Physical Mechanisms of the Storage Stability of Fast Pyrolysis Bio-oils. NREL/SR-570-27613.
11 Baldwin, R.M. and Felik, C.J. (2013). Bio-oil stabilization and upgrading by hot gas filtration. *Energy Fuels* 27: 3224–3238.
12 Elkasabi, Y., Mullen, C.A., and Boateng, A.A. (2014). Distillation and isolation of commodity chemicals from bio-oil made by tail-gas reactive pyrolysis. *ACS Sustainable Chem. Eng.* 2: 2042–2052.
13 Elliott, D.C., Wang, H., French, R. et al. (2014). Hydrocarbon liquid production from biomass via hot-vapor-filtered fast pyrolysis and catalytic hydroprocessing of the bio-oil. *Energy Fuels* 28: 5909–5917.
14 Zacher, A.H., Olarte, M.V., Santosa, D.M. et al. (2104). A review and perspective of recent bio-oil hydrotreating research. *Green Chem.* 16: 491–515.

15 Elliott, D.C., Hart, T.R., Neuenshawander, G.G. et al. (2012). Catalytic hydroprocessing of fast pyrolysis bio-oil from pine sawdust. *Energy Fuels* 26: 3891–3896.
16 Elliott, D.C., Biller, P., Ross, A.B. et al. (2015). Hydrothermal liquefaction of biomass: developments from batch to continuous process. *Bioresour. Technol.* 178: 146–156.
17 Linck, M., Felix, L., Marker, T., and Roberts, M. (2014). Integrated biomass hydropyrolysis and hydrotreating: a brief review. *Wiley Interdiscip. Rev.: Energy Environ.* 3: 575–581.
18 Dutta, A., Sahir, A., Tan, E., et al. (2015). Process Design and Economics for the Conversion of Lignocellulosic Biomass to Hydrocarbon Fuels – Thermochemical Research Pathways with in situ and ex situ Upgrading of Fast Pyrolysis Vapors. NREL/TP-5100-62455.
19 Coombs, D.S., Alberti, A., Armbruster, T. et al. (1998). Recommended nomenclature for zeolite minerals: report of the subcommittee on zeolites of the international mineralogical association, commission on new minerals and mineral names. *Mineral. Mag.* 62: 533–571.
20 Weitkamp, J. (2000). Zeolites and catalysis. *Solid State Ionics* 131: 175–188.
21 Davis, R.J. (2003). New perspectives on basic zeolites as catalysts and catalyst supports. *J. Catal.* 216: 396–405.
22 Mihalcik, D.J., Mullen, C.A., and Boateng, A.A. (2011). Screening acidic zeolites for catalytic fast pyrolysis of biomass and its components. *J. Anal. Appl. Pyrolysis* 90: 197–203.
23 Aho, A., Kumar, N., Eränen, K. et al. (2008). Catalytic pyrolysis of woody biomass in a fluidized bed reactor: influence of the zeolite structure. *Fuel* 87: 2493–2501.
24 Mullen, C.A., Boateng, A.A., Mihalcik, D.J., and Goldberg, N.M. (2011). Catalytic fast pyrolysis of white oak wood in a bubbling fluidized bed. *Energy Fuels* 25: 5444–5451.
25 Jae, J., Tompsett, G.A., Foster, A.J. et al. (2011). Investigation into the shape selectivity of zeolite catalysts for biomass conversion. *J. Catal.* 279: 257–268.
26 Liu, C., Wang, H., Karim, A.M. et al. (2014). Catalytic fast pyrolysis of lignocellulosic biomass. *Chem. Soc. Rev.* 43: 7549–7623.
27 Olsbye, U., Svelle, S., Lillerud, K.P. et al. (2015). The formation and degradation of active species during methanol conversion of protonated zeotype catalysts. *Chem. Soc. Rev.* 44: 7155–7176.
28 Serrano, D.P., Aguado, J., and Escola, J.M. (2012). Developing advanced catalysts for the conversion of polyolefinic waste plastics into fuels and chemicals. *ACS Catal.* 2: 1924–1941.
29 Zhang, H., Cheng, Y.-T., Vispute, T.P. et al. (2011). Catalytic conversion of biomass-derived feedstocks into olefins and aromatics with ZSM-5: the hydrogen to carbon effective ratio. *Energy Environ. Sci.* 4: 2297–2307.
30 Lei, X., Jockusch, S., Ottaviani, M.F., and Turro, N.J. (2003). In situ EPR investigation of the addition of persistent benzyl radicals to acrylates on ZSM-5 zeolites. Direct spectroscopic detection of the initial steps in a supramolecular photopolymerization. *Photochem. Photobiol. Sci.* 2: 1095–1100.

31 Jae, J.H., Tompsett, G.A., Lin, Y. et al. (2010). Depolymerization of lignocellulosic biomass to fuel precursors: maximizing carbon efficiency by combining hydrolysis with pyrolysis. *Energy Environ. Sci.* 3: 358–365.

32 Mullen, C.A. and Boateng, A.A. (2010). Catalytic pyrolysis-GC/MS of lignin from several sources. *Fuel Process. Technol.* 91: 1446–1458.

33 Carlson, T.R., Jae, J., Lin, Y.–.C. et al. (2010). Catalytic fast pyrolysis of glucose with HZSM-5: the combined homogenous and heterogeneous reactions. *J. Catal.* 270: 110–124.

34 Lin, Y.C., Cho, J., Topsett, G.A. et al. (2009). Kinetics and mechanism of cellulose pyrolysis. *J. Phys. Chem. C* 113: 20097–20107.

35 Kawamoto, H., Murayama, M., and Saka, S. (2003). Pyrolysis behavior of levoglucosan as an intermediate in cellulose pyrolysis: polymerization into polysaccharide as a key reaction to carbonized product formation. *J. Wood Sci.* 49: 469–473.

36 Cheng, Y.–.T. and Huber, G.W. (2011). Chemistry of furan conversion into aromatics and olefins over HZSM-5: a model biomass conversion reaction. *ACS Catal.* 1: 611–628.

37 Horne, P.A. and Williams, P.T. (1994). Premium quality fuels and chemicals from the fluidized bed pyrolysis of biomass with zeolite catalyst upgrading. *Renewable Energy* 5: 810–812.

38 Zhang, H., Xaio, R., Huang, H., and Xaio, G. Comparison of non-catalytic and catalytic fast pyrolysis of corncob in a fluidized bed reactor. *Bioresour. Technol.* 100: 1428–1434.

39 Carlson, T.R., Cheng, Y.-T., Jae, J., and Huber, G.W. (2011). Production of green aromatics and olefins by catalytic fast pyrolysis of wood sawdust. *Energy Environ. Sci.* 4: 145–161.

40 Foster, A.J., Jae, J., Cheng, Y.-T. et al. (2012). Optimizing the aromatic yield and distribution from catalytic fast pyrolysis of biomass over ZSM-5. *Appl. Catal., A* 423-424: 154–161.

41 Mullen, C.A. and Boateng, A.A. (2013). Accumulation of inorganic impurities on HZSM-5 zeolites during catalytic fast pyrolysis of switchgrass. *Ind. Eng. Chem. Res.* 52: 17156–17161.

42 Paasikallio, V., Lindfors, C., Kuoppala, E. et al. (2014). Product quality and catalyst deactivation in a four day catalytic fast pyrolysis production run. *Green Chem.* 16: 3549–3559.

43 Jae, J., Coolman, R., Mountziaris, T.J., and Huber, G.W. (2014). Catalytic fast pyrolysis of lignocellulosic biomass in a process development unit with continual catalyst addition and removal. *Chem. Eng. Sci.* 108: 33–46.

44 Wang, K., Kim, K.H., and Brown, R.C. (2014). Catalytic pyrolysis of individual components of lignocellulosic biomass. *Green Chem.* 16: 727–735.

45 Mante, O.D. and Agblevor, F.A. (2014). Catalytic pyrolysis for the production of refinery-ready biocrude oils from six different biomass sources. *Green Chem.* 16: 3364–3377.

46 Karanjkar, P.U., Coolman, R.J., Huber, G.W. et al. (2014). Production of aromatics by catalytic fast pyrolysis of cellulose in a bubbling fluidized bed reactor. *AlChE J.* 60: 1320–1335.

47 Paasikallio, V., Kalogiannis, K., Lappas, A. et al. (2017). Catalytic fast pyrolysis: influencing bio-oil quality with the catalyst-to-biomass ratio. *Energy Technol.* 5: 94–103.

48 Iisa, K., French, R.J., Orton, K.A. et al. (2017). Production of low-oxygen bio-oil via ex situ catalytic fast pyrolysis and hydrotreating. *Fuel* 207: 413–422.

49 Starace, A.K., Black, B.A., Lee, D.D. et al. (2017). Characterization and catalytic upgrading of aqueous stream carbon from catalytic fast pyrolysis of biomass. *ACS Sustainable Chem. Eng.* 5: 11761–11769.

50 Mukarakate, C., Zhang, X., Stanton, A.R. et al. (2014). Real-time monitoring of the deactivation of HZSM-5 during upgrading of pine pyrolysis vapors. *Green Chem.* 16: 1444–1461.

51 Mullen, C.A., Boateng, A.A., Dadson, R.B., and Hashem, F.M. (2014). Biological mineral range effects on biomass conversion to aromatic hydrocarbon via catalytic pyrolysis over HZSM-5. *Energy Fuels* 28: 7014–7024.

52 Wan, S., Waters, C., Stevens, A. et al. (2015). Decoupling HZSM-5 catalyst activity from deactivation during upgrading of pyrolysis oil vapors. *ChemSusChem* 8: 552–559.

53 Hemelsoet, K., Van der Mynsbrugge, J., De Wispelaere, K. et al. (2013). Unraveling the reaction mechanisms governing methanol-to-olefins catalysis by theory and experiment. *ChemPhysChem* 14: 1526–1545.

54 Gayubo, A.G., Aguayo, A.T., Atutxa, A. et al. (2004). Transformation of oxygenate components of biomass pyrolysis oil on a HZSM-5 zeolite. I. Alcohols and phenols. *Ind. Eng. Chem. Res.* 43: 2610–2618.

55 Mullen, C.A., Dorado, C., and Boateng, A.A. (2018). Catalytic co-pyrolysis of switchgrass and polyethylene over HZSM-5: catalyst deactivation and coke formation. *J. Anal. Appl. Pyrolysis* 129: 195–203.

56 Mullen, C.A., Tarves, P.C., Raymundo, L.M. et al. (2018). Fluidized bed catalytic pyrolysis of eucalyptus over HZSM-5: effect of acid density and gallium modification on catalyst deactivation. *Energy Fuels* 32: 1771–1778.

57 Mukarakate, C., McBrayer, J.D., Evans, T.J. et al. (2015). Catalytic fast pyrolysis of biomass: the reactions of water and aromatic intermediates produces phenols. *Green Chem.* 17: 4217–4227.

58 Hoang, T.Q., Zhu, Z., Sooknoi, T. et al. (2010). A comparison of the relativities of propanal and proplylene on HZSM-5. *J. Catal.* 271: 201–208.

59 Mullen, C.A., Tarves, P.C., and Boateng, A.A. (2017). Role of potassium exchange in catalytic pyrolysis of biomass over ZSM-5: formation of alkyl phenols and furans. *ACS Sustainable Chem. Eng.* 5: 2154–2162.

60 Shanshan, S., Zhang, H., Xiao, R. et al. (2018). Controlled regeneration of ZSM-5 catalysts in the combined oxygen and steam atmosphere for catalytic pyrolysis of biomass-derivates. *Energy Convers. Manage.* 155: 175–181.

61 Huber, G.W., Cheng, Y.-T., Carlson, T., et al. (2012). Catalytic Pyrolysis of Solid Biomass and Related Biofuels, Aromatic, and Olefin Compounds. US Patent 8,277,643 B2, filed 3 March, 2009 and issued 2 October 2012.

62 Wang, L., Lei, H., Bu, Q. et al. (2014). Aromatic hydrocarbons production from ex situ catalysis of pyrolysis vapor over zinc modified ZSM-5 in a

packed-bed catalysis coupled with microwave pyrolysis reactor. *Fuel* 129: 78–85.

63 Vichaphund, S., Aht-ong, D., Sricharoenchaikul, V., and Atong, D. (2015). Production of aromatic compounds from catalytic fast pyrolysis of Jatropha residues using metal/HZSM-5 prepared by ion-exchange and impregnation methods. *Renewable Energy* 79: 28–37.

64 Huang, Y., Wei, L., Crandall, Z. et al. (2013). Combining Mo–Cu/HZSM-5 with a two-stage catalytic pyrolysis system for pine sawdust thermal conversion. *Fuel* 150: 656–663.

65 Du, Z., Ma, X., Li, Y. et al. (2013). Production of aromatic hydrocarbons by catalytic pyrolysis of microalgae with zeolites: catalyst screening in a pyroprobe. *Bioresour. Technol.* 139: 397–401.

66 French, R. and Cernik, S. (2010). Catalytic pyrolysis of biomass for biofuels production. *Fuel Process. Technol.* 91: 25–32.

67 Mullen, C.A. and Boateng, A.A. (2015). Production of aromatic hydrocarbons via catalytic pyrolysis of biomass over Fe-modified HZSM5 zeolites. *ACS Sustainable Chem. Eng.* 3: 1623–1631.

68 Cheng, Y.-T., Jae, J., Fan, W., and Huber, G.W. (2012). Production of renewable aromatic hydrocarbons by catalytic fast pyrolysis of lignocellulosic biomass with bifunctional Ga/ZSM-5 catalysts. *Angew. Chem. Int. Ed.* 51: 1387–1390.

69 Park, H.J., Dong, J.I., Jeon, J.K. et al. (2007). Conversion of the pyrolytic vapor of radiata pine over zeolites. *J. Ind. Eng. Chem.* 13: 182–189.

70 Kelkar, S., Saffron, C.M., Li, Z. et al. (2014). Aromatics from biomass pyrolysis vapor using a bifunctional mesoporous catalyst. *Green Chem.* 16: 803–812.

71 Shultz, E.L., Mullen, C.A., and Boateng, A.A. (2017). Aromatic hydrocarbon production from eucalyptus urophylla pyrolysis over several metal-modified ZSM-5 catalysts. *Energy Technol.* 5: 196–204.

72 Kim, J.W., Park, S.H., Jung, J. et al. (2013). Catalytic pyrolysis of mandarin residue from the mandarin juice processing industry. *Bioresour. Technol.* 136: 431–436.

73 Yung, M.M., Stanton, A.R., Iisa, K. et al. (2016). Multiscale evaluation of catalytic upgrading of biomass pyrolysis vapors on Ni- and Ga-modified ZSM-5. *Energy Fuels* 30: 9471–9479.

74 Iliopoulou, E.F., Stefanidis, S., Kalogiannis, K. et al. (2014). Pilot-scale validation of co-ZSM-5 catalyst performance in the catalytic upgrading of biomass pyrolysis vapors. *Green Chem.* 16: 662–674.

75 Yung, M.M., Starace, A.K., Mukarakate, C. et al. (2016). Biomass catalytic pyrolysis on Ni/ZSM-5: effects of nickel pretreatment and loading. *Energy Fuels* 30: 5259–5268.

76 Nowak, I., Quartararo, J., Derouane, E.G., and Védrine, J.C. (2003). Effect of H_2–O_2 pre-treatments on the state of gallium in Ga/H-ZSM-5 propane aromatisation catalysts. *Appl. Catal., A* 251: 107–120.

77 Li, S., Lepore, A.W., Salazar, M.F. et al. (2017). Selective conversion of bio-derived ethanol to renewable BTX over Ga-ZSM-5. *Green Chem.* 19: 4344–4352.

78 Lee, E.H., Park, R., Kim, H. et al. (2016). Hydrodeoxygenation of guaiacol over Pt loaded zeolitic materials. *J. Ind. Eng. Chem.* 27: 18–21.
79 Yu, Y., Li, X., Su, L. et al. (2012). The role of shape selectivity in catalytic fast pyrolysis of lignin with zeolite catalysts. *Appl. Catal., A* 447–448: 115–123.
80 Iliopolou, E.F., Antonakou, E.V., Karakoulia, S.A. et al. (2007). Catalytic conversion of biomass pyrolysis products by mesoporous materials: effect of steam stability and acidity of Al-MCM-41 catalysts. *Chem. Eng. J.* 134: 51–57.
81 Li, Q., Li, W., Zhang, D., and Xi-feng, Z. (2009). Analytical pyrolysis-gas chromatography/mass spectrometry (Py-GC/MS) of sawdust with Al/SBA-15 catalysts. *J. Anal. Appl. Pyrolysis* 84: 131–138.
82 Li, J., Li, X., Zhou, G. et al. (2014). Catalytic fast pyrolysis of biomass with mesoporous ZSM-5 zeolites prepared by desilication with NaOH solutions. *Appl. Catal., A* 470: 115–122.
83 Hoff, T.C., Gardner, D.W., Thiladkaratne, R. et al. (2017). Elucidating the effect of desilication on aluminum-rich ZSM-5 zeolite and its consequences on biomass catalytic fast pyrolysis. *Appl. Catal., A* 529: 68–78.
84 Gamliel, D.P., Cho, H.J., Fan, W., and Valla, J.A. (2016). On the effectiveness of tailored mesoporous MFI zeolites for biomass catalytic fast pyrolysis. *Appl. Catal., A* 522: 109–119.
85 Hoff, T.C., Gardner, D.W., Thilakaratne, R. et al. (2016). Tailoring ZSM-5 zeolites for the fast pyrolysis of biomass to aromatic hydrocarbons. *ChemSusChem* 9: 1473–1482.
86 Lee, H.W., Park, S.H., Jeon, J. et al. (2014). Upgrading of bio-oil derived from biomass constituents over hierarchical unilamellar mesoporous MFI zeolites. *Catal. Today* 232: 119–126.
87 Xu, M., Mukarakate, C., Iisa, K. et al. Deactivation of multilayered MFI nanosheet zeolite during upgrading of biomass pyrolysis vapors. *ACS Sustainable Chem. Eng.* 5: 5477–5484.
88 Mante, O.D., Dayton, D.C., Carpenter, J.R. et al. (2018). Pilot-scale catalytic fast pyrolysis of loblolly pine over γ-Al_2O_3 catalyst. *Fuel* 214: 569–579.
89 Agblevor, F.A., Elliott, D.C., Santosa, D.M. et al. (2016). Red mud catalytic pyrolysis of pinyon juniper and single-stage hydrotreatment of oils. *Energy Fuels* 30: 7947–7958.
90 Case, P.A., Wheeler, M.C., and Desisto, W.J. (2014). Formate assisted pyrolysis of pine sawdust for in-situ oxygen removal and stabilization of bio-oil. *Bioresour. Technol.* 173: 177–184.
91 Stefanidis, S.D., Karakoulia, S.A., Kalogiannis, K.G. et al. (2016). Natural magnesium oxide (MgO) catalysts: a cost-effective sustainable alternative to acid zeolites for the in situ upgrading of biomass fast pyrolysis oil. *Appl. Catal., B* 196: 155–173.
92 Stefanidis, S.D., Kalogiannis, K.G., Iliopulou, E.F. et al. (2011). In-situ upgrading of biomass pyrolysis vapors: catalyst screening on a fixed bed reactor. *Bioresour. Technol.* 102: 8261–8267.
93 Case, P.A., Truong, C., Wheeler, M.C., and DeSisto, W.J. (2015). Calcium-catalyzed pyrolysis of lignocellulosic biomass components. *Bioresour. Technol.* 192: 247–252.

94 Kim, P., Rials, T., Labbé, N., and Chmely, S.C. (2016). Screening of mixed-metal oxide species for catalytic ex situ vapor-phase deoxygenation of cellulose by Py-GC/MS coupled with multivariate analysis. *Energy Fuels* 30: 3167–3174.

95 Nolte, M.W. and Shanks, B.H. (2017). A perspective on catalytic strategies for deoxygenation in biomass pyrolysis. *Energy Technol.* 5: 7–18.

96 Murugappan, K., Mukarakate, C., Budhi, S. et al. (2016). Supported molybdenum oxides as effective catalysts for the catalytic fast pyrolysis of lignocellulosic biomass. *Green Chem.* 18: 5548–5557.

97 Nolte, M.W., Zhang, J., and Shanks, B.H. (2016). Ex situ hydrodeoxygenation in biomass pyrolysis using molybdenum oxide and low pressure hydrogen. *Green Chem.* 18: 134–138.

98 Wang, K., Dayton, D.C., Peters, J.E., and Mante, O.D. (2017). Reactive catalytic fast pyrolysis of biomass to produce high-quality bio-crude. *Green Chem.* 19: 3243–3251.

99 Lee, W., Kumar, A., Wang, Z., and Bhan, A. (2015). Chemical titration and transient kinetic studies of site requirements in Mo_2C-catalyzed vapor phase anisole hydrodeoxygenation. *ACS Catal.* 5: 4104–4114.

100 Sullivan, M.M. and Bhan, A. (2016). Acetone hydrodeoxygenation over bifunctional metallic-acidic molybdenum carbide catalysts. *ACS Catal.* 6: 1145–1152.

101 Prasomsri, T., Nimmanwudipong, T., and Román-Leshkov, Y. (2013). Effective hydroxeoxygenation of biomass-derived oxygenates into unsaturated hydrocarbons by MoO_3 using low H_2 pressures. *Energy Environ. Sci.* 6: 1732–1738.

102 Wang, Z., Lu, Q., Zhu, X., and Zhang, Y. (2011). Catalytic fast pyrolysis of cellulose to prepare levoglucosenone using sulfated zirconia. *ChemSusChem* 4: 79–84.

103 Ye, X., Lu, Q., Wang, X. et al. (2017). Catalytic fast pyrolysis of cellulose and biomass to selectively produce levglucosenone using activated carbon catalyst. *ACS Sustainable Chem. Eng.* 5: 10815–10825.

104 Fabbri, D., Torri, C., and Mancini, I. (2007). Pyrolysis of cellulose catalyzed by nanopowder metal oxides: production and chacterisation of a chiral hydroxylactone and its role as building block. *Green Chem.* 9: 1374–1379.

105 Lu, Q., Zhang, Z., Ye, X. et al. (2017). Selective production of 4-ethyl guaiacol from catalytic fast pyrolysis of softwood biomass using Pd/SBA-15 catalyst. *J. Anal. Appl. Pyrolysis* 123: 237–243.

106 Lu, Q., Ye, X., Zhang, Z. et al. (2016). Catalytic fast pyrolysis of bagasse using activated carbon catalyst to selectively produce 4-ethyl phenol. *Energy Fuels* 30: 10618–10626.

107 Perego, C., Bosetti, A., Ricci, M., and Millini, R. (2017). Zeolite materials for biomass conversion to biofuel. *Energy Fuels* 31: 7721–7733.

108 Adkins, B., Stamires, D., Bartek, R. et al. (2012). Improved catalyst for thermocatalytic conversion of biomass to liquid fuels and chemicals, WO Patent 2012/142490 A1.

2

The Upgrading of Bio-Oil via Hydrodeoxygenation

Adetoyese O. Oyedun, Madhumita Patel, Mayank Kumar, and Amit Kumar

University of Alberta, 10-263 Donadeo Innovation Centre for Engineering, Department of Mechanical Engineering, Edmonton, Alberta T6G 1H9, Canada

2.1 Introduction

The production of liquid transportation fuels from biomass has received increased attention in the last two decades due to environmental concerns [1–4]. Bio-oil, a product of fast pyrolysis or hydrothermal liquefaction (HTL), is a promising renewable energy carrier and a means of producing liquid transportation fuels from biomass [2, 5–9]. Bio-oil is a dark brown liquid, and its physical properties are a function of its composition (various oxygenated organic compounds) [10]. The advantages of using bio-oil as a liquid fuel include the abundance of biomass, its renewability, negligible nitrogen and sulfur content, and nearly zero net-carbon emissions [2, 11, 12]. However, bio-oil produced from fast pyrolysis has several challenges that need to be addressed as the bio-oil can significantly impact the supply chain of liquid transportation fuels. The main challenge is the oxygen content of bio-oil, which can be up to 50% for bio-oil from fast pyrolysis, and leads to chemical and thermal instabilities, polymerization, and storage difficulties; in addition, bio-oil with high oxygen content has a low heating value and is immiscible with crude oil from fossil sources [2, 9, 13, 14]. The polymerization tendency of bio-oil causes it to quickly form a solid coke-like product when distilled, and heavier compounds are formed by polymerization and condensation reactions during prolonged storage [1]. Since upgrading bio-oil through hydrodeoxygenation (HDO) partially or totally eliminates the oxygenated compounds, it is essential to stabilize the bio-oil through this process before it can be widely accepted as a conventional liquid transportation fuel.

Several approaches, broadly categorized as physical and chemical, have been developed and tested for bio-oil upgrading and stabilization [15]. Emulsification and hot vapor filtration are examples of physical methods [16–18]; chemical methods include hydrogen processing, steam reforming, solvent addition, catalytic pyrolysis, and rapid thermal processing [13, 14, 19–22]. Hydrogen processing involves the elimination of heteroatoms such as oxygen (O), nitrogen (N), and sulfur (S) via HDO, hydrodenitrogenation (HDN), and hydrodesulfurization

Chemical Catalysts for Biomass Upgrading, First Edition. Edited by Mark Crocker and Eduardo Santillan-Jimenez.

(HDS), respectively [13, 23]. The HDO of bio-oil is considered to be the most effective method for bio-oil upgrading and involves the hydrotreating of bio-oils with high-pressure H_2 gas at moderate temperatures (300–600 °C) in the presence of catalysts [13] to convert oxygen-containing compounds to oxygen-free liquid transportation fuels. As bio-oil contains a number of oxygen-containing compounds such as phenols, aldehydes, alcohols acids, esters, and ketones [24–28], it is challenging to fully understand all the reaction pathways for the HDO reaction [29]. According to a review by Patel and Kumar [5], most oxygen compounds found in the bio-oil produced through fast pyrolysis are water (20–30%), lignin-derived components (10–15%), acids (13–15%), phenols (10–18%), aldehydes (8–10%), esters (2–5%), and ketones (8–10%). The quantity and the quality of the bio-oil depend on the chemical composition of biomass, thermochemical conversion approach – fast pyrolysis, intermediate pyrolysis, catalytic pyrolysis, or HTL – reactor type, and operating conditions. As bio-oil is made up of multifunctional compounds, the process conditions and catalysts for HDO vary depending on the bio-oil reactivity. Table 2.1, which lists the attributes of the bio-crudes obtained from HTL and pyrolysis, shows that HTL is superior to fast pyrolysis in most regards, except for sulfur content [30]. Indeed, HTL-derived bio-oil has a lower O (oxygen) and moisture content, giving it a higher heating value than pyrolytic oils. Bio-oils have low pH values of 2–3 as a result of their high amounts of carboxylic acids, such as acetic and formic acids, and this strong acidity makes the oils extremely unstable [31]. The condensation reactions that accelerate bio-oil aging and cause its properties to decline are promoted by the acidity of bio-oil, which also causes bio-oil immiscibility with petro-fuels [5]. Strong acidity also makes bio-oil very corrosive and extremely unstable at elevated temperatures. Therefore, stringent requirements are needed in the construction of the vessels used for both storage and the upgrading processes applied before bio-oil can be used as transportation fuels [31].

This chapter presents recent trends in the HDO of bio-oil with an emphasis on catalyst design issues for HDO such as structure–activity relationships, selectivity, stability, and catalyst deactivation.

Table 2.1 Comparison of bio-crude obtained from the HTL and the pyrolysis of biomass.

Attributes	HTL	Pyrolysis
C (wt%)	73	58
H (wt%)	8	6
O (wt%)	16	36
S (ppm)	<45	29
High heating value (HHV) (MJ/kg)	35.7	22.6
Moisture	5.1	24.8

Source: Gollakota et al. 2018 [30]. Reproduced with permission of Elsevier.

2.2 Hydrodeoxygenation (HDO)

HDO is a process in which carbon–oxygen bonds are cleaved in the presence of hydrogen and a catalyst resulting in the removal of oxygen atoms from oxygen-containing compounds. This process is very similar to HDS, which removes sulfur atoms from petroleum crude oil to form stable petro-fuels. The efficiency with which oxygen is removed and upgraded oil is stabilized depends on the catalyst type, hydrogen pressure, temperature, and bio-oil quality.

HDO is divided into two broad categories: low temperature HDO (mild or partial HDO) and high temperature HDO (deep or complete HDO). In mild HDO, extremely unstable compounds (acidic oxygen compounds) are stabilized, and dissolved water is separated to enhance the energy density. The temperature is maintained between 175 and 250 °C and the pressure is typically >100 bar of hydrogen in the presence of a catalyst [5]. In contrast, high temperature HDO involves the complete conversion of oxygen-containing compounds into both high molecular-weight alkanes and aromatic hydrocarbons. The conversion occurs at a temperature of 350–400 °C and a pressure of >200 bar of hydrogen in the presence of a catalyst [5]. In this chapter, major types of oxygenated

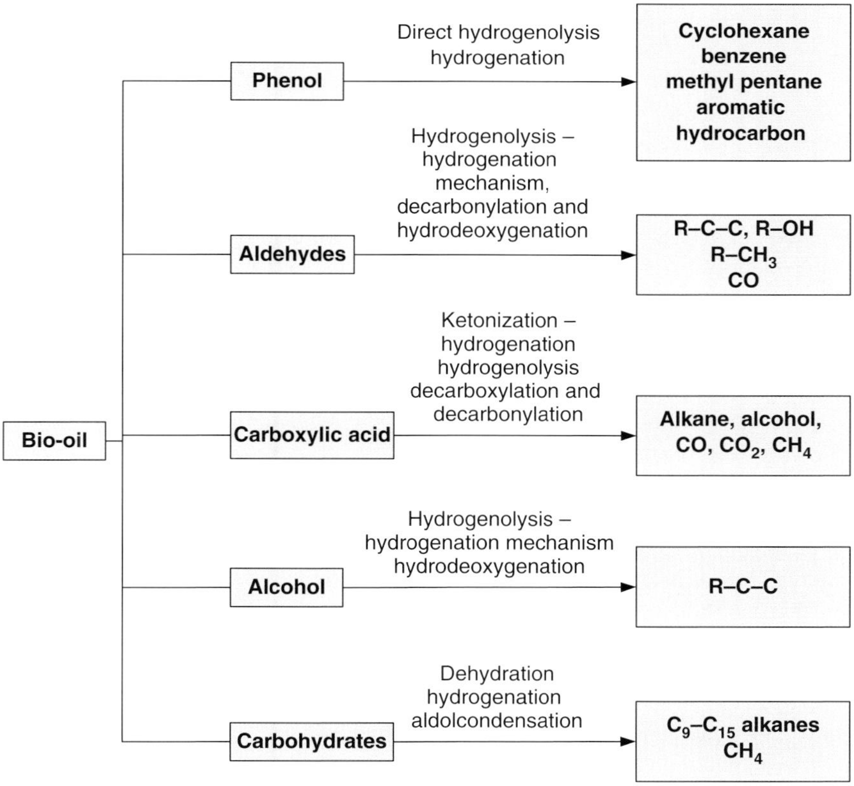

Figure 2.1 Hydrodeoxygenation pathways for major oxygenate compounds in bio-oil.

compounds are considered including phenol, aldehyde, carboxylic acid, alcohol, and carbohydrate model compounds (see Figure 2.1).

2.2.1 Hydrodeoxygenation of Phenol as a Model Compound

In general, different phenol-derived compounds present in bio-oil range between 10 and 18 wt%; however, because of phenol's standard structure, it is considered a key compound in bio-oil and is the model compound most often used during HDO studies. Therefore, it is essential to understand the detailed HDO reaction mechanism of phenol. Theoretically, phenol degradation occurs through three routes:

(1) Direct HDO to sequentially form benzene, cyclohexane, and methyl pentane in the presence of hydrogen.
(2) A combination of hydrogenation and hydrogenolysis of phenol to form an intermediate (cyclohexanol), which further undergoes dehydration to form cyclohexane.
(3) A combination of direct HDO and hydrogenation of phenol to form cyclohexanone followed by dehydration and hydrogenation reactions to form methyl pentane.

The reaction mechanism is largely driven by the catalyst employed. The performance of different types of catalysts, from commercial to noble, has been investigated for the upgrading of phenol-derived compounds [5, 32–38]. On commercial catalysts (Ni–Mo or Co–Mo on alumina), phenol is upgraded through two independent reactions: direct deoxygenation of phenol to aromatics, which is the primary product, and deoxygenation through aromatic hydrogenation to form naphtha compounds [33–35]. On the surface of Ni–W catalysts supported on active carbon, phenol conversion follows two routes: the first is the hydrogenolysis reaction to form benzenes by breaking of the C—O bond and the second route is a combination of hydrogenation and hydrogenolysis to form cyclohexanol, which is converted to cyclohexane and methyl cyclopentane via further hydrogenation [32]. However, noble metal catalysts such as Pd supported on carbon follow a different pathway than typical commercial catalysts [38]. The first step in this route is a metal-catalyzed aromatic ring hydrogenation, which is followed by naphthenic alcohol dehydration and metal-catalyzed cycloalkene hydrogenation [37, 38].

While a number of studies have considered phenol compounds in general as the model compounds used to investigate the HDO mechanism, others have focused on phenol-derived compounds – such as guaiacol, anisole, and cresole – to probe the HDO mechanism and have assessed specific catalysts in the HDO of each model compound.

2.2.1.1 HDO of Phenolic (Guaiacol) Model Compounds

Guaiacol, with its one methoxy functional group (–OCH_3) and one phenolic functional group (–OH), is considered a representative model compound for bio-oil originating from high lignin content biomass [29, 39, 40]. During the HDO of guaiacol (shown in Figure 2.2), the oxygen atom of the phenolic

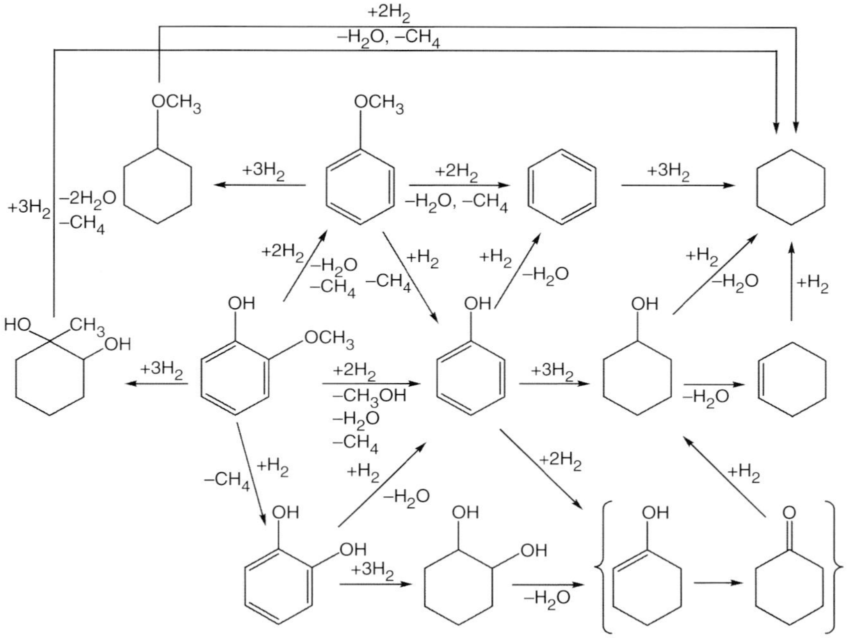

Figure 2.2 Guaiacol HDO conversion pathways. Source: Bykova et al. 2012 [1]. Reproduced with permission of Elsevier.

functional group is removed through two different reaction routes: (i) the hydrogenation of the aromatic ring followed by the elimination of the –OH group and (ii) the cleavage of the bond between the oxygen atom of the phenol group and the carbon atom of the aromatic ring [1, 29, 41]. The methoxy functional group is also eliminated through two reactions: (i) demethoxylation, which results from the rupture of the bond between the oxygen atom and the carbon in the aromatic ring to form phenol and methanol as byproducts, and (ii) demethylation, which comprises the cleavage of C—O bond of the $-OCH_3$ group to form catechol and methane as products [1, 41].

The effect of the catalyst on reaction selectivity in the HDO of guaiacol over noble metal catalysts – including Pt, Rh, Ru, and Pd supported on $SiO_2-Al_2O_3$, Al_2O_3, and nitric acid-treated carbon black – was studied by Lee et al. [42]. The reaction was conducted in a batch reactor at 40 bar of hydrogen pressure and 250 °C, and the acid-site-measurement-dependent catalyst results were used to elucidate the catalytic roles of acidic supports and metal nanoparticles. The results show that acidic supports were indispensable to the deoxygenation of oxygenates and that the metals were responsible for the hydrogenation of aromatic rings. The highest cyclohexane yield was obtained over $Ru/SiO_2-Al_2O_3$ and $Rh/SiO_2-Al_2O_3$ among the various combinations of metals and supports [42].

Lee et al. [39] used platinum-loaded HY zeolite (Pt/HY) catalysts with different Si/Al molar ratios for the HDO upgrading of guaiacol to various hydrocarbons. The yield of the main product, cyclohexane, increased with decreasing Si/Al molar ratio and results suggested that the ratio of the platinum metal particles to Brønsted acid sites must be optimized to realize maximum production of alkylated cyclic compounds and cyclohexane with high octane number.

Bykova et al. [1] carried out the HDO of guaiacol using a series of Ni-based catalysts with different stabilizing components in an autoclave at 17 MPa H_2 and 320 °C. Their results show that Ni-based catalysts prepared by a sol–gel method and stabilized using ZrO_2 and SiO_2 are the most active catalysts in the HDO of guaiacol and the main products formed are cyclohexanone, 1-methylcyclohexane-1,2-diol, and cyclohexane. These authors also investigated the effect of temperature on catalyst activity and product distribution in the HDO process and found that the degree of hydrodeoxygenation increases with temperature, while guaiacol conversion decreases due to catalyst coking at elevated temperatures [1]. The authors used a two-stage bio-oil HDO approach to prevent intense coking of the catalysts. The first stage involved the hydrogenation and partial deoxygenation of unsaturated oxygen-containing organics in the low temperature range of 175–250 °C, while the second stage comprised the hydrotreatment of the bio-oil at elevated temperatures in the range of 350–400 °C.

2.2.1.2 HDO of Phenolic (Anisole) Model Compounds

Anisole consists of an aromatic ring with a single methoxy group ($-OCH_3$) and its hydrodeoxygenation follows two reaction pathways [29, 43], as shown in Figure 2.3:

1. Transalkylation to toluene, cresols, and xylenols and demethylation of anisole to phenol.
2. Ring hydrogenation and hydrogenation of phenol to cyclohexane and benzene.

Of all the phenol-containing model compounds in bio-oil, anisole has been studied the least even though it has a similar structure to the main products of lignin depolymerization during the fast pyrolysis of wood [43]. The direct cleavage of the methoxy group in anisole is weak because the $C_{aromatic}$—O bond is stronger than the C_{methyl}—O bond [43]. The hydrogenolysis and hydrocracking of the $C_{aromatic}$–O–C_{methyl} in the anisole model compound required a bifunctional catalyst consisting of a metal and an acid [44].

Shi et al. [46] investigated the adsorption and reaction of anisole on Pt and Pt/Zn catalysts using both high surface area and model single crystal-supported metal catalysts. The authors demonstrated that the bonding configuration of anisole on Pt facilitates hydrogenation of the aromatic ring, while the bonding configuration of anisole on Pt/Zn limits this hydrogenation. However, the Zn-modified Pt exhibits high selectivity for C—O bond scission in anisole. The authors concluded that the Pt/Zn catalyst is the most effective catalyst for the hydrodeoxygenation of lignin-derived aromatic oxygenates, especially since this formulation displays low activity for ring hydrogenation.

2.2.1.3 HDO of Phenolic (Cresol) Model Compounds

The HDO mechanism of cresol depends on the position of the methyl group, the three isomers of the cresol molecule being *p*-cresol, *m*-cresol, and *o*-cresol. The two main reaction pathways in the HDO of cresol are hydrogenolysis of the bond

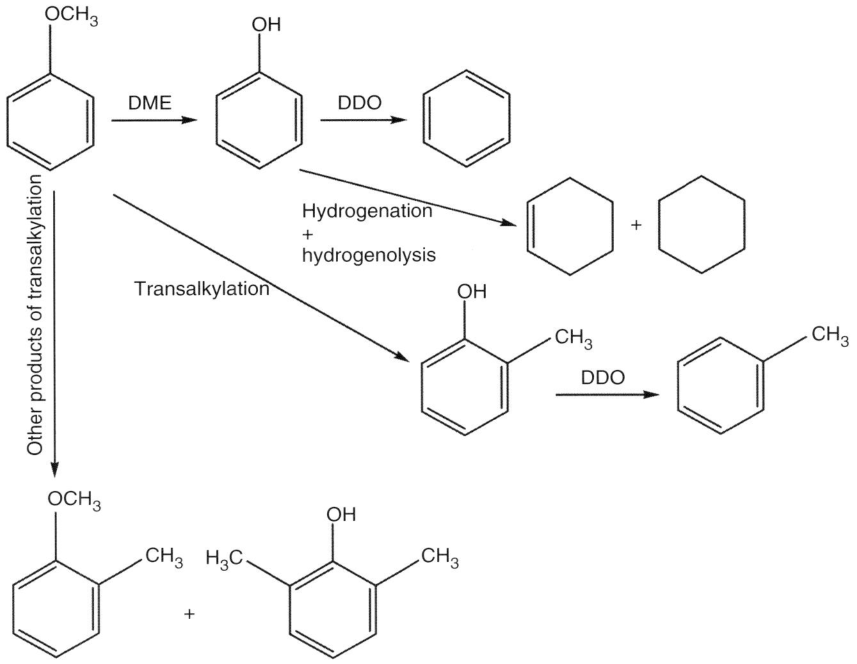

Figure 2.3 Anisole HDO conversion pathways. DME, Demethylation; DDO, Direct deoxygenation. Source: Viljava et al. 2000 [45]. Reproduced with permission of Elsevier.

between the aromatic ring and the –OH group and the rapid dehydration of cresol followed by ring hydrogenation, which produce toluene and methylcyclohexane as the two main products, respectively [35]. These two products have high octane numbers due to the presence of the methyl group, which favors the production of aromatics in the HDO process relative to the phenol molecule [47, 48]. The HDO conversion pathway for *p*-cresol is presented in Figure 2.4.

Foster et al. [49] investigated the HDO of *m*-cresol using Pt catalysts supported on γ-Al_2O_3 and SiO_2 at 0.5 atm H_2 and 260 °C to produce methylcyclohexane and toluene. They found that the HDO reaction involves the combination of acid-catalyzed dehydration and Pt-catalyzed hydrogenation reactions, the dominant reaction pathway to toluene involving the formation of a partially hydrogenated oxygenated intermediate. Increasing the hydrogen pressure increases the selectivity to methylcyclohexane, and modifying the support material had a dramatic impact on the rate of *m*-cresol conversion.

2.2.2 Hydrodeoxygenation of Aldehyde Model Compounds

The HDO of aldehydes involves four major routes [50–53]: (i) direct hydrogenolysis to form alkanes, (ii) hydrogenation followed by hydrogenolysis to yield alkanes, (iii) direct hydrogenation followed by dehydration and additional hydrogenation, and (iv) decarbonylation to afford CO and alkanes.

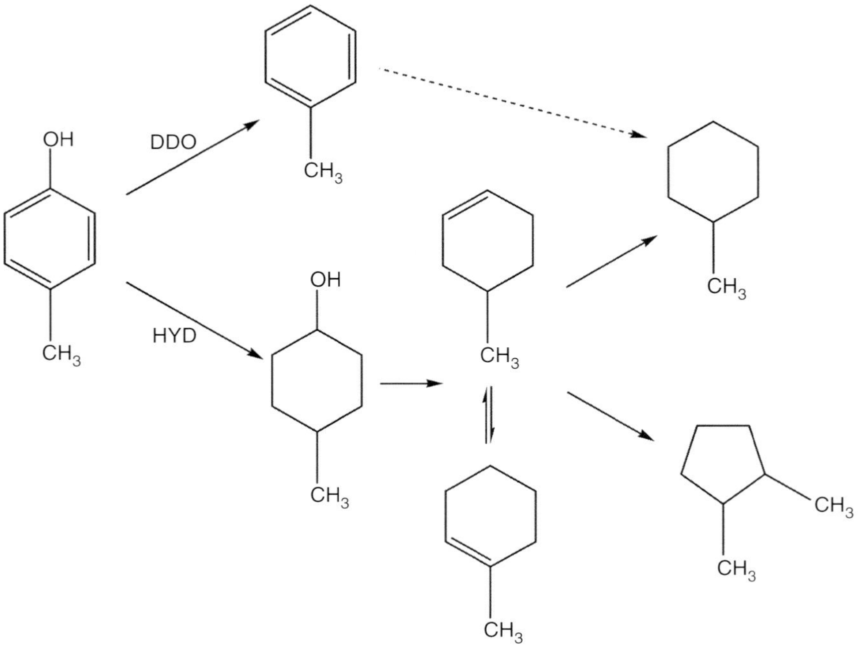

Figure 2.4 *p*-Cresol HDO conversion pathways. HYD, Hydrogenation; DDO, Direct deoxygenation. Source: Laurent and Delmon 1993 [35]. Reproduced with permission of American Chemical Society.

The hydrodeoxygenation of benzaldehyde was studied by Procházková et al. [54] over Pd/ZSM-5 in an autoclave using a temperature range of 30–130 °C and pressure ranging from 1 to 6 MPa. The reaction scheme for the HDO of benzaldehyde is shown in Figure 2.5. Benzaldehyde (A) forms toluene (C) through direct hydrogenolysis – with benzyl alcohol (B) as intermediate – or via the direct cleavage of the carbonyl bond (C—O) in a hexane environment. Toluene can also be produced from HDO of benzaldehyde in a methanol environment through several intermediates including dimethylacetal benzaldehyde (D), benzyl alcohol (B), and benzylmethyl ether (E) [54].

Saadi et al. [52] investigated copper-based catalysts supported on Al_2O_3, SiO_2, TiO_2, CeO_2, and ZrO_2 for the hydrogenation of benzaldehyde at atmospheric pressure and 100–350 °C. These authors found that the production of benzyl alcohol, toluene, and benzene depend on the nature of the support and the reaction temperature. Hydrogenolysis and hydrogenation are the dominating reactions in the upgrading of aldehydes to alkanes.

Zhang et al. [55] studied the HDO mechanism of acetaldehyde in a high pressure stainless steel fixed bed reactor over bimetallic Ni–Mo carbide supported on SiO_2. The authors concluded that the HDO of acetaldehyde can proceed via a number of reaction mechanisms including the following:

(1) Direct hydrogenation to form alcohol.
(2) Decarbonylation to form methane.

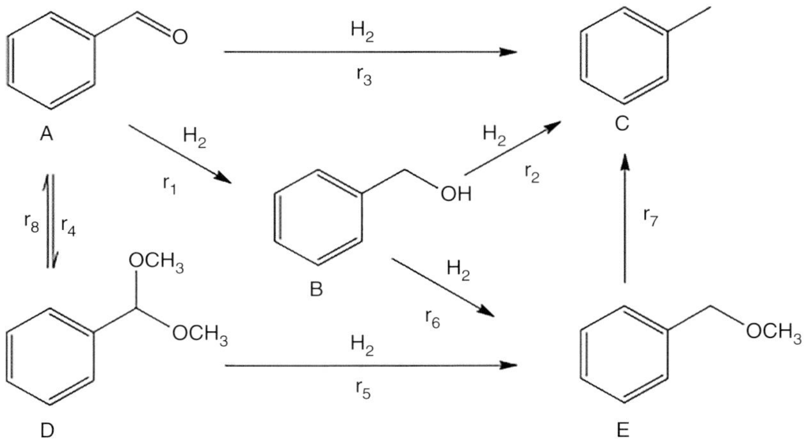

Figure 2.5 Hydrodeoxygenation of benzaldehyde. Source: Procházková et al. 2007 [54]. Reproduced with permission of Elsevier.

(3) Self-aldol condensation of acetaldehyde (side reaction) on the acid sites to form aldol alcohol, which undergoes dehydration and hydrogenation to yield butyl aldehydes.

Saliently, Snell et al. [56] investigated in detail the aldol condensation reaction of acetaldehyde, which takes place during the HDO reaction over a bifunctional aluminophosphate catalyst. The products of this condensation are as follows:

(1) Crotonaldehyde through self-aldol condensation of acetaldehyde with acetaldehyde.
(2) 3-Penten-2-one through cross-aldol condensation of acetaldehyde and acetone.
(3) 3-Methyl-3-penten-one and 4-hexen-3-one through cross-condensation of acetaldehyde and methyl ethyl ketone.

2.2.3 Hydrodeoxygenation of Carboxylic Acid Model Compounds

Carboxylic acids (acetic acid, formic acid, benzoic acid, etc.) present in bio-oil are the most unstable compounds and are responsible for the high total acid number (TAN) number and low pH value of bio-oil. The HDO of carboxylic acid takes place in three major steps, as several authors have described [57–60]:

(1) The dissociation of the carboxylic acid to carboxylate on the catalyst surface, followed by hydrogenation to alcohol.
(2) The decomposition of the C—C bond (through decarbonylation and decarboxylation) to form CO, CO_2, and alkanes with one less carbon atom (with further hydrogenation of CO and CO_2 to CH_4).
(3) The conversion of the alcohol formed in step 1 to an alkane through the dehydration–hydrogenation mechanism.

Serrano-Ruiz and Dumesic [59] studied the HDO of lactic acid over Pt/Nb_2O_5 at 573 K and 57 bar. The authors proposed that two major intermediates are formed: primary intermediates – such as propanoic acid (through the removal of the hydroxyl group following the cleavage of the C—O bond) and acetaldehyde along with CO and CO_2 (through decarbonylation and decarboxylation) – and the secondary product, ethanol (through direct hydrogenolysis). These intermediates are further upgraded to alkanes, ketones, and carbonyls through dehydration–hydrogenation, ketonization, and aldol condensation, respectively. Chen et al. [61] investigated the kinetics of aqueous phase HDO of a mixture of lactic acid and propionic acid in a three-phase stirred batch reactor at 343–423 K and 3.4–10.3 MPa, ultimately developing a two-site Langmuir–Hinshelwood (L–H) kinetic model for the hydrogenation of the feed to produce alcohols.

Olcay et al. [58] studied the aqueous phase hydrogenation of acetic acid over different types of transition metal catalysts (Ru/C, Pt/C, Pd/C, Rh/C, Ir/Al_2O_3, Raney Ni, and Raney Cu) over a temperature range of 110–290 °C and a pressure of 5.17 MPa. They proposed that on Ru and Ni, acetate converts to acetyl and oxygen; on Rh, Pt, Ru, and Ir, acetic acid converts to acetyl and a hydroxyl group; and on Cu, acetic acid converts to ethane-1-ol-1-olate.

2.2.4 Hydrodeoxygenation of Alcohol Model Compounds

Alcohol HDO has three independent routes, as summarized by Peng et al. [62]:

1. Dehydrogenation of alcohol to ketones followed by dehydration.
2. Direct hydrogenolysis to alkanes.
3. Dehydration followed by hydrogenation.

Wawrzetz et al. [63] discussed the HDO of glycerol in a high pressure tubular reactor over bifunctional Pt/Al_2O_3 at 498 K and a pressure of 29 bar. In carrying out the HDO of glycerol, they found that the presence of a catalyst comprising acid–base and metal functions is required. The initiation steps for the HDO of glycerol are dehydrogenation and dehydration. The dehydrogenation of the hydroxyl group is followed by decarbonylation and subsequent water–gas shift or disproportionation to form a carboxylic acid or alcohol. In the second route, glycerol is dehydrated followed by hydrogenation and decarboxylation reactions to form alkanes. There was no evidence of hydrogenolysis of the C—C or C—O bond on the surface of the Pt catalyst. However, the Pt catalyst favors the hydrogenation reaction. Davda et al. [64] studied the aqueous phase reforming of ethylene glycol over Ni, Pd, Pt, Ru, Rh, and Ir supported on silica catalysts at temperatures of 483 and 498 K and a pressure of 22 bar. The authors also observed that for the upgrading of ethylene glycol, there are two pathways other than the reforming process:

(1) The combination of hydrogenation and dehydration to form alcohols.
(2) The dehydrogenation of ethylene glycol followed by rearrangement to form an aldehyde, which undergoes hydrogenolysis to form alkanes.

2.2.5 Hydrodeoxygenation of Carbohydrate Model Compounds

The carbohydrates present in bio-oil are upgraded to alkanes through acid-catalyzed dehydration followed by aldol condensation to form large organic compounds that are further processed through a combination of hydrogenation and dehydration [65, 66].

Wildschut et al. [67] studied the hydrogenation of three different carbohydrate fractions (D-glucose, D-cellobiose, and D-sorbitol) present in pyrolysis oil using a Ru/C catalyst at a temperature of 250 °C and a pressure of 100 bar. They proposed two pathways for the HDO reaction: (i) hydrogenation of carbohydrates followed by hydrogenolysis to form small polyols (glycols, propanediol) and (ii) an undesired thermal pathway to form 5-hydroxymethyl furfural/levulinic acid/lumen (minor pathway). Verma and Gehlawat [68] studied the kinetics of the hydrogenation of D-glucose to sorbitol in a high pressure, stirred Parr reactor using a Raney Ni catalyst. They found that the reaction between dissolved hydrogen and D-glucose was very slow and the reaction followed homogenous kinetics with a reaction order of 1 with respect to both hydrogen and D-glucose.

Yan et al. [69] investigated the one-step conversion of cellobiose to alcohols using four transition metal catalysts (Ru, Rh, Pt, and Pd nanocluster) and a Parr autoclave at 120 °C and 4 MPa H_2. Of all the catalysts, Ru clusters could selectively cleave the C—O bond to accomplish one-step hydrogenation to form C_1-alcohols.

2.3 Chemical Catalysts for the HDO Reaction

The catalyst employed has an important role in HDO reactions. The reaction mechanisms discussed for different model compounds are completely dependent on the catalyst, reaction conditions (temperature, pressure, space velocity), and feedstock composition. A catalyst basically consists of two components: active phase and support.

Table 2.2 summarizes the catalyst, reactor type and dimensions, and temperature and pressure, as well as the degree of HDO achieved in the reactor during the upgrading of fast pyrolysis oil produced from different biomass feedstocks. HDO technology is similar to crude oil refinery upgrading via HDS, which removes sulfur as H_2S. To do so, sulfided Ni–Mo or Co–Mo supported on γ-alumina are used, these commercial catalysts also being used as the reference catalyst for the HDO reaction to remove oxygen in the form of water. Besides commercial catalysts, researchers have investigated both noble metal and base metal catalysts on various supports. Of the noble metal catalysts, Rh, Ru, and Pd are widely used. In the transition metal category, Ni, Co, W, and Cu are employed. The noble metal catalysts provide a higher oil yield than the transition metal catalysts. The reason might be due to the bonding between the reactant and the active sites during the reaction phase, which enhances reactant conversion to the desired product. Supports also play an important role in the HDO reaction. Most contain oxygen, which reduces catalyst deactivation, provide sites for the active metal to settle on, enhance the reactivity of the reaction, etc. Alumina, silica, carbon, titania, zirconia, and ceria are generally used as catalyst supports in this reaction. Support selection is determined by the desired end product and the nature of the reactant. Ahmadi et al. [70] compared transition metal catalysts on a number of supports (Co–Mo/SBA-15, Co–Mo/MCM–41, Co–Mo/γ-Al_2O_3, Co–Mo/HZSM-5,

Table 2.2 Catalysts, operating conditions, and HDO yields for HDO processing of bio-oil.

Feedstock	Reactor type	Reactor dimensions	Catalyst used	Temperature (°C)	Pressure (bar)	HDO (wt%)	References
Pine sawdust	Autoclave reactor	500 ml, Parr 4575	Co/SiO_2, $Fe–Co/SiO_2$, Fe/SiO_2, Co/HZSM-5	300	3.45 MPa	15–35	[72]
Loblolly pine	Stainless steel, high-pressure autoclave reactor	1.8 l Parr batch	Nickel/silica–alumina catalyst	340	1000 psig	31.4–35.6	[73]
Pine wood chips	Fixed bed continuous reactor	ID 1″, height 30″	$CoMo/\gamma\text{-}Al_2O_3$	375–400	10.34 MPa	35	[74]
Raw pine sawdust	Autoclave reactor	500 ml, Parr 4575	$Ni–Zn/Al_2O_3$	400	5000 psig	20–45	[75]
Raw pine sawdust	Autoclave reactor	500 ml, Parr 4575	Zinc powder with zero valence and 5.00 wt% Pd/C	200, 250, and 300	1–4.45 MPa	20–25	[76]
Raw pine sawdust	Autoclave reactor	500 ml, Parr 4575	Zinc metal with zero valency and zinc oxide	250, 300, 350, and 400	3.45 MPa	10–15	[77]
Raw pine sawdust	Stainless steel high pressure reactor	500 ml, Parr 4575	Zn–Pd/C	150, 200, and 250	1.38, 2.76, and 4.14 MPa	35–45	[78]
Dried cornstalk	High pressure stainless steel airtight pilot-scale autoclave	24 l	Bimetallic ammonium nickel molybdate	280, 310, 340, and 370	4 MPa	26.7–16.3	[79]
Pyrolysis oil	Stirred autoclave reactor (Parr)	Nominal internal volume of 500 ml	CoMo/SBA-15, CoMo/MCM-41, $CoMo/\gamma\text{-}Al_2O_3$, CoMo/HZSM-5, CoMo/C-Pellet, and CoMo/C-Powder	300 and 350	20.7 and 22.5 MPa,	45–65 (light oil and heavy oil)	[70]

Pyrolysis oil from lignocellulosic biomass	Batch autoclave	100 ml	Ru/C, Ru/TiO_2, Ru/Al_2O_3, Pt/C, and Pd/C Sulfided NiMo/Al_2O_3 and CoMo/Al_2O_3	250 and 350	100 and 200 bar	Up to 60	[80]
Miscanthus sinensis	Autoclave		Ru/C and Pt/C	250, 300, or 350	3 MPa	60–80	[81]
Yellow poplar	Stainless steel (SUS316) autoclave reactor	200 ml	Ni/C, Ni/SBA-15, and Ni/Al-SBA-15	300	3 MPa	55–75	[82]
Switchgrass, *Eucalyptus benthamii*, and equine manure feedstock	Parr bench-top reactor	100 ml	Pt, Ru, or Pd on carbon supports	320	2100 psi		[83]

Co–Mo/C-pellet, and Co–Mo/C-powder) with commercial Ru/C catalysts. Of all the catalysts, the noble metal catalyst Ru/C performed better than any other because of its high activity. However, Ru is very expensive and difficult to regenerate after the HDO reaction. Less coke formed on the surface of SBA-15 and MCM-41 due to their mesoporous nature, and the activity of fresh and regenerated MCM-41 was the same in the HDO reaction. The coke deposition rate depends on the support's acidity, which was highest for HZSM-5 followed by alumina, MCM-41, and SBA-15.

Venderbosch et al. [71] studied the HDO mechanism for the production of biofuel from pyrolysis oil in the presence of a Ru/C catalyst. In the first step, bio-oil is processed in the first stage hydrotreater to produce stabilized bio-oil by removing the unstable, polar, and viscous oxygen compounds through the hydrodeoxygenation reaction. The stabilized oil from the first hydrotreater contains high molecular weight compounds that need to be processed in a high temperature hydrotreater. In this second stage hydrotreater, complete hydrodeoxygenation takes place to convert high molecular weight stabilized compounds to low molecular weight compounds that can be hydrocracked to form gasoline and diesel range compounds. In general, bio-oil consists of different oxygenated compounds; therefore, most researchers study individual model compounds and their reaction mechanisms to understand their conversion in the hydrotreater.

2.3.1 Catalyst Promoters for HDO

In petroleum hydroprocessing technology, metal sulfide catalyst activity is modified through promoting MoS_2-supported Al_2O_3 with cobalt or nickel [84]. Chemical catalyst promoters affect the reaction selectivity through chemical properties such as the type of active site and physical properties such as particle size, pore volume, and catalyst surface area [29]. In short, catalyst promoters influence the performance of chemical catalysts during the HDO process largely through the structure of the catalyst and its active sites [29].

Jacobs et al. [85] investigated a number of supports – such as Al_2O_3, SiO_2, and ZrO_2 – and several promoters including metal cations and noble metals. The promoters are employed in hydrodeoxygenation, which allows easy access to the active sites available for desired reactions. Some noble metal promoters, including Pt and Ru, undergo reduction at lower temperatures compared with cobalt oxides [86, 87]. Also, textural promoters including certain catalyst supports or support modifiers enhance cluster dispersion, improve resistance due to attrition, and improve S tolerance [85]. Romero et al. [27] studied the influence of nickel and cobalt promoters on Mo/Al_2O_3 activity during the hydrodeoxygenation of 2-ethylphenol in a fixed-bed reactor at 7 MPa and 340 °C in the presence of hydrogen sulfide in order to attain sulfidation. In their study, support acidity generated oxygenated compounds and both Ni and Co improved the rate of deoxygenation; however, Ni was found to promote the hydrodeoxygenation pathway, while Co influenced both direct deoxygenation and hydrodeoxygenation, signifying that route selectivity was affected by the catalyst used. Yoosuk et al. [84] studied the activity of Ni–Mo sulfides and found them to be more active than Ni or Mo sulfides alone, with the maximum activity achieved at an atomic ratio of

0.3. Bui et al. [88] determined the influence of Co as a promoter for MoS_2 catalysts in the HDO of guaiacol and reported enhanced activity through improved direct deoxygenation, the major products being methyl-substituted compounds such as toluene, while methylcyclohexene was produced with unpromoted catalysts. Hence, the addition of promoters to sulfided catalysts has an effect on HDO, although their effect on non-sulfided catalysts is still under investigation [40].

2.3.2 Catalyst Supports for HDO

The material used as catalyst support has a great influence on catalyst performance in general and product selectivity in particular during the HDO process. Extensive studies have been performed using sulfide-supported Co–Mo and Ni–Mo catalysts by changing the catalyst's surface chemistry [89–93]. In this regard, modifiers such as potassium and platinum along with supports including silica as well as activated carbon have been used as a replacement for Al_2O_3 supports [90–92]. The results show that catalyst design has an impact on HDO chemistry. For instance, the incorporation of surface O on activated carbon improves the interaction of the support with the active sites, thereby improving activity and selectivity [94, 95].

Pinheiro et al. [95] and Bui et al. [96] investigated sulfided Co–Mo/Al_2O_3 with the aim of coprocessing straight-run gas oil with model bio-oil compounds at a hydrotreating facility and showed the feasibility of treating oxygenates and other hydrocarbon molecules during HDO and hydrodesulfurization. Zakzeski et al. [97] reviewed and reported various studies showing that sulfided Co–Mo and Ni–Mo with Al_2O_3 supports are active at temperatures ranging from 500 to 700 K at ~3–7 MPa, the main reactions involving six-membered ring structures such as phenol and benzene. A few studies have used expensive noble metal-based catalysts, which have considerable advantages over sulfided catalysts such as higher reactivity and energy efficiency during HDO, low S stripping, and flexibility in catalyst design [98, 99]. Lin et al. [100] studied the catalytic HDO of guaiacol on a supported Rh-based catalyst and found it to have excellent HDO activity, irrespective of its ability to saturate benzene rings during hydrodeoxygenation. Moreover, other supported catalysts such as sulfided Co–Mo and Ni–Mo led to aromatics but also formed coke [100]. When sulfided catalysts are used instead of supported noble metals, the hydrogenation activity is considerably improved [40]. For instance, Şenol et al. [101] reported high hydrogenation activity using a Ni–Mo catalyst, which increased the yield of cyclohexane from phenol.

2.3.3 Catalyst Selectivity for HDO

Catalyst selectivity, which depends on the active phase, support, and promoter type, greatly influences the hydrodeoxygenation mechanism of bio-oil. Hence, an ideal HDO catalyst should display higher selectivity toward C—O cleavage, instead of C—C scission, and must show high stability under HDO conditions [102]. Although most existing HDO catalysts result in C—C cleavage, Mo_2C is found to have relatively higher selectivity toward C—O scission with almost negligible C—C cleavage, thereby producing many stable products including unsaturated compounds with C—C bonds [102]. Sitthisa and Resasco [103] studied furfural hydrodeoxygenation over metal catalysts such as Cu, Ni, and Pd and found higher selectivity using Cu/SiO_2 during furfural alcohol formation. These authors also reported an increase in selectivity to decarbonylation with rising temperature. In general, selective hydrodeoxygenation enhances furfural stability in the formation of gasoline derivatives [103]. Other studies have shown Cu catalysts to be highly selective in C—O hydrogenation during aldehyde formation, these catalysts favoring the reduction of the carbonyl group when supported on acidic SiO_2 [52, 104]. Supported Co–Mo catalysts showed a decrease in the selectivity of oxygenated molecules with an increase in temperature [105]. In bimetallic Ni–Mo catalysts, the selectivity to hydrocarbons was higher than that for Mo/Al_2O_3 and Ni/Al_2O_3 catalysts, which could be attributed to a Ni–Mo synergy in such bimetallic catalysts [106]. In addition, the Ni/Al_2O_3 catalyst formed products through a decarboxylation pathway, while the Ni–Mo/Al_2O_3 catalyst resulted in products formed via hydrogenation. Another study involving the HDO of phenol on a Ni–Mo amorphous catalyst also showed a synergistic effect from the active sites of Ni and Mo [107]. However, a zeolite-supported Pt catalyst functions as a bifunctional catalyst, wherein molecules formed via hydrogenation led to enhanced activity and selectivity toward monocyclic and bicyclic compounds, which is relevant for phenolics conversion during the upgrading of bio-oils in the hydrodeoxygenation process [108].

2.3.4 Catalyst Deactivation During HDO

A detailed understanding of the deactivation mechanism for chemical catalysts used in the HDO process is essential to design and develop a more effective system and to guarantee a high yield of desired products. The main reasons for catalyst deactivation are catalyst poisoning due to the presence of sulfur in the bio-oil, coking/fouling through carbon deposition, and thermal degradation of supports and active sites due to the high temperature and pressure of the reaction [5, 109]. Shafaghat et al. [29] recently reviewed the deactivation of HDO catalysts. Simply put, catalyst deactivation refers to the loss of catalytic activity in the hydrodeoxygenation process. Such deactivation occurs as a result of catalyst poisoning from the chemisorption of compounds such as S- and/or N-containing molecules onto active sites, fouling from coke deposition of the catalyst surface, thermal disintegration of the catalyst support, vapor formation, undesirable reactions taking place on supports and/or promoters, and pore loss through attrition [109, 110]. However, the most common form of catalyst deactivation is fouling, which results from coke deposition [5]. Catalysts may facilitate the Boudouard reaction, wherein CO is dissociated into carbon on the surface of catalysts. Coke also forms as a result of the decomposition of tungstic acid on the surface active sites, which blocks pores [32]. Coke deposition is a function of the biomass and the reaction conditions. Centeno et al. [90] used sulfided Co–Mo supported on γ-alumina and reported the formation of coke due to feed interaction with the catalyst support. This phenomenon is acidity related through the interaction of the Lewis sites and the donation of protons by the Brønsted sites. Other supports such as activated C and ZrO_2 are comparatively inert, resulting in less coking [111]. Although Ni acts as a promoter to enhance Mo activity, it has the tendency to produce coke at high temperatures – as illustrated by Ardiyanti et al. [111] – and increases in viscosity due to the release of high molecular weight hydrocarbons. For noble metal catalysts, coking is not a major issue, although it leads to sintering. In the case of oil derived from microalgae, S- and N-containing compounds act as impurities. The chemisorption of S compounds is known to be irreversible; hence, S molecules cause the deactivation of HDO catalysts [5, 112].

The composition of lignocellulosic biomass, reflected in bio-oil, also affects catalyst deactivation. For instance, as bio-oil is known to have ~40–50 wt% O-containing molecules, the compounds produced therefrom are unstable and rapidly polymerize to coke [13]. In general, avoiding coke formation requires careful consideration of the catalyst, support, and reaction conditions.

2.4 Research Gaps

The catalytic HDO of biomass-derived bio-oil offers considerable challenges during the production of transportation fuels. There is considerable research on the implementation of continuous systems at the industrial scale informed by techno-economic analyses. However, there are gaps with respect to the use of appropriate catalyst systems (including supports and promoters) to obtain end products with enhanced quality and quantity that meet fuel requirements and standards compared with petroleum derivatives. To understand the use of promoters and the characterization of the intrinsic activity of catalysts, characterization studies in terms of adsorption and infrared (IR) spectroscopy are needed [27]. Advances in the understanding of HDO mechanisms and surface chemistry at an experimental and theoretical scale would help researchers in both industry and academia design catalysts to target desired products [113]. Moreover, supports and promoters play a crucial role in HDO mechanisms, which are affected by process conditions. When Co–Mo-based catalysts on several supports were used for the HDO reaction, it was observed that increase of the temperature lowered the selectivity with respect to oxygenated products, irrespective of the type of support employed [105]. In addition, ash content during a thermochemical process impacts bio-oil quality and quantity; ash that comes in contact with the upgrading catalyst leads to deactivation. Hence, there is a need to study the influence of HDO pathways with varying ash quantities [114]. Other contaminants such as solids, char, and excess water, as well as reactor materials, can cause deposits and corrosion, thereby negatively impacting the catalyst during the HDO process [115, 116].

Catalyst fouling has been observed during HDO and has an impact on the overall economics of the process [117]. The influence of high pressure thermal treatment on the lifetime of catalysts is still unclear, and the development of this area is the key to understanding catalyst stability and selectivity at optimal reaction conditions [114]. Most studies have been done with model compounds that rely on the structure and chemistry of a particular representative molecule; however, real feedstock poses challenges in terms of bulk effects due to complex molecular interactions. The latter are linked with the adsorption of particular molecules on the active sites of catalysts, leading to the inhibition of other molecules that show considerably weaker active site–support interactions [118–120]. Research with respect to catalyst structure includes the development of an isomerization catalyst function as well as process optimization [106]. Desirable catalysts are expected to have sufficient acidity to allow isomerization and minimize cracking, thereby obtaining high product yields. From an economic and environmental perspective, reducing unnecessary H_2 consumption during HDO is vital, as the H_2 consumption is sensitive to catalyst structure [106].

Most of the studies on chemical catalysts for hydrodeoxygenation have focused on catalyst screening tests without considering catalyst deactivation. Catalyst deactivation for the HDO process should receive more attention in order to fast-track the commercialization of bio-oil upgrading via catalytic hydrodeoxygenation. Technological gaps in the HDO of bio-oils pose additional challenges attributed to low bio-oil quality (i.e. high oxygen content, impurities, and coking), these challenges requiring catalyst systems with effective promoters and supports. The continued development of catalysts with significant economic and environmental perspectives would help improve HDO technology for renewable fuel production.

2.5 Conclusions

The hydrodeoxygenation reaction for bio-oil upgrading involves different reaction mechanisms and pathways to form desired products. The reaction mechanism for different model compounds is dependent on the chemical catalysts used during the upgrading process.

In this chapter, we discussed the upgrading of different model compounds for bio-oil and the impact of various catalysts on the HDO mechanism of these model compounds. The selectivity toward various chemicals was discussed, as well as the effect of promoters for different catalysts. The loss of catalyst activity in the HDO process, or catalyst deactivation, was also reviewed, and we conclude that this is an important research gap that needs to be filled for most catalysts.

Acknowledgments

The authors are grateful to the NSERC/Cenovus/Alberta Innovates Associate Industrial Research Chair Program in Energy and Environmental Systems Engineering and the Cenovus Energy Endowed Chair in Environmental Engineering at the University of Alberta for financial support for this research. As a part of the University of Alberta's Future Energy Systems (FES) research initiative, this research was made possible in part thanks to funding from the Canada First

Research Excellence Fund (CFREF). Astrid Blodgett is thanked for editing this chapter.

References

1 Bykova, M., Ermakov, D.Y., Kaichev, V. et al. (2012). Ni-based sol–gel catalysts as promising systems for crude bio-oil upgrading: guaiacol hydrodeoxygenation study. *Appl. Catal., B* 113: 296–307.
2 Wang, W., Yang, Y., Luo, H. et al. (2011). Amorphous Co–Mo–B catalyst with high activity for the hydrodeoxygenation of bio-oil. *Catal. Commun.* 12 (6): 436–440.
3 Kumar, M., Oyedun, A.O., and Kumar, A. (2018). A review on the current status of various hydrothermal technologies on biomass feedstock. *Renewable Sustainable Energy Rev.* 81 (Part 2): 1742–1770.
4 Oyedun, A.O., Kumar, A., Oestreich, D. et al. (2018). The development of the production cost of oxymethylene ethers as diesel additives from biomass. *Biofuels, Bioprod. Biorefin.* 12 (4): 694–710.
5 Patel, M. and Kumar, A. (2016). Production of renewable diesel through the hydroprocessing of lignocellulosic biomass-derived bio-oil: a review. *Renewable Sustainable Energy Rev.* 58: 1293–1307.
6 Patel, M., Zhang, X., and Kumar, A. (2016). Techno-economic and life cycle assessment on lignocellulosic biomass thermochemical conversion technologies: a review. *Renewable Sustainable Energy Rev.* 53: 1486–1499.
7 Zhao, C., Kou, Y., Lemonidou, A.A. et al. (2009). Highly selective catalytic conversion of phenolic bio–oil to alkanes. *Angew. Chem. Int. Ed.* 121 (22): 4047–4050.
8 Kumar, M., Oyedun, A.O., and Kumar, A. (2017). Hydrothermal liquefaction of biomass for the production of diluents for bitumen transport. *Biofuels, Bioprod. Biorefin.* 11: 811–829.
9 Patel, M., Oyedun, A.O., Kumar, A., and Gupta, R. (2018). A techno-economic assessment of renewable diesel and gasoline production from aspen hardwood. *Waste Biomass Valorization*: 1–16.
10 Xiu, S. and Shahbazi, A. (2012). Bio-oil production and upgrading research: a review. *Renewable Sustainable Energy Rev.* 16 (7): 4406–4414.
11 Chheda, J.N., Huber, G.W., and Dumesic, J.A. (2007). Liquid–phase catalytic processing of biomass–derived oxygenated hydrocarbons to fuels and chemicals. *Angew. Chem. Int. Ed.* 46 (38): 7164–7183.
12 Huber, G.W., Iborra, S., and Corma, A. (2006). Synthesis of transportation fuels from biomass: chemistry, catalysts, and engineering. *Chem. Rev.* 106 (9): 4044–4098.
13 Furimsky, E. (2000). Catalytic hydrodeoxygenation. *Appl. Catal., A* 199 (2): 147–190.
14 Maggi, R. and Delmon, B. (1994). Characterization and upgrading of bio-oils produced by rapid thermal processing. *Biomass Bioenergy* 7 (1): 245–249.

15 Lee, H., Kim, Y.-M., Lee, I.-G. et al. (2016). Recent advances in the catalytic hydrodeoxygenation of bio-oil. *Korean J. Chem. Eng.* 33 (12): 3299–3315.
16 Chen, T., Wu, C., Liu, R. et al. (2011). Effect of hot vapor filtration on the characterization of bio-oil from rice husks with fast pyrolysis in a fluidized-bed reactor. *Bioresour. Technol.* 102 (10): 6178–6185.
17 Jiang, X. and Ellis, N. (2009). Upgrading bio-oil through emulsification with biodiesel: mixture production. *Energy Fuels* 24 (2): 1358–1364.
18 Jiang, X. and Ellis, N. (2010). Upgrading bio-oil through emulsification with biodiesel: thermal stability. *Energy Fuels* 24 (4): 2699–2706.
19 Chen, T., Wu, C., and Liu, R. (2011). Steam reforming of bio-oil from rice husks fast pyrolysis for hydrogen production. *Bioresour. Technol.* 102 (19): 9236–9240.
20 Garcia, L.A., French, R., Czernik, S., and Chornet, E. (2000). Catalytic steam reforming of bio-oils for the production of hydrogen: effects of catalyst composition. *Appl. Catal., A* 201 (2): 225–239.
21 Hew, K., Tamidi, A., Yusup, S. et al. (2010). Catalytic cracking of bio-oil to organic liquid product (OLP). *Bioresour. Technol.* 101 (22): 8855–8858.
22 Wildschut, J., Melian-Cabrera, I., and Heeres, H. (2010). Catalyst studies on the hydrotreatment of fast pyrolysis oil. *Appl. Catal., B* 99 (1–2): 298–306.
23 Peters, J.E., Carpenter, J.R., and Dayton, D.C. (2015). Anisole and guaiacol hydrodeoxygenation reaction pathways over selected catalysts. *Energy Fuels* 29 (2): 909–916.
24 Akhtar, J. and Amin, N.A.S. (2011). A review on process conditions for optimum bio-oil yield in hydrothermal liquefaction of biomass. *Renewable Sustainable Energy Rev.* 15 (3): 1615–1624.
25 Fisk, C.A., Morgan, T., Ji, Y. et al. (2009). Bio-oil upgrading over platinum catalysts using in situ generated hydrogen. *Appl. Catal., A* 358 (2): 150–156.
26 Joshi, N. and Lawal, A. (2012). Hydrodeoxygenation of pyrolysis oil in a microreactor. *Chem. Eng. Sci.* 74: 1–8.
27 Romero, Y., Richard, F., and Brunet, S. (2010). Hydrodeoxygenation of 2-ethylphenol as a model compound of bio-crude over sulfided Mo-based catalysts: promoting effect and reaction mechanism. *Appl. Catal., B* 98 (3–4): 213–223.
28 Romero, Y., Richard, F., Renème, Y., and Brunet, S. (2009). Hydrodeoxygenation of benzofuran and its oxygenated derivatives (2,3-dihydrobenzofuran and 2-ethylphenol) over NiMoP/Al_2O_3 catalyst. *Appl. Catal., A* 353 (1): 46–53.
29 Shafaghat, H., Rezaei, P.S., and Daud, W.M.A.W. (2015). Effective parameters on selective catalytic hydrodeoxygenation of phenolic compounds of pyrolysis bio-oil to high-value hydrocarbons. *RSC Adv.* 5 (126): 103999–104042.
30 Gollakota, A., Kishore, N., and Gu, S. (2018). A review on hydrothermal liquefaction of biomass. *Renewable Sustainable Energy Rev.* 81 (Part 1): 1378–1392.
31 Zhang, Q., Chang, J., Wang, T., and Xu, Y. (2007). Review of biomass pyrolysis oil properties and upgrading research. *Energy Convers. Manage.* 48 (1): 87–92.

32 Echeandia, S., Arias, P.L., Barrio, V.L. et al. (2010). Synergy effect in the HDO of phenol over Ni–W catalysts supported on active carbon: effect of tungsten precursors. *Appl. Catal., B* 101 (1–2): 1–12.

33 Gevert, B., Otterstedt, J., and Massoth, F. (1987). Kinetics of the HDO of methyl-substituted phenols. *Appl. Catal.* 31 (1): 119–131.

34 Yang, Y., Gilbert, A., and Xu, C. (2009). Hydrodeoxygenation of bio-crude in supercritical hexane with sulfided CoMo and CoMoP catalysts supported on MgO: a model compound study using phenol. *Appl. Catal., A* 360 (2): 242–249.

35 Laurent, E. and Delmon, B. (1993). Influence of oxygen-, nitrogen-, and sulfur-containing compounds on the hydrodeoxygenation of phenols over sulfided cobalt-molybdenum/.Gamma.-alumina and nickel-molybdenum/.Gamma.-alumina catalysts. *Ind. Eng. Chem. Res.* 32 (11): 2516–2524.

36 Feng, J., Hse, C.Y., Wang, K. et al. (2017). Directional liquefaction of biomass for phenolic compounds and in situ hydrodeoxygenation upgrading of phenolics using bifunctional catalysts. *Energy* 135: 1–13.

37 Wang, L., Zhang, J., Yi, X. et al. (2015). Mesoporous ZSM-5 zeolite-supported Ru nanoparticles as highly efficient catalysts for upgrading phenolic biomolecules. *ACS Catal.* 5 (5): 2727–2734.

38 Zhao, C., He, J., Lemonidou, A.A. et al. (2011). Aqueous-phase hydrodeoxygenation of bio-derived phenols to cycloalkanes. *J. Catal.* 280 (1): 8–16.

39 Lee, H., Kim, H., Yu, M.J. et al. (2016). Catalytic hydrodeoxygenation of bio-oil model compounds over Pt/HY catalyst. *Sci. Rep.* 6: 28765.

40 Saidi, M., Samimi, F., Karimipourfard, D. et al. (2014). Upgrading of lignin-derived bio-oils by catalytic hydrodeoxygenation. *Energy Environ. Sci.* 7 (1): 103–129.

41 Bykova, M., Bulavchenko, O., Ermakov, D.Y. et al. (2011). Guaiacol hydrodeoxygenation in the presence of Ni-containing catalysts. *Catal. Ind.* 3 (1): 15–22.

42 Lee, C.R., Yoon, J.S., Suh, Y.-W. et al. (2012). Catalytic roles of metals and supports on hydrodeoxygenation of lignin monomer guaiacol. *Catal. Commun.* 17: 54–58.

43 Loricera, C., Pawelec, B., Infantes-Molina, A. et al. (2011). Hydrogenolysis of anisole over mesoporous sulfided CoMoW/SBA-15 (16) catalysts. *Catal. Today* 172 (1): 103–110.

44 Huuska, M.K. (1986). Effect of catalyst composition on the hydrogenolysis of anisole. *Polyhedron* 5 (1): 233–236.

45 Viljava, T.-R., Komulainen, R., and Krause, A. (2000). Effect of H_2S on the stability of $CoMo/Al_2O_3$ catalysts during hydrodeoxygenation. *Catal. Today* 60 (1–2): 83–92.

46 Shi, D., Arroyo-Ramírez, L., and Vohs, J.M. (2016). The use of bimetallics to control the selectivity for the upgrading of lignin-derived oxygenates: reaction of anisole on Pt and PtZn catalysts. *J. Catal.* 340: 219–226.

47 Massoth, F., Politzer, P., Concha, M. et al. (2006). Catalytic hydrodeoxygenation of methyl-substituted phenols: correlations of kinetic parameters with molecular properties. *J. Phys. Chem. B* 110 (29): 14283–14291.

48 Wan, H., Chaudhari, R.V., and Subramaniam, B. (2012). Catalytic hydroprocessing of *p*-cresol: metal, solvent and mass-transfer effects. *Top. Catal.* 55 (3–4): 129–139.

49 Foster, A.J., Do, P.T., and Lobo, R.F. (2012). The synergy of the support acid function and the metal function in the catalytic hydrodeoxygenation of *m*-cresol. *Top. Catal.* 55 (3–4): 118–128.

50 Keane, M.A. (1997). Gas phase hydrogenation/hydrogenolysis of benzaldehyde and *o*-tolualdehyde over Ni/SiO_2. *J. Mol. Catal. A: Chem.* 118 (2): 261–269.

51 Vannice, M.A. and Poondi, D. (1997). The effect of metal-support interactions on the hydrogenation of benzaldehyde and benzyl alcohol. *J. Catal.* 169 (1): 166–175.

52 Saadi, A., Rassoul, Z., and Bettahar, M. (2000). Gas phase hydrogenation of benzaldehyde over supported copper catalysts. *J. Mol. Catal. A: Chem.* 164 (1–2): 205–216.

53 Haffad, D., Kameswari, U., Bettahar, M.M. et al. (1997). Reduction of benzaldehyde on metal oxides. *J. Catal.* 172 (1): 85–92.

54 Procházková, D., Zámostný, P., Bejblová, M. et al. (2007). Hydrodeoxygenation of aldehydes catalyzed by supported palladium catalysts. *Appl. Catal., A* 332 (1): 56–64.

55 Zhang, W., Zhang, Y., Zhao, L., and Wei, W. (2010). Catalytic activities of NiMo carbide supported on SiO_2 for the hydrodeoxygenation of ethyl benzoate, acetone, and acetaldehyde. *Energy Fuels* 24 (3): 2052–2059.

56 Snell, R.W., Combs, E., and Shanks, B.H. (2010). Aldol condensations using bio-oil model compounds: the role of acid–base bi-functionality. *Top. Catal.* 53 (15): 1248–1253.

57 Chen, L., Zhu, Y., Zheng, H. et al. (2011). Aqueous-phase hydrodeoxygenation of carboxylic acids to alcohols or alkanes over supported Ru catalysts. *J. Mol. Catal. A: Chem.* 351: 217–227.

58 Olcay, H., Xu, L., Xu, Y., and Huber, G.W. (2010). Aqueous-phase hydrogenation of acetic acid over transition metal catalysts. *ChemCatChem* 2 (11): 1420–1424.

59 Serrano-Ruiz, J.C. and Dumesic, J.A. (2009). Catalytic processing of lactic acid over Pt/Nb_2O_5. *ChemSusChem* 2 (6): 581–586.

60 Zhang, Z., Jackson, J.E., and Miller, D.J. (2008). Effect of biogenic fermentation impurities on lactic acid hydrogenation to propylene glycol. *Bioresour. Technol.* 99 (13): 5873–5880.

61 Chen, Y., Miller, D.J., and Jackson, J.E. (2007). Kinetics of aqueous-phase hydrogenation of organic acids and their mixtures over carbon supported ruthenium catalyst. *Ind. Eng. Chem. Res.* 46 (10): 3334–3340.

62 Peng, B., Zhao, C., Mejía-Centeno, I. et al. (2012). Comparison of kinetics and reaction pathways for hydrodeoxygenation of C_3 alcohols on Pt/Al_2O_3. *Catal. Today* 183 (1): 3–9.

63 Wawrzetz, A., Peng, B., Hrabar, A. et al. (2010). Towards understanding the bifunctional hydrodeoxygenation and aqueous phase reforming of glycerol. *J. Catal.* 269 (2): 411–420.

64 Davda, R.R., Shabaker, J.W., Huber, G.W. et al. (2003). Aqueous-phase reforming of ethylene glycol on silica-supported metal catalysts. *Appl. Catal., B* 43 (1): 13–26.

65 Huber, G.W., Chheda, J.N., Barrett, C.J., and Dumesic, J.A. (2005). Production of liquid alkanes by aqueous-phase processing of biomass-derived carbohydrates. *Science* 308 (5727): 1446–1450.

66 West, R.M., Tucker, M.H., Braden, D.J., and Dumesic, J.A. (2009). Production of alkanes from biomass derived carbohydrates on bi-functional catalysts employing niobium-based supports. *Catal. Commun.* 10 (13): 1743–1746.

67 Wildschut, J., Arentz, J., Rasrendra, C.B. et al. (2009). Catalytic hydrotreatment of fast pyrolysis oil: model studies on reaction pathways for the carbohydrate fraction. *Environ. Prog. Sustainable Energy* 28 (3): 450–460.

68 Verma, R. and Gehlawat, J.K. (1989). Kinetics of hydrogenation of D–glucose to sorbitol. *J. Chem. Technol. Biotechnol.* 46 (4): 295–301.

69 Yan, N., Zhao, C., Luo, C. et al. (2006). One-step conversion of cellobiose to C_6-alcohols using a ruthenium nanocluster catalyst. *J. Am. Chem. Soc.* 128 (27): 8714–8715.

70 Ahmadi, S., Yuan, Z., Rohani, S., and Xu, C. (2016). Effects of nano-structured CoMo catalysts on hydrodeoxygenation of fast pyrolysis oil in supercritical ethanol. *Catal. Today* 269: 182–194.

71 Venderbosch, R., Ardiyanti, A., Wildschut, J. et al. (2010). Stabilization of biomass–derived pyrolysis oils. *J. Chem. Technol. Biotechnol.* 85 (5): 674–686.

72 Cheng, S., Wei, L., Julson, J., and Rabnawaz, M. (2017). Upgrading pyrolysis bio-oil through hydrodeoxygenation (HDO) using non-sulfided Fe–Co/SiO_2 catalyst. *Energy Convers. Manage.* 150: 331–342.

73 Luo, Y., Guda, V.K., Hassan, E.B. et al. (2016). Hydrodeoxygenation of oxidized distilled bio-oil for the production of gasoline fuel type. *Energy Convers. Manage.* 112: 319–327.

74 Luo, Y., Guda, V.K., Steele, P.H., and Wan, H. (2016). Hydrodeoxygenation of oxidized and hydrotreated bio-oils to hydrocarbons in fixed-bed continuous reactor. *BioResources* 11 (2): 4415–4431.

75 Cheng, S., Wei, L., Julson, J. et al. (2017). Hydrocarbon bio-oil production from pyrolysis bio-oil using non-sulfide Ni–Zn/Al_2O_3 catalyst. *Fuel Process. Technol.* 162: 78–86.

76 Cheng, S., Wei, L., Alsowij, M.R. et al. (2018). In situ hydrodeoxygenation upgrading of pine sawdust bio-oil to hydrocarbon biofuel using Pd/C catalyst. *J. Energy Inst.* 91 (2): 163–171.

77 Cheng, S., Wei, L., Julson, J. et al. (2017). Hydrodeoxygenation upgrading of pine sawdust bio-oil using zinc metal with zero valency. *J. Taiwan Inst. Chem. Eng.* 74: 146–153.

78 Huang, Y., Wei, L., Zhao, X. et al. (2016). Upgrading pine sawdust pyrolysis oil to green biofuels by HDO over zinc-assisted Pd/C catalyst. *Energy Convers. Manage.* 115: 8–16.

79 Yang, T., Jie, Y., Li, B. et al. (2016). Catalytic hydrodeoxygenation of crude bio-oil over an unsupported bimetallic dispersed catalyst in supercritical ethanol. *Fuel Process. Technol.* 148: 19–27.

80 Wildschut, J., Mahfud, F.H., Venderbosch, R.H., and Heeres, H.J. (2009). Hydrotreatment of fast pyrolysis oil using heterogeneous noble-metal catalysts. *Ind. Eng. Chem. Res.* 48 (23): 10324–10334.
81 Oh, S., Hwang, H., Choi, H.S., and Choi, J.W. (2015). The effects of noble metal catalysts on the bio-oil quality during the hydrodeoxygenative upgrading process. *Fuel* 153: 535–543.
82 Oh, S., Choi, H.S., Choi, I.-G., and Choi, J.W. (2017). Evaluation of hydrodeoxygenation reactivity of pyrolysis bio-oil with various Ni-based catalysts for improvement of fuel properties. *RSC Adv.* 7 (25): 15116–15126.
83 Elkasabi, Y., Mullen, C.A., Pighinelli, A.L.M.T., and Boateng, A.A. (2014). Hydrodeoxygenation of fast-pyrolysis bio-oils from various feedstocks using carbon-supported catalysts. *Fuel Process. Technol.* 123: 11–18.
84 Yoosuk, B., Tumnantong, D., and Prasassarakich, P. (2012). Amorphous unsupported Ni–Mo sulfide prepared by one step hydrothermal method for phenol hydrodeoxygenation. *Fuel* 91 (1): 246–252.
85 Jacobs, G., Das, T.K., Zhang, Y. et al. (2002). Fischer–Tropsch synthesis: support, loading, and promoter effects on the reducibility of cobalt catalysts. *Appl. Catal., A* 233 (1): 263–281.
86 Iglesia, E., Soled, S.L., Fiato, R.A., and Via, G.H. (1993). Bimetallic synergy in cobalt ruthenium Fischer-Tropsch synthesis catalysts. *J. Catal.* 143 (2): 345–368.
87 Bruce, L.A., Hoang, M., Hughes, A.E., and Turney, T.W. (1993). Ruthenium promotion of Fischer–Tropsch synthesis over coprecipitated cobalt/ceria catalysts. *Appl. Catal., A* 100 (1): 51–67.
88 Bui, V.N., Laurenti, D., Afanasiev, P., and Geantet, C. (2011). Hydrodeoxygenation of guaiacol with CoMo catalysts. Part I: promoting effect of cobalt on HDO selectivity and activity. *Appl. Catal., B* 101 (3–4): 239–245.
89 Laurent, E. and Delmon, B. (1994). Study of the hydrodeoxygenation of carbonyl, carylic and guaiacyl groups over sulfided CoMo/γ-Al_2O_3 and NiMo/γ-Al_2O_3 catalysts: I. Catalytic reaction schemes. *Appl. Catal., A* 109 (1): 77–96.
90 Centeno, A., Laurent, E., and Delmon, B. (1995). Influence of the support of CoMo sulfide catalysts and of the addition of potassium and platinum on the catalytic performances for the hydrodeoxygenation of carbonyl, carboxyl, and guaiacol-type molecules. *J. Catal.* 154 (2): 288–298.
91 Ferrari, M., Maggi, R., Delmon, B., and Grange, P. (2001). Influences of the hydrogen sulfide partial pressure and of a nitrogen compound on the hydrodeoxygenation activity of a CoMo/carbon catalyst. *J. Catal.* 198 (1): 47–55.
92 Ferrari, M., Delmon, B., and Grange, P. (2002). Influence of the impregnation order of molybdenum and cobalt in carbon-supported catalysts for hydrodeoxygenation reactions. *Carbon* 40 (4): 497–511.
93 Ferrari, M., Lahousse, C., Centeno, A. et al. (1998). Influence of the impregnation order of molybdenum and cobalt in carbon supported catalysts for hydrodeoxygenation reactions. *Stud. Surf. Sci. Catal.* 118: 505–515.

94 De la Puente, G., Gil, A., Pis, J., and Grange, P. (1999). Effects of support surface chemistry in hydrodeoxygenation reactions over CoMo/activated carbon sulfided catalysts. *Langmuir* 15 (18): 5800–5806.
95 Pinheiro, A., Hudebine, D., Dupassieux, N., and Geantet, C. (2009). Impact of oxygenated compounds from lignocellulosic biomass pyrolysis oils on gas oil hydrotreatment. *Energy Fuels* 23 (2): 1007–1014.
96 Bui, V.N., Toussaint, G., Laurenti, D. et al. (2009). Co-processing of pyrolysis bio oils and gas oil for new generation of bio-fuels: hydrodeoxygenation of guaïacol and SRGO mixed feed. *Catal. Today* 143 (1–2): 172–178.
97 Zakzeski, J., Bruijnincx, P.C., Jongerius, A.L., and Weckhuysen, B.M. (2010). The catalytic valorization of lignin for the production of renewable chemicals. *Chem. Rev.* 110 (6): 3552–3599.
98 Elliott, D.C. and Hart, T.R. (2008). Catalytic hydroprocessing of chemical models for bio-oil. *Energy Fuels* 23 (2): 631–637.
99 Gutierrez, A., Kaila, R., Honkela, M. et al. (2009). Hydrodeoxygenation of guaiacol on noble metal catalysts. *Catal. Today* 147 (3–4): 239–246.
100 Lin, Y.-C., Li, C.-L., Wan, H.-P. et al. (2011). Catalytic hydrodeoxygenation of guaiacol on Rh-based and sulfided CoMo and NiMo catalysts. *Energy Fuels* 25 (3): 890–896.
101 Şenol, O., Ryymin, E.-M., Viljava, T.-R., and Krause, A. (2007). Effect of hydrogen sulphide on the hydrodeoxygenation of aromatic and aliphatic oxygenates on sulphided catalysts. *J. Mol. Catal. A: Chem.* 277 (1–2): 107–112.
102 Ren, H., Yu, W., Salciccioli, M. et al. (2013). Selective hydrodeoxygenation of biomass–derived oxygenates to unsaturated hydrocarbons using molybdenum carbide catalysts. *ChemSusChem* 6 (5): 798–801.
103 Sitthisa, S. and Resasco, D.E. (2011). Hydrodeoxygenation of furfural over supported metal catalysts: a comparative study of Cu, Pd and Ni. *Catal. Lett.* 141 (6): 784–791.
104 Chambers, A., Jackson, S.D., Stirling, D., and Webb, G. (1997). Selective hydrogenation of cinnamaldehyde over supported copper catalysts. *J. Catal.* 168 (2): 301–314.
105 Kubička, D., Šimáček, P., and Žilková, N. (2009). Transformation of vegetable oils into hydrocarbons over mesoporous-alumina-supported CoMo catalysts. *Top. Catal.* 52 (1–2): 161–168.
106 Choudhary, T.V. and Phillips, C.B. (2011). Renewable fuels via catalytic hydrodeoxygenation. *Appl. Catal., A* 397 (1): 1–12.
107 Wang, W.-Y., Yang, Y.-Q., Bao, J.-G., and Luo, H.-A. (2009). Characterization and catalytic properties of Ni–Mo–B amorphous catalysts for phenol hydrodeoxygenation. *Catal. Commun.* 11 (2): 100–105.
108 Hong, D.-Y., Miller, S.J., Agrawal, P.K., and Jones, C.W. (2010). Hydrodeoxygenation and coupling of aqueous phenolics over bifunctional zeolite-supported metal catalysts. *Chem. Commun.* 46 (7): 1038–1040.
109 Bartholomew, C.H. (2001). Mechanisms of catalyst deactivation. *Appl. Catal., A* 212 (1): 17–60.
110 Forzatti, P. and Lietti, L. (1999). Catalyst deactivation. *Catal. Today* 52 (2–3): 165–181.

111 Ardiyanti, A., Khromova, S., Venderbosch, R. et al. (2012). Catalytic hydrotreatment of fast pyrolysis oil using bimetallic Ni–Cu catalysts on various supports. *Appl. Catal., A* 449: 121–130.

112 Ardiyanti, A., Gutierrez, A., Honkela, M. et al. (2011). Hydrotreatment of wood-based pyrolysis oil using zirconia-supported mono-and bimetallic (Pt, Pd, Rh) catalysts. *Appl. Catal., A* 407 (1–2): 56–66.

113 Vlachos, D.G. and Caratzoulas, S. (2010). The roles of catalysis and reaction engineering in overcoming the energy and the environment crisis. *Chem. Eng. Sci.* 65 (1): 18–29.

114 Zacher, A.H., Olarte, M.V., Santosa, D.M. et al. (2014). A review and perspective of recent bio-oil hydrotreating research. *Green Chem.* 16 (2): 491–515.

115 Diebold, J.P. (1999). *A Review of the Chemical and Physical Mechanisms of the Storage Stability of Fast Pyrolysis Bio-Oils*. Golden, CO: National Renewable Energy Lab.

116 Elliott, D.C. (1994). Water, alkali and char in flash pyrolysis oils. *Biomass Bioenergy* 7 (1–6): 179–185.

117 Elliott, D.C., Lee, S.-J., and Hart, T.R. (2012). Stabilization of Fast Pyrolysis Oil: Post Processing. Final Report, Pacific Northwest National Laboratory (PNNL), Richland, WA, PNNL-21549.

118 Girgis, M.J. and Gates, B.C. (1991). Reactivities, reaction networks, and kinetics in high-pressure catalytic hydroprocessing. *Ind. Eng. Chem. Res.* 30 (9): 2021–2058.

119 Choudhary, T.V. (2007). Structure – reactivity – mechanistic considerations in heavy oil desulfurization. *Ind. Eng. Chem. Res.* 46 (25): 8363–8370.

120 Choudhary, T.V., Parrott, S., and Johnson, B. (2008). Understanding the hydrodenitrogenation chemistry of heavy oils. *Catal. Commun.* 9 (9): 1853–1857.

3

Upgrading of Bio-oil via Fluid Catalytic Cracking

Idoia Hita, Jose Maria Arandes, and Javier Bilbao

University of the Basque Country UPV/EHU, Faculty of Science and Technology, Department of Chemical Engineering, PO Box 644, 48080 Bilbao, Spain

3.1 Introduction

The petroleum industry is currently at a crossroads due to different reasons, like the uncertain situation of its traditional markets and the evolution of the exploitation of other conventional fossil resources (i.e. coal and natural gas). Furthermore, environmental protection policies are boosting the exploitation of renewable energy sources including biomass, solar energy, wind energy, and tidal energy. However, the transition toward economies alternative to that based on petroleum faces both inertia and resistance from economic interests affected by changes in the energy model. Furthermore, this transition requires the technological development of new processes whose implementation is not immediate. In many cases, this implementation is not even foreseeable in the mid- and long term, and therefore, a transition period is expected in which the petroleum industry will continue to play a key role in satisfying the global energy demand. In this transition period of competitive fossil and renewable energy markets, the petroleum industry must deal not only with an increasing demand for fuel and raw materials but also with increasingly strict policies concerning the quality of their products (such as automotive fuels with minimal amounts of sulfur, nitrogen, olefins, and aromatics).

As a response to these challenges, the petroleum industry is making strong investments in the reconditioning of existing – and the installation of new – units, to make them more versatile and efficient, both in the valorization of petroleum fractions and in meeting stricter quality standards. As part of this changing scenario, an interest for integrating new feeds (derived from alternative energy sources like biomass, or societal residues like tires or waste plastics) into traditional petroleum refineries has emerged. Due to their capacity and versatility, the refinery units presenting the greatest potential for the integration of new feeds are fluidized catalytic cràcking (FCC), hydrotreating (HDT) and coker units.

The development of novel routes and the adaptation of existing ones for the valorization of waste streams can be framed within the R&D platforms of the

Chemical Catalysts for Biomass Upgrading, First Edition. Edited by Mark Crocker and Eduardo Santillan-Jimenez.

biorefinery and the waste refinery, depending on the nature of the processed feed. The technological development of biorefineries, conceived as the set of routes for the valorization of biomass in general – and lignocellulosic biomass in particular – for the production of fuels and raw materials, is motivated by environmental policies and the need to reduce CO_2 emissions [1, 2]. Furthermore, in developing countries with high cropping potential, the valorization of biomass offers a great opportunity for coordinating and synergizing both energy and agricultural policies as well as developing homegrown technologies.

The main routes for the valorization of biomass at the industrial scale are (i) fermentation, (ii) gasification, and (iii) fast pyrolysis. The resulting products are bioethanol, synthesis gas, and bio-oil, respectively, which in turn present distinct valorization and upgrading options. The potential of bio-ethanol is well established as a fuel (mainly by mixing it with either gasoline or diesel) [3–5], as a raw material for the production of olefins and other hydrocarbons [6–8], and as a hydrogen vector [9–11]. However, the economic viability of the production of bioethanol from lignocellulosic biomass at the industrial scale requires additional advances on several fronts, such as (i) enzymatic engineering to efficiently depolymerize cellulose and hemicellulose into fermentable sugars [12], (ii) metabolic engineering to develop more active and selective fermentation microorganisms, and (iii) stage integration to minimize the energy requirements of the global process and increase its productivity [13–15]. The production of synthesis gas from biomass through gasification has also received much attention, mainly linked to the development of a rising number of gas-to-liquid (GTL) plants dedicated to the synthesis of methanol or dimethyl ether (DME) and to hydrocarbon production via Fischer–Tropsch [16, 17], in which the catalysts used play a pivotal role [18]. As mentioned earlier, bio-oil can be valorized through several alternative routes, which are explained in detail in Sections 3.2.2, 3.2.3 and 3.3 within this chapter. Among these, mostly due to its promising prospects for implementation at the industrial scale, coprocessing of bio-oil and petroleum feedstock in traditional refineries in general – and in FCC and HDT units in particular – represents one of the most interesting approaches [19] (Figure 3.1).

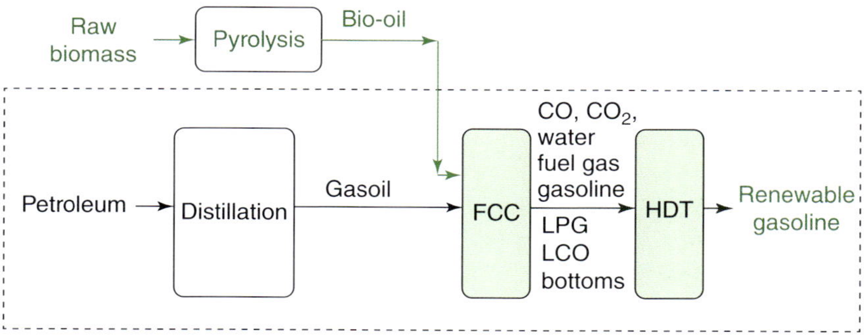

Figure 3.1 Simplified proposed scheme for coprocessing bio-oil in an existing refinery. Source: Pinho et al. 2014 [19]. Reprinted with permission of Walter de Gruyter GmbH.

3.2 Bio-oil

3.2.1 Bio-oil Production via Fast Pyrolysis

Fast pyrolysis of biomass offers several advantages in comparison to other biorefinery conversion routes (such as fermentation and gasification) not only because it has a limited environmental impact but also because it has great potential for the valorization of the resulting products – particularly the liquid fraction or bio-oil – at an industrial scale [20].

Basically, pyrolysis is a thermal decomposition process in which organic components are degraded in an oxygen-deprived environment. Conventional fast pyrolysis conditions include temperatures between 450 and 550 °C, heating rates ranging from 10^3 to 10^4 K/s, and residence times of less than one second for volatile compounds, along with fast cooling of the pyrolyzates exiting the reactor. The products formed comprise a liquid fraction known as bio-oil, a solid residue denominated char and a fraction of noncondensable gases. Typical yields of these three fractions fall between 60 and 75 wt% for bio-oil, 15–25 wt% for char, and 10–20 wt% for gas products. Notably, Carpenter et al. [2] have reviewed the different techniques available for the characterization of the main biomass feeds and products associated with pyrolysis processes.

The fundamental aspects of fast pyrolysis and the readiness level of different technologies have also been summarized in a number of recent reviews [21–24]. Briefly, pyrolysis reactors are mostly (i) bubbling fluidized beds [25–27]; (ii) fluidized circulating or dragging beds [28]; (iii) ablative reactors (either flat wall, rotating, or cyclonic) [29]; (iv) conical spouted bed reactors (CSBRs) [30, 31]; (v) screw reactors; or (vi) vacuum reactors [32, 33]. Saliently, while vacuum operation is known to reduce the energy required by the process and also favor bio-oil collection [34], the process can also be carried out in an autothermal regime without sacrificing bio-oil quality [35].

In short, bio-oil is envisioned as a liquid fuel intermediate of great promise due to its renewable nature, its low sulfur and nitrogen content, and its potential to be carbon neutral in terms of the CO_2 emissions stemming from its combustion. Against this backdrop, the main features of different valorization approaches (with an emphasis on FCC routes) will be discussed in detail in Sections 3.2.2, 3.2.3, and 3.3 that follow.

3.2.2 General Characteristics, Composition, and Stabilization of Bio-oil

Bio-oil is a complex mixture of water (15–30 wt%) and a plethora of organic oxygenates. A comparison of its properties with those of a conventional fuel oil (see Table 3.1) [2, 36–38], readily illustrates that bio-oil has a lower heating value, is immiscible with petroleum-derived fuels as well as being both chemically and thermally unstable (as a consequence of its high oxygen content), while also being highly viscous (40–100 cP at 50 °C) and corrosive ($2 < pH < 4$). Notably, Gollakota et al. [39] recently reported the properties of a wide range of pyrolysis oils derived from different biomass sources and obtained using various approaches and process conditions.

Table 3.1 Typical bio-oil and conventional fuel oil properties.

Property	Bio-oil	Conventional fuel oil
Water content (wt%)	15–30	0.1
pH	2.5	–
Density (kg/m^3)	1200	240
Elemental composition (wt%)		
C	54–58	85
H	5.5–7	11
O	35–40	1.0
N	0–0.2	0.3
Ash (wt%)	0–0.2	0.1
Higher heating value (MJ/kg)	16–19	42
Viscosity (cP at 50 °C)	40–100	180
Solids (wt%)	0.2–1	1
Distillation residue (wt%)	<50	1
Turbine emissions (g/MJ)		
NO_X	<0.7	1.4
SO_X	0	0.3

In short, bio-oil can be considered as a microemulsion in which a continuous phase – consisting of an aqueous solution of products evolved from the decomposition of holocellulose (the carbohydrate fraction of lignocellulose comprising both cellulose and hemicellulose) and lignin – stabilizes a discontinuous phase mainly consisting of macromolecules originating from lignin pyrolyzates, which tend to repolymerize and form pyrolytic lignin. The stabilization of this microemulsion occurs through hydrogen bonding and the formation of nano- and micromicelles [40]. As mentioned earlier, bio-oil contains a wide array of compounds that have their origin in the depolymerization and fragmentation of the three main components in biomass: cellulose, hemicellulose, and lignin [41]. These compounds can be classified into five main groups: (i) hydroxyaldehydes, (ii) hydroxyketones, (iii) sugars, (iv) carboxylic acids, and (v) phenolic components, albeit a more detailed classification would distinguish phenols, ketones, acids, esters, aldehydes, alcohols, furans, anhydrous-sugars, nitrogen-containing compounds, hydrocarbons, and carboxylic acids [42]. The average molecular weight of bio-oil is in the 370–1000 g/mol range, although bio-oil contains a fraction of oligomeric species (derived from lignin and cellulose) with molecular weights of up to 5000 g/mol that cannot be identified through GC–MS methodologies and that require advanced analysis techniques (i.e. HPLC/ES-MS, FTIR, GPC, or NMR) [2, 43, 44]. The amount of water that can be dissolved in bio-oil ranges between 22 and 31 wt%, the actual value depending on the characteristics of the starting biomass [45]. By adding extra water, the bio-oil separates into two different phases (aqueous and organic), allowing for the independent valorization of each fraction. Molecular distillation is known to facilitate the separation in several phases, affording an organic fraction with a high calorific value [46, 47].

The main bio-oil properties that must be taken into account in order for bio-oil to be used either as feedstock or fuel in refinery units are its oxygen, water, ash, and metal content, as well as its viscosity, pH, volatility, and corrosiveness [39]. Bio-oil pretreatments can be carried out either in the pyrolysis process itself (to improve bio-oil composition via dehydration-cracking reactions) or through bio-oil stabilization approaches.

3.2.2.1 Adjustment of Bio-oil Composition Through Pyrolytic Strategies

The oxygen content in bio-oil can be reduced through dehydration-cracking reactions either by using an acid catalyst (typically a zeolite) in the pyrolysis reactor (*in situ*) [48] or by placing such catalyst in a reactor connected in series (*ex situ*), as schematized in Figure 3.2 [49–51]. Fluidized bed reactors in general, and CSBRs in particular, allow for the acid catalysts to be used *in situ* while also avoiding segregation issues [52]. During catalytic pyrolysis, the oxygen content of bio-oil can be decreased through the formation of hydrocarbons, aromatics, and polyaromatics [48]. A two-step approach (low temperature pyrolysis followed by catalytic pyrolysis) has also recently been reported by Zhang et al. [53] as a suitable strategy to remove undesirable compounds (e.g. coke precursors) prior to the catalytic pyrolysis stage to ultimately enhance the production of hydrocarbons.

Through their work using a pilot plant equipped with an FCC reactor, Lappas et al. proved that there are significant synergies between the thermal and catalytic cracking that take place during catalytic pyrolysis, which results not only in a reduction in the oxygen content of bio-oil but also in changes in the composition of the latter, leading to a decrease in the molecular weight of the heavier

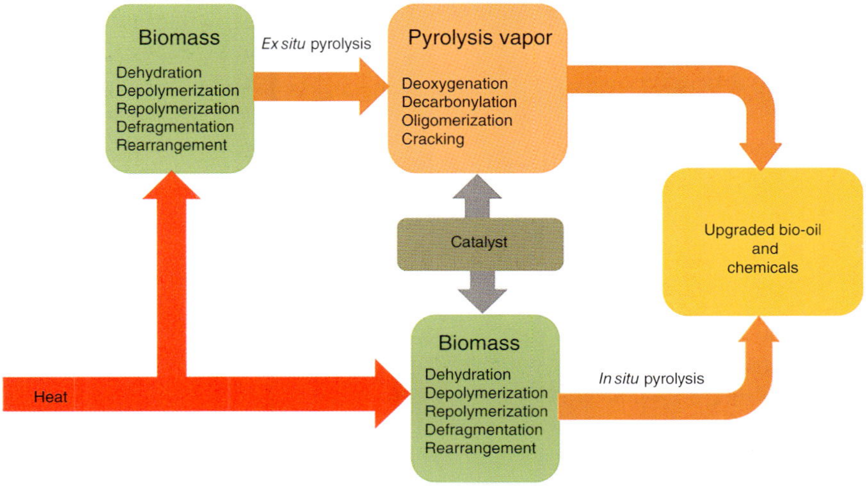

Figure 3.2 *In situ* and *ex situ* catalytic pyrolysis strategies for the production of pyrolysis oil. Source: Kabir and Hameed 2017 [51]. Reprinted with permission of Elsevier.

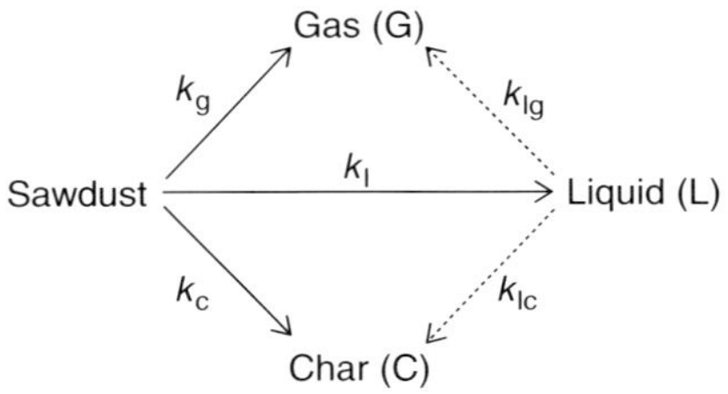

Figure 3.3 Kinetic scheme for the catalytic pyrolysis of biomass using a HZSM-5 catalyst. Source: Atutxa et al. 2005 [55]. Reproduced with permission of American Chemical Society.

fractions [54]. Atutxa et al. [55], using a CSBR, established a kinetic scheme for the *in situ* catalytic pyrolysis of sawdust using a HZSM-5 catalyst (see Figure 3.3) and determined the kinetic parameters of the thermal and catalytic stages, which are, respectively, represented by the continuous and dotted arrows in Figure 3.3. These authors also demonstrated the different reactivity of several bio-oil components, acids, and alcohols displaying high conversion values and ketones affording lower conversion numbers, these results having been more recently confirmed by Aho et al. [56].

Other authors have reported on the noteworthy behavior of mesoporous catalysts like MCM-41 and Al-MCM-41 [57–59], which are becoming increasingly applied as pyrolysis catalysts due to their ability to facilitate the internal diffusion of heavy volatile components (such as methoxyphenols), enhancing the yields of aromatic compounds and partially eliminating the coking responsible for the deactivation of zeolite catalysts [51]. Similarly, in their work using a pilot plant equipped with an FCC reactor, Lappas et al. [60] investigated the advantages of catalytic pyrolysis (with an FCC catalyst) over thermal pyrolysis, which included an improvement in the gasoline and diesel yields. These results are in line with those obtained by Zhang et al. [61], who also reported on the good performance of FCC catalysts tested for the same purpose. Interestingly, recent reports show that spent FCC catalysts display similar activity to a HZSM-5 zeolite in the catalytic pyrolysis of empty fruit bunches in a fixed bed reactor while also affording higher selectivity toward (alkyl)phenolic compounds [62].

Naturally, different types of zeolites (e.g. HY, USY, HZSM-5, or β-zeolite) show distinct performance in catalytic pyrolysis, which is mostly due to differences in pore structure and shape selectivity, as shown in Figure 3.4 [63, 64]. Since an extensive discussion on this topic falls outside of the scope of this chapter, the reader interested in further information in this regard is referred to Chapter 1, as well as to a recent review by Kabir and Hameed [51] containing the latest and most relevant developments in the catalytic pyrolysis of lignocellulosic biomass.

3.2.2.2 Bio-oil Stabilization

The high reactivity of the oxygenated compounds in bio-oil render the latter highly unstable. As a consequence, stored bio-oil undergoes a series of physicochemical transformations known as "aging" that increase its viscosity and water content and depress its volatility. In this process, the bio-oil microemulsion breaks, and chemical reactions (predominantly esterification and condensation) of aldehydes, ketones, and other (mostly oxygenated) components take

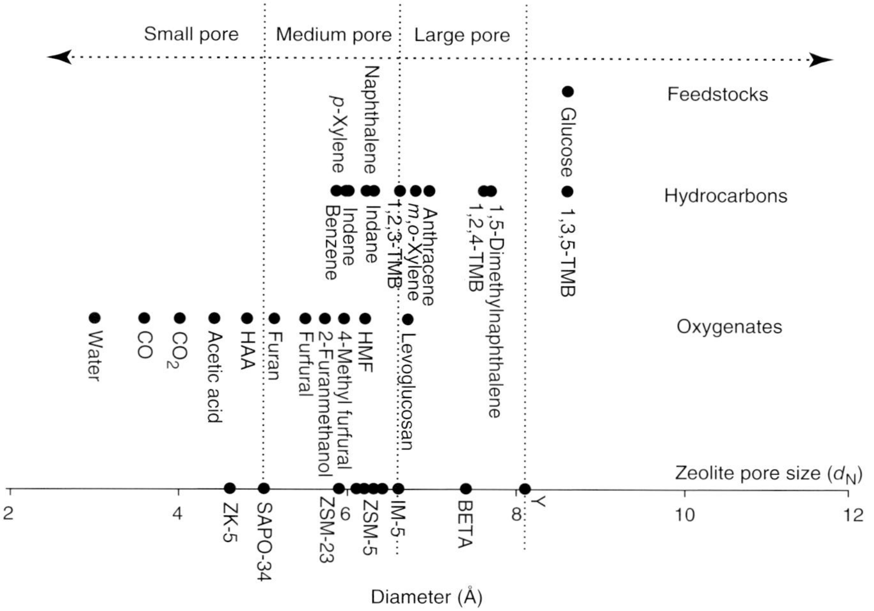

Figure 3.4 Comparison of zeolite pore diameter and the kinetic diameter of pyrolysis feedstock and products. Source: Jae et al. 2011 [65]. Reprinted with permission of Elsevier.

place, leading to the formation of heavier compounds [66–68]. As reported by Talmadge et al. [69], bio-oil aging occurs more rapidly during the first three months of storage (see Figure 3.5). It should also be noted that although the content of solids in bio-oil is typically <1%, the removal of these particles via filtration is necessary in order to curb the aging process [67].

A schematic representation of the approach commonly employed to valorize untreated bio-oil through the production of biofuels and chemicals is shown in Figure 3.6. Prior to catalytic valorization (for which the main routes will be extensively discussed later in this chapter with an emphasis in catalytic cracking through FCC) – and in order to improve both the storage stability and the properties of bio-oil to make it more amenable to upgrading – different treatments (mostly based on physical methods) can be applied, the most common being thermal aging, the addition of solvents, ash and char removal, emulsification, and the addition of antioxidants [67].

The addition of methanol represents an effective method of bio-oil stabilization, with an addition of 10 wt% methanol decreasing the rate of accelerated aging by a factor of 20 [70]. These effects are the consequence of (i) physical dilution, (ii) molecular dilution and the change in the microstructure of the emulsion, and (iii) reactions between methanol and a number of components in bio-oil hindering detrimental chain growth [67].

In line with the fact that temperature can greatly affect the reactions between bio-oil oxygenates, thermal aging also appears to be a viable strategy for bio-oil

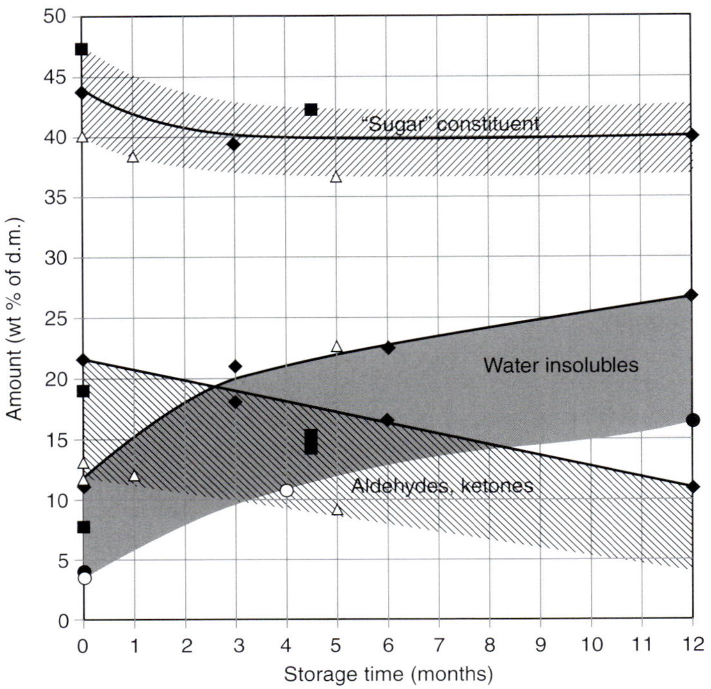

Figure 3.5 Aging trends of some of the main components in bio-oil. Source: Talmadge et al. 2014 [69]. Reprinted with permission of Elsevier.

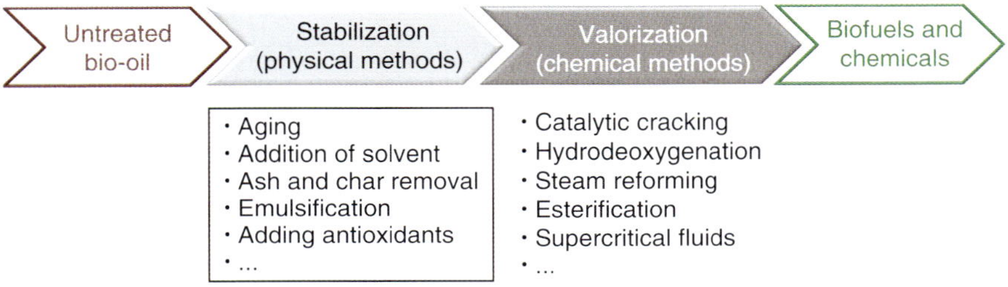

Figure 3.6 Strategies for the valorization of untreated bio-oil to valuable biofuels and chemicals with a previous stabilization stage.

stabilization. In this regard, it is interesting to note that the treatment of bio-oil in the 200–300 °C temperature range has been investigated at high pressure (200 bar) in an effort to decrease its oxygen and water content [71]. Notably, Bertero et al. [72] reported that thermal treatment in the 350–550 °C range decreases the concentration of phenolics and other compounds of high molecular weight. Given that thermal aging affects a highly complex reaction network in which significant synergistic and otherwise interdependent transformations occur, establishing a protocol that leads to reproducible results is of great difficulty [73, 74].

3.2.3 Valorization Routes for Bio-oil

Among the valorization routes available for bio-oil to be upgraded to biofuels, three main approaches have been explored intensively: catalytic cracking, hydrodeoxygenation (HDO), and steam reforming. More recently, esterification and more novel methods involving supercritical fluids have also started attracting attention [39]. The catalyst formulations and operation conditions typically employed in the three main bio-oil upgrading routes mentioned earlier are displayed in Figure 3.7, along with the main products that can be obtained from each process. In short, while catalytic cracking predominantly yields bio-gasoline (C_5–C_{12}), liquified petroleum gases (LPGs), and light cycle oil (LCO), hydroprocessing routes mostly produce diesel fraction products (C_{12}–C_{20}), and steam reforming is the most suitable route for the production of hydrogen.

The fundamental aspects of these three routes are discussed in the following Sections 3.2.3.1, 3.2.3.2, and 3.3, particular emphasis being placed on catalytic cracking routes mainly due to the relevance of FCC units, which is also highlighted in the discussion.

3.2.3.1 Hydroprocessing

Bio-oil hydroprocessing aims for producing higher quality fuels (with increased hydrogen and decreased oxygen content) without significantly affecting the boiling point (bp) range of the feed processed [75]. The prospect of improving the properties of bio-oil to enable its use as a fuel or as a feedstock in FCC refinery units has been studied not only over conventional hydrotreatment catalysts – namely, sulfided Ni–Mo and Co–Mo functions supported on SiO_2 and SiO_2—Al_2O_3 – but also over supported Pt, Pd, and Ru formulations,

Catalyst	Conditions	Products
Catalytic cracking		
Predominantly HZSM-5 zeolites. Others: HY, mordenite, FCC equilibrium catalyst	Temperature: 300–500 °C Pressure: atmospheric	C_1–C_4 hydrocarbons Biogasoline
Hydroprocessing		
NiMo, CoMo and noble metals supported on Al_2O_3 and activated carbon mostly. Other supports: ZrO_2, TiO_2, zeolites	Temperature: 300–600 °C Pressure: 20–80 bar	Renewable diesel
Steam reforming		
Predominantly Ni, but also Co, Fe, and noble metals (Pt, Ir, Rh, Ru) over Al_2O_3. Other supports: CeO_2–ZrO_2, HZSM-5, CNT	Temperature: 550–850 °C Pressure: atmospheric	Hydrogen

Figure 3.7 Main catalytic valorization routes and product fraction obtained from bio-oil.

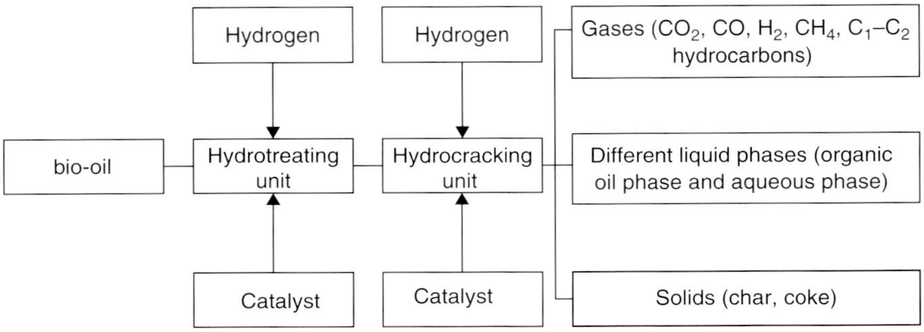

Figure 3.8 General scheme for hydroprocessing stages. Source: Patel and Kumar 2016 [77]. Reprinted with permission of Elsevier.

which provide higher activity and stability in reaction media with high water content [76].

The general flow diagram for hydroprocessing proposed by Patel and Kumar [77] is shown in Figure 3.8. The goal of the first hydrotreating unit is the removal of heteroatoms (i.e. sulfur and nitrogen) and unsaturated hydrocarbons from the feed, while in the subsequent hydrocracking unit cracking, dealkylation, isomerization, and hydrodecyclization (ring opening) reactions are predominant. Additionally, decarboxylation, decarbonylation, water–gas shift, methanation, and coke formation reactions can occur in both units.

Hydroprocessing is typically carried out at temperatures in the 300–500 °C range, with the following stoichiometry:

$$-(CH_2O)- + H_2 \rightarrow -(CH_2)- + H_2O$$

where $-(CH_2O)-$ represents the carbohydrate fraction in bio-oil.

For further insights on the topic of bio-oil upgrading through hydroprocessing, its potential, and technical feasibility, the reader is referred to Chapter 2 and also the reviews authored by Saidi et al. [76] and Patel and Kumar [77].

3.2.3.2 Steam Reforming

The availability of biomass and the advantages that producing bio-oil through pyrolysis offer relative to the production of other biomass-derived oxygenates (e.g. methanol, DME, ethanol) also render bio-oil an interesting vector for the large-scale production of hydrogen via steam reforming, which is also a suitable strategy to avoid separating water from bio-oil [78, 79]. Admittedly, although the study of the steam reforming of oxygenates has provided great insights for the development of efficient catalysts and the identification of the most propitious operating conditions, work in this area has been challenged by the complex composition of bio-oil and its instability, particularly in the case of raw bio-oil. The presence of phenolic compounds (derived from the pyrolysis of the lignin fraction in biomass) in bio-oil exacerbates the aforementioned problems due to their polymerization when bio-oil is heated [80]. Moreover, polymerization can take place prior to bio-oil vaporization as well as later on in the reforming reactor. As a consequence, most reports in the literature focus on the reforming of model oxygenates [79, 81–83], little work having been reported either on the reforming of the aqueous phase [84, 85] or on the use of raw bio-oil [86, 87].

3.2.3.3 Extraction of Valuable Components from Bio-oil

There is great interest in the extraction of a number of valuable components from bio-oil, including phenolics (for the formulation of resins), organic volatile acids (for the synthesis of antifreeze solutions), levoglucosan, hydroxyacetaldehyde, and other additives (for the pharmaceutical industry, the synthesis of fertilizers, and aromatics for the food industry) [88, 89]. Saliently, Wang et al. [90] have proposed a multistep separation approach – using an acid solution, an alkaline solution, or organic solvents – that allows for the recovery of a phenolic-enriched bio-oil fraction with a concentration of phenolics higher than 94%. Furthermore, raw bio-oil and the organic fraction thereof have also been applied for the formulation of phenol-formaldehyde resins [91]. Finally, levoglucosan (an important component in bio-oil) can also be used as feedstock in the production of chiral reactants, polysaccharides, and antibiotics. However, it can also be hydrolyzed to afford glucose, thus providing a route for the facile production of bio-ethanol [92, 93].

3.3 Catalytic Cracking of Bio-oil: Fundamental Aspects

The catalytic cracking of bio-oil represents a versatile strategy for the production of automotive fuels and chemicals – including light olefins, Benzene + Toluene + Xylenes (BTX), and aromatics – either in dedicated units designed for that purpose and operating at optimal conditions to achieve specific production goals or in existing refinery FCC units in which bio-oil is coprocessed with traditional FCC feeds.

3.3.1 The FCC Unit

The traditional function of FCC units is to transform heavy petroleum fractions (mainly atmospheric and vacuum gas oil, VGO) into high octane gasoline, light gasoil, and olefin-rich light gases. However, FCC units are also a key component of the emerging sustainable refinery concept mainly due to their remarkable versatility, which makes them capable of processing nonconventional feeds including (i) secondary refinery streams [94, 95], (ii) external streams derived from the consumer society such as plastic waste [96] and waxes produced from the pyrolysis of polyolefin plastics [97], and (iii) bio-based materials like vegetable oil or bio-oil [98–100]. In fact, FCC is the bio-oil coprocessing strategy most widely investigated to date [101].

Figure 3.9 schematizes the four main sections of the FCC unit, namely, the pneumatic transport reactor (riser) section, the gas–solid separation section, the catalyst steam-stripping section, and the catalyst regeneration section (where regeneration is accomplished via combusting the coke deposits on the surface of the spent catalyst).

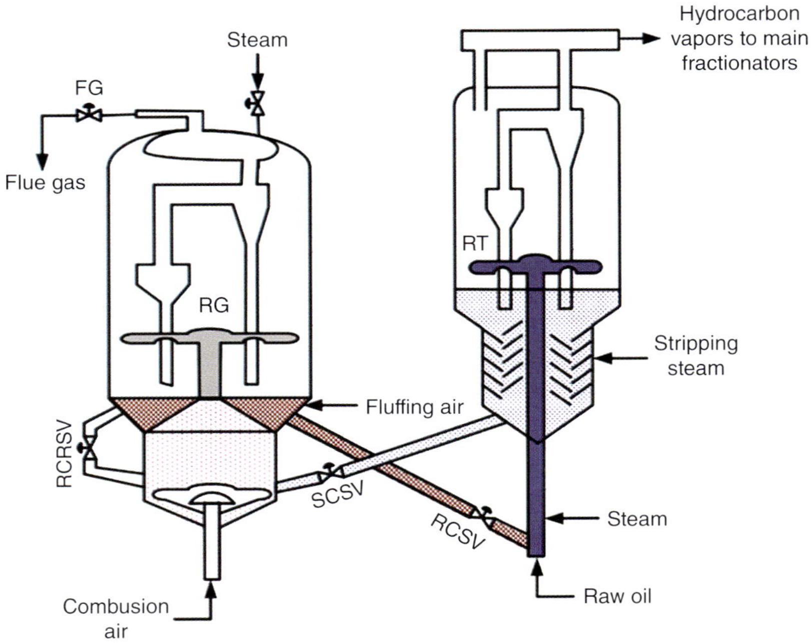

Figure 3.9 Sections of a FCC unit. FG, flue gas; RG, regenerator; RCRSV, regenerated catalyst recirculation standpipe valve; RCSV, regenerated catalyst standpipe valve; SCSV, spent catalyst standpipe valve. Source: Jarullah et al. 2017 [102]. Reprinted with permission of Elsevier.

During FCC operation, the preheated feed – VGO or recirculation gasoil with a boiling point (bp) > 344 °C – is vaporized in contact with steam at a temperature of 350–450 °C and made to converge at the bottom of the reactor with the stream of regenerated catalyst at 650–700 °C in such a way that the resulting temperature of the mix is ca. 530 °C. As the vaporized feed (482–538 °C) drags the catalyst through the riser, which has a length of 25–40 m and a diameter of 0.6–1.8 m, the gas undergoes considerable expansion due to cracking and the gas velocity typically reaches values in the 5–15 m/s range. The fluid-dynamic regime corresponds to a dense phase pneumatic transport bed due to the high catalyst-to-oil ratio (C/O = 4–9) with a residence time for the gas and catalyst of four to eight seconds. At the upper part of the reactor, when cracking reactions are completed, the spent catalyst is rapidly separated from the products in high-efficiency cyclones to prevent secondary reactions. Vapors leave the reactor through the reactor head and are directed toward the fractionation and concentration systems. Typical FCC yields are 5 wt% dry gases (H_2, C_1–C_2), 20 wt% light gases (LPG, C_3–C_4), 45 wt% gasoline (C_5–C_{12}), 15 wt% LCO, 10 wt% heavy cycle oil (HCO), and 5 wt% coke. The deactivated catalyst circulates toward the stripping section – in which the interstitial hydrocarbons adsorbed on the catalyst are separated using a countercurrent stream of steam (2–5 kg/1000 kg of catalyst) – prior to being directed toward a fractionation column. The catalyst (containing 0.4–2.5 wt% coke) flows to the bubbling fluidized bed regeneration reactor (10–15 m diameter) where coke is removed through combustion with air at 620–845 °C. In the regenerator, the catalyst is both reactivated and made to acquire sensitive heat by increasing its temperature by 150–200 °C, which makes it able to satisfy the thermal requirements of the unit.

The catalyst inventory of a FCC unit is approximately 270–300 ton for a unit that processes 40 000 barrels/day and, given the high residence time of the catalyst within the regeneration reactor (6–11 minutes) in comparison to the residence time in the riser (3–8 seconds), the catalyst spends most of the time within the regenerator describing 100–400 cycles/day. The gas at the exit of the regeneration reactor contains fine particles formed due to catalyst attrition, which need to be removed and substituted by a new catalyst stream (approximately 1 kg of new catalyst per 600 kg of spent catalyst).

To date, a number of different units have been used for studying and simulating FCC units, which are being intensively applied for exploring new catalysts and potential feeds [103]. The use of spent (coked) catalysts has also been leveraged in the conversion of heavier feedstock with decreased selectivity toward coke formation, which also improves the quality of the gasoline and LCO fractions.

3.3.2 Cracking Reactions and Mechanisms

The mechanism through which catalytic cracking proceeds involves a series of carbocations, which are formed on the acid sites of the catalyst through two different routes, leading to the formation of two distinct types of ions:

- Carbenium ions, $R{-}C_nH_{2n}{}^+$, which are formed either through the abstraction of a hydride ion from paraffinic molecules on a Lewis acid site or through the addition of a proton on a Brønsted acid site to an olefin:

$$R{-}C_nH_{(2n+1)} + L \rightarrow R{-}C_nH_{2n}{}^+ + LH^-$$
$$R{-}C_nH_{(2n-1)} + BH \rightarrow R{-}C_nH_{2n}{}^+ + B^-$$

- Carbonium ions, $R{-}C_nH_{(2n+2)}{}^+$, which are formed by the addition of a proton to a paraffin:

$$R{-}C_nH_{(2n+1)} + BH \rightarrow R{-}C_nH_{(2n+2)}{}^+ + B^-$$
$$R{-}C_nH_{(2n+2)}{}^+ \rightarrow R{-}C_nH_{2n}{}^+ + H_2$$

However, carbonium ions are unstable (their formation requires very highly acidic sites) and have a tendency to form carbenium ions via disproportionation. Therefore, FCC proceeds mainly through the route resulting in the formation of carbenium ions, whose stability order follows the trend tertiary > secondary > primary. In addition, carbocations are very reactive species that are involved in the following reactions:

- *β-Scission reactions*. These reactions are unimolecular, endothermic, and irreversible. The C—C bond including the β but not the α C holding the positive charge is the position cleaved since it requires less energy, leading to the formation of an olefin and a paraffin:

$$R_1{-}\alpha CH^+\beta CH_2CH_2CH_2R_2 \rightarrow R_2{-}CH_2{-}CH_2{}^+ + H_2C{-}CH{-}R_1$$

- *Isomerization reactions.* The isomerization of the carbenium ion (more frequent in tertiary ions, due to their higher stability) is a reversible endothermic reaction that occurs via hydrogen transposition or a carbocation:

$$\mathrm{R{-}C_nH_{2n}CH_2^+ \leftrightarrow R{-}C_n^{\;+}H_{(2n-1)}CH_3}$$

- *Hydrogen transfer reactions.* Proton transfer reactions are bimolecular and occur between a saturated species and a carbenium ion. Typically, an olefin and a naphthene (which are adsorbed in adjacent sites) will react to form a paraffin and an aromatic compound:

$$\mathrm{R_1{-}CH^+{-}R_2 + R_3H \rightarrow R_1CH_2R_2 + R_3^{\;+}}$$

- *Condensation reactions.* Carbocations with a double bond have the ability to cyclize, affording naphthenes and aromatic compounds.

3.3.3 Cracking of Oxygenated Compounds

Insights into bio-oil cracking have been gathered through the study and understanding of the mechanisms and reaction stages undergone by its individual components. For instance, Gayubo et al. [104, 105] have studied the cracking of different bio-oil model components over HZSM-5 zeolite catalysts at 400 °C. Among their results, the similarity of the kinetic schemes for alcohol cracking reactions (see Figure 3.10a for propanol and Figure 3.10b for butanol) as well as those of well-known transformations of methanol (methanol to gasoline, MTG; and methanol to olefins, MTO) and ethanol (bioethanol to gasoline, BTG; and bioethanol to olefins, BTO) are particularly noteworthy.

Another notable result deals with the heterogeneity in the reactivity of different oxygenates, among which the limited reactivity of phenolic compounds stands out, along with the high instability of 2-methoxyphenol (a precursor to considerable amounts of coke of thermal origin). The latter suggests that the removal of phenolic components prior to the catalytic valorization of bio-oil could be an

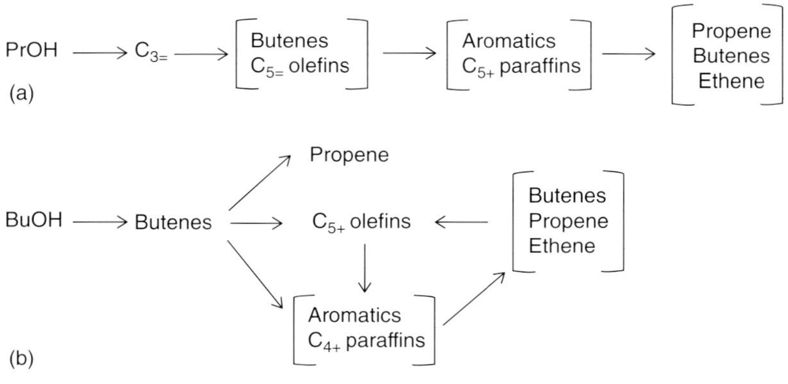

Figure 3.10 Reaction scheme for the catalytic cracking of (a) 1- and 2-propanol and (b) butanol using a HZSM-5 catalyst. Source: Gayubo et al. 2004 [104]. Reproduced with permission of American Chemical Society.

effective strategy. The challenges associated with opening phenolic rings has also been observed over HY zeolite catalysts by Graça et al. [106] who concluded that the high adsorption capacity of these compounds on the outside of the zeolite crystals can also lead to pronounced catalyst deactivation. Moreover, the same authors also reported that while process temperature plays a key role in catalyst deactivation when HY zeolites are used (due to the effect of phenol diffusion within the pore structure), the effect of the temperature on deactivation is greatly palliated when HZSM-5 zeolites are employed [107]. Further, these authors proved that mixing both HY and HZSM-5 zeolites represents a suitable approach to confer upon HZSM-5 a higher resistance to deactivation [108].

Acetone is another reactive compound – albeit less so than alcohols – presenting a reaction scheme (summarized in Figure 3.11a) in which isobutene intermediates the formation of light olefins and aromatics, the production of the former being favored at temperatures above 450 °C due to C_{5+} olefin and C_{4+} paraffin cracking. Heavier ketones including butanone present similar schemes (see Figure 3.11b). Unsurprisingly, while acetic acid conversion is intermediated by acetone, the conversion of heavier acids proceeds via heavier ketone intermediates. In studies on the conversion of acetone into gasoline-range products, nanocrystalline HZSM-5 zeolites have been found to have suitable features for the production of an enhanced low-benzene, toluene-rich, and xylene-rich product pool with a high octane number [109].

Gayubo et al. [105] reported low selectivity toward light olefins during the catalytic cracking of acetaldehyde, where only C_{6+} olefins and the trimethyltrioxane trimer are produced and approximately 50% of the acetaldehyde transforms into coke, which suggests that removing acetaldehyde along with the phenolic fraction prior to the valorization of bio-oil represents an advisable approach. Moreover,

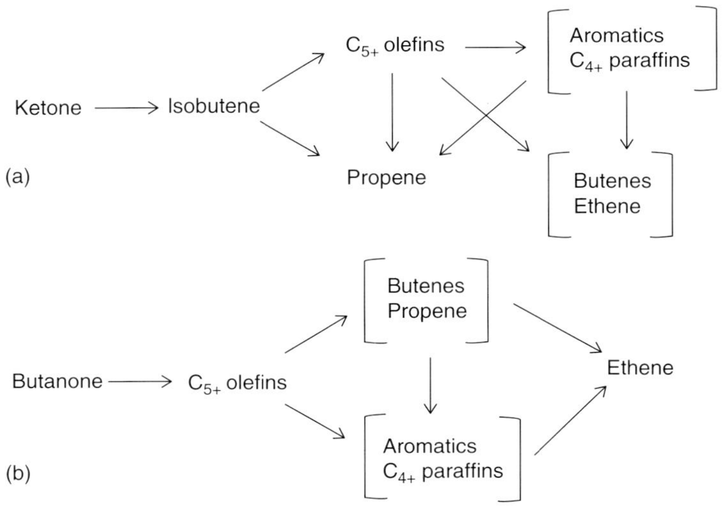

Figure 3.11 Reaction scheme for the catalytic cracking of (a) acetone and (b) butanone using a HZSM-5 catalyst. Source: Gayubo et al. 2004 [105]. Reproduced with permission of American Chemical Society.

through a comparative study in which the transformation of oxygenated compounds representative of bio-oil was investigated collectively and individually, it was determined that acetaldehyde, 2-methoxyphenol, and furfural not only present low reactivity but are also responsible for the formation of coke on the exterior of the catalyst particles [110].

Several authors have also observed synergies between oxygenates within bio-oil during catalytic cracking transformations. Wang et al. [111] observed that a high gasoline yield (about 30%) could be obtained in a fixed bed reactor (operated at 400 °C and 2 MPa) by co-feeding ethanol and reactive bio-oil model compounds (such as acetic acid, cyclopropanone, and hydroxypropanone). On the other hand, synergies between hydrocarbons resulting from the cracking of the FCC feed (e.g. naphthenes and alkanes) and oxygenates representative of bio-oil have also been proven. Finally, in work performed by Graça et al. [106, 108, 112] below 450 °C, the preferred adsorption of phenolics (particularly over the HZSM-5 zeolite commonly used as an additive in FCC catalysts) was observed along with their tendency to exacerbate coke formation, once again highlighting the importance of limiting the amount of phenolics in the feed.

3.3.4 Cracking of Bio-oil

Several authors have compared the performance of different acid catalysts in the cracking of bio-oil, Adjaye and Bakhshi [113] observing the following reactivity order in a fixed bed reactor operated between 330 and 410 °C: HZSM-5 > H-mordenite > HY > SiO_2–Al_2O_3 > silicate.

Zeolites HZSM-5 and HY tested in the 290–500 °C range provide hydrocarbon yields between 6% and 26%, product mixtures consisting mostly of aromatic compounds and paraffins when HZSM-5 and HY are employed, respectively [114, 115]. The yields of oxygenated compounds – which are mainly phenolics, alcohols, acids, ketones, and furans – is in the 6–18% range, high coke yields of 10–30% being observed (particularly when using the HY zeolite). In the aforementioned contributions, as well as in others where the effect of temperature on the yield of cracking products was studied [39, 114], it has been observed that at elevated temperatures the main reaction pathways are dehydration, cracking, and decarboxylation and decarbonylation reactions, which predominate above 500 °C.

It should be noted that different components in bio-oil display distinct reactivity; for instance, aldehydes are more susceptible to decarbonylation than ketones.

The studies of Gayubo et al. [116] emphasize the similarity between two kinetic schemes: (i) that of the cracking of the aqueous fraction of bio-oil over a HZSM-5 catalyst in the 400–500 °C range and (ii) that of the transformation of methanol and ethanol to form olefins, paraffins, and aromatics, highlighting the possibility of co-feeding bio-oil and methanol in the MTO process. However, some challenges need to be addressed for this to become a viable approach, including the high coke deposition (>9%, mainly deposited on the external surface of the zeolite crystals) and the irreversible HZSM-5 zeolite deactivation

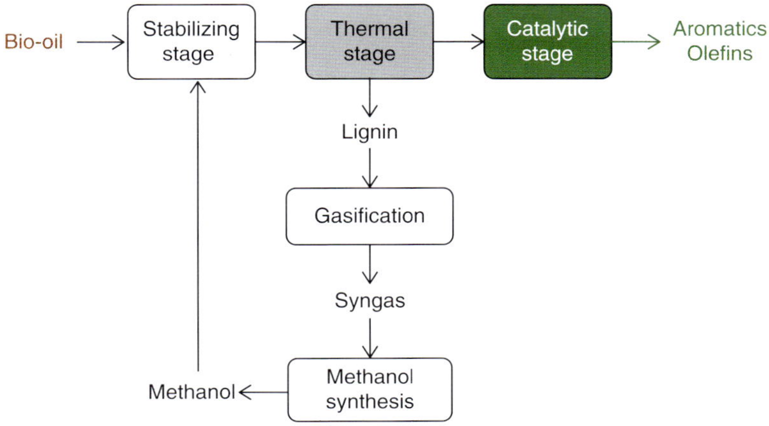

Figure 3.12 Two-step process for the catalytic transformation of raw bio-oil. Source: Gayubo et al. 2010 [117]. Adapted with permission of John Wiley and Sons.

due to dealuminization above 450 °C, which is a consequence of the high water content of the reaction media and requires more efficient catalyst stabilization.

The desire to achieve efficient raw bio-oil cracking while avoiding the fast catalyst deactivation caused by (pyrolytic lignin-derived) coke deposition, motivated Gayubo et al. [117] to propose the two-step thermocatalytic process schematized in Figure 3.12. In the first – thermal (*sans* catalyst) – step, pyrolytic lignin is deposited, the remaining volatiles being catalytically transformed in line in the second step using a fluidized bed reactor and a HZSM-5 zeolite catalyst. The composition and acidic properties of the zeolite play a key role in determining the selectivity of the process, which can be directed either toward the production of an aromatic BTX fraction [118] or olefins [119]. Indeed, moderating zeolite acidity curbs the occurrence of secondary reactions, thus favoring the production of olefins as primary products.

Figure 3.12 also considers the possibility of integrating the valorization of pyrolytic lignin (through gasification and methanol synthesis) in the process. As previously stated, methanol is a bio-oil stabilizing agent, so co-feeding methanol and bio-oil can mitigate coke deposition on the catalyst by increasing the H/C ratio in the feed [120]. Moreover, using a fluidized bed reactor can also help limit catalyst deactivation relative to fixed bed reactors due to the homogenization of catalyst activity throughout the bed.

Wang et al. [111] have observed synergies during the catalytic cracking of bio-oil over a HZSM-5 catalyst in a fixed bed reactor operated at 400 °C and 2 MPa. Briefly, by processing bio-oil (obtained from molecular distillation with a water content of 4.8 wt%) and ethanol in a 2 : 3 ratio, these authors obtained a high gasoline yield of 25.9 wt% and a coke yield of 3.2 wt%, liquid products consisting of 98.3 wt% hydrocarbons (mostly BTX) with CO_2, CO, and propylene as the main gas products.

3.4 Bio-oil Cracking in the FCC Unit

Despite a number of limitations to mimic the industrial FCC conditions in terms of catalyst features and process conditions (e.g. temperature, contact time), laboratory studies to date have achieved high bio-oil conversion, identified important synergies between the cracking of hydrocarbons and oxygenates, and determined the effect of the oxygenated fraction composition on product distribution.

3.4.1 Cracking of Model Oxygenates

Bertero et al. [121, 122] have compared the distribution of products stemming from the thermal and catalytic cracking – performed in fixed bed reactors constituting micro activity test (MAT) units at 500 °C using an equilibrated FCC catalyst – of individual model compounds representative of the different chemical groups in bio-oil. In contrast with results obtained at 400 °C over HZSM-5 zeolites [104, 105], oxygenates displayed very high overall reactivity over the FCC catalyst, significant differences being observed in the reactivity and product distribution for different compounds. Further, the importance of thermal cracking was found to be small compared with that of catalytic cracking, a notable exception being 1,2,4-trimethoxybenzene, which also presents the highest tendency toward coke formation. These results confirm the conclusions obtained by Valle et al. [123] regarding the origin of coke during the coprocessing of bio-oil and methanol mixtures.

3.4.2 Coprocessing of Oxygenates and Their Mixtures with Vacuum Gas Oil (VGO)

The effect of coprocessing pure oxygenates and oxygenate mixtures with VGO in FCC units has been studied by several authors [124–127], using the MAT units and conditions typically used for the evaluation of FCC catalysts. When assessing the results obtained from coprocessing studies, two main facts must be considered: (i) the transfer of hydrogen from hydrocarbons to oxygenates and (ii) the competitive adsorption of the feed on the acidic sites of the catalyst, which is considered to be stronger for the hydrocarbons in VGO [106, 112]. Coprocessing increases the yield of dry gases (C_1–C_2) and decreases that of C_3–C_4 hydrocarbons. The yields of the gasoline and LCO fractions are not significantly affected, although the concentration of aromatics in the gasoline fraction increases along with the concentration of certain oxygenated compounds (e.g. alkylphenols). Coke yield increases – as much as 30% relative to the cracking of VGO alone – and high yields of H_2O, CO, and CO_2 are also observed. Additionally, the presence of HZSM-5 as an additive within the FCC catalyst has been observed to increase the conversion of oxygenates.

Naik et al. [128] have compared the distribution of products obtained through the cracking of VGO and its mixtures with acetic acid (6%) and guaiacol (12.4%) in a MAT unit operated at 530 °C and using an equilibrated FCC catalyst-to-oil (C/O) ratio in the 3–9 range. This work resulted in several noteworthy observations, including (i) the complete conversion of acetic acid and its capacity for enhancing olefin yields; (ii) the co-feeding of guaiacol resulting in the formation of phenol and the increase in the yield of aromatics and coke, which is also favored by increases in the C/O ratio; and (iii) conditions resulting in high water yields also attenuating coke formation.

Corma et al. [125] studied the cracking reactivity of two model compounds – namely, glycerol and sorbitol – in a MAT unit using different formulations, including a fresh and an equilibrated FCC catalyst. In runs involving glycerol, HZSM-5 zeolite displayed an outstandingly low selectivity toward coke along with the highest olefin (ethylene, propylene, butene) and aromatic yields, while USY zeolite showed the highest overall activity for glycerol conversion. On the other hand, runs involving sorbitol afforded a similar product selectivity (over the same catalysts) to that observed when glycerol was employed, but with a higher CO formation. Dehydration and hydrogen transfer reactions led to the formation of olefins, paraffins, and coke, while aromatic compounds were mostly formed through Diels–Alder reactions. Coprocessing glycerol and VGO (in a volumetric ratio of 2 : 1) did not greatly affect product selectivity, albeit coke formation increased significantly causing a drop in conversion. Saliently, the relevance of hydrogen-producing and hydrogen-consuming reactions in the catalytic cracking of biomass-derived oxygenates was highlighted by Huber and Corma [129], who proposed the reaction network in Figure 3.13. In short, these authors suggested that the catalytic transformation of model biomass oxygenates occurred through five main types of processes: (i) dehydration reactions, (ii) cracking of large oxygenated molecules, (iii) hydrogen-producing reactions, (iv) hydrogen-consuming reactions, and (v) synthesis of larger molecules through reactions forming C—C bonds.

3.4.3 Cracking of Bio-oil and Its Mixtures with VGO

Al-Sawabi et al. [130] reviewed the literature available at the time on the cracking of oxygenates and bio-oil – both alone and co-fed with VGO – explaining some contradictory results by invoking the fact that studies are commonly performed using different experimental conditions, catalysts, and feeds, as well as the fact that comparison of product distributions are often made at different conversion values. The main insights put forward by Al-Sawabi et al. can be summarized as follows: (i) coprocessing results in lower conversion than that obtained with hydrocarbons constituting traditional FCC feeds; (ii) coprocessing also results in higher yield of dry gases; (iii) synergies lower olefin yields and increase aromatic and paraffin yields as a consequence of the hydrogen transfer from hydrocarbons; (iv) catalyst deactivation due to coke formation is attenuated due to the nature of the coke formed being different (although coke yields are typically higher); (v) synergies in the coke formation process are notable, the coke content showing no correlation with the amount of oxygenates in the feed; (vi) the gasoline fraction contains oxygenates such as (alkyl)phenols; and (vii) oxygen in the feed is eliminated as water.

Based on the results they obtained processing pure VGO and coprocessing a mixture of 80% VGO and 20% hydrogenated bio-oil (see entry 1 in Table 3.2),

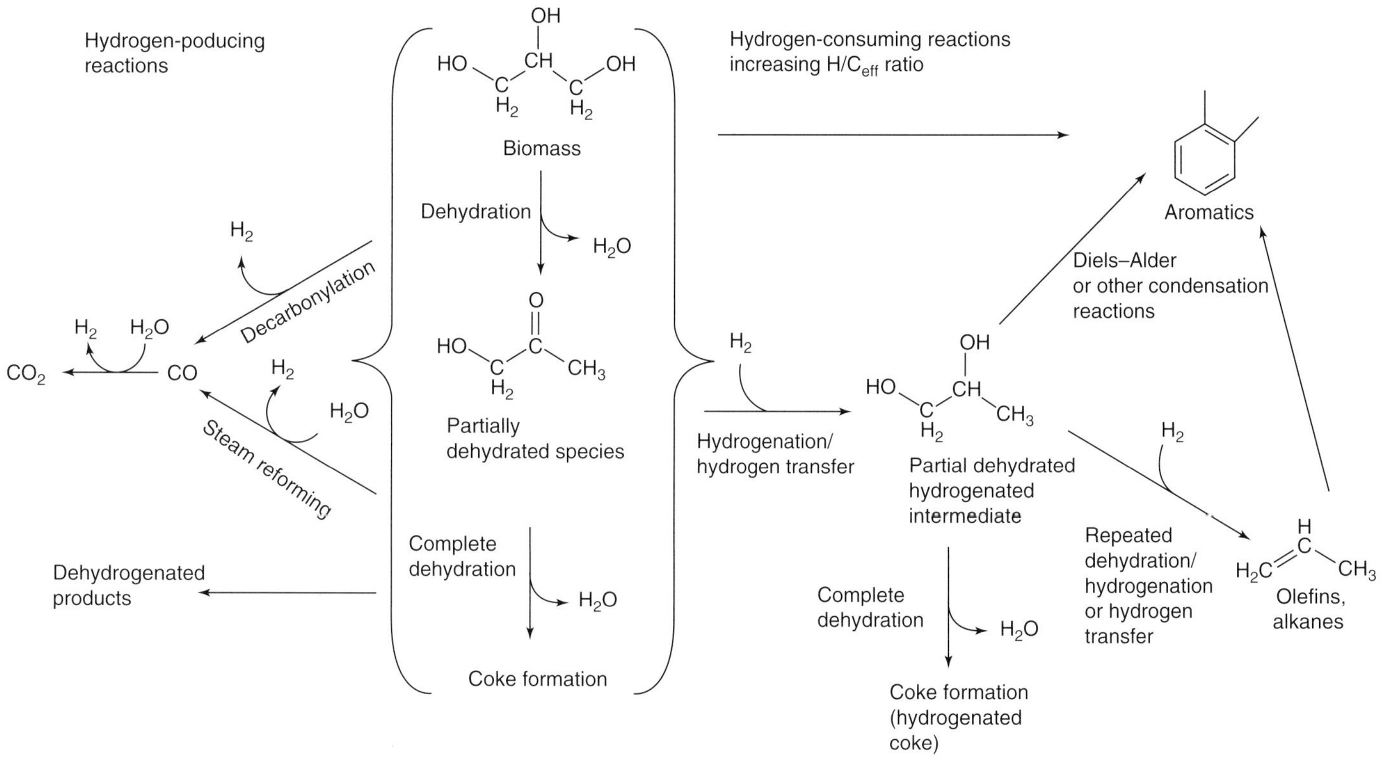

Figure 3.13 Reaction pathways for the catalytic cracking of oxygenated compounds derived from biomass. Source: Huber and Corma 2007 [129]. Reprinted with permission of Elsevier.

Fogassy et al. [126] proposed a scheme for the cracking of hydrocarbons and oxygenated compounds over an FCC catalyst involving five main sets of reactions that is very similar to that proposed by Huber and Corma [129] displayed in Figure 3.13. At different conversion levels (upon varying C/O ratios), analogous gasoline yields were achieved by processing pure VGO and coprocessing VGO with the hydrotreated bio-oil, along with comparable yields of LCO and bottom fractions. However, the presence of bio-oil significantly increased coke yields, the amount of coke being almost double (at 80% conversion) than that obtained cracking VGO alone.

Bertero and Sedran [131] used an FCC catalyst in a MAT unit operated at 500 °C to compare the product distribution obtained from untreated bio-oil, treated (thermally aged) bio-oil, and a synthetic mixture of eight oxygenates representative of bio-oil. Notably, coke yields were high (9–14%) irrespective of the feed employed. However, the synthetic bio-oil led to lower hydrocarbon and coke yields and provided a product distribution that highly differs from that of real bio-oil, which evidences the importance of synergies between the oxygenates in bio-oil. The aging treatment given to bio-oil was observed to have a positive effect, as it increased hydrocarbon yields ca. 25% while decreasing the oxygenates and coke yields 55% and 20%, respectively.

The results of Thegarid et al. [132] are also noteworthy, since they found that the results obtained by cracking VGO with a previously hydrotreated bio-oil are very similar to those obtained by coprocessing VGO with an untreated bio-oil derived from a catalytic pyrolysis process, which implies lower operational costs (see entry 2 in Table 3.2). Furthermore, the authors discuss the prospect of improving HZSM-5 catalysts so that the produced bio-oil (from the catalytic pyrolysis) can be fed directly into FCC units without any other necessary improvement. Also of note is the work of Gueudré et al. [133], who provided extensive information on the nature and location of coke in the aforementioned process. Indeed, these authors determined that coke was formed via both conventional VGO cracking pathways (producing hard coke) and reactions involving the oxygenated compounds in bio-oil (i.e. sugars, furans, phenolics) that led to a softer kind of coke or "bio-coke."

Catalytic pyrolysis oil has been reported to have great potential for coprocessing in pilot scale FCC units, this approach having afforded very similar selectivity to gasoline to the cracking of VGO alone in blends of up to 10% bio-oil, as observed by Wang et al. [134] (see entry 3 in Table 3.2). However, the production of hydrogen and light alkanes is negatively affected due to the hydrogen transfer reactions that take place between bio-oil and VGO. Moreover, catalytic pyrolysis is preferable over thermal pyrolysis as a method for bio-oil production that avoids the formation of water in the subsequent catalytic cracking stage.

Pinho et al. [100] coprocessed bio-oil and VGO blends (containing 10 and 20 wt% bio-oil) in a pilot plant (150 kg/h) at 540 and 560 °C (see entry 4 in Table 3.2). Similar gasoline yields of 39–42 wt% were reported when processing only VGO or blends containing up to 10 wt% bio-oil. However, gasoline yields decreased to 36–38 wt% when blends with a higher bio-oil content were employed, although higher conversion levels (+2–3%) were observed under certain conditions. Due to the oxygen content of bio-oil, oxygen-bearing compounds – such as CO, CO_2, and H_2O – are produced in higher proportions (up to 10 wt%) through coprocessing. Vis-à-vis the conversion of the carbon contained in the bio-oil to liquid products, carbon efficiency was estimated at around 30%. In more recent work performed in a larger unit processing 200 kg/h, Pinho et al. [135] explored the effect of using aged bio-oil during the coprocessing of VGO and bio-oil blends (containing 5% and 10% bio-oil) at 540 °C (see entry 5 in Table 3.2). While 9-month-old bio-oil could be processed without technical difficulties, 21-month-old bio-oil hindered FCC operation and led to the overcracking of the gasoline fraction. Gasoline yields were highest at conversions of 68–69%, these maximum values being similar for all the processed feeds.

In recent work, Gueudre et al. [136] studied the effect of coprocessing a mixture of 90% VGO (74% hydrotreated VGO and 26% decant oil) and 10% bio-oil at 560 °C over a stabilized USY zeolite (see entry 6 in Table 3.2). The bio-oil used had been previously hydrotreated using a Ni-based catalyst and two different C/O ratios, namely, 1.5 and 6. Enhanced gasoline yields were achieved at a C/O ratio of 6, along with higher coke and LPG yields and lower yields of LCO and bottoms. Moreover, the produced gasoline obtained through these conditions comprised 50–55% aromatics and 25% isoparaffins, having a research octane number (RON) of ca. 87.5. These authors also determined that the hydrotreatment of bio-oil should be restricted to hydrogen consumption levels below 200 NL/kg in order to prevent excessive coking and produce high quality gasoline.

The influence of coprocessing VGO with dry bio-oil, fast pyrolysis oil, and bio-oil pretreated via HDO in a MAT unit was explored by Lindfors et al. [137] (see entry 7 in Table 3.2). VGO and dry bio-oil mixtures afforded the lowest yield of liquid product, while the difference in the yields obtained using the VGO and pyrolysis oil mixtures and the VGO and HDO bio-oil mixtures were relatively small. Moreover, the composition of the liquid products recovered did not differ significantly when coprocessing VGO with either fast pyrolysis oil or HDO bio-oil. The authors also concluded that the proportion of bio-oil had to be kept low ($\leq$20%) in order to avoid coking issues, the lighter nature and higher H/C ratio of HDO bio-oil leading to limited coke formation.

The remarkable extent of the synergies between the cracking of hydrocarbons in VGO and that of oxygenates in bio-oil was explored by Ibarra et al. [138]. Indeed, these authors observed that the conversion of the blend (with 20 wt% bio-oil) was higher than that of VGO alone over the whole temperature range investigated (500–560 °C), which they attributed to synergistic effects and joint cracking (see Figure 3.14). Relatively high conversion values (49–67%) were reached at a reaction time of three seconds, evidencing the speed at which the reaction takes place. The aforementioned synergies also had a great impact on the distribution of products obtained from the different feeds (VGO, raw bio-oil, and their blends) as well as on the composition of the gasoline fraction, as shown in Figure 3.15. The yields of dry gas, LPG, gasoline, and LCO (Figure 3.15a) obtained with the blend fell between those obtained using the individual feeds. These results indicate that coprocessing has an attenuating effect on gasoline overcracking, the presence of H_2O potentially being a contributing factor. A similar effect was observed for decarbonylation and

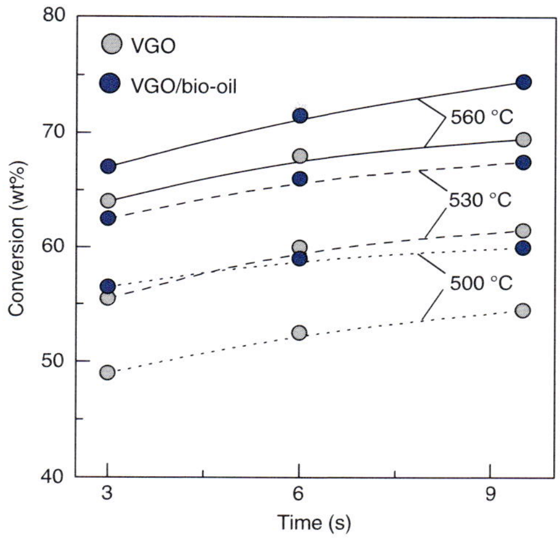

Figure 3.14 Evolution of the conversion with the reaction time in the catalytic cracking of VGO and VGO + bio oil (20 wt%) blends at different reaction temperatures. Source: Ibarra et al. 2016 [138]. Adapted with permission of American Chemical Society.

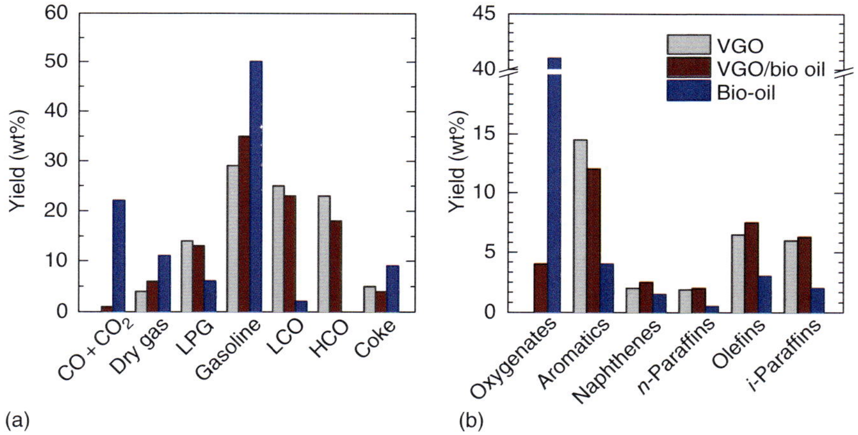

Figure 3.15 (a) Product fraction distribution and (b) gasoline fraction composition in the cracking of VGO, raw bio-oil, and a VGO/bio-oil (20 wt%) blend at 500 °C and six seconds. Source: Ibarra et al. 2016 [138]. Adapted with permission of American Chemical Society.

decarboxylation reactions, based on the low CO and CO_2 yields obtained when processing the blend. Furthermore, coprocessing can also curb coke formation and have a significant impact on the composition of the gasoline produced (see Figure 3.15b). Enhanced cracking of oxygenated compounds led to a lower presence of oxygenates in the gasoline fraction, while the yields of paraffins (both linear and branched), naphthenes, and olefins were higher than expected.

In a complementary study, Ibarra et al. [99] carried out extensive characterization studies on the coke formed during the catalytic cracking of a VGO, raw bio-oil, and a VGO/bio-oil (20 wt%) mixture under FCC conditions at 500 °C (see entry 8 in Table 3.2). Through this work, the authors established that two main types of coke are formed: (i) ordered, polycondensed aromatic structures formed from heavy hydrocarbons and (ii) disordered, more aliphatic coke formed from

Table 3.2 Overview of experimental conditions and product yields obtained from the coprocessing of bio-oil and VGO blends in FCC conditions.

#	References	VGO/ bio-oil (wt%)	Temperature (°C)	Conversion (%)	Yields (wt%)								
					Dry gas	LPG	Gasoline	LCO	Bottoms	Coke	CO	CO_2	H_2O
1	[126][a)]	100/0	500	64–85	0.9–2.2	15–29	43–47	12–32	2–5.5	2.2–3.7			
	[126][a)]	80/20	500	74–84	1.9–2.9	20–26	45–46	14–24	1.7–3	4–7			
2	[132][b)]	100/0	500	63–87	1–2.2	16–30	46–48	15–34	2–5	2.5–4.5		0	
	[132][b)]	90/10	500	74–87	1.9–2.8	20–26	46–48	16–23	2–3	4–7		0.2–0.3	
3	[135][c)]	100/0	525	66–73	3.8–5.1	13.6–17.5	40–42	16–17.9	11.7–14.9	7.1–8.2	0	0.1	
	[135][c)]	90/10	525	69–74	2.4–3.2	12.5–13.7	41–42	15.6–17.8	11.2–13.6	7.5–8.4	0.1–1.4	0.2–0.5	0
	[135][c)]	90/10	500	78–87	1.5–2.2	15.20	55–57	15–20	2–3	4–6		0.2–0.3	0.8–0.9
4	[100][d)]	100/0	540	67–69	4	15–16	40–42	17.5–18.5	13–15	7–8	0	0	0
	[100][d)]	100/0	560	72.5–75.5	5–6	18–21	39–42	14.5–16	10–11.5	8–10	0	0.1	0
	[100][d)]	90/10	540	68.5	3	13	40.5	17.5	14	7.5	2	0.5	2
	[100][d)]	80/20	540	70–72	2–3	8–12	37–38	15–16	12–14	8.5–10	2.7–3.4	0.7–0.9	6–10
	[100][d)]	80/20	560	75–76	3.5	14–15	36–36.5	14	10.5	9	3.5–4	1	8
5	[135][e)]	100/0	540	63–70	3.5–5	15.5–18	40–41.5	17–19.5	12.5–17	6–7	0	0	0
	[135][e)]	95/5	540	63–71	2.5–3.5	12–19	39–42	39–42	12–17	5.7–7	1–2	0–0.5	1–4
	[135][e)]	90/10	540	64–70	2–3	10–14	37–41	37–41	13–17	6–7	1–2	0.5	3–5.5
6	[136][f)]	90/10	560	35–41	0.2–0.9	2–5	29–31	23–25	35–41	1–2.2	0.2–0.7[g)]		
	[136][f)]	90/10	560	64–69	0.9–2.0	12–15	37–41	16–19	15–18	4–5.5	0.3–1[g)]		

7	[137][h)]	100/0	482	16	1	2	11	10	64	2	
	[137][h)]	5/95	482	16	2	3	8	10	63	3	
	[137][h)]	10/90	482	18	3	2	9	10	60	5	
	[137][h)]	20/80	482	24	5	4	10	12	43	6	
	[137][h)]	30/70	482	35	7	5	10	11	33	14	
	[137][i)]	100/0	482	30	2	8	16	15	54	5	
	[137][i)]	80/20	482	41	4	9	17	16	37	10	
	[137][i)]	80/20	482	40	3	9	19	17	38	10	
	[137][i)]	80/20	482	38	3	9	18	16	42	8	
8	[99, 138][j)]	100/0	500	55	4	14	29	25	23	5	0[g)]
	[99, 138][j)]	80/20	500	61	6	13	35	23	18	4	1[g)]
	[99, 138][j)]	0/100	500	80	11	6	50	2	0	9	2[g)]

a) Fixed bed reactor (ID = 12 mm, L = 340 mm), 1 g catalyst, C/O = 2.9, 4.5, 5.9.
b) Fixed bed reactor (ID = 12 mm, L = 340 mm), 2 g catalyst, (i) pure VGO, C/O = 3.1, 6.0, 8.9; (ii) 10% HDO oil + VGO, C/O = 3, 4, 6; (iii) 10% bio-oil + VGO, C/O = 2.9, 3.2, 5.5.
c) Pilot scale FCC riser (ID = 7 mm, height = 9 m), C/O = 5.1, 6.3, 7.2, 8.7.
d) Demonstration scale FCC unit; feed rate = 150 kg/h.
e) Demonstration scale FCC unit; feed rate = 200 kg/h.
f) Micro activity test (MAT) unit with a quartz tube reactor (ID = 14 mm, L = 452 mm), 3–7 g catalyst.
g) $CO + CO_2$.
h) Micro activity test (MAT) unit with a quartz tube reactor (ID = 14 mm, L = 318 mm), 10 g catalyst; C/O = 1.
i) Micro activity test (MAT) unit with a quartz tube reactor (ID = 14 mm, L = 318 mm), 10 g catalyst; C/O = 3.
j) Riser simulator, C/O = 6 (for VGO and mix), 7.6 (for bio-oil).

oxygenated compounds. The formation of the first type of coke was attenuated by the presence of bio-oil in the feed as a consequence of the powerful synergistic effects that arise during bio-oil coprocessing.

A kinetic model based on work involving a reaction scheme with 10 real bio-oil components and 40 bio-oil pseudocomponents was proposed by Cruz et al. [139], the model proving to be robust enough to accurately predict both the behavior of VGO and hydrotreated bio-oil (20%) blends in FCC units and the temperature profile in the reactor.

In short, albeit coprocessing of bio-oils with traditional feedstock in FCC units presents great potential, technical difficulties (e.g. corrosion issues) need to be addressed, particularly if previously untreated bio-oils with a high oxygen content are to be employed [140].

Finally, although VGO is the most common feed for FCC units, it is not the only feedstock that has been tested in bio-oil coprocessing studies. For instance, de Miguel Mercader et al. [124] compared the cracking of the aqueous fraction of hydroprocessed bio-oil (20%) with a long residue (heaviest fraction from the crude oil distillation process) at 520 °C using an FCC catalyst in a MAT unit comprising a fluidized bed reactor. The results were similar to those obtained by cracking the residue alone in terms of conversion (60%), gasoline yield (44–46%), and LCO yield (23–25%). However, the authors also concluded that it is necessary to pretreat the bio-oil in order to avoid high coke and dry gas formation during the cracking of the undiluted bio-oil.

More recently, Errekatxo et al. [127] explored the use of glycerol and VGO blends (1/4 wt/wt) as a potential feedstock for FCC performed over a HY zeolite at 550 °C. The conversion of the blend was higher than that of VGO alone and produced less gasoline and diesel due to a higher yield of gases and coke, albeit gasoline with a higher RON was produced as a consequence of a higher amount of oxygenated moieties. Coke characterization studies suggested that coke formation occurs sequentially during coprocessing, oxygenated coke being formed first and its evolution resulting in a more aromatic type of coke, which is in agreement with previous reports by Corma et al. [125].

The potential of hydrotreated vegetable oils as a co-feed for FCC units has also been explored. As concluded by Al Sabawi et al. in their review [130], coprocessing vegetable oils with petroleum feedstock typically leads to the production of less gasoline and diesel along with higher octane and cetane numbers. These authors also concluded that HZSM-5 zeolites (either as an additive or as catalyst) enhance the production of both organic distillates and fuels.

3.5 Conclusions and Critical Discussion

Coprocessing bio-oil with the traditional VGO feed in FCC refinery units holds great potential for the valorization of biomass at an industrial scale, increasing the raw materials available for the production of fuels and added-value products (platform chemicals), and decreasing net CO_2 emissions. Interest in this approach is based on the positive outlook and technological development of biomass fast pyrolysis. Indeed, bio-oil can be produced from very diverse biomass sources and in a delocalized way using mobile fast pyrolysis units, which can be placed in temporary locations in which agricultural and forestry activities are being carried out. In this regard, the fast pyrolysis of shrubbery, herbaceous plants, and subproducts from the wood industry is particularly interesting, since the valorization of the resulting bio-oil would contribute to improving the economy of agricultural and sparsely populated regions. Moreover, the valorization of these residual materials can also help preserve areas threatened by fires.

The FCC unit is suitable for bio-oil valorization mainly due to its great versatility, which allows for processing diverse feeds and, given its great capacity, could process bio-oil produced in – and transported – from different locations. Therefore, advances in the standardization of raw bio-oil composition and in its storage and transport stability are key, particularly considering that the feed composition is a crucial parameter in the operation of FCC units.

The results reported to date in the literature regarding the coprocessing of mixtures of raw bio-oil and traditional refinery feeds (such as VGO) are encouraging but rather scarce, few studies having been performed using experimental conditions resembling those of an industrial riser reactor. However, numerous academic studies have demonstrated the various advantages of coprocessing oxygenated bio-oil components with hydrocarbon VGO components, which have been attributed to synergies in the reaction mechanisms and to the attenuation in coke formation relative to the coke formed when processing hydrocarbons alone. Given the magnitude of the FCC unit, it is necessary to boost research with realistic feeds (including VGO), with commercial catalysts and with pilot plant scales. Among the aspects on which further research should focus, the effect of bio-oil co-feeding on catalyst deactivation and regeneration is paramount, mainly due to its strong impact on the energetic balance of FCC units consisting of a reactor and an interconnected regenerator. Another important aspect to be explored further is the role of (unreacted) oxygenated compounds in the gasoline and diesel product streams. Likewise, progress in the kinetic modeling of the cracking of mixed feeds is necessary to attain precise control of the unit and for establishing optimal reaction conditions in the reaction–regeneration sections.

Besides representing a technological challenge, coprocessing bio-oil in refinery units has economic, social, and ethical implications for traditional petroleum refineries. Given the average FCC unit capacity (50 000 barrels per day), it is reasonable to assume that bio-oil coprocessing would be limited to only a moderate fraction of that total amount. Consequently, the economic impact of incorporating this new feed in already amortized refinery units would also be relatively moderate. However, the coprocessing refinery could adopt the "sustainable refinery" role and become the recipient of economic stimuli according to their contribution to the reduction in CO_2 emissions. Furthermore, the social impact of this approach is of great importance, as it integrates several economic and social interests, helping the petroleum industry contribute not only to the remission of net CO_2 emissions but also to boosting the economy of other sectors (i.e. agricultural, forestry, eco-industry). In summary, by leveraging the potential of the coprocessing approach, refineries could progressively contribute to a bio-based economy for green fuels jointly produced and commercialized with petroleum-derived products.

References

1 Huber, G.W., O'Connor, P., and Corma, A. (2007). Processing biomass in conventional oil refineries: production of high quality diesel by hydrotreating vegetable oils in heavy vacuum oil mixtures. *Appl. Catal., A* 329: 120–129.

2 Carpenter, D., Westover, T.L., Czernik, S., and Jablonski, W. (2014). Biomass feedstocks for renewable fuel production: a review of the impacts of feedstock and pretreatment on the yield and product distribution of fast pyrolysis bio-oils and vapors. *Green Chem.* 16: 384–406.

3 Hansen, A.C., Zhang, Q., and Lyne, P.W.L. (2005). Ethanol-diesel fuel blends – a review. *Bioresour. Technol.* 96: 277–285.

4 Aydın, F. and Öğüt, H. (2017). Effects of using ethanol-biodiesel-diesel fuel in single cylinder diesel engine to engine performance and emissions. *Renewable Energy* 103: 688–694.

5 Lapuerta, M., Armas, O., and Herreros, J.M. (2008). Emissions from a diesel-bioethanol blend in an automotive diesel engine. *Fuel* 87: 25–31.

6 Gayubo, A.G., Alonso, A., Valle, B. et al. (2010). Hydrothermal stability of HZSM-5 catalysts modified with Ni for the transformation of bioethanol into hydrocarbons. *Fuel* 89: 3365–3372.

7 Gayubo, A.G., Alonso, A., Valle, B. et al. (2010). Selective production of olefins from bioethanol on HZSM-5 zeolite catalysts treated with NaOH. *Appl. Catal., B* 97: 299–306.

8 Goto, D., Harada, Y., Furumoto, Y. et al. (2010). Conversion of ethanol to propylene over HZSM-5 type zeolites containing alkaline earth metals. *Appl. Catal., A* 383: 89–95.

9 Garbarino, G., Riani, P., Lucchini, M.A. et al. (2013). Cobalt-based nanoparticles as catalysts for low temperature hydrogen production by ethanol steam reforming. *Int. J. Hydrogen Energy* 38: 82–91.

10 Vicente, J., Montero, C., Ereña, J. et al. (2014). Coke deactivation of Ni and co catalysts in ethanol steam reforming at mild temperatures in a fluidized bed reactor. *Int. J. Hydrogen Energy* 39: 12586–12596.

11 Montero, C., Remiro, A., Benito, P.L. et al. (2018). Optimum operating conditions in ethanol steam reforming over a $Ni/La_2O_3{-}\alpha Al_2O_3$ catalyst in a fluidized bed reactor. *Fuel Process. Technol.* 169: 207–216.

12 Banerjee, S., Mudliar, S., Sen, R. et al. (2010). Commercializing lignocellulosic bioethanol: technology bottlenecks and possible remedies. *Biofuels, Bioprod. Biorefin.* 4: 77–93.

13 Sánchez, O.J. and Cardona, C.A. (2005). Biotechnological production of fuel ethanol. I: obtaining from various raw materials. *Interciencia* 30: 22–47.

14 Sánchez, O.J. and Cardona, C.A. (2005). Biotechnological production of fuel ethanol. II: integration of processes. *Interciencia* 30: 48–64.

15 Van Zyl, W.H., Lynd, L.R., Den Haan, R., and McBride, J.E. (2007). Consolidated bioprocessing for bioethanol production using *Saccharomyces cerevisiae*. *Adv. Biochem. Eng./Biotechnol.* 108: 205–235.

16 Zhang, W. (2010). Automotive fuels from biomass via gasification. *Fuel Process. Technol.* 91: 866–876.

17 Bulushev, D.A. and Ross, J.R.H. (2011). Catalysis for conversion of biomass to fuels via pyrolysis and gasification: a review. *Catal. Today* 171: 1–13.

18 De Lasa, H., Salaices, E., Mazumder, J., and Lucky, R. (2011). Catalytic steam gasification of biomass: catalysts, thermodynamics and kinetics. *Chem. Rev.* 111: 5404–5433.

19 Pinho, A.D.R., De Almeida, M.B.B., Mendes, F.L., and Ximenes, V.L. (2014). Production of lignocellulosic gasoline using fast pyrolysis of biomass and a conventional refining scheme. *Pure Appl. Chem.* 86: 859–865.

20 Huber, G.W., Iborra, S., and Corma, A. (2006). Synthesis of transportation fuels from biomass: chemistry, catalysts, and engineering. *Chem. Rev.* 106: 4044–4098.

21 Meier, D., Van De Beld, B., Bridgwater, A.V. et al. (2013). State-of-the-art of fast pyrolysis in IEA bioenergy member countries. *Renewable Sustainable Energy Rev.* 220: 619–641.

22 Guedes, R.E., Luna, A.S., and Torres, A.R. (2018). Operating parameters for bio-oil production in biomass pyrolysis: a review. *J. Anal. Appl. Pyrolysis* 129: 134–149.

23 Bridgwater, A.V. (2012). Review of fast pyrolysis of biomass and product upgrading. *Biomass Bioenergy* 38: 68–94.

24 Isahak, W.N.R.W., Hisham, M.W.M., Yarmo, M.A., and Yun Hin, T.Y. (2012). A review on bio-oil production from biomass by using pyrolysis method. *Renewable Sustainable Energy Rev.* 16: 5910–5923.

25 DeSisto, W.J., Hill, N., Beis, S.H. et al. (2010). Fast pyrolysis of pine sawdust in a fluidized-bed reactor. *Energy Fuels* 24: 2642–2651.

26 Matta, J., Bronson, B., Gogolek, P.E.G. et al. (2017). Comparison of multi-component kinetic relations on bubbling fluidized-bed woody biomass fast pyrolysis reactor model performance. *Fuel* 210: 625–638.

27 Dong, N.H., Luo, K.H., and Wang, Q. (2017). Modeling of biomass pyrolysis in a bubbling fluidized bed reactor: impact of intra-particle heat conduction. *Fuel Process. Technol.* 161: 199–203.

28 Van de Velden, M., Baeyens, J., and Boukis, I. (2008). Modeling CFB biomass pyrolysis reactors. *Biomass Bioenergy* 32: 128–139.

29 Luo, G., Chandler, D.S., Anjos, L.C.A. et al. (2017). Pyrolysis of whole wood chips and rods in a novel ablative reactor. *Fuel* 194: 229–238.

30 Alvarez, J., Lopez, G., Amutio, M. et al. (2014). Bio-oil production from rice husk fast pyrolysis in a conical spouted bed reactor. *Fuel* 128: 162–169.

31 Arregi, A., Lopez, G., Amutio, M. et al. (2018). Role of operating conditions in the catalyst deactivation in the in-line steam reforming of volatiles from biomass fast pyrolysis. *Fuel* 216: 233–244.

32 Zheng, Y., Zhang, Y., Xu, J. et al. (2017). Co-pyrolysis behavior of fermentation residues with woody sawdust by thermogravimetric analysis and a vacuum reactor. *Bioresour. Technol.* 245: 449–455.

33 Yang, X., Yuan, C., Xu, J., and Zhang, W. (2014). Co-pyrolysis of Chinese lignite and biomass in a vacuum reactor. *Bioresour. Technol.* 173: 1–5.

34 Amutio, M., Lopez, G., Aguado, R. et al. (2011). Effect of vacuum on lignocellulosic biomass flash pyrolysis in a conical spouted bed reactor. *Energy Fuels* 25: 3950–3960.

35 Amutio, M., Lopez, G., Artetxe, M. et al. (2012). Influence of temperature on biomass pyrolysis in a conical spouted bed reactor. *Resour. Conserv. Recycl.* 259: 23–31.

36 Czernik, S. and Bridgwater, A.V. (2004). Overview of applications of biomass fast pyrolysis oil. *Energy Fuels* 18: 590–598.

37 Mohan, D., Pittman, C.U. Jr.,, and Steele, P.H. (2006). Pyrolysis of wood/biomass for bio-oil: a critical review. *Energy Fuels* 20: 848–889.

38 Lu, Q., Li, W.Z., and Zhu, X.F. (2009). Overview of fuel properties of biomass fast pyrolysis oils. *Energy Convers. Manage.* 50: 1376–1383.

39 Gollakota, A.R.K., Reddy, M., Subramanyam, M.D., and Kishore, N. (2016). A review on the upgradation techniques of pyrolysis oil. *Renewable Sustainable Energy Rev.* 58: 1543–1568.

40 Fratini, E., Bonini, M., Oasmaa, A. et al. (2006). SANS analysis of the microstructural evolution during the aging of pyrolysis oils from biomass. *Langmuir* 22: 306–312.

41 Dhyani, V. and Bhaskar, T. (2018). A comprehensive review on the pyrolysis of lignocellulosic biomass. *Renewable Energy* 129: 695–716.

42 Lazzari, E., Schena, T., Marcelo, M.C.A. et al. (2018). Classification of biomass through their pyrolytic bio-oil composition using FTIR and PCA analysis. *Ind. Crops Prod.* 111: 856–864.

43 Ozagac, M., Bertino-Ghera, C., Uzio, D. et al. (2018). Catalytic hydroconversion of pyrolytic bio-oil: understanding and limiting macromolecules formation. *Biomass Bioenergy* 108: 501–510.

44 Keleş, S., Kar, T., Akgün, M., and Kaygusuz, K. (2017). Catalytic fast pyrolysis of hazelnut cupula: characterization of bio-oil. *Energy Sources Part A* 39: 2216–2225.

45 Oasmaa, A. and Czernik, S. (1999). Fuel oil quality of biomass pyrolysis oils – state of the art for the end users. *Energy Fuels* 13: 914–921.

46 Chen, J.-H., Lu, L., and Wang, S.-R. (2017). Mild hydrogenation of simulated bio-oil based on molecular distillation. *J. Fuel Chem. Technol.* 45: 1056–1063.

47 Wang, Y., Wang, S., Leng, F. et al. (2015). Separation and characterization of pyrolytic lignins from the heavy fraction of bio-oil by molecular distillation. *Sep. Purif. Technol.* 152: 123–132.

48 Aho, A., Salmi, T., and Murzin, D.Y. (2013). Catalytic pyrolysis of lignocellulosic biomass, Chapter 5. In: *The Role of Catalysis for the Sustainable Production of Bio-Fuels and Bio-Chemicals*, 137–159. Elsevier.

49 Iliopoulou, E.F., Stefanidis, S., Kalogiannis, K. et al. (2014). Pilot-scale validation of co-ZSM-5 catalyst performance in the catalytic upgrading of biomass pyrolysis vapours. *Green Chem.* 16: 662–674.

50 Naqvi, S.R. and Naqvi, M. (2018). Catalytic fast pyrolysis of rice husk: influence of commercial and synthesized microporous zeolites on deoxygenation of biomass pyrolysis vapors. *Int. J. Energy Res.* 42: 1352–1362.

51 Kabir, G. and Hameed, B.H. (2017). Recent progress on catalytic pyrolysis of lignocellulosic biomass to high-grade bio-oil and bio-chemicals. *Renewable Sustainable Energy Rev.* 70: 945–967.

52 Olazar, M., Aguado, R., Bilbao, J., and Barona, A. (2000). Pyrolysis of sawdust in a conical spouted-bed reactor with a HZSM-5 catalyst. *AlChE J.* 46: 1025–1033.

53 Zhang, H., Shao, S., Jiang, Y. et al. (2017). Improving hydrocarbon yield by two-step pyrolysis of pinewood in a fluidized-bed reactor. *Fuel Process. Technol.* 159: 19–26.

54 Lappas, A.A., Samolada, M.C., Iatridis, D.K. et al. (2002). Biomass pyrolysis in a circulating fluid bed reactor for the production of fuels and chemicals. *Fuel* 81: 2087–2095.

55 Atutxa, A., Aguado, R., Gayubo, A.G. et al. (2005). Kinetic description of the catalytic pyrolysis of biomass in a conical spouted bed reactor. *Energy Fuels* 19: 765–774.

56 Aho, A., Kumar, N., Eränen, K. et al. (2008). Catalytic pyrolysis of woody biomass in a fluidized bed reactor: influence of the zeolite structure. *Fuel* 87: 2493–2501.

57 Iliopoulou, E.F., Antonakou, E.V., Karakoulia, S.A. et al. (2007). Catalytic conversion of biomass pyrolysis products by mesoporous materials: effect of steam stability and acidity of Al-MCM-41 catalysts. *Chem. Eng. J.* 134: 51–57.

58 Li, X., Dong, L., Zhang, J. et al. (2017). In-situ catalytic upgrading of biomass-derived vapors using HZSM-5 and MCM-41: effects of mixing ratios on bio-oil preparation. *J. Energy Inst.* https://doi.org/10.1016/j.joei.2017.10.015.

59 Torri, C., Reinikainen, M., Lindfors, C. et al. (2010). Investigation on catalytic pyrolysis of pine sawdust: catalyst screening by Py-GC-MIP-AED. *J. Anal. Appl. Pyrolysis* 88: 7–13.

60 Lappas, A.A., Dimitropoulos, V.S., Antonakou, E.V. et al. (2008). Design, construction, and operation of a transported fluid bed process development unit for biomass fast pyrolysis: effect of pyrolysis temperature. *Ind. Eng. Chem. Res.* 47: 742–747.

61 Zhang, H., Xiao, R., Wang, D. et al. (2009). Catalytic fast pyrolysis of biomass in a fluidized bed with fresh and spent fluidized catalytic cracking (FCC) catalysts. *Energy Fuels* 223: 6199–6206.

62 Ro, D., Kim, Y.M., Lee, I.G. et al. (2018). Bench scale catalytic fast pyrolysis of empty fruit bunches over low cost catalysts and HZSM-5 using a fixed bed reactor. *J. Cleaner Prod.* 176: 298–303.

63 Du, S., Gamliel, D.P., Valla, J.A., and Bollas, G.M. (2016). The effect of ZSM-5 catalyst support in catalytic pyrolysis of biomass and compounds abundant in pyrolysis bio-oils. *J. Anal. Appl. Pyrolysis* 122: 7–12.

64 Jia, L.Y., Raad, M., Hamieh, S. et al. (2017). Catalytic fast pyrolysis of biomass: superior selectivity of hierarchical zeolites to aromatics. *Green Chem.* 19: 5442–5459.

65 Jae, J., Tompsett, G.A., Foster, A.J. et al. (2011). Investigation into the shape selectivity of zeolite catalysts for biomass conversion. *J. Catal.* 279: 257–268.

66 Zhu, L., Li, K., Ding, H., and Zhu, X. (2018). Studying on properties of bio-oil by adding blended additive during aging. *Fuel* 211: 704–711.
67 Chen, D., Zhou, J., Zhang, Q., and Zhu, X. (2014). Evaluation methods and research progresses in bio-oil storage stability. *Renewable Sustainable Energy Rev.* 40: 69–79.
68 Meng, J., Moore, A., Tilotta, D. et al. (2014). Toward understanding of bio-oil aging: accelerated aging of bio-oil fractions. *ACS Sustainable Chem. Eng.* 2: 2011–2018.
69 Talmadge, M.S., Baldwin, R.M., Biddy, M.J. et al. (2014). A perspective on oxygenated species in the refinery integration of pyrolysis oil. *Green Chem.* 16: 407–453.
70 Diebold, J.P. and Czernik, S. (1997). Additives to lower and stabilize the viscosity of pyrolysis oils during storage. *Energy Fuels* 11: 1081–1091.
71 de Miguel Mercader, F., Groeneveld, M.J., Kersten, S.R.A. et al. (2010). Pyrolysis oil upgrading by high pressure thermal treatment. *Fuel* 89: 2829–2837.
72 Bertero, M., De La Puente, G., and Sedran, U. (2011). Effect of pyrolysis temperature and thermal conditioning on the coke-forming potential of bio-oils. *Energy Fuels* 25: 1267–1275.
73 Alsbou, E. and Helleur, B. (2014). Accelerated aging of bio-oil from fast pyrolysis of hardwood. *Energy Fuels* 28: 3224–3235.
74 Ben, H. and Ragauskas, A.J. (2012). In situ NMR characterization of pyrolysis oil during accelerated aging. *ChemSusChem* 5: 1687–1693.
75 Xiu, S. and Shahbazi, A. (2012). Bio-oil production and upgrading research: a review. *Renewable Sustainable Energy Rev* 16: 4406–4414.
76 Saidi, M., Samimi, F., Karimipourfard, D. et al. (2014). Upgrading of lignin-derived bio-oils by catalytic hydrodeoxygenation. *Energy Environ. Sci.* 7: 103–129.
77 Patel, M. and Kumar, A. (2016). Production of renewable diesel through the hydroprocessing of lignocellulosic biomass-derived bio-oil: a review. *Renewable Sustainable Energy Rev.* 58: 1293–1307.
78 Nabgan, W., Tuan Abdullah, T.A., Mat, R. et al. (2017). Renewable hydrogen production from bio-oil derivative via catalytic steam reforming: an overview. *Renewable Sustainable Energy Rev.* 79: 347–357.
79 Ma, Z., Xiao, R., and Zhang, H. (2017). Catalytic steam reforming of bio-oil model compounds for hydrogen-rich gas production using bio-char as catalyst. *Int. J. Hydrogen Energy* 42: 3579–3585.
80 Bai, X., Kim, K.H., Brown, R.C. et al. (2014). Formation of phenolic oligomers during fast pyrolysis of lignin. *Fuel* 128: 170–179.
81 González-Gil, R., Chamorro-Burgos, I., Herrera, C. et al. (2015). Production of hydrogen by catalytic steam reforming of oxygenated model compounds on Ni-modified supported catalysts. Simulation and experimental study. *Int. J. Hydrogen Energy* 40: 11217–11227.
82 Yadav, A.K. and Vaidya, P.D. (2017). Kinetic investigation on butanol steam reforming over Ru/Al_2O_3 catalyst. *Int. J. Hydrogen Energy* 42: 25203–25212.

83 Mei, Y., Wu, C., and Liu, R. (2016). Hydrogen production from steam reforming of bio-oil model compound and byproducts elimination. *Int. J. Hydrogen Energy* 41: 9145–9152.

84 Resende, K.A., Ávila-Neto, C.N., Rabelo-Neto, R.C. et al. (2015). Thermodynamic analysis and reaction routes of steam reforming of bio-oil aqueous fraction. *Renewable Energy* 80: 166–176.

85 Bimbela, F., Ábrego, J., Puerta, R. et al. (2017). Catalytic steam reforming of the aqueous fraction of bio-oil using Ni–Ce/Mg–Al catalysts. *Appl. Catal., B* 209: 346–357.

86 Valle, B., Aramburu, B., Olazar, M. et al. (2018). Steam reforming of raw bio-oil over Ni/La_2O_3–αAl_2O_3: influence of temperature on product yields and catalyst deactivation. *Fuel* 216: 463–474.

87 Arandia, A., Remiro, A., Oar-Arteta, L. et al. (2017). Reaction conditions effect and pathways in the oxidative steam reforming of raw bio-oil on a Rh/CeO_2–ZrO_2 catalyst in a fluidized bed reactor. *Int. J. Hydrogen Energy* 42: 29175–29185.

88 Effendi, A., Gerhauser, H., and Bridgwater, A.V. (2008). Production of renewable phenolic resins by thermochemical conversion of biomass: a review. *Renewable Sustainable Energy Rev.* 12: 2092–2116.

89 Pang, B., Yang, S., Fang, W. et al. (2017). Structure-property relationships for technical lignins for the production of lignin-phenol-formaldehyde resins. *Ind. Crops Prod.* 108: 316–326.

90 Wang, S., Wang, Y., Cai, Q. et al. (2014). Multi-step separation of monophenols and pyrolytic lignins from the water-insoluble phase of bio-oil. *Sep. Purif. Technol.* 122: 248–255.

91 Xu, J., Jiang, J., Lv, W. et al. (2010). Rice husk bio-oil upgrading by means of phase separation and the production of esters from the water phase, and novolac resins from the insoluble phase. *Biomass Bioenergy* 34: 1059–1063.

92 Bennett, N.M., Helle, S.S., and Duff, S.J.B. (2009). Extraction and hydrolysis of levoglucosan from pyrolysis oil. *Bioresour. Technol.* 100: 6059–6063.

93 Wang, J.-Q., Zheng, J.-L., Wang, J.-T., and Lu, Z.-M. (2018). A separation and quantification method of levoglucosan in biomass pyrolysis. *Ind. Crops Prod.* 113: 266–273.

94 Arandes, J.M., Torre, I., Azkoiti, M.J. et al. (2008). Effect of atmospheric residue incorporation in the fluidized catalytic cracking (FCC) feed on product stream yields and composition. *Energy Fuels* 22: 2149–2156.

95 Arandes, J.M., Torre, I., Azkoiti, M.J. et al. (2009). HZSM-5 zeolite as catalyst additive for residue cracking under FCC conditions. *Energy Fuels* 23: 4215–4223.

96 Passamonti, F.J. and Sedran, U. (2012). Recycling of waste plastics into fuels. LDPE conversion in FCC. *Appl. Catal., B* 125: 499–506.

97 Arandes, J.M., Torre, I., Azkoiti, M.J. et al. (2008). Effect of catalyst properties on the cracking of polypropylene pyrolysis waxes under FCC conditions. *Catal. Today* 133-135: 413–419.

98 Bielansky, P., Weinert, A., Schönberger, C., and Reichhold, A. (2011). Catalytic conversion of vegetable oils in a continuous FCC pilot plant. *Fuel Process. Technol.* 92: 2305–2311.
99 Ibarra, A., Veloso, A., Bilbao, J. et al. (2016). Dual coke deactivation pathways during the catalytic cracking of raw bio-oil and vacuum gasoil in FCC conditions. *Appl. Catal., B* 182: 336–346.
100 Pinho, A.D.R., de Almeida, M.B.B., Mendes, F.L. et al. (2015). Co-processing raw bio-oil and gasoil in an FCC unit. *Fuel Process. Technol.* 131: 159–166.
101 Castello, D. and Rosendahl, L. (2018). Coprocessing of pyrolysis oil in refineries, Chapter 9. In: *Direct Thermochemical Liquefaction for Energy Applications*, 293–317. Woodhead Publishing.
102 Jarullah, A.T., Awad, N.A., and Mujtaba, I.M. (2017). Optimal design and operation of an industrial fluidized catalytic cracking reactor. *Fuel* 206: 657–674.
103 Corma, A. and Sauvanaud, L. (2013). FCC testing at bench scale: new units, new processes, new feeds. *Catal. Today* 218-219: 107–114.
104 Gayubo, A.G., Aguayo, A.T., Atutxa, A. et al. (2004). Transformation of oxygenate components of biomass pyrolysis oil on a HZSM-5 zeolite. I. Alcohols and phenols. *Ind. Eng. Chem. Res.* 43: 2610–2618.
105 Gayubo, A.G., Aguayo, A.T., Atutxa, A. et al. (2004). Transformation of oxygenate components of biomass pyrolysis oil on a HZSM-5 zeolite. II. Aldehydes, ketones, and acids. *Ind. Eng. Chem. Res.* 43: 2619–2626.
106 Graça, I., Comparot, J.D., Laforge, S. et al. (2009). Effect of phenol addition on the performances of H-Y zeolite during methylcyclohexane transformation. *Appl. Catal., A* 353: 123–129.
107 Graça, I., Comparot, J.D., Laforge, S. et al. (2009). Influence of phenol addition on the H-ZSM-5 zeolite catalytic properties during methylcyclohexane transformation. *Energy Fuels* 23: 4224–4230.
108 Graça, I., Lopes, J.M., Ribeiro, M.F. et al. (2012). *N*-heptane cracking over mixtures of HY and HZSM-5 zeolites: influence of the presence of phenol. *Fuel* 94: 571–577.
109 Viswanadham, N. and Saxena, S.K. (2013). Enhanced performance of nano-crystalline ZSM-5 in acetone to gasoline (ATG) reaction. *Fuel* 105: 490–495.
110 Gayubo, A.G., Aguayo, A.T., Atutxa, A. et al. (2005). Undesired components in the transformation of biomass pyrolysis oil into hydrocarbons on an HZSM-5 zeolite catalyst. *J. Chem. Technol. Biotechnol.* 80: 1244–1251.
111 Wang, S., Wang, Y., Cai, Q., and Guo, Z. (2014). Production of bio-gasoline by co-cracking of acetic acid in bio-oil and ethanol. *Chin. J. Chem. Eng.* 22: 98–103.
112 Graça, I., Ribeiro, F.R., Cerqueira, H.S. et al. (2009). Catalytic cracking of mixtures of model bio-oil compounds and gasoil. *Appl. Catal., B* 90: 556–563.
113 Adjaye, J.D. and Bakhshi, N.N. (1995). Production of hydrocarbons by catalytic upgrading of a fast pyrolysis bio-oil. Part I: conversion over various catalysts. *Fuel Process. Technol.* 45: 161–183.

114 Adjaye, J.D. and Bakhshi, N.N. (1995). Production of hydrocarbons by catalytic upgrading of a fast pyrolysis bio-oil. Part II: comparative catalyst performance and reaction pathways. *Fuel Process. Technol.* 45: 185–202.

115 Williams, P.T. and Horne, P.A. (1994). Characterisation of oils from the fluidised bed pyrolysis of biomass with zeolite catalyst upgrading. *Biomass Bioenergy* 7: 223–236.

116 Gayubo, A.G., Aguayo, A.T., Atutxa, A. et al. (2004). Deactivation of a HZSM-5 zeolite catalyst in the transformation of the aqueous fraction of biomass pyrolysis oil into hydrocarbons. *Energy Fuels* 18: 1640–1647.

117 Gayubo, A.G., Valle, B., Aguayo, A.T. et al. (2010). Pyrolytic lignin removal for the valorization of biomass pyrolysis crude bio-oil by catalytic transformation. *J. Chem. Technol. Biotechnol.* 85: 132–144.

118 Valle, B., Gayubo, A.G., Aguayo, A.T. et al. (2010). Selective production of aromatics by crude bio-oil valorization with a nickel-modified HZSM-5 zeolite catalyst. *Energy Fuels* 24: 2060–2070.

119 Valle, B., Gayubo, A.G., Alonso, A. et al. (2010). Hydrothermally stable HZSM-5 zeolite catalysts for the transformation of crude bio-oil into hydrocarbons. *Appl. Catal., B* 100: 318–327.

120 Gayubo, A.G., Valle, B., Aguayo, A.T. et al. (2009). Attenuation of catalyst deactivation by cofeeding methanol for enhancing the valorisation of crude bio-oil. *Energy Fuels* 23: 4129–4136.

121 Bertero, M. and Sedran, U. (2013). Upgrading of bio-oils over equilibrium FCC catalysts. Contribution from alcohols, phenols and aromatic ethers. *Catal. Today* 212: 10–15.

122 Bertero, M., de la Puente, G., and Sedran, U. (2013). Products and coke from the conversion of bio-oil acids, esters, aldehydes and ketones over equilibrium FCC catalysts. *Renewable Energy* 60: 349–354.

123 Valle, B., Castaño, P., Olazar, M. et al. (2012). Deactivating species in the transformation of crude bio-oil with methanol into hydrocarbons on a HZSM-5 catalyst. *J. Catal.* 285: 304–314.

124 de Miguel Mercader, F., Groeneveld, M.J., Kersten, S.R.A. et al. (2010). Production of advanced biofuels: co-processing of upgraded pyrolysis oil in standard refinery units. *Appl. Catal., B* 96: 57–66.

125 Corma, A., Huber, G.W., Sauvanaud, L., and O'Connor, P. (2007). Processing biomass-derived oxygenates in the oil refinery: catalytic cracking (FCC) reaction pathways and role of catalyst. *J. Catal.* 247: 307–327.

126 Fogassy, G., Thegarid, N., Toussaint, G. et al. (2010). Biomass derived feedstock co-processing with vacuum gas oil for second-generation fuel production in FCC units. *Appl. Catal., B* 96: 476–485.

127 Errekatxo, A., Ibarra, A., Gutierrez, A. et al. (2017). Catalytic deactivation pathways during the cracking of glycerol and glycerol/VGO blends under FCC unit conditions. *Chem. Eng. J.* 307: 955–965.

128 Naik, D.V., Kumar, V., Prasad, B. et al. (2014). Catalytic cracking of pyrolysis oil oxygenates (aliphatic and aromatic) with vacuum gas oil and their characterization. *Chem. Eng. Res. Des.* 92: 1579–1590.

129 Huber, G.W. and Corma, A. (2007). Synergies between bio- and oil refineries for the production of fuels from biomass. *Angew. Chem. Int. Ed.* 46: 7184–7201.
130 Al-Sabawi, M., Chen, J., and Ng, S. (2012). Fluid catalytic cracking of biomass-derived oils and their blends with petroleum feedstocks: a review. *Energy Fuels* 26: 5355–5372.
131 Bertero, M. and Sedran, U. (2013). Conversion of pine sawdust bio-oil (raw and thermally processed) over equilibrium FCC catalysts. *Bioresour. Technol.* 135: 644–651.
132 Thegarid, N., Fogassy, G., Schuurman, Y. et al. (2014). Second-generation biofuels by co-processing catalytic pyrolysis oil in FCC units. *Appl. Catal., B* 145: 161–166.
133 Gueudré, L., Thegarid, N., Burel, L. et al. (2015). Coke chemistry under vacuum gasoil/bio-oil FCC co-processing conditions. *Catal. Today* 257 (Part 2): 200–212.
134 Wang, C., Li, M., and Fang, Y. (2016). Coprocessing of catalytic-pyrolysis-derived bio-oil with VGO in a pilot-scale FCC riser. *Ind. Eng. Chem. Res.* 55: 3525–3534.
135 Pinho, A.D.R., de Almeida, M.B.B., Mendes, F.L. et al. (2017). Fast pyrolysis oil from pinewood chips co-processing with vacuum gas oil in an FCC unit for second generation fuel production. *Fuel* 188: 462–473.
136 Gueudré, L., Chapon, F., Mirodatos, C. et al. (2017). Optimizing the bio-gasoline quantity and quality in fluid catalytic cracking co-refining. *Fuel* 192: 60–70.
137 Lindfors, C., Paasikallio, V., Kuoppala, E. et al. (2015). Co-processing of dry bio-oil, catalytic pyrolysis oil, and hydrotreated bio-oil in a micro activity test unit. *Energy Fuels* 29: 3707–3714.
138 Ibarra, A., Rodríguez, E., Sedran, U. et al. (2016). Synergy in the cracking of a blend of bio-oil and vacuum gasoil under fluid catalytic cracking conditions. *Ind. Eng. Chem. Res.* 55: 1872–1880.
139 Cruz, P.L., Montero, E., and Dufour, J. (2017). Modelling of co-processing of HDO-oil with VGO in a FCC unit. *Fuel* 196: 362–370.
140 Brady, M.P., Keiser, J.R., Leonard, D.N. et al. (2017). Corrosion of stainless steels in the riser during co-processing of bio-oils in a fluid catalytic cracking pilot plant. *Fuel Process. Technol.* 159: 187–199.

4

Stabilization of Bio-oil via Esterification

Xun Hu

University of Jinan, School of Material Science and Engineering, 36, NanXinzhuang West Road, Jinan, 250022, China

4.1 Introduction

Biomass is an organic matter formed by the capture of pockets of solar energy with chlorophyll, which has provided us with food, clothes, and energy from the very beginning of human history. Until the late 1800s, biomass still provided more than 90% of world energy supply, after which biomass use for energy began to decrease as fossil fuels became the preferred energy resource [1]. However, the non-renewable nature of fossil fuels and the pollution caused by their use have brought a number of enormous social issues (i.e. constant fighting between countries for controlling oil resources) and environmental issues (i.e. CO_2 emissions) [2]. In this context, the energy stored in renewable, sustainable, and carbon-neutral biomass has again attracted great interest in recent years [3, 4]. It is predicted that bioenergy (including traditional and modern uses) will become the most important renewable energy source for the world, providing 10–15% of the world's energy demand by 2035.

There are several ways for extracting energy from biomass. The most traditional one is direct combustion. Biomass can also be gasified to produce syngas, which can be used as gaseous fuel or be used as a feedstock for chemicals production via Fischer–Tropsch (F–T) synthesis [5, 6]. Another important route is the pyrolysis of biomass to produce gaseous products, condensable liquid (bio-oil), and solid bio-char [7, 8]. The gaseous products are a mixture of hydrogen, methane, and carbon oxides, which can be burned to provide the energy for biomass pyrolysis. Bio-char contains the inorganic nutrients gathered by plants during growth, which can be returned to the field and be utilized by plants again.

Bio-oil is the liquid product from pyrolysis of biomass. It is generally viscous and has a dark brown color [9, 10]. Complexity is a major feature of bio-oil. This is because bio-oil is a very complex mixture of water and organic compounds including sugars, sugar monomers, furans, aldehydes, ketones, carboxylic acids, phenolics, pyrolytic lignin, and so on [11, 12], as shown in Table 4.1 [13]. Most of these components contain some oxygen-containing functionalities, which

Chemical Catalysts for Biomass Upgrading, First Edition. Edited by Mark Crocker and Eduardo Santillan-Jimenez.

Table 4.1 Selected compounds in the water-like phase and in the paste-like phase of bio-oil produced from pyrolysis of mallee bark.

Compounds	Retention time (min)	In water-like phase (area %)	In paste-like phase (area %)
Acids			
Acetic acid	8.09	11.76	4.70
Propanoic acid	9.38	1.04	1.07
Butyric acid	10.51	0.25	0.35
Levulinic acid	17.59	0.37	0.25
n-Hexadecanoic acid	22.30	0.07	1.14
Ketones and aldehydes			
2-Propanone, 1-hydroxy-	5.59	6.29	2.11
Acetaldehyde, hydroxy-	5.86	1.72	0.51
3-Pentanone, 2-methyl-	9.31	0.29	0.26
1-Hydroxy-2-butanone	6.93	1.07	0.59
Cyclopentenones			
2-Cyclopenten-1-one	6.55	1.41	1.58
2-Cyclopenten-1-one, 2-methyl-	6.81	0.45	0.73
2-Cyclopenten-1-one, 3-methyl-	9.07	0.75	1.03
1,2-Cyclopentanedione	12.11	2.63	1.88
2-Cyclopenten-1-one, 2-hydroxy-3-methyl-	12.78	2.44	2.74
2-Cyclopenten-1-one, 3-ethyl-2-hydroxy-	13.49	0.27	0.36
Esters			
1,2-Ethanediol, diacetate	8.44	1.56	1.30
2-Pentenoic acid, 4-methyl-, methyl ester	12.98	0.41	0.71
Alcohols			
1,2-Ethanediol	10.44	0.76	0.54
Furans			
Furfural	8.34	4.54	5.85
Ethanone, 1-(2-furanyl)-	8.91	0.29	0.38
2-Furancarboxaldehyde, 5-methyl-	9.82	0.90	2.16
2-Furanmethanol	10.88	0.86	0.96
2(5*H*)-Furanone, 5-methyl-	11.08	0.43	0.64
2(5*H*)-Furanone, 3-methyl-	11.52	0.28	0.66

Table 4.1 (Continued)

Compounds	Retention time (min)	In water-like phase (area %)	In paste-like phase (area %)
2(5*H*)-Furanone	11.92	0.90	0.61
4-Methyl-5*H*-furan-2-one	13.41	0.34	0.46
2-Furancarboxaldehyde, 5-(hydroxymethyl)-	19.04	2.32	1.94
Sugars			
1,4:3,6-Dianhydro-.alpha.-D-glucopyranose	18.17	1.43	1.19
Levoglucosan	25.54	27.77	7.06
Aromatics			
Benzene, 1-methoxy-4-methyl-	7.96	0	0.31
Phenol, 2-methoxy-	13.10	2.41	5.05
Phenol, 2-methoxy-4-methyl-	13.97	0.43	1.21
2-Methoxy-5-methylphenol	14.11	1.01	2.65
Phenol	14.59	2.81	5.19
Phenol, 4-ethyl-2-methoxy-	14.86	0.24	1.18
Phenol, 3-methyl-	15.42	1.61	1.14
2-Methoxy-4-vinylphenol	16.43	0.66	2.38
Phenol, 2,6-dimethoxy-	17.06	4.98	7.66
1,2,4-Trimethoxybenzene	17.82	2.10	5.62
Benzene, 1,2,3-trimethoxy-5-methyl-	18.33	1.47	3.49
Phenol, 2,6-dimethoxy-4-(2-propenyl)-	19.41	0.41	1.53
Ethanone, 1-(3,4-dimethoxyphenyl)-	19.67	1.47	5.88
Phenol, 2,6-dimethoxy-4-(2-propenyl)-	20.91	1.28	5.35
Ethanone, 1-(4-hydroxy-3,5-dimethoxyphenyl)-	22.83	1.03	1.64
Hydroquinone	23.00	2.18	2.52
Benzaldehyde, 4-hydroxy-3,5-dimethoxy-	22.46	0.54	1.03
Phenol, 3,4,5-trimethoxy-	23.28	1.05	1.24
3,5-Dimethoxy-4-hydroxycinnamalde	27.42	0.72	1.17

Source: Hu et al. 2012 [13]. Reprinted with permission of Elsevier.

make the components very reactive. Instability is a major issue not only in upgrading of bio-oil at elevated temperatures but also in the storage of bio-oil at room temperature [14, 15]. During pyrolysis, especially the fast pyrolysis process, the residence time for bio-oil in the reactor is generally very short (i.e. a few seconds). The organic compounds produced from the cleavage of the chemical bonds of biomass do not have enough time to react with each other before quenching with coolant to bio-oil. Thus, bio-oil as a product does not reach "thermodynamic equilibrium" during its formation from pyrolysis. As a result, even during storage under ambient conditions, reactions between the components of bio-oil continue, which generally lead to the condensation of the compounds [16, 17]. Such a condensation process can be remarkably accelerated at elevated temperatures [18].

The condensation of bio-oil is a major issue for the application of bio-oil. Bio-oil can be potentially used as a fuel for transportation. Nevertheless, bio-oil has a low heating value due to the high water content. Thus, bio-oil has to be upgraded via processes such as hydrotreatment to reduce the oxygen content [19]. Unfortunately, the hydrotreatment of bio-oil involves heating bio-oil to elevated temperatures (i.e. from 150 to 400 °C) for the hydrogenation and/or hydrodeoxygenation reactions to take place [20, 21]. During this process, the polymerization or cracking of bio-oil proceeds, leading to serious coking [22, 23]. The coke formed can cover the surface of the catalyst and result in rapid catalyst deactivation or lead to blockage of the reactor, making the continuous hydrotreatment of bio-oil difficult [24].

To address the high propensity of bio-oil toward polymerization, a methodology of two-stage hydrotreatment of bio-oil has been developed [25, 26]. The process includes as a first step the hydrogenation of bio-oil with generally noble metal catalysts under mild conditions (low temperature) to stabilize bio-oil via saturation of the reactive functionalities such as carbonyl groups and as a second step the hydrodeoxygenation of some of the oxygen-containing functionalities under harsh conditions (high temperatures) to reduce the oxygen content of bio-oil and to increase the heating value of the product as fuel [26]. Nevertheless, although the two-stage hydrotreating process can alleviate coking to some extent, coking is still an issue. Other methods still need to be developed to tackle the high instability of bio-oil.

As mentioned earlier, the multiple type of organics with varied functionalities in bio-oil are the main reason for its instability. The water content in bio-oil is typically ca. 30% and strongly depends on the pyrolysis process and the origin of the biomass [27, 28]. The rest of the oil mainly consists of organics. The high concentration of organics in bio-oil increases the chance for polymerization reactions and also creates the high viscosity of bio-oil. Dilution of the organics with organic solvents, such as alcohols, acetone, esters, etc., can drastically lower the viscosity of bio-oil and improve its flow properties [29, 30]. This process is called emulsification and can suppress the polymerization of bio-oil during the storage (aging) to some extent [31]. Nevertheless, the emulsification process cannot solve the issue of the bio-oil instability. This is because even when some solvents such as methanol can react with some of the aldehydes and convert them to acetals, the majority of the reactive functionalities in the bio-oil still remain. Emulsification can improve the physical properties of bio-oil such as the flowability, but cannot significantly impact the composition of bio-oil [32], which is the origin of its high instability.

Bio-oil not only contains the organics that are propone to polymerization but also contains the catalysts, namely, carboxylic acids, which catalyze the polymerization reactions [33]. These carboxylic acids include formic acid, acetic acid, propionic acid, butyric acid, levulinic acid, etc. The aqueous solution of these acids is acidic, while polymerization of many of the organics in bio-oil requires the involvement of hydrogen ions to accelerate the process [33]. The naturally present carboxylic acids in bio-oil thus play the role of catalysts for the polymerization of the organics in bio-oil. Our previous work has confirmed that the carboxylic acids catalyze the polymerization or the cross-polymerization of the main components of bio-oil such as sugars, furans, and phenolics [33]. The impact of the carboxylic acids on polymerization of bio-oil is also related to the acidity of the carboxylic acids. Formic acid has a lower pK_a than acetic acid and thus is superior in catalyzing the polymerization reactions. In bio-oil, the content of formic acid is generally lower than that of acetic acid [11]. The concentration of acetic acid in bio-oil ranges from 5% to 10%, the acetic acid mainly originating from dissociation of the acetyl group in hemicellulose during biomass pyrolysis [34, 35]. In addition to these carboxylic acids that can be detected and quantified with gas chromatography-mass spectrometry (GC–MS), there is also a significant amount of heavy carboxylic acids in bio-oil, which cannot be detected with GC–MS. In a previous study [36], we developed nonaqueous potentiometric titration to quantify the total acid number of the heavy acids in bio-oil. The results showed that the GC–MS detectable carboxylic acids account for only 29–45% (mole basis) of all the carboxylic acids in bio-oil prepared from pyrolysis of mallee wood. The structures of these heavy acids in bio-oil remain to be elucidated, but it is most likely that carboxylic groups are attached to the benzene rings in pyrolytic lignin. The light carboxylic acids plus the heavy acids in bio-oil comprise a very important fraction in bio-oil. This fraction significantly affects the properties of bio-oil, not only acting as catalysts for the polymerization of other organics in bio-oil but also making bio-oil corrosive.

Esterification is a reaction between an alcohol and carboxylic acid, via which the carboxylic acids in bio-oil are converted into neutral esters [37]. Esterification can effectively remove the majority of the acids (not all due to the reaction equilibrium of esterification reactions), which alleviates the corrosiveness of bio-oil. Nevertheless, from a survey of the current literature, the esterification of bio-oil involves far more than just esterification reactions between the carboxylic acids and the alcohols. A number of other reactions take place in parallel, including the conversion and transformation of sugars, aldehydes, furans, and aromatics. The esterification of bio-oil is more accurately described as the acid-catalyzed conversion of the organics in an alcohol-rich medium. Many reactions, which are catalyzed by an acid catalyst, can take place. Indeed, as the reaction medium is alcohol-rich, many components of the bio-oil undergo reactions with the alcohol. As a consequence, the reaction network under esterification conditions is very complex. The composition of bio-oil changes significantly during esterification, contributing to the bio-oil stabilization. In the Sections 4.2 and 4.3, the detailed reaction routes of the major bio-oil components during esterification and the effect on bio-oil stability are discussed in detail.

4.2 Reactions of the Main Components of Bio-Oil Under Esterification Conditions

4.2.1 Sugars

Sugars, including sugar monomers and oligomers, are one of the major components of bio-oil and are produced from the degradation of cellulose and hemicellulose during biomass pyrolysis [38, 39]. In general, levoglucosan is one of the most abundant detectable anhydrate sugars in bio-oil, while other sugars such as glucose, galactose, xylose, etc. are also present. The reactivity of sugars under the conditions of acid treatment is broadly similar to that in acid-catalyzed conversion of model sugars [40, 41]. Nevertheless, bio-oil is a very complex reaction medium. Other components in bio-oil may be involved in the conversion of sugars, a topic that needs further investigation. The C6 sugars and the C5 sugars behave quite differently during acid-catalyzed conversions [42, 43]. In a methanol-rich medium, levoglucosan can hydrolyze to glucose, while it can also be converted directly to methyl glucopyranosides via alcoholysis [42], as shown in Figure 4.1. The methyl glucopyranosides can be further converted to methyl levulinate in methanol-rich medium with the formation of 5-hydroxymethylfurfural (HMF) and the ether or acetal of HMF as the intermediate products. The further conversion of methyl glucopyranosides strongly depends on the experimental conditions. All of the C6 sugars such as galactose, glucose, and fructose have similar reaction pathways during acid-catalyzed conversion in alcohol-rich medium [44].

The methanol added for esterification of bio-oil and the water in bio-oil have distinct effects on the propensity of the C6 sugars toward polymerization [42], as shown in Figure 4.2. In water or methanol medium, the reaction intermediates involved are quite different, which is the main reason for the different tendencies of the sugars toward polymerization. In water medium with levoglucosan as the starting reactant, glucose and HMF are the main reaction intermediates. In comparison, in methanol medium methyl glucopyranoside and the ether or acetal of HMF are the main reaction intermediates, as shown in Figure 4.1. The main difference is that the reactive functionalities in glucose and HMF intermediates are protected in methanol, which is proposed to be the main reason for the suppression of polymerization reactions. The effects of protecting the functionalities of HMF on their condensation pathways is shown in Figure 4.3 [42]. Other alcohols like ethanol can also suppress the polymerization of the C6 sugars [45].

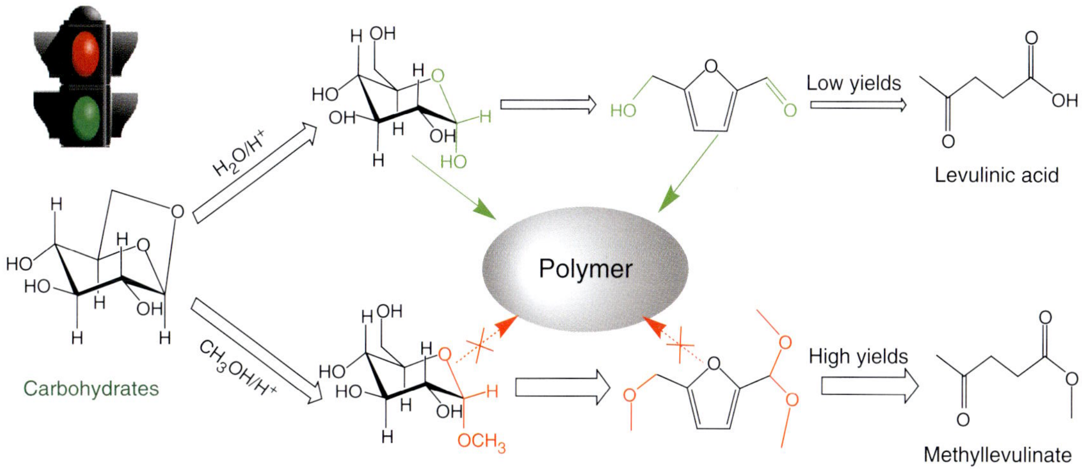

Figure 4.1 Reaction pathways of levoglucosan in a water-rich medium and a methanol-rich medium. Source: Hu and Li 2011 [42]. Reprinted with permission of Royal Society of Chemistry.

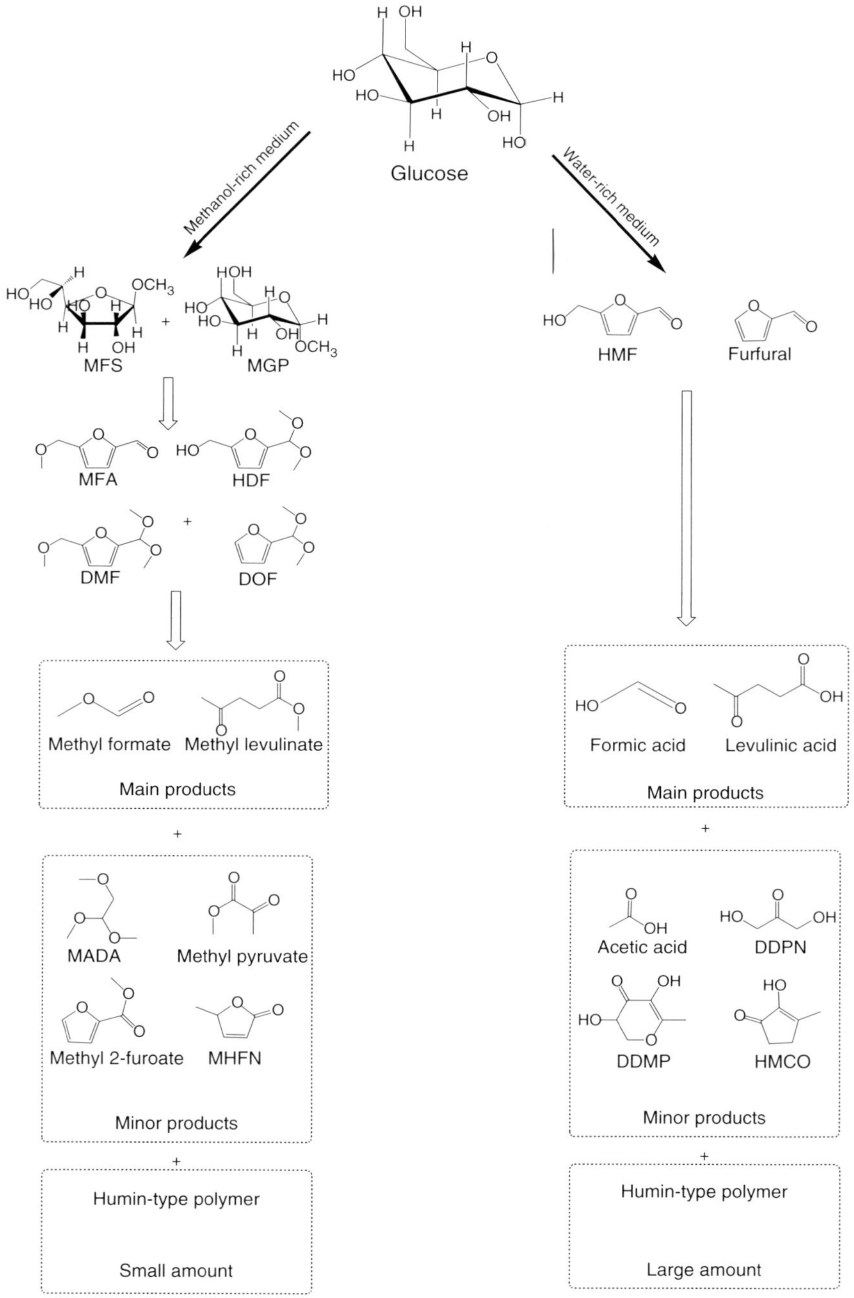

Figure 4.2 Reaction pathways of glucose in methanol-rich medium and water-rich medium. Source: Hu et al. 2011 [41]. Reprinted with permission of Elsevier.

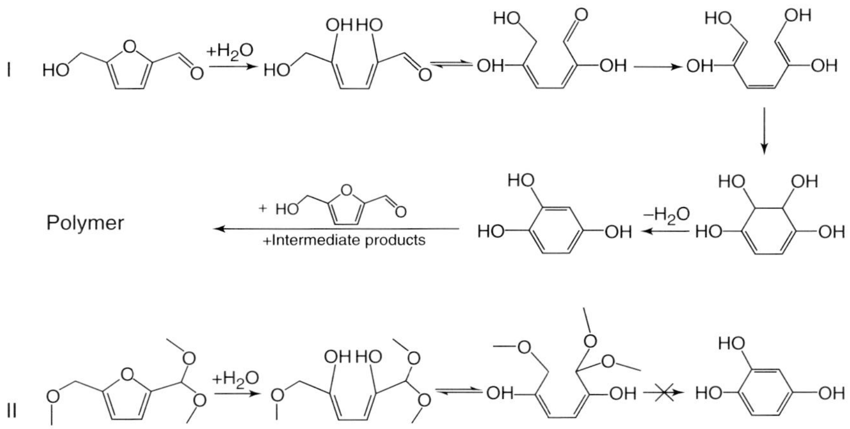

Figure 4.3 Humin formation from HMF (Route I) and the molecular modification of HMF to 2-(dimethoxymethyl)-5-(methoxymethyl)furan (DMMF) to suppress humin formation (Route II). Source: Hu and Li 2011 [42]. Reprinted with permission of Royal Society of Chemistry.

C5 sugars behave quite differently in terms of polymerization [43]. Xylose in methanol-rich medium can be converted into methyl xylosides, as shown in Figure 4.4. The further dehydration of the methyl xylosides produces furfural. Furfural has a very high tendency toward polymerization whether in methanol or in water medium, as shown in Figures 4.5 and 4.6, in which the reaction routes for the polymerization of furfural are proposed. Furfural is the main product from the dehydration of C5 sugars, while levulinic acid is the main product from C6 sugars. In the presence of an acid catalyst in water, levulinic acid is relatively stable, while furfural mainly polymerizes [43]. This is the main reason for the high propensity of C5 sugars toward polymerization. After esterification of bio-oil with methanol, a significant amount of methyl levulinate is formed, while furfural disappears in the products [13]. The polymerization of C5 sugars and furfural during esterification reactions negatively impacts the activity of the esterification catalyst. This will be discussed later.

There are several sugar monomers in bio-oil, including levoglucosan, glucose, galactose, and fructose. Hu et al. tried to elucidate the effects of the molecular structure of the sugars on their conversion in ethanol/water [38]. It was found that the sugar structure significantly affects the conversion, as shown in Figure 4.7. The yields of levulinic acid/ester followed the order: fructose > glucose > levoglucosan ≈ galactose. The behavior of sugar oligomers such as sucrose, maltose, raffinose, and beta-cyclodextrins was similar to their basic units. Nevertheless, the sugar oligomers are relatively large in size, which creates some steric hindrance when using Amberlyst 70, a solid acidic resin, as the catalyst.

In addition to the simple sugars mentioned, there are also sugar oligomers present in bio-oil. The exact structures of these sugar oligomers are still unknown. Li and coworkers performed the esterification of bio-oil at low temperatures (<90 °C) and found that levoglucosan could be produced [47].

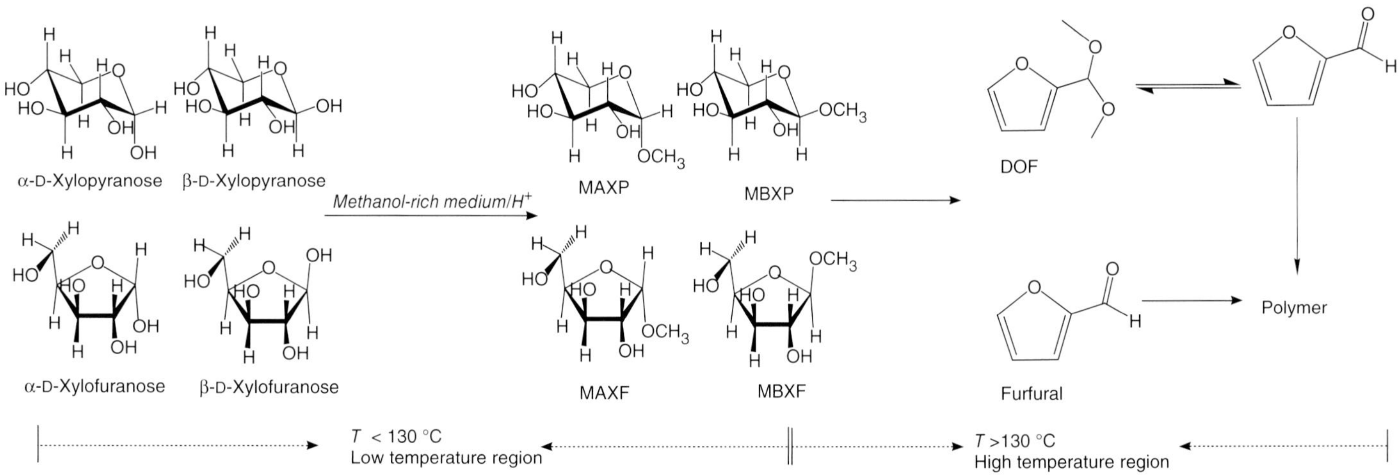

Figure 4.4 The main reaction pathways of xylose in methanol-rich medium. Source: Hu et al. 2012 [43]. Reprinted with permission of John Wiley and Sons.

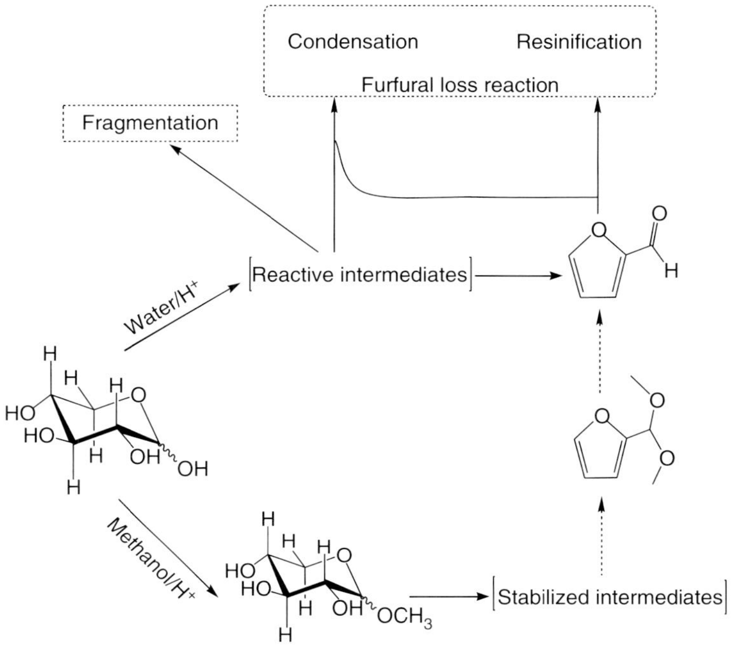

Figure 4.5 The polymerization of xylose in water medium and methanol medium. Source: Hu et al. 2012 [43]. Reprinted with permission of John Wiley and Sons.

It was assumed that levoglucosan was produced from decomposition of sugar oligomers. Other than this work, few studies have investigated the reaction behavior of the heavy sugar oligomers in bio-oil. These sugar oligomers not only have a complex structure but also have a large molecular size. Like the conversion of beta-cyclodextrins over Amberlyst 70 [38], steric hindrance must affect the conversion of the heavy sugar oligomers. During the esterification of bio-oil, a significant amount of coke can be formed [48]. How the heavy components of bio-oil such as the heavy sugar oligomers contribute to polymerization reactions requires further investigation.

Given that the C6 sugars and furans such as HMF and furfuryl alcohol can be converted to methyl levulinate during esterification of bio-oil in methanol-rich medium [13], as shown in Figure 4.8, it is possible to produce methyl levulinate from bio-oil via esterification as reported by Hu et al. [49]. The process includes a first step to extract or separate the sugars from bio-oil via extraction with water and subsequent extraction with $CHCl_3$ solvent, as shown in Figures 4.9 and 4.10. The extraction with water aims to separate the polar sugars from less polar components. Nevertheless, some phenolics are also extracted. Thus, extraction with $CHCl_3$ was further performed to separate the sugars from the phenolics. Subsequent acid treatment of the extracted sugars produced levulinic acid/ester. The separation of levulinic acid and methyl levulinate was also achieved via extraction with CH_2Cl_2 or $CHCl_3$, as shown in Figure 4.10. The method developed

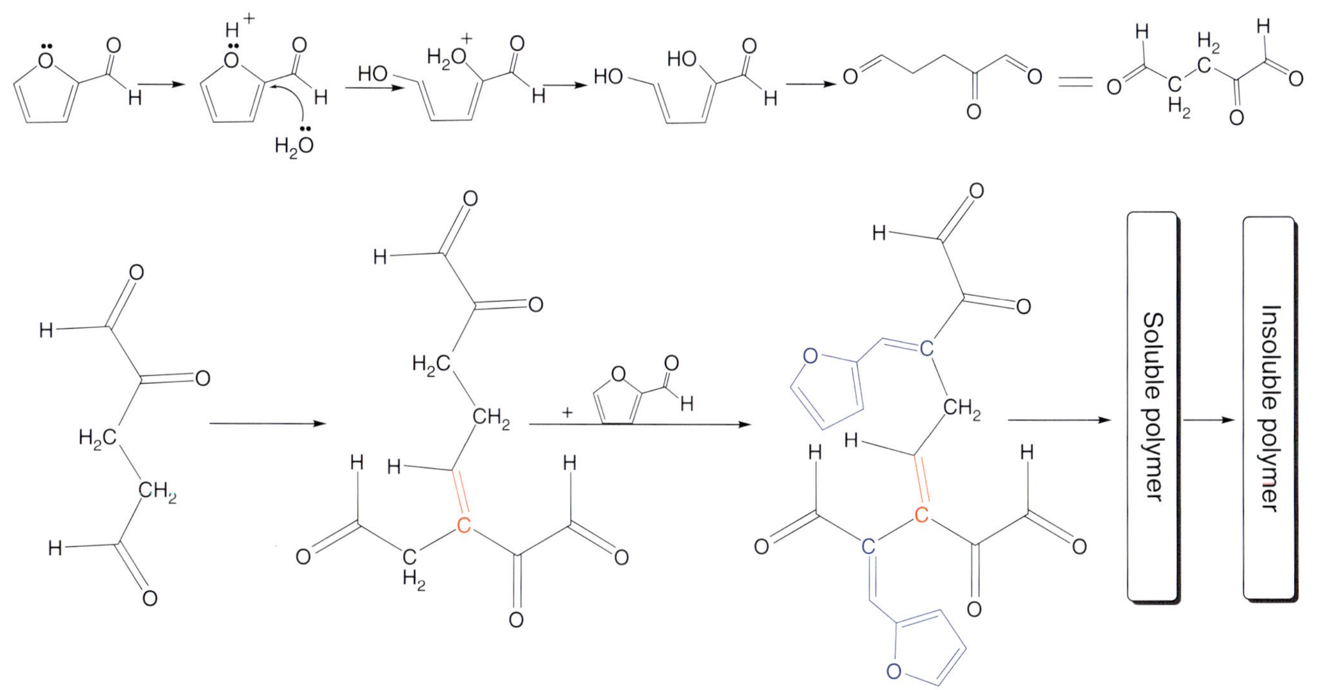

Figure 4.6 Proposed reaction pathways for opening of the furan ring in furfural and the following cross-polymerization. Source: Hu et al. 2014 [46]. Reprinted with permission of American Chemical Society.

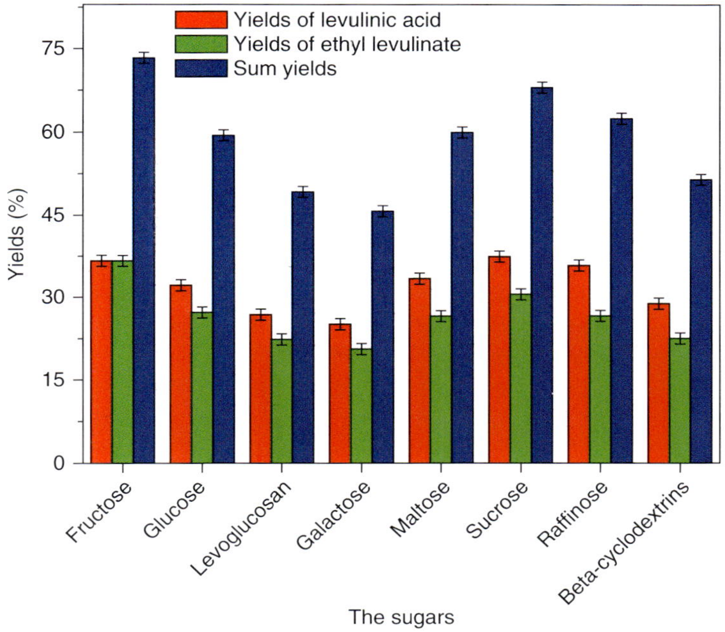

Figure 4.7 The total yields of levulinic acid and ethyl levulinate from the acid treatment of mono- and poly-sugars. Source: Hu et al. 2013 [38]. Reprinted with permission of Elsevier.

here transforms the "cumbersome" sugars in bio-oil into value-added chemicals. Similar methodology should be developed to extract or to produce certain other chemicals or groups of chemicals from bio-oil.

4.2.2 Carboxylic Acids

The initial aim of esterification is to transform the corrosive carboxylic acids in bio-oil into neutral esters [50] via the reaction of carboxylic acids with various alcohols, as shown in Figure 4.11. However, in general the carboxylic acids cannot be completely converted, due to the reaction equilibrium between alcohol and carboxylic acid on one side and the ester and water on the other side. Bio-oil contains water and, during esterification, other reactions such as sugar dehydration and etherification reactions between alcohols also produce water. Owing to the significant amount of water present, the carboxylic acids cannot be completely converted into esters. In principle, the water in the system can be removed *in situ* via distillation. However, there are many light compounds in bio-oil, and the alcohols used in esterification such as methanol also have a low boiling point. How to remove the water in bio-oil and the water formed during esterification is an issue to be resolved. Zhang et al. made an attempt to solve this problem by employing an olefin as the reactant for bio-oil esterification [51, 52]. The olefin can hydrolyze to produce an alcohol, while the alcohol can react with carboxylic acids to form esters. Such a process would consume the water originally present in the bio-oil and could drive the esterification of the carboxylic acids to completion. However,

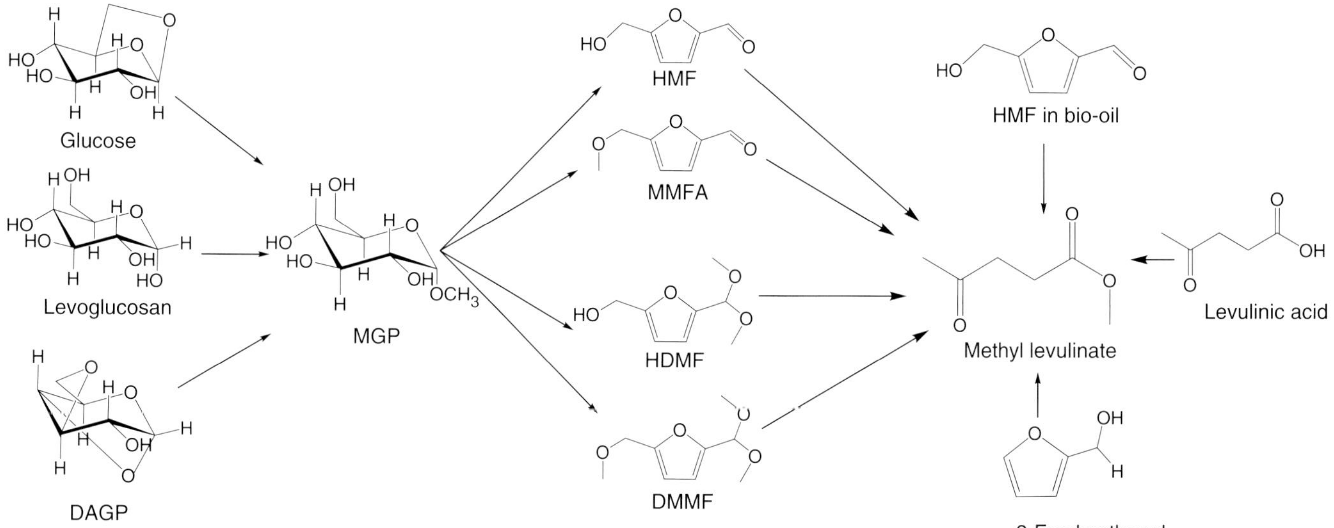

Figure 4.8 The reaction network for the formation of methyl levulinate in bio-oil esterification. Source: Hu et al. 2012 [13]. Reprinted with permission of Elsevier.

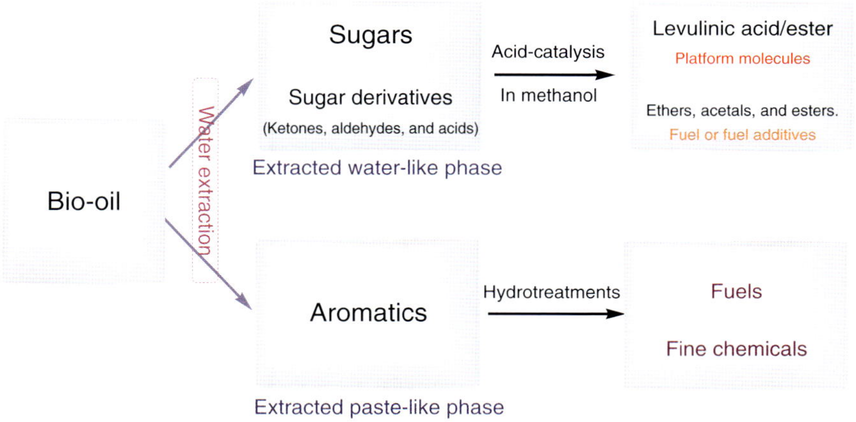

Figure 4.9 Utilization of different fractions of bio-oil via water extraction. Source: Hu et al. 2012 [49]. Reprinted with permission of Royal Society of Chemistry.

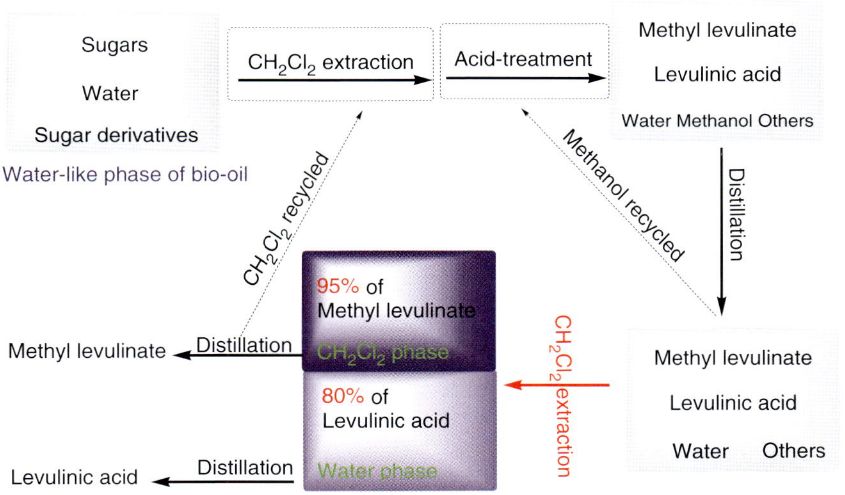

Figure 4.10 The production and separation of levulinic acid/ester via acid catalysis and extraction. Source: Hu et al. 2012 [49]. Reprinted with permission of Royal Society of Chemistry.

phase separation is an issue given that olefins possess low polarity, while bio-oil contains water and many polar products. The resulting phase separation would significantly decrease the reaction efficiency. Alcohol thus still needs to be added to the reaction medium to tackle the phase separation issue. The presence of an alcohol could also suppress polymerization reactions. In addition, the olefin can also react with phenolic compounds, forming alkylated phenols and making the bio-oil less hydrophilic.

In addition to the externally added alcohols, the carboxylic acids in bio-oil can also react with the phenolics present, which is also a type of esterification reaction [53]. However, the phenolics are not as reactive as the small alcohols like

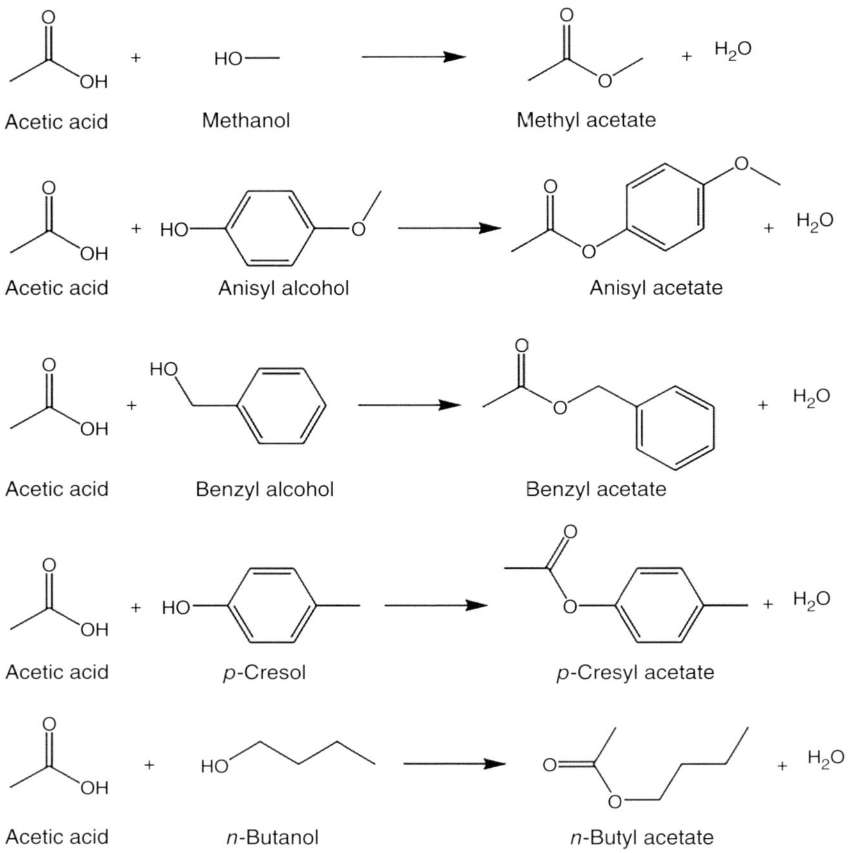

Figure 4.11 Bio-oil upgrading via esterification of acetic acid with various alcohols. Source: Osatiashtiani et al. 2017 [50]. Reprinted with permission of Springer Nature. (http://creativecommons.org/licenses/by/4.0/).

methanol. Research on the esterification of the carboxylic acids in bio-oil with phenolics is still very limited and further effort is required to verify the feasibility of this reaction route. In addition, one-step hydrogenation/esterification of bio-oil has been proposed by several researchers [54–56]. The philosophy of the process is to hydrogenate the furans like furfural and the aldehydes and the ketones to produce alcohols that are used for esterification of the carboxylic acids in bio-oil. The alcohols produced from hydrogenation of furans and aldehydes are large in size and are generally less reactive toward esterification. Hence, harsh reaction conditions need to be employed to achieve reasonable conversion of the carboxylic acids. How to optimize the process to facilitate the reaction between carboxylic acids and the larger alcohols is a question deserving further attention.

As mentioned earlier, there are significant amounts of heavy carboxylic acids in bio-oil that cannot be detected with GC–MS, as shown in Table 4.2 [36]. The reactivity behavior of these heavy carboxylic acids during esterification of bio-oil remains unknown. The conversion of the heavy carboxylic acids may

Table 4.2 The percentages of heavy carboxylic acids in the total carboxylic acids in wood oil, bark oil, and leaf oil at various pyrolysis temperatures.

Pyrolysis temperatures (°C)	Heavy carboxylic acids (% mole basis)		
	Wood oil	Bark oil	Leaf oil
350	45	33	78
375	40	—	—
400	38	34	76
450	33	45	74
500	29	40	76
550	—	30	76
580	29	23	75

Source: Wu et al. 2014 [36]. Reprinted with permission of Elsevier.

suffer from steric hindrance, which may negatively affect their conversion. Up to now, few studies have investigated the behavior of these heavy carboxylic acids in bio-oil. Understanding their reactivity during bio-oil esterification deserves further attention.

4.2.3 Furans

Some furans, such as furfural, HMF, and furfuryl alcohol, are present in bio-oil as a consequence of their formation during biomass pyrolysis. The dehydration of sugars in bio-oil during esterification also produces furans like furfural and HMF [57]. The furans in bio-oil are generally unstable. During bio-oil esterification, furfural mainly polymerizes, while furfuryl alcohol and HMF are converted to levulinic acid/ester [58]. The reaction medium also significantly affects the conversion of the furans, except for furfural. In methanol, furfuryl alcohol and HMF are selectively converted to methyl levulinate [58]. In water, furfuryl alcohol and HMF polymerize to a significant extent. Furfural polymerizes in either methanol or water medium [58]. Furan does not have a carbonyl or hydroxyl group, the polymerization of which could be suppressed in methanol medium, as shown in Figure 4.12. The conversion of the C6 and C5 sugars, the dehydration of sugars, and the further conversion of the furans can all lead to coke formation, and the sugars/furans may also cross-polymerize. The contribution of sugars and furans toward polymerization remains unclear. Hu et al. investigated this issue and found that during the conversion of glucose to levulinate the yield of coke was 24% [60]. The dehydration of glucose to 5-hydroxymethylfurfural (HMF) contributes ca. 7% out of the 24% of the coke, while the conversion of HMF to levulinic acid contributes to ca. 17% of the coke. In comparison, in the dehydration of xylose to furfural, negligible amounts of coke form, while almost all furfural polymerizes to coke, as shown in Figure 4.13. The furans are the main source of coke during acid-catalyzed conversion of the sugars [60]. The cross-polymerization

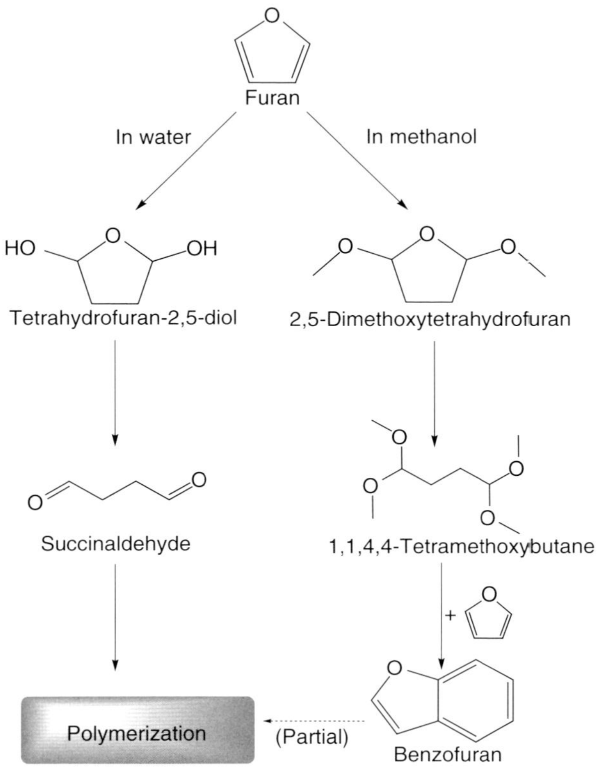

Figure 4.12 The distribution of the main products during the conversion of furan in water and in methanol. Source: Hu et al. 2016 [59]. Reprinted with permission of Royal Society of Chemistry.

between HMF and glucose as well as the cross-polymerization between furfural and xylose may have some limited effect on the structure of the soluble polymer, but do not have appreciable impact on coke formation.

4.2.4 Aldehydes and Ketones

Aldehydes and ketones are important fractions of bio-oil, which are mainly derived from the degradation of cellulose and hemicellulose during biomass pyrolysis. Hydroxyaldehyde and hydroxyacetone are the typical light aldehyde/ketone in bio-oil. Hydroxyaldehyde can react with methanol via acetalization and etherification, forming 1,1,2-trimethoxyethane [13]. This product is unstable under esterification conditions and is further converted to other products. Hydroxyacetone is also converted to some extent, but it is difficult to identify the corresponding product due to the complex composition of the products. Model experiments show that under acidic conditions, hydroxyacetone can cross-polymerize with guaiacol [33]. In bio-oil, there are a number of phenolic compounds structurally similar to guaiacol. Hydroxyacetone probably condenses with them. Most aldehydes and ketones in bio-oil

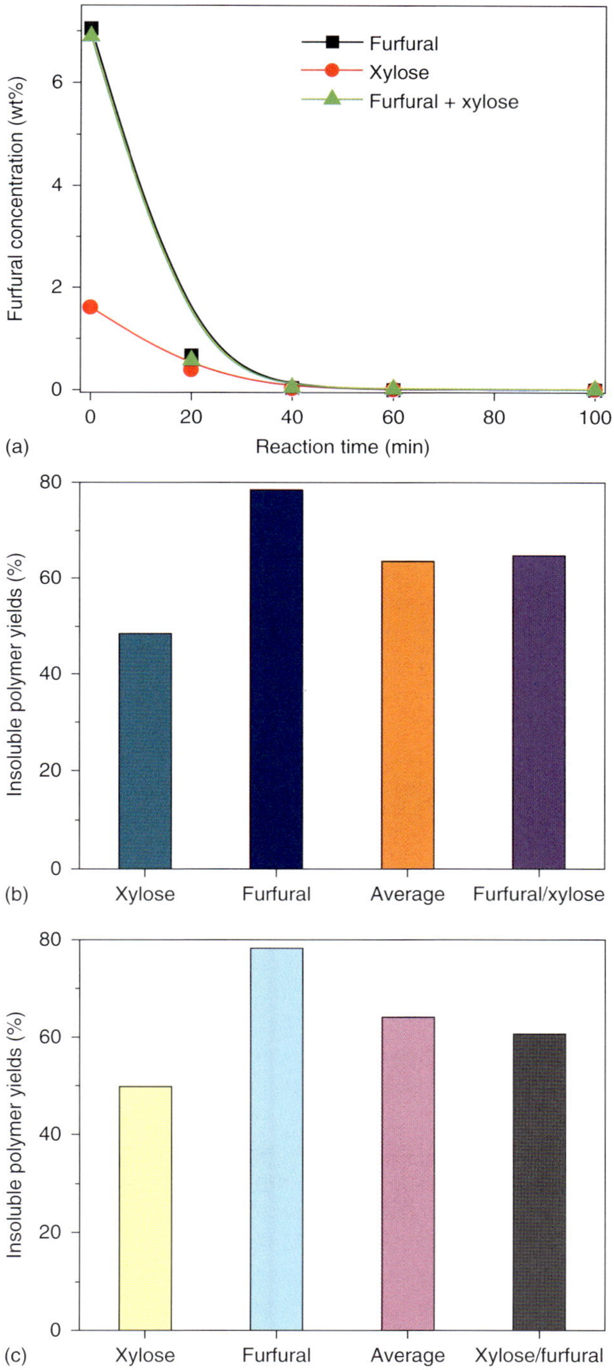

Figure 4.13 Product/reactant distribution in acid-catalyzed conversion of furfural, xylose and furfural/xylose. Reaction conditions for experiments in (a) and (b): Furfural and/or xylose loaded each: 6 g; water: 60 ml; T = 190 °C; Reaction time: 100 min; Amberlyst 70 loaded: 18 g. Reaction conditions for the experiments in (c): Furfural and/or xylose loaded each: 15 g; water: 60 ml; T = 180 °C; Reaction time: 100 min; Amberlyst 70 loaded: 15 g. Source: Hu et al. 2015 [60]. Reprinted with permission of Elsevier.

contain both a carbonyl group and an α-H, which are very reactive toward polymerization via aldol condensation with other compounds. In principle, the carbonyl group in aldehydes and ketones can be converted to alcohols via hydrogenation. The alcohols could be used to esterify the carboxylic acids in bio-oil, i.e. hydrogenation can be coupled with esterification. To achieve the effective esterification of carboxylic acids with the alcohol produced via hydrogenation of aldehydes/ketones, the amount of aldehydes/ketones in bio-oil has to match the amount of carboxylic acids. This needs to be considered during optimization of the hydrogenation/esterification process.

4.2.5 Phenolics

Phenolic compounds are one of the most important fractions of bio-oil and are mainly derived from the degradation of lignin. The proportion of phenolics detectable with GC–MS is quite small. Li and coworkers employed a nonaqueous potentiometric titration method and found that the light phenolic compounds detected with GC–MS account for only 3% (mole basis) of the total phenolics [36], as shown in Table 4.3. The majority of the phenolics are heavy components in bio-oil, the structure of which is largely unclear. Furthermore, it is also not clear how these heavy phenolics behave under esterification conditions. Hu et al. investigated the reactivity of the light phenolic monomers during esterification of bio-oil obtained from mallee bark [13]. They found that aromatics possessing a carbonyl group, like 3,4-dimethoxyacetophenone, could be converted, as shown in Figure 4.14. However, it was difficult to identify the corresponding product to clarify the reaction pathway. From model compound experiments, it was shown that guaiacol can react with hydroxyacetone under esterification conditions [13]. Polymerization is probably one of the reaction pathways for the conversion of the phenolics in bio-oil. In addition, as mentioned before, phenolics contain a

Table 4.3 The percentages of heavy phenolic components in the total phenolic groups in wood oil, bark oil, and leaf oil at various pyrolysis temperatures.

Pyrolysis temperatures (°C)	Heavy phenolic components (wt% mole basis)		
	Wood oil	Bark oil	Leaf oil
350	98	97	94
375	97	—	—
400	97	97	94
450	97	94	95
500	97	93	91
550	—	94	92
580	98	87	92

Source: Wu et al. 2014 [36]. Reprinted with permission of Elsevier.

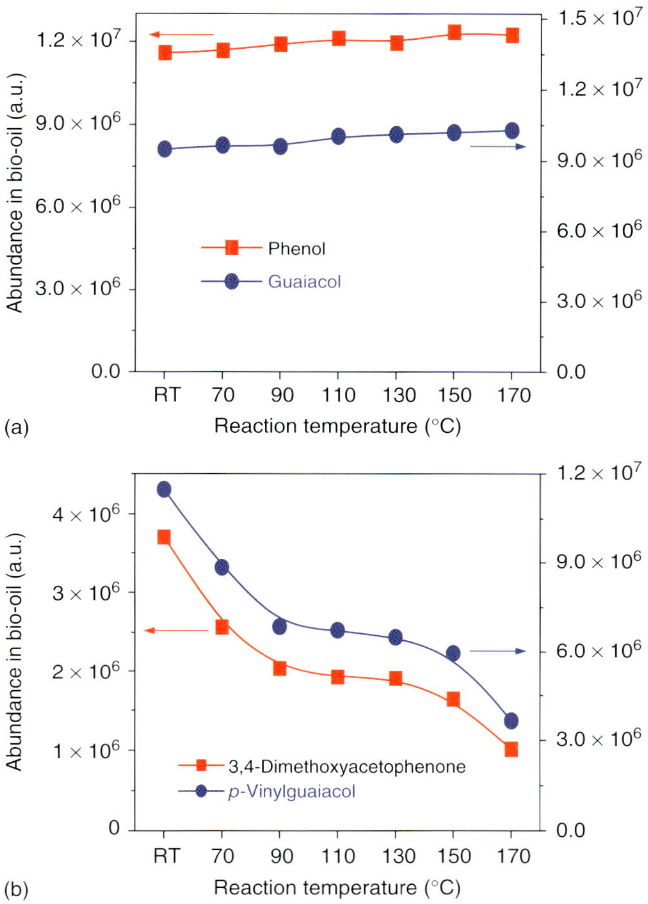

Figure 4.14 Behavior of typical aromatics including (a) phenol and guaiacol and (b) 3,4-dimethoxyacetophenone and *p*-vinylguaiacol in bio-oil during esterification. Source: Hu et al. 2012 [13]. Reprinted with permission of Elsevier.

hydroxyl group, which could be used to esterify the carboxylic acids in bio-oil. More effort should be devoted to this research area.

4.2.6 Other Components

Bio-oil is very complex in composition, and the exact composition depends on the biomass properties and pyrolysis conditions. In addition to the components discussed earlier, there are other important fractions in bio-oil, including esters, terpenoids, N-containing organics, metal species, etc. Esterification of the carboxylic acids in bio-oil produces esters, while in bio-oil there are also some heavy esters. Under esterification conditions, these heavier esters undergo transesterification reactions, producing small esters [13, 61–63], as shown in Figure 4.15. The small esters are more volatile, which improves the properties of bio-oil for use as

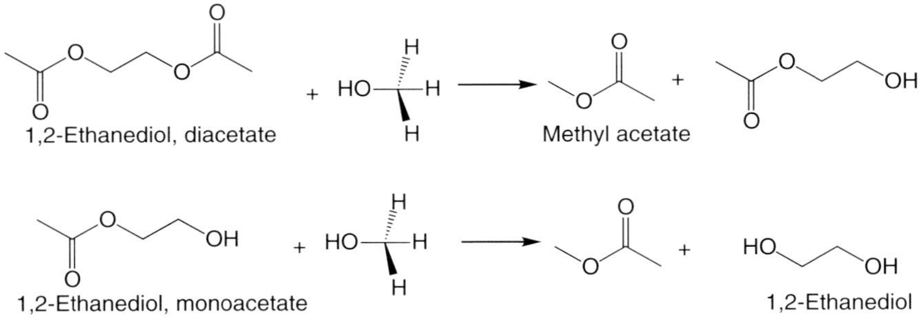

Figure 4.15 Transesterification reactions in bio-oil esterification. Source: Hu et al. 2012 [13]. Reprinted with permission of Elsevier.

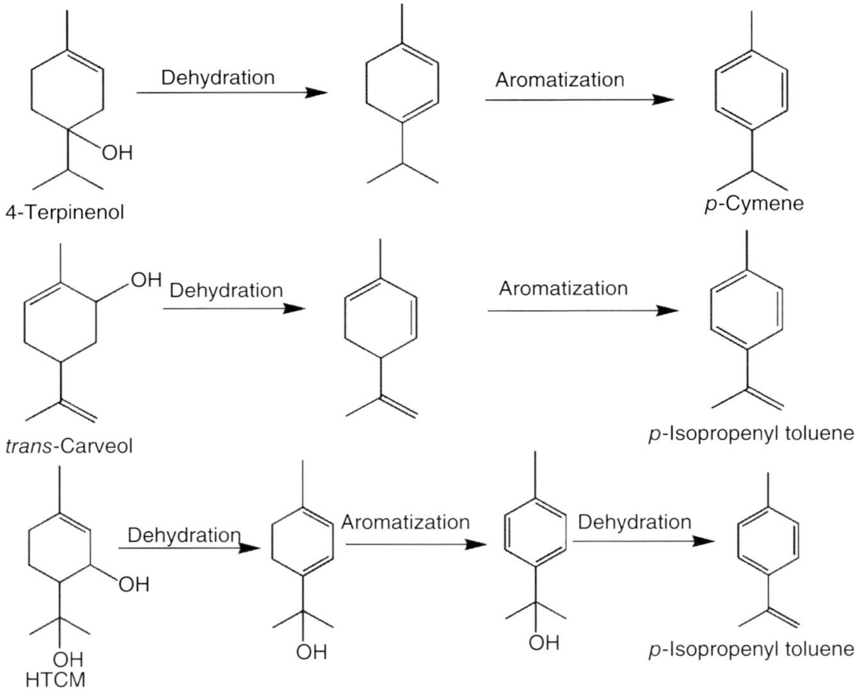

Figure 4.16 Proposed reaction pathways for the aromatization of terpenoids. HTCM, 2-hydroxy-α,α,4-trimethyl-3-cyclohexene-1-methanol. Source: Hu et al. 2012 [64]. Reprinted with permission of Elsevier.

fuel. Terpenoids and eucalyptol are found in bio-oil produced from pyrolysis of mallee leaves [64]. The concentration of eucalyptol in mallee leaves bio-oil can be up to 10 wt% [64]. The terpenoids and eucalyptol can be converted to aromatic hydrocarbons via a series of steps including ring-opening, dehydration, isomerization and aromatization steps, as shown in Figure 4.16 [64]. The conversion of the terpenoids and eucalyptol to aromatic hydrocarbons decreases the oxygen content of the bio-oil and also improves the properties of the bio-oil for use as

Table 4.4 Elemental composition of bio-oil produced from the mallee leaves and from mallee wood from Western Australia.

Bio-oils	Elements percentage (%)				
	C	H	N	S	O[a)]
Leaves bio-oil produced at 450 °C	60.9	8.8	1.3	0.08	28.92
Leaves bio-oil produced at 500 °C	52.6	9.1	1.1	0.04	37.16
Leaves bio-oil produced at 550 °C	58.6	9.2	1.3	0.05	30.85
Wood bio-oil produced at 500 °C	45.0	7.3	<0.1	0.01	47.69

a) The percentage of oxygen was calculated by difference.
Source: Hu et al. 2013 [65]. Reprinted with permission of Elsevier.

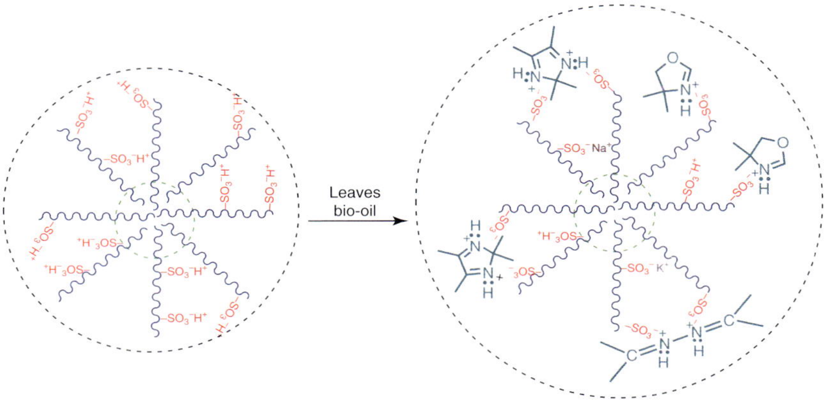

Figure 4.17 Poisoning of Amberlyst 70 catalyst by N-containing organics in bio-oil produced from leaves. Source: Hu et al. 2013 [65]. Reprinted with permission of Elsevier.

fuel. Nitrogen is present in the chlorophyll in leaves. The pyrolysis of leaves transfers some of the nitrogen to the resulting bio-oil, as shown in Table 4.4. Some N-containing organics in bio-oil were found to be strong bases, which can react with solid acid catalysts such as Amberlyst 70 even during mixing of the bio-oil with the catalysts at room temperature [65], as shown in Figure 4.17. This can result in rapid deactivation of the catalyst. Nevertheless, the deactivation of the solid acid catalyst is reversible. The deactivated catalyst can be regenerated by ion exchange with concentrated sulfuric acid, as shown in Figure 4.18.

In addition to the N-containing organics, inorganic species are also transferred to bio-oil during biomass pyrolysis. Separation of the volatiles and the solid bio-char particles formed depends on the performance of the cyclone separators employed. Under some experimental conditions char particles can be transferred to the bio-oil. Bio-oil contains many carboxylic acids, which may react with and dissolve some metal species. In addition, the solid acid catalyst might also react with char particles, leading to the occurrence of ion exchange reactions [65]. This can deactivate the catalyst, as shown in Table 4.5 and Figure 4.19. The deactivation of Amberlyst 70 by metal species is also reversible, as shown in

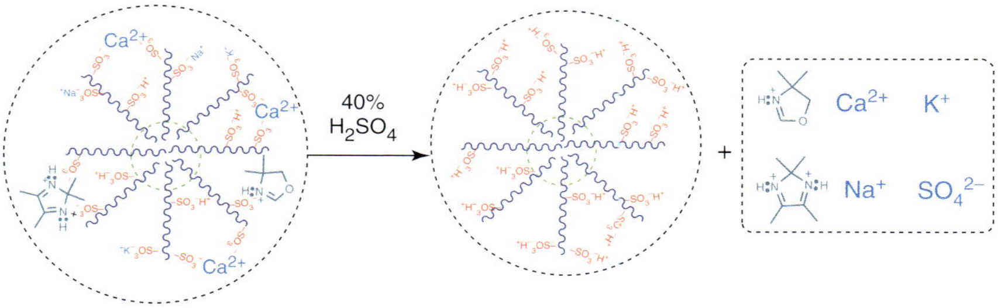

Figure 4.18 Regeneration of deactivated Amberlyst 70 catalyst via processing in concentrated sulfuric acid. Source: Hu et al. 2013 [65]. Reprinted with permission of Elsevier.

Table 4.5 Abundance of metal ions in catalyst after bark bio-oil esterification at 90 °C.

Metal ions	Metal ions on deactivated catalyst (μg/g catalyst)		
	1% catalyst	3% catalyst	5% catalyst
Ca	4886.1	1867.7	1205.9
K	559.9	452.8	261.6
Na	193.3	921.5	593.4
Mg	478.9	136.4	83.9
Fe	0	56.8	34.0
Al	0	0	0

Source: Hu et al. 2013 [65]. Reprinted with permission of Elsevier.

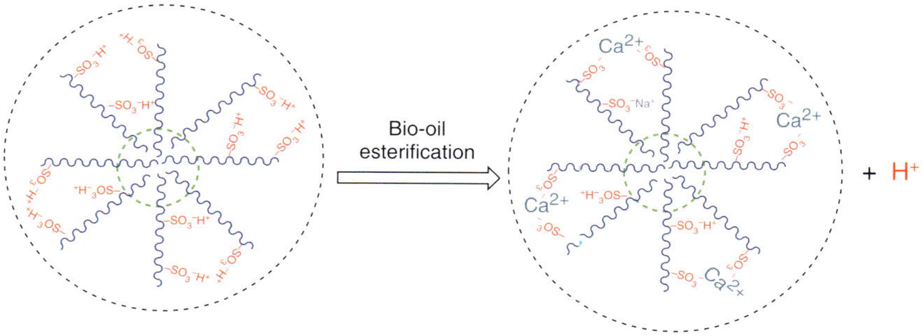

Figure 4.19 Poisoning of Amberlyst 70 catalyst with metal ions in bio-oil. Source: Hu et al. 2013 [65]. Reprinted with permission of Elsevier.

Figure 4.18. Immersing the deactivated catalyst in a concentrated acid solution replaces the metal ions with hydrogen ions, restoring the activity of the catalyst.

Bio-oil is very complex and contains a number of components with distinct properties. The different components impact the esterification process and the catalyst in different ways, making the reaction network for bio-oil esterification rather complex. During the design of any process for bio-oil esterification, several factors need to be considered. The water in bio-oil can affect the reaction equilibrium between esters and acids. The organic components such as sugars, furans, phenolics, and the heavy components can polymerize, affecting the activity of the catalyst. The N-containing components and metal species can also negatively affect catalyst activity. Corresponding measures should be taken to tackle these challenges in bio-oil esterification.

4.3 Processes for Esterification of Bio-oil

4.3.1 Esterification of Bio-oil Under Subcritical or Supercritical Conditions

The basic process for bio-oil esterification uses an alcohol medium and an acid catalyst, aiming to convert the carboxylic acids to esters to decrease the corrosiveness of the bio-oil. Various alcohols such as methanol, ethanol, butanol, and even the phenolic compounds in bio-oil can be used to esterify the carboxylic acids. In most studies, methanol is used as the esterifying agent. This is because methanol has a small molecular size and relatively high reactivity for esterification reactions. Nevertheless, methanol has a low boiling point, and operating the esterification experiment at high reaction temperatures generates significant autogenous pressure, requiring the use of appropriate reactor materials.

Some researchers have even increased the pressure for bio-oil esterification to higher ranges, i.e. have investigated bio-oil esterification in alcohols under supercritical conditions [66, 67]. Esterification under supercritical conditions can not only esterify acetic acid but also other components such as the phenolics. This is because the phenolics become more reactive at the high temperatures at which supercritical esterification is performed (250 °C). The supercritical esterification of bio-oil involves a complex reaction network. The reaction pathways of the main components remain unclear. At high reaction temperature, the polymerization of bio-oil is also more extensive [68]. In general, the benefit of esterification under supercritical esterification is not significant. The esterification of bio-oil under subcritical conditions is preferable due to the relatively milder operating conditions and the lower cost.

4.3.2 Removal of the Water in Bio-oil to Enhance Conversion of Carboxylic Acids

The esterification of the carboxylic acids in bio-oil with alcohols generates water. The water in bio-oil, together with the water produced from esterification reactions and the dehydration of sugars, shifts the reaction equilibrium to the acid direction. Hence, the presence of significant amounts of water in the reaction medium negatively affects the conversion of carboxylic acids. To achieve the complete or near complete conversion of acetic acid, Qin et al. developed a reactor with a knockout trap configuration for esterification to achieve the online removal of water [69]. 1-Butanol and sulfuric acid were used as the reactant and catalyst, respectively. The continuous removal of water helped to achieve almost

100% conversion of acetic acid. Azeotropic distillation and vacuum distillation have also been developed to remove the water in bio-oil before esterification [70]. In this way, the water in the reaction medium is mainly derived from esterification, which enables higher conversions of carboxylic acids to be achieved. Reactive distillation has also been proposed for continuous water removal [71]. In general, a high boiling-point alcohol is used [72]. The main disadvantage of the process is that after esterification the removal of the high boiling-point alcohol would be energy intensive. Furthermore, during the distillation, most of the compounds with a boiling point lower than that of water are also removed together with the water. These light organics, like ethers and esters, are good fuel additives. Moreover, their presence can help to lower the viscosity of the bio-oil.

To avoid the removal of water in bio-oil and to simultaneously achieve the high conversion of carboxylic acids, esterification of bio-oil with olefins/alcohols was developed [51, 52]. As mentioned before, olefins hydrolyze in bio-oil, which consumes water in the bio-oil and produces an alcohol for the esterification of the carboxylic acids present. The main issues are that olefins cannot homogeneously disperse in bio-oil and that it is still necessary to add an alcohol to the reaction medium. Another issue is that serious coking can occur if the concentration of added alcohol is too low. Bio-oil has a very high tendency toward polymerization. The added alcohols play an important role in suppressing polymerization of the bio-oil. Typical reaction networks for the conversion of phenolics and acetic acid are shown in Figures 4.20 and 4.21, respectively.

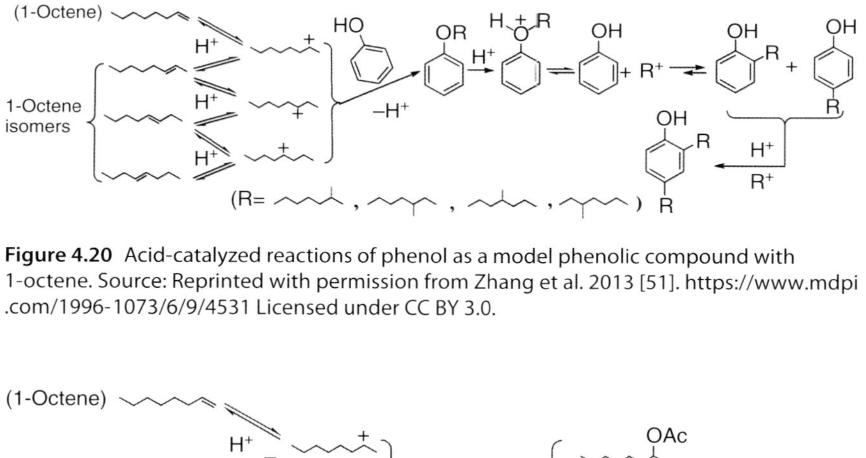

Figure 4.20 Acid-catalyzed reactions of phenol as a model phenolic compound with 1-octene. Source: Reprinted with permission from Zhang et al. 2013 [51]. https://www.mdpi.com/1996-1073/6/9/4531 Licensed under CC BY 3.0.

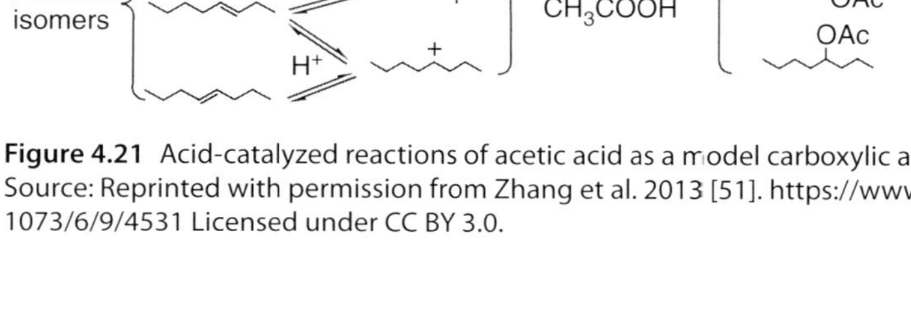

Figure 4.21 Acid-catalyzed reactions of acetic acid as a model carboxylic acid with 1-octene. Source: Reprinted with permission from Zhang et al. 2013 [51]. https://www.mdpi.com/1996-1073/6/9/4531 Licensed under CC BY 3.0.

4.3.3 In-line Esterification of Bio-oil

The previously mentioned methods are applied *ex situ* for the esterification of bio-oil. In-line esterification of bio-oil was proposed by Hilten et al. [73]. In the process, esterification is coupled with biomass pyrolysis. The bio-oil vapor is directly quenched in alcohols that act as coolant. The biggest advantage of such a process is that the carboxylic acids and the aldehydes in the bio-oil can be converted *in situ* and that the process can be operated in a continuous mode. The integration of biomass pyrolysis and bio-oil esterification, as well as continuous operation, would significantly lower the cost of the process. The main issue with the in-line esterification conducted by Hilten et al. is that the conversion of acetic acid does not reach 50%. This is because no acid catalyst was used in the process. Future studies might integrate a solid acid catalyst in the process to accelerate the conversion of acetic acid. In the literature, many studies have reported the esterification of bio-oil in batch reactors. The kinetics for the conversion of acetic acid and the polymerization of bio-oil in a batch reactor and in a continuous fixed-bed reactor might not be the same, and hence this topic needs further investigation.

4.3.4 Esterification Coupled with Oxidation

Aside from the esterification of bio-oil, pretreatment of bio-oil has also been studied by some researchers. Xu et al. used ozone to partially oxidize the aldehydes in bio-oil into acids [74]. The carboxylic acids produced from the oxidation and the carboxylic acids originally present in bio-oil are subsequently esterified with alcohols. Tanneru et al. also investigated the partial oxidation of the aldehydes in bio-oil with ozone or H_2O_2 and found that the combination of ozone/H_2O_2 performed much better than the use of a single oxidant [75]. Aldehydes in bio-oil are generally reactive. The conversion of aldehydes to carboxylic acids and the subsequent esterification increases the amount of ester formed, which improves the properties of bio-oil as fuel, while the simultaneous removal of carboxylic acids and aldehydes contributes to the stabilization of bio-oil. In addition to aldehydes, the oxidation process should also impact the other bio-oil components, which deserves further attention. Understanding this is very important to evaluate the feasibility of the oxidation plus esterification process on the properties of bio-oil.

4.3.5 Esterification Coupled with Hydrogenation

In addition to coupling with oxidation, esterification can also be coupled with hydrogenation. Esterification can effectively remove the carboxylic acids in bio-oil and can also transform some aldehydes into acetals. However, esterification cannot remove the carbonyl groups in ketones and other unsaturated functionalities such as carbon–carbon double bonds. Hydrogenation targets these functionalities, and, depending on reaction conditions and hydrogenation catalysts, these unsaturated functionalities may be removed via hydrogenation. More importantly, hydrogenation of the aldehydes and ketones produces alcohols, which can be used to esterify the carboxylic acids in bio-oil, as

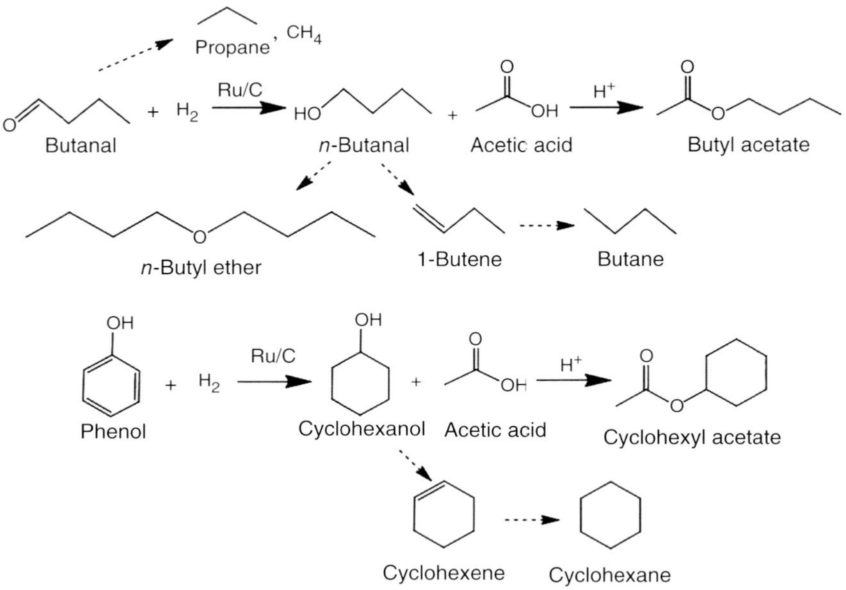

Figure 4.22 Reaction network for the hydrogenation–esterification of phenol, *n*-butanal, and acetic acid. Solid line arrows represent desired reaction pathways. Dash line arrows represent undesired reaction pathways. Source: Chen et al. 2014 [76]. Reprinted with permission of Elsevier.

shown in Figure 4.22 [76]. Another advantage of coupling hydrogenation and esterification is that polymerization reactions can be suppressed, due to the conversion of the unsaturated functionalities in the reactive components such as aldehydes/ketones. In this process, the development of a bifunctional catalyst with both acidic sites for esterification and metallic sites for hydrogenation is the key to achieve the simultaneous hydrogenation of the aldehydes/ketones and the esterification of the carboxylic acids.

4.3.6 Steric Hindrance in Bio-oil Esterification

Steric hindrance is an issue that needs to be addressed, based on the following considerations. Bio-oil contains a range of compounds with varied sizes and the molecular weight distribution varies from hundreds to thousands. The conversion of these molecules, especially the heavy ones, could be impeded by steric hindrance. The steric hindrance becomes more significant if a solid acid catalyst is used to esterify bio-oil. Frequently, solid acid catalysts are used to achieve the easy separation of catalysts from products. Solid acid catalysts such as Amberlyst 70 can swell in polar solvents, making the inner surface available to reactants [65]. However, the expansion of Amberlyst 70 catalyst has a limitation, which also depends on the polarity of the solvent. Furthermore, the hydrogen ions mainly disperse in local vicinity of the catalyst. The reactants have to diffuse into the inner structure of the catalyst to access the acidic sites. Hu et al. found that the carboxylic acids in bio-oil produced from pyrolysis of mallee leaf have different

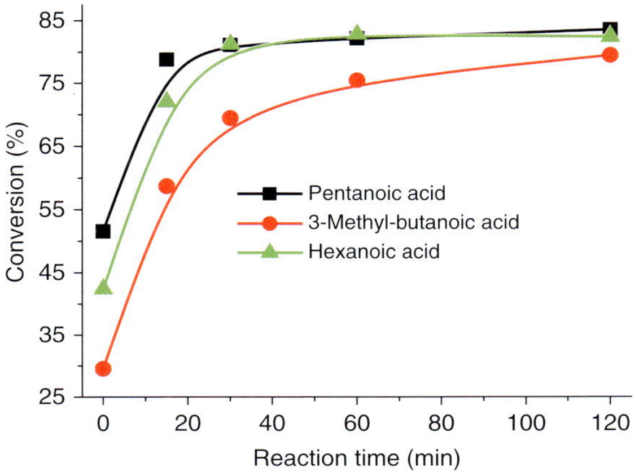

Figure 4.23 Effect of carbon chain branching on the esterification of carboxylic acids. Source: Hu et al. 2013 [77]. Reprinted with permission of Elsevier.

reactivity over Amberlyst 70 catalyst [77], as shown in Figure 4.23. It took much longer to reach the reaction equilibrium for the conversion of carboxylic acids with long carbon chains, which is a typical effect of steric hindrance [77].

In addition to carboxylic acids possessing long carbon chains, bio-oil also contains many other heavy components. For example, the phenolics detectable with GC–MS account for only 3% of the total phenolic components (on a molar basis) [36]. The conversion of these heavy phenolic components may also encounter steric hindrance. The use of mineral acid catalysts such as sulfuric acid might reduce the effect of steric hindrance from the catalyst sites. Wu et al. investigated the esterification of bio-oil using Amberlyst 70 and sulfuric acid as the catalysts and compared the kinetics for the conversion of the carboxylic acids and phenolics [78]. The results showed that under the same reaction conditions, the conversion rate for the carboxylic acids was much faster in the presence of sulfuric acid compared with Amberlyst 70. Similarly, the conversion of the phenolic components was also faster using the sulfuric acid catalyst, as shown in Figure 4.24. Clearly, steric effects significantly affect the conversion of the organic components in bio-oil, which should be kept in mind during the development of catalysts for bio-oil esterification.

4.3.7 Coking in Esterification of Bio-oil

Coking is one of the biggest issues during hydrotreatment of bio-oil and is also a serious issue in bio-oil esterification. Hu et al. investigated the mechanism of Amberlyst 70 deactivation during bio-oil esterification [65]. Several factors contributed to deactivation of the catalyst. Amberlyst 70 is an acidic resin, and the swelling of the catalyst in alcohols exposes the acid sites, as shown in Figure 4.25. Ion exchange between the $-SO_3H$ groups and the metal ions in bio-oil may decrease the acidity of the catalyst, while N-containing organics can react with the acidic sites of the catalyst. Furthermore, the formation of coke on the catalyst can fill the pores of the catalyst. All of these factors can lead to catalyst deactivation. Nevertheless, the deactivation induced by ion exchange or the N-containing organics is reversible, while the deactivation induced by coking

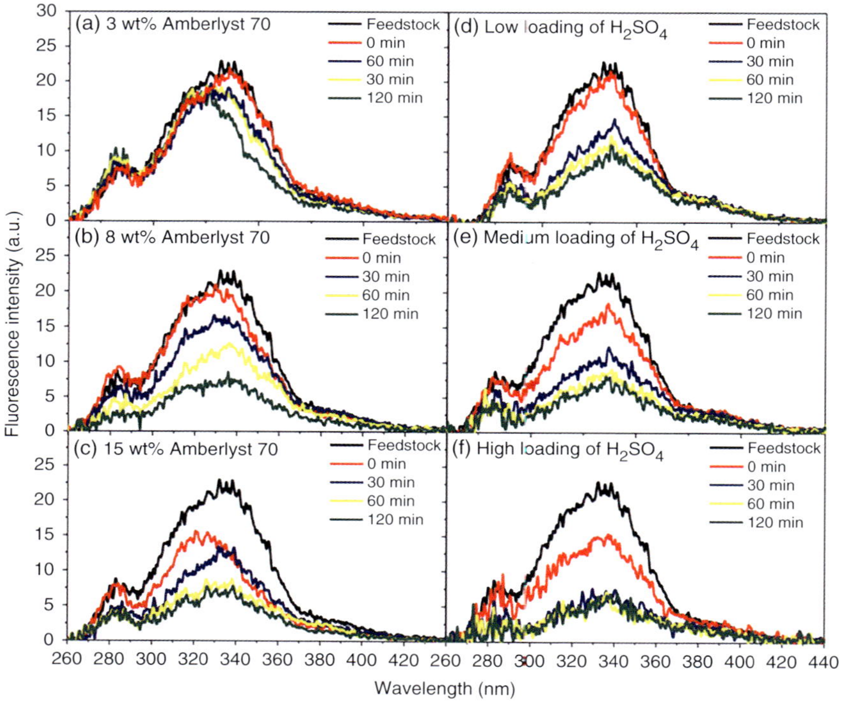

Figure 4.24 Fluorescence spectra of bio-oil feedstock and the products from acid treatment of bio-oil in methanol at 170 °C with a stirring rate of 500 rpm as a function of reaction time with Amberlyst 70 (a–c) or sulfuric acid (d– f) as the catalysts. The fluorescence intensity is on bio-oil basis. Low, medium, and high loadings of H_2SO_4 mean the hydrogen ions applied were equal to the amounts of hydrogen ions in the cases of 3, 8, and 15 wt% Amberlyst 70, respectively. Source: Wu et al. 2018 [78]. Reprinted with permission of Elsevier.

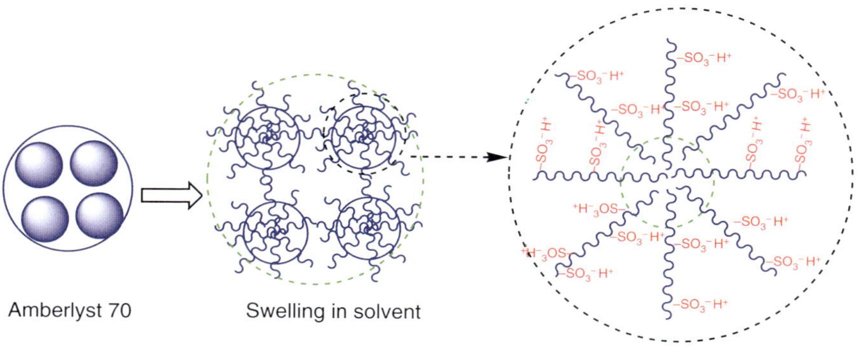

Figure 4.25 Swelling of Amberlyst 70 catalyst with solvent. Source: Hu et al. 2013 [65]. Reprinted with permission of Elsevier.

cannot be reversed. This is because Amberlyst 70 is a carbon-based solid acidic resin, and cannot be regenerated via coke combustion, unlike metal oxide-based catalysts.

Wu et al. investigated coke formation during esterification of bio-oil obtained from mallee biomass [48]. It was found that a solid deposit, which was not defined as coke, can be formed on the surface of the catalyst even after mixing bio-oil and catalyst at room temperature, as shown in Table 4.6. The properties, origin, and mechanism of formation of the solid deposit obtained at room temperature remain unclear. The reaction medium significantly affects coke formation. Compared with a water-rich medium, an alcohol-rich medium can suppress coke formation to some extent. The use of phenol as the solvent can also suppress coke formation but to a limited extent. In the previous model compound study, it was observed that alcohols can suppress coke formation from the dehydration of C6 sugars, HMF, and furfuryl alcohol, but cannot suppress the polymerization of xylose and certain other reactive components in bio-oil [43]. Cross-polymerization of the main components of bio-oil was also

Table 4.6 Variation of coke yield with different upgrading parameters.

Reaction medium	Mass ratio[a)]	Bio-oil (g)	Reaction time (min)	Reaction temperature (°C)	Catalyst loading[b)] (wt%)	Coking yield[c)] (wt%)
Water	1 : 2	25	120	170	15	27.6
Methanol	1 : 2	25	120	RT[d)]	15	2.9
Methanol	1 : 2	25	120	170	15	15.4
Ethanol	1 : 2	25	120	170	15	15.4
1-Propanal	1 : 2	25	120	170	15	15.5
1-Butanol	1 : 2	25	120	170	15	15.6
Ethylene glycol	1 : 2	25	120	170	15	18.0
Methanol	1 : 2	25	120	90	15	3.2
Methanol	1 : 2	25	120	130	15	7.9
Methanol	1 : 2	25	120	170	5	10.4
Methanol	1 : 2	25	120	170	10	14.8
Methanol	1 : 2	25	30	170	15	11.5
Methanol	1 : 2	25	360	170	15	16.2
Methanol	1 : 1	40	120	170	15	15.4
Methanol	1 : 1	40	120	170	15	20.0[e)]
Phenol	1 : 2	20	120	170	15	19.6

a) The mass ratio refers to the ratio of bio-oil to reaction medium.
b) The catalyst loading was based on the mass of bio-oil and solvent.
c) The coke yield was based on the mass of bio-oil.
d) Reaction occurred at room temperature.
e) The mass of coke included soluble polymeric material and insoluble polymeric, while other experiments only referred to insoluble polymeric material.

Source: Wu et al. 2016 [48]. Reprinted with permission of the Royal Society of Chemistry.

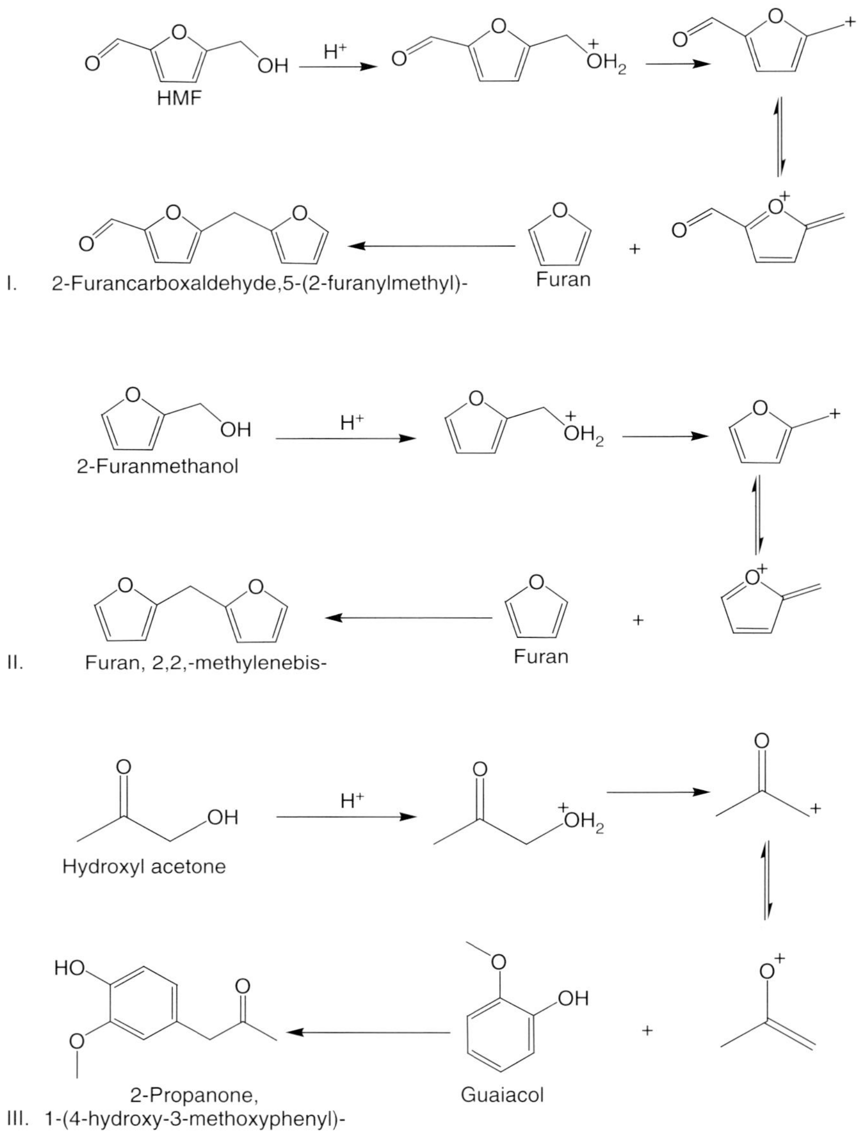

Figure 4.26 Typical condensation reactions in acid treatment of model compounds. HMF, 5-(hydroxymethyl)furfural. Source: Hu et al. 2013 [33]. Reprinted with permission of John Wiley and Sons.

observed, as shown in Figure 4.26. Coking can also take place in an alcohol-rich medium. In addition, it was found that the reaction parameters significantly affected coke formation. Higher reaction temperature and higher loading of the acid catalyst promoted coke formation. Characterization of the coke showed that it contains aromatic ring structures and carbonyl functionalities. However, due to the complicated composition of bio-oil, the mechanism of coke formation during esterification remains unclear. Further effort is required to understand the specific contribution of the bio-oil components toward polymerization and to develop effective measures to suppress coke formation.

4.3.8 Effects of Bio-oil Esterification on the Subsequent Hydrotreatment

Esterification significantly modifies the composition of bio-oil, which may impact subsequent hydrotreatment. Westerhof et al., compared the effects of bio-oil stabilization via hydrogenation and via esterification on subsequent hydrotreatment of bio-oil at 400 °C [79]. The process design is shown in Figure 4.27. Basically, the pretreatment of bio-oil via stabilization with hydrogen or esterification with methanol or ethanol showed similar effects on the yields and the properties of the resulting biofuels. One of the appreciable differences is that the yields of the gaseous products were relatively higher during hydrotreatment of the esterified bio-oil, which was related to the formation of ethers and esters during esterification. The authors also calculated the energy balance and mass balance of the process, as shown in Figure 4.28. In general, the difference in the energy balance and the mass balance of the biofuel products was not very significant. Hence, esterification of bio-oil with methanol or ethanol can achieve similar effects to the stabilization of bio-oil with hydrogenation over noble metal catalysts.

The advantage of stabilizing bio-oil with alcohols is that the overall hydrogen consumption is low as the esterification does not require the involvement of hydrogen. The process can be operated at mild conditions and a hydrogenation catalyst is not required. An issue to be addressed is how best to connect the esterification and the hydrotreatment processes, given that esterification in most current literature is operated in batch mode, while hydrotreatment is commonly operated in continuous mode. In addition, the esterification of bio-oil is a relatively slow process, while hydrotreatment is relatively fast. Matching the kinetics of the two processes should therefore also be addressed. Furthermore, bio-oil esterification generally uses an excess of alcohol to achieve high conversion of acetic acid. How to recycle the alcohols also needs to be addressed. In comparison, stabilization of bio-oil via hydrogenation under mild conditions offers the advantages that the process can be operated in a continuous mode and the reaction rate for hydrogenation is fast. Moreover, bio-oil stabilization via hydrogenation can be integrated into the hydrotreatment process. The main disadvantages are that noble metal catalysts are typically used and the consumption of hydrogen is high. Westerhof et al., have attempted to understand how the esterification and hydrogenation of bio-oil under mild conditions affects the subsequent hydrotreatment of bio-oil under harsh conditions [79]. Nevertheless, the effect of esterification and mild hydrogenation on coke formation during hydrotreatment remains unclear and requires further study.

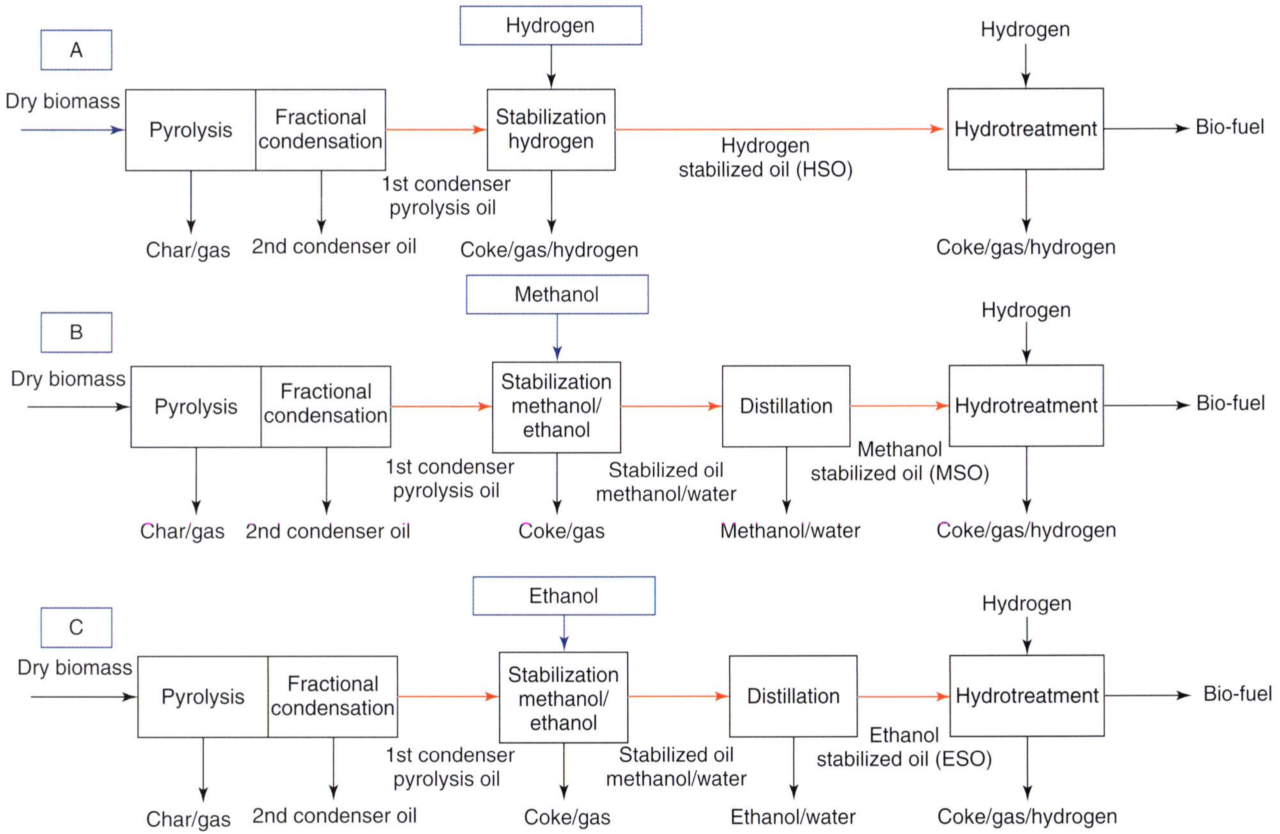

Figure 4.27 A summary of the concepts discussed herein. The pyrolysis temperature was around 500 °C and the condenser temperature was 80 °C. Temperature stabilization with methanol and ethanol was 150 or 170 °C for two or four hours, respectively, with Amberlyst 70 catalyst. Temperature stabilization with hydrogen was 250 °C for four hours with Ru/C catalyst. The hydrotreatment step was performed at 400 °C by using NiMo/Al_2O_3 catalyst. Source: Westerhof et al. 2016 [79]. Reprinted with permission of John Wiley and Sons.

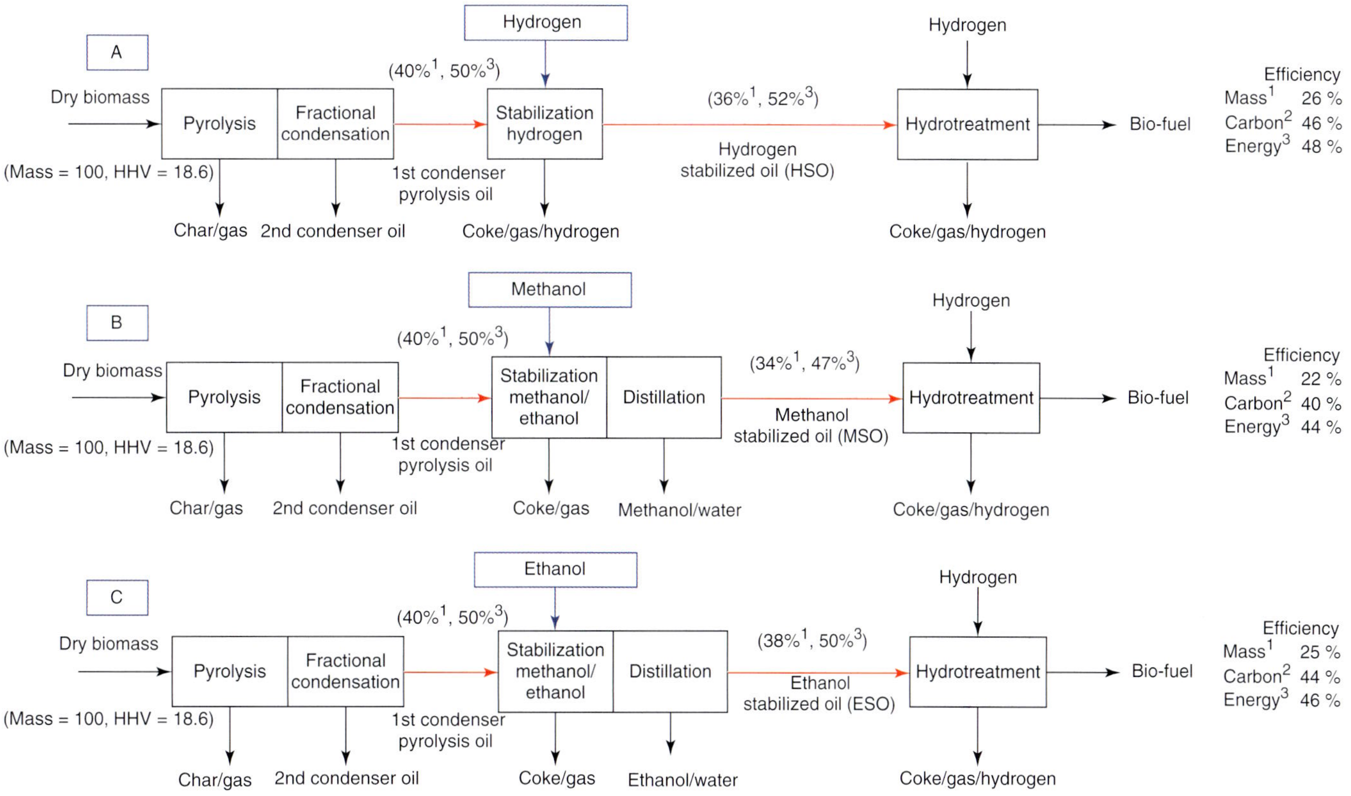

Figure 4.28 Overall evaluation of efficiency of the different processes for the conversion of biomass to biofuel. The pyrolysis temperature was around 500 °C and the condenser temperature was 80 °C. Temperature stabilization with methanol and ethanol was 150 or 170 °C for two or four hours, respectively, with Amberlyst 70 catalyst. Temperature stabilization with hydrogen was 250 °C for four hours with Ru/C catalyst. The hydrotreatment step was performed at 400 °C with NiMo/Al_2O_3 catalyst. [1]Mass efficiency, in kg of alcohol consumed and/or of hydrogen consumed; [2]Carbon content of alcohol consumed; [3]energy efficiency, in energy from methanol or ethanol and/or hydrogen. Source: Westerhof et al. 2016 [79]. Reprinted with permission of Wiley.

4.4 Catalysts

Bio-oil contains carboxylic acids. If the reaction time is long enough and the reaction temperature is high enough, esterification of the carboxylic acids in bio-oil may occur to some extent even in the absence of an acid catalyst. Nevertheless, a catalyst is generally required to accelerate the esterification reaction rate. A number of catalysts have been developed for the esterification of bio-oil and/or the typical components of bio-oil, as shown in Table 4.7. Both base catalysts like K_2CO_3/Al_2O_3–NaOH and acid catalysts like Amberlyst 70 can catalyze esterification reactions [94]. However, base catalysts generally show lower activity than acid catalysts [94]; hence, research has focused mainly on acid catalysts. Both mineral acids and solid acid catalysts can be used for bio-oil esterification. Mineral acid catalysts are very simple and are readily dispersed in bio-oil, minimizing steric effects. However, separation of the mineral acid from the liquid products is difficult. Furthermore, mineral acids are more corrosive. Solid acid catalysts do not have the separation issue and are generally regarded as noncorrosive. However, in our work concerning acid treatment of sugars with Amberlyst 70, conducted in an autoclave made of SS316, we observed that the Amberlyst 70 catalyst caused the SS316 material to corrode (unpublished results). Hence, the corrosiveness of solid acidic resins needs to be taken into consideration during use.

There have been a number of solid acid catalysts developed or tested in bio-oil esterification, as shown in Table 4.7. Solid acidic resin catalysts like Amberlyst 70, Amberlyst 15, Amberlyst 35, Amberlyst 36, and NKC-9-type ion-exchange resins have been widely employed in esterification reactions [92, 95, 96]. These catalysts generally have a high surface acid density, although the operating temperature for these catalysts is relatively low. For example, the maximum operating temperature for Amberlyst 70 is 190 °C. Hence, this type of catalyst cannot be used in the esterification of bio-oil under supercritical conditions. In addition, leaching is an issue when using these catalysts in bio-oil esterification. Our experience shows that solid acid resin catalysts like Amberlyst 70 can leach even at room temperature (unpublished results). Possibly some sulfuric acid is not completely washed away during manufacture of the catalyst. Leaching is also observed after bio-oil esterification at elevated temperatures (e.g. 170 °C). Leaching of $-SO_3H$ groups leads to the loss of acidic sites from the catalyst and also renders the esterified bio-oil corrosive again.

The steric effects encountered during conversion of the heavy components in bio-oil has been discussed earlier. The majority of the acidic sites in the solid acidic resin are dispersed in the local area of the flexible branches of the catalyst, and not in the reaction medium [96]. This makes it difficult for bulky molecules to access the acidic sites. Nevertheless, a high concentration of hydrogen ions exists in the spatially restricted inner structure of the catalysts. Given that most of the acid-catalyzed reactions, including polymerization reactions, take place over these acidic sites, some products – such as those formed from the polymerization reactions – might not be able to escape from the inner pores, which would inevitably lead to pore blockage and catalyst deactivation. Unfortunately, solid resin catalysts are carbon-based materials and cannot be regenerated by burning off the coke deposits. Sulfonated carbon materials have also been tried by many

Table 4.7 Typical catalysts used in bio-oil esterification.

Entry	Methods	Catalyst	T (°C)	References
1	Esterification of acetic acid	SO_3H–SBA-15/Pr(10)	50	[80]
2	Esterification of fatty acids	Macroporous cation exchange resin	70	[81]
3	Esterification of levulinic acid	ZrO_2–TiO_2–SO_3H	110	[82]
4	Esterification of acetic acid or acetaldehyde	Silica sulfuric acid	140	[83]
5	Esterification of acetic acid or acetaldehyde	SBA-15–SO_3H	100	[84]
6	Esterification of acetic acid	ZSM-5, mordenite, etc.	220	[53]
7	Esterification of phenol, acetic acid, acetaldehyde, hydroxyacetone, glucose	Dowex50WX2, Amberlyst15, Amberlyst36	120	[85]
8	Esterification/hydrogenation of acetaldehyde, acetic acid, or butyl aldehyde	Pt/$Al_2(SiO_3)_3$, Pt/HZSM-5	150	[86]
9	Esterification/hydrogenation of furfural or acetic acid	Pd/Al–SBA-15	150	[87]
10	Esterification/hydrogenation of acetic acid or acetaldehyde	Pt/SBA15–PrSO_3H	150	[88]
11	Esterification of bio-oil by Miscanthus	SO_4^{2-}/ZrO_2–TiO_2	100	[89]
12	Esterification of bio-oil	SO_4^{2-}/ZrO_2–TiO_2, etc.	90	[90]
13	Esterification of water phase of bio-oil	ZSM–5	135	[91]
14	Esterification of bio-oil by woods	Nafion SAC-13, Amberlyst 15, Sulfuric acid	74	[92]
15	Esterification of bio-oil by wood	No catalyst	80	[93]
16	Esterification of bio-oil	SiO_2/TiO_2–SO_4^{2-}, K_2CO_3/Al_2O_3–NaOH		[94]
17	Esterification of bio-oil from sawdust	Amberlyst 35	120	[95]
18	Esterification of bio-oil by wood chips	NKC-9-type ion-exchange resins	70	[96]
19	Esterification of water-soluble fraction	H_2SO_4		[69]
20	Esterification under reduced pressure of bio-oil	KOH, H_2SO_4, HCl, Nafion SAC-13	80	[71]
21	Esterification of bio-oil by sawdust	*p*-Toluene sulfonic acid/Carbon	80	[72]
22	Esterification with water removal of bio-oil by forest thinnings	H_2SO_4, Amberlyst-15, Amberlyst-36, etc.	120	[97]

(*continued*)

Table 4.7 (Continued)

Entry	Methods	Catalyst	*T* (°C)	References
23	Esterification of bio-oil from pine sawdust	Sulfidized SiO_2 and TiO_2	90	[98]
24	Esterification of bio-oil by pine wood	Ni/Si–Al	225	[99]
25	Esterification of bio-oil by pine chips	Silica sulfuric acid	120	[51]
26	Esterification of bio-oil by pine chips	Dowex50WX2, Amberlyst 70, Amberlyst 15, etc.	150	[52]
27	In-line esterification of biomass by pine wood	No catalyst	127	[73]
28	Ozone oxidation followed by esterification of bio-oil	$NaHSO_4$	116	[74]
29	Pre-oxidation/esterification of bio-oil by pine wood	Ru/Al_2O_3, K_2CO_3	340	[75]
30	Super-critical esterification of bio-oil by rice husk	Acidic aluminum silicate or HZSM-5	260	[66, 67]
31	Hydrogenation/esterification of bio-oil by rice husk	Pt/C, Pd/C, Ru/C, Ru/HZSM-5	300	[100]
32	Hydrogenation/esterification of bio-oil by *Pinus sylvestris* L.	SO_4^{2-}/ZrO_2/SBA-15	260	[101]
33	Hydrogenation/esterification of bio-oil by rice husk	$Ni/SiO_2/ZrO_2$	280	[55]
34	Hydrogenation/esterification of bio-oil by pine sawdust	Ru/C	300	[102]
35	Hydrogenation/esterification of bio-oil by rice husk	$Pt/Al_2(SiO_3)_3$, Pt/SiO_2, PtNi/MgO	290	[103]
36	Hydrogenation/esterification of low-boiling fraction	$Pt/Al_2(SiO_3)_3$, Pt/MgO, Pt/C	250	[104]
37	Hydrogenation/esterification of bio-oil from rice husk	$Pt/SO_4^{2-}/ZrO_2$/SBA-15	300	[54]
38	Conversion of sugars under conditions of esterification	Amberlyst 70	90–170	[46, 59, 105–108]

researchers [72]. However, these carbon-based acid catalysts have the same issues as those of the resin catalysts.

Sulfuric acid-treated metal oxides are another type of solid acid catalyst and contain both Brønsted acid sites and Lewis acid sites [82, 89, 90, 94, 101]. They also suffer from similar issues to those of resin catalysts, including steric hindrance and leaching of the acidic sites. Leaching from SO_4^{2-}/MO_x catalysts is even more serious, especially in aqueous medium. Hydrolysis of the surface sulfate leads to the loss of catalyst activity. It is worth noting that metal species can also be loaded to make bifunctional catalysts such as $Pt/SO_4^{2-}/ZrO_2$/SBA-15 [54]. These bifunctional catalysts are mainly used for the coupled hydrogenation and esterification of bio-oil.

In addition to catalyst stability and acid site density, the hydrophilic or hydrophobic properties of solid acid catalysts have also attracted attention [80]. Research in this area is mainly focused on SBA-15-based catalysts. The hydrophilic or hydrophobic properties of catalysts can be regulated via grafting organic functionalities [80]. Miao and Shanks performed esterification of acetic acid with methanol using propylsulfonic acid-functionalized SBA-15 materials as the catalyst [80]. They found that the catalyst showed a comparable site activity to H_2SO_4. They also investigated the water tolerance of the catalyst. Results showed that SO_3H-SBA-15 exhibited less reaction inhibition due to water than sulfuric acid, which was due to the hydrophobic propyl groups bonded to the $-SO_3H$ group. Further grafting of the catalyst with hydrophobic propyl groups further increased the tolerance toward water. Ding et al. employed a SBA-15 catalyst grafted with a heteropoly acid for acetic acid esterification with *n*-butanol [109]. They found that the conversion of acetic acid surpassed the limit of the chemical equilibrium. This is mainly due to the rejection of the hydrophobic *n*-butyl acetate from the catalytic sites. As a consequence, the reverse reaction – *n*-butyl acetate hydrolysis – was suppressed, achieving a high conversion of acetic acid. Wilson et al. prepared a propylsulfonic acid derivatized solid acid catalyst via post-modification of mesoporous SBA-15 silica with mercaptopropyltrimethoxysilane (MPTMS) for the esterification of acetic acid. Results showed that catalytic activity was promoted upon cofunctionalization with hydrophobic octyl chains. Catalyst characterization with inverse gas chromatography showed that the increased activity was related to the increased hydrophobicity of the catalyst surface [50, 110]. The effect of the hydrophilic/hydrophobic properties of a faujasite catalyst on its activity in acetic acid on esterification is shown in Figure 4.29. Overall, these studies show that the hydrophilic/hydrophobic properties of the acidic sites significantly affect the activity of acid catalysts, which should be taken into consideration in the development of catalysts for esterification of water-rich bio-oil.

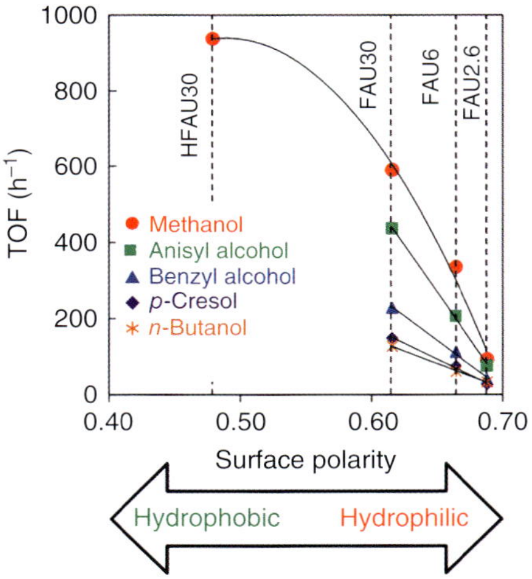

Figure 4.29 Turnover frequency (TOFs) as a function of FAU zeolite surface polarity. (Reaction conditions: 5 mmol acetic acid, acid:alcohol = 1 : 5 mol:mol, 10 ml toluene, 100 mg catalyst, 100 °C.) Source: Osatiashtiani e al. 2017 [50]. Reprinted with permission of Springer Nature. (http://creativecommons.org/licenses/by/4.0/).

4.5 Summary and Outlook

Bio-oil is a very complex mixture of organics, which makes the esterification of bio-oil a complicated process. The main objective of bio-oil esterification is to transform the acidic carboxylic acids into neutral esters. However, the actual reaction network is far more complex than the esterification of carboxylic acids. Esterification of bio-oil in alcohols is better described as the acid treatment of bio-oil in an alcohol-rich medium. Not only acetic acid but also the sugars, furans, aldehydes, and phenolics react with water, alcohols, or other components in some way, which significantly modifies the composition and the properties of the bio-oil. Various aspects of bio-oil esterification have been investigated, including the reaction network, process parameters, reactors, and catalysts. Nevertheless, a number of issues still need to be addressed. Some of the major issues remaining are listed as follows:

1. An evaluation of the effects of bio-oil esterification on the properties and subsequent applications of the bio-oil should be conducted. Esterification is employed to remove the carboxylic acids in bio-oil. Indeed, typically the majority of the light carboxylic acids are transformed or removed. Nevertheless, the behavior of the heavy acids remains to be elucidated. Furthermore, bio-oil esterification significantly affects its composition. However, how this change in composition impacts the stability of the bio-oil needs further investigation. Moreover, the effect of esterification on subsequent bio-oil upgrading via hydrotreatment also deserves further attention.
2. The reaction network during bio-oil esterification needs to be further clarified. From the existing literature, there is some understanding concerning the reaction pathways of the light components. For the heavy components, up to now little is known about their structures or their detailed reaction pathways during esterification. The heavy components are very important fractions of bio-oil, which significantly affect its further processing. How to handle or process the heavy components of bio-oil is a major problem and one of the biggest challenges facing bio-oil utilization.
3. Coking is one of the key issues that should be addressed. Coke formation leads to catalyst deactivation and consumes the organics in bio-oil. Furthermore, coking is the main reason for the irreversible deactivation of carbon-based catalysts. Coking involves a complex reaction network. The manner in which the main bio-oil components contribute to polymerization needs to be clarified. Corresponding measures need to be taken to suppress coke formation.
4. Catalysts specifically designed for bio-oil esterification should be developed. Steric hindrance is a problem for conversion of especially the heavy components of bio-oil. Easy accessibility of the active sites to molecules with varied molecular size should be a major feature of such catalysts. In addition, bio-oil is prone to polymerization. The catalyst must be amenable to burn-off the coke deposits to recover its activity.
5. In current studies, bio-oil esterification is generally performed in batch reactors. Batch reactors are convenient for achieving good mass balances and for investigating the reaction network of the compounds present.

However, performing bio-oil esterification in continuous reactors could lower the cost of the process due to the higher throughput. The reaction rates for acid-catalyzed reactions such as carboxylic acid esterification and the conversion of sugars are generally not high at low reaction temperatures. Long residence times are thus required. How to configure continuous flow reactors to meet the reaction kinetics of the major components in bio-oil during esterification need to be further investigated.

6. Alcohols are generally used in excess in bio-oil esterification. Recovery of the unconsumed alcohols needs to be considered. The simplest approach to recover the alcohol is distillation, which, however, is energy intensive. Furthermore, the light components of the bio-oil and the alcohols cannot be separated easily by distillation. If the alcohols in the esterified bio-oil are not separated, the effect of the alcohols on subsequent processing of the bio-oil needs to be understood. In addition, etherification reactions between the alcohols during bio-oil esterification also consume the alcohols, producing ethers and water. How to best suppress the etherification reaction to avoid excessive consumption of alcohols is also a question deserving of attention.
7. The last, and the simplest question, is the cost of the process. Is the input and output of the esterification of bio-oil balanced in terms of the economics and energy balance?

Acknowledgments

This work was supported by the National Natural Science Foundation of China (No. 51876080), the Program for Taishan Scholars of Shandong Province Government, the Recruitment Program of Global Young Experts (Thousand Youth Talents Plan), the Natural Science Fund of Shandong Province (ZR2017BB002), and the Key R&D Program of Shandong Province (2018GSF116014).

References

1 Huber, G.W., Iborra, S., and Corma, A. (2006). Synthesis of transportation fuels from biomass: chemistry, catalysts, and engineering. *Chem. Rev.* 106: 4044–4098.

2 Kim, S. and Dale, B.D. (2005). Environmental aspects of ethanol derived from no-tilled corn grain: nonrenewable energy consumption and greenhouse gas emissions. *Biomass Bioenergy* 28: 475–489.

3 Zhang, Q., Chang, J., Wang, T., and Xu, Y. (2007). Review of biomass pyrolysis oil properties and upgrading research. *Energy Convers. Manage.* 48: 87–92.

4 Hiloidhari, M., Das, D., and Baruah, D.C. (2014). Bioenergy potential from crop residue biomass in India. *Renewable Sustainable Energy Rev.* 32: 504–512.

5 Sansaniwal, S.K., Pal, K., Rosen, M.A., and Tyagi, S.K. (2017). Recent advances in the development of biomass gasification technology: a comprehensive review. *Renewable Sustainable Energy Rev.* 72: 363–384.
6 Shahbaz, M., Yusup, S., Inayat, A. et al. (2017). The influence of catalysts in biomass steam gasification and catalytic potential of coal bottom ash in biomass steam gasification: a review. *Renewable Sustainable Energy Rev.* 73: 468–476.
7 Wang, S., Dai, G., Yang, H., and Luo, Z. (2017). Lignocellulosic biomass pyrolysis mechanism: a state-of-the-art review. *Prog. Energy Combust. Sci.* 62: 33–86.
8 Guedes, R.E., Luna, A.S., and Torres, A.R. (2018). Operating parameters for bio-oil production in biomass pyrolysis: a review. *J. Anal. Appl. Pyrolysis* 129: 134–149.
9 Sharma, R.K. and Bakhshi, N.N. (1991). Upgrading of wood-derived bio-oil over HZSM-5. *Bioresour. Technol.* 35: 57–66.
10 Adjaye, J.D., Sharma, R.K., and Bakhshi, N.N. (1992). Characterization and stability analysis of wood-derived bio-oil. *Fuel Process. Technol.* 31: 241–256.
11 Oasmaa, A., Solantausta, Y., Arpiainen, V. et al. (2010). Fast pyrolysis bio-oils from wood and agricultural residues. *Energy Fuels* 24: 1380–1388.
12 Hossain, A.K. and Davies, P.A. (2013). Pyrolysis liquids and gases as alternative fuels in internal combustion engines–a review. *Renewable Sustainable Energy Rev.* 21: 165–189.
13 Hu, X., Gunawan, R., Mourant, D. et al. (2012). Acid-treatment of the bio-oil from mallee bark in methanol-rich medium: reaction pathways of the main components in bio-oil. *Fuel* 97: 512–522.
14 Tzanetakis, T., Ashgriz, N., James, D.F., and Thomson, M.J. (2008). Liquid fuel properties of a hardwood-derived bio-oil fraction. *Energy Fuels* 22: 2725–2733.
15 Garcia-Perez, M., Shen, J., Wang, X.S., and Li, C.Z. (2010). Production and fuel properties of fast pyrolysis oil/bio-diesel blends. *Fuel Process. Technol.* 91: 296–305.
16 Alsbou, E. and Helleur, B. Accelerated aging of bio-oil from fast pyrolysis of hardwood. *Energy Fuels* 28: 3224–3235.
17 Zhu, L., Li, K., Ding, H., and Zhu, X. (2018). Studying on properties of bio-oil by adding blended additive during aging. *Fuel* 211: 704–711.
18 Chen, D., Zhou, J., Zhang, Q., and Zhu, X. (2014). Evaluation methods and research progresses in bio-oil storage stability. *Renewable Sustainable Energy Rev.* 40: 69–79.
19 Wang, H., Male, J., and Wang, Y. (2013). Recent advances in hydrotreating of pyrolysis bio-oil and its oxygen-containing model compounds. *ACS Catal.* 3: 1047–1070.
20 Zacher, A.H., Olarte, M.V., Santosa, D.M. et al. (2014). A review and perspective of recent bio-oil hydrotreating research. *Green Chem.* 16: 491–515.
21 Zhang, L., Hu, G., Hu, S. et al. (2018). Hydrogenation of fourteen biomass-derived phenolics in water and in methanol: their distinct reaction behaviours. *Sustainable Energy Fuels* 2: 751–758.

22 Li, X., Gunawan, R., Wang, Y. et al. (2014). Upgrading of bio-oil into advanced biofuels and chemicals. Part III. Changes in aromatic structure and coke forming propensity during the catalytic hydrotreatment of a fast pyrolysis bio-oil with Pd/C catalyst. *Fuel* 116: 642–649.

23 Xiu, S. and Shahbazi, A. (2012). Bio-oil production and upgrading research: a review. *Renewable Sustainable Energy Rev.* 16: 4406–4414.

24 French, R.J., Stunkel, J., and Baldwin, R.M. (2011). Mild hydrotreating of bio-oil: effect of reaction severity and fate of oxygenated species. *Energy Fuels* 25: 3266–3274.

25 Elliott, D.C. (2007). Historical developments in hydroprocessing bio-oils. *Energy Fuels* 21: 1792–1815.

26 Elliott, D.C., Hart, T.R., Neuenschwander, G.G. et al. (2012). Catalytic hydroprocessing of fast pyrolysis bio-oil from pine sawdust. *Energy Fuels* 26: 3891–3896.

27 Mohan, D., Pittman, C.U., and Steele, P.H. (2006). Pyrolysis of wood/biomass for bio-oil: a critical review. *Energy Fuels* 20: 848–889.

28 WanIsahak, W.N.R., Hisham, M.W.M., Yarmo, M.A., and Hin, T.Y. (2012). A review on bio-oil production from biomass by using pyrolysis method. *Renewable Sustainable Energy Rev.* 16: 5910–5923.

29 Ikura, M., Stanciulescu, M., and Hogan, E. (2003). Emulsification of pyrolysis derived bio-oil in diesel fuel. *Biomass Bioenergy* 24: 221–232.

30 Garcia-Perez, M., Adams, T.T., Goodrum, J.W. et al. (2007). Production and fuel properties of pine chip bio-oil/biodiesel blends. *Energy Fuels* 21: 2363–2372.

31 Kim, T.-S., Kim, J.-Y., Kim, K.-H. et al. (2012). The effect of storage duration on bio-oil properties. *J. Anal. Appl. Pyrolysis* 95: 118–125.

32 Boucher, M.E., Chaala, A., Pakdel, H., and Roy, C. (2000). Bio-oils obtained by vacuum pyrolysis of softwood bark as a liquid fuel for gas turbines. Part II: stability and ageing of bio-oil and its blends with methanol and a pyrolytic aqueous phase. *Biomass Bioenergy* 19: 351–361.

33 Hu, X., Wang, Y., Mourant, D. et al. (2013). Polymerization on heating up of bio-oil: a model compound study. *AIChE J.* 59: 888–900.

34 Antal, M.J. Jr., (1983). Biomass pyrolysis: a review of the literature part 1—carbohydrate pyrolysis. In: *Advances in Solar Energy*, 61–111. Boston, MA: Springer.

35 Bridgwater, A.V. (2012). Review of fast pyrolysis of biomass and product upgrading. *Biomass Bioenergy* 38: 68–94.

36 Wu, L., Hu, X., Mourant, D. et al. (2014). Quantification of strong and weak acidities in bio-oil via non-aqueous potentiometric titration. *Fuel* 115: 652–657.

37 Hu, X., Gunawan, R., Mourant, D. et al. (2017). Upgrading of bio-oil via acid-catalyzed reactions in alcohols – a mini review. *Fuel Process. Technol.* 155: 2–19.

38 Hu, X., Wu, L., Wang, Y. et al. (2013). Acid-catalyzed conversion of mono- and poly-sugars into platform chemicals: effects of molecular structure of sugar substrate. *Bioresour. Technol.* 133: 469–474.

39 Asadullah, M., Rahman, M.A., Ali, M.M. et al. (2007). Production of bio-oil from fixed bed pyrolysis of bagasse. *Fuel* 86: 2514–2520.

40 Girisuta, B., Janssen, L.P.B.M., and Heeres, H.J. (2006). Green chemicals: a kinetic study on the conversion of glucose to levulinic acid. *Chem. Eng. Res. Des.* 84: 339–349.

41 Hu, X., Lievens, C., Larcher, A., and Li, C.-Z. (2011). Reaction pathways of glucose during esterification: effects of reaction parameters on the formation of humin type polymers. *Bioresour. Technol.* 102: 10104–11013.

42 Hu, X. and Li, C.Z. (2011). Levulinic esters from the acid-catalysed reactions of sugars and alcohols as part of a bio-refinery. *Green Chem.* 13: 1676–1679.

43 Hu, X., Lievens, C., and Li, C.Z. (2012). Acid-catalyzed conversion of xylose in methanol-rich medium as part of biorefinery. *ChemSusChem* 5: 1427–1434.

44 Hu, X., Wang, S., Wu, L. et al. (2014). Acid-treatment of C5 and C6 sugar monomers/oligomers: insight into their interactions. *Fuel Process. Technol.* 126: 315–323.

45 Hu, X., Wu, L., Wang, Y. et al. (2012). Mediating acid-catalyzed conversion of levoglucosan into platform chemicals with various solvents. *Green Chem.* 14: 3087–3098.

46 Hu, X., Westerhof, R.J.M., Dong, D. et al. (2014). Acid-catalyzed conversion of xylose in 20 solvents: insight into interactions of the solvents with xylose, furfural, and the acid catalyst. *ACS Sustainable Chem. Eng.* 2: 2562–2575.

47 Gunawan, R.R., Li, X., Larcher, A.V. et al. (2012). Hydrolysis and glycosidation of sugars during the esterification of fast pyrolysis bio-oil. *Fuel* 95: 146–151.

48 Wu, L., Hu, X., Wang, S. et al. (2016). Formation of coke during the esterification of pyrolysis bio-oil. *RSC Adv.* 6: 86485–86493.

49 Hu, X., Mourant, D., Gunawan, R. et al. (2012). Production of value-added chemicals from bio-oil via acid catalysis coupled with liquid–liquid extraction. *RSC Adv.* 2: 9366–9370.

50 Osatiashtiani, A., Puértolas, B., Oliveira, C.C.S. et al. (2017). On the influence of Si:Al ratio and hierarchical porosity of FAU zeolites in solid acid catalysed esterification pretreatment of bio-oil. *Biomass Convers. Biorefin.* 7: 331–342.

51 Zhang, Z., Pittman, C.U. Jr.,, Sui, S. et al. (2013). Catalytic upgrading of bio-oil by reacting with olefins and alcohols over solid acids: reaction paths via model compound studies. *Energies* 6: 1568–1589.

52 Zhang, Z., Wang, Q., Tripathi, P., and Pittman, C.U. Jr., (2011). Catalytic upgrading of bio-oil using 1-octene and 1-butanol over sulfonic acid resin catalysts. *Green Chem.* 13: 940–949.

53 Milina, M., Mitchell, S., and Pérez-Ramírez, J. (2014). Prospectives for bio-oil upgrading via esterification over zeolite catalysts. *Catal. Today* 235: 176–183.

54 Dang, Q., Luo, Z., Zhang, J. et al. (2013). Experimental study on bio-oil upgrading over Pt/SO_4^{2-}/ZrO_2/SBA-15 catalyst in supercritical ethanol. *Fuel* 103: 683–692.

55 Zhang, X., Chen, L., Kong, W. et al. (2015). Upgrading of bio-oil to boiler fuel by catalytic hydrotreatment and esterification in an efficient process. *Energy* 84: 83–90.

56 Zhou, M., Liu, P., Wang, K. et al. (2016). Catalytic hydrogenation and one step hydrogenation-esterification to remove acetic acid for bio-oil upgrading: model reaction study. *Catal. Sci. Technol.* 6: 7783–7792.

57 Zhao, H., Holladay, J.E., Brown, H., and Zhang, Z.C. (2007). Metal chlorides in ionic liquid solvents convert sugars to 5-hydroxymethylfurfural. *Science* 316: 1597–1600.

58 Hu, X., Westerhof, R.J.M., Wu, L. et al. (2015). Upgrading biomass-derived furans via acid-catalysis/hydrogenation: the remarkable difference between water and methanol as the solvent. *Green Chem.* 17: 219–224.

59 Hu, X., Jiang, S., Kadarwati, S. et al. (2016). Effects of water and alcohols on the polymerization of furan during its acid-catalyzed conversion into benzofuran. *RSC Adv.* 6: 40489–40501.

60 Hu, X., Kadarwati, S., Wang, S. et al. (2015). Biomass-derived sugars and furans: which polymerize more during their hydrolysis? *Fuel Process. Technol.* 137: 212–219.

61 Demirbas, A. (2008). Comparison of transesterification methods for production of biodiesel from vegetable oils and fats. *Energy Convers. Manage.* 49: 125–130.

62 Leung, D.Y.C. and Guo, Y. (2006). Transesterification of neat and used frying oil: optimization for biodiesel production. *Fuel Process. Technol.* 87: 883–890.

63 He, M., Mourant, D., Gunawan, R. et al. (2012). Yield and properties of bio-oil from the pyrolysis of mallee leaves in a fluidised-bed reactor. *Fuel* 102: 506–513.

64 Hu, X., Gunawan, R., Mourant, D. et al. (2012). Esterification of bio-oil from mallee leaves with a solid acid catalyst: conversion of the cyclic ether and terpenoids into hydrocarbons. *Bioresour. Technol.* 123: 249–255.

65 Hu, X., Lievens, C., Mourant, D. et al. (2013). Investigation of deactivation mechanisms of a solid acid catalyst during esterification of the bio-oils from mallee biomass. *Appl. Energy* 111: 94–103.

66 Peng, J., Chen, P., Lou, H., and Zheng, X. (2008). Upgrading of bio-oil over aluminum silicate in supercritical ethanol. *Energy Fuels* 22: 3489–3492.

67 Peng, J., Chen, P., Lou, H., and Zheng, X. (2009). Catalytic upgrading of bio-oil by HZSM-5 in sub- and super-critical ethanol. *Bioresour. Technol.* 100: 3415–3418.

68 Valle, B., Castaño, P., Olazar, M. et al. (2012). Deactivating species in the transformation of crude bio-oil with methanol into hydrocarbons on a HZSM-5 catalyst. *J. Catal.* 285: 304–314.

69 Qin, F., Cui, H., Yi, W., and Wang, C. (2014). Upgrading the water-soluble fraction of bio-oil by simultaneous esterification and acetalation with online extraction. *Energy Fuels* 28: 2544–2553.

70 Sundqvist, T., Oasmaa, A., and Koskinen, A. (2015). Upgrading fast pyrolysis bio-oil quality by esterification and azeotropic water removal. *Energy Fuels* 29: 2527–2534.

71 Mahfud, F.H., Melia'n-Cabrera, I., Manurung, R., and Heeres, H.J. (2007). Upgrading of flash pyrolysis oil by reactive distillation using a high boiling alcohol and acid catalysts. *Process Saf. Environ. Prot.* 85: 466–472.

72 Wang, C., Hu, Y., Chen, Q. et al. (2013). Bio-oil upgrading by reactive distillation using *p*-toluene sulfonic acid catalyst loaded on biomass activated carbon. *Biomass Bioenergy* 56: 405–411.

73 Hilten, R.N., Bibens, B.P., Kastner, J.R., and Das, K.C. (2010). In-line esterification of pyrolysis vapor with ethanol improves bio-oil quality. *Energy Fuels* 24: 673–682.

74 Xu, J., Jiang, J., Dai, W. et al. (2011). Bio-oil upgrading by means of ozone oxidation and esterification to remove water and to improve fuel characteristics. *Energy Fuels* 25: 1798–1801.

75 Tanneru, S.K., Parapati, D.R., and Steele, P.H. (2014). Pretreatment of bio-oil followed by upgrading via esterification to boiler fuel. *Energy* 73: 214–220.

76 Chen, W., Luo, Z., Yu, C. et al. (2014). Catalytic transformations of acids, aldehydes, and phenols in bio-oil to alcohols and esters. *Fuel* 135: 55–62.

77 Hu, X., Mourant, D., Wang, Y. et al. (2013). Acid-catalysed treatment of the mallee leaf bio-oil with methanol: effects of molecular structure of carboxylic acids and esters on their conversion. *Fuel Process. Technol.* 106: 569–576.

78 Wu, L., Hu, X., Wang, S. et al. (2018). Acid-treatment of bio-oil in methanol: the distinct catalytic behaviours of a mineral acid catalyst and a solid acid catalyst. *Fuel* 212: 412–421.

79 Westerhof, R.J.M., Oudenhoven, S.R.G., Hu, X. et al. (2016). Biofuel and methyl levulinate from biomass-derived fractional condensed pyrolysis oil and alcohol. *Energy Technol.* 4: 1–12.

80 Miao, S. and Shanks, B.H. (2009). Esterification of biomass pyrolysis model acids over sulfonic acid-functionalized mesoporous silicas. *Appl. Catal., A* 359: 113–120.

81 Fu, J., Chen, L., Lv, P. et al. (2015). Free fatty acids esterification for biodiesel production using self-synthesized macroporous cation exchange resin as solid acid catalyst. *Fuel* 154: 1–8.

82 Li, Z., Wnetrzak, R., Kwapinski, W., and Leahy, J.J. (2012). Synthesis and characterization of sulfated TiO_2 nanorods and ZrO_2/TiO_2 nanocomposites for the esterification of biobased organic acid. *ACS Appl. Mater. Interfaces* 4: 4499–4505.

83 Ye, J., Liu, C., Fu, Y. et al. (2014). Upgrading bio-oil: simultaneous catalytic esterification of acetic acid and alkylation of acetaldehyde. *Energy Fuels* 28: 4267–4272.

84 Lohitharn, N. and Shanks, B.H. (2009). Upgrading of bio-oil: effect of light aldehydes on acetic acid removal via esterification. *Catal. Commun.* 11: 96–99.

85 Zhang, Z., Sui, S., Tan, S. et al. (2013). Catalytic conversion of bio-oil to oxygen-containing fuels by simultaneous reactions with 1-butanol and 1-octene over solid acids: model compound studies and reaction pathways. *Bioresour. Technol.* 130: 789–792.

86 Tang, Y., Yu, W., Mo, L. et al. (2008). One-step hydrogenation-esterification of aldehyde and acid to ester over bifunctional Pt catalysts: a model reaction as novel route for catalytic upgrading of fast pyrolysis bio-oil. *Energy Fuels* 22: 3484–3488.

87 Yu, W., Tang, Y., Mo, L. et al. (2011). Bifunctional Pd/Al-SBA-15 catalyzed one-step hydrogenation–esterification of furfural and acetic acid: a model reaction for catalytic upgrading of bio-oil. *Catal. Commun.* 13: 35–39.

88 Tang, Y., Miao, S., Shanks, B.H., and Zheng, X. (2010). Bifunctional mesoporous organic–inorganic hybrid silica for combined one-step hydrogenation/esterification. *Appl. Catal., A* 375: 310–317.

89 Liu, Y., Li, Z., Leahy, J.J., and Kwapinski, W. (2015). Catalytically upgrading bio-oil via esterification. *Energy Fuels* 29: 3691–3698.

90 Song, M., Zhong, Z., and Dai, J. (2010). Different solid acid catalysts influence on properties and chemical composition change of upgrading bio-oil. *J. Anal. Appl. Pyrolysis* 89: 166–170.

91 Wei, Y., Lei, H., Zhu, L. et al. (2015). Oxygen-containing fuels from high acid water phase pyrolysis bio-oils by ZSM−5 catalysis: kinetic and mechanism studies. *Energies* 8: 5898–5915.

92 Moens, L., Black, S.K., Myers, M.D., and Czernik, S. (2009). Study of the neutralization and stabilization of a mixed hardwood bio-oil. *Energy Fuels* 23: 2695–2699.

93 Udomsap, P., Yeinn, Y.H., Hui, J.T.H. et al. (2011). Towards stabilization of bio-oil by addition of antioxidants and solvents, and emulsification with conventional hydrocarbon fuels. In: *2011 International Conference and Utility Exhibition on Power and Energy Systems: Issues & Prospects for Asia (ICUE)*, Pattaya City, Thailand (28–30 September 2011), 1–5. IEEE.

94 Zhang, Q., Chang, J., Wang, T., and Xu, Y. (2006). Upgrading bio-oil over different solid catalysts. *Energy Fuels* 20: 2717–2720.

95 Reyhanitash, E., Tymchyshyn, M., Yuan, Z. et al. (2014). Upgrading fast pyrolysis oil via hydrodeoxygenation and thermal treatment: effects of catalytic glycerol pretreatment. *Energy Fuels* 28: 1132–1138.

96 Wang, J.J., Chang, J., and Fan, J. (2010). Upgrading of bio-oil by catalytic esterification and determination of acid number for evaluating esterification degree. *Energy Fuels* 24: 3251–3255.

97 Sundqvist, T., Oasmaa, A., and Koskinen, A. (2015). Upgrading fast pyrolysis bio-oil quality by esterification and a zeotropic water removal. *Energy Fuels* 29: 2527–2534.

98 Guo, Z., Wang, S., Xu, G. et al. (2011). Upgrading of bio-oil molecular distillation fraction with solid acid catalyst. *Bioresources* 6: 2539–2550.

99 Tao, J., Penmetsa, V.K., and Steele, P.H. (2015). Addition of gaseous olefin to produce boiler fuel and stability studies. *Environ. Prog. Sustainable Energy* 34: 933–941.

100 Chen, W., Luo, Z., Yu, C. et al. (2014). Upgrading of bio-oil in supercritical ethanol: catalysts screening, solvent recovery and catalyst stability study. *J. Supercrit. Fluids* 95: 387–393.

101 Zhang, J., Luo, Z., Dang, Q. et al. (2012). Upgrading of bio-oil over bifunctional catalysts in supercritical monoalcohols. *Energy Fuels* 26: 2990–2995.

102 Xu, X., Zhang, C., Zhai, Y. et al. (2014). Upgrading of bio-oil using supercritical 1-butanol over a Ru/C heterogeneous catalyst: role of the solvent. *Energy Fuels* 28: 4611–4621.

103 Li, W., Pan, C., Sheng, L. et al. (2011). Upgrading of high-boiling fraction of bio-oil in supercritical methanol. *Bioresour. Technol.* 102: 9223–9228.

104 Li, W., Pan, C., Zhang, Q. et al. (2011). Upgrading of low-boiling fraction of bio-oil in supercritical methanol and reaction network. *Bioresour. Technol.* 102: 4884–4889.

105 Hu, X., Jiang, S., Wu, L. et al. (2017). One-pot conversion of the biomass-derived xylose and furfural into levulinate esters via acid catalysis. *Chem. Commun.* 53: 2938–2941.

106 Hu, X., Kadarwati, S., Song, Y., and Li, C.Z. (2016). Simultaneous hydrogenation and acid-catalyzed conversion of the biomass-derived furans in solvents with distinct polarities. *RSC Adv.* 6: 4647–4656.

107 Hu, X., Wang, S., Westerhof, R.J.M. et al. (2015). Acid-catalyzed conversion of C6 sugar monomer/oligomers to levulinic acid in water, tetrahydrofuran and toluene: importance of the solvent polarity. *Fuel* 141: 56–63.

108 Hu, X., Song, Y., Wu, L. et al. (2013). One-pot synthesis of levulinic acid/ester from C5 carbohydrates in a methanol medium. *ACS Sustainable Chem. Eng.* 1: 1593–1599.

109 Liu, H., Xue, N., Peng, L. et al. (2009). The hydrophilic/hydrophobic effect of porous solid acid catalysts on mixed liquid phase reaction of esterification. *Catal. Commun.* 10: 1734–1737.

110 Manayil, J.C., Santos, V.C., Jentoft, F.C. et al. (2017). Octyl Co-grafted $PrSO_3H$/SBA-15: tunable hydrophobic solid acid catalysts for acetic acid esterification. *ChemCatChem* 9: 2231–2238.

5

Catalytic Upgrading of Holocellulose-Derived C_5 and C_6 Sugars

Xingguang Zhang[1], Zhijun Tai[2], Amin Osatiashtiani[2], Lee Durndell[3], Adam F. Lee[4], and Karen Wilson[4]

[1] *Nanjing Forestry University, College of Chemical Engineering, Jiangsu Key Lab for the Chemistry and Utilization of Agricultural and Forest Biomass, Nanjing, 210037, China*
[2] *Aston University, Aston Triangle, European Bioenergy Research Institute, Birmingham B4 7ET, UK*
[3] *School of Geography, Earth and Environmental Sciences, University of Plymouth, Drake Circus, Plymouth PL4 8AA, UK*
[4] *RMIT University, School of Science, Applied Chemistry and Environmental Science, 124 La Trobe Street, Melbourne, VIC 3000, Australia*

5.1 Introduction

The quest for sustainable resources to meet the demands of a rapidly rising global population while mitigating the risks of rising CO_2 emissions and associated climate change represents a grand challenge for humanity. Global CO_2 emissions are predicted to increase from 6.6 to 11.0 GtC/yr by 2050 [1] due to increasing energy demands. Despite significant growth in proven and predicted fossil fuel reserves over the next two decades, e.g. heavy crude oil, tar sands, deep water wells, and shale oil and gas, there are great uncertainties in the economics of their exploitation through current extraction methodologies, and crucially, an increasing proportion of such carbon resources (estimates vary between 65% and 80% [2, 3]) cannot be burned without breaching targets from Paris COP21 for a <1.5 °C increase in mean global temperature relative to preindustrial levels [4–7]. Realizing these ambitious targets will require a transition from traditional fossil fuel resources to renewables, notably nuclear, solar, wind and wave power, and biomass [8]. Biomass is a generic term describing biogenic organic matter, typically formed through photosynthesis, with the potential to replace fossil feedstocks for the production of bio-based transportation fuels, heat, power, and biomaterials, in addition to underpinning new chemical platforms. The US Department of Energy identified a range of sugar-derived "platform chemicals," accessible by chemical or biochemical transformation of lignocellulosic biomass, which are potential targets for production in biorefineries [9]. Akin to petroleum refining, biorefineries integrate biomass conversion processes so that biofuels, chemicals, and energy are coproduced to maximize waste biomass valorization; an approach analogous to current petroleum refineries that deliver high volume/low value (fuels and commodity chemicals) and low volume/high

Chemical Catalysts for Biomass Upgrading, First Edition. Edited by Mark Crocker and Eduardo Santillan-Jimenez.

value (fine/speciality chemicals) products in tandem, the latter supporting the economic success of the industry [10, 11]. Catalytic technologies have played a critical role in the economic development of the petrochemical industry and modern society, underpinning 90% of chemical manufacturing processes and contributing to over 20% of all industrial products [12, 13]. In a post-petroleum era, catalysis will support analogous biorefinery technology, and researchers will need to rise to the challenge of synthesizing chemical intermediates, advanced functional materials, and fuels from nonpetroleum feedstocks.

Polysaccharides are attractive potential feedstocks for bio-refineries, with the transformation of C_5 and C_6 sugars offering routes to the production of fine, bulk, and platform chemicals typically produced via petrochemical routes. While starch $(C_6H_{10}O_5)_n$ derived from wheat, corn, tuberous plants, sugar cane [14], or sorghum [15], is easily processed to glucose via hydrolysis, such food crops are not considered sustainable feedstocks. In this respect, lignocellulose derived from waste agricultural or forestry materials (e.g. logging or mill and manufacturing residues) or perennial herbaceous plants and short rotation woody crops (e.g. miscanthus, eucalyptus, or willow), is considered a particularly viable option for sustainable fuels and chemicals production [16, 17]. Lignocellulose comprises cellulose (30–50% of total lignocellulosic dry mass) and hemicellulose (20–40% of total dry mass), assembled from C_6 and C_5 sugars, respectively, such as glucose, xylose, and amylose, and bound together by poly-phenolic lignin that makes up the remaining 15–25% of dry biomass and imparts rigidity to plants and trees. The use of agricultural waste biomass residues [18], such as bagasse, rice husk, and straw, offers an attractive means to valorize waste materials that would otherwise be left to decompose, or in the case of rice straw, simply burned, thereby releasing atmospheric CO_2 [19]. Provided that sustainable land management practices are implemented to safeguard soil quality, biodiversity, and water use, it is claimed that sustainably sourced waste residues could supply 16% of road transport fuel by 2030, delivering greenhouse gas savings >60% across the European Union [20].

Isolation of sugars from lignocellulose is particularly challenging and requires extensive preprocessing of the raw materials [21]. Fractionation to separate cellulose, lignin, and hemicellulose is typically performed via acid or base hydrolysis, steam explosion [22], or organosolv treatments [23, 24]. Once separated, the hemicellulose and cellulose fractions are typically hydrolyzed to C_5 and C_6 sugars, respectively, for processing into fuels and/or chemicals via enzymatic [25] or chemical (acid or base) [26, 27], supercritical water [28], or more recently ionic liquid (IL) [29] treatments. Mechanocatalysis using H_2SO_4-impregnated lignocellulosic substrates is another approach showing enormous promise for clean and efficient depolymerization of lignocellulose in a one-pot process. The application of mechanical forces alongside acid catalysts can convert dried lignocellulose into water-soluble products comprising oligosaccharides and lignin fragments [30, 31].

5.2 Catalytic Transformation of C_5–C_6 Sugars

The transformation of biomass-derived building blocks will require different processes to those currently employed in the petrochemical industry, e.g. reverse chemical transformations wherein highly functional biomolecules are deoxygenated to their target product, instead of being oxygenated as is usual when starting from crude oil [32]. New classes of catalysts are urgently required that are compatible with hydrophilic, bulky substrates to facilitate the move away from existing short-chain hydrocarbon supplies. Improvements and innovations in catalyst and process design are necessary to deliver high-value chemicals from biomass-derived building blocks. Low temperature (<200 °C) thermochemical aqueous phase processing of sugars is of particular interest, as this affords highly functional intermediates via selective deoxygenation and functionalization to produce hydrocarbons, aldehydes, alcohols, acids, and esters [33–36]. Figure 5.1 illustrates the diverse products that can be synthesized from glucose via such transformations [37]. There is a growing interest in aqueous phase processing of (hemi)cellulosic-derived sugars that are formed through chemical or enzymatic hydrolysis. Here we review recent developments in catalyst design for the most common transformations to deoxygenate C_5–C_6 sugars: isomerization, dehydration, hydrogenation, and hydrogenolysis. Oxidation of sugars is another important transformation that has been employed to produce high value chemical products and has been extensively reviewed elsewhere [38]. The focus of this chapter will be on catalytic transformations in aqueous solvents, complementing excellent reviews on the application of alternative solvents, particularly ILs [39–41] and supercritical CO_2 [42] elsewhere.

5.2.1 Isomerization Catalysts

Hydrolysis of hemicellulose and cellulose predominately yields C_5 (xylose and arabinose) and C_6 sugars (glucose) that exist in three conformers in water: a pyranose ring, a furanose ring, or open-chain structures [43, 44]. Isomerization reactions of sugars, particularly xylose → xylulose, and glucose → fructose, represent important transformations in the production of platform molecules such as 5-hydroxymethylfurfural (5-HMF) and furfural [45]. The earliest studies of hexose isomerization by Lobry de Bruyn in 1895 employed homogeneous alkaline catalysts [46]; however, significant effort has been directed since to develop efficient heterogeneous catalysts for sugar isomerization, notably zeolitic solid Lewis acids [47–49], hydrotalcite (HT) solid bases [50], and metallosilicate solid bases [51].

Selective isomerization is hampered by side reactions including the epimerization of xylose to lyxose and of glucose to mannose, as well as the retroaldolization of fructose to glyceraldehyde and dihydroxyacetone and of glucose to glycolaldehyde and erythrose (Figure 5.2) [43, 44, 52].

Glucose is also susceptible to condensation and mutarotation [53, 54], and the mechanism of its isomerization to fructose and associated active sites remains debated [55–57]. Formation of solid humins and degradation products that are intractable to valorization is also problematic at elevated temperatures (>100 °C). Xylose and glucose isomerization are equilibrium-limited, so care should be taken to ensure that reaction kinetics are determined away from equilibrium conversion.

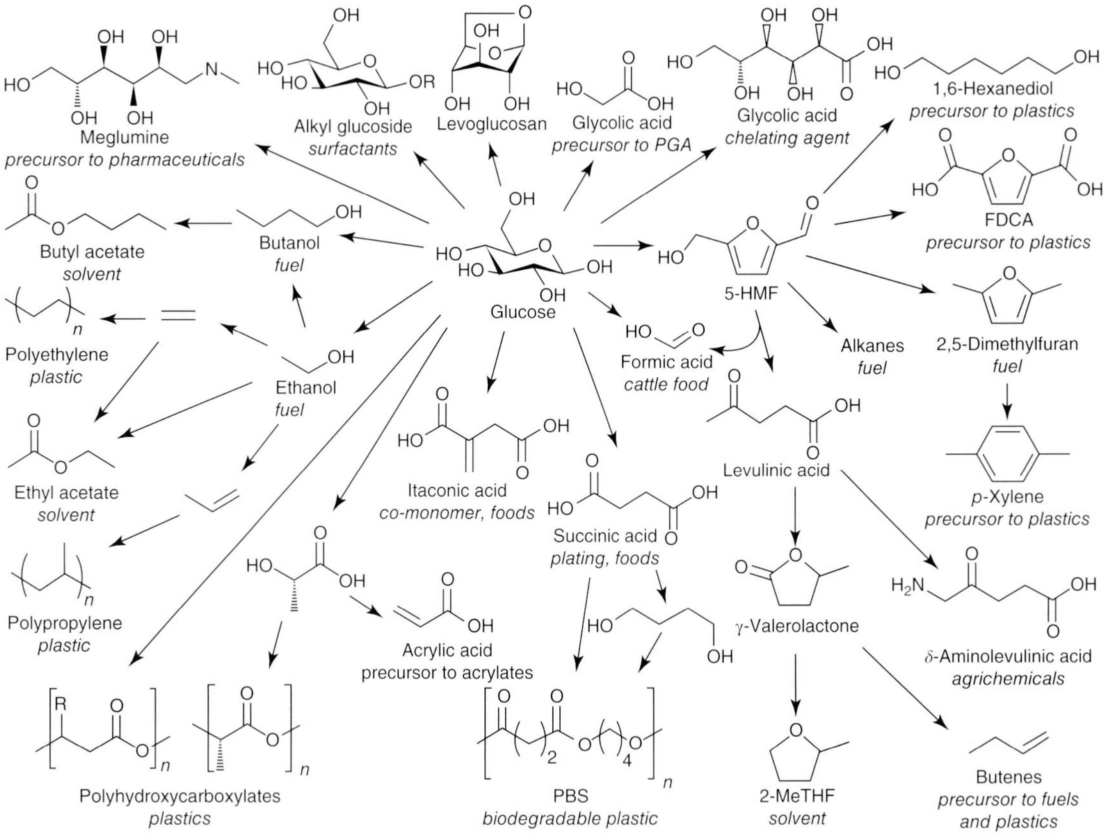

Figure 5.1 Chemical products derived from the catalytic transformation of glucose. Source: Kobayashi and Fukuoka [37]. Reproduced with permission of Royal Society of Chemistry.

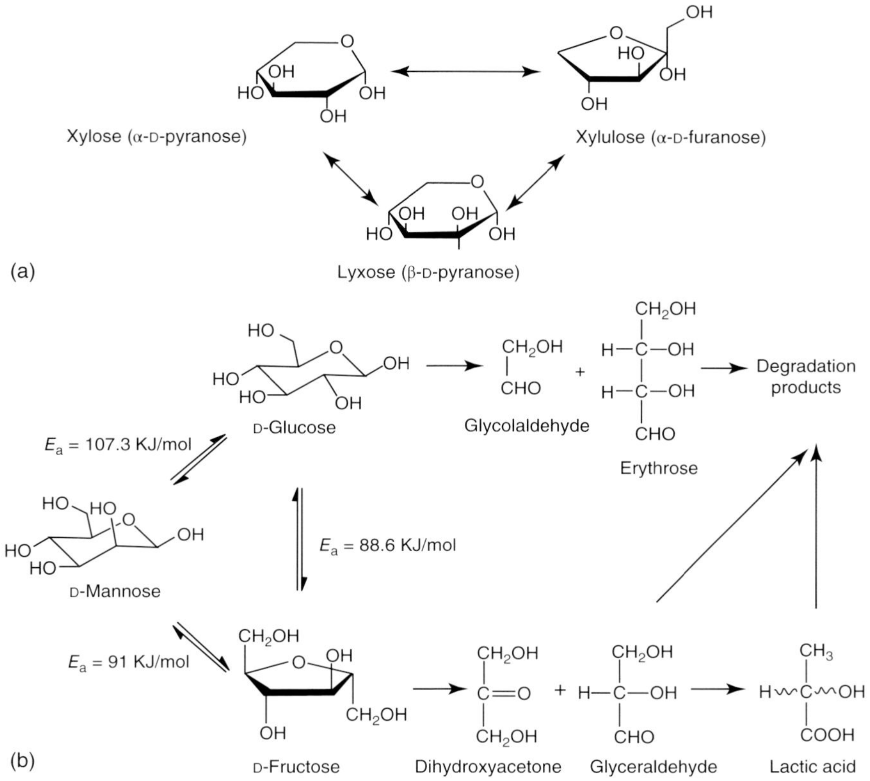

Figure 5.2 Principal reaction pathways for the isomerization of (a) xylose over Sn-beta zeolites. Source: Reprinted with permission from Choudhary et al. [43]. Copyright 2011, American Chemical Society. and (b) glucose over Mg-Al hydrotalcites and Sn-BEA zeolites. Source: Delidovich 2014 [52] and Sweeney 2014 [44]. Adapted with permission of The Royal Society of Chemistry.

5.2.1.1 Zeolites

Zeolites are frequently employed to convert carbohydrates into platform molecules due to their well-defined and controllable physicochemical properties, such as acid and base character [58]. Sn-Beta zeolite is a promising Lewis acid catalyst for carbohydrate isomerization [45], although it delivers a broad distribution of monosaccharides, illustrated for a dealuminated Sn-Beta (Sn/de-Al-β) Lewis acid catalyst in Table 5.1 [59].

Product selectivity depends on the relative energy barriers for the different isomerization pathways. For glucose, the energy barrier to fructose is ~17% lower than that to mannose, hence fructose is the dominant isomeric product under appropriate kinetic control [44]. Over Sn/de-Al-β at 110 °C, 70% selectivity to fructose is observed at equilibrium conversion, equating to a fructose yield of ~40% [59]. Separate studies of xylose isomerization over Sn-β-zeolite at 110 °C report a maximum xylulose yield of 27%, with a corresponding lyxose yield of 11%, at 60% xylose conversion [43].

Table 5.1 Product distribution following isomerization of C_6 and C_5 sugars over Sn/de-Al-β zeolite.

Reagent	Monosaccharide yield (%)	Monosaccharide distribution (%)		
		Glucose	Fructose	Mannose
Glucose	92.6	59.4	32.4	8.2
Fructose	83.9	19.9	61.4	18.7
Mannose	85.5	15.6	55.1	29.3
		Galactose	Tagatose	Talose
Galactose	89.5	69.4	28.3	2.3
		Ribose	Ribulose	Arabinose
Ribose	76.2	40.5	26.3	33.3
		Xylose	Xylulose	Lyxose
Xylose	78.0	49.8	4.6	45.5

Source: Dijkmans et al. 2013 [59]. Reproduced with permission of Royal Society of Chemistry.

Glucose isomerization over Sn-Beta at 110 °C has also been compared with Ti-Beta of similar pore dimensions and mesoporous Sn-MCM-41 and Ti-MCM-41. Sn and Ti-Beta are more efficient in terms of fructose yield (32% and 14%, respectively) from a 10 wt% glucose solution. The catalytic activity of these zeolites is partly attributed to the unique large pore environment of β-zeolites. Sn-containing MCM-41 exhibited significantly lower isomerization activity, most likely due to a lower degree of Sn incorporation into the MCM framework. Since $SnCl_4 \cdot 5H_2O$ and SnO_2 alone were inactive, it was concluded that Sn atoms within the zeolite frameworks are the active sites. Glucose isomerization can be influenced by framework or extra-framework active sites, the choice of solvent, and the nature of exchanged cations. According to Davis and workers [45], framework tin sites in Sn-Beta catalyze the isomerization of glucose to fructose via a Lewis acid-mediated intramolecular hydride shift in water (Figure 5.3a), whereas the former epimerization into mannose occurs by a Lewis acid-mediated intramolecular carbon shift. Extra-framework Sn located within the hydrophobic zeolite channels isomerized glucose to fructose in water and methanol through a base-catalyzed H^+ transfer mechanism. Isotope labeling (^2H and ^{13}C) of glucose highlights that isomerization products are formed through either hydrogen or carbon shifts, which are catalyst and solvent dependent, as summarized in Figure 5.3b. Another factor influencing isomerization activity is the synergy between Lewis acidic Sn centers and proximate hydroxyl groups derived from coadsorbed water molecules or internal silanols. A density functional theory (DFT) study by Hensen and coworkers [57], demonstrated that, independent of the zeolite topology, the rate-controlling H-shift elementary step is strongly favored when the aldehyde C1—O1 moiety of *o*-Glu is preactivated by a proton donor in the vicinity of the Lewis acidic Sn site. This cooperative action of the lattice Sn and hydrogen bond donor allows for efficient stabilization of

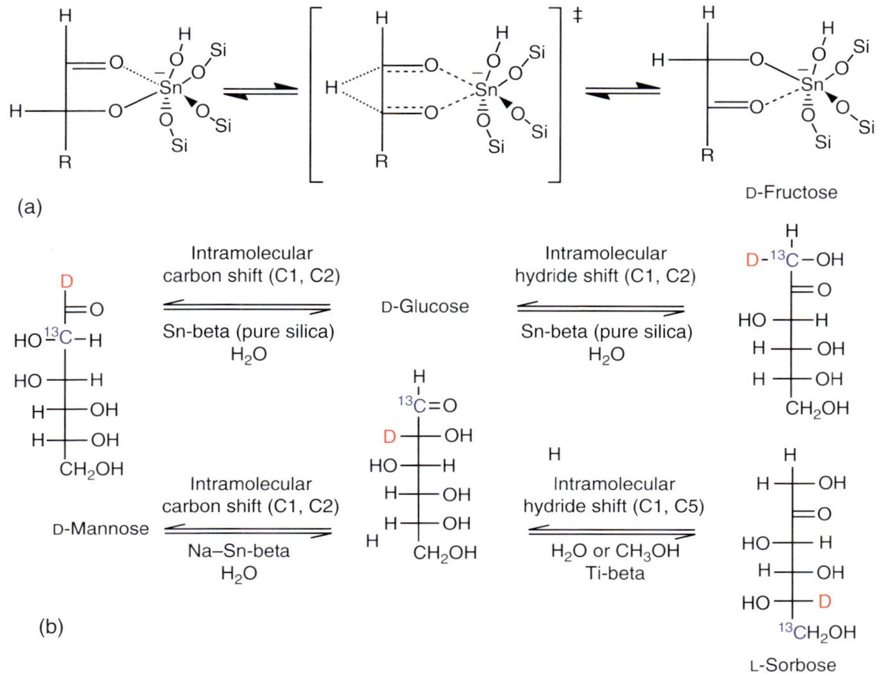

Figure 5.3 (a) Proposed glucose coordination to Sn during isomerization to fructose. Source: Adapted with permission from Moliner et al. 2010 [45]. Copyright 2010, National Academy of Sciences. and (b) Principal reaction pathways for glucose isomerization catalyzed by Sn-beta or Ti-beta in water or methanol. Source: Adapted with permission from Bermejo-Deval et al. 2014 [47], Bermejo-Deval et al. 2012 [55], Gounder and Davis [60]. Copyright 2014, 2012, and 2013, respectively, American Chemical Society.

the negative charge by redistributing it between the O2 and O1 moieties of the adsorbed carbohydrate during the H-shift reaction.

Cation exchanged zeolites X, Y, and A have also been employed in glucose isomerization [49]. For example, NaX, KX, and CsY possessing moderate basicity achieved 86% fructose selectivity; however, glucose conversion remains <25% and is potentially accompanied by cation leaching. Other alkaline earth metal cations such as Ca- and Ba-exchanged X, Y, and A zeolites exhibit lower selectivity toward fructose.

5.2.1.2 Hydrotalcites

HTs of general formula $[M(II)_{1-x}M(III)_x(OH)_2]^{x+}(A_{n-x/n})\cdot mH_2O$ are a subset of microporous-layered double hydroxides (LDHs) [61–63], which comprise brucite-like hydroxide sheets containing octahedrally coordinated M^{2+} (e.g. Mg^{2+}) and M^{3+} (e.g. Al^{3+}) cations, separated by interlayer A^{n-} charge-balancing anions, where x is the fraction of M^{3+} (typically $0.17 < x < 0.34$ for aluminum) and m denotes the water of crystallization. HTs are readily synthesized via coprecipitation from nitrates [64]; however the use of Na or K hydroxide/carbonate solutions for precipitation is ill-advised for liquid phase catalysis, due to leaching of entrained alkali residue during reaction [65, 66]. This issue can be circumvented through use of an alkali-free precipitation route employing $NH_4OH/(NH_4)_2CO_3$ and/or urea, and this provides more reliable materials to explore structure–reactivity correlations within thermally activated and rehydrated Mg–Al HTs [62]. HTs exhibit an unusual "memory effect," whereby their calcination results in high surface area mixed metal oxides whose original double hydroxide structure can be restored upon rehydration. HTs can also exhibit strong Lewis basicity, with the strength dependent on the M(II):M(III) ratio within the HT sheets, enabling facile tuning of their physicochemical and catalytic properties.

Various commercial Mg–Al HTs, prepared as the hydroxide, carbonate, and mixed carbonate-hydroxide anionic forms, have been explored for glucose isomerization [67]. Hydroxide containing structures were both more active and selective than the carbonate form; however, all deactivated after only 15% glucose conversion, concomitant with a rapid fall in their initial high (90%) selectivity to fructose. The initial activity and selectivity were claimed to return on recycling, presumably due to the removal of reversibly adsorbed products during catalyst regeneration. The effect of calcination (HT_C) and calcination/rehydration (HT_R) of a parent Mg–Al HT (HT_A) (Mg:Al molar ratio = 3) also impacts the performance for glucose isomerization [50]. Rehydration improved the activity of the Mg–Al HT (HT_R) catalyst for fructose conversion, which was attributed to the high abundance of weak base sites (Figure 5.4).

The hydrophobic/hydrophilic nature of commercial Mg–Al HTs was also investigated in batch and continuous aqueous phase glucose isomerization [52]. Hydrophobic materials exhibited lower conversion but superior selectivity to fructose, exceeding 92% at 30% glucose conversion. The main by-products were glyceraldehyde and dihydroxyacetone from the retroaldol reaction of fructose (and glucose conversion to glycolaldehyde and erythrose) (Figure 5.2b); catalyst

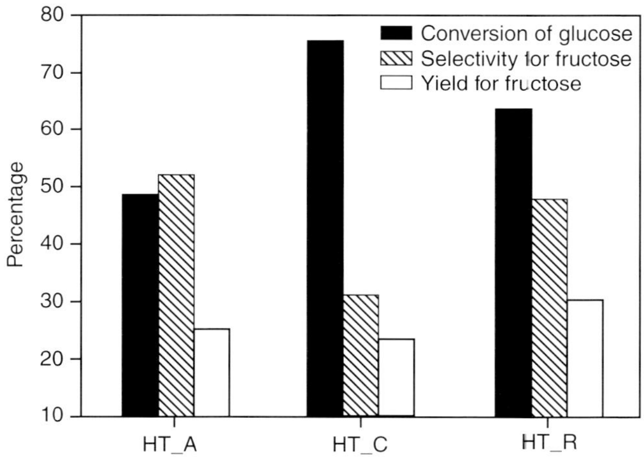

Figure 5.4 Catalytic performance of hydrotalcite catalysts (HT_A, HT_C, and HT_R) in the isomerization of glucose to fructose at 100 °C after five hours. Source: Yu et al. 2012 [50]. Reprinted with permission of Elsevier.

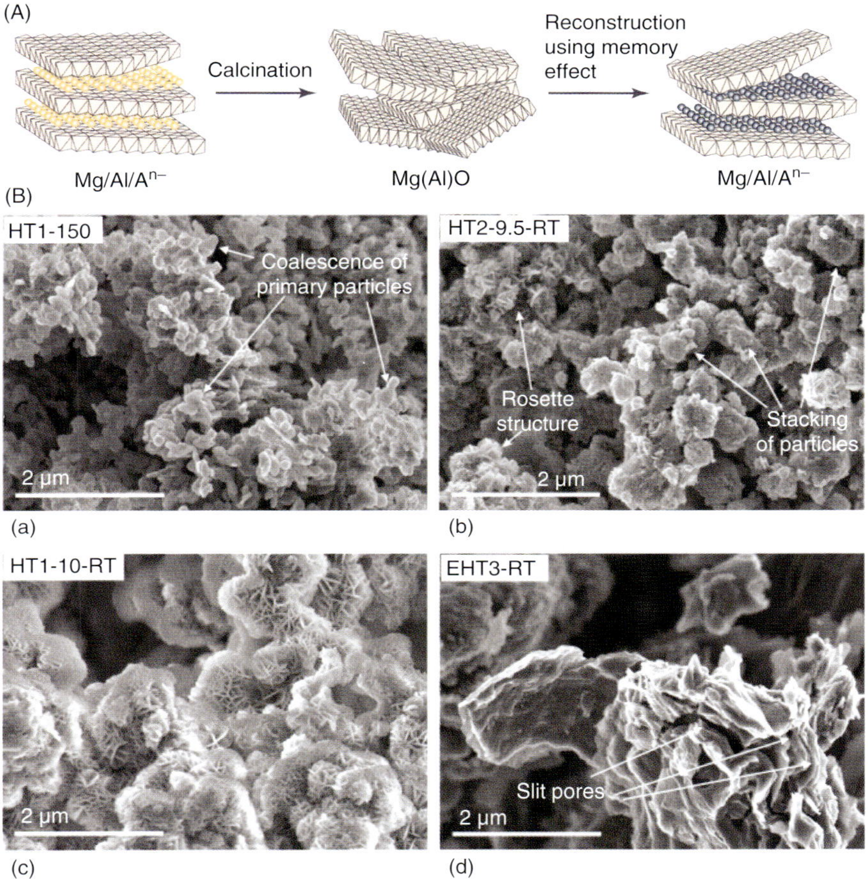

Figure 5.5 (A) Hydrotalcite structure and memory effect. Source: Tichit and Coq 2003 [69]. Reprinted with permission of Springer Nature. (B) SEM micrographs of selected hydrotalcites with different morphologies and structures: (a) HT1-150 [$Mg_{0.78}Al_{0.22}(OH)_2(CO_3)_{0.11} \cdot 0.57H_2O$], (b) HT2-9.5-RT [$Mg_{0.76}Al_{0.24}(OH)_2(CO_3)_{0.12} \cdot 0.85H_2O$], (c) HT1-10-RT [$Mg_{0.75}Al_{0.25}(OH)_2(CO_3)_{0.125} \cdot 0.71H_2O$], and (d) EHT3-RT [$Mg_{0.69}Al_{0.31}(OH)_2(CO_3)_{0.155} \cdot 0.79H_2O$]. Source: Delidovich and Palkovits 2015 [68]. Adapted with permission of Elsevier.

deactivation was ascribed to strong adsorption of these by-products. Acidic degradation products such as lactic acid can also neutralize HTs and induce magnesium leaching; however hot filtration tests suggest that isomerization occurs exclusively via heterogeneous catalysis. Structure–activity relations for glucose isomerization have been examined over Mg–Al HTs with different Mg:Al molar ratios, textural properties, and morphology (Figure 5.5) [68].

Glucose conversion was proportional to base site density, with weak base sites associated with interlayer anions at the edges of primary particles and defects on the basal planes of the HT identified as the active sites. The dimensions of coherent crystallographic domains and primary particles, and their agglomeration, had a significant impact on the accessibility of base sites and resulting catalytic activity. HT prepared in an aqueous medium at pH 10 exhibited the optimal sand rose structure, which may be responsible for the improved activity and selectivity of the porous $Mg_{0.75}Al_{0.25}(OH)_2(CO_3)_{0.125}{\cdot}0.71H_2O$ (HT1-10-RT in Figure 5.5). Since these catalysts were prepared using alkali precipitants, the possibility of homogeneous catalysis by leached Na^+ cannot be discounted.

5.2.1.3 Other Solid Catalysts

Other materials that can be employed for glucose isomerization include metallosilicates, niobium, tantalum, and zirconium compounds that possess Lewis acidity or basicity. A variety of metallosilicates such as $Na_9Si_{12}Ti_5O_{38}(OH)$ were tested in glucose isomerization in aqueous media, yielding 20–40% fructose [51]. Catalytic activities were comparable or superior to those achievable with a commercial NaX zeolite or NaOH under these reaction conditions, although catalysts deactivated after two recycles due to a combination of sodium/potassium leaching, loss of crystallinity, and surface passivation. Zirconium salts such as zirconium carbonate and zirconium phosphate are also active for glucose isomerization, with the carbonate performing best with a maximum conversion of 45% and fructose selectivity of 76% at 120 °C, and excellent stability over five recycles. Zirconia, synthesized both in bulk form [70] and as ultrathin conformal monolayers over mesoporous SBA-15 [71], exhibits modest glucose conversions of up to 20%, but high fructose selectivity >82–84% at 100 °C, with the latter thin film system also displaying excellent hydrothermal stability.

5.2.2 Dehydration Catalysts

Acid-catalyzed dehydration of sugars and their derivatives offers a convenient means to deoxygenate biomass-derived molecules [72]. Furfural and 5-HMF both feature in the US DOE top 10 list of valuable platform chemicals as revised by Bozell and Petersen [9], whose formation from C_5 and C_6 sugars requires a dehydration step. Conventionally, catalytic dehydration of C_5–C_6 sugars is performed using liquid acids such as H_2SO_4, HCl, H_3PO_4, oxalic and maleic acids, and water as solvent [73, 74]. Other homogeneous Lewis acid catalysts such as metal halides (e.g. $CrCl_3$) have also been explored in conjunction with a liquid Brønsted acid (e.g. HCl) [75, 76]. In common with most homogeneous acid-catalyzed processes, corrosion, toxicity, and product separation and recovery remain problematic. Solid acid catalysts offer promising alternatives for sugar dehydration.

5-HMF is a key intermediate for the production of linear alkanes in diesel and jet fuels, and valuable chemical building blocks including furan derivatives and linear C_6 diols [77–79]. However, while glucose is a preferential feedstock for 5-HMF production, since it can be directly obtained from cellulose, this route is challenging due to the (slow) Lewis acid or base-catalyzed isomerization that must precede subsequent Brønsted acid-catalyzed fructose dehydration (Figure 5.6) [80]. New catalysts are sought for low temperature aqueous glucose conversion to 5-HMF that optimize both steps [81].

The analogous dehydration of xylose to furfural is another important process for the production of furfuryl alcohol, THFA (tetrahydrofurfuryl alcohol), and C_5

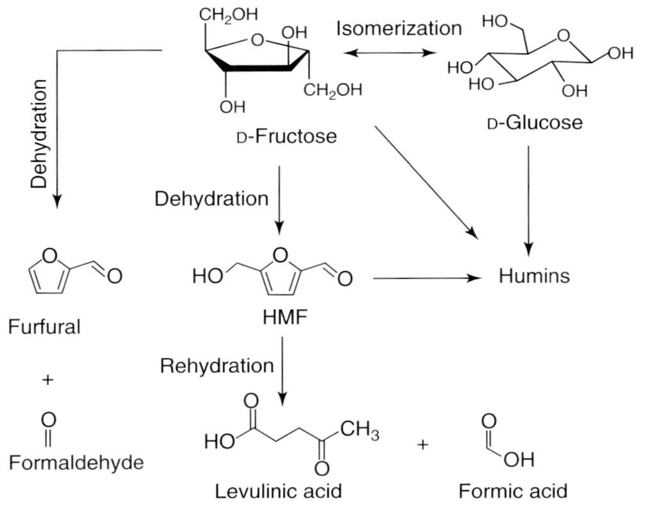

Figure 5.6 Reaction network for glucose/fructose dehydration catalyzed by solid acids/bases. Source: Dornath and Fan 2014 [80]. Adapted with permission of Elsevier.

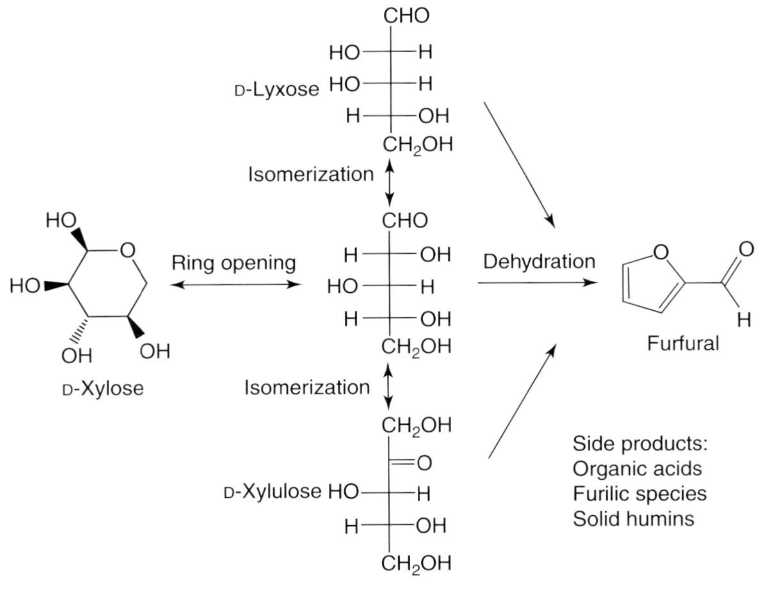

Figure 5.7 Reaction network of xylose dehydration catalyzed by solid acids/bases. Source: Sairanen et al. 2013 [83]. Adapted with permission of Springer Nature.

diols. Unfortunately, poor selectivity to furfural is often observed as the temperatures employed for xylose dehydration (120–200 °C) are significantly higher than those required for xylose isomerization (100–120 °C) [80, 82]; xylulose and lyxose formation therefore compete with furfural formation (Figure 5.7) [83]. High temperature catalytic sugar transformations also result in poor carbon mass balances due to furfural fragmentation into organic acids (formic and levulinic acid), oligomerization into furilic species, and humins formation. Active and selective catalysts able to operate at lower temperature can limit humin production.

5.2.2.1 Zeolitic and Mesoporous Brønsted Solid Acids

Brønsted acidic proton-exchanged zeolites have been widely investigated for sugar dehydration [84, 85]. The impact of zeolite structure on catalytic performance in xylose dehydration to furfural has been reported for H-form ferrierite, ZSM-5, modernite, beta, and Y zeolites possessing similar acid site densities and external/specific areas [86]. Xylose conversion and selectivity to furfural correlated with pore channel size, with conversion linearly proportional to channel diameter; H-ZSM-5 exhibited the highest furfural selectivity due to matching of its channel size to the molecular dimensions. Dealumination and desilication also improved the activity of H-ZSM-5, attributed to an increased Brønsted acid density [86]. However, mass balances, which are critical for systems where slow in-pore diffusion may favor side reactions and humin formation, were not reported. Quantitative comparison of catalyst selectivity must be undertaken at iso-conversion for processes where side reactions are important.

The effect of Si:Al ratio on the catalytic performance of H-MCM-22 in xylose dehydration was also investigated [87]. Catalyst activity was inversely proportional to Si:Al ratio; however, a constant 54% furfural yield was obtained indicating a concomitant decrease in selectivity. ITQ-2, the delaminated analogue of H-MCM-22, exhibited comparable performance for samples with Si:Al = 24, despite possessing a significantly higher external surface area, reflecting a similar acid loading. ITQ-2 was however easier to regenerate, suggesting deactivation by coking was suppressed, and the formation of different carbonaceous deposits over the two zeolites.

The catalytic performance of hierarchical USY, beta, and ZSM-5 zeolites [88] prepared by alkaline pretreatment reveals that both acid loading and strength influence fructose dehydration (Figure 5.8). Modified zeolites offer higher 5-HMF selectivity than their parent materials, possibly due to improved accessibility of acid sites in the hierarchically porous materials.

A detailed study of H-Beta-18 (Si:Al ratio of 18) [89] revealed that fructose dehydration was accompanied by isomerization to glucose and decomposition of reactively formed 5-HMF to formic and levulinic acids. Isomerization between fructose and glucose is mainly promoted by octahedral Al atoms at lattice defect sites and extra-framework Al atoms, which act as Lewis acids. Organic acids produced by 5-HMF rehydration may induce leaching of aluminosilicate species from the zeolite, with dissolved aluminosilicate species, promoting undesired side reactions. Changing the Si:Al ratio in aluminosilicate or zeolite catalysts, and hence Lewis:Brønsted character and concentration of extra-framework Al, is therefore anticipated to impact 5-HMF selectivity.

The impact of extra-framework Al on selectivity during fructose dehydration was also observed for Al-SBA-15 [90], wherein the density of non-framework Al species increased with overall Al loading. The acid site loading could be increased from 0.147 to 0.267 mmol/g; however, for Si:Al $>$ 10, octahedral Al species, rather than the desired tetrahedral species, were favored. According

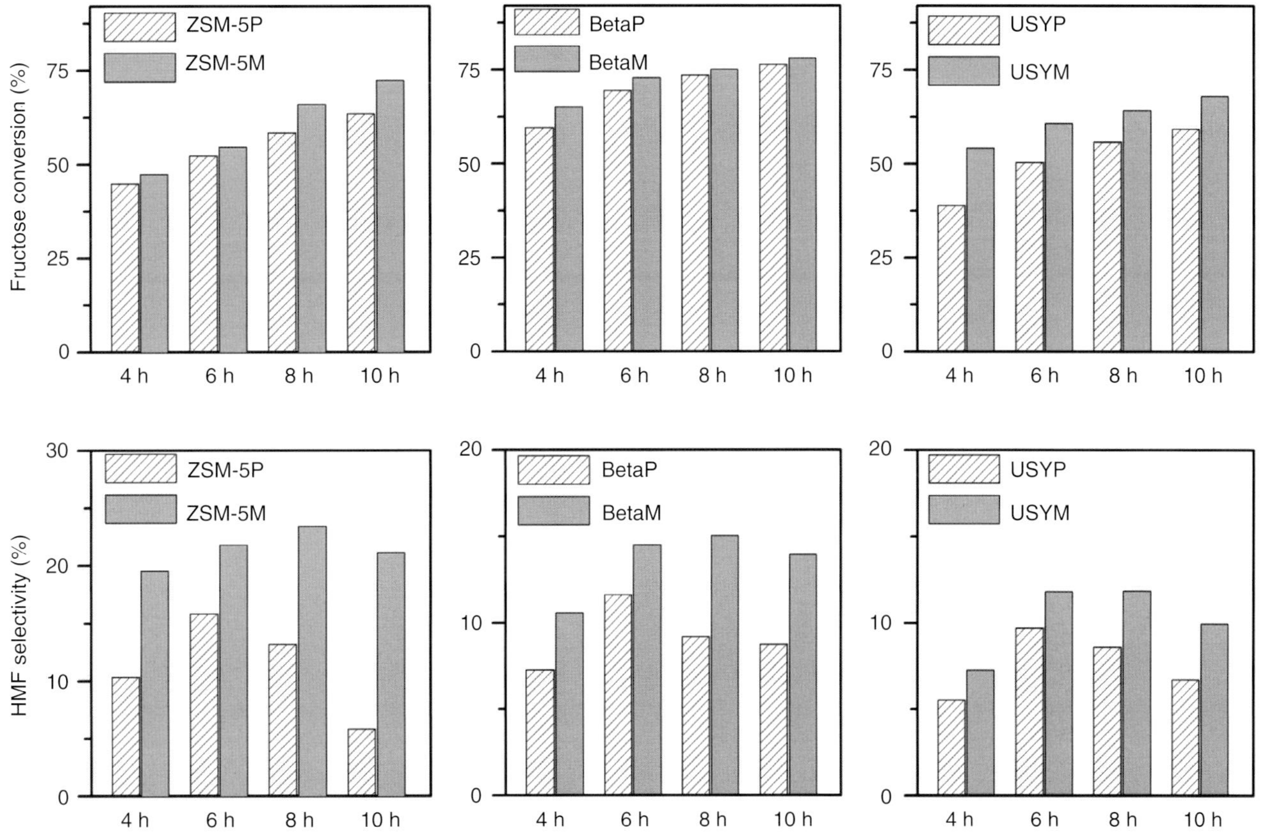

Figure 5.8 Fructose conversion and 5-HMF selectivity of parent (P) and modified (M) zeolites. Source: Rac et al. 2014 [88]. Adapted with permission of Elsevier.

to literature [91–94], only tetrahedral Al sites introduce the Brønsted acidity required for the selective dehydration of fructose to 5-HMF. Non-framework Al cations are proposed to promote humin formation; Al-SBA-15 (Si:Al = 10) that possessed the highest proportion of non-framework Al was only 59% selective to 5-HMF at 68% fructose conversion. In contrast, Al-SBA-15 (Si:Al = 40) that possessed mainly tetrahedral Al sites was 89% selective to 5-HMF, albeit at a slightly reduced fructose conversion.

The yield and selectivity of 5-HMF from fructose is often low in the aqueous phase due to competing side reactions; hence, process optimization often explores mixed organic-water solvents to facilitate the extraction of reactively formed 5-HMF [95]. Nijhuis and coworkers [96] studied the utility of organic solvents to extract 5-HMF from fructose dehydration over zeolite catalysts including MOR, ZSM-5, BEA, and amorphous aluminosilicate. MOR zeolite exhibited the highest selectivity, with methyl isobutyl ketone (MIBK) addition significantly increasing the initial selectivity to 5-HMF. It was postulated that MIBK fills the zeolite pores and interacts strongly with the acid sites, thereby displacing reactively formed 5-HMF from the catalyst surface and preventing its oligomerization. Silylated zeolites were also investigated for fructose dehydration and afforded superior selectivity at higher conversion in a biphasic system through the deactivation of (unselective) external acid sites.

The effect of pore structure on the selective dehydration of fructose to 5-HMF has also been demonstrated by Moreau et al. [97] over microporous H-type zeolites, including H-Y faujasites, H-mordenites, H-beta, and H-ZSM-5 at 165 °C using a water and MIBK (to extract 5-HMF) solvent mix. Fructose conversion and selectivity to 5-HMF were dependent on the acidity and pore structure, with conversion over H-Y faujasites and H-mordenites increasing as the Si:Al ratio decreased from 15 to 11, respectively, albeit at the expense of lower selectivity to 5-HMF. The zeolite structure also influenced 5-HMF selectivity, with the bimodal channel structures of H-mordenites favoring 5-HMF formation relative to the trimodal channel structure in H-Y faujasites, H-beta, and H-ZSM-5. A particularly high selectivity of >90–95% to 5-HMF was reported over H-mordenites, which was attributed to the shape selective properties. Further optimization using dealuminated H-mordenites was also attempted [98]; however, as both porosity and Si:Al ratio changed in concert, it was difficult to evaluate factors critical for high selectivity to 5-HMF. Although some zeolites, such as H-mordenites and H-Y faujasites, showed a high selectivity to 5-HMF (>90%), potential structural collapse of zeolites must be considered when these materials are used for aqueous phase biomass conversion. Moreover, water-soluble zeolitic species can homogeneously catalyze reactions of fructose and 5-HMF [89, 99]. Lercher and coworkers [84] studied the stability of zeolite Y and ZSM-5 in hot water (150–200 °C) finding that the hydrothermal stability was strongly dependent on the framework type. ZSM-5 was stable at 150 and 200 °C, whereas Y-type zeolites were damaged under these conditions. When the Si:Al ratio was increased to 41, zeolite Y lost all structural order after hydrothermal treatment at 200 °C for six hours, attributed to the hydrolysis of siloxane bonds (Si—O—Si) with OH^- from hot water, rather than dealumination.

5.2.2.2 Sulfonic Acid Functionalized Hybrid Organic–Inorganic Silicas

The strong acidity of sulfonic acids (SO_3H) makes them attractive Brønsted solid acid catalyst replacements for H_2SO_4 in acid catalysis [100–105]. Acidic ion-exchange resins were an early class of such catalysts to be utilized for fructose dehydration to 5-HMF [106, 107]. However, strong acid resin catalysts are considered unstable in water above 100 °C, and hence often require solvent mixtures [108]. A mixed acetone/water solvent was employed for the catalytic dehydration of fructose into 5-HMF over a strong acid cation-exchange resin (Dowex® 50wx8) under microwave heating; a 73.4% 5-HMF yield and 95.1% fructose conversion was obtained at 150 °C for an acetone:H_2O volume ratio of 7 : 3. However, even 30 vol% water promoted 5-HMF degradation to levulinic and formic acid. The inherent instability of commercial polymer resins under the conditions used for hydrothermal processing of sugars [109] means they are conventionally employed for fructose dehydration in nonaqueous solvents, including ILs and dimethyl sulfoxide (DMSO) solvent mixtures [110].

While solvation of $-SO_3H$ sites at elevated temperatures (>150 °C) in water limits the hydrothermal stability of sulfonic acid catalysts, hybrid organic–inorganic silica and carbon-based analogues offer greater stability than resins. Framework substituted, periodic mesoporous organosilicas (PMOs) synthesized by condensation of organic-linker-bridged disilanes, $(RO)_3Si{-}R{-}Si(OR)_3$, offer improved stability when subsequently sulfonated [111, 112]. The stability and activity of mesoporous SBA-15 sulfonic acid catalysts and hybrid organosilica catalysts prepared via co-condensation and grafting functionalization methods was studied in detail by Hensen and coworkers [113]. Sulfur leaching was investigated at 120 °C under hydrothermal conditions over seven days and revealed that sulfonic functional groups on co-condensed materials were much more stable than those introduced by post-synthetic grafting. Although fructose conversion and 5-HMF selectivity in aqueous solvents were poor, they could be significantly improved by the use of a DMSO solvent. The highest activity was observed over the co-condensed hybrid organosilica-based catalyst SBA-C_2Ph-coc containing ethylphenylsulfonic acid groups, which afforded 88% 5-HMF at 99% fructose conversion.

Dumesic and coworkers [114] compared propylsulfonic acid-functionalized ethane-bridged PMOs containing either 90 or 45 mol% BTME (1,2-bis(trimethoxysilyl)-ethane). E90 and E45 variants possessed hexagonally ordered mesoporous structures, high surface areas (548 and 559 m^2/g), and well-defined mesopores channels of 4.2 and 4.5 nm diameter, respectively. Fructose dehydration at 130 °C under continuous flow showed that all these ordered mesoporous catalysts (E0, E45, and E90) exhibited higher activity and stability than commercial sulfonic acids, with initial fructose conversion >80% and 5-HMF selectivity of 65–75%. Selectivity to 5-HMF remained >60% even after 24 hours reaction at 130 °C in steam, with E45 exhibiting superior stability to E90 with a much lower deactivation rate.

Further improvements in 5-HMF selectivity from fructose dehydration are often reported using biphasic systems to immediately remove the 5-HMF product from the aqueous phase to minimize side reactions and levulinic and formic acid formation. Jérôme and coworkers [115] adopted this strategy for sulfonic

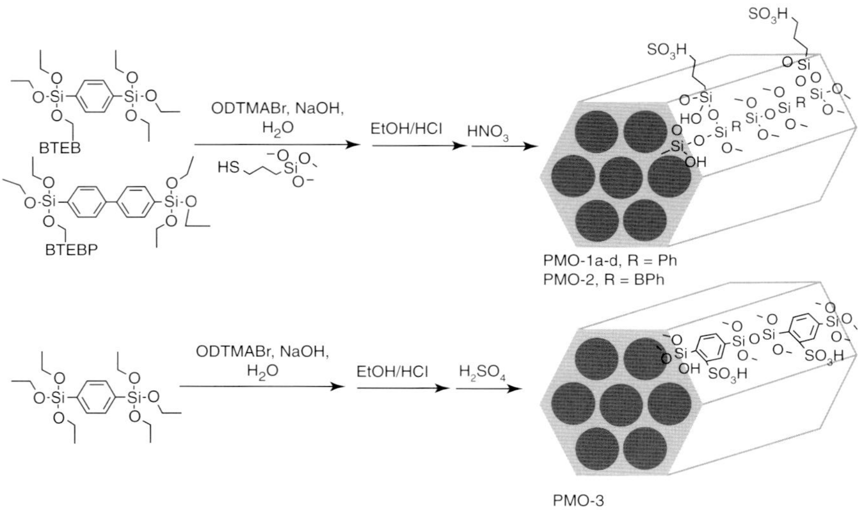

Figure 5.9 Synthetic route to sulfonic acid functionalized PMO materials. Source: [115]. Reprinted with permission of Royal Society of Chemistry.

acid-functionalized PMO catalysts for fructose dehydration. The synthesis of these catalysts is shown in Figure 5.9 and resulted in catalysts with high acid loadings between 0.25 and 1.11 mmol/g.

PMO-1 with acid site loading of 0.36 mmol/g produced the highest turnover frequency (TOF) of 945 h^{-1} for fructose dehydration in water-MIBK/2-butanol at 160 °C. Jérôme et al. attributed the highest activity to a combination of (i) phenyl rings on the pore wall creating a micro hydrophobic environment that prevents the solvation of sulfonic sites by water, (ii) improved anchorage of sulfonic sites on a propyl chain, and (iii) a moderate sulfonic site loading. Overall PMO materials offer improved stability than Amberlyst-15 or purely siliceous acid-functionalized materials such as SBA–SO_3H and are promising materials for selectively converting fructose to 5-HMF.

Alternative organic linkers for the production of organic sulfonic groups have been explored through the synthesis of 3-(butylthio) propane-1-sulfonic acid (BTPSA)-functionalized SiO_2 catalysts, TAA-A380, and TAA-SBA-15 by Scott and coworkers [116] who used post-synthetic grafting and co-condensation methods, illustrated in Figure 5.10 for TAA-SBA-15.

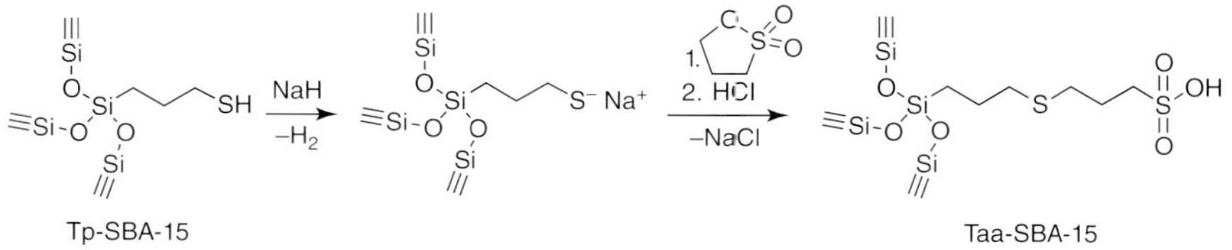

Figure 5.10 Proposed synthetic scheme for a bifunctional acid catalyst supported on silica. Source: Crisci et al. 2010 [116]. Reprinted with permission of Springer Nature.

The sulfur loading of post-synthetic grafted TAA-A380 material was only 0.38 mmol/g, which was much lower than that of TAA-SBA-15 (2.3 mmol/g). Acid groups incorporated by co-condensation were also more robust under harsh reaction conditions of 180 °C than those introduced by post-synthetic grafting. The activity of both catalysts was investigated for fructose dehydration to 5-HMF in a biphasic solvent system (water-MIBK/2-butanol), wherein TAA-SBA-15 afforded 74% selectivity to 5-HMF at 66% fructose conversion. However, the co-condensed SBA-15 still underwent eventual collapse of the ordered mesoporous structure and loss of acid function at these elevated temperatures. To address this [117], an alternative one-pot synthetic route was developed to prepare bifunctional silicas (TESAS-SBA-15 and SSA-SBA-15) with higher acid loadings in which the silane TESAS (3-((3-(trimethoxysilyl)propyl)thio)propane-1-sulfonic acid) was utilized in the initial sol–gel step. The resulting TESAS-SBA-15 and SSA-SBA-15 materials exhibited excellent structural order with pore diameters of ~4.7 nm and high acidic site loadings of 1.25 and 0.82 mmol/g, respectively. Fructose dehydration at 130 °C in water-MIBK/2-butanol gave fructose conversion and 5-HMF selectivity as high as 84% and 71% over TESAS-SBA-15, and 81% fructose conversion and 65% selectivity to 5-HMF over SSA-SBA-15, both higher than that obtained for a commercial Amberlyst-70 catalyst.

Silica nanoparticles with a mesoporous shell silica bead and dense silica core (MSHS) have also been developed for the catalytic dehydration of xylose to furfural in water. These were modified with Brønsted acidic sulfonic acid groups (SO_3H–MSHS) or with Lewis acidic Al^{3+} centers (Al–MSHS) (Figure 5.11) [82]. SO_3H–MSHS outperformed Al–MSHS in respect of selectivity to furfural due to the latter driving xylose isomerization to lyxose. The hydrothermal stability of SO_3H–MSHS was superior to other mesoporous silicas such as MCM-41 at 170–190 °C; however, humin formation under these conditions was not monitored. The application of MSHS materials is limited by their relatively low surface areas and consequent low acid site densities, and they are prone to irreversible deactivation of sulfonic acid groups through hydrolysis at elevated temperatures [118].

The issue of low active site loadings was addressed by "hairy" solid acid catalysts designed by outward growth of poly(sodium 4-styrenesulfonate) brushes from the surface of silica core particles (Figure 5.12). Subsequent acidification resulted

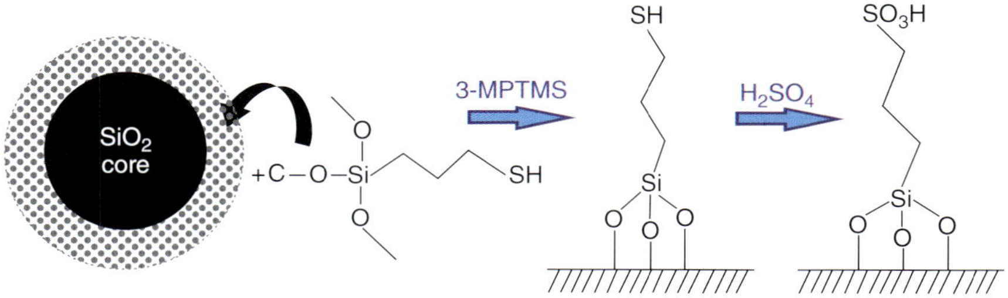

Figure 5.11 Synthesis of MSHS catalysts modified by SO_3H groups. Source: Jeong et al. 2011 [82]. Adapted with permission of Elsevier.

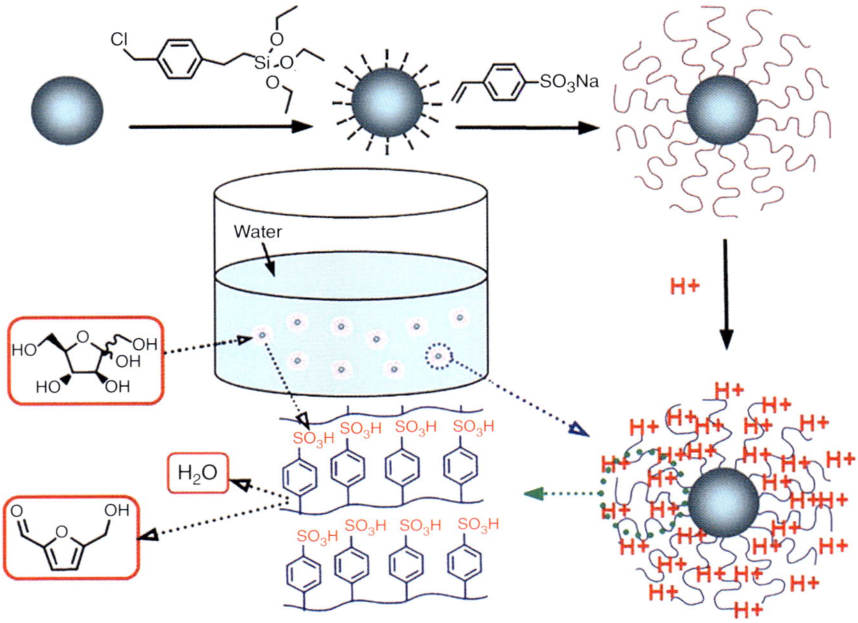

Figure 5.12 Synthesis of poly(4-styrenesulfonic acid) silica particles for fructose dehydration to 5-HMF. Source: Tian et al. 2013 [119]. Reproduced with permission of Royal Society of Chemistry.

in poly-(4-styrenesulfonic acid) (PSSH) brushes [119], which exhibit improved 5-HMF yields from fructose dehydration in water when compared to the free homopolymer acid. This was attributed to a unique solvation micro-environment formed by the densely grafted PSSH chains in the brush structures.

In order to readily separate the solid catalyst and minimize handling losses, Fu and coworkers [120] prepared a novel magnetic SBA-15–SO_3H material by incorporating magnetic Fe_3O_4 nanoparticles into the mesoporous SiO_2 wall. The average pore diameter of Fe_3O_4–SBA–SO_3H was 4.8 nm, the average pore volume was 0.49 cm^3/g, and the Brunauer–Emmett–Teller (BET) surface area was 464 m^2/g. The Fe_3O_4–SBA-15–SO_3H magnetic catalyst exhibited good activity when used in conjunction with a K–OMS-2 oxidation catalyst for the one pot-conversion of fructose into 2,5-diformylfuran, delivering 80% 2,5-diformylfuran at 99% fructose conversion.

Sulfonation of incompletely carbonized biomass, such as glucose, starch, and cellulose, by concentrated or fuming sulfuric acid, can be used to produce carbon materials with –SO_3H, –COOH, and phenolic functions as solid catalysts for the hydrolysis of cellulose and dehydration of fructose to 5-HMF [121]. Despite comparatively low sulfonic acid densities (relative to –COOH and phenolic functions), these carbonaceous catalysts were more active for 5-HMF production than conventional solid acids such as HZSM-5, SO_4^{2-}/TiO_2–ZrO_2, Amberlyst-15, and even H_2SO_4. At 140 °C a 41% yield of 5-HMF at 60% fructose conversion was obtained in only 30 minutes using the cellulose-derived catalyst in [BMIM]Cl; however, this catalyst rapidly deactivated and required subsequent thermal regeneration and sulfonation. Hu et al. [121] also proposed that [BMIM]Cl could facilitate the direct synthesis of 5-HMF from glucose.

Sulfonated graphene and graphene oxides have also been tested for xylose dehydration [122]. Sulfonated graphene oxide proved an active and water-tolerant solid acid catalyst even at very low catalyst loadings of ~0.5 wt% (relative to xylose), showing negligible deactivation over 12 reuses at 200 °C, with an average furfural yield of 61% vs. 44% for the uncatalyzed system.

5.2.2.3 Metal–Organic Frameworks

Metal–organic frameworks (MOFs) have emerged as one of most exciting areas of materials science [123]. MOFs comprise metal ions or clusters interconnected by organic linkers across two or three dimensions [124]. Owing to their porous structure and high surfaces areas, MOFs have found early application in heterogeneous catalysis. However, in contrast to zeolites, the introduction of an organic component to MOFs offers greater structural and chemical diversity, although MOFs (historically) could not compete with zeolites in terms of thermochemical stability. Many MOFs possess permanent microporosity, while others collapse on desolvation; however, this is not problematic for liquid phase catalysis [125].

Zhang et al. [126] employed phosphotungstic acid (PTA) encapsulated in MIL-101Cr in the dehydration of fructose to 5-HMF. This high surface area (1352 m^2/g) PTA3.0@MIL-101Cr catalyst afforded 79% 5-HMF at after 2.5 hours reaction in [EMIM]Cl solvent at 80 °C and was also active in DMSO, yielding 63% 5-HMF after 30 minutes at 130 °C. Hu et al. [127] synthesized hierarchically porous hafnium MOFs named NUS-6(Hf) via the modulated hydrothermal (MHT) approach at 80 °C (Figure 5.13). The resulting material showed good stability up to 150 °C, and quantitative conversion of fructose (>99%) with extremely high chemoselectivity (98%) to 5-HMF in DMSO at

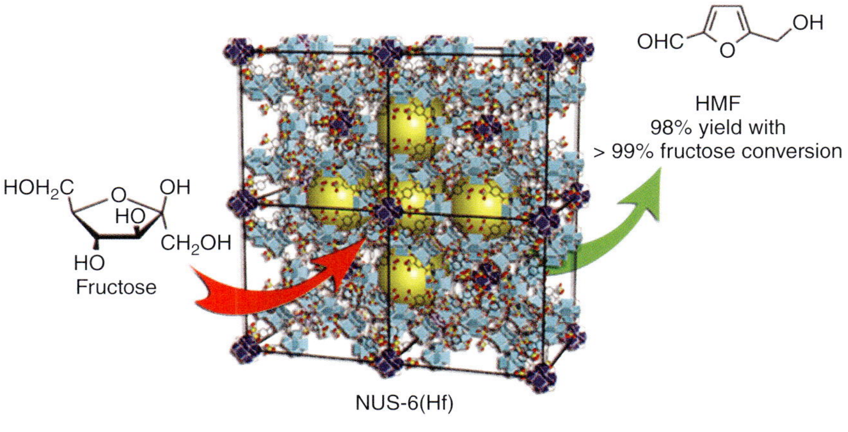

Figure 5.13 A 1 × 2 × 2 unit supercell of NUS-6(Hf) with mesopores indicated by yellow spheres. Source: Reprinted with permission from Hu et al. 2016 [127]. Copyright 2016, American Chemical Society.

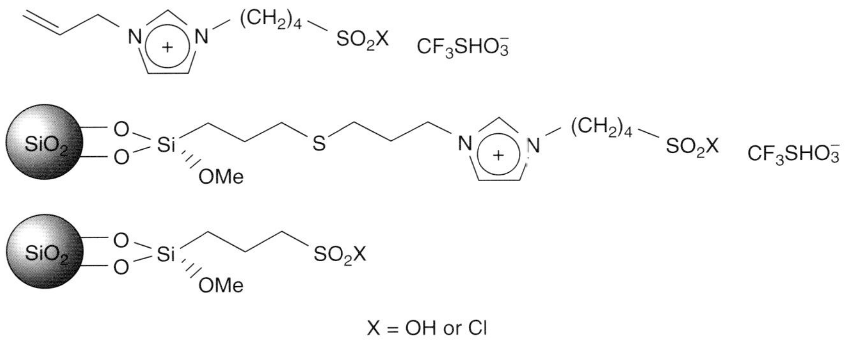

Figure 5.14 Structure of SiO_2 supported acidic IL catalysts. Source: Bao et al. 2008 [128]. Reprinted with permission of Elsevier.

100 °C. The excellent catalytic activity of NUS-6(Hf) was attributed to its strong Brønsted acidity and optimal pore size that inhibited side reactions.

5.2.2.4 Supported Ionic Liquids

Acidic ILs are considered promising catalysts and solvents for fructose dehydration to 5-HMF; however, as solvents their high cost is to date prohibitive and hinders the recovery and isolation of 5-HMF (necessitating ion-exchange resins to recover the IL). [29] The development of heterogeneous (supported) acidic IL catalysts has therefore sparked interest, although such systems have typically been exploited for 5-HMF production in nonaqueous solvents.

Silica gel immobilized acidic IL catalysts [128] ILIS–SO_3H and ILIS–SO_2Cl (Figure 5.14) offered 70% and 67% yields of 5-HMF, respectively at 100% fructose conversion in DMSO following four minutes microwave irradiation. Sulfated ILs have also been immobilized on nanosized amorphous silica particles using 1-(tri-ethoxy silyl-propyl)-3-methyl-imidazolium hydrogen sulfate (IL-HSO_4) precursors [129]. Catalysts with different particle sizes (214–504 nm) and IL loading were prepared, with >60% 5-HMF obtained at 99.9% fructose conversion in DMSO solvent at 130 °C. Fructose conversion and 5-HMF yield increased with silica nanoparticle size, possibly reflecting the higher IL loading of larger particles. This performance was competitive with protonated ZSM-5 and β zeolites, and SiO_2–SO_3H, with no IL leaching or deactivation even after six recycles.

In addition to mono-functionalized catalysts, bifunctional mesoporous SiO_2 nanoparticles comprising both sulfonic acid (SO_3H) and IL functionalities were also explored [130]. Using DMSO solvent a 72.5% 5-HMF yield was obtained at 97% fructose conversion. This bifunctional catalyst could be recycled four times without deactivation, although some activity loss occurred on the fifth run. Quantitative benchmarking of such catalysts is hindered by the catalytic properties of DMSO alone for 5-HMF production from fructose under these conditions.

5.2.3 Catalysts for Tandem Isomerization and Dehydration of C_5–C_6 Sugars

One-pot cascade reactions are an important route to improve process efficiency during chemical synthesis by minimizing product isolation steps [131, 132] and are well suited to converting sugars to platform chemicals [133, 134]. However, such cascades require the design of multifunctional catalysts in which the spatial distribution of active sites for each step is precisely controlled. Glucose conversion to 5-HMF exemplifies such a cascade, involving the Lewis acid- or base-catalyzed isomerization of glucose to fructose, followed by Brønsted acid-catalyzed dehydration of fructose to 5-HMF, and hence has attracted significant interest [77].

5.2.3.1 Bifunctional Zeolites and Mesoporous Solid Acids

Cascade processes for the production of 5-HMF or furfural directly from glucose or xylose, respectively, have been explored using framework-substituted Sn-beta zeolites [43, 47] in conjunction with HCl as a Brønsted acid catalyst. In this system, catalyst stability under harsh acidic environments is problematic, and the use of corrosive mineral acids undesirable from a green chemistry perspective. Wang et al. [135] prepared a Sn-montmorillonite bifunctional catalyst by ion-exchange of natural calcium montmorillonite with an aqueous tin tetrachloride solution. The resulting material possessed both Lewis and Brønsted acid sites of weak to moderate strength. A 54% 5-HMF yield was obtained for 98% glucose conversion at 160 °C in a biphasic THF/DMSO solvent (70 : 30 volume ratio). The synergistic effect of tin ions (Lewis acid sites for isomerization) and Sn–OH (Brønsted acid sites for dehydration) was held responsible for the catalyst performance (Figure 5.15).

The impact of alternative solvents in sugar dehydration has also been explored, with ILs [136–139] or deep eutectic solvents (DESs) showing promise due to their ability to dissolve lignin or cellulose selectively during lignocellulose fractionation [140–143]. Such routes to catalytically convert sugars would be important for IL or DES based biorefineries.

Liu and coworkers [144] investigated the activity of various commercial zeolite catalysts, including HY-zeolite, H-mordenite, Hβ-zeolite, and HZSM-5 for the conversion of glucose to 5-HMF in a ([BMIM]Cl) IL. Hβ-zeolite with Si:Al molar ratio of 25 exhibited the best catalytic activity among the previously mentioned zeolites. At 150 °C, 81% glucose conversion and 50% 5-HMF yield was obtained. Synergy between surface Brønsted acid and Lewis acid sites over Hβ-zeolite were claimed essential for good activity (Figure 5.16). While the Hβ-zeolite catalyst exhibited decreased 5-HMF yield upon reuse, regeneration was possible by calcination. However, in general the isolation of 5-HMF from ILs is challenging [145], with the high temperatures required to distill the product often leading to 5-HMF degradation, or necessitating ion-exchange resins or solid phase extraction [146] to recover the IL.

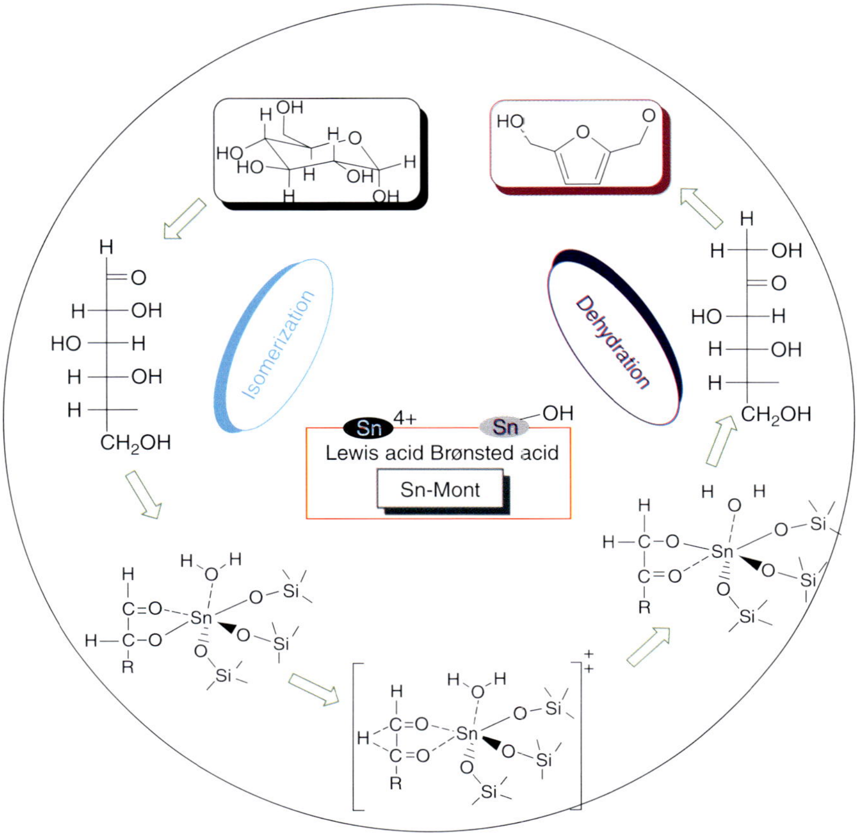

Figure 5.15 Proposed reaction mechanism for the conversion of glucose to 5-HMF catalyzed over Sn-Mont catalyst. Source: Wang et al. 2012 [135]. Reprinted with permission of Royal Society of Chemistry.

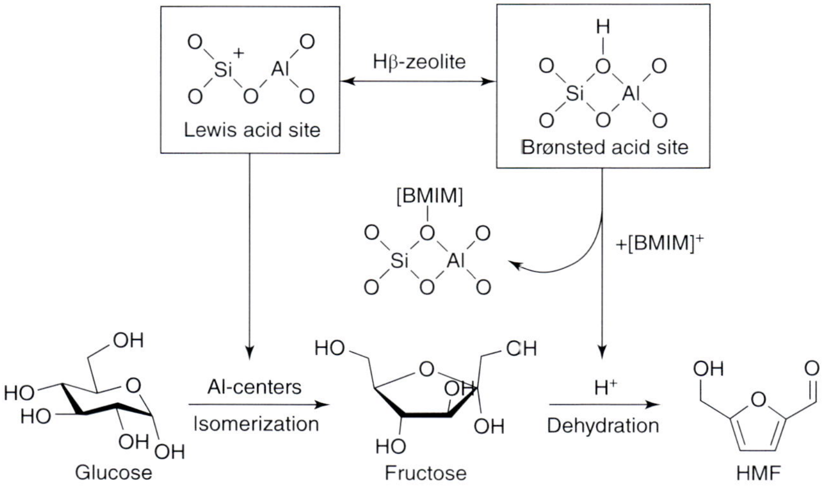

Figure 5.16 Proposed mechanism for the conversion of glucose into 5-HMF over Hβ-zeolite catalyst in [BMIM]Cl. Source: Hu et al. 2014 [144]. Reprinted with permission of Elsevier.

5.2.3.2 Metal Oxides, Sulfates, and Phosphates

Metal oxides are extensively used in catalysis as active components or supports and have been investigated in various reaction media for C_5–C_6 sugar dehydration. The hydrothermal stability and amphoteric and Lewis acid character of titanium, niobium, tungsten, and tantalum oxides [147, 148], makes them particularly interesting candidates for aqueous phase biomass processing [118]. Judicious doping through formation of mixed metal oxides or by sulfation/phosphorylation can also introduce Brønsted acidity for catalytic cascades [70].

Hot-compressed water, including hydrothermal and supercritical water ($T_c = 374.2\,°C$, $P_c = 22.1\,MPa$, and $\rho_c = 0.323\,g/cm^3$), is considered a green solvent for biomass refining. Early studies by Watanabe et al. [149] used TiO_2 and ZrO_2 catalysts for glucose conversion in hot-compressed water (200 °C). Anatase TiO_2 directed both isomerization and dehydration steps to produce the highest 5-HMF yield, suggesting the presence of base and acid sites. Mixed monoclinic/tetragonal ZrO_2 catalysts promote glucose isomerization to fructose, attributed to their base character [150], whereas rutile TiO_2 (r-TiO_2) is inactive due to a low density of acid and base sites. Glucose isomerization to fructose is promoted over ZrO_2 and TiO_2 catalysts under microwave irradiation [151]. In the absence of catalyst, only 7% 5-HMF is obtained in hot-compressed water. Given neither catalyst exhibits Brønsted acidity *ex situ*, the enhanced 5-HMF formation observed over anatase TiO_2 suggests the catalyst's acidic properties are modified by the increased ionic product of subcritical water or supercritical water [150, 152]. H_2SO_4 or H_3PO_4 promoted metal oxides, such as SO_4^{2-}/ZrO_2 and SO_4^{2-}/Al_2O_3, have more strong Brønsted acidic sites and controlled base sites than pure metal oxides and hence should be promising catalysts for 5-HMF production from glucose [153]. Bifunctional SO_4^{2-}/ZrO_2–Al_2O_3 catalysts, prepared by impregnation of $Zr(OH)_4$ and $Zr(OH)_4$–$Al(OH)_3$ with chlorosulfonic acid, possess both Brønsted acid and base sites, but no Lewis acidity. Addition of Al slightly decreases the acid density at the expense of increased (moderate strength) basicity. Such a CSZA-3 catalyst possessing a Zr:Al molar ratio of 1 : 1 achieved a 48% 5-HMF yield at 99% glucose conversion in DMSO at 130 °C. Although the authors attributed this performance to a good balance of acid and base sites, homogeneous catalysis by the solvent cannot be discounted.

Bifunctional sulfated zirconia (SZ) catalysts with tunable physicochemical properties were also investigated for the one-pot production of 5-HMF from glucose [70]. A comparison of acidic properties and reactivity toward both glucose and fructose revealed that submonolayer SO_4 coverages offer the optimal balance of base sites on the exposed zirconia surface for glucose isomerization to fructose, and polydentate Brønsted acid sulfoxy species coordinated to the underlying zirconia support active for fructose dehydration to 5-HMF (Figure 5.17). TOFs for fructose dehydration evidenced that this step of the cascade was catalyzed solely by Brønsted acidic sulfoxy groups. However, such catalysts possessed low surface areas, and hence sulfation of higher area ZrO_2/SBA-15 was explored, in which conformal ZrO_2 monolayers were coated over a mesoporous SBA-15 template [71]. The resulting SZ/SBA-15 bilayer catalyst exhibited excellent hydrothermal stability, and a threefold enhancement in 5-HMF productivity from either glucose or fructose, compared with nonporous SZ analogues.

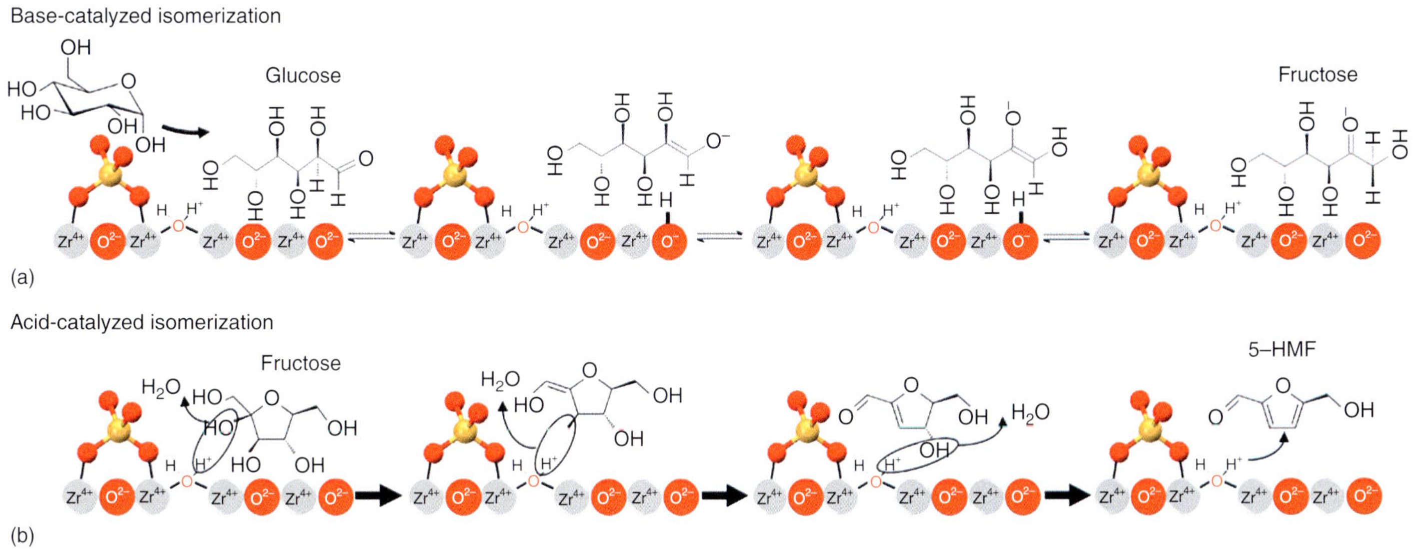

Figure 5.17 Proposed bifunctional, surface-catalyzed mechanism of direct glucose conversion to 5-HMF: (a) base-catalyzed isomerization of glucose to fructose (note O^{2-} sites of monoclinic ZrO_2 provide the basic sites, and the Lewis acidic Zr^{4+} may stabilize enolate intermediates). (b) Brønsted acid-catalyzed dehydration of fructose to 5-HMF over sulfoxy species. Source: Osatiashtiani et al. 2014 [70]. Reproduced with permission of Royal Society of Chemistry.

SZ/SBA-15 is also a promising catalyst for the one-pot cascade synthesis of alkyl levulinates from glucose [154]. The excellent hydrothermal stability of this material is particularly attractive for aqueous phase biomass processing.

Metal phosphates are widely studied in academia and industry as heterogeneous catalysts, with niobium and tantalum phosphates of interest for biomass conversion [155, 156]. Early studies revealed niobium phosphate as active for the aqueous phase dehydration of fructose to 5-HMF [155], although little correlation was reported between acidity and performance, despite good conversions between 30% and 60% and high initial 5-HMF selectivity, albeit selectivity declined sharply over the course of the reaction to unidentified (presumably polymeric) secondary products. Hara and coworkers [157] demonstrated a striking promotion of H_3PO_4-treated $Nb_2O_5{\cdot}nH_2O$ for 5-HMF production from glucose in water at 120 °C. Whereas the parent $Nb_2O_5{\cdot}nH_2O$ only achieved 12% selectivity to 5-HMF at 100% glucose conversion, H_3PO_4–$Nb_2O_5{\cdot}nH_2O$ offered 52% 5-HMF at almost complete conversion. Phosphoric acid treated niobic acid (NA-p) catalysts were also investigated by Yang et al. [158, 159] for the dehydration of glucose at 160 °C, affording 49% 5-HMF in a water/2-butanol solvent. In comparison, phosphoric acid-treated hydrated tantalum oxide (TA-p) exhibited superior activity and stability to NA-p, resulting in 58% yield of 5-HMF from fructose in a water/2-butanol mixture. Bulk niobia possesses a low surface area, and its calcination induces sintering, dehydration, and crystallization of orthorhombic niobia nanoparticles, and a concomitant loss of acid sites and Brønsted character. Tapia Reche et al. [160] therefore synthesized a high surface area niobic acid functionalized SBA-15 analogue through peptization of a niobic acid sol with H_2O_2. The resulting Nb_2O_5/SBA-15 exhibited pure Brønsted acidity independent of Nb loading. Niobic acid nanoparticles were active for the aqueous phase isomerization of glucose to fructose, and subsequent fructose dehydration to 5-HMF under mild reaction conditions in water. Pure Brønsted acidic niobic acid nanoparticles dispersed over SBA-15 showed enhanced TOFs for fructose dehydration to 5-HMF but were unable to isomerize glucose due to a lack of Lewis acidity.

Carbon-supported niobia was also explored for glucose conversion to 5-HMF in a biphasic system. Although this catalyst exhibited improved stability to pure niobia [161], the carbon catalyst partitioned into the organic phase due to its hydrophobic nature, resulting in a low activity toward glucose. In light of this, Shanks and coworkers [162] prepared three kinds of niobia/carbon black catalysts (Nb/CB-1-DP, Nb/CB-2-DP, and Nb/CS-HT) via deposition–precipitation and deposition–precipitation–carbonization. The resulting control over surface hydrophobicity led to these catalysts respectively residing in the organic phase, aqueous phase, or at the solvent interface. As expected, the hydrophilic Nb/CB-2-DP was most active and afforded 20% 5-HMF in a sec-butyl phenol/water mix at 170 °C. Although these carbon-supported catalysts demonstrated different reactivity, a lack of characterization of the acid sites and niobia particle sizes, as well as the range of reaction conditions employed, hinder insight into any corresponding quantitative structure–activity relationships.

Mixed metal oxides offer interesting opportunities to tune Brønsted and Lewis acidity relative to their oxide constituents. CeO_2–Nb_2O_5 mixed oxides prepared by co-precipitation were tested for fructose dehydration [163]. No crystalline mixed oxide phases were observed; however, the CeO_2:Nb_2O_5 ratio influenced the density of strong acid sites and associated conversion and selectivity. Higher Nb_2O_5 loadings favored fructose dehydration and 5-HMF selectivity, although there was no clear evidence of any synergy between oxide components, with pure niobia outperforming any of the CeO_2–Nb_2O_5 composites.

Doped and oxide-supported vanadyl phosphates ($VOPO_4{\cdot}2H_2O$) with Brønsted and Lewis acid sites have also been studied for fructose dehydration. Doped vanadyl phosphates were prepared by isomorphic substitution of VO^{3+} groups with trivalent metals Fe^{3+}, Cr^{3+}, Ga^{3+}, Mn^{3+}, and Al^{3+} [164]. In the case of supported vanadyl phosphates, the support acidity was also important in promoting dehydration. For doped samples, Fe–$VOPO_4{\cdot}2H_2O$ offered the best activity and selectivity, converting concentrated aqueous fructose solutions with a 5-HMF productivity of 376 mmol/g_{cat} h at 80 °C while avoiding forming insoluble polymers or 5-HMF rehydration. Fructose dehydration over copper phosphates depended on their surface acidity (Brønsted/Lewis character) and morphology [165]. Heat treatment of $CuHPO_4{\cdot}H_2O$ nanoneedles produced α-$Cu_2P_2O_7$ nanocrystals at 600 °C and rodlike nanostructures at 900 °C (Figure 5.18). These thermally processed phosphates exhibited weaker acidity ($+3.3 \leq H_0 \leq +4.8$) but enhanced productivity compared with H_3PO_4, with the α-$Cu_2P_2O_7$-900 achieving 36% 5-HMF yield.

Calcium and α-strontium phosphates have been investigated for glucose, fructose, and cellulose conversion in hot-compressed water [166]. These catalysts were prepared by a modified coprecipitation method, resulting in wormlike morphologies observed by SEM. Despite their lack of porosity and very low (0.5 g/m^2) surface areas, high activity was observed for glucose and fructose dehydration, and the hydrolysis and dehydration of cellulose, over CaP_2O_6 and α-$Sr(PO_3)_2$ phosphate that promote 5-HMF formation from both monosaccharides (Figure 5.19).

In comparison with the preceding niobium, vanadium, copper and alkaline earth phosphates, zirconium phosphates ($ZrPO_x$) are more widely studied.

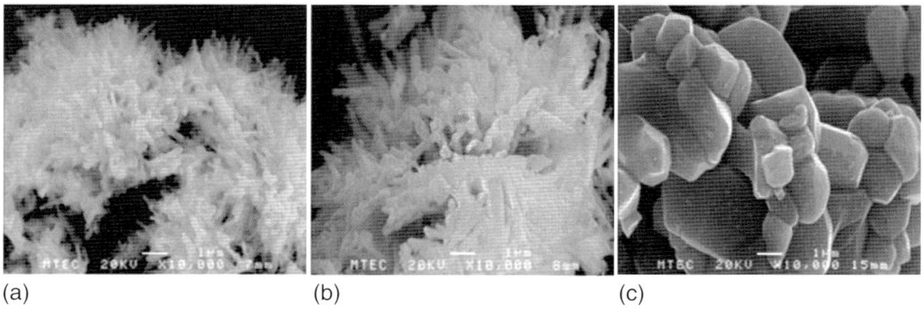

Figure 5.18 SEM images of (a) $CuHPO_4{\cdot}H_2O$ (parent copper phosphates), (b) α-$Cu_2P_2O_7$-600 (calcined at 600 °C), and (c) α-$Cu_2P_2O_7$-900 (calcined at 900 °C). Source: Khemthong et al. 2012 [165]. Adapted with permission of Elsevier.

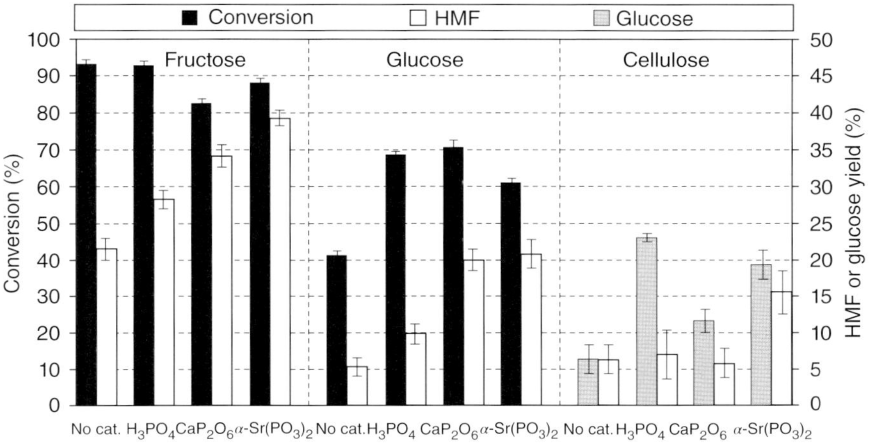

Figure 5.19 Fructose, glucose, and cellulose conversion, and 5-HMF yields for H_3PO_4, CaP_2O_6, or α-$Sr(PO_3)_2$. Source: Daorattanachai et al. 2012 [166]. Reproduced with permission of Elsevier.

Crystalline $ZrPO_x$, precipitated from $ZrCl_2$, was explored for fructose dehydration in subcritical water at 240 °C, achieving around 80% fructose conversion after a mere 120 seconds with 61% selectivity to 5-HMF [167]. However, these aggressive conditions resulted in a high background dehydration rate. Metal phosphates of aluminum ($AlPO_x$), titanium ($TiPO_x$), zirconium ($ZrPO_x$), and niobium ($NbPO_x$) have also been compared for glucose dehydration [168], with a view to examining the effect of different phosphate species and associated acidity. Acid strength and activity both increased with decreasing electronegativity of the metal cation, with $NbPO_x > ZrPO_x > TiPO_x > AlPO_x$. The Brønsted:Lewis acid site distribution also influenced 5-HMF selectivity, with excess Lewis acidity driving unselective glucose transformation to humins. Silylation of $NbPO_x$ and $ZrPO_x$ eliminated unselective Lewis acid sites, thereby dramatically increasing 5-HMF selectivity that reached 60% at 135 °C. Porous metal phosphates have also attracted interest. Cheng et al. [169] reported mesoporous zirconium phosphates obtained by hydrothermal synthesis from organic amine (dodecylamine and hexadecylamine) templates that exhibited high conversion (up to 96%) for xylose dehydration in water, attaining furfural yields of 52%. This promising performance was attributed to their open internal structures and abundant Brønsted/Lewis acid sites; mesoporous $ZrPO_x$ showed good stability and was easily regenerated by calcination.

While (transition) metal phosphates show promise for C_5–C_6 sugar dehydration, control over their textural properties is limited due to restructuring under high temperature calcinations necessary to remove structure-directing agents [170, 171]; mild calcination or solvent extraction (e.g. acid and ethanol) are insufficient to completely remove organic templates. Synthesis of porous metal phosphates is also currently restricted to a limited range of metals: zirconium, titanium, and niobium [172]. Continued effort should be devoted to preparing porous and ordered metal phosphates with improved active site accessibility.

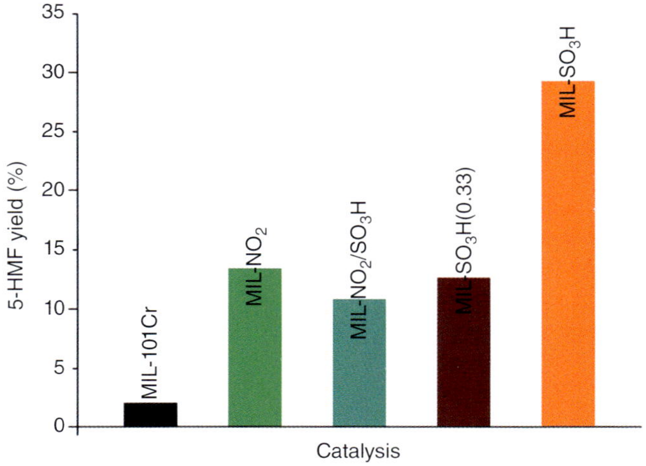

Figure 5.20 Conversion of glucose to 5-HMF over functionalized MIL-X (catalyst 5.22×10^{-5} mol, 223 mg glucose, 130 °C, 5 ml THF:H_2O (v:v) 39 : 1, 24 hours). Source: Herbst and Janiak 2016 [173]. Reproduced with permission of The Royal Society of Chemistry.

5.2.3.3 Metal–Organic Frameworks

The diverse chemical functionality of MOFs has also been exploited for the cascade glucose transformation to 5-HMF. Herbst and Janiak [173] compared the activity of MIL-101Cr, its nitro, sulfonic acid, and mixed nitro/sulfonic acid derivatives, respectively, MIL-101Cr, MIL-101Cr-NO_2, MIL-101Cr-NO_2/SO_3H, MIL-101Cr-SO_3H(33%), and MIL-101Cr-SO_3H(100%), for glucose conversion to 5-HMF. This was the first successful catalytic application of a bifunctional MOF, and utilized a biphasic THF:H_2O (39 : 1 volume ratio) solvent. MIL-101Cr-SO_3H yielded 29% 5-HMF yield after 24 hours at 130 °C, (Figure 5.20); however, it exhibited poor reusability due to pore blockage.

Many MOFs exhibit limited hydrothermal stability, a problem addressed for the bifunctional Lewis and Brønsted acidic MIL-101(Cr) by coating with silylating agents for furfural production from xylose [174]. When MIL-101(Cr) was employed in a water:toluene (3 : 7 volume ratio) solvent at 170 °C, 49% furfural was obtained at 86% xylose conversion; however, the subsequent yield decreased to only 10% after four recycles due to structural collapse. In contrast, MIL-101(Cr) functionalized with the hydrophobic silylating agent octadecyltrichlorosilane (OTS), exhibited improved hydrothermal stability and remained stable even after eight recycles. OTS functionalized MIL also exhibited improved activity (92% xylose conversion) and furfural yield (56%) at 170 °C, due to the superior resistance of active sites toward poisoning.

5.2.4 Catalysts for the Hydrogenation of C_5–C_6 Sugars

Hydrogenation is one of the most common reaction classes in chemical synthesis [175] and of great importance in transforming unsaturated functional groups in biomass and sugar derivatives [36, 176] to high value products for chemical and food sectors. In particular, the development of bifunctional, metal doped solid acid or base hydrogenation catalysts is attracting interest for the low temperature one-pot hydrodeoxygenation or selective deoxygenation of carbohydrates [72, 177–179].

There are many examples of industrially significant processes based around C_5 and C_6 sugar hydrogenation. The hydrogenation of xylose to xylitol yields an artificial sweetener whose sweetening capacity exceeds that of sucrose, making it ideally suited for those diagnosed with diabetes owing to its low insulin demand. Xylitol also possesses interesting anti-carcinogenic properties, and its demand has increased within the cosmetic, pharmaceutical, and food sectors [180–182]. Sorbitol, the C_6 analogue produced from glucose hydrogenation, finds application as a cosmetic thickener or sweetener and can be further valorized through dehydration to sorbitan and isosorbide, or by hydrogenolysis to other commercially relevant polyols. Sorbitan and isosorbide are used in medicinal therapies, and their corresponding esters are frequently employed as food stabilizers and emulsifiers. Isosorbide can be further deoxygenated to yield light alkanes ($<C_6$) for the production of second-generation biofuels [183, 184].

Aqueous phase hydrogenation of glucose, fructose, and xylose has been extensively studied over supported metal catalysts [185, 186], with early studies revealing ruthenium, platinum, cobalt, copper, rhodium, and nickel as promising candidates for such transformations. Selectivity is often compromised by competing polymerization, degradation, isomerization, and dehydration side reactions of sugars under the elevated temperatures (100–140 °C) employed (Figure 5.21). There is growing interest in the development of catalysts able to selectively hydrogenate C_5 and C_6 sugars under mild conditions, either through modifying the support structure and/or electronic properties, or using promoters/cocatalysts to control the reaction pathway, sugar stability, and final product selectivity.

The source of hydrogen for sustainable processes is also much debated, although options are emerging for the production of "green hydrogen" [187] from nonfossil sources, via aqueous phase reforming [188, 189], steam reforming/gasification [190–192] of biomass derivatives, or water splitting by electrolysis [193, 194] or photocatalysis [195, 196]. Here we summarize recent advances in the development of hydrogenation catalysts for C_5 and C_6 sugars.

5.2.4.1 Ni Catalysts

Raney Ni was historically employed for the hydrogenation of sugar compounds due to its low cost. Typically, Raney Ni is prepared from Ni–Al alloys, with variable amounts of aluminum removal through controlled alkaline leaching to yield a porous architecture with tunable Ni loading. Promoters such as Cr, Mo, Fe, and Sn (M), have been utilized in $Ni_{40-x}Al_{60}M_x$ Raney-catalyzed glucose hydrogenation to sorbitol [197], with catalytic activity enhanced sevenfold for Fe-promoted Raney Ni (Figure 5.22). Boron and phosphorous containing Raney Ni amorphous alloys also display enhanced catalytic activity, vs. a parent Raney nickel, in glucose hydrogenation [198–200]. It has been postulated that enhanced Ni dispersion, in conjunction with electronic perturbation of metallic Ni by surface promoter species, induces stronger C—O adsorption to Ni atoms. Low valent promoters act as Lewis acid sites, binding glucose through the oxygen of the C—O group,

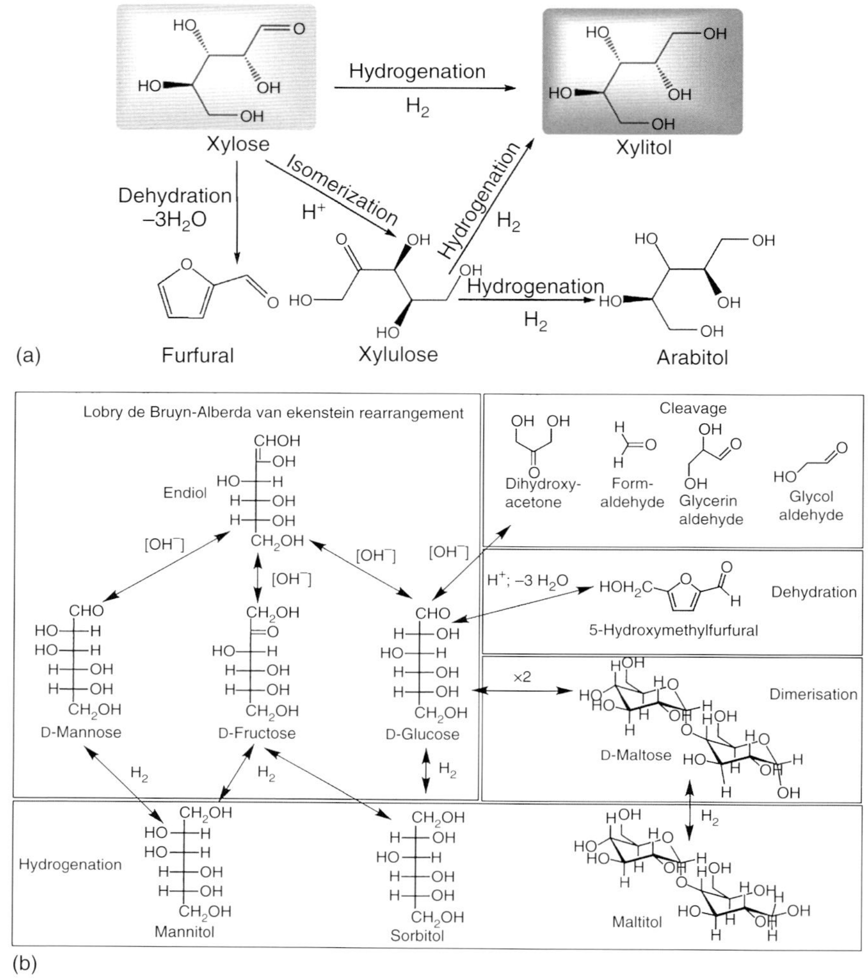

Figure 5.21 (a) Hydrogenation of xylose with possible side reactions. Source: Mishra et al. 2013 [185]. Reproduced with permission of Elsevier. (b) Hydrogenation of glucose with possible side reactions over oxides-supported Ni and Ru catalysts. Source: Kusserow et al. 2003 [186]. Reproduced with permission of John Wiley & Sons.

thereby polarizing this bond to favor hydrogenation by surface hydrogen. Addition of Cr, Mo, and W promoters further enhances activity (Figure 5.22), which passes through a maximum with respect to promoter surface concentration, due to high concentrations site-blocking active Ni sites. The promoter type, distribution, and oxidation state all influence the resulting performance enhancement for Ni-catalyzed glucose hydrogenation.

The impact of preparation method on the aqueous phase hydrogenation of glucose in batch and flow mode over supported Ni nanoparticles has also been

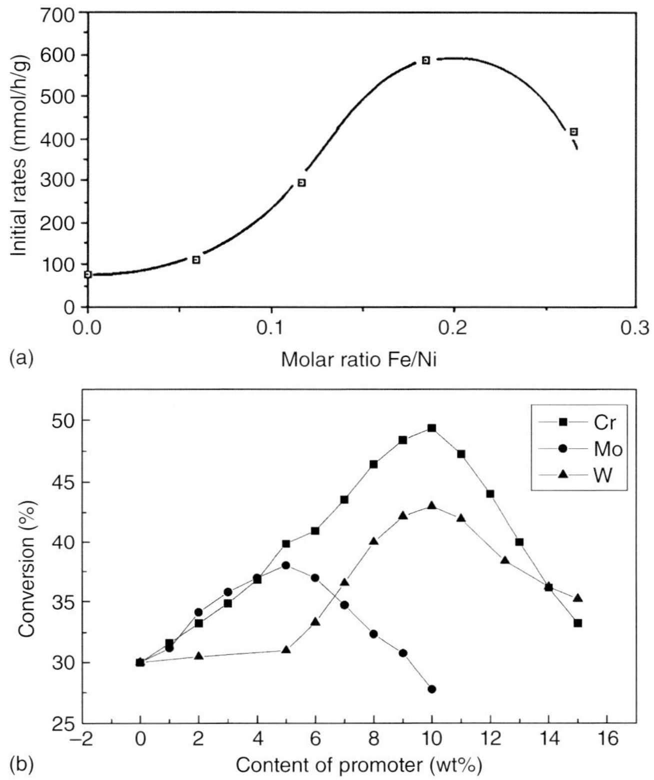

Figure 5.22 (a) Initial rate of glucose hydrogenation over fresh Raney $Ni_{40-x}Al_{60}M_x$ catalyst as a function of Fe:Ni ratio. Source: Gallezot et al. 1994 [197]. Reproduced with permission of Elsevier. (b) Glucose conversion over doped Ni-B/SiO_2 as a function of promoter concentration during hydrogenation. Source: Li et al. 2002 [199]. Adapted with permission of Elsevier.

explored (Figure 5.23) [186]. Supported Ni catalysts exhibit higher catalytic activity than Raney Ni, with Ni impregnated on alumina demonstrating the highest glucose conversion, comparable to a commercial Ni68T catalyst, despite containing only 5 wt% vs. 68 wt% Ni, and also delivered 90% sorbitol selectivity. Specific sorbitol productivities decreased with oxide support in the order $Al_2O_3 > TiO_2 \gg SiO_2 > C$, suggesting that Ni dispersion and support texture and acidity play important roles in catalytic performance. Catalyst pretreatment also influences catalytic performance. Schimpf et al. [200] demonstrated that calcination prior to reduction of a Ni/SiO_2 catalyst increased glucose conversion and sorbitol selectivity by fully decomposing the nickel precursor. Zhang et al. [201, 202] developed Ni/Cu/Al HT precursors and Ni/Cu/Al/Fe HT-like catalysts for the hydrogenation of glucose and fructose, respectively. Elevated reduction temperatures in the catalyst activation step also enhanced activity and sorbitol selectivity for $Ni_{1.85}Cu_1Al_{1.15}$-catalyzed glucose hydrogenation and activity for $Ni_{4.63}Cu_1Al_{1.82}Fe_{0.79}$ HT-catalyzed fructose hydrogenation.

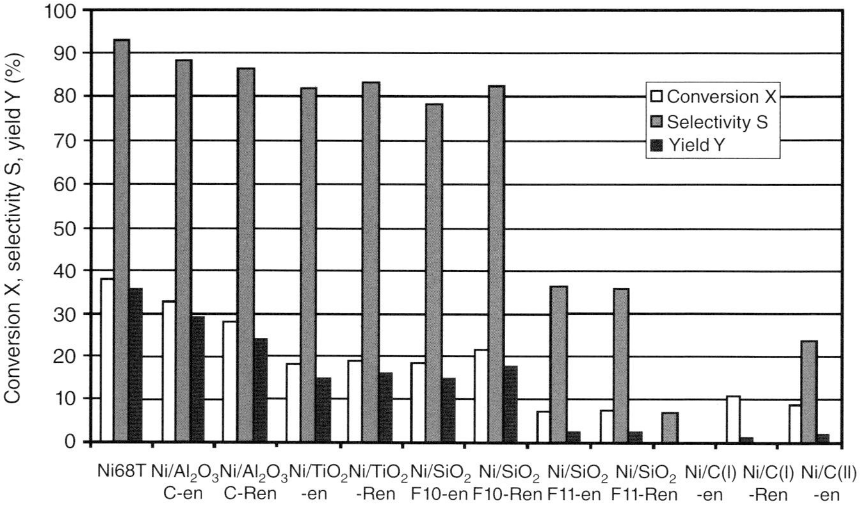

Figure 5.23 Catalytic performance of Ni systems in glucose hydrogenation: "en" = impregnation; "Ren" = incipient wetness. Source: Kusserow et al. 2003 [186]. Reproduced with permission of John Wiley & Sons.

5.2.4.2 Ru Catalysts

Ruthenium catalysts are more widely investigated than their Ni analogues for the hydrogenation of sugars due to their greater efficiency for aqueous phase carbonyl hydrogenations [203]. Ru nanoparticles on various supports demonstrate excellent activity and selectivity for glucose and fructose hydrogenation in batch and continuous flow reactors [186, 204]. Glucose hydrogenation over Ru supported on activated carbon in a continuous trickle bed reactor (100 °C, 80 bar H_2) revealed that sorbitol selectivity was inversely proportional to residence time, with longer residence times favoring sorbitol epimerization to mannitol. Continuous flow operation offered significant selectivity improvements over batch experiments, and crucially the catalyst was stable for several weeks on-stream with negligible metal leaching. Ru/C catalysts are comparable to, or more active, selective, and stable than Raney Ni catalysts, which suffer from leaching of both Ni and promoters [204].

Liu et al. screened Ru nanoparticles embedded on mesoporous carbon microfibers (Ru/C-NFs) for aqueous phase glucose hydrogenation [205]. The open, mesoporous support structure facilitated significant activity and stability enhancements, attributed to the unique microfiber morphology and hydrogen spillover between the carbon support and Ru nanoparticles. N-doping of the support improved hydrogen adsorption and Ru wettability, further augmenting catalytic performance.

Murzin and coworkers [206, 207] attempted to ascertain structure–reactivity relationships for Ru/C catalysts in glucose hydrogenation to sorbitol (and galactose/arabinose hydrogenation [207]) for Ru nanoparticles spanning from 1.2 to 10 nm. All three sugars demonstrated structure sensitivity, with the highest TOFs observed for ~3 nm Ru nanoparticles. Dehydration to glycerol was favored over Ru nanoparticles ≥10 nm, with a maximum sorbitol yield of 96% obtained for <3 nm particles. Crezee et al. [208] investigated the kinetics of D-glucose hydrogenation to D-sorbitol over 5 wt% Ru/C in a semi-batch stirred autoclave (40–75 bar H_2). Hydrogenation was first order with respect to hydrogen and D-glucose for concentrations <0.3 M but zero order at higher concentrations. Modeling implicated Langmuir–Hinshelwood–Hougen–Watson (LHHW) kinetics wherein the surface reaction was rate-determining; however this study was unable to distinguish between: (i) noncompetitive (molecular or dissociative) adsorption of H_2 and D-glucose, (ii) competitive adsorption of molecular H_2 and D-glucose, or (iii) competitive adsorption of dissociatively chemisorbed hydrogen and D-glucose.

Several studies have been conducted on the incorporation of Ru nanoparticles within zeolites [185, 209, 210]. The preparative route to Ru/ZSM-5 catalysts greatly influenced their behavior in glucose hydrogenation. Guo et al. [210] screened Ru/ZSM-5 obtained by addition of a Ru precursor to the parent ZSM-5 synthetic mixture (Ru/ZSM-5-TF), vs. loading of Ru nanoparticles onto a commercial microporous ZSM-5 (Ru/ZSM-5-MS) or alkali-treated mesoporous ZSM-5 (Ru/ZSM-5-AT) by incipient wetness impregnation. All Ru/ZSM-5 catalysts outperformed Ru/C. Ru/ZSM-5-TF exhibited uniform and well-dispersed Ru particles throughout the ZSM-5 framework, which resulted in the highest conversion and sorbitol selectivity.

Mechanistic aspects of Ru-catalyzed hydrogenation of bio-derived carbonyl compounds (albeit not the C_5–C_6 sugars) offer some insight into the active species [203]. Jae et al. [211] used X-ray photoelectron spectroscopy (XPS), X-ray absorption spectroscopy (XAS), and high resolution transmission electron microscopy (HRTEM) to investigate Ru-catalyzed 5-HMF hydrogenation to 2,5-DM; fresh Ru nanoparticles were dominated by surface RuO_x that underwent *in situ* reduction to the metal and concomitant deactivation. RuO_x was proposed to behave as a Lewis acid, promoting a Meerwein–Ponndorf–Verley (MPV) pathway via interhydride transfer from 2-propanol to 5-HMF to form 2,5-bis(hydroxymethyl)furan. Prereduced Ru/C exhibited moderate hydrogenolysis activity toward 5-HMF and ~30% selectivity to 2,5-DMF. A synergy between physically mixed (partially) oxidized and metal phases increased 2,5-DMF selectivity to 70%. DFT calculations and experimental studies of Ru-catalyzed levulinic acid [20, 212], furfural [213], L-arabinose [214], and 2-butanone [215] hydrogenation have highlighted the crucial role of surface H_2O in regulating substrate activity, Ru surface oxidation state, and reactivity toward adsorbed intermediates. Cooperative interactions between C—O functions in reactive intermediates bound to Ru with (coadsorbed) neighboring water molecules appear to decrease the barrier for levulinic acid hydrogenation to γ-valerolactone (GVL); higher experimental rates were observed as the hydrogen-bond donor capability of the solvent increased [20]. Computational studies suggest that water undergoes dissociative adsorption over Ru metal to produce ionic surface species and atomic hydrogen [216], the latter participating in hydrogenation.

5.2.4.3 Pt Catalysts

Mono- and bimetallic Pt catalysts have been the subject of theoretical, experimental, and industrial hydrogenation studies due to platinum's high activity, selectivity, and stability for C—O hydrogenation [72, 217–221]. However, there are fewer studies of the Pt-catalyzed selective transformation of xylose (glucose) to xylitol (sorbitol) than for Ni and Ru analogues. Pt on activated carbon cloth (ACC) was investigated for glucose hydrogenation (Figure 5.24) [212, 222]. In comparison with traditional powder or granular carbon supports, ACC exhibits a high surface area and narrow micropore size distribution, favoring efficient liquid phase mass transport. Pt introduced through cation exchange after oxidation of the ACC support yielded a homogeneous dispersion of 2–3 nm Pt nanoparticles within the carbon fibers. Pt/ACC catalysts were highly active for glucose hydrogenation, resistant to acidic and basic media, and delivered >99.5% sorbitol, superior to conventional Pt/C catalysts. Activated charcoal also shows promise as a support for Pt-catalyzed glucose hydrogenation [223] and fructose hydrogenation to mannitol [224]. However, in common with many studies utilizing carbon supports, the chemical composition and distribution of surface functions on the carbon were poorly characterized, limiting their reproducibility.

Structure–activity relationships for the aqueous phase hydrogenation of D-glucose to D-sorbitol were undertaken using a systematic family of Pt nanoparticles supported on a low surface area silica and high surface area SBA-15 mesoporous silica. Sorbitol productivity was structure insensitive for Pt particles between 3 and 17 nm, whereas competing D-glucose isomerization to D-fructose was favored over small, electron deficient Pt particles (<3 nm, Figure 5.25) [217].

Bifunctional Pt catalysts employing acidic γ-Al_2O_3 and basic HT supports have been examined for xylose and glucose hydrogenation [225]. A combined sorbitol and mannitol yield of 68% was obtained from glucose over Pt/γ-Al_2O_3 in combination with HT, with an alkaline medium increasing the concentration of open-form glucose that is more readily hydrogenated. The same combination achieved 99% xylose conversion and 82% xylitol yield. Unfortunately, these catalysts lost significant activity after only two runs at 90 °C possibly due to poor hydrothermal stability and coking.

5.2.4.4 Other Hydrogenation Catalysts

Cu/SiO_2 and Cu/ZnO/Al_2O_3 catalysts have been investigated for fructose hydrogenation to mannitol [226]. Acoustic treatment of the catalyst–substrate mixture enhanced activity but had negligible impact on selectivity over Cu/SiO_2 and Raney Ni catalysts. Such irradiation is proposed to deform the catalyst surface and hence (somehow) hinder catalyst deactivation; strong deactivation occurred without sonication. The commercial CuO–ZnO (61 wt% CuO and 39 wt% ZnO) methanol synthesis catalyst was also active for fructose hydrogenation [227]. In both cases, mannitol was the major hydrogenation product over copper catalysts, with some sorbitol arising from fructose isomerization to glucose. The highest reported mannitol selectivity was approximately 70%.

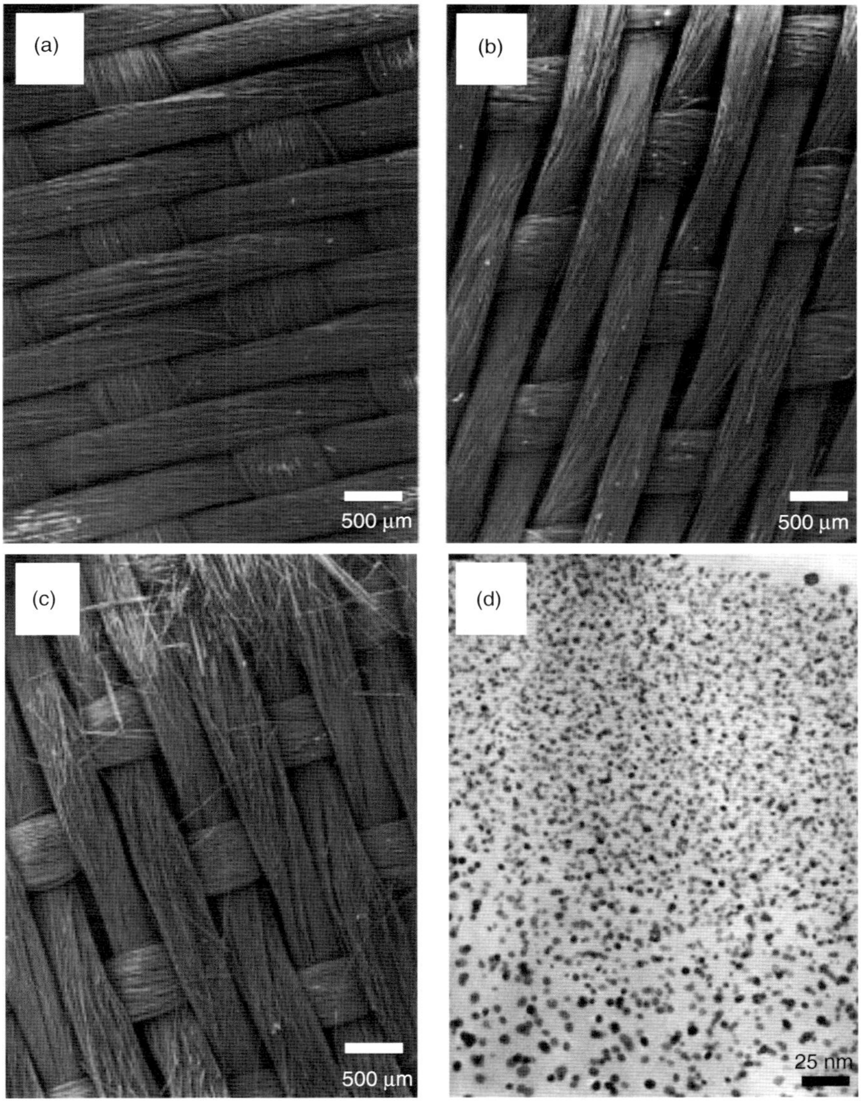

Figure 5.24 Scanning electron micrographs of activated carbon cloth at different preparation stages: (a) parent carbon, (b) after NaClO oxidation, (c) 10 wt% Pt/ACC, and (d) TEM image of 10 wt% Pt/ACC catalyst. Source: Perrard et al. 2007 [222]. Adapted with permission of Elsevier.

5.2.5 Hydrogenolysis Catalysts

In its broadest definition, hydrogenolysis is the rupture of chemical bonds via their interaction with hydrogen [228]. Although this is viewed as a versatile transformation to cleave C—C and C—O bonds to produce polyols from carbohydrates or transform other biomass-derived molecules [34, 229], reaction

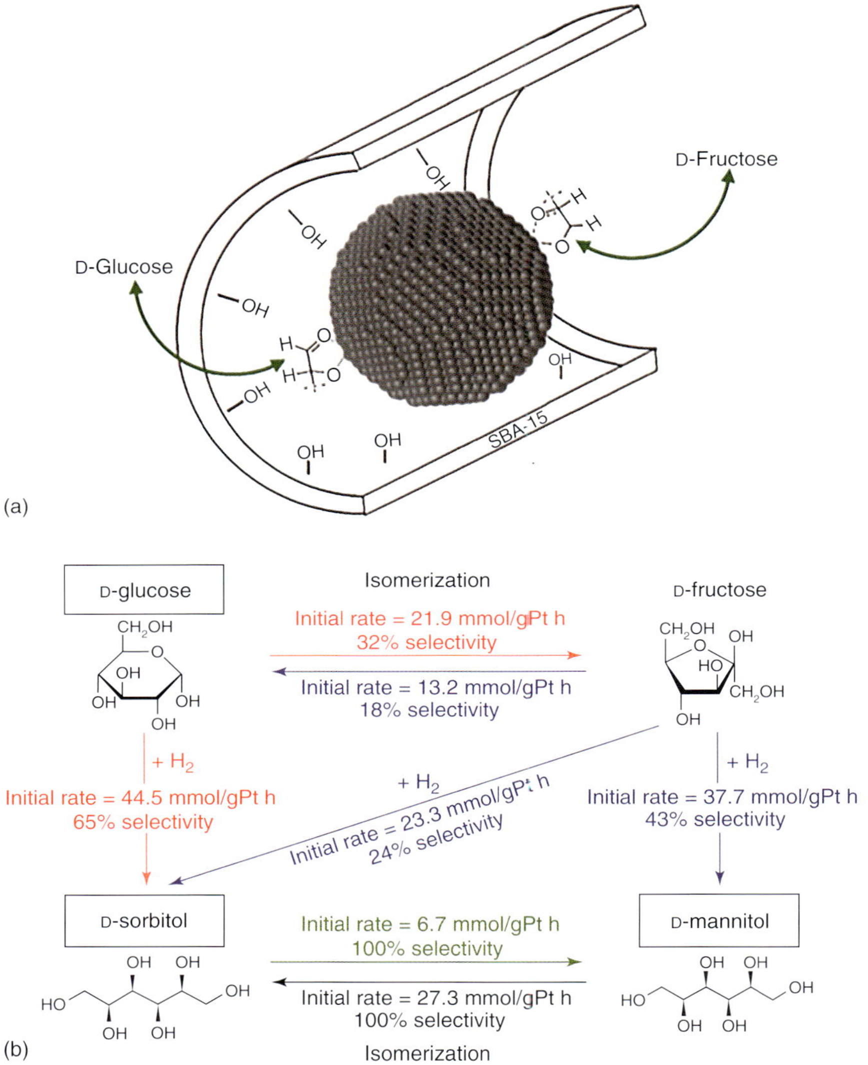

Figure 5.25 (a) Proposed isomerization of D-glucose to D-fructose over low coordination sites of Pt nanoparticles. (b) Reaction network for D-glucose and D-fructose hydrogenation at 140 °C and 40 bar H_2 over 1.5 wt% Pt/SBA-15 catalysts. Source: Zhang et al. [217]. Reproduced with permission of American Chemical Society.

conditions tend to be more forcing than hydrogenation reactions. Competing reaction networks involving C–O or C–C cleavage can operate during sugar hydrogenolysis, and early work employing Ni, Cu, Co, Fe, and Pt catalysts at elevated temperature and pressure (240 °C, 100 bar H_2), resulted in poor product selectivity with a broad distribution spanning glycerol, propylene glycol, higher polyhydric alcohols, ethane, and methanol. For example, glucose hydrogenolysis features competing hydrogenation pathways to sorbitol,

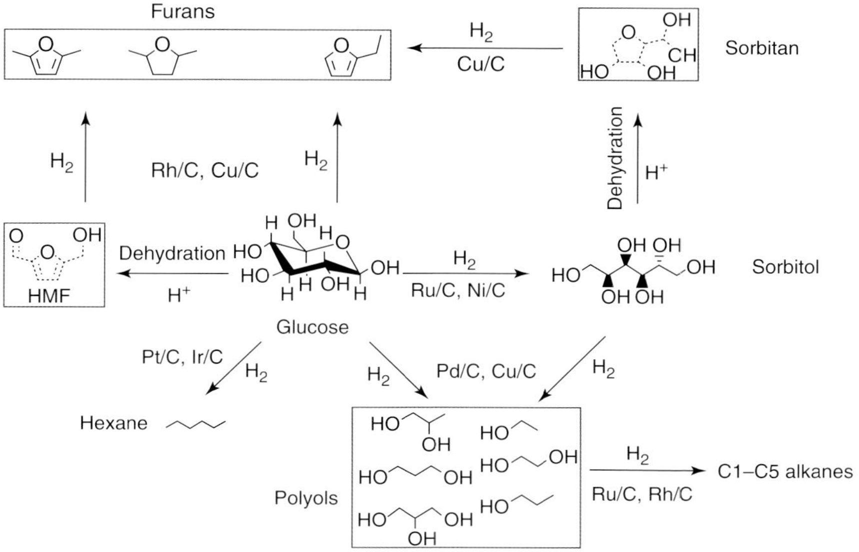

Figure 5.26 Reaction network for glucose hydrogenolysis. Source: Liu et al. 2015 [233]. Adapted with permission of Elsevier.

dehydration to 5-HMF, carbon chain fragmentation, and subsequent cascade reactions (Figure 5.26). Selective hydrogenolysis is thus especially challenging [34, 230] and requires multifunctional catalysts, or combinations of different catalyst classes [231, 232] able to promote the entire desired cascade under mild conditions.

Bifunctional catalysts were first observed to improve selectivity during cellulose and sugar hydrogenolysis. Ni is most widely used for sugar hydrogenolysis, and in conjunction with basic promotors, including NaOH, $Ca(OH)_2$, and $NaCO_3$, it exhibits enhanced activity for sorbitol production from carbohydrates owing to faster rates of hydrogenation vs. hydrogenolysis [34]. Application of Ni catalysts for hydrogenolysis is limited by metal leaching under the harsh conditions typically employed [234]. $Mg(OH)_2$-promoted Cu/Al_2O_3 has also been reported [34], and indeed many doped Cu systems have been explored such as $Cu–CeO_2$ [235, 236], CuO–ZnO [227], $Cu/CaO/Al_2O_3$ [237], and Ni-doped CuO–ZnO [238]. Direct hydrogenolysis of cellulose using $Cu–ZnO/Al_2O_3$ at 245 °C and 50 bar H_2 produces mainly ethylene glycol, 1,2-propane diol, and glycerol [239]; however, the C_3:C_2 product distribution from sugar hydrogenolysis is sensitive to the chosen substrate. Hydrogenation of cellulose end units yields sorbitol that undergoes subsequent hydrogenolysis. Basic promotors are beneficial for Ru/Al_2O_3- or Ru/TiO_2-catalyzed hydrogenolysis of sorbitol and xylitol to polyols including glycerol, propylene glycol, and ethylene glycol [240]. In basic media, sorbitol hydrogenolysis proceeds mainly via reverse aldolization to mixtures of two or three carbon atom molecules. Under neutral conditions, hydrogenolysis predominantly involves the middle of the molecular carbon chain, favoring glycerol and 1,2-propanediol [241].

Catalytic processes for biomass hydrogenolysis have been reviewed elsewhere in detail [34], with the mechanisms of dehydration, retro-aldol, retro-Claisen, and retro-Michael steps leading to fragmentation considered. Acid and base promoters exert a significant impact on product selectivity, which varies depending on how the C–O hydrogenolytic cleavage step is initiated. Lewis acids are postulated to bind polyols during base-catalyzed steps (Figure 5.27), highlighting the complex cooperative surface structures that need to be designed into hydrogenolysis catalysts. Acid–base pairs on metal oxide supports are often invoked in hydrogenolysis and are also implicated in ketonization and aldol condensation [242–244].

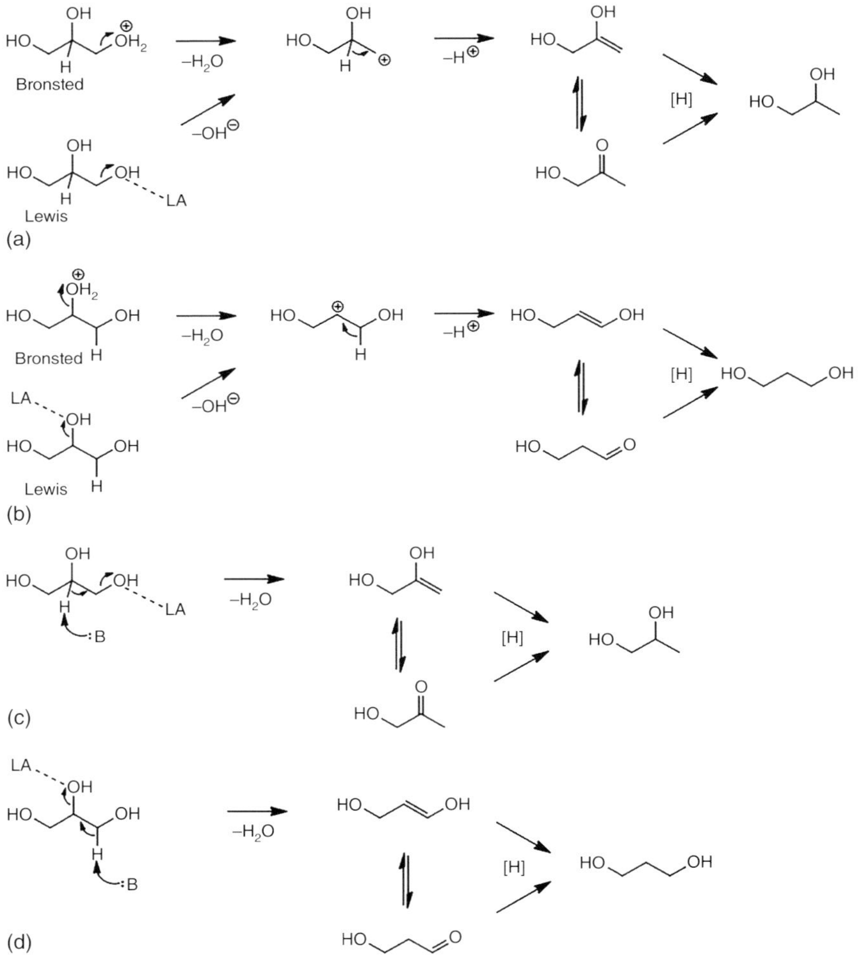

Figure 5.27 Role of support acid or base properties in polyol hydrogenolysis over multifunctional catalysts. Brønsted or Lewis acid-assisted dehydration of (a) primary and (b) secondary alcohols. Base-assisted hydride abstraction with concerted reaction via a Lewis acid site stabilized alkoxide anion for (c) primary alcohol dehydration, and (d) secondary alcohol dehydration. Source: Ruppert Agnieszka et al. 2012 [34]. Reproduced with permission of John Wiley and Sons.

5.2.6 Other Reactions

Sugars can also be converted into sugar fatty acid esters through enzymatic processes [245] or base-catalyzed transesterification with triglycerides [246] for pharmaceutical, cosmetic, and food applications [247] and the formation of biodegradable sugar ester surfactants [248]. Alkyl lactate production from sugars has been explored by Sels and coworkers [249] using bifunctional Sn-doped carbon–silica catalyst (Sn–Si–CSM, carbon modified Sn–Si–MCM-41). Sn–Si–CSM materials exhibit both Brønsted and Lewis acid sites, with the latter originating from Sn(IV) sites, while the weak Brønsted acid groups are associated with O-containing functional groups in the carbon framework. These bifunctional materials are active for the catalytic esterification of mono- and disaccharides with methanol to methyl lactate, using methanol as both solvent and reactant. The highest methyl lactate yield from glucose was 17% over a Sn–Si–CSM-773-20.4 catalyst prepared via carbonization at 500 °C and possessing ~20 wt% carbon.

There is also growing interest in hydrogen transfer reactions [250] for hydrogenation of bio-derived molecules using formic acid [251–253] or secondary alcohols [254, 255] as the hydrogen source. Although care must be taken to ensure a sustainable hydrogen donor is identified [256], such hydrogen transfer reactions offer significant safety advantages over the use of high pressure molecular hydrogen. Cascade hydrogen transfer reactions of xylose to GVL (Figure 5.28) exemplifies this strategy, in which an acid-catalyzed dehydration of xylose to furfural is followed by a hydrogenation step to form furfuryl alcohol. Reactively formed furfuryl alcohol in turn undergoes an acid-catalyzed hydration to levulinic acid, which subsequently hydrogenates to hydroxypentanoic acid

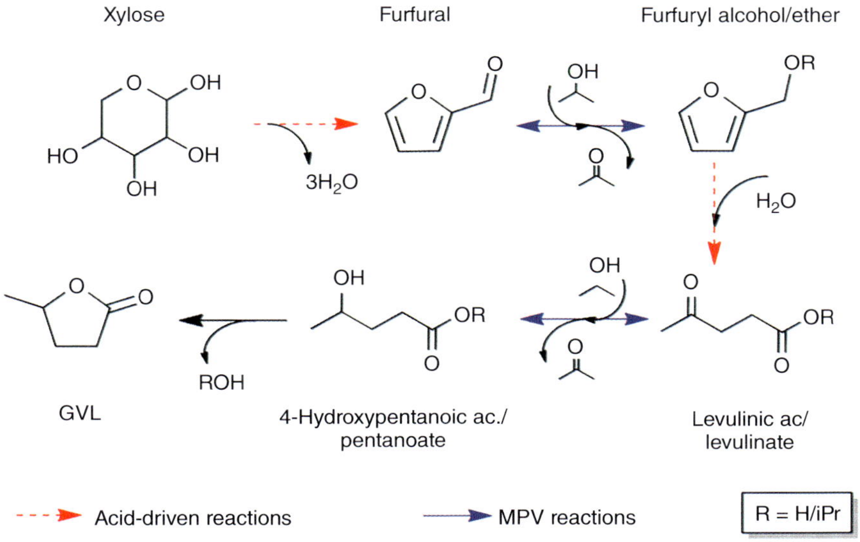

Figure 5.28 Cascade transformation of xylose to GVL through acid-catalyzed steps and MPV hydrogen transfer process. Source: Hernandez et al. 2016 [257]. Reproduced with permission of Royal Society of Chemistry.

prior to a final lactonization step to yield GVL. The direct production of GVL through this one-pot process offers significantly reduced operating costs by minimizing the need to isolate and purify reaction intermediates. Bifunctional Zr- and Al-containing beta [257, 258] and USY [259] zeolite catalysts, and physical mixtures of zeolite HY(2.6) with superparamagnetic $FeZrO_x$ nanoparticles show promise for this process [260].

5.3 Conclusions and Future Perspectives

Significant progress has been made in the development of heterogeneous catalysts and associated processes for the transformation of C_5 and C_6 sugars related to biorefinery applications. Waste-derived sugars are a promising feedstock for the production of renewable chemicals and advanced transportation fuels through low temperature, predominantly hydrothermal routes that exploit advances in materials and surface chemistry to engineer tailored inorganic and hybrid inorganic–organic solid catalysts with tunable acid/base character and/or electronic/geometric properties. However, as with all areas of catalysis, efforts are required to standardize reaction conditions, feedstock composition, reactor designs, and performance indicators to permit quantitative comparison of disparate catalytic systems. Biomass pretreatments such as steam explosion [22] or enzymatic [25] and chemical (acid or base) [26, 27] routes to fractionate and hydrolyze cellulose will ultimately produce aqueous sugar streams with mixed impurities. Aqueous phase sugar processing presents new challenges for heterogeneous catalysis related to the variable nature of the feed stream (in terms of component composition and concentration), solubility of sugars and reaction products, and the dissolution of reactive gases [96, 261, 262]. For industry to adopt processes based around these outputs, either the catalytic process must be adaptable to regular changes in feedstock composition, or a well-defined, mixed output stream must be agreed upon for targeted fractionation (akin to petroleum refinery operation). Biphasic media and cosolvents may overcome some of these issues and facilitate continuous processing and integrated product separation; however, solvent selection must adhere to general green chemistry and sustainable technology principles [263–266], being nontoxic, safe to handle, nonflammable, and noncorrosive. Such aspects are particularly important as water itself is considered a finite resource whose use in product isolation and purification must be minimized.

From a practical and environmental perspective, the continued development of hydrothermally stable catalysts able to operate in water should be encouraged. Support materials based on niobia, titania, zirconia, or tungstates that exhibit excellent stability under hydrothermal conditions, and catalysts employing such oxides as supports, will increasingly feature in high temperature sugar transformations. Templated porous carbons with controlled surface functionality will also likely come to the fore in aqueous phase sugar conversion due to their stability under acidic and alkaline environments [267].

A major challenge to selective catalytic sugar transformations is the prevalence of competing side reactions. For example, isomerization often competes with dehydration, and hydrogenation at high reaction temperatures is hindered by site-blocking by insoluble side products such as humins. Several strategies may offer improved selectivity: (i) passivation of catalyst supports for hydrogenation and oxidation to minimize support contributions to reactions [178, 268, 269], (ii) tunable Brønsted/Lewis acid character and strength [270, 271], (iii) tunable catalyst hydrophobicity to minimize water inhibition and promote organic reactant adsorption in the aqueous phase [272, 273], and (iv) control over oxide and metal surface terminations to favor specific crystal facets that have desirable adsorptive and/or catalytic properties [178, 274, 275]. Advanced nanomaterials with hierarchical macro-mesoporous or meso-microporous architectures also offer a means to improve mass transport compared with microporous materials through superior in-pore accessibility [276]. Hierarchical porous materials are also effective in stabilizing highly dispersed active species. Catalyst deactivation during hydrogenation or hydrogenolysis is often attributed to metal leaching (particularly for Ni). However, coke formation and attendant pore-blocking and/or active site-blocking is also problematic. Catalyst development must therefore address improved mass transport through the use of high surface area and/or hierarchical porous supports. Promoting strong metal-support interactions offers one route to increase resistance to metal leaching *in situ*, and associated electronic perturbation of the active site may also improve product selectivity [277]. Encapsulated nanoparticles [278], such as those developed by Joo et al. [279], in which Pt nanoparticles are enclosed within porous silica to form a Pt core@porous silica-shell structure have demonstrated exceptional thermal stability for CO oxidation and ethene hydrogenation, preventing metal sintering or loss, and may show promise for metal-catalyzed C_5–C_6 sugar hydrogenation.

Atomic layer deposition (ALD) is becoming a mature technology for catalyst preparation that is underpinning breakthroughs in catalyst development [280]. ALD employs self-limiting chemical reactions between gaseous precursors and a solid surface to deposit thin films of one or more metals [281] or oxides. Precursors can infiltrate mesoporous materials, producing highly uniform, conformal thin film coatings on surfaces with atomic scale precision [282, 283]. The development of a diverse range of coated support materials with desirable properties such as improved hydrothermal stability and/or acidity/basicity for sugar conversion should be possible by the application of ALD to coating tailored porous architectures generated by dual templating methods. In this context, single atom (site) catalysts offer exciting opportunities to bridge the gap between homogeneous and heterogeneous catalysis in sugar transformations, with the added benefit of precious metal thrifting [284]. While the concept of single site catalysts is common in framework solids or immobilized organometallic complexes [285], single site catalysts based around atoms anchored to graphene is also attracting attention [286, 287]. N-doped graphene encapsulated Fe_2O_3 nanoparticles generated via pyrolysis of well-defined amine-ligated metal complexes [288] offers an interesting route to nonprecious metal, chemoselective hydrogenation catalysts, with nitrogen ligand selection expected to influence catalyst performance. Such iron-derived catalysts are active for transfer hydrogenation using formic acid [289] and may find future applications in sugar conversion.

Pt, Ru, and Ni heterogeneous catalysts continue to dominate the carbohydrate hydrogenation literature; however, quantitative catalyst comparisons are limited due to a lack of detailed kinetic analyses regarding the identification, and elimination, of external and in-pore diffusion limitations. Such information is critical for catalyst development, which is currently hampered by misleading claims of high selectivity in sugar hydrogenation, despite reactions being performed in regimes promoting glucose and xylose isomerization, and wherein carbon mass balances are often neglected. Process optimization through flow chemistry should also improve selectivity by allowing better control over reactant/product residence times and removing primary products from the reaction zone before competitive side reactions initiate. Aqueous phase glucose hydrogenation to sorbitol in a trickle-bed reactor is enhanced at short residence times [234, 290], which avoids slower sorbitol epimerization to mannitol.

Catalytic C_5–C_6 sugar reforming often involves multistep cascade reactions that are desirably performed in a one-pot fashion. Examples include: 5-HMF production through glucose isomerization to fructose over a solid base/Lewis acid and subsequent fructose dehydration over Brønsted acid sites [70]; the stepwise production of polyols or alkanes via hydrogenation of glucose to sorbitol over metal (Ni, Pt, or Ru) sites and subsequent hydrogenolysis over metal-acid bifunctional catalysts [230]; metal-promoted solid acids for the production of isosorbide (an important intermediate for the synthesis of pharmaceuticals, chemicals, and polymers) from glucose via hydrogenation and dehydration; and the production of GVL from xylose via Brønsted acid catalysis coupled with Lewis acid- or base-catalyzed MPV reactions using isopropanol as a hydrogen donor [291–294]. Cascades are advantageous in respect to atom economy, process time, labor and resource management, and waste generation [295]. One-pot catalytic synthetic processes are highly desirable in industry but require multifunctional catalysts to become industrially viable processes [131]. New concepts for catalyst design through spatially orthogonal chemical functionalization of hierarchically porous silica frameworks are set to allow unprecedented control over reaction sequences [132]. Ultimately, the development of new processes for biomass conversion to drop-in and next-generation chemicals and fuels requires collaboration between catalytic chemists, chemical engineers, and experts in multiscale simulations to exploit innovative reactor designs.

References

1 Lewis, N.S. and Nocera, D.G. (2006). Powering the planet: chemical challenges in solar energy utilization. *Proc. Natl. Acad. Sci. U.S.A.* 103 (43): 15729.
2 Authority C.C. (2013). Reducing Australia's Greenhouse Gas Emissions – Targets and Progress Review Draft Report. Commonwealth of Australia, ISBN: 978-0-9873828-6-3.
3 Secretariat C.C. (2013). The Critical Decade 2013 Climate Change Acience, Risks and Responses. Commonwealth of Australia, ISBN: 978-1-925006-21-6.

4 United Nations Framework Convention on Climate Change (UNFCCC) (2015). *The Paris Agreement*. New York: UNFCCC.

5 Agency I.E. (2013) Redrawing the Energy Climate Map, World Energy Outlook Special Report.

6 Jakob, M. and Hilaire, J. (2015). Climate science: unburnable fossil-fuel reserves. *Nature* 517 (7533): 150–152.

7 Gerrard, E. and Mansour, A. (2015). COP21 and the Paris agreement. *Aust. Environ. Law Dig.* 2 (4): 3.

8 Jacobson, M.Z. and Delucchi, M.A. (2011). Providing all global energy with wind, water, and solar power, part I: technologies, energy resources, quantities and areas of infrastructure, and materials. *Energy Policy* 39 (3): 1154–1169.

9 Bozell, J.J. and Petersen, G.R. (2010). Technology development for the production of biobased products from biorefinery carbohydrates-the US Department of Energy's "Top 10" revisited. *Green Chem.* 12 (4): 539–554.

10 Lipinsky, E.S. (1981). Chemicals from biomass: petrochemical substitution options. *Science* 212 (4502): 1465–1471.

11 James, H.C., Fabien, E.I.D., and Thomas, J.F. (2008). The integration of green chemistry into future biorefineries. *Biofuels, Bioprod. Biorefin.* 3 (1): 72–90.

12 Thomas, J.M. (2012). The societal significance of catalysis and the growing practical importance of single-site heterogeneous catalysts. *Proc. R. Soc. London, Ser. A* 468 (2143): 1884–1903.

13 Somorjai, G.A., Frei, H., and Park, J.Y. (2009). Advancing the frontiers in nanocatalysis, biointerfaces, and renewable energy conversion by innovations of surface techniques. *J. Am. Chem. Soc.* 131 (46): 16589–16605.

14 Kim, M. and Day, D.F. (2011). Composition of sugar cane, energy cane, and sweet sorghum suitable for ethanol production at Louisiana sugar mills. *J. Ind. Microbiol. Biotechnol.* 38 (7): 803–807.

15 Tew, T.L., Cobill, R.M., and Richard, E.P. (2008). Evaluation of sweet sorghum and sorghum x sudangrass hybrids as feedstocks for ethanol production. *Bioenergy Res.* 1 (2): 147–152.

16 Cherubini, F. (2010). The biorefinery concept: using biomass instead of oil for producing energy and chemicals. *Energy Convers. Manage.* 51 (7): 1412–1421.

17 Menon, V. and Rao, M. (2012). Trends in bioconversion of lignocellulose: biofuels, platform chemicals & biorefinery concept. *Prog. Energy Combust. Sci.* 38 (4): 522–550.

18 Hoogwijk, M., Faaij, A., van den Broek, R. et al. (2003). Exploration of the ranges of the global potential of biomass for energy. *Biomass Bioenergy* 25 (2): 119–133.

19 Kadam, K.L., Forrest, L.H., and Jacobson, W.A. (2000). Rice straw as a lignocellulosic resource: collection, processing, transportation, and environmental aspects. *Biomass Bioenergy* 18 (5): 369–389.

20 Michel, C., Zaffran, J., Ruppert, A.M. et al. (2014). Role of water in metal catalyst performance for ketone hydrogenation: a joint experimental and theoretical study on levulinic acid conversion into gamma-valerolactone. *Chem. Commun.* 50 (83): 12450–12453.

21 Mosier, N., Wyman, C., Dale, B. et al. (2005). Features of promising technologies for pretreatment of lignocellulosic biomass. *Bioresour. Technol.* 96 (6): 673–686.
22 Glasser, W.G. and Wright, R.S. (1998). Steam-assisted biomass fractionation. II. Fractionation behavior of various biomass resources. *Biomass Bioenergy* 14 (3): 219–235.
23 Zhao, X., Cheng, K., and Liu, D. (2009). Organosolv pretreatment of lignocellulosic biomass for enzymatic hydrolysis. *Appl. Microbiol. Biotechnol.* 82 (5): 815–827.
24 Fatih Demirbas, M. (2009). Biorefineries for biofuel upgrading: a critical review. *Appl. Energy* 86 (Supplement 1 (0)): S151–S161.
25 Alvira, P., Tomás-Pejó, E., Ballesteros, M., and Negro, M.J. (2010). Pretreatment technologies for an efficient bioethanol production process based on enzymatic hydrolysis: a review. *Bioresour. Technol.* 101 (13): 4851–4861.
26 Mok, W.S., Antal, M.J., and Varhegyi, G. (1992). Productive and parasitic pathways in dilute acid-catalyzed hydrolysis of cellulose. *Ind. Eng. Chem. Res.* 31 (1): 94–100.
27 Rinaldi, R. and Schüth, F. (2009). Acid hydrolysis of cellulose as the entry point into biorefinery schemes. *ChemSusChem* 2 (12): 1096–1107.
28 Sasaki, M., Kabyemela, B., Malaluan, R. et al. (1998). Cellulose hydrolysis in subcritical and supercritical water. *J. Supercrit. Fluids* 13 (1–3): 261–268.
29 Binder, J.B. and Raines, R.T. (2010). Fermentable sugars by chemical hydrolysis of biomass. *Proc. Natl. Acad. Sci. U.S.A.* 107: 4516–4521.
30 Kaufman Rechulski, M.D., Käldström, M., Richter, U. et al. (2015). Mechanocatalytic depolymerization of lignocellulose performed on hectogram and kilogram scales. *Ind. Eng. Chem. Res.* 54 (16): 4581–4592.
31 Carrasquillo-Flores, R., Kaeldstroem, M., Schueth, F. et al. (2013). Mechanocatalytic depolymerization of dry (ligno)cellulose as an entry process for high-yield production of furfurals. *ACS Catal.* 3 (5): 993–997.
32 Schlaf, M. (2006). Selective deoxygenation of sugar polyols to alpha,omega-diols and other oxygen content reduced materials – a new challenge to homogeneous ionic hydrogenation and hydrogenolysis catalysis. *Dalton Trans.* (39): 4645–4653.
33 Fernando, S., Adhikari, S., Chandrapal, C., and Murali, N. (2006). Biorefineries: current status, challenges, and future direction. *Energy Fuels* 20 (4): 1727–1737.
34 Ruppert, A.M., Weinberg, K., and Palkovits, R. (2012). Hydrogenolysis goes bio: from carbohydrates and sugar alcohols to platform chemicals. *Angew. Chem. Int. Ed.* 51 (11): 2564–2601.
35 Chheda, J.N., Huber, G.W., and Dumesic, J.A. (2007). Liquid-phase catalytic processing of biomass-derived oxygenated hydrocarbons to fuels and chemicals. *Angew. Chem. Int. Ed.* 46 (38): 7164–7183.
36 Zhang, X., Wilson, K., and Lee, A.F. (2016). Heterogeneously catalyzed hydrothermal processing of C5–C6 sugars. *Chem. Rev.* 116 (19): 12328–12368.

37 Kobayashi, H. and Fukuoka, A. (2013). Synthesis and utilisation of sugar compounds derived from lignocellulosic biomass. *Green Chem.* 15 (7): 1740–1763.
38 Climent, M.J., Corma, A., and Iborra, S. (2011). Converting carbohydrates to bulk chemicals and fine chemicals over heterogeneous catalysts. *Green Chem.* 13 (3): 520–540.
39 Brandt, A., Grasvik, J., Hallett, J.P., and Welton, T. (2013). Deconstruction of lignocellulosic biomass with ionic liquids. *Green Chem.* 15 (3): 550–583.
40 Wang, H., Block, L.E., and Rogers, R.D. (2014). Catalytic conversion of biomass in ionic liquids, Chapter 1. In: *Catalysis in Ionic Liquids: From Catalyst Synthesis to Application*, 1–19. The Royal Society of Chemistry https://doi.org/10.1039/9781849737210-00001.
41 Yang, Y., Zhang, C., and Zhang, Z.C. (2018). Advances in catalytic transformations of carbohydrates and lignin in ionic liquids and mechanistic studies. *Wiley Interdiscip. Rev.: Energy Environ.* 7 (3): e284.
42 Morais, A.R., da Costa Lopes, A.M., and Bogel-Łukasik, R. (2014). Carbon dioxide in biomass processing: contributions to the green biorefinery concept. *Chem. Rev.* 115 (1): 3–27.
43 Choudhary, V., Pinar, A.B., Sandler, S.I. et al. (2011). Xylose isomerization to xylulose and its dehydration to furfural in aqueous media. *ACS Catal.* 1 (12): 1724–1728.
44 Sweeney, M.D. and Xu, F. (2012). Biomass converting enzymes as industrial biocatalysts for fuels and chemicals: recent developments. *Catalysts* 2 (2): 244.
45 Moliner, M., Román-Leshkov, Y., and Davis, M.E. (2010). Tin-containing zeolites are highly active catalysts for the isomerization of glucose in water. *Proc. Natl. Acad. Sci. U.S.A.* 107 (14): 6164–6168.
46 de Bruyn, C.A.L. and van Ekenstein, W.A. (1895). Action des alcalis sur les sucres, II. Transformation réciproque des uns dans les autres des sucres glucose, fructose et mannose. *Recl. Trav. Chim. Pays-Bas* 14 (7): 203–216.
47 Bermejo-Deval, R., Orazov, M., Gounder, R. et al. (2014). Active sites in Sn-beta for glucose isomerization to fructose and epimerization to mannose. *ACS Catal.* 4 (7): 2288–2297.
48 Li, Y.-P., Head-Gordon, M., and Bell, A.T. (2014). Analysis of the reaction mechanism and catalytic activity of metal-substituted beta zeolite for the isomerization of glucose to fructose. *ACS Catal.* 4 (5): 1537–1545.
49 Moreau, C., Durand, R., Roux, A., and Tichit, D. (2000). Isomerization of glucose into fructose in the presence of cation-exchanged zeolites and hydrotalcites. *Appl. Catal., A* 193 (1–2): 257–264.
50 Yu, S., Kim, E., Park, S. et al. (2012). Isomerization of glucose into fructose over Mg–Al hydrotalcite catalysts. *Catal. Commun.* 29: 63–67.
51 Lima, S., Dias, A.S., Lin, Z. et al. (2008). Isomerization of D-glucose to D-fructose over metallosilicate solid bases. *Appl. Catal., A* 339 (1): 21–27.
52 Delidovich, I. and Palkovits, R. (2014). Catalytic activity and stability of hydrophobic Mg–Al hydrotalcites in the continuous aqueous-phase isomerization of glucose into fructose. *Catal. Sci. Technol.* 4 (12): 4322–4329.

53 Liu, D., Nimlos, M.R., Johnson, D.K. et al. (2010). Free energy landscape for glucose condensation reactions. *J. Phys. Chem. A* 114 (49): 12936–12944.
54 Qian, X. and Wei, X. (2012). Glucose isomerization to fructose from ab Initio molecular dynamics simulations. *J. Phys. Chem. B* 116 (35): 10898–10904.
55 Bermejo-Deval, R., Gounder, R., and Davis, M.E. (2012). Framework and extraframework tin sites in zeolite beta react glucose differently. *ACS Catal.* 2 (12): 2705–2713.
56 Zhao, H., Holladay, J.E., Brown, H., and Zhang, Z.C. (2007). Metal chlorides in ionic liquid solvents convert sugars to 5-hydroxymethylfurfural. *Science* 316 (5831): 1597–1600.
57 Li, G., Pidko, E.A., and Hensen, E.J.M. (2014). Synergy between Lewis acid sites and hydroxyl groups for the isomerization of glucose to fructose over Sn-containing zeolites: a theoretical perspective. *Catal. Sci. Technol.* 4 (8): 2241–2250.
58 Smit, B. and Maesen, T.L.M. (2008). Towards a molecular understanding of shape selectivity. *Nature* 451 (7179): 671–678.
59 Dijkmans, J., Gabriels, D., Dusselier, M. et al. (2013). Productive sugar isomerization with highly active Sn in dealuminated [small beta] zeolites. *Green Chem.* 15 (10): 2777–2785.
60 Gounder, R. and Davis, M.E. (2013). Titanium-beta zeolites catalyze the stereospecific isomerization of d-glucose to l-sorbose via intramolecular C_5–C_1 hydride shift. *ACS Catal.* 3 (7): 1469–1476.
61 Mills, S.J., Christy, A.G., Génin, J.M.R. et al. (2012). Nomenclature of the hydrotalcite supergroup: natural layered double hydroxides. *Mineral. Mag.* 76 https://doi.org/10.1180/minmag.2012.076.5.10.
62 Cantrell, D.G., Gillie, L.J., Lee, A.F., and Wilson, K. (2005). Structure-reactivity correlations in MgAl hydrotalcite catalysts for biodiesel synthesis. *Appl. Catal., A* 287 (2): 183–190.
63 Di Serio, M., Ledda, M., Cozzolino, M. et al. (2006). Transesterification of soybean oil to biodiesel by using heterogeneous basic catalysts. *Ind. Eng. Chem. Res.* 45 (9): 3009–3014.
64 Xi, Y. and Davis, R.J. (2009). Influence of textural properties and trace water on the reactivity and deactivation of reconstructed layered hydroxide catalysts for transesterification of tributyrin with methanol. *J. Catal.* 268 (2): 307–317.
65 Fraile, J.M., Garcia, N., Mayoral, J.A. et al. (2009). The influence of alkaline metals on the strong basicity of Mg–Al mixed oxides: the case of transesterification reactions. *Appl. Catal., A* 364 (1–2): 87–94.
66 Cross, H.E. and Brown, D.R. (2010). Entrained sodium in mixed metal oxide catalysts derived from layered double hydroxides. *Catal. Commun.* 12 (3): 243–245.
67 Lecomte, J., Finiels, A., and Moreau, C. (2002). Kinetic study of the isomerization of glucose into fructose in the presence of anion-modified hydrotalcites. *Starch – Stärke* 54 (2): 75–79.

68 Delidovich, I. and Palkovits, R. (2015). Structure–performance correlations of Mg–Al hydrotalcite catalysts for the isomerization of glucose into fructose. *J. Catal.* 327: 1–9.
69 Tichit, D. and Coq, B. (2003). Catalysis by hydrotalcites and related materials. *Cattech* 7 (6): 206–217.
70 Osatiashtiani, A., Lee, A.F., Brown, D.R. et al. (2014). Bifunctional SO_4/ZrO_2 catalysts for 5-hydroxymethylfufural (5-HMF) production from glucose. *Catal. Sci. Technol.* 4 (2): 333–342.
71 Osatiashtiani, A., Lee, A.F., Granollers, M. et al. (2015). Hydrothermally stable, conformal, sulfated zirconia monolayer catalysts for glucose conversion to 5-HMF. *ACS Catal.* 5 (7): 4345–4352.
72 Taylor, M.J., Durndell, L.J., Isaacs, M.A. et al. (2016). Highly selective hydrogenation of furfural over supported Pt nanoparticles under mild conditions. *Appl. Catal., B* 180: 580–585.
73 Salak Asghari, F. and Yoshida, H. (2006). Acid-catalyzed production of 5-hydroxymethyl furfural from D-fructose in subcritical water. *Ind. Eng. Chem. Res.* 45 (7): 2163–2173.
74 Hansen, T.S., Woodley, J.M., and Riisager, A. (2009). Efficient microwave-assisted synthesis of 5-hydroxymethylfurfural from concentrated aqueous fructose. *Carbohydr. Res.* 344 (18): 2568–2572.
75 Choudhary, V., Sandler, S.I., and Vlachos, D.G. (2012). Conversion of xylose to furfural using Lewis and Brønsted acid catalysts in aqueous media. *ACS Catal.* 2 (9): 2022–2028.
76 Choudhary, V., Mushrif, S.H., Ho, C. et al. (2013). Insights into the interplay of Lewis and Brønsted acid catalysts in glucose and fructose conversion to 5-(hydroxymethyl)furfural and levulinic acid in aqueous media. *J. Am. Chem. Soc.* 135 (10): 3997–4006.
77 van Putten, R.-J., van der Waal, J.C., de Jong, E. et al. (2013). Hydroxymethylfurfural, a versatile platform chemical made from renewable resources. *Chem. Rev.* 113 (3): 1499–1597.
78 Tong, X., Ma, Y., and Li, Y. (2010). Biomass into chemicals: conversion of sugars to furan derivatives by catalytic processes. *Appl. Catal., A* 385 (1–2): 1–13.
79 Zakrzewska, M.E., Bogel-Łukasik, E., and Bogel-Łukasik, R. (2011). Ionic liquid-mediated formation of 5-hydroxymethylfurfural—a promising biomass-derived building block. *Chem. Rev.* 111 (2): 397–417.
80 Dornath, P. and Fan, W. (2014). Dehydration of fructose into furans over zeolite catalyst using carbon black as adsorbent. *Microporous Mesoporous Mater.* 191: 10–17.
81 Qian, X., Johnson, D.K., Himmel, M.E., and Nimlos, M.R. (2010). The role of hydrogen-bonding interactions in acidic sugar reaction pathways. *Carbohydr. Res.* 345 (13): 1945–1951.
82 Jeong, G.H., Kim, E.G., Kim, S.B. et al. (2011). Fabrication of sulfonic acid modified mesoporous silica shells and their catalytic performance with

dehydration reaction of D-xylose into furfural. *Microporous Mesoporous Mater.* 144 (1–3): 134–139.
83 Sairanen, E., Vilonen, K., Karinen, R., and Lehtonen, J. (2013). Functionalized activated carbon catalysts in xylose dehydration. *Top. Catal.* 56 (9–10): 512–521.
84 Ravenelle, R.M., Schüßler, F., D'Amico, A. et al. (2010). Stability of zeolites in hot liquid water. *J. Phys. Chem. C* 114 (46): 19582–19595.
85 Lourvanij, K. and Rorrer, G.L. (1993). Reactions of aqueous glucose solutions over solid-acid Y-zeolite catalyst at 110-160 °C. *Ind. Eng. Chem. Res.* 32 (1): 11–19.
86 You, S.J. and Park, E.D. (2014). Effects of dealumination and desilication of H-ZSM-5 on xylose dehydration. *Microporous Mesoporous Mater.* 186: 121–129.
87 Antunes, M.M., Lima, S., Fernandes, A. et al. (2012). Aqueous-phase dehydration of xylose to furfural in the presence of MCM-22 and ITQ-2 solid acid catalysts. *Appl. Catal., A* 417–418: 243–252.
88 Rac, V., Rakić, V., Stošić, D. et al. (2014). Hierarchical ZSM-5, beta and USY zeolites: acidity assessment by gas and aqueous phase calorimetry and catalytic activity in fructose dehydration reaction. *Microporous Mesoporous Mater.* 194: 126–134.
89 Kruger, J.S., Choudhary, V., Nikolakis, V., and Vlachos, D.G. (2013). Elucidating the roles of zeolite H-BEA in aqueous-phase fructose dehydration and HMF rehydration. *ACS Catal.* 3 (6): 1279–1291.
90 Lucas, N., Kokate, G., Nagpure, A., and Chilukuri, S. (2013). Dehydration of fructose to 5-hydroxymethyl furfural over ordered AlSBA-15 catalysts. *Microporous Mesoporous Mater.* 181: 38–46.
91 Kuehl, G.H. and Timken, H.K.C. (2000). Acid sites in zeolite beta: effects of ammonium exchange and steaming. *Microporous Mesoporous Mater.* 35–36: 521–532.
92 de Lucas, A., Canizares, P., Durán, A., and Carrero, A. (1997). Dealumination of HZSM-5 zeolites: effect of steaming on acidity and aromatization activity. *Appl. Catal., A* 154 (1): 221–240.
93 Poh, N.E., Nur, H., Muhid, M.N.M., and Hamdan, H. (2006). Sulphated AlMCM-41: mesoporous solid Brønsted acid catalyst for dibenzoylation of biphenyl. *Catal. Today* 114 (2): 257–262.
94 Weingarten, R., Tompsett, G.A., Conner, W.C. Jr.,, and Huber, G.W. (2011). Design of solid acid catalysts for aqueous-phase dehydration of carbohydrates: the role of Lewis and Brønsted acid sites. *J. Catal.* 279 (1): 174–182.
95 Swift, T.D., Bagia, C., Nikolakis, V. et al. (2013). Reactive adsorption for the selective dehydration of sugars to furans: modeling and experiments. *AlChE J.* 59 (9): 3378–3390.
96 Ordomsky, V.V., van der Schaaf, J., Schouten, J.C., and Nijhuis, T.A. (2012). The effect of solvent addition on fructose dehydration to 5-hydroxymethylfurfural in biphasic system over zeolites. *J. Catal.* 287: 68–75.

97 Moreau, C., Durand, R., Pourcheron, C., and Razigade, S. (1994). Preparation of 5-hydroxymethylfurfural from fructose and precursors over H-form zeolites. *Ind. Crops Prod.* 3 (1): 85–90.

98 Moreau, C., Durand, R., Razigade, S. et al. (1996). Dehydration of fructose to 5-hydroxymethylfurfural over H-mordenites. *Appl. Catal., A* 145 (1): 211–224.

99 Kruger, J.S., Nikolakis, V., and Vlachos, D.G. (2014). Aqueous-phase fructose dehydration using Brønsted acid zeolites: catalytic activity of dissolved aluminosilicate species. *Appl. Catal., A* 469: 116–123.

100 Zhou, Y., Huang, R., Ding, F. et al. (2014). Sulfonic acid-functionalized α-zirconium phosphate single-layer nanosheets as a strong solid acid for heterogeneous catalysis applications. *ACS Appl. Mater. Interfaces* 6 (10): 7417–7425.

101 Nabae, Y., Liang, J., Huang, X. et al. (2014). Sulfonic acid functionalized hyperbranched poly(ether sulfone) as a solid acid catalyst. *Green Chem.* 16 (7): 3596–3602.

102 Russo, P.A., Antunes, M.M., Neves, P. et al. (2014). Solid acids with SO_3H groups and tunable surface properties: versatile catalysts for biomass conversion. *J. Mater. Chem. A* 2 (30): 11813–11824.

103 Toda, M., Takagaki, A., Okamura, M. et al. (2005). Green chemistry: biodiesel made with sugar catalyst. *Nature* 438 (7065): 178–178.

104 Hara, M., Yoshida, T., Takagaki, A. et al. (2004). A carbon Material as a strong protonic acid. *Angew. Chem. Int. Ed.* 43 (22): 2955–2958.

105 Chung, P.-W., Charmot, A., Olatunji-Ojo, O.A. et al. (2014). Hydrolysis catalysis of miscanthus xylan to xylose using weak-acid surface sites. *ACS Catal.* 4 (1): 302–310.

106 Kei-ichi, S., Yoshihisa, I., and Hitoshi, I. (2001). Catalytic activity of lanthanide(III) ions for the dehydration of hexose to 5-hydroxymethyl-2-furaldehyde in water. *Bull. Chem. Soc. Jpn.* 74 (6): 1145–1150.

107 Yoshio, N. and Shunichi, M. (1980). The dehydration of D-Fructose to 5-hydroxymethyl-2-furaldehyde. *Bull. Chem. Soc. Jpn.* 53 (12): 3705–3706.

108 Qi, X., Watanabe, M., Aida, T.M., and Smith, J.R.L. (2008). Catalytic dehydration of fructose into 5-hydroxymethylfurfural by ion-exchange resin in mixed-aqueous system by microwave heating. *Green Chem.* 10 (7): 799–805.

109 Wang, L. and Xiao, F.-S. (2015). Nanoporous catalysts for biomass conversion. *Green Chem.* 17 (1): 24–39.

110 Lansalot-Matras, C. and Moreau, C. (2003). Dehydration of fructose into 5-hydroxymethylfurfural in the presence of ionic liquids. *Catal. Commun.* 4 (10): 517–520.

111 Wang, W., Lofgreen, J.E., and Ozin, G.A. (2010). Why PMO? Towards functionality and utility of periodic mesoporous organosilicas. *Small* 6 (23): 2634–2642.

112 Burleigh, M.C., Markowitz, M.A., Jayasundera, S. et al. (2003). Mechanical and hydrothermal stabilities of aged periodic mesoporous organosilicas. *J. Phys. Chem. B* 107 (46): 12628–12634.

113 van der Graaff, W.N.P., Olvera, K.G., Pidko, E.A., and Hensen, E.J.M. (2014). Stability and catalytic properties of porous acidic (organo)silica materials for conversion of carbohydrates. *J. Mol. Catal. A: Chem.* 388–389: 81–89.

114 Tucker, M.H., Crisci, A.J., Wigington, B.N. et al. (2012). Acid-functionalized SBA-15-type periodic mesoporous organosilicas and their use in the continuous production of 5-hydroxymethylfurfural. *ACS Catal.* 2 (9): 1865–1876.

115 Bispo, C., De Oliveira Vigier, K., Sardo, M. et al. (2014). Catalytic dehydration of fructose to HMF over sulfonic acid functionalized periodic mesoporous organosilicas: role of the acid density. *Catal. Sci. Technol.* 4 (8): 2235–2240.

116 Crisci, A.J., Tucker, M.H., Dumesic, J.A., and Scott, S.L. (2010). Bifunctional solid catalysts for the selective conversion of fructose to 5-hydroxymethylfurfural. *Top. Catal.* 53 (15): 1185–1192.

117 Crisci, A.J., Tucker, M.H., Lee, M.-Y. et al. (2011). Acid-functionalized SBA-15-type silica catalysts for carbohydrate dehydration. *ACS Catal.* 1 (7): 719–728.

118 Xiong, H., Pham, H.N., and Datye, A.K. (2014). Hydrothermally stable heterogeneous catalysts for conversion of biorenewables. *Green Chem.* 16 (11): 4627–4643.

119 Tian, C., Bao, C., Binder, A. et al. (2013). An efficient and reusable “hairy” particle acid catalyst for the synthesis of 5-hydroxymethylfurfural from dehydration of fructose in water. *Chem. Commun.* 49 (77): 8668–8670.

120 Yang, Z.-Z., Deng, J., Pan, T. et al. (2012). A one-pot approach for conversion of fructose to 2,5-diformylfuran by combination of Fe_3O_4-SBA-SO_3H and K-OMS-2. *Green Chem.* 14 (11): 2986–2989.

121 Hu, L., Zhao, G., Tang, X. et al. (2013). Catalytic conversion of carbohydrates into 5-hydroxymethylfurfural over cellulose-derived carbonaceous catalyst in ionic liquid. *Bioresour. Technol.* 148: 501–507.

122 Lam, E., Chong, J.H., Majid, E. et al. (2012). Carbocatalytic dehydration of xylose to furfural in water. *Carbon* 50 (3): 1033–1043.

123 Herbst, A. and Janiak, C. (2017). MOF catalysts in biomass upgrading towards value-added fine chemicals. *CrystEngComm* 19 (29): 4092–4117.

124 Li, Y.S., Liang, F.Y., Bux, H. et al. (2010). Molecular sieve membrane: supported metal–organic framework with high hydrogen selectivity. *Angew. Chem.* 122 (3): 558–561.

125 Lee, J., Farha, O.K., Roberts, J. et al. (2009). Metal-organic framework materials as catalysts. *Chem. Soc. Rev.* 38 (5): 1450–1459.

126 Zhang, Y., Degirmenci, V., Li, C., and Hensen, E.J.M. (2011). Phosphotungstic acid encapsulated in metal–organic framework as catalysts for carbohydrate dehydration to 5-hydroxymethylfurfural. *ChemSusChem* 4 (1): 59–64.

127 Hu, Z., Peng, Y., Gao, Y. et al. (2016). Direct synthesis of hierarchically porous metal–organic frameworks with high stability and strong Brønsted acidity: the decisive role of hafnium in efficient and selective fructose dehydration. *Chem. Mater.* 28 (8): 2659–2667.

128 Bao, Q., Qiao, K., Tomida, D., and Yokoyama, C. (2008). Preparation of 5-hydroymethylfurfural by dehydration of fructose in the presence of acidic ionic liquid. *Catal. Commun.* 9 (6): 1383–1388.

129 Sidhpuria, K.B., Daniel-da-Silva, A.L., Trindade, T., and Coutinho, J.A.P. (2011). Supported ionic liquid silica nanoparticles (SILnPs) as an efficient and recyclable heterogeneous catalyst for the dehydration of fructose to 5-hydroxymethylfurfural. *Green Chem.* 13 (2): 340–349.

130 Lee, Y.-Y. and Wu, K.C.W. (2012). Conversion and kinetics study of fructose-to-5-hydroxymethylfurfural (HMF) using sulfonic and ionic liquid groups bi-functionalized mesoporous silica nanoparticles as recyclable solid catalysts in DMSO systems. *Phys. Chem. Chem. Phys.* 14 (40): 13914–13917.

131 Climent, M.J., Corma, A., Iborra, S., and Sabater, M.J. (2014). Heterogeneous catalysis for tandem reactions. *ACS Catal.* 4 (3): 870–891.

132 Parlett Christopher, M.A., Isaacs Mark, A., Beaumont Simon, K. et al. (2015). Spatially orthogonal chemical functionalization of a hierarchical pore network for catalytic cascade reactions. *Nat. Mater.* 15: 178.

133 Yamaguchi, S. and Baba, T. (2016). A novel strategy for biomass upgrade: cascade approach to the synthesis of useful compounds via C–C bond formation using biomass-derived sugars as carbon nucleophiles. *Molecules* 21 (7): 937.

134 Li, H., Bhadury, P.S., Riisager, A., and Yang, S. (2014). One-pot transformation of polysaccharides via multi-catalytic processes. *Catal. Sci. Technol.* 4 (12): 4138–4168.

135 Wang, J., Ren, J., Liu, X. et al. (2012). Direct conversion of carbohydrates to 5-hydroxymethylfurfural using Sn-Mont catalyst. *Green Chem.* 14 (9): 2506–2512.

136 Mora-Pale, M., Meli, L., Doherty, T.V. et al. (2011). Room temperature ionic liquids as emerging solvents for the pretreatment of lignocellulosic biomass. *Biotechnol. Bioeng.* 108 (6): 1229–1245.

137 da Costa Lopes, A.M., João, K.G., Morais, A.R.C. et al. (2013). Ionic liquids as a tool for lignocellulosic biomass fractionation. *Sustainable Chem. Processes* 1 (1): 1–31.

138 Swatloski, R.P., Spear, S.K., Holbrey, J.D., and Rogers, R.D. (2002). Dissolution of cellose with ionic liquids. *J. Am. Chem. Soc.* 124 (18): 4974–4975.

139 Brandt, A., Hallett, J.P., Leak, D.J. et al. (2010). The effect of the ionic liquid anion in the pretreatment of pine wood chips. *Green Chem.* 12 (4): 672–679.

140 Francisco, M., van den Bruinhorst, A., and Kroon, M.C. (2013). Low-transition-temperature mixtures (LTTMs): a new generation of designer solvents. *Angew. Chem. Int. Ed.* 52 (11): 3074–3085.

141 Xia, S., Baker, G.A., Li, H. et al. (2014). Aqueous ionic liquids and deep eutectic solvents for cellulosic biomass pretreatment and saccharification. *RSC Adv.* 4 (21): 10586–10596.

142 Chheda, J.N., Roman-Leshkov, Y., and Dumesic, J.A. (2007). Production of 5-hydroxymethylfurfural and furfural by dehydration of biomass-derived mono- and poly-saccharides. *Green Chem.* 9 (4): 342–350.

143 Selva, M., Gottardo, M., and Perosa, A. (2013). Upgrade of biomass-derived levulinic acid via Ru/C-catalyzed hydrogenation to γ-valerolactone in

aqueous–organic–ionic liquids multiphase systems. *ACS Sustainable Chem. Eng.* 1 (1): 180–189.
144 Hu, L., Wu, Z., Xu, J. et al. (2014). Zeolite-promoted transformation of glucose into 5-hydroxymethylfurfural in ionic liquid. *Chem. Eng. J.* 244: 137–144.
145 Ståhlberg, T., Fu, W., Woodley, J.M., and Riisager, A. (2011). Synthesis of 5-(hydroxymethyl)furfural in ionic liquids: paving the way to renewable chemicals. *ChemSusChem* 4 (4): 451–458.
146 Liu, W., Zheng, F., Li, J., and Cooper, A. (2013). An ionic liquid reaction and separation process for production of hydroxymethylfurfural from sugars. *AIChE J.* 60 (1): 300–314.
147 Tanabe, K., Sumiyoshi, T., Shibata, K. et al. (1974). A new hypothesis regarding the surface acidity of binary metal oxides. *Bull. Chem. Soc. Jpn.* 47 (5): 1064–1066.
148 Auroux, A. and Gervasini, A. (1990). Microcalorimetric study of the acidity and basicity of metal oxide surfaces. *J. Phys. Chem.* 94 (16): 6371–6379.
149 Watanabe, M., Aizawa, Y., Iida, T. et al. (2005). Catalytic glucose and fructose conversions with TiO_2 and ZrO_2 in water at 473 K: relationship between reactivity and acid–base property determined by TPD measurement. *Appl. Catal., A* 295 (2): 150–156.
150 Watanabe, M., Aizawa, Y., Iida, T. et al. (2005). Glucose reactions with acid and base catalysts in hot compressed water at 473 K. *Carbohydr. Res.* 340 (12): 1925–1930.
151 Qi, X., Watanabe, M., Aida, T.M., and Smith, R.L. Jr., (2008). Catalytical conversion of fructose and glucose into 5-hydroxymethylfurfural in hot compressed water by microwave heating. *Catal. Commun.* 9 (13): 2244–2249.
152 Bröll, D., Kaul, C., Krämer, A. et al. (1999). Chemistry in supercritical water. *Angew. Chem. Int. Ed.* 38 (20): 2998–3014.
153 Yan, H., Yang, Y., Tong, D. et al. (2009). Catalytic conversion of glucose to 5-hydroxymethylfurfural over SO_4^{2-}/ZrO_2 and $SO_4^{2-}/ZrO_2–Al_2O_3$ solid acid catalysts. *Catal. Commun.* 10 (11): 1558–1563.
154 Morales, G., Osatiashtiani, A., Hernandez, B. et al. (2014). Conformal sulfated zirconia monolayer catalysts for the one-pot synthesis of ethyl levulinate from glucose. *Chem. Commun.* 50 (79): 11742–11745.
155 Armaroli, T., Busca, G., Carlini, C. et al. (2000). Acid sites characterization of niobium phosphate catalysts and their activity in fructose dehydration to 5-hydroxymethyl-2-furaldehyde. *J. Mol. Catal. A: Chem.* 151 (1–2): 233–243.
156 Lin, R. and Ding, Y. (2013). A review on the synthesis and applications of mesostructured transition metal phosphates. *Materials* 6 (1): 217.
157 Nakajima, K., Baba, Y., Noma, R. et al. (2011). $Nb_2O_5{\cdot}nH_2O$ as a heterogeneous catalyst with water-tolerant lewis acid sites. *J. Am. Chem. Soc.* 133 (12): 4224–4227.
158 Yang, F., Liu, Q., Bai, X., and Du, Y. (2011). Conversion of biomass into 5-hydroxymethylfurfural using solid acid catalyst. *Bioresour. Technol.* 102 (3): 3424–3429.

159 Yang, F., Liu, Q., Yue, M. et al. (2011). Tantalum compounds as heterogeneous catalysts for saccharide dehydration to 5-hydroxymethylfurfural. *Chem. Commun.* 47 (15): 4469–4471.
160 Reche, M.T., Osatiashtiani, A., Durndell, L.J. et al. (2016). Niobic acid nanoparticle catalysts for the aqueous phase transformation of glucose and fructose to 5-hydroxymethylfurfural. *Catal. Sci. Technol.* 6 (19): 7334–7341.
161 Xiong, H., Pham, H.N., and Datye, A.K. (2013). A facile approach for the synthesis of niobia/carbon composites having improved hydrothermal stability for aqueous-phase reactions. *J. Catal.* 302: 93–100.
162 Xiong, H., Wang, T., Shanks, B.H., and Datye, A.K. (2013). Tuning the location of niobia/carbon composites in a biphasic reaction: dehydration of D-glucose to 5-hydroxymethylfurfural. *Catal. Lett.* 143 (6): 509–516.
163 Stošić, D., Bennici, S., Rakić, V., and Auroux, A. (2012). CeO_2–Nb_2O_5 mixed oxide catalysts: Preparation, characterization and catalytic activity in fructose dehydration reaction. *Catal. Today* 192 (1): 160–168.
164 Carlini, C., Patrono, P., Galletti, A.M.R., and Sbrana, G. (2004). Heterogeneous catalysts based on vanadyl phosphate for fructose dehydration to 5-hydroxymethyl-2-furaldehyde. *Appl. Catal., A* 275 (1–2): 111–118.
165 Khemthong, P., Daorattanachai, P., Laosiripojana, N., and Faungnawakij, K. (2012). Copper phosphate nanostructures catalyze dehydration of fructose to 5-hydroxymethylfufural. *Catal. Commun.* 29: 96–100.
166 Daorattanachai, P., Khemthong, P., Viriya-empikul, N. et al. (2012). Conversion of fructose, glucose, and cellulose to 5-hydroxymethylfurfural by alkaline earth phosphate catalysts in hot compressed water. *Carbohydr. Res.* 363: 58–61.
167 Asghari, F.S. and Yoshida, H. (2006). Dehydration of fructose to 5-hydroxymethylfurfural in sub-critical water over heterogeneous zirconium phosphate catalysts. *Carbohydr. Res.* 341 (14): 2379–2387.
168 Ordomsky, V.V., Sushkevich, V.L., Schouten, J.C. et al. (2013). Glucose dehydration to 5-hydroxymethylfurfural over phosphate catalysts. *J. Catal.* 300: 37–46.
169 Cheng, L., Guo, X., Song, C. et al. (2013). High performance mesoporous zirconium phosphate for dehydration of xylose to furfural in aqueous-phase. *RSC Adv.* 3 (45): 23228–23235.
170 Sanz, F., Parada, C., Rojo, J.M., and Ruiz-Valero, C. (1999). Crystal structure, magnetic properties, and ionic conductivity of a new mixed-anion phosphate $Na_4Ni_5(PO_4)_2(P_2O_7)_2$. *Chem. Mater.* 11 (10): 2673–2679.
171 Luo, Q., Shen, S., Lu, G. et al. (2009). Synthesis of cubic ordered mesoporous YPO_4:Ln^{3+} and their photoluminescence properties. *J. Mater. Chem.* 19 (43): 8079–8085.
172 Jiménez-Jiménez, J., Maireles-Torres, P., Olivera-Pastor, P. et al. (1998). Surfactant-assisted synthesis of a mesoporous form of zirconium phosphate with acidic properties. *Adv. Mater.* 10 (10): 812–815.
173 Herbst, A. and Janiak, C. (2016). Selective glucose conversion to 5-hydroxymethylfurfural (5-HMF) instead of levulinic acid with MIL-101Cr MOF-derivatives. *New J. Chem.* 40 (9): 7958–7967.

174 Chatterjee, A., Hu, X., and Lam, F.L.-Y. (2018). Towards a recyclable MOF catalyst for efficient production of furfural. *Catal. Today* 314: 129–136.
175 Irfan, M., Glasnov, T.N., and Kappe, C.O. (2011). Heterogeneous catalytic hydrogenation reactions in continuous-flow reactors. *ChemSusChem* 4 (3): 300–316.
176 Besson, M., Gallezot, P., and Pinel, C. (2014). Conversion of biomass into chemicals over metal catalysts. *Chem. Rev.* 114 (3): 1827–1870.
177 Huber, G.W., Cortright, R.D., and Dumesic, J.A. (2004). Renewable alkanes by aqueous-phase reforming of biomass-derived oxygenates. *Angew. Chem.* 116 (12): 1575–1577.
178 Gao, J., Zhang, X., Yang, Y. et al. (2013). Surface-passivated SBA-15-supported gold nanoparticles: highly improved catalytic activity and selectivity toward hydrophobic substrates. *Chem. Asian J.* 8 (5): 934–938.
179 Zhang, Q., Jiang, T., Li, B. et al. (2012). Highly selective sorbitol hydrogenolysis to liquid alkanes over Ni/HZSM-5 catalysts modified with pure silica MCM-41. *ChemCatChem* 4 (8): 1084–1087.
180 Chen, X., Jiang, Z.-H., Chen, S., and Qin, W. (2010). Microbial and bioconversion production of D-xylitol and its detection and application. *Int. J. Biol. Sci* 6 (7): 834–844.
181 Wisniak, J., Hershkowitz, M., and Stein, S. (1974). Hydrogenation of xylose over platinum group catalysts. *Ind. Eng. Chem. Prod. Res. Dev.* 13 (4): 232–236.
182 da Silva, S.S. and Chandel, A.K. (2012). *D-Xylitol: Fermentative Production, Application and Commercialization*, 2012e. New York, NY: Springer Group, Berlin.
183 Zhang, H., Haba, M., Okumura, M. et al. (2013). Novel formation of Ag/Au bimetallic nanoparticles by physical mixture of monometallic nanoparticles in dispersions and their application to catalysts for aerobic glucose oxidation. *Langmuir* 29 (33): 10330–10339.
184 Vilcocq, L., Cabiac, A., Especel, C. et al. (2013). Transformation of sorbitol to biofuels by heterogeneous catalysis: chemical and industrial considerations. *Oil Gas Sci. Technol. – Rev. IFP Energies nouvelles* 68 (5): 841–860.
185 Mishra, D.K., Dabbawala, A.A., and Hwang, J.-S. (2013). Ruthenium nanoparticles supported on zeolite Y as an efficient catalyst for selective hydrogenation of xylose to xylitol. *J. Mol. Catal. A: Chem.* 376: 63–70.
186 Kusserow, B., Schimpf, S., and Claus, P. (2003). Hydrogenation of glucose to sorbitol over nickel and Ruthenium catalysts. *Adv. Synth. Catal.* 345 (1–2): 289–299.
187 Dincer, I. (2012). Green methods for hydrogen production. *Int. J. Hydrogen Energy* 37 (2): 1954–1971.
188 Vaidya, P.D. and Lopez-Sanchez, J.A. (2017). Review of hydrogen production by catalytic aqueous-phase reforming. *ChemistrySelect* 2 (22): 6563–6576.
189 Davda, R.R., Shabaker, J.W., Huber, G.W. et al. (2005). A review of catalytic issues and process conditions for renewable hydrogen and alkanes by aqueous-phase reforming of oxygenated hydrocarbons over supported metal catalysts. *Appl. Catal., B* 56 (1): 171–186.

190 Haryanto, A., Fernando, S., Murali, N., and Adhikari, S. (2005). Current status of hydrogen production techniques by steam reforming of ethanol: a review. *Energy Fuels* 19 (5): 2098–2106.
191 Czernik, S., French, R., Feik, C., and Chornet, E. (2002). Hydrogen by catalytic steam reforming of liquid byproducts from biomass thermoconversion processes. *Ind. Eng. Chem. Res.* 41 (17): 4209–4215.
192 Saxena, R.C., Seal, D., Kumar, S., and Goyal, H.B. (2008). Thermo-chemical routes for hydrogen rich gas from biomass: a review. *Renewable Sustainable Energy Rev.* 12 (7): 1909–1927.
193 Wang, M., Wang, Z., Gong, X., and Guo, Z. (2014). The intensification technologies to water electrolysis for hydrogen production – a review. *Renewable Sustainable Energy Rev.* 29: 573–588.
194 Carmo, M., Fritz, D.L., Mergel, J., and Stolten, D. (2013). A comprehensive review on PEM water electrolysis. *Int. J. Hydrogen Energy* 38 (12): 4901–4934.
195 Ni, M., Leung, M.K.H., Leung, D.Y.C., and Sumathy, K. (2007). A review and recent developments in photocatalytic water-splitting using TiO_2 for hydrogen production. *Renewable Sustainable Energy Rev.* 11 (3): 401–425.
196 Chen, X., Shen, S., Guo, L., and Mao, S.S. (2010). Semiconductor-based photocatalytic hydrogen generation. *Chem. Rev.* 110 (11): 6503–6570.
197 Gallezot, P., Cerino, P.J., Blanc, B. et al. (1994). Glucose hydrogenation on promoted raney-nickel catalysts. *J. Catal.* 146 (1): 93–102.
198 Li, H., Wang, W., and Fa Deng, J. (2000). Glucose hydrogenation to sorbitol over a skeletal Ni–P amorphous alloy catalyst (Raney Ni–P). *J. Catal.* 191 (1): 257–260.
199 Li, H., Li, H., and Deng, J.-F. (2002). Glucose hydrogenation over Ni–B/SiO_2 amorphous alloy catalyst and the promoting effect of metal dopants. *Catal. Today* 74 (1–2): 53–63.
200 Schimpf, S., Louis, C., and Claus, P. (2007). Ni/SiO_2 catalysts prepared with ethylenediamine nickel precursors: influence of the pretreatment on the catalytic properties in glucose hydrogenation. *Appl. Catal., A* 318: 45–53.
201 Zhang, J., Wu, S., Liu, Y., and Li, B. (2013). Hydrogenation of glucose over reduced Ni/Cu/Al hydrotalcite precursors. *Catal. Commun.* 35: 23–26.
202 Zhang, J., Xu, S., Wu, S., and Liu, Y. (2013). Hydrogenation of fructose over magnetic catalyst derived from hydrotalcite precursor. *Chem. Eng. Sci.* 99: 171–176.
203 Michel, C. and Gallezot, P. (2015). Why is ruthenium an efficient catalyst for the aqueous-phase hydrogenation of biosourced carbonyl compounds? *ACS Catal.* 5 (7): 4130–4132.
204 Hoffer, B.W., Crezee, E., Mooijman, P.R.M. et al. (2003). Carbon supported Ru catalysts as promising alternative for Raney-type Ni in the selective hydrogenation of D-glucose. *Catal. Today* 79–80: 35–41.
205 Liu, J., Bai, P., and Zhao, X.S. (2011). Ruthenium nanoparticles embedded in mesoporous carbon microfibers: preparation, characterization and catalytic properties in the hydrogenation of d-glucose. *Phys. Chem. Chem. Phys.* 13 (9): 3758–3763.

206 Aho, A., Roggan, S., Simakova, O.A. et al. (2015). Structure sensitivity in catalytic hydrogenation of glucose over ruthenium. *Catal. Today* 241 (Part B): 195–199.

207 Simakova, I.L., Demidova, Y.S., Murzina, E.V. et al. (2016). Structure sensitivity in catalytic hydrogenation of galactose and arabinose over Ru/C catalysts. *Catal. Lett.* 146 (7): 1291–1299.

208 Crezee, E., Hoffer, B.W., Berger, R.J. et al. (2003). Three-phase hydrogenation of D-glucose over a carbon supported ruthenium catalyst—mass transfer and kinetics. *Appl. Catal., A* 251 (1): 1–17.

209 Mishra, D.K., Dabbawala, A.A., Park, J.J. et al. (2014). Selective hydrogenation of D-glucose to D-sorbitol over HY zeolite supported ruthenium nanoparticles catalysts. *Catal. Today* 232: 99–107.

210 Guo, X., Wang, X., Guan, J. et al. (2014). Selective hydrogenation of D-glucose to D-sorbitol over Ru/ZSM-5 catalysts. *Chin. J. Catal.* 35 (5): 733–740.

211 Jae, J., Zheng, W., Karim, A.M. et al. (2014). The role of Ru and RuO_2 in the catalytic transfer hydrogenation of 5-hydroxymethylfurfural for the production of 2,5-dimethylfuran. *ChemCatChem* 6 (3): 848–856.

212 Tan, J., Cui, J., Deng, T. et al. (2015). Water-promoted hydrogenation of levulinic acid to γ-valerolactone on supported ruthenium catalyst. *ChemCatChem* 7 (3): 508–512.

213 Jenness, G.R. and Vlachos, D.G. (2015). DFT study of the conversion of furfuryl alcohol to 2-methylfuran on RuO_2 (110). *J. Phys. Chem. C* 119 (11): 5938–5945.

214 Cortese, R., Duca, D., Sifontes Herrera, V.A., and Murzin, D.Y. (2012). l-Arabinose conformers adsorption on ruthenium surfaces: a DFT study. *J. Phys. Chem. C* 116 (28): 14908–14916.

215 Wan, H., Vitter, A., Chaudhari, R.V., and Subramaniam, B. (2014). Kinetic investigations of unusual solvent effects during Ru/C catalyzed hydrogenation of model oxygenates. *J. Catal.* 309: 174–184.

216 Feibelman, P.J. (2002). Partial dissociation of water on Ru (0001). *Science* 295 (5552): 99–102.

217 Zhang, X., Durndell, L.J., Isaacs, M.A. et al. (2016). Platinum-catalyzed aqueous-phase hydrogenation of D-glucose to D-sorbitol. *ACS Catal.* 6 (11): 7409–7417.

218 Durndell, L.J., Parlett, C.M., Hondow, N.S. et al. (2015). Selectivity control in Pt-catalyzed cinnamaldehyde hydrogenation. *Sci. Rep.* 5: 9425.

219 Durndell, L.J., Wilson, K., and Lee, A.F. (2015). Platinum-catalysed cinnamaldehyde hydrogenation in continuous flow. *RSC Adv.* 5 (97): 80022–80026.

220 Zheng, R., Porosoff, M.D., Weiner, J.L. et al. (2012). Controlling hydrogenation of C=O and C=C bonds in cinnamaldehyde using silica supported Co–Pt and Cu–Pt bimetallic catalysts. *Appl. Catal., A* 419–420: 126–132.

221 Delbecq, F. and Sautet, P. (1995). Competitive C=C and C=O adsorption of α-β-unsaturated aldehydes on Pt and Pd surfaces in relation with the selectivity of hydrogenation reactions: a theoretical approach. *J. Catal.* 152 (2): 217–236.

222 Perrard, A., Gallezot, P., Joly, J.-P. et al. (2007). Highly efficient metal catalysts supported on activated carbon cloths: a catalytic application for the hydrogenation of D-glucose to D-sorbitol. *Appl. Catal., A* 331: 100–104.

223 Ahmed, M. (2012). Kinetics studies of D-glucose hydrogenation over activated charcoal supported platinum catalyst. *Heat Mass Transfer* 48 (2): 343–347.

224 Ahmed, M.J. and Kadhum, A.A.H. (2011). Hydrogenation of D-fructose over activated charcoal supported platinum catalyst. *J. Taiwan Inst. Chem. Eng.* 42 (1): 114–119.

225 Tathod, A., Kane, T., Sanil, E.S., and Dhepe, P.L. (2014). Solid base supported metal catalysts for the oxidation and hydrogenation of sugars. *J. Mol. Catal. A: Chem.* 388–389: 90–99.

226 Toukoniitty, B., Kuusisto, J., Mikkola, J.-P. et al. (2005). Effect of ultrasound on catalytic hydrogenation of D-fructose to D-mannitol. *Ind. Eng. Chem. Res.* 44 (25): 9370–9375.

227 Kuusisto, J., Mikkola, J.-P., Casal, P.P. et al. (2005). Kinetics of the catalytic hydrogenation of D-fructose over a CuO–ZnO catalyst. *Chem. Eng. J.* 115 (1–2): 93–102.

228 Sinfelt, J.H. (1970). Catalytic hydrogenolysis over supported metals. *Catal. Rev.* 3 (1): 175–205.

229 ten Dam, J. and Hanefeld, U. (2011). Renewable chemicals: dehydroxylation of glycerol and polyols. *ChemSusChem* 4 (8): 1017–1034.

230 Vilcocq, L., Cabiac, A., Especel, C. et al. (2014). New insights into the mechanism of sorbitol transformation over an original bifunctional catalytic system. *J. Catal.* 320: 16–25.

231 Lazaridis, P.A., Karakoulia, S., Delimitis, A. et al. (2015). D-Glucose hydrogenation/hydrogenolysis reactions on noble metal (Ru, Pt)/activated carbon supported catalysts. *Catal. Today* 257 (Part 2): 281–290.

232 Liu, C., Zhang, C., Hao, S. et al. (2016). WO_x modified Cu/Al_2O_3 as a high-performance catalyst for the hydrogenolysis of glucose to 1,2-propanediol. *Catal. Today* 261: 116–127.

233 Liu, C., Zhang, C., Liu, K. et al. (2015). Aqueous-phase hydrogenolysis of glucose to value-added chemicals and biofuels: a comparative study of active metals. *Biomass Bioenergy* 72: 189–199.

234 Gallezot, P., Nicolaus, N., Flèche, G. et al. (1998). Glucose hydrogenation on ruthenium catalysts in a trickle-bed reactor. *J. Catal.* 180 (1): 51–55.

235 Dar, B.A., Khalid, S., Wani, T.A. et al. (2015). Ceria-based mixed oxide supported CuO: an efficient heterogeneous catalyst for conversion of cellulose to sorbitol. *Green Sustainable Chem.* 5 (01): 15.

236 Zhu, S., Gao, X., Zhu, Y., and Li, Y. (2016). Tailored mesoporous copper/ceria catalysts for the selective hydrogenolysis of biomass-derived glycerol and sugar alcohols. *Green Chem.* 18 (3): 782–791.

237 Jin, X., Shen, J., Yan, W. et al. (2015). Sorbitol hydrogenolysis over hybrid $Cu/CaO–Al_2O_3$ catalysts: tunable activity and selectivity with solid base incorporation. *ACS Catal.* 5 (11): 6545–6558.

238 Wang, X., Wu, F., Yao, S. et al. (2012). Ni–Cu/ZnO-catalyzed hydrogenolysis of cellulose for the production of 1,2-alkanediols in hot compressed water. *Chem. Lett.* 41 (5): 476–478.

239 Tajvidi, K., Hausoul Peter, J.C., and Palkovits, R. (2014). Hydrogenolysis of cellulose over Cu-based catalysts—analysis of the reaction network. *ChemSusChem* 7 (5): 1311–1317.
240 Chopade S.P., Miller D.J., Jackson J.E., Werpy T.A., Frye Jr, J.G., Zacher A.H. (2001) Catalysts and Process for Hydrogenolysis of Sugar Alcohols to Polyols. US Patent US6291725B1.
241 Sohounloue, D.K., Montassier, C., and Barbier, J. (1983). Catalytic hydrogenolysis of sorbitol. *React. Kinet. Catal. Lett.* 22 (3): 391–397.
242 Pham, T.N., Sooknoi, T., Crossley, S.P., and Resasco, D.E. (2013). Ketonization of carboxylic acids: mechanisms, catalysts, and implications for biomass conversion. *ACS Catal.* 3 (11): 2456–2473.
243 Lewis Jennifer, D., Van de Vyver, S., and Román-Leshkov, Y. (2015). Acid–base pairs in Lewis acidic zeolites promote direct aldol reactions by soft enolization. *Angew. Chem. Int. Ed.* 54 (34): 9835–9838.
244 Jahangiri, H., Osatiashtiani, A., Bennett, J.A. et al. (2018). Zirconia catalysed acetic acid ketonisation for pre-treatment of biomass fast pyrolysis vapours. *Catal. Sci. Technol.* 8 (4): 1134–1141.
245 Gumel, A.M., Annuar, M.S.M., Heidelberg, T., and Chisti, Y. (2011). Lipase mediated synthesis of sugar fatty acid esters. *Process Biochem.* 46 (11): 2079–2090.
246 Deshpande, P.S., Deshpande, T.D., Kulkarni, R.D., and Mahulikar, P.P. (2013). Synthesis of sucrose–coconut fatty acids esters: reaction kinetics and rheological analysis. *Ind. Eng. Chem. Res.* 52 (43): 15024–15033.
247 Pérez, B., Anankanbil, S., and Guo, Z. (2017). Synthesis of sugar fatty acid esters and their industrial utilizations A2, Chapter 10. In: *Fatty Acids* (ed. M.U. Ahmad), 329–354. AOCS Press https://doi.org/10.1016/B978-0-12-809521-8.00010-6.
248 Baker, I.J.A., Matthews, B., Suares, H. et al. (2000). Sugar fatty acid ester surfactants: structure and ultimate aerobic biodegradability. *J. Surfactants Deterg.* 3 (1): 1–11.
249 de Clippel, F., Dusselier, M., Van Rompaey, R. et al. (2012). Fast and selective sugar conversion to alkyl lactate and lactic acid with bifunctional carbon–silica catalysts. *J. Am. Chem. Soc.* 134 (24): 10089–10101.
250 Johnstone, R.A., Wilby, A.H., and Entwistle, I.D. (1985). Heterogeneous catalytic transfer hydrogenation and its relation to other methods for reduction of organic compounds. *Chem. Rev.* 85 (2): 129–170.
251 Bulushev, D.A. and Ross, J.R.H. (2011). Vapour phase hydrogenation of olefins by formic acid over a Pd/C catalyst. *Catal. Today* 163 (1): 42–46.
252 Tuteja, J., Choudhary, H., Nishimura, S., and Ebitani, K. (2013). Direct synthesis of 1,6-hexanediol from HMF over a heterogeneous Pd/ZrP catalyst using formic acid as hydrogen source. *ChemSusChem* 7 (1): 96–100.
253 Thananatthanachon, T. and Rauchfuss, T..B. (2010). Efficient production of the liquid fuel 2,5-dimethylfuran from fructose using formic acid as a reagent. *Angew. Chem.* 122 (37): 6766–6768.
254 Alonso, F., Riente, P., and Yus, M. (2011). Nickel nanoparticles in hydrogen transfer reactions. *Acc. Chem. Res.* 44 (5): 379–391.

255 Scholz, D., Aellig, C., and Hermans, I. (2013). Catalytic transfer hydrogenation/hydrogenolysis for reductive upgrading of furfural and 5-(hydroxymethyl)furfural. *ChemSusChem* 7 (1): 268–275.

256 Enthaler, S., von Langermann, J., and Schmidt, T. (2010). Carbon dioxide and formic acid—the couple for environmental-friendly hydrogen storage? *Energy Environ. Sci.* 3 (9): 1207–1217.

257 Hernandez, B., Iglesias, J., Morales, G. et al. (2016). One-pot cascade transformation of xylose into [gamma]-valerolactone (GVL) over bifunctional Bronsted-Lewis Zr–Al-beta zeolite. *Green Chem.* 18 (21): 5777–5781.

258 Melero, J.A., Morales, G., Iglesias, J. et al. (2017). Efficient one-pot production of [gamma]-valerolactone from xylose over Zr–Al-beta zeolite: rational optimization of catalyst synthesis and reaction conditions. *Green Chem.* 19 (21): 5114–5121.

259 López-Aguado, C., Paniagua, M., Iglesias, J. et al. (2018). Zr-USY zeolite: efficient catalyst for the transformation of xylose into bio-products. *Catal. Today* 304: 80–88.

260 Li, H., Fang, Z., and Yang, S. (2016). Direct conversion of sugars and ethyl levulinate into γ-valerolactone with superparamagnetic acid–base bifunctional $ZrFeO_x$ nanocatalysts. *ACS Sustainable Chem. Eng.* 4 (1): 236–246.

261 Huber, G.W., Chheda, J.N., Barrett, C.J., and Dumesic, J.A. (2005). Production of liquid alkanes by aqueous-phase processing of biomass-derived carbohydrates. *Science* 308 (5727): 1446–1450.

262 Román-Leshkov, Y., Chheda, J.N., and Dumesic, J.A. (2006). Phase modifiers promote efficient production of hydroxymethylfurfural from fructose. *Science* 312 (5782): 1933–1937.

263 Anastas, P. and Eghbali, N. (2010). Green chemistry: principles and practice. *Chem. Soc. Rev.* 39 (1): 301–312.

264 Sheldon, R.A. (2005). Green solvents for sustainable organic synthesis: state of the art. *Green Chem.* 7 (5): 267–278.

265 Prat, D., Hayler, J., and Wells, A. (2014). A survey of solvent selection guides. *Green Chem.* 16 (10): 4546–4551.

266 Welton, T. (2015). Solvents and sustainable chemistry. *Proc. R. Soc. London, Ser. A* 471 (2183): 20150502.

267 Titirici, M.-M., White, R.J., Brun, N. et al. (2015). Sustainable carbon materials. *Chem. Soc. Rev.* 44 (1): 250–290.

268 Mishra, D.K., Lee, J.-M., Chang, J.-S., and Hwang, J.-S. (2012). Liquid phase hydrogenation of D-glucose to D-sorbitol over the catalyst (Ru/NiO–TiO_2) of ruthenium on a NiO-modified TiO_2 support. *Catal. Today* 185 (1): 104–108.

269 Lou, L.-L., Du, H., Shen, Y. et al. (2014). Chiral Ir(I) complex supported on external surface passivated SBA-15 as heterogeneous catalyst for asymmetric transfer hydrogenation of aromatic ketones. *Microporous Mesoporous Mater.* 187: 94–99.

270 Dabbawala, A.A., Mishra, D.K., Huber, G.W., and Hwang, J.-S. (2015). Role of acid sites and selectivity correlation in solvent free liquid phase dehydration of sorbitol to isosorbide. *Appl. Catal., A* 492: 252–261.

271 Nolan M.R. (2014). Selective dehydration of polyols over solid acids and metal-acid bifunctional catalysts: Towards a catalytic toolbox for rational

catalyst design. Graduate theses and dissertations, Iowa State University, Ames, Iowa.

272 Zhao, S., Cheng, M., Li, J. et al. (2011). One pot production of 5-hydroxymethylfurfural with high yield from cellulose by a Bronsted–Lewis-surfactant-combined heteropolyacid catalyst. *Chem. Commun.* 47 (7): 2176–2178.

273 Richter, F.H., Pupovac, K., Palkovits, R., and Schüth, F. (2013). Set of acidic resin catalysts to correlate structure and reactivity in fructose conversion to 5-hydroxymethylfurfural. *ACS Catal.* 3 (2): 123–127.

274 Zang, W., Li, G., Wang, L., and Zhang, X. (2015). Catalytic hydrogenation by noble-metal nanocrystals with well-defined facets: a review. *Catal. Sci. Technol.* 5 (5): 2532–2553.

275 Pham, T.C.T., Kim, H.S., and Yoon, K.B. (2011). Growth of uniformly oriented silica MFI and BEA zeolite films on substrates. *Science* 334 (6062): 1533–1538.

276 Parlett, C.M.A., Wilson, K., and Lee, A.F. (2013). Hierarchical porous materials: catalytic applications. *Chem. Soc. Rev.* 42 (9): 3876–3893.

277 Matsubu, J.C., Zhang, S., DeRita, L. et al. (2017). Adsorbate-mediated strong metal–support interactions in oxide-supported Rh catalysts. *Nat. Chem.* 9 (2): 120.

278 Mao, H., Peng, S., Yu, H. et al. (2014). Facile synthesis of highly stable heterogeneous catalysts by entrapping metal nanoparticles within mesoporous carbon. *J. Mater. Chem. A* 2 (16): 5847–5851.

279 Joo, S.H., Park, J.Y., Tsung, C.-K. et al. (2009). Thermally stable Pt/mesoporous silica core–shell nanocatalysts for high-temperature reactions. *Nat. Mater.* 8 (2): 126–131.

280 O'Neill, B.J., Jackson, D.H.K., Lee, J. et al. (2015). Catalyst design with atomic layer deposition. *ACS Catal.* 5 (3): 1804–1825.

281 Christensen, S.T., Feng, H., Libera, J.L. et al. (2010). Supported Ru–Pt bimetallic nanoparticle catalysts prepared by atomic layer deposition. *Nano Lett.* 10 (8): 3047–3051.

282 Libera, J.A., Elam, J.W., and Pellin, M.J. (2008). Conformal ZnO coatings on high surface area silica gel using atomic layer deposition. *Thin Solid Films* 516 (18): 6158–6166.

283 George, S.M. (2010). Atomic layer deposition: an overview. *Chem. Rev.* 110 (1): 111–131.

284 Yang, X.-F., Wang, A., Qiao, B. et al. (2013). Single-atom catalysts: a new Frontier in heterogeneous catalysis. *Acc. Chem. Res.* 46 (8): 1740–1748.

285 Thomas, J.M., Raja, R., and Lewis, D.W. (2005). Single-site heterogeneous catalysts. *Angew. Chem. Int. Ed.* 44 (40): 6456–6482.

286 Sun, S., Zhang, G., Gauquelin, N. et al. (2013). Single-atom catalysis using Pt/graphene achieved through atomic layer deposition. *Sci. Rep.* 3: 1775.

287 Yan, H., Cheng, H., Yi, H. et al. (2015). Single-atom Pd1/graphene catalyst achieved by atomic layer deposition: remarkable performance in selective hydrogenation of 1,3-butadiene. *J. Am. Chem. Soc.* 137 (33): 10484–10487.

288 Jagadeesh, R.V., Surkus, A.-E., Junge, H. et al. (2013). Nanoscale Fe_2O_3-based catalysts for selective hydrogenation of nitroarenes to anilines. *Science* 342 (6162): 1073–1076.

289 Jagadeesh, R.V., Natte, K., Junge, H., and Beller, M. (2015). Nitrogen-doped graphene-activated iron-oxide-based nanocatalysts for selective transfer hydrogenation of nitroarenes. *ACS Catal.* 5 (3): 1526–1529.

290 Déchamp, N., Gamez, A., Perrard, A., and Gallezot, P. (1995). Kinetics of glucose hydrogenation in a trickle-bed reactor. *Catal. Today* 24 (1–2): 29–34.

291 Sun, P., Long, X., He, H. et al. (2013). Conversion of cellulose into isosorbide over bifunctional ruthenium nanoparticles supported on niobium phosphate. *ChemSusChem* 6 (11): 2190–2197.

292 Xi, J., Zhang, Y., Ding, D. et al. (2014). Catalytic production of isosorbide from cellulose over mesoporous niobium phosphate-based heterogeneous catalysts via a sequential process. *Appl. Catal., A* 469: 108–115.

293 Op de Beeck, B., Geboers, J., Van de Vyver, S. et al. (2013). Conversion of (ligno)cellulose feeds to isosorbide with heteropoly acids and Ru on carbon. *ChemSusChem* 6 (1): 199–208.

294 Yamaguchi, A., Sato, O., Mimura, N., and Shirai, M. (2015). One-pot conversion of cellulose to isosorbide using supported metal catalysts and ion-exchange resin. *Catal. Commun.* 67: 59–63.

295 Nicolaou, K.C., Edmonds, D.J., and Bulger, P.G. (2006). Cascade reactions in total synthesis. *Angew. Chem. Int. Ed.* 45 (43): 7134–7186.

6

Chemistry of C—C Bond Formation Reactions Used in Biomass Upgrading: Reaction Mechanisms, Site Requirements, and Catalytic Materials

Tuong V. Bui, Nhung Duong, Felipe Anaya, Duong Ngo, Gap Warakunwit, and Daniel E. Resasco

University of Oklahoma, Chemical, Biological, and Materials Engineering Department, 100 East Boyd Street, Norman, OK 73019, USA

6.1 Introduction

Biomass, particularly lignocellulosic biomass, is a promising platform for the production of renewable fuels and chemicals. When life cycle analysis and differences in geographic location and extent of development – industrialized vs. developing countries – are considered in the global techno-economic analysis, significant advantages over other energy sources may be predicted. Yet, the process of biomass utilization has proven to be a complicated challenge. Different technological approaches have been investigated for biomass deconstruction; among them, thermal conversion is a techno-economically attractive option that can be scaled up and may allow for full utilization of the main components of lignocellulosic biomass. These components are biopolymers composed of distinctive building blocks that control their thermal stability and subsequent chemical reactivity. For example, hemicellulose is the most reactive component, comprising an amorphous and highly branched polymer of various sugar units that can be depolymerized very easily. In contrast, cellulose is a linear polymer of many glucose units, connected via acetal beta linkages. It contains crystalline parts that H-bond together into fibers that give cellulose its unique mechanical properties and chemical stability. On the other hand, lignin is a highly branched and interconnected polymer of phenyl propane units, which are interconnected by several types of C—O and C—C bond linkages.

The depolymerization of biomass involves the cleavage of a number of different bonds that produces a wide range of oxygenated monomers and oligomers that include partially dehydrated saccharides, furanic, and phenolic compounds with oxygen functionalities such as carboxyl, carbonyl, and methoxy groups, ketones, etc. These monomers and oligomers are exceedingly reactive and tend to agglomerate and repolymerize unless a catalyst is strategically interposed to guide the conversion in the direction of valuable compounds instead of amorphous humins. That is, an appropriate catalyst can selectively capitalize on the high reactivity imparted by the oxygenated groups (COOH, CO, Ar–O–C,

Chemical Catalysts for Biomass Upgrading, First Edition. Edited by Mark Crocker and Eduardo Santillan-Jimenez.

Ar–OH, etc.) to form products with molecular weights in the range of fuels and chemicals, as opposed to uncontrollably formed humins.

In this chapter, we explore the different C—C bond forming reactions that can be utilized to optimize the upgrading process and maximize the yield in the desirable molecular weight range for the specific production of gasoline, diesel, or chemicals. Some of these C–C coupling reactions include aldol condensation, alkylation, hydroxyalkylation, acylation, and ketonization. The goal of the chapter is to critically evaluate the different ideas presented in the literature concerning the reaction mechanisms for each of these reactions – structure/function relationships being analyzed – with a focus on the topological and chemical requirements of the active sites (acid, base, acid–base pairs, both of Brønsted and Lewis types) – along with some of the specific materials (oxides, zeolites) that are particularly effective for each task.

6.2 Mechanisms and Site Requirements of C–C Coupling Reactions

6.2.1 Aldol Condensation: Mechanism and Site Requirement

6.2.1.1 Base-Catalyzed Aldol Condensation

6.2.1.1.1 Mechanism The most widely accepted mechanism for base-catalyzed aldol condensation on solid surfaces includes three main elementary steps: (i) deprotonation; (ii) carbon–carbon bond formation, and (iii) reprotonation. First, the basic sites abstract the α-H from the adsorbed ketones/aldehydes to produce a surface-enolate specie (deprotonation or enolization step). The enolate then reacts with the carbonyl carbon of adjacent adsorbed ketones/aldehydes (Langmuir–Hinshelwood) or gas phase ones (Eley–Rideal) to generate an aldol adduct (C–C coupling step) whose O atom consecutively accepts the proton back-transferred from the surface (reprotonation step). The aldol product is then dehydrated to generate an α,β-unsaturated ketone (final aldol condensation product) (Scheme 6.1).

6.2.1.1.2 Site Requirement In principle, all surface oxide basic sites (O^{2-}, Me^{n+}–O^{2-} pairs, and OH^- sites) could be active for H-abstraction. Those sites are identified/quantified by the different modes of CO_2 adsorbed on the active sites by Fourier-transform infrared spectroscopy (FTIR). There are three adsorption modes with different frequencies that correspond to three types of basic sites: high-strength (O^{2-} sites), medium-strength (Me^{n+}–O^{2-} pairs), and low-strength (OH^- sites) (Figure 6.1) [3–8].

Ordóñez et al. [9] have suggested that the aldol condensation of furfural and acetone requires acid–base pair sites and a high density of strong basic sites. Faba et al. [10] have proposed that medium-strength basic sites are responsible for promoting the enolate formation, which is the rate determining step of the reaction. Frolich and coworkers [11] have studied hydrotalcite-based Zn(Mg)Al oxides for the gas phase aldol condensation of furfural and acetone, finding a

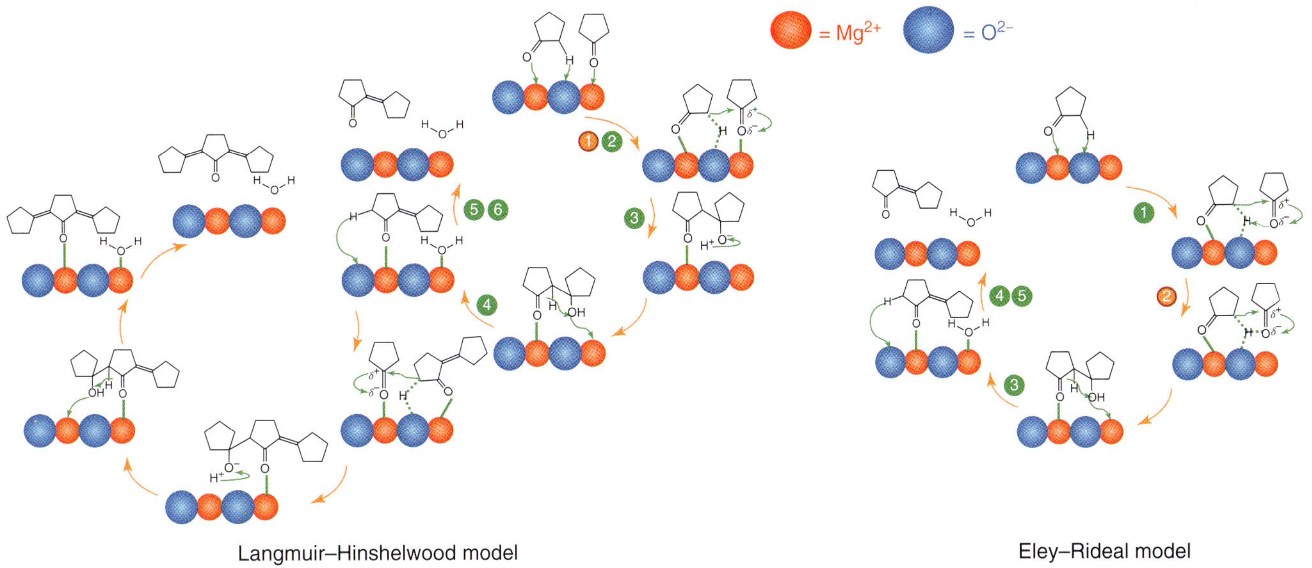

Scheme 6.1 Proposed mechanism for the self-condensation of cyclopentanone on MgO [1, 2].

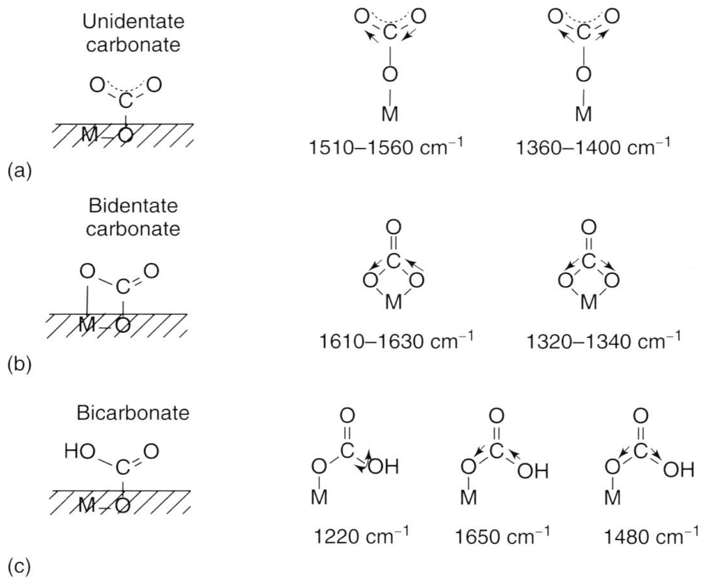

Figure 6.1 IR bands of adsorbed CO_2 surface species. (a) High-strength basic site, (b) medium-strength basic site, and (c) low-strength basic site. Source: Di Cosimo et al. 1998 [6]. Reproduced with permission of Elsevier.

linear relationship between the condensation rate and the total amount of basic sites, which are dominated by the medium-strength sites (Mg^{2+}–O^{2-} pair) (and confirmed by CO_2 FTIR) (Figure 6.2). However, the authors have also stated that other types of active sites such as O^{2-} and OH^- might also be involved. In contrast, there is no apparent trend observed between acid density and catalytic activity (Figure 6.2), which would indicate that Lewis acid sites are not crucial for the reaction. However, the linear relationship between furfural conversion and basic/acid sites ratio suggests a combined contribution of both types of site in the measured rate. It is possible that acid sites could be required as the adsorption sites for stabilizing the enolate in the H-abstraction step. Nevertheless, the actual role of acid sites has not been clarified. Even so, one of the most obvious effects of the acid sites is the fast dehydration of the alcohol. A good correlation is observed between the increase in formation of α,β-unsaturated ketone with increasing acid density [3, 12].

Wang et al. [13] have proposed a novel method for identification/quantification of relevant active sites under aldol condensation reaction conditions by introducing different amounts of deactivation agents into the reaction system including CO_2 [14], 2,6-di-*tert*-butylpyrydine [15], pyridine [16], and propanoic acid [16]. The addition of pyridine and propanoic acid decreased the measured rate, which implies that weak Lewis acid sites and Ti^{2+}–O^{2-} pairs are the active sites (Figure 6.3a). The mechanism of aldol condensation on TiO_2 surfaces is very similar to that proposed on MgO (Scheme 6.2). That is, the weak Lewis acid Ti^{4+} cation helps to stabilize the enolate species generated by α-H abstraction on the adjacent lattice O^{2-} basic site.

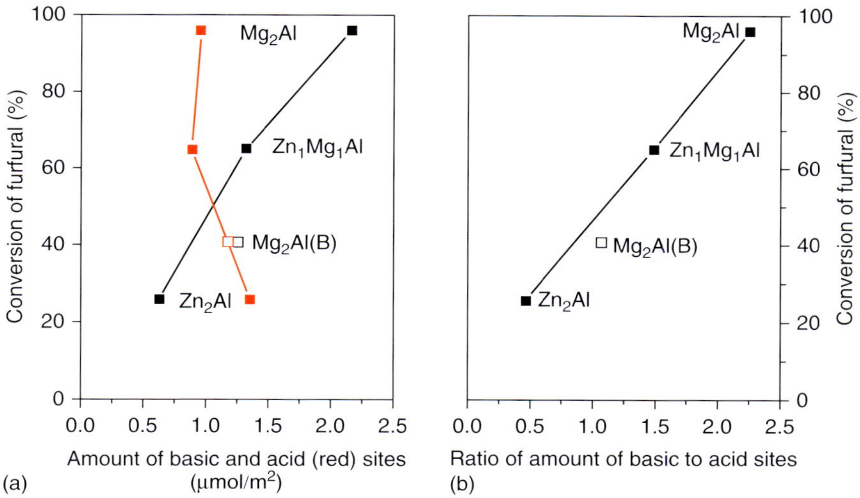

Figure 6.2 Dependence of the furfural conversion on the amount of basic or acid sites per square meter of catalyst (a) and on the ratio of amount of basic to acid sites (b). Zn/Mg/Al molar ratio of Zn_2Al, Zn_1Mg_1Al, Mg_2Al, and $Mg_2Al(B)$ are 1.9/0/1, 0.5/0.7/1, 0/1.9/1, and 0/1.9/1, respectively. Sample Mg_2Al and $Mg_2Al(B)$ were synthesized with different concentrations of M^{2+} (Mg^{2+} and Zn^{2+}) and Al^{3+} in the synthesis solution (0.15 and 1 M for Mg_2Al and $Mg_2Al(B)$, respectively). Source: Smoláková et al. 2017 [11]. Reproduced with permission of American Chemical Society.

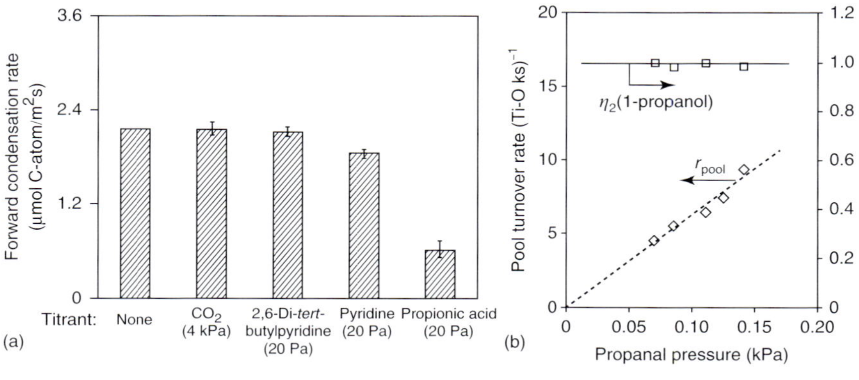

Figure 6.3 (a) Effect of different titrants on areal forward acetone condensation rate on TiO_2 (P25) + 20 wt% Cu/SiO_2 (1 : 1 mass) at 523 K. (b) Effect of propanol pressure on turnover rates for the conversion of 1-propanol-propanal equilibrated reactions TiO_2 (P25) + 20 wt% Cu/SiO_2 (1 : 1 mass) at 523 K. Dashed lines and solid line represent the regressed fits to the functional form of rate expression $r_{enolate}/[L]_{TiO_2}$ (presented in the main text) and a linear regression, respectively. Source: Wang et al. 2016 [13]. Reproduced with permission of Elsevier.

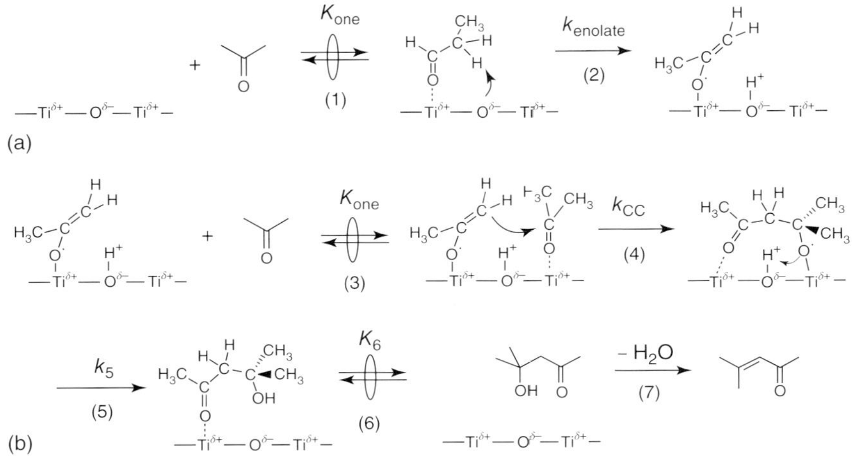

Scheme 6.2 Aldol condensation pathway of acetone on bifunctional TiO_2 catalysts; k_i and K_i are kinetic constant of the forward step and equilibrium constant for step *i*, respectively. (a) Enolate formaton form acetone. (b) Aldol condensation reaction between enolate and acetone. Source: Reproduced with permission from Wang et al. [13].

It has been observed that the consumption rate of 1-propanal (or 1-propanol/1-propanal pool) increases proportionally with 1-propanal pressure, indicating that deprotonation is a kinetically relevant step. This mechanism is consistent with the observed first-order rate dependency on 1-propanal pressure (Figure 6.3b) [13]. A similar concept has been applied for the self- and cross-aldol condensation of acetone and formaldehyde on TiO_2 [2]. If the rate-limiting step is H-abstraction, the total rate would be controlled by the formation of enolate, which should respond to a first-order rate expression with respect to the acetone pressure (Figure 6.4a). Since formaldehyde does not have an α-H, the enolate can only be formed by the deprotonation of acetone. Therefore, the total condensation rate remains constant with increasing HCHO pressure (when acetone pressure was kept constant), confirming the hypothesis (Figure 6.4b). On the other hand, the selectivity of cross-/self-products is determined by the difference in activation energy between cross- and self-coupling steps and the adsorption constants of HCHO and acetone on the TiO_2 surface [2]. From kinetic fitting of reported data, the activation energy of enolate formation on $Ti^{2+}–O^{2-}$ pairs is estimated to be ~124 kJ/mol. The barrier for C—H bond cleavage is higher for ketones compared with aldehydes and long-alkyl chain aldehydes compared with shorter ones, due to steric effects [2].

Wang et al. have also conducted DFT calculations to model this reaction using a $Ti_5O_{19}H_{18}$ cluster derived from the (101) plane of the TiO_2 anatase surface consisting of $Ti_{5c}–Ti_{6c}–O_{2c}–Ti_{5c}$ and $Ti_{5c}–O_{3c}–Ti_{5c}$ sites [2]. A shorter $Ti_{5c}—O_{2c}$ bond on the $Ti_{5c}–Ti_{6c}–O_{2c}–Ti_{5c}$ site compared with the $O_{3c}—Ti_{5c}$ bond on the $Ti_{5c}–O_{3c}–Ti_{5c}$ site leads to greater stability for the enolate and C–C coupling on $Ti_{5c}–Ti_{6c}–O_{2c}–Ti_{5c}$. Enolate formation is reported to be the kinetically relevant step on $Ti_{5c}–O_{3c}–Ti_{5c}$ sites. On the other hand, the long distance between

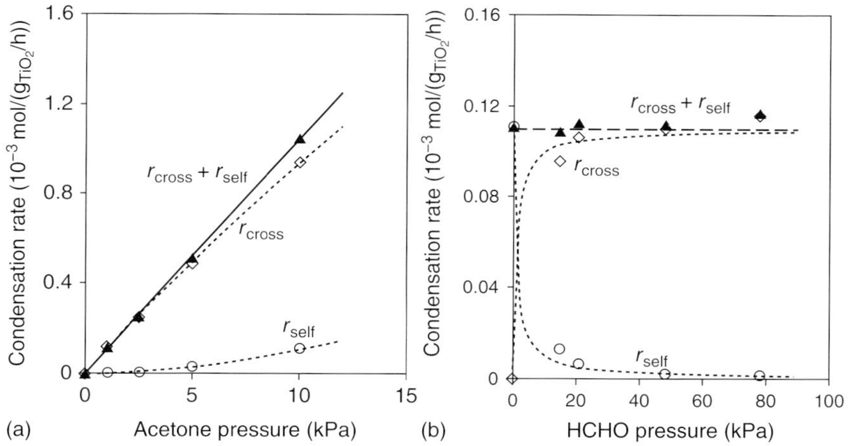

Figure 6.4 Self-condensation (r_{self}, ○) and cross-condensation (r_{cross}, ◇) rate of acetone–HCHO reactions and their sum (◆) as a function of (a) acetone and (b) HCHO pressure (TiO_2; 20% Ag/SiO_2; 1 : 8 mass ratio; 773 K, 10 kPa H_2, 0–10 kPa acetone, 5 kPa methanol for (a); 1 kPa acetone and 0–10 kPa methanol for (b); (<5% acetone conversion; HCHO pressure measued in effluent stream). Dashed lines are regression fits to the functional form of the rate expression. Source: Wang and Iglesia 2016 [2]. Reproduced with permission of American Chemical Society.

Ti_{5c}–O_{2c} and the second Ti_{5c} adsorption site on Ti_{5c}–Ti_{6c}–O_{2c}–Ti_{5c} increases the activation barrier for the reprotonation (129 kJ/mol), which makes it the rate-determining step. On the Ti_{5c}–O_{3c}–Ti_{5c} site, the activation energy for the reprotonation step on this surface is much lower (50 kJ/mol). Consequentially, the total barrier for Ti_{5c}–Ti_{6c}–O_{2c}–Ti_{5c} sites is 129 kJ/mol, which is 7 kJ/mol higher than that on Ti_{5c}–O_{3c}–Ti_{5c} sites. As a result, a higher aldol condensation rate was observed on Ti_{5c}–O_{3c}–Ti_{5c} sites, which are prevalent on the anatase surface. A longer Ti—O bond and a stronger strength of the acid–base pair of Ti_{5c}–Ti_{6c}–O_{2c}–Ti_{5c} on the rutile surface are responsible for the low activity for aldol condensation (10^{13}-fold lower rate than anatase in the studied conditions).

DFT calculations for aldol condensation of acetaldehyde on $Mg_{LC}{}^{2+}$–$O_{L'C}{}^{2-}$ (where LC and L′C stand for the coordination number of Mg^{2+} and O^{2-}, respectively) have shown that a low-coordinated O atom with strong basicity increases the barrier for the reprotonation step, which would prevent the alcohol adduct from being formed and desorbed [17]. Equally important, the adsorption of acetaldehyde on Mg^{2+}–O^{2-} pairs is determined by the coordination environment of Mg^{2+}. A stronger interaction with Mg^{2+} can stabilize the enolate transition state (TS) and reduce the barrier for deprotonation by pulling the acetaldehyde molecules closer to the basic site. However, it should be noticed that too strong bonding with Mg^{2+} would possibly require higher energy for desorption of the products, which could lead to fast deactivation. In conclusion, an optimum relative strength of Lewis acid M^{n+} and Brønsted basic O^{2-} sites is desirable to achieve the highest condensation rate (Figure 6.5). That is, the strength of basic O^{2-} sites and acidic Mg^{2+} sites should be sufficient to promote enolization but not so high as to impede the reprotonation/desorption step.

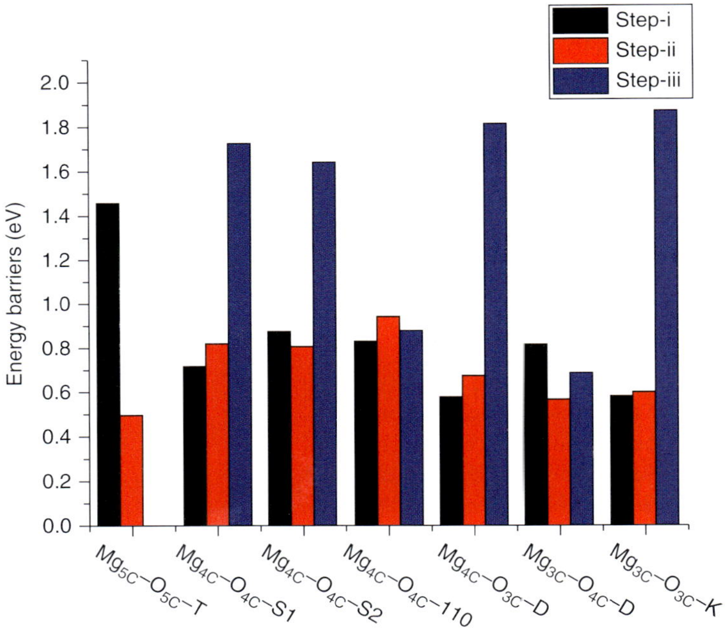

Figure 6.5 The energy barriers of Step i (deprotonation), Step ii (C–C coupling), Step iii (reprotonation) on Mg^{2+}–O^{2-} pairs with different coordination numbers. Source: Fan et al. 2017 [17]. Reproduced with permission of Royal Society of Chemistry.

6.2.1.2 Acid-Catalyzed Aldol Condensation: Mechanism and Site Requirement

6.2.1.2.1 Mechanism The acid-catalyzed aldol condensation of a wide range of ketones and aldehydes has been reported on zeolites [18–22], Amberlyst [23, 24], acid-functionalized mesoporous materials [25, 26], $CoAlPO_4$-5 [27], and MOF [28]. The mechanism for acid-catalyzed aldol condensation is initiated by simultaneous tautomerization and protonation. The former step can be activated by both Lewis and Brønsted acid sites, while the latter can only happen on Brønsted acids via proton transfer from active sites to the O atom of carbonyl groups. Herrmann and Iglesia [29] have proposed that the Brønsted acid sites on large pore zeolites (BEA, FAU, MCM41) are responsible for activating adsorbed acetone via tautomerization. The acetone is first adsorbed on the surface via the interaction between the O atom of the carbonyl group with the surface proton (Scheme 6.3). The adsorbed species then undergoes tautomerization with the concerted configuration on the Brønsted acid site. The concerted mechanism is in agreement with several previous studies [30, 31]. DFT calculations show that the proton does not completely transfer tc the adsorbed acetone but instead stabilizes the molecule via H-bonding, which agrees with a ^{13}C NMR study for

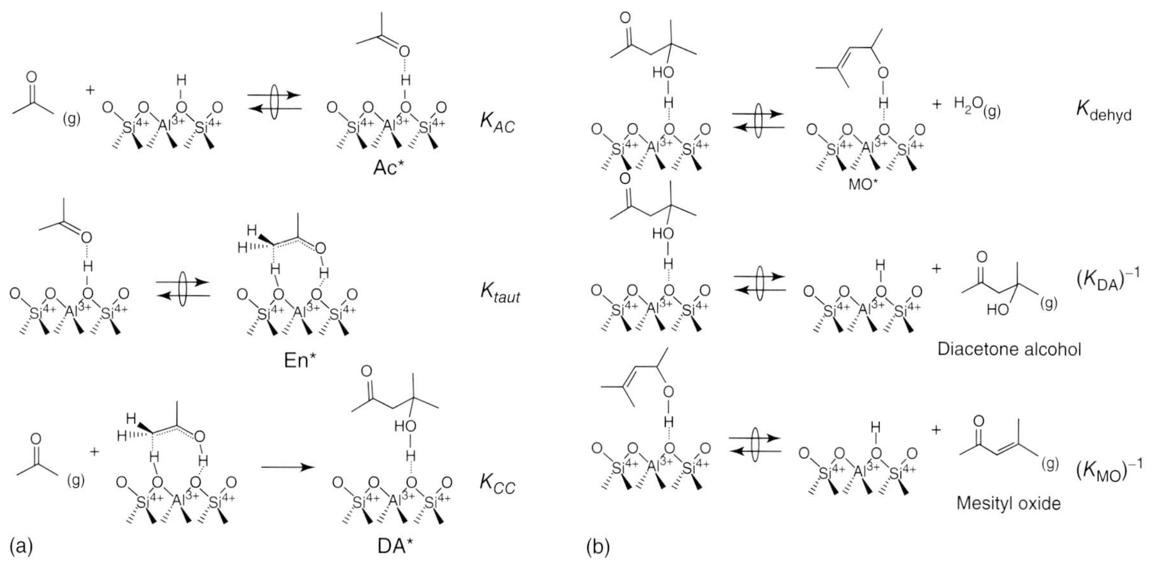

Scheme 6.3 Proposed mechanism of acid-catalyzed aldol condensation of acetone on a zeolite surface. Source: Reproduced with permission from Herrmann and Iglesia 2017 [29].

the adsorption of acetaldehyde on HZSM-5 [32, 33]. The former enol undergoes an electrophilic attack on the C atom of the gas phase acetone (Eley–Rideal mechanism) to produce an aldol adduct [29]. In the second step, the C—C bond is formed simultaneously with proton transfer in which the enol donates an α-proton to the O atom of the zeolite surface while accepting a proton from the Brønsted acid site to the O atom of the carbonyl group, as also suggested by Solans-Monfort et al. [31]. The diacetone alcohol is quickly dehydrated to produce mesityl oxide, which can be further isomerized to isomethyl oxide (Scheme 6.3). In contrast, other authors [30, 34–36] have reported that the most stable surface species is a silyl-ether complex, which is the protonated adsorbed acetone at the TS. The formed enol is bound to the Brønsted site via a weak π interaction [30]. The questions as to whether a Brønsted or a Lewis acid site is essential for the reaction and which step is kinetically relevant (tautomerization or C–C coupling) have been addressed repeatedly in the literature.

6.2.1.2.2 Site Requirement

Activation (Enolization) by Brønsted Sites Kikhtyanin et al. [37, 38] have proposed that the Brønsted acid sites on zeolites (HZSM-5, HBEA, HMOR, and HSDUSY) play a crucial role in aldol condensation of furfural/acetone. It has been shown that regardless of the zeolite crystal framework structure, the H-form of the catalysts are all active for the reaction [37], whereas the NH_4-form exhibits poor activity. Moreover, two H-BEA catalysts with similar Brønsted acidity but different Lewis acidity perform similarly, suggesting that the Brønsted acid sites are more important. The authors have also pointed out that the dimerization of two aldol adducts could only be seen on acid catalysts. This result has been used in several studies to distinguish the contribution of acid or basic sites in aldol condensation reactions [39].

Herrmann and Iglesia [29] have reported a similar value of rate constants calculated from the first-order rate expression for several MFI and BEA zeolites with different Si/Al ratios (MFI: 17–168, BEA: 12–43) and H^+/Al ratios (MFI: 0.65–1.03, BEA: 0.22–1.04) (Figure 6.6). There are three main points that back up their conclusion of an Eley–Rideal mechanism with C–C coupling as the kinetically relevant step: (i) the surface was nearly saturated with H-bonded acetone as confirmed by FTIR and DFT calculations under the reaction conditions; (ii) the acetone condensation rate was first order in the partial pressure of acetone in the gas phase; and (iii) no isotope effect was observed when acetone-d_0/H_2 vs. acetone-d_6/D_2 were used, indicating that the kinetically relevant step does not involve a proton-transfer process (i.e. formation/breaking of C—H/O—H bonds).

On the other hand, some DFT studies on the acetone/formaldehyde system using cluster models have shown that the tautomerization of acetone is the rate-limiting step on different zeolites: HZSM-5 [30, 31, 40], HY [40], H-FER [30], and HMCM-22 [30], which is in agreement with an experimental gas phase study [41].

Activation (Enolization) by Lewis Sites Dumitriu et al. [18] have found out that silicalite-1 is also active for aldol condensation of a mixture of acetaldehyde and formaldehyde. This material does not have any strong Brønsted acid sites, except

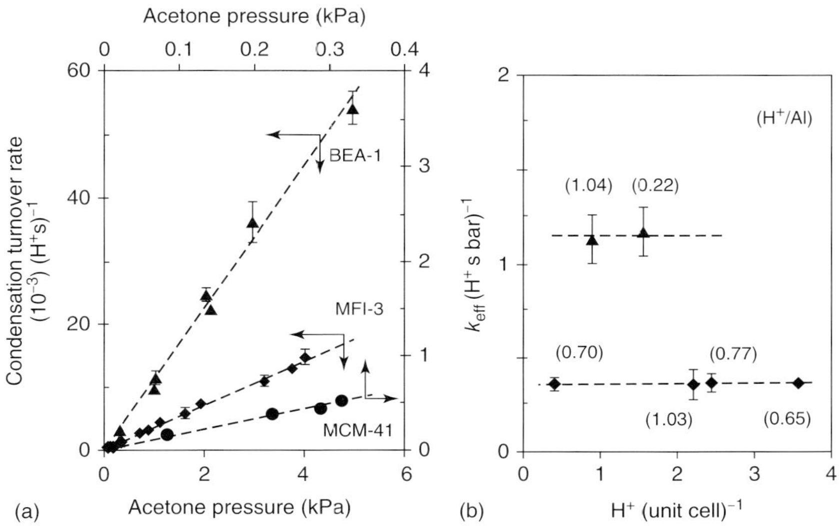

Figure 6.6 (a) Condensation turnover rate as a function of acetone pressure on Pt/BEA-1 + Pt/SiO_2 (1 wt% intracrystalline Pt and 1 wt% extracrystalline Pt; triangle), MFI-3 + Pt/SiO_2 (1 wt% extra-crystalline Pt; diamonds), and MCM-41 + Pt/SiO_2 (0.66 wt% extracrystalline Pt; circle) (473 K, 27 kPa H_2) (b) Effective first-order rate constants as a function of proton density (per unit cell) on Pt-containing samples of MFI and BEA samples with different H^+/Al ratios (in parentheses) at 473 K. Source: Herrmann and Iglesia 2017 [29]. Reproduced with permission of Elsevier.

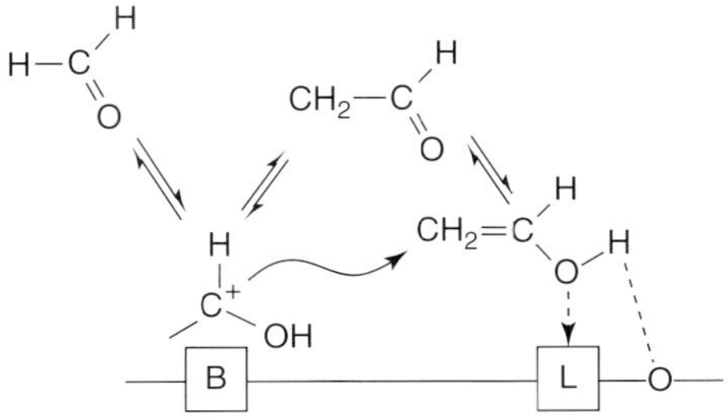

Scheme 6.4 Proposed mechanism of cooperative catalysis of Brønsted and Lewis acid sites. Source: Reproduced with permission from Dumitriu et al. 2001 [18].

for some weak acidity from terminal Si–OH groups. The authors have attributed the catalytic activity to the catalytic action of Lewis acid sites in combination with silanol groups, acting as the Brønsted acid sites (Scheme 6.4). The role of the Brønsted acid sites is to protonate the acetone to render the C of the carbonyl group electrophilic, while that of the Lewis acid is to facilitate tautomerization of the adsorbed ketone. However, the nature of the Lewis sites on this material was not mentioned. Biaglow et al. [42] have proposed that acetone condensation could be catalyzed by non-framework Al species. The cooperation effect of both types of acid sites has also been invoked to explain the high activity observed in heteroatom modified-MFI zeolites (Fe, Al, Ge, B), which have both Brønsted and

Lewis sites. The Lewis acid site can bind to acetone and enhance the electrophilicity of the carbonyl group's C atom, which is very similar to the protonation effect caused by Brønsted acid sites [42]. However, this hypothesis has not been tested experimentally.

Panov and Fripiat [43] have utilized FTIR to investigate the adsorption/condensation of acetone on alumina and USY, HY, and HZSM-5 zeolites at 105 °C. They observed that on alumina, which possesses only Lewis acid sites, the reaction occurred at a much faster rate than on zeolites (containing both Brønsted and Lewis acid sites). In contrast, an HZSM-5 zeolite sample without non-framework aluminum exhibited zero activity. A linear relationship between the initial rate and density of Lewis acid sites can be seen in Figure 6.7. In contrast, no clear trend was observed with respect to Brønsted site density, suggesting that the Lewis acid sites are kinetically relevant. From FTIR spectra, it was shown that acetone could adsorb on both Lewis and Brønsted sites characterized by different stretching vibration frequencies of the carbonyl group (1650–1690 cm^{-1} for Brønsted sites and 1690–1705 cm^{-1} for Lewis sites). This observation suggests a combined role for both types of active sites. In this manner, the Lewis sites would tautomerize the acetone while the Brønsted sites would adsorb the coreactant for the C–C coupling step. If this is the case, close proximity between Brønsted and Lewis sites would be a crucial requirement for the cooperative effect. The lack of adjacent sites could lead to a low condensation rate, even when there is a high density of Brønsted sites available. On alumina, the absence of Brønsted sites led to an Eley–Rideal mechanism, in which the tautomerized acetone picks up the coreactant from the gas phase. The C–C

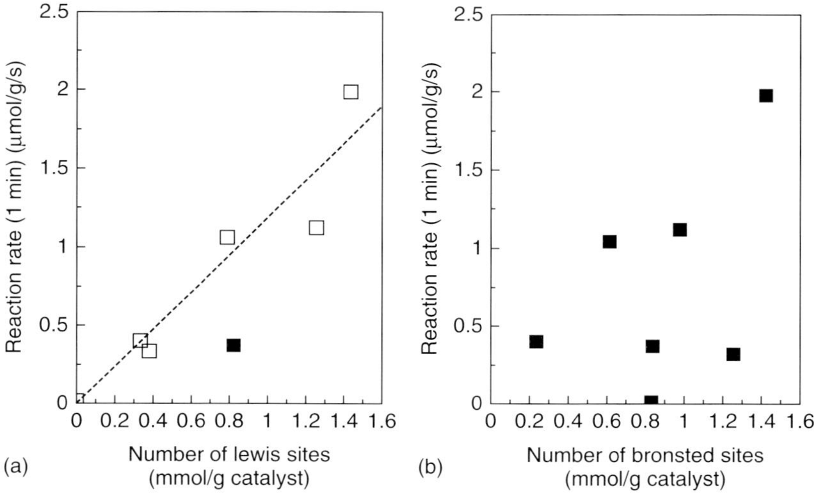

Figure 6.7 Initial reaction rate R_0 (after one minute of reaction) on zeolite vs. the number of Lewis sites (a) or Brønsted sites (b). The black square in (a) is the reaction rate observed for two experiments on *H*-mordenite, dealuminated at 700 °C. Source: Panov and Fripiat 1998 [43]. Reproduced with permission from of Elsevier.

coupling was proposed as the rate-limiting step without further explanation [43].

The activation of aldehydes by Lewis acid sites has also been reported in other catalysts such as $CoAlPO_4$-5 [27] and MOF [28]. Kikhtyanin et al. [28] have studied the role of Lewis acid sites in MOF materials for the aldol condensation of furfural and acetone. MOF catalysts investigated included Cu-BTC (Basikite C300) and Fe-BTC (Basolite F300). They showed opposite catalytic behavior even though both materials have a similar density of Lewis acid sites (~2.3–2.5 mmol/g). While Cu-BTC showed poor activity for aldol condensation, Fe-BTS exhibited 10-fold higher activity. The authors proposed that the Fe-BTS catalyst possesses weak Brønsted acidity originating from the promoting effect of water. However, the nature of the active sites was not reported. Hypothetically, they suggested that Fe^{3+} can hydrolyze water molecules in their first coordination sphere to generate a metal aqua complex that could donate a proton to activate the ketone. The poor activity of Cu-BTC indicates that Lewis acid sites alone are not sufficient to initiate the aldol condensation reaction.

There are two issues that have been widely discussed in the literature, but without a clear conclusion. In fact, conflicting proposals have often been reported:

- The Brønsted acid site could participate in either tautomerization or protonation. The former renders the activated species nucleophilic (enol formation), while the latter makes the carbonyl carbon more electrophilic toward attack by the enol [32, 44]. The question is which pathway would be more favorable than the other if a carbonyl compound is introduced to a surface that has mostly Brønsted acid sites. In fact, the difference in activation energy between these steps has never been compared in the literature. This is important, especially for the systems having more than two carbonyl groups, since the reaction pathway and product selectivity between cross and self aldol condensation could be different. An extensive investigation of this matter is necessary.
- Similarly, the roles of Lewis acid sites are still unclear. Compared with Brønsted acid-catalyzed aldol condensation, the Lewis acid-catalyzed case is less well understood. Some questions to be addressed include whether Lewis sites alone can catalyze aldol condensation, whether the Lewis site requires the assistance or cooperative catalysis of an adjacent Lewis or Brønsted site, and what is the essential requirement for an active Lewis acid catalyst.

6.2.2 Alkylation: Mechanism and Site Requirement

6.2.2.1 Lewis Acid-Catalyzed Alkylation Mechanism

In industry, Friedel–Crafts Lewis acid catalysts such as BF_3 and $AlCl_3$ have been utilized in alkylation processes [45, 46]. Other Lewis acid catalysts, such as ionic liquids [47–51] and phosphonium ionic liquids immobilized on SiO_2 [52] have been investigated in the literature. Alkylation reactions catalyzed by homogeneous Lewis acids have been proposed to proceed via the formation of a polarized complex of the alkylating agent and the catalyst [45, 46].

6.2.2.2 Brønsted Acid-Catalyzed Alkylation Mechanism

6.2.2.2.1 Mechanism Many solid acid catalysts have been investigated for alkylation, including metal phosphates (e.g. $AlPO_4$, rare earth phosphates, BPO_4) [53–61], alkali/alkaline earth metal ion doped γ-Al_2O_3 [62–71], acidic zeolites [57, 72–84], etc. On Brønsted acid sites, the mechanism of the alkylation reaction between an alcohol and a nucleophilic substrate is generally thought to follow the surface-alkoxide pathway. As an example, the alkylation of phenol by ethanol is shown in Scheme 6.5. The ethanol is first activated by a proton to form a surface-ethoxide cation with water as a by-product. This ethoxide species can either attack phenol to produce protonated ethyl phenol or undergo deprotonation to form ethylene. The activated ethyl phenol then releases the surface proton and desorbs. Diethyl phenol can also be generated from secondary alkylation of ethyl phenol. The alkylating agent prefers to attack at the *ortho*- and *para*-positions of the phenol ring, due to the higher electron density at these positions, caused by the electron donating effect of the OH group [85]. An alternative alkylation mechanism that involves the formation of oxonium and carbenium ions has also been proposed [86]. In the presence of Brønsted acid sites, the alcohol is first protonated to form an oxonium ion. Dehydration of the oxonium ion generates a carbenium ion, which can desorb to produce an olefin. The olefin can also readsorb from the bulk phase to reform a carbenium ion. Either the oxonium or carbenium ion can function as an electrophile that would alkylate a substrate to form coupling products. In this case, the alkylating agents are the alcohol/alkene without the formation of an intermediate alkoxide. For alkylation involving phenolics (e.g. phenol, *m*-cresol), an alkylating agent can

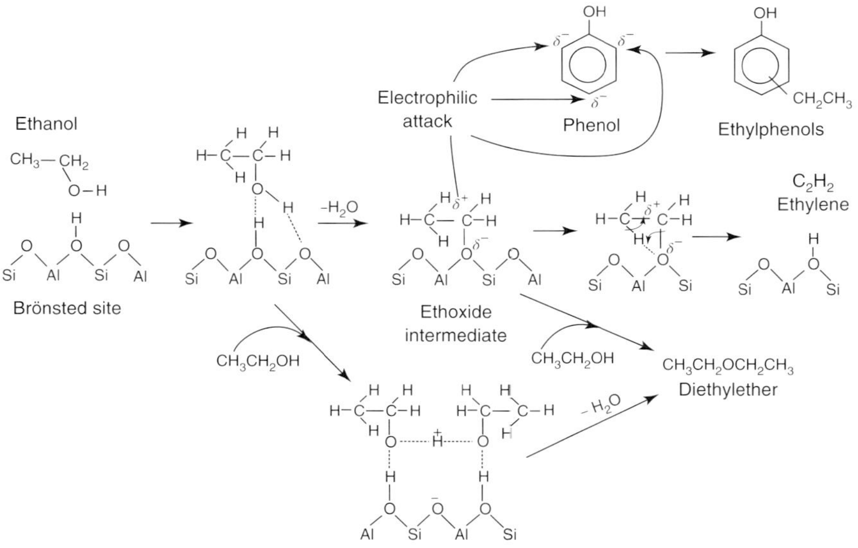

Scheme 6.5 Ethanol conversion mechanism via the formation of surface-bound ethoxide intermediate for phenol alkylation, dehydration to ethylene, and dehydration/condensation to diethylether. Source: Reproduced with permission from Sad et al. 2014 [85].

attack either the C-aromatic ring or the *O*-hydroxyl to generate C-alkylation and O-alkylation products, respectively [86]. The selectivity between the two types of alkylation depends on several factors, such as the nature of the active sites (acid or basic and strength), temperature, and residence time [71, 76, 87–90]. In general, O-alkylation is more favorable in milder conditions (i.e. low temperatures and weaker acid sites [72, 76, 82, 89, 91–96]), whereas C-alkylation requires stronger acid sites and higher temperatures [57, 76, 80, 97, 98].

Two important questions have been discussed in recent literature [99]: (i) How many elementary steps are involved in the reaction mechanism and how do they proceed along the reaction coordinate? (ii) What is the nature of the alkylating agent? Regarding the first question, two different mechanisms have been proposed for the alkylation reaction: a stepwise (consecutive or dissociative) and a direct (concerted or associative) pathway. Both mechanisms have been investigated theoretically and experimentally [100–102]. While the former involves the formation of a stable (alkoxide) intermediate on the surface with two distinct steps – alkoxide formation and C–C coupling – the latter proceeds in a single step, considering the alkoxide as a TS rather than an intermediate. An illustration of the two different mechanisms is shown in Figure 6.8. In the stepwise mechanism, methanol first adsorbs on a Brønsted acid site in an "end-on" mode through two hydrogen bonds, this adsorption mode being in agreement with other studies [96, 97, 103–107]. Water is then released to generate a methoxide species on the surface. In the next step, benzene approaches and reacts with the methoxide to form a new C—C bond [101]. The calculated activation barriers are 221 kJ/mol for methanol dehydration and 186 kJ/mol for methylation (at 400 K) [101]. Other studies using cluster modeling and the own n-layer integrated molecular orbital molecular mechanics (ONIOM) method have reported a similar value for the intrinsic energy barrier for methanol dehydration of ~200 kJ/mol [108, 109], which appears exceedingly high. However, recent DFT calculations have shown much lower barriers (i.e. 100–150 kJ/mol) for this step on a mordenite zeolite. Such moderate values for the reaction energy barrier are closer to experimentally observed activation energies [110]. On HZSM-11 and HZSM-5 zeolites, the alkoxide formation is also demonstrated as the kinetically relevant step [111–113]. In the direct mechanism, there is a single barrier to go from the adsorbed reactants to the products. The calculated activation barrier for the single step mechanism is ~195 kJ/mol [101]. Lower energy barriers have also been reported (i.e. 129 and 149 kJ/mol for benzene methylation on HZSM-11 and HZSM-5, respectively [111]).

Even though the existence of surface-bound methoxide has been confirmed experimentally by *in situ* FTIR and NMR analysis [99, 100, 114–116], some kinetic measurements/experimental data are in agreement with the direct pathway [99, 114, 117]. Theoretical calculations have not reached a definitive answer concerning the dominant pathway [118]. Several kinetic modeling studies have also been attempted, supporting the stepwise pathway [86, 119, 120]. Both the Eley–Rideal and Langmuir–Hinshelwood models have been shown to fit experimental data from the alkylation of *iso*-propanol/*m*-cresol on H-BEA zeolite well [86]. The rate-limiting step was assumed to be C—C bond formation [86]. A Langmuir–Hinshelwood mechanism has also been proposed for

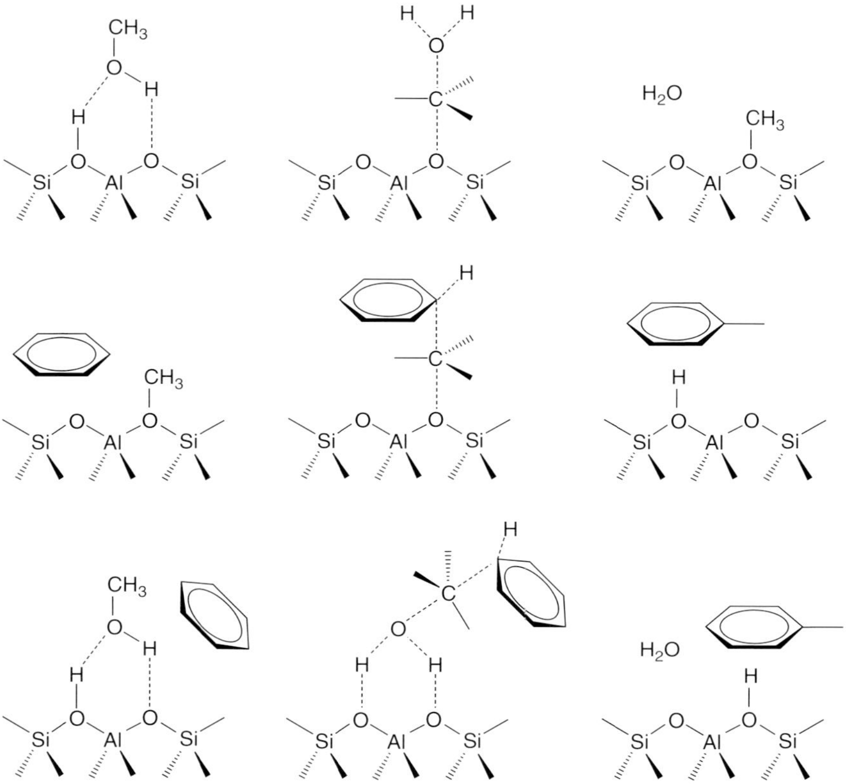

Figure 6.8 Schematic representation of the stepwise (top and middle row) and direct (bottom row) methylation mechanism for benzene. Source: Svelle et al. 2011 [99]. Reproduced with permission of Springer Nature.

the ethanol/toluene system on an MFI catalyst [119]. Other systems such as propene/benzene on zeolites [121, 122], cyclohexene/phenol on sulfonic styrene divinylbenzene resins [120], and cyclohexene/phenol on different solid acids [92] have been proposed to follow an Eley–Rideal mechanism. On the other hand, Eckstein et al. [123] have observed no difference in the activation barrier between O-alkylation and C-alkylation in ethanol/phenol system, and also for C-alkylation of ethanol with different aromatic compounds (phenol or toluene) on MFI zeolites. Moreover, no kinetic isotope effect was detected, suggesting that the first step is the most relevant step kinetically.

Regarding the second question, it has been proposed that five different species could be generated from an alcohol that are all possible alkylating agents, including (i) alkoxide: an alcohol is dehydrated on the acid site to form a surface-bound alkoxide species; (ii) alcohol: an alcohol is directly protonated, producing a $C_nH_{2n+1}{-}OH^{2+}$ species (oxonium ion) at the TS; (iii) alkene: an alkene, the dehydration product of an alcohol, is protonated, producing a carbenium ion in the TS; (iv) intramolecular rearrangement of O-alkylated ethers to form C-alkylated products; and (v) dimethyl ether (DME): self-etherification of an alcohol to generate an ether that can interact with the acid site on the surface and act as an alkylating agent.

While the alkoxide pathway has been widely reported in the literature, the involvement of alcohols or alkenes as alkylating agents has also been addressed [86, 122, 124–126]. Some studies have shown that ethanol can quickly undergo dehydration to produce ethylene with very high selectivity (>98%) at relatively low temperatures over steamed ZSM-5 zeolite [117, 127, 128], which consequentially results in no significant difference being observed when using ethanol or ethylene as the alkylating agents [129, 130]. This result suggests that the actual alkylating agent is the alkene. On the other hand, González-Borja and Resasco [86] have reported that both the alcohol and the alkene could be involved distinctively as the alkylating agents, based on two main points. First, under the same reaction conditions, the use of *iso*-propanol and propylene produced comparable total (C- and O-alkylated) product yields (Figure 6.9a). However, it should be noted that when *iso*-propanol was used as the feed, ether formation was more favorable than when a propylene feed was used; different product distributions for the two cases indicated distinct mechanisms (Figure 6.9a). Second, if the alkene pathway was dominant, an alcohol had to be dehydrated to an alkene before the alkylation reaction could proceed. Therefore, the use of either *n*-propanol or *iso*-propanol would both result in *iso*-propyl alkylation via the coupling of a secondary carbenium ion with the substrate. However, when *n*-propanol was used as the feed, a high selectivity toward *n*-propyl alkylation was experimentally observed instead of *iso*-propyl alkylation (Figure 6.9b). This result suggests that *iso*-propanol or *n*-propanol could be directly activated by a proton [86]. Theoretical calculations for the alkylation of *tert*-butanol and phenolics have also revealed that the activation barrier for C—C bond formation by using olefins is very close to that found with alcohols [131].

Some papers in the literature have reported that the C-alkylated products can also originate from the intramolecular rearrangement of O-alkylated products, which would make the ether the alkylating agent [85, 86, 88, 95, 132–135]. The

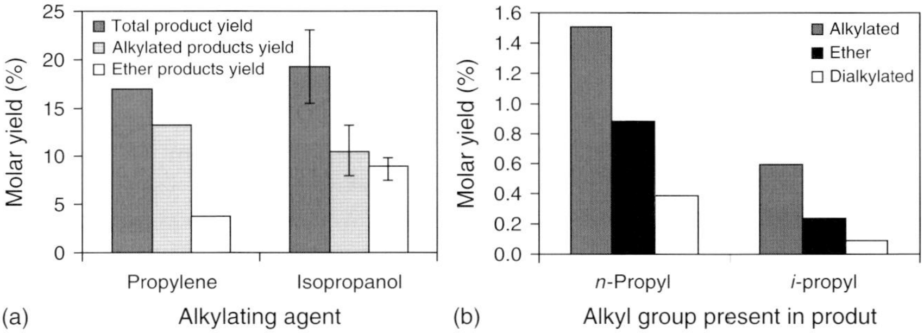

Figure 6.9 (a) Alkylated and ether product yields for alkylation of *m*-cresol with propylene and *iso*-propanol (Conditions: isopropene/*m*-cresol = 3.8, T = 473 K, P = 4.8 MPa, catalyst mass = 20 mg HY, reaction time = one hour). (b) Product yields for alkylation of *m*-cresol with *n*-propanol (Conditions: *n*-propanol/*m*-cresol = 3.7, T = 473 K, P = 4.8 MPa, catalyst mass = 50 mg HY, reaction time = one hour). Source: González-Borja and Resasco 2015 [86]. Reproduced with permission of John Wiley and Sons.

conversion of ether to alkylated products could proceed via an ionic pathway [133] (Scheme 6.6) or via the decomposition of the ether back to an olefin and a phenolic compound, which are then recoupled [86, 126].

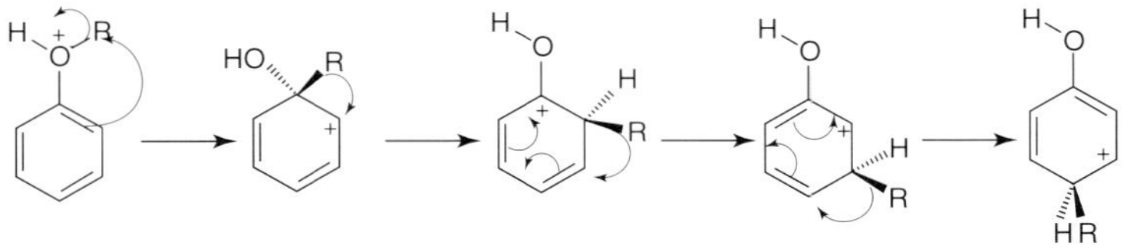

Scheme 6.6 Five possible protonated configurations, p-ether, p-*ipso*, p-*ortho*, p-*meta*, and p-*para*, for migrations of the alkyl group in the ionic pathway. Source: Reproduced with permission from Ma et al. 2006 [133].

In principle, DME can act as the alkylating agent [54, 69, 70, 100, 109, 112, 136–138]. In fact, some authors have proposed that DME is a more effective alkylating agent than methanol for benzene methylation based on both experimental and theoretical calculations [136]. When DME/benzene was used, the alkylation rate was higher than when using methanol/benzene on all the catalysts tested (HZSM-5, HSSZ-24, HSAPO-5) in the temperature range 250–350 °C.

6.2.2.2.2 Site Requirement Sad et al. [139] have investigated the catalytic performance of HBEA, HZSM5, HMCM22, HY, SiO_2–Al_2O_3, and tungstophosphoric acid (HPA) on SiO_2 for the methylation of phenol at 200 °C. The pure Brønsted acid HPD/SiO_2 catalyst exhibited very high activity toward O-alkylation compared with C-alkylation, indicating that Brønsted acidity is much more selective toward O alkylation [92, 139]. Similar catalytic behavior has been reported on Brønsted super-acid catalysts such as Nafion-H, a perfluorinated sulfonic acid resin, over which the selectivity to anisole was over 80% for 35% phenol conversion at 473 K [62]. It was suggested that the selectivity toward either O-alkylation or C-alkylation depends on the adsorption mode of the nucleophilic substrate. The aromatic ring of phenol adsorbs perpendicularly to the surface on the protonic sites of heteropoly acids (HPA)/SiO_2, providing easy accessibility to the oxygen of the hydroxyl group for the attack of an activated alkylating agent. Zeolites, such as HBEA and HY, which contain high Lewis and Brønsted acid density, display high yields of C-alkylated products. Since the simultaneous effect of Brønsted and Lewis sites is required to promote C-alkylation [87], the phenol should adsorb in a flat mode on the surface-Lewis acid sites, so the activated alkylating agent on the Brønsted site can approach the aromatic C [62].

The moderately strong Brønsted and Lewis acid sites on catalysts such as SiO_2–Al_2O_3 or Al-MCM-41 [139] have been shown to be insufficient to catalyze alkylation reactions at the conditions studied (200 °C). Indeed, higher temperatures are necessary [59, 62, 74, 140]. The poor activity of Lewis solid acid γ-Al_2O_3 has also been reported for liquid phase alkylation of cyclohexene and phenol at 200 °C [126]. Moreover, stronger solid acid catalysts such as HZSM-5 and HMCM22 zeolites encounter another problem, i.e. the slow intra-crystalline diffusion due to their small pore size can outweigh their high density of strong Brønsted and Lewis sites. Another problem associated with microporous catalysts for alkylation reactions is the isomerization of *p*-cresol to *o*- and *m*-cresol, which is catalyzed by strong Brønsted acid sites [129, 139].

6.2.2.3 Base-Catalyzed Alkylation: Mechanism and Site Requirement

Alkylation can also be catalyzed by basic catalysts such as MgO, alkali cation-exchanged β, Y, ZSM-5 [141–143], MgAl-CHT, and CuAl-CHT [90]. For example, the base-catalyzed side-chain methylation of toluene with methanol produces styrene via attachment of a methyl group from methanol into the methyl side-chain of the aromatic ring. The reaction pathway has been proposed to include three elementary steps (Scheme 6.7). In the first step, methanol is dehydrogenated on a strong basic site to a formaldehyde intermediate [144–146]. The interaction between formaldehyde and the surface polarizes the carbonyl C of formaldehyde. In a parallel step, the adsorption of toluene (an electron pair donor) on the cation (a Lewis acid site) promotes the electron withdrawing effect of the aromatic ring, inducing polarization of the methyl C–H and giving the carbon atom a partial negative charge. The polarization effect is further enhanced via the interaction of the methyl group with a basic surface-O atom. In Step 2, C—C bond formation takes place, followed by dehydration (Step 3) and formation of an adduct (Scheme 6.7). The C–C coupling step is likely to occur in a concerted way to minimize the lifetime of the ionic species [141, 145]. Alternatively, the side-chain alkylation of toluene with methanol has been proposed to occur via a free radical pathway [89, 147].

A cooperative effect of acid–base sites has been suggested for the reaction [145, 148, 149]. The distance between the acid–base sites influences the polarization of the aromatic ring and the activation barrier for C—C bond formation [150]. It has been suggested that there are three crucial requirements for a catalyst to be active in the side-chain alkylation reaction [146]: (i) sufficient basic strength to dehydrogenate methanol to formaldehyde, (ii) enough capacity to stabilize toluene on the surface and activate the methyl group (Lewis acidity), and (iii) appropriate stoichiometric ratio of reactants and appropriate distance between adsorbed methanol and toluene. For example, Cs-Y, MgO, and hydrotalcite are not suitable catalysts for side-chain alkylation because Cs-Y does not have sufficient basic strength for the dehydrogenation, and MgO and hydrotalcite do not have cations large enough for appropriate accessibility, adsorption, and polarization of toluene. Cs-X zeolite, on the other hand, satisfies all three conditions, resulting in good catalytic performance.

6.2.3 Hydroxyalkylation: Mechanism and Site Requirement

Several catalyst systems have been studied for the hydroxyalkylation reaction including homogeneous Brønsted acids (HCOOH, CH_3COOH, H_3PO_4, H_2SO_4 [151], *para*-toluene sulfonic acid (PTA) [152]); heterogeneous Brønsted solid acids (Amberlyst 15, Nafion 212 [152], acidic resin [153]); homogeneous Lewis acids (BF_3, $BeCl_2$, $TiCl_4$, $SbCl_5$, $SnCl_4$); super acids ($HF{\cdot}SbF_5$ and $HSO_3F{\cdot}SbF_5$ [154–156]); solid Lewis acids (Al-MCM-41 [157], Sn/Si-MCM41 [158]); and Brønsted and Lewis acid solid catalysts (USY [159], HY [159, 160], HZSM-5

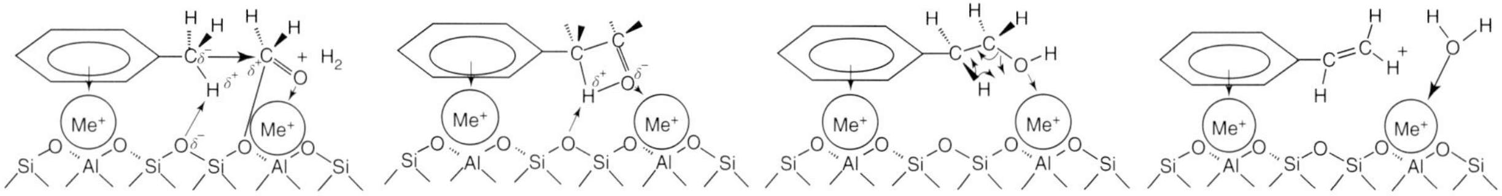

Scheme 6.7 Proposed mechanism for based-catalyzed side-chain alkylation. Source: Reproduced with permission from Palomares et al. 1997 [145].

[157, 159, 160], HBEA [24, 157, 159, 160], Al-MCM-41 [157], heteropolyacids deposited on solid supports [161–163], montmorillonite clay, $NbOPO_4$ [164], and MOFs [165]).

6.2.3.1 Brønsted Acid-Catalyzed Mechanism

An example of a proposed Brønsted acid-catalyzed mechanism for the hydroxyalkylation of ketones/aldehydes with aromatics on a solid surface [166–168] is shown in Scheme 6.8. In the first step, formaldehyde (1) is protonated to form a hydroxymethyl carbocation (2), which then undergoes electrophilic substitution at the *ortho*/*para* carbon atom of the phenol molecule. This C–C coupling step includes several elementary steps regarding the rearrangement of C—C, C—H, and C—O bonds inside the aromatic ring to form a hydroxyl oxygen cation (3). This unstable species then returns the proton to the surface to form 4-hydroxy benzyl alcohol (4). The alcohol product could be reprotonated to generate a carbocation intermediate (5), which continues to react with another phenol molecule to yield a trimer intermediate and water as a by-product (6). Afterward, the intermediate species (6) is deprotonated to produce bisphenol F (7) [169–172]. Accordingly, it should be noticed that the electron density of the substrate plays an essential role. Barthel et al. [173] have shown that the consumption rate of substituted benzenes (phenol, anisole, and toluene) for hydroxyalkylation with chloral on zeolites follows the order: anisole ($X = OCH_3$) > phenol ($X = OH$) > toluene ($X = CH_3$). This order matches the corresponding σ_{p+} substituent constant, which demonstrates the strength of the electron-donating effect of the ring substituent. In this study, it is likely that the kinetically relevant step is associated with C—C bond formation since a higher electron density in the aromatic ring results in a higher condensation rate. Other mechanisms have also been proposed for the reaction such as a Lewis acid-catalyzed mechanism on BF_3, $BeCl_2$, $AlBr_3$, $TiCl_4$, $SbCl_5$, $SnCl_4$ [154–156, 174–180], or super-electrophiles [154, 181–188]. Lewis acids activate the carbonyl compounds by polarizing the C—O bond of the carbonyl group via interaction with the oxygen lone pair, making the C-carbonyl more susceptible to nucleophilic attack [155, 174–180].

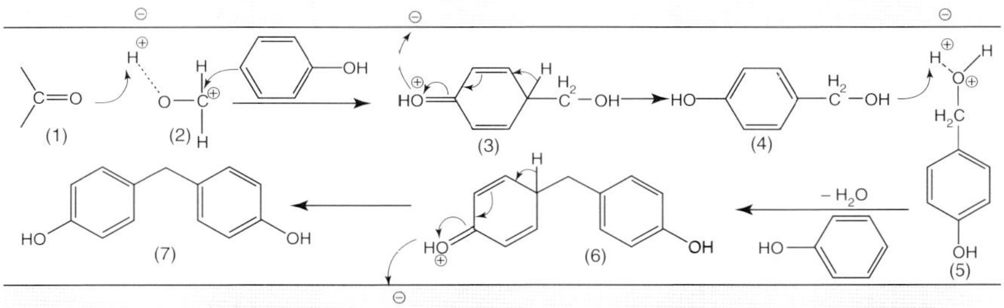

Scheme 6.8 Mechanism of the hydroxyalkylation of formaldehyde/phenol. Source: Reproduced with permission from Wei et al. 2016 [169].

6.2.3.2 Site Requirement

Brønsted and Lewis acids both activate carbonyl compounds by rendering the carbonyl C electrophilic. In heterogeneous systems, there is lack of comprehensive studies related to the mechanism of Lewis acid-catalyzed hydroxyalkylation on solid surfaces. This is important, especially when one has to deal with a complex catalyst system that might contain both types of acid sites (Brønsted and Lewis). Several studies on different solid catalysts have not clearly differentiated the contribution of each site type to the observed activity [163, 169, 189]. For example, Garade et al. [171] have investigated the production of bisphenol F from the hydroxyalkylation of phenol and formaldehyde on dodecatungstophosphoric acid (DTP) impregnated on fumed silica. They have shown that pure SiO_2 shows negligible activity due to the presence of only weak Lewis acid sites. The deposition of DTP on SiO_2 significantly increases the density of strong sites as well as the total acid density (both Brønsted and Lewis sites), resulting in an increase in product yields. However, no further explanation about the role of each site type or their contribution has been reported.

Garade et al. [190] have shown that the conversion of formaldehyde as well as the trimer selectivity on DTP/montmorillonite K10 catalyst increased with increasing Brønsted/Lewis ratios, suggesting that the Brønsted sites are more crucial in formaldehyde activation [190–193]. On the other hand, Bai et al. [194] have reported that zeolites modified with oxalic acid exhibited a significant increase in Lewis acid site density and a decrease in Brønsted acid site density (as confirmed by pyridine-IR). This resulted in an enhancement of catalytic performance/stability for the hydroxyalkylation of anisole and chloral. They attributed the improvement of the turnover number to the higher Lewis site density, which suggests that the Lewis sites are more effective for this reaction. Separately, a study of the *p*-cresol/formaldehyde system has demonstrated that a 70% yield of the coupling product could be achieved on Sn/Si-MCM41 catalyst at 90 °C for two hours. The Lewis acid site corresponded to Sn^{4+} in tetrahedral coordination in the silica framework [158].

Wu et al. [195] have proposed a synergy between both types of acid sites in their study, using mesoporous Al-incorporated silica-pillared clay interlayer catalysts for the production of bisphenol F from phenol/formaldehyde. The weak and moderately strong acid sites were proposed to originate from OH groups bonded to the pillars' Al ions (Al–OH), while the strong acid sites were associated with the OH groups bonded to the tetrahedrally coordinated Al ions. Bands at 1447, 1544, and 1490 cm^{-1} in FTIR-pyridine spectra corresponded to pyridine adsorbed on Lewis acid sites, Brønsted acid sites and pyridine associated with both Lewis and Brønsted acid sites, respectively [196–198]. They have proposed that while the Brønsted site activates formaldehyde by donating a proton (H^+) to the *O*-carbonyl, the Lewis site (Al^{3+}) abstracts a H^- from the carbonyl C atom [199]. Increase of the Al content leads to an increase in Lewis/Brønsted site ratio as well as an enhancement in activity. The activation of formaldehyde by Lewis acid sites has been reported to require a higher activation energy than that by Brønsted acid sites. Therefore, the higher condensation rate observed at high Lewis/Brønsted ratios suggests that the reaction might be catalyzed by the combination (synergy effect) of both site types rather than by just a single one [195]. However, the nature of this synergy effect is not fully addressed.

The type of active sites also influences the selectivity toward the products. While Brønsted acid sites promote trimer production, Lewis sites with weaker strength favor dimer generation. For instance, high selectivity toward the trimer bisphenol F was obtained with a high density of strong acid sites [161, 190]. Lewis acid sites with moderate activity (low temperature-programmed desorption (TPD) desorption temperature range) were responsible for the highest selectivity toward the dimer of *p*-cresol/formaldehyde hydroxyalkylation on MFI structured molecular sieves of SnO_2/Al_2O_3 [200]. Increasing the Al content led to an increase of Brønsted acid density, resulting in enhanced formaldehyde conversion as well as trimer yield. Tan et al. [201] have reported that Brønsted acid sites favor the formation of the 4,4′-substituted product, whereas Lewis sites favor the production of 2,4′ and 2,2′-isomers in the formaldehyde/phenol system [201].

6.2.4 Acylation: Mechanism and Site Requirement

Acylation can be catalyzed by different catalysts, including homogeneous acid catalysts such as $AlCl_3$, $ZnCl_3$, $FeCl_3$, HF, etc. and heterogeneous catalysts such as zeolites [202–207], metal oxides [208–210], heteropolyacids [211–213], or supported sulfonic acids [205, 214], etc. In this reaction, the molecule that donates the acyl species is called the acylating agent, while the molecule that receives the acyl group is referred to as the substrate. Different compounds can be used as acylating agents such as carboxylic acids, aromatic esters, acyl halides, or acyl anhydrides, while aromatic or furanic molecules can act as the substrate. The reaction occurs via a two-step mechanism: (1) formation of the acyl species from the acylating agent and (2) C—C bond formation between the acyl species and the substrate. Figure 6.10 illustrates an example of this two-step mechanism for the acylation of anisole by acetyl chloride over a zeolite catalyst [215].

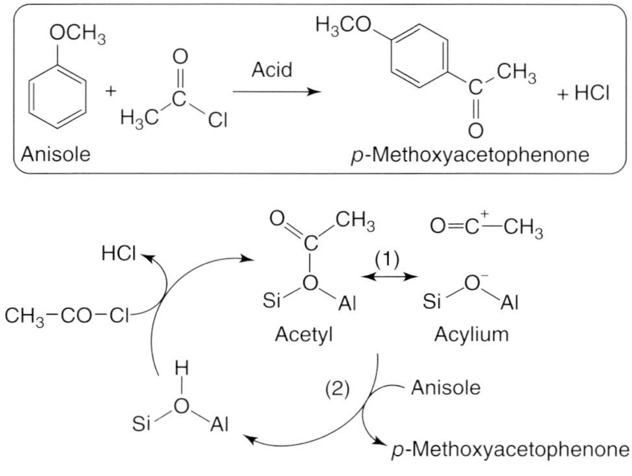

Figure 6.10 Friedel–Crafts acylation of anisole over a zeolite. Source: Lezcano-Gonzalez et al. 2013 [215]. Reproduced with permission of John Wiley and Sons.

Acylation has been extensively investigated in organic chemistry with different combinations of acylating agents, substrates, and catalysts. Maggi et al. [205, 214] have reviewed in great detail the acylation of different substrates including arenes, aromatic ethers, aromatic thioethers, phenolic substrates, and heterocycles. The reactivity of each substrate family has been compared with different types of solid acid catalysts including zeolites, clays, metal oxides, acid-treated metal oxides, HPA, and Nafion. In another work, Bejblová et al. [216] have reviewed this topic, focusing on the relationship between catalyst structure and the corresponding acylation activity for different molecular sieves, including zeolites, isomorphously substituted mesoporous silica materials, MCM-41, and SBA-15. Herein, we will highlight some fundamental aspects that can be considered crucial and relevant in the context of biomass upgrading. Recently published work on the acylation of furanic molecules, these molecules being an important product fraction of pyrolyzed biomass, has been extensively discussed elsewhere [205, 214, 216, 217].

6.2.4.1 Mechanistic Aspects of Acylation Reactions

Acylation happens via a two-step mechanism and there is not a definite rate-limiting step; rather, the rate depends on the relative energy barriers of each step. A recent study [218] concerning liquid phase acylation of lignin-derived aromatic molecules has shown that, when a less effective acylating agent (e.g. acetic acid) is used with an effective substrate (e.g. *m*-cresol or anisole), the rate-limiting step is the formation of the acyl species. However, when an effective acylating agent (aromatic ester) and an ineffective aromatic substrate (toluene) are used, the rate-limiting step becomes C—C bond formation [203, 218, 219]. Clearly, oxygen-containing substituents, such as hydroxyl (–OH) and alkoxy (–OCH_3), activate the aromatic ring for the C–C coupling step with the acyl species, enhancing their substrate efficiency. On the other hand, aromatic compounds with only alkyl substituents are poor aromatic substrates [218–221]. Regarding the activity of different acylating agents, the following order has been established: acyl halides, acyl anhydrides > aromatic esters > carboxylic acids, alkyl esters [218, 222–224].

Over zeolites, the nature of the active acylating intermediate was determined by Corma and coworkers to be a covalent acyl-zeolite complex, rather than an acylium cation as usually assumed before [215, 225]. In another study, Kresnawahjuesa et al. [222] have shown that over H-ZSM5 at temperatures below 400 K, acetyl chloride and acetic anhydride react with the Brønsted acid sites to create acetyl intermediates, while acetic acid forms a hydrogen-bonded complex with the Brønsted acid sites. This means that acyl halide and acyl anhydride are more active than carboxylic acids, agreeing with reported experimental results [223, 224].

At the same time, the acylation activity of carboxylic acids strongly depends on the carbon chain length [226, 227]. A study from Chiche et al. has reported this dependence using toluene and *p*-xylene as substrates with CeY as the catalyst and linear C_2–C_{22} carboxylic acids as acylating agents [227]. As shown in Figure 6.11, the acylation yield significantly increases when the C chain length

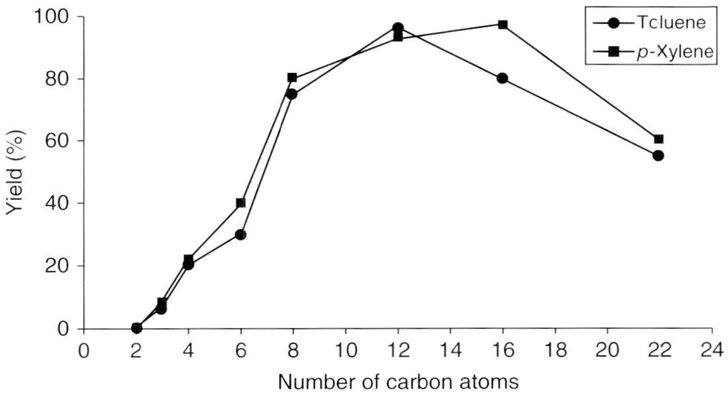

Figure 6.11 Relationship between acylation yield and the number of carboxylic acid C atoms, with toluene and *p*-xylene as substrates. Source: Reproduced with permission from Ref. [205, 227].

increases from two to eight, does not change much in the range C_8–C_{16}, and finally decreases up to C_{22}.

Pyrolysis bio-oil derived from biomass contains a significant amount of short-chain carboxylic acids (C_2–C_3) [228], which are inherently weak acylating agents [218, 227]. Therefore, in order to apply acylation chemistry to bio-oil upgrading, ways must be found to enhance their activity. For example, they can be converted to acyl halides [229–231] or acyl anhydrides [232, 233] using different types of homogeneous catalysts. These processes are feasible but in general require extra steps, which results in increased costs stemming from the need for additional reaction and separation units. Another method is to convert carboxylic acids into aromatic esters via esterification with phenol or *m*-cresols, which are directly derived from biomass [218]. This chemistry can occur without a catalyst or *in situ* with acylation being catalyzed by acids [218].

Over large pore zeolites such as beta or Y, acetic acid has been reported to be an ineffective acylating agent that is unable to produce acyl species at an appreciable rate [218, 227]. Recently, Crossley and coworkers have reported that by using a small pore size zeolite such as H-ZSM5, acetic acid can directly generate acyl species. Furthermore, using methyl furan as the substrate can help stabilize the acyl species and reduce the barrier for C—C bond formation [234]. H-ZSM5 has been reported to have lower activity than larger pore zeolites such as HY or HBEA for acylation with large aromatic substrates such as arenes or phenolic compounds [202, 218, 223, 235]. However, when using methyl furan as a substrate, H-ZSM5 exhibits higher activity. A smaller pore zeolite will have a lower energy barrier for acyl formation from short-chain carboxylic acids [234], but in the case of large substrate molecules such as arenes or phenolic compounds, the small pores will present a disadvantage for the desorption of the final acyl products, which leads to increased catalyst deactivation and lower overall yields [236]. That is, the overall acylation activity depends not only on the nature of the acylating agents and substrates but also on the acidity and pore size of the catalyst materials and desorption of the final acylation products.

Another aspect with important consequences for industrial applications is catalyst deactivation, which can be caused by heavy compounds or coke formation, competitive adsorption of ketone products, or ketene formation. It has been found that the acylation yield decreases rapidly after a short period of time on stream (TOS) [218, 219, 237–240]. Analysis of the catalyst after reaction by different techniques (NMR, IR, extraction with different solvents, thermogravimetric analysis (TGA), etc.) has revealed that a significant fraction of the acylated products remains on the catalyst, suggesting that the aromatic ketone products adsorb strongly on the surface and compete with the reactants [219, 241–243]. Other reports have reached similar conclusions. That is, the adsorption of the aromatic ketone product is much stronger than that of the reactants, leading to the negative impact of the acylation products [203, 240, 241, 244]. This competitive adsorption is the primary reason for the quick deactivation of the catalyst. A small fraction of coke is derived from the di- and tri-acylated products that cause pore blockage, leading to the long-term deactivation of the catalysts [241, 243, 245]. The deactivation due to ketene formation is not significant in this case [246]. The zeolite catalysts can be regenerated and initial activity can be restored by calcination [247].

Catalyst deactivation during acylation can be mitigated by adjusting the reaction conditions, e.g. using a high molar ratio of substrate to acylating agent [238, 241, 248], using solvents with moderate polarity [223, 242, 247, 249], or using a liquid phase reactor system that can continuously extract the ketone products [202, 242]. Furthermore, reducing the diffusion path helps to enhance the accessibility of the acid site to reactants and the desorption of the ketone products, which as a result helps to reduce deactivation. This enhancement can be achieved by modifying the zeolite structure, which can be done using nanocrystalline zeolites [219, 250], zeolites supported on SiC foam monoliths [251, 252], or mesoporous zeolites [216, 253–257].

In addition to the previously mentioned modifications, the acylation activity of zeolites can also be improved by introducing metals, especially rare earth metals [224, 227, 258–260]. Gauthier et al. have investigated the activity of different metal-exchanged Y zeolites for the acylation of toluene with octanoic acid, rare earth metal exchanged zeolites being the most active [258]. As shown in Figure 6.12, the acylation activity follows the order Cr^{3+}, $Zr^{4+} < Mg^{2+}$, Cu^{2+}, $Co^{2+} \ll H^+ \ll Pr^{3+}$, La^{3+}, Gd^{3+}, Yb^{3+}, Ce^{3+} [258]. The degree of cation exchange is not linearly correlated with activity and there is a threshold amount below which there is no effect on the activity whatsoever [258]. Enhancement effects observed by addition of rare earth metals have been reported with zeolite Y [227, 255, 261], beta [224, 255, 259], and ZSM-5 [260].

6.2.4.2 Role of Brønsted vs. Lewis Acid in Acylation Over Zeolites

The role of Brønsted vs. Lewis acid sites in zeolites has been investigated to determine the active site for acylation activity. The method normally used to distinguish the role of each type of acid site is to observe the effects of varying the Brønsted/Lewis ratio on catalytic activity. However, manipulating the ratio of the two types of acidity inevitably induces changes in the zeolite structure, which might create defects and eventually influence acylation activity. Therefore, there

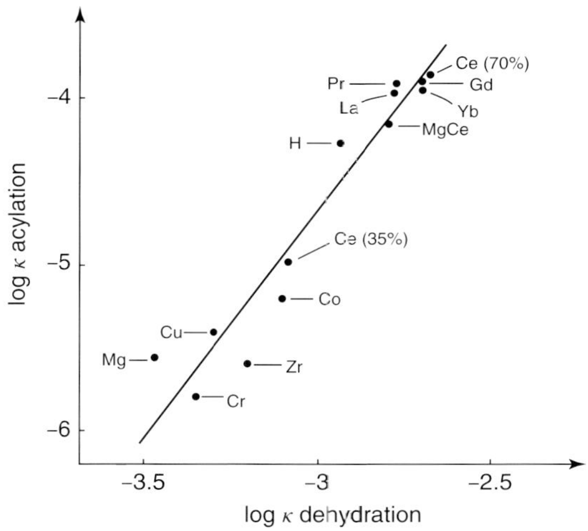

Figure 6.12 Correlation between rate of toluene acylation and cyclohexanol dehydration of different metal-exchange Y zeolites. Source: Gauthier et al. 1989 [258]. Reproduced with permission of Elsevier.

have been contradictory conclusions about the actual role of Brønsted [204, 238, 259] and Lewis acid sites [225].

For example, Gauthier et al. have reported a linear correlation between the rate of toluene acylation and the rate of cyclohexanol dehydration, as shown in Figure 6.12 [258]. Cyclohexanol dehydration is a Brønsted acid-catalyzed reaction, suggesting that acylation is also catalyzed by Brønsted acids. A linear correlation with cyclohexanol dehydration has also been observed in another study of toluene acylation by acetic anhydride over metal-exchanged zeolite beta [259]. In another study, zeolites were analyzed by pyridine-FTIR before and after the acylation reaction, and the results show that after reaction the Brønsted acidity was reduced more significantly than Lewis acidity [238]. The decrease in Brønsted acidity aligns with a significant drop in the catalyst activity, indicating that Brønsted acid sites are responsible for the reaction [238]. Similar conclusions about the important role of Brønsted acidity were also derived from a study of anisole acylation with acetic anhydride over zeolite beta [262]. The Brønsted acid sites were selectively poisoned by an organic base (2,4-dimethylquinoline), while the Lewis acid sites were not affected. The corresponding acylation rate was linearly correlated with the Brønsted acid site density. Interestingly, Koehle et al. have studied the acylation of methyl furan over Lewis acidic Sn-beta zeolite and found that the reaction is more favorable on the Brønsted acidic silanol groups of the hydrolyzed open site than on the Lewis acidic Sn metal center [263].

On the other hand, studies from Bejblova et al. have shown that on metal-exchanged zeolite beta and USY, the most active catalysts also have the highest Lewis acid density, suggesting that Lewis acidity controls the chemistry [255]. In other work on anisole acylation with propionic acid, zeolite Y was steamed at different conditions and tested for activity; the zeolites with higher Lewis acid concentrations exhibited higher activities [225]. Interestingly, work by Bigi et al. has suggested that a delicate balance between both types of acid is needed to maximize acylation activity [264]. The authors tested HY zeolites with different Lewis/Brønsted acid ratios for acylation of dimethoxybenzene with different linear acyl chlorides, a zeolite with a balanced amount of Lewis/Brønsted acid sites and strength being more active than zeolites with either weaker or stronger acidity, regardless of the acylating agent.

6.2.5 Ketonization: Mechanism and Site Requirement

Ketonization is a well-known reaction that converts two carboxylic acids into a ketone, with water and carbon dioxide as by-products:

$$R_1COOH + R_2COOH \rightarrow R_1COR_2 + H_2O + CO_2$$

The reaction has gained noticeable attention, not only in the industrial production of acetone but also for biofuel applications due to its ability to reduce the oxygen content and increase the energy density of bio-oil. In the literature, extensive research has been performed on the mechanism of ketonization, bulk ketonization and surface ketonization being widely accepted mechanisms [265]. While the bulk ketonization mechanism involves the decomposition of a carboxylate salt, the surface ketonization mechanism requires the abstraction of an α-hydrogen and the formation of several surface intermediates – namely, ketene, acyl carbonium ions, β-keto acids, and acid anhydride, all of which occur on the surface of the catalyst. Pestman et al. have reported that the mechanism of acetic acid ketonization on different oxides such as TiO_2, Al_2O_3, and ZnO_2 follows different pathways [266–268]. Oxides with low lattice energy such as MgO, CaO, BaO, SrO, and CdO strongly interact with carboxylic acids to form acetate salts, which decompose into acetone under thermal treatment (bulk ketonization mechanism). As for the oxides with high lattice energy such as TiO_2, ZrO_2, CeO_2, etc., the surface mechanism dominates. However, in some cases, both mechanisms may occur simultaneously. Snell and Shanks have shown that CeO_2 can catalyze ketonization reactions through either the bulk or the surface pathway depending on the reaction temperature [269]. A discussion of the mechanisms and catalysts for ketonization of carboxylic acid can be found in a review authored by Pham et al. [265]. In this contribution, only surface ketonization will be discussed.

6.2.5.1 Mechanism of Surface Ketonization

The mechanism of surface ketonization has been the center of controversy for many years. Several studies have proposed that a carboxylate intermediate adsorbed on a coordinatively unsaturated metal site can decompose into a surface ketene intermediate that consequentially couples with an adsorbed carboxylate species to form the ketone [270–273]. However, according to recent studies, the ketene is a side product rather than the ketonization reactive intermediate [274–276]. Alternatively, a mechanism involving the formation of a β-keto acid intermediate has been proposed [274, 277–281]. An example of the proposed mechanism of ethanoic acid ketonization on TiO_2 (anatase) is shown in Scheme 6.9. The gas phase ketonization of ethanoic acid is initiated by the dissociation of the acid on a Ti–O pair to form a monodentate carboxylate species (AcO*) (Step 1). The second oxygen atom of the carboxylate species can also bind to an adjacent Ti site, while the α-H is abstracted by a lattice O-atom to form a bidentate carboxylate species (*AcO*) (Step 2). Under the studied reaction

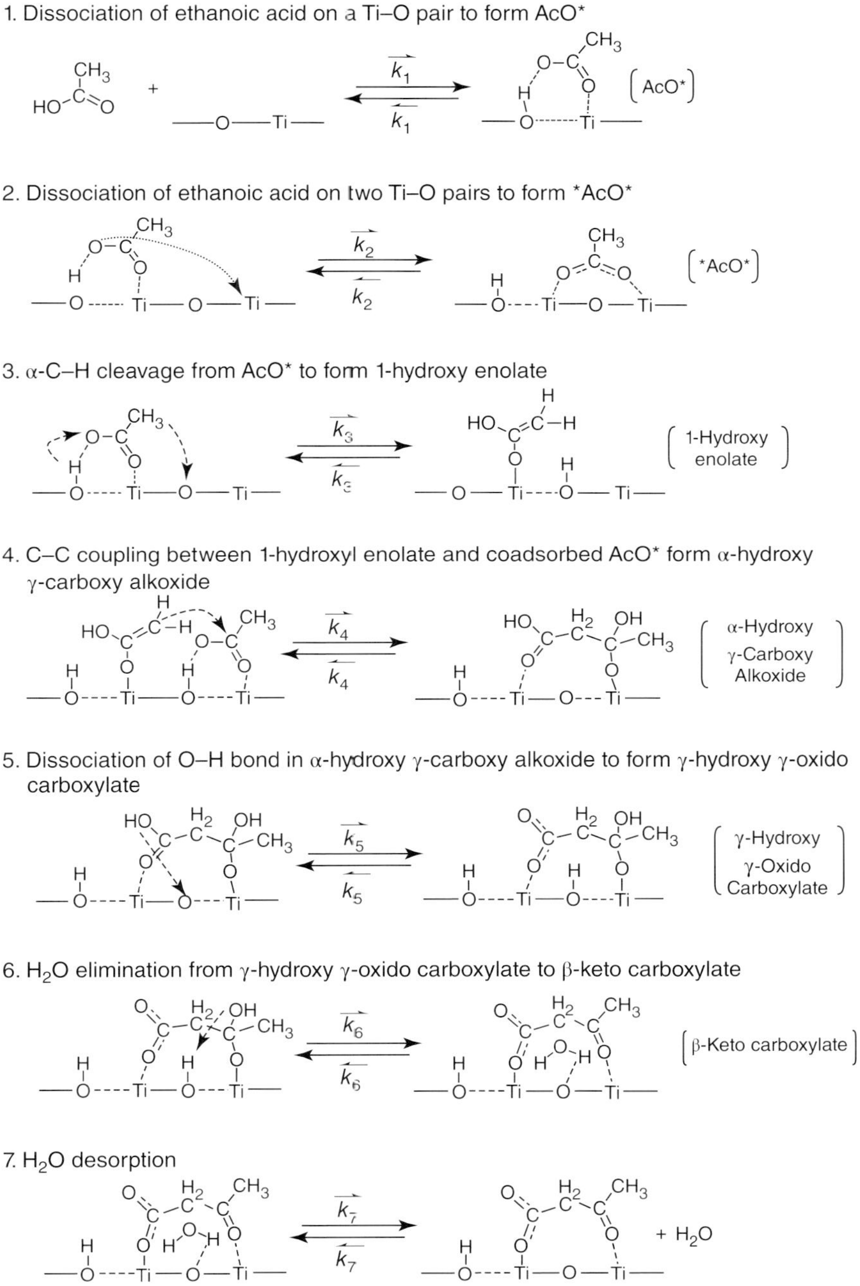

Scheme 6.9 Proposed elementary steps for the ketonization of ethanoic acid on acid–base pairs of TiO_2 anatase. Source: Reproduced with permission from Wang and Iglesia 2017 [274].

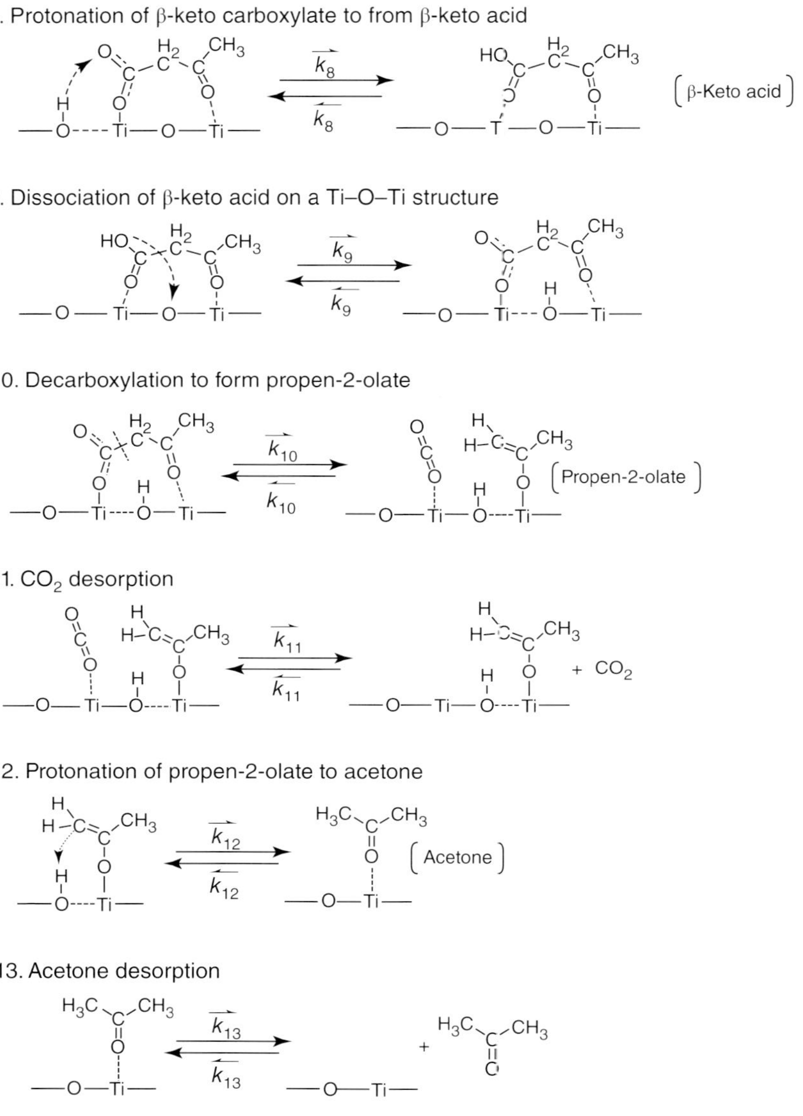

Scheme 6.9 (*Continued*)

conditions, infrared spectra show that the AcO* is the most abundant surface species, which participates in ketonization as a reactive intermediate. A similar result has been reported on several other catalysts including TiO_2 [282–284], ZrO_2 (monoclinic) [280, 285], CeO_2 [278, 284, 285], and $ZnO–Cr_2O_3$ [286]. On the other hand, the bidentate carboxylate (*AcO*) acts as a spectator, leading to catalyst deactivation. The bidentate species can be slowly dehydrated to form a gaseous ketene in trace amounts [274]. In the next step, the monodentate carboxylate is deprotonated at the α-H by the O-atom of the second Ti–O pair to form the 1-hydroxy enolate (Step 3). The enolate then couples with the carboxyl C-atom of an adjacent AcO* species to form α-hydroxy γ-carboxy alkoxide (Step 4). The dehydration of this species on a vicinal lattice O-atom produces a β-keto carboxylate (Steps 5–7), which is then protonated to form a β-keto acid (Step 8). The decomposition of the β-keto acid generates a C_3 enolate and CO_2. Finally, the C_3 enolate is reprotonated to form acetone. No H/D isotope effect has been reported, which rules out the involvement of the formation/cleavage of bonds containing an H-atom in the kinetically relevant step (Steps 3, 5, 6, 8, 9, and 12). The C–C coupling step (Step 4) is proposed to be rate-limiting on both TiO_2 anatase and rutile, consistent with the lack of an H/D isotopic effect, kinetic measurements, and DFT calculations [274].

On acidic zeolites, the ketonization of carboxylic acids involves the formation of a β-keto acid originating from the coupling of an adsorbed carboxylic acid and a surface acyl species. The acyl species is generated via the protonation of a carboxylic acid, followed by dehydration as depicted in Scheme 6.10. The β-keto acid then decomposes to yield the ketone, water, and CO_2. Recent work from Crossley and coworkers has demonstrated that the ketonization of acetic acid on ZSM-5 zeolite follows a second-order kinetic dependence with respect to acetic acid, suggesting that the rate-limiting step could be the C—C bond formation between the acyl and another acid to form the β-keto acid [281].

One of the key factors in the ketonization reaction is the need for an α-hydrogen in at least one of the carboxylic acids participating in the reaction [265, 277, 280, 281, 287]. The role of the α-hydrogen is to facilitate either enolate formation on oxide surfaces (Scheme 6.9) or the activation (tautomerization) (Scheme 6.10) of a second carboxylate on zeolite surfaces, which facilitates the C–C coupling step to form a β-keto acid [265]. Notably, a study published in 1962 illustrated that ketonization of a tertiary carboxylic acid is also feasible. It was stated that 2,2,5,5-tetramethyladipic acid can be converted into the corresponding ketone with 52–72% yield in the presence of KF or BaO [288]. This result is inconsistent with the β-keto acid mechanism, which requires the involvement of an α-hydrogen. Recently, Oliver-Tomas et al. have reported

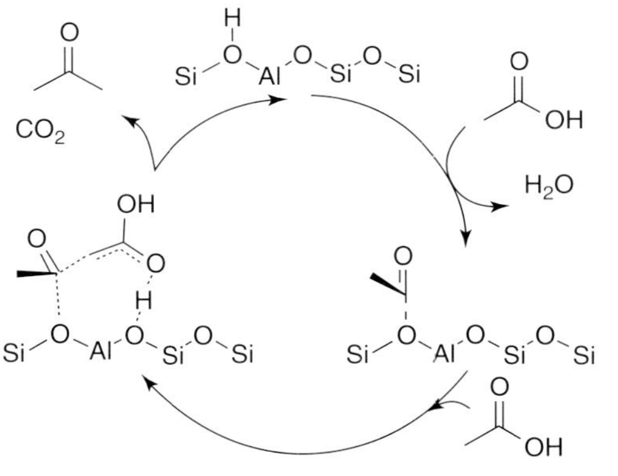

Scheme 6.10 Proposed mechanism of acetic acid ketonization on ZSM-5 zeolites. Source: Reproduced with permission from Gumidyala et al. 2016 [281].

that pivalic acid – a carboxylic acid without α-hydrogens – could be converted into a mixture of ketones, with the exception of the symmetrical ketonic decarboxylation product, i.e. 2,2,4,4-tetramethyl-3-pentanone, at 550 °C, over ZrO_2 [289]. They pointed out that the ketonization of the tertiary carboxylic acid (pivalic acid) at such high temperature occurs via a completely different pathway involving the transformation of the acid itself into other carboxylic acids that bear α-hydrogen atoms. This observation strongly confirms once again the β-keto acid pathway as the universal mechanism for the decarboxylation of carboxylic acids [289].

6.2.5.2 Site Requirement

The ketonization of acetic acid has been studied on different catalysts including TiO_2 anatase, TiO_2 rutile, and TiO_2 P25 [270, 274, 279, 283, 290], ZrO_2 [274, 289, 291], CeO_2 [271, 272, 292, 293], supported Co–Mo [294], Fe_3O_4/SiO_2 [295], etc. According to the mechanisms proposed earlier, the two critical requirements for the active sites for ketonization are (i) coordinatively unsaturated surface metal cations/oxygen anions for the initial deprotonation of the carboxylic acid forming a surface carboxylate and its subsequent enolization (or secondary deprotonation of the carboxylate) and (ii) nearby surface cation sites that can adsorb other carboxylates, which allow the adsorbate–adsorbate interaction that eventually leads to the formation of a reactive intermediate such as the β-keto acid.

These characteristics can be described in terms of acid–base pairs or redox properties [265]. According to Wang and Iglesia [274], the acid–base pairs of TiO_2 and ZrO_2 are the active sites for the ketonization of C_2–C_4 carboxylic acids. For example, on the TiO_2 surface, the coordinately unsaturated Ti site is responsible for the adsorption of the carboxylic acid, while the lattice O-atom is responsible for dissociating the adsorbed acid to form a surface carboxylate species or deprotonating the carboxylate into an enolate. The authors have reported that the turnover rate on TiO_2 anatase is much higher than that of TiO_2 rutile due to the differences in the surface structure. That is, the distance between Ti and O sites in Ti–O pairs and the distance between the Ti centers of two adjacent Ti–O pairs on anatase are different compared with the rutile surface. In addition, the closer Ti–Ti centers on the rutile surface stabilize the binding of two O-atoms on the two Ti centers, resulting in the dominance of bidentate species on the surface. The bidentate species has been shown to be a spectator that deactivates the catalyst, leading to a lower condensation rate. A shorter distance between Ti centers also induces steric repulsion between neighboring monodentate species adsorbed on adjacent Ti–O pairs, leading to the destabilization of the monodentate adsorption mode on the rutile surface. The longer Ti–Ti distance on anatase, in contrast, favors the formation of the monodentate species, which is the reactive intermediate for the ketonization reaction, resulting in a higher turnover rate compared with that on the rutile surface. Finally, the kinetically relevant C–C coupling TS on anatase is more stable due to the stabilization effect via H-bonding between the TS and the vicinal surface OH species derived from the dissociation of co-adsorbed carboxylic acids. This stabilization is lacking on the rutile surface due to the longer distance between Ti and O sites in Ti–O pairs, which renders the TS and surface OH species farther apart.

In this mechanism, the exposed metal cations act as Lewis acid sites, while the oxygen anions function as Brønsted basic sites. However, the affinity of the carboxylic acid oxygen for the adsorption sites is not necessary due to the acidity of the site. Instead, it could also be associated with the oxophilicity of cations or oxygen vacancies. Therefore, ketonization activity may not be directly related to the acid–base properties of amphoteric oxides [265]. Studies by Resasco and coworkers on the ketonization of acetic acid on Ru/TiO_2 have shown that coordinatively unsaturated Ti^{3+} sites are crucial for the reaction and that under H_2 pre-treatment, Ru metal assists to increase the density of Ti^{3+} sites by enhancing the reducibility of TiO_2, resulting in an increase in rate [282]. Tosoni and Pacchioni have reported that on reduced zirconia, Zr^{3+} centers facilitate acyl radical formation from acetate ions. Moreover, the surface, which is rich in oxygen vacancies, helps to stabilize enolate and anion acyl species, which consequentially reduces the activation barrier for each step [291].

6.3 Optimization and Design of Catalytic Materials for C–C Bond Forming Reactions

6.3.1 Oxides

6.3.1.1 Magnesia (MgO)

Magnesium oxide is a typical basic oxide with the basic sites originating from low coordination surface oxygen atoms. Depending on the location of sites on different surface positions (edges, kinks, or corners) the basic strength of the sites can be different, basicity following the order: ${O_{5c}}^{2-}$ (terrace surface) < ${O_{4c}}^{2-}$ (edges) < ${O_{3c}}^{2-}$ (corners) (Figure 6.13a) [296]. MgO also manifests weak Lewis acidity due to the presence of Mg^{2+}. FTIR analysis on parent MgO has demonstrated three different CO_2 adsorption species with the following basic strength: unidentate carbonate (O^{2-} anions) > bidentate carbonate ($Mn^+–O^{2-}$) pairs > bicarbonate (OH^- groups) [6].

Commercial MgO has a very low surface area (~20 m^2/g) and concentration of basic sites, which limit its application in many fields such as adsorption, separation, and catalysis. Recent research has focused on developing synthesis or modification methods to control the structural/morphological properties of MgO and tailor them to specific applications, such as higher surface area for adsorption or higher acidity–basicity for activity enhancement in catalysis [299]. The modification of MgO can be conducted in two main ways: (i) direct modification on MgO via control of the synthesis conditions [300–315] or (ii) modification by incorporating foreign matter – either inorganic (such as Al_2O_3 [6, 316–318], ZrO_2 [10, 319, 320], B_2O_3 [321–325], TiO_2 [326, 327], SiO_2 [328–332], and alkali metals including Li, Na, K, and Cs [333, 334]) or organic (organosilanes, ethylene glycol, etc. [335]) – into MgO to produce hybrid materials with controllable properties.

6.3.1.1.1 Modification of MgO via Control of Synthesis Conditions One of the conventional techniques to synthesize high surface area MgO involves the thermal decomposition of appropriate precursors, which can be commercially available

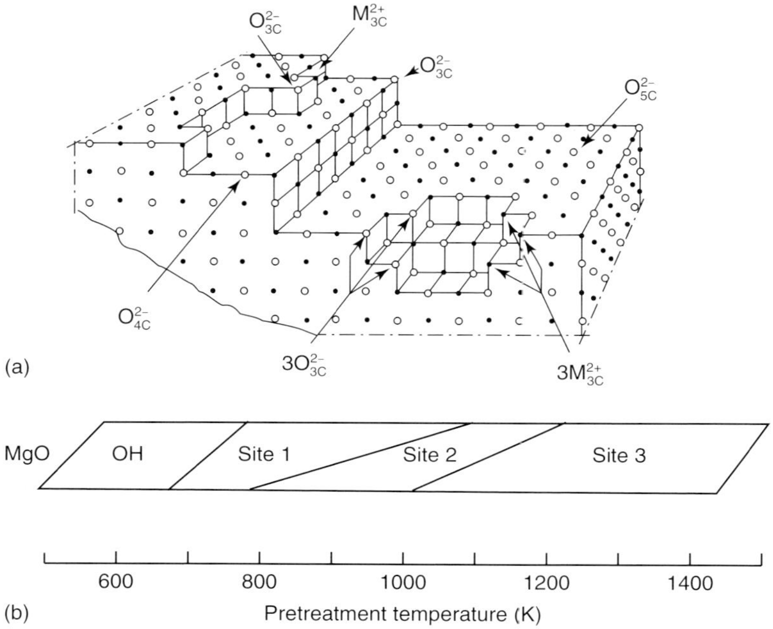

Figure 6.13 (a) Surface model of MgO. (b) Appearance of four site-types on alkaline earth oxides. Source: Reproduced with permission from Hattori 2004 [297] and Hattori 1988 [298].

or synthesized by precipitation or sol–gel methods [6, 10, 336]. The conditions of the thermal treatment (temperature, time, starting materials, etc.) tremendously affect the structure, the density of active surface species and surface area, and the catalytic properties of the oxide [298, 337]. For example, at temperatures in the range 400–600 °C, catalytically active magnesia (caustic phase) is produced from $Mg(OH)_2$ decomposition, while at higher temperatures, the inactive (burnt) form of MgO is formed. At temperatures above 1800 °C, magnesium oxide exists in the fully burnt form, which is industrially used in fire-resistant materials [337]. According to the MgO surface model [296], active sites with different coordination numbers can be generated at different thermal treatment temperatures. The weakest active site ($Mg^{2+}_{4C}O^{2-}_{4C}$ ion pair at the edge of the (100) plane; Site 1) is formed at the lowest temperature range (Figure 6.13b). At higher temperatures, the medium-strength site ($Mg^{2+}_{3C}O^{2-}_{4C}$ ion pair; Site 2) and the highest-strength site ($Mg^{2+}_{3C}O^{2-}_{3C}$; Site 3) can be created.

While the low surface area and density of basic sites of commercial MgO can be improved via a rehydration method [302, 312, 338, 339], several efforts have also been made to improve the basicity of MgO by controlling either particle morphology [340] or particle size [299]. For example, Climent et al. [299] have demonstrated that the initial rate of aldol condensation of acetone/aldehyde increases exponentially with decreasing crystal size. Nano-crystallite MgO (3 nm, 670 m^2/g) is reported to produce a 98% yield of aldol adducts with 100% selectivity. The change of MgO basicity with particle size is ascribed to the increasing number of low-coordinated sites (corner or step) as the particles get smaller. Nanoparticle MgO has been synthesized by thermal decomposition [308], sol–gel [311, 320, 341], sol–gel/surfactant [315, 342], wet chemical [343], and precipitation/surfactant [344] methods. The synthesis of MgO nanoparticles with different morphologies such as cube and hexagonal plate, spherical, random nanoflakes, and arranged nanoflakes toward flower and house-of-cards morphologies has proved to substantially increase the number of sites at edges, steps, and corners, which leads to high activity toward the aldol condensation of acetophenone and benzaldehyde compared with bulk MgO [345].

The synthesis of mesoporous MgO is also a promising approach to expand the surface area and basicity of the parent material. The larger pore size and more open porosity of this material can significantly promote the formation/diffusion of bulky compounds normally produced from C–C formation reactions, especially in the field of biomass conversion. Even though the application of mesoporous MgO in the field of catalysis is still largely unnoticed in the literature, its potential is enormous. Notably, many methods have been proposed for introducing mesoporosity into parent MgO [339, 341, 346–363].

6.3.1.1.2 Modification of MgO via Incorporation of Inorganic Materials into Parent MgO

The textural and acid–base properties of MgO can be controlled by incorporating different inorganic dopants into the parent MgO, including secondary oxides – SiO_2 [328], Al_2O_3 [10, 318, 364], ZrO_2 [10] – metallic ions [365–367], or alkali metals (Li, Na, K) [334]. The most common incorporation methods are coprecipitation [10, 319, 321, 328, 334, 336] and impregnation [318, 319, 364]. As discussed previously, depending on the type of C–C coupling reaction or the desired selectivity toward a specific product, the requirement for the active sites is different. For example, while side-chain alkylation demands strong basic sites, aldol condensation requires medium strength basic sites [10, 11, 145]. However, the existence of strong basic sites on the MgO surface in aldol condensation would promote the generation of strongly adsorbed polycyclic products, which rapidly deactivate the catalyst [17, 368]. The introduction of mixed oxides can also increase the acidity of the parent MgO, which has proven beneficial in some cases, e.g. facilitating the dehydration of aldol adducts [3, 12] or participating in catalytically cooperative effects between acidic and basic sites [18].

Mixed Oxides

MgO–Al_2O_3 and MgO–ZrO_2 Tsuji et al. [318] have investigated two different methods to incorporate γ-Al_2O_3 into MgO via impregnation and decomposition of hydrotalcite precursors. The addition of γ-Al_2O_3 has been shown to reduce the density of strong basic sites (O^{2-}), which normally exist on the surface of MgO [6, 316, 317]. Di Cosimo et al. [6] and Kikhtyanin et al. [3] have stated that for the MgO–Al_2O_3 system, at low alumina content, the density of the strong basic sites dramatically decreased with increasing Al% due to the coverage of amorphous AlO_y or isolated Al centers on the surface that block the strong basic sites (at values of Mg/Al > 5). At higher Al contents (5 > Mg/Al > 1), the basicity increased due to the emergence of more defect sites (low-coordinated oxygen). This is the result of the incorporation of Al^{3+} ions into the MgO matrix that generates charge compensation within the MgO lattice [6, 369]. Excess loading of Al (Mg/Al < 1) forms the bulk $MgAl_2O_4$ spinel phase and decreases basic site density. Additionally, the introduction of Al_2O_3 into MgO also generates Lewis acid sites, including strong Lewis sites coming from the Al^{3+} and weaker Lewis sites originating from Mg^{2+} in the MgO phase or mixed phase $MgAl_2O_4$ [6, 318, 369, 370]. This enhancement in Lewis acidity has shown to be important in controlling the cross selectivity of aldol condensation reactions and facilitating dehydration of aldol adducts [3, 12, 364].

The incorporation of Zr has been shown to exert similar effects on the acid–base properties of the parent MgO [10, 320, 371, 372]. Sádaba et al. have shown that the mixed oxide MgO–ZrO_2 has different active sites on the surface, depending on the Zr% loading. The site strengths are sorted in the order: Mg–O–Zr on MgO > Mg–O–Mg on MgO ≈ Mg–O–Zr on $Mg_xZr_{1-x}O_{2-x}$, the most active catalyst being the one that has the highest density of Mg–O–Zr on MgO (or Zr^{4+} species) [371]. Moreover, MgO–ZrO_2 has been dispersed on a support with the purpose of modifying its basicity relative to the bulk material (total density and type of basic sites) [319]. For example, on non-microporous carbonaceous materials such as high surface area graphite, the support helps to increase the dispersion of active sites, which consequentially increases the density of medium-strength sites. The intrinsic activity per site (bidentate sites) is also enhanced compared with the parent MgO as a result of the interaction with the support [319]. Another example of this approach can be found in the work of Faba et al. [373].

MgO–B_2O_3 MgO–B_2O_3 mixed oxide can be synthesized by either coprecipitation or impregnation [321–324]. Nevertheless, conventional methods tend to result in low surface area and pore connectivity, as well as in phase segregation [325]. Recently, Resasco and coworkers have suggested a novel method to introduce boron oxide (B_2O_3) into MgO by the combustion of citric acid at high temperature [325]. Compared with conventional techniques (coprecipitation and impregnation), this method can produce higher surface area metal oxide foam with a hierarchical meso- and macroporous structure. The catalyst surface includes two distinct phases: small MgO clusters enriched with trigonal coordinated boron species (BO_3) and a flat surface containing tetrahedral boron species (BO_4). The high mobility of B_2O_3 at high combustion temperature allows the formation of a glassy B_2O_3–metal oxide interface, which lessens the interfacial tension at the vapor–glass interface and stabilizes the large voids inside the porous structure. A surface covered by large clusters of Mg-enriched B–MgO, a smooth surface composed of glassy B_2O_3–MgO, and large cavities interconnected by mesoporous channels can be seen from TEM images (Figure 6.14a,i–iii). Homogeneous distribution of B and macroscopic pores can also be observed in SEM analysis (Figure 6.14b,i–iii). X-ray photoelectron spectroscopy (XPS) analysis shows that the addition of B shifted the Mg_{2p} peak to higher binding energy. This is because the presence of B^{3+} cations in MgO delocalizes the electronic cloud around the Mg^{2+}cations and makes the latter more electron deficient. At the same time, the O_{1s} peak also

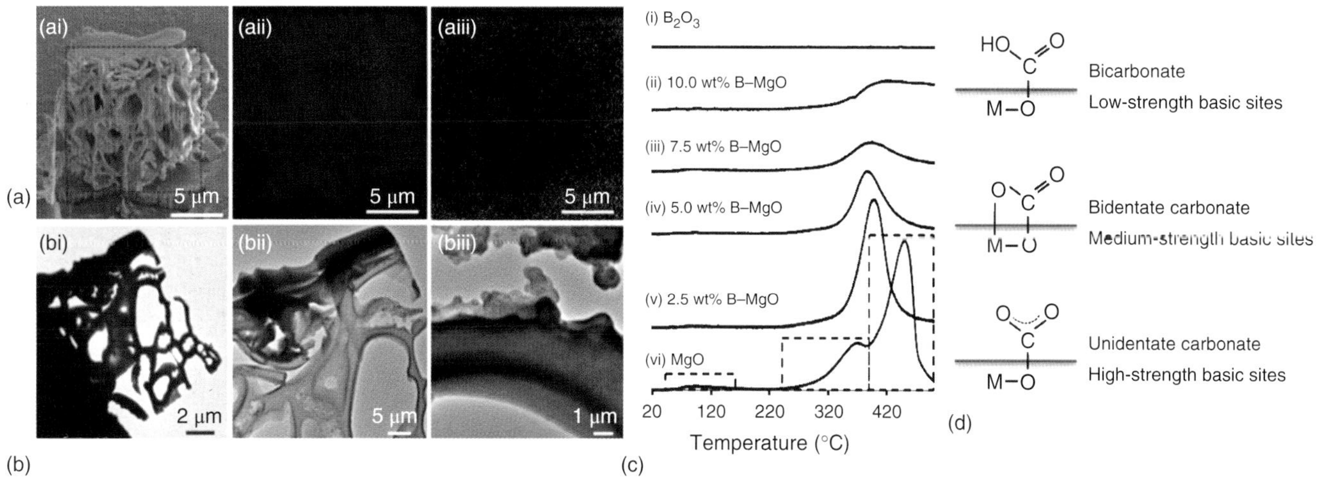

Figure 6.14 Cross-sectional characterization of the internal structure of B–MgO. (a) (i) SEM analysis of the sample during focused ion beam (FIB) etching. (ii, iii) Energy-dispersive X-ray spectroscopy (EDX) mapping of the cross section demonstrating the microscopic distribution of Mg (red, ii) and boron (yellow, iii). (b) (i–iii) TEM analysis of the cross section at three different magnifications. (c) TPD profiles of CO_2 over MgO, B_2O_3, and mixed B_2O_3–MgO catalysts. (d) Different CO_2 adsorption modes on MgO [379]. Source: Pham et al. 2016 [325]. Reproduced with permission of John Wiley and Sons.

shifted to higher binding energy, indicating that O^{2-} ions become less electron rich. As a consequence, the basicity of $MgO-B_2O_3$ decreased as shown in Figure 6.14c. The incorporation of B substantially reduces the unidentate sites while increasing the density of bidentate sites (2.5 wt% B–MgO; Figure 6.14c). The basicity of B–MgO continued to decrease with increasing B loading. At the same time, a slow increase in acidity was observed at high B loadings (according to pyridine–TPD measurements). The catalyst was tested for aldol condensation/Meerwein–Ponndorf–Verley (MPV) reactions of ethanol and acetaldehyde in the liquid phase at 250 °C. By controlling the proper loading of B, the catalyst could be specifically designed for the target reaction. For example, the highest yield toward the desired products (C_4 aldol and C_4 MPV products) could be achieved at 7.5 wt% B on MgO, which manifests the proper balance of acid–base properties, meso-/macroporosity, and surface area.

Other MgO-based binary oxide systems such as $MgO-SiO_2$ (sol–gel), MgO supported on SiO_2 (impregnation) [328–332], $MgO-TiO_2$ (sol–gel) [326, 327], MgO–supported on mesoporous silica, $MgAl_2O_4$ [374], and ZnMgAl [11] – have also been reported in the literature [375]. Besides conventional synthesis methods, other techniques have been introduced to synthesize MgO-based mixed oxides with controllable textural/basic properties, including a peroxo-route [376], coprecipitation with ethanol solvent [377], and coprecipitation combined with sonication (ultrasound) [378].

Alkali-Metal Díez et al. [333] have studied the catalytic performance of MgO doped with alkali metal (denoted as A) such as Li, Na, K, and Cs for the aldol condensation of citral and acetone. The basicity of alkali metal oxides increases when going from Li to Cs due to increasing negative charge on the oxygen atom ($-q_O$). However, as the atomic weight of A increases, the corresponding A_2O oxide becomes increasingly bulky. For alkali metals with cationic radii larger than Mg^{2+}, the bulky A_2O phase tends to block the catalyst pores and basic sites. As a result, lower citral conversion and pseudoionones (PS) yield (cross aldol adduct) were obtained. Indeed, Li/MgO has been shown to be the most active catalyst for the reaction. According to CO_2–TPD, the basicity increased monotonically with Li loading up to 0.5 wt%. This enhancement is caused by the spreading of the more basic Li_2O phase ($-q_O = 0.8$) on the MgO surface ($-q_O = 0.5$), leading to the increased electron-donating property of the oxygen atoms. At Li loadings of >0.5 wt%, the base site density decreased due to particle agglomeration and the formation of a stable lithium carbonate phase that blocks the active surface of MgO [19], resulting in the decrease of the PS formation rate [333]. Doping with alkali metals (Li, Na, K) has also been conducted on mixed oxide systems such as $MgO-Al_2O_3$ hydrotalcite [334]. However, this type of catalyst suffers from severe leaching of alkalinity into the reaction medium, as has been reported by Abelló et al. [334]. This might be a critical problem in terms of practical applications.

6.3.1.1.3 Modification of MgO via Incorporation of Organic Materials into Parent MgO The common purpose of integrating organic compounds with MgO is to create a hydrophobic surface, which can improve the interface compatibility between inorganic and organic phases [380]. These types of materials have been applied in polymer fillers, adsorbents [335, 337, 381, 382], and chromatography [335]. Recently, Resasco and coworkers proposed a novel method to synthesize hydrophobic MgO with octadecyltrichlorosilane (OTS), which is then applied as a catalyst for the aldol condensation of cyclopentanone [383]. The parent MgO was directly functionalized with OTS or coated with a thin layer of silica to create a core–shell structure before being functionalized with OTS. The mesoporous silica layer provides not only a high density of surface silanols for the functionalization process but also easy accessibility for the reactants to reach internal active sites (Scheme 6.11). CO_2–TPD analysis has shown a decrease in strong basic site density as well as in overall basicity on both functionalized catalysts. It is possible that during the functionalization, the OTS selectively titrates and blocks the most basic sites. However, the functionalized catalysts were more resistant to the deactivation caused by either water or heavy side products, thanks to the appropriate basicity and the hydrophobic linkers. This modification technique has been proven to be a potential way of enhancing catalyst stability against water, which is found in abundance in bio-oil.

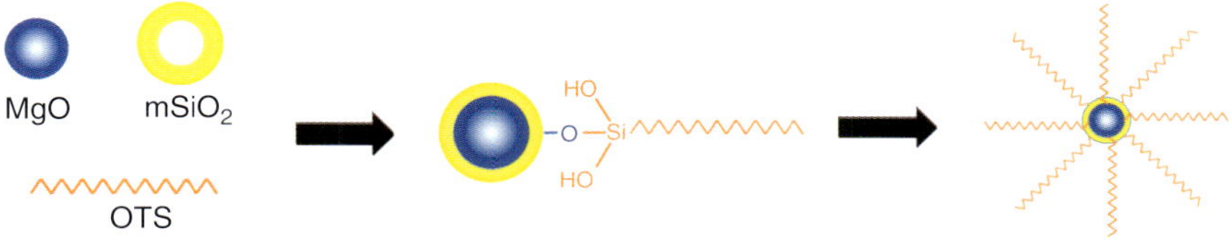

Scheme 6.11 Illustration of core–shell MgO@SiO_2–OTS catalyst system.

6.3.1.2 Zirconia (ZrO_2)

Zirconium(IV) oxide (ZrO_2), also known as zirconium dioxide or zirconia, is one of the most commonly used materials in catalysis. Due to its excellent physicochemical properties such as high melting point (over 2973 K) and high thermochemical stability under reductive and oxidative environments, ZrO_2 has been extensively utilized as a catalyst support, being even better than γ-alumina or silica under some circumstances [384]. The coexistence of Lewis acid and Brønsted base site pairs (Zr^{4+}–O^{2-}) also allows ZrO_2 to function as a catalyst itself for various chemical reactions including CO hydrogenation, Fischer–Tropsch synthesis, aldol condensation, acylation, and alkylation [385, 386].

There are currently three known phases of ZrO_2 (Figure 6.15): monoclinic ZrO_2 at temperatures below 1443 K, tetragonal ZrO_2 from 1443 to 2643 K, and cubic ZrO_2 at temperatures beyond 2643 K [387, 388]. Tetragonal ZrO_2 has been shown to possess high catalytic activity, but it is usually transformed into monoclinic ZrO_2 in the common range of calcination and reaction temperatures, a phenomenon associated with the transition of the crystalline structure into a less ordered state due to changes in volume and detrimental cracks within the lattice structure induced by heating or cooling processes [389–395]. Moreover, ZrO_2 is also prone to deactivation by water as a consequence of sintering of the active sites [396, 397]. Such issues have motivated studies on surface

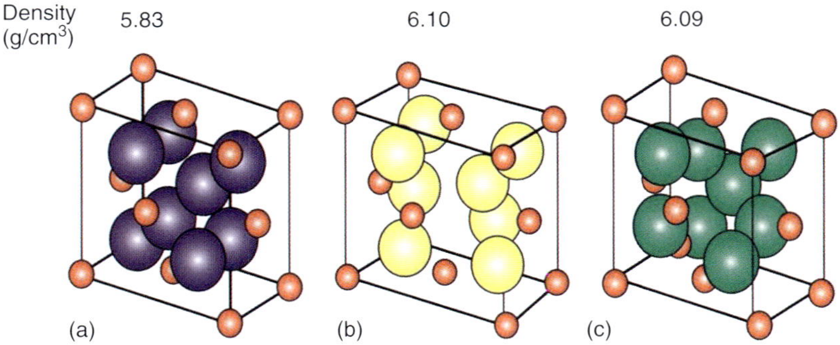

Figure 6.15 Crystal structure of monoclinic (a), tetragonal (b), and cubic zirconia (c). Red spheres represent O atoms, while the other colors represent Zr atoms. Source: Volpato et al. 2011 [388]. Reproduced with permission of Intechopen.

modification to not only improve the water resistance but also preserve the valuable tetragonal phase during reactions at elevated temperatures. The surface properties of zirconia (acidic/basic, hydrophobic/hydrophilic) can be adjusted through manipulation of synthesis conditions [398–400], incorporation of either inorganic materials (e.g. metal oxides, sulfate anions) [401] or organic materials (e.g. perfluoroalkylsilanes) [402–404], or even photo-excitation [405]. To gain more insights into these aspects, this section will address the manner by which typical dopants such as MgO, ZnO, and CeO_2 can be used to tune the overall acidity–basicity of ZrO_2 in industrially applicable aldol condensations, followed by the use of carbon to stabilize tetragonal ZrO_2 in aqueous media [406]. The modification of ZrO_2 with MgO has been discussed previously in Section 6.3.1.1.

6.3.1.2.1 Modification of ZrO_2 via Incorporation of Inorganic Materials into Parent ZrO_2

Zinc Oxide (ZnO) The acid–base properties and catalytic activity of ZrO_2 can be modified by the incorporation of ZnO. Sun et al. [407] proposed the addition of ZnO into ZrO_2 to passivate strong Lewis acid sites and weaken Brønsted acid sites while improving the basicity required for the direct conversion of bio-ethanol to isobutene (ETIB), a valuable intermediate for the production of fuels and chemicals. Diffuse reflectance infrared Fourier transform spectroscopy (DRIFTS) measurements of adsorbed pyridine show that pure ZrO_2 has both Brønsted and strong Lewis acid sites, while pure ZnO has only weak Lewis acid sites in low concentration. The incorporation of ZnO into ZrO_2 through the hard template method largely neutralizes both Brønsted and strong Lewis acid sites on ZrO_2 but does not generate new acid sites. In addition, CO_2–TPD spectra indicate that pure ZrO_2 has both weak Lewis basic oxygen atoms, corresponding to surface hydroxyl groups, and strong Lewis acid–Brønsted base site pairs (Zr^{4+}–O^{2-}), while pure ZnO only has a smaller amount of the latter type. The addition of ZnO into ZrO_2 also attenuates both types of basic sites without forming new desorption peaks. Results from DRIFTS of adsorbed pyridine and CO_2–TPD suggest that a $Zn_xZr_yO_z$ catalyst mostly contains weak Lewis acid sites along with moderate amounts of surface hydroxyl groups on Zr^{4+} and acid–base site pairs. Another example can be found in the work of Crisci et al. [408].

Ceria (CeO_2) CeO_2–ZrO_2 has also been used as a catalyst to assist C–C coupling reactions [409, 410]. For instance, Kumar et al. [411] prepared CeO_2–ZrO_2 (32.2% CeO_2) via the sol–gel method and tested it for activity in the aldol condensation of heptanal and benzaldehyde to produce jasminaldehyde. In Ce-doped ZrO_2, X-ray diffraction (XRD) peaks of fluorite CeO_2 emerge along with those of tetragonal ZrO_2, but there is no trace of monoclinic ZrO_2. The disappearance of monoclinic ZrO_2 in the mixed oxide suggests that Ce^{4+} species effectively stabilize the tetragonal ZrO_2 structure. Brunauer–Emmett–Teller (BET) and TPD results show that the incorporation of the more basic CeO_2 into the more acidic ZrO_2 not only increases the surface area but also optimizes the acidic and basic densities. While weak acidic and basic sites dominate in pure ZrO_2 and CeO_2, Ce-doped ZrO_2 possesses greater proportions of medium-strength acidic and basic sites. Under optimal reaction conditions, CeO_2–ZrO_2 is much more active and more selective to jasminaldehyde than pure CeO_2 and pure ZrO_2.

Carbonization Apart from doping with various metal oxides, carbonization has been considered a promising approach to enhance the catalytic activity of ZrO_2. Wu et al. [412] have shown that relative to ZrO_2, carbonized ZrO_2/C catalysts display an enhanced yield of acetone from acetic acid ketonization along with superior stability in aqueous reaction media. ZrO_2 solids prepared via the Zr-based MOF route and sol–gel method have well-dispersed surface C–Zr species existing at the edge between ZrO_2 and graphite, which correlates with remarkably high activities in aqueous-phase ketonization. The authors proposed that the strong interaction between graphite and ZrO_2 enhances ketonization activity. Moreover, for the carbonized ZrO_2 catalysts, these carbon species are sufficiently exposed to the bulk phase and readily capture acetic acid to enrich its surface concentration, thus benefiting the acid chemisorption on finely dispersed tetragonal ZrO_2 islets and C–Zr boundaries. Under this circumstance, the competitive adsorption of water as either a solvent or a side product is significantly weakened due to hydrophobicity (as depicted in Scheme 6.12).

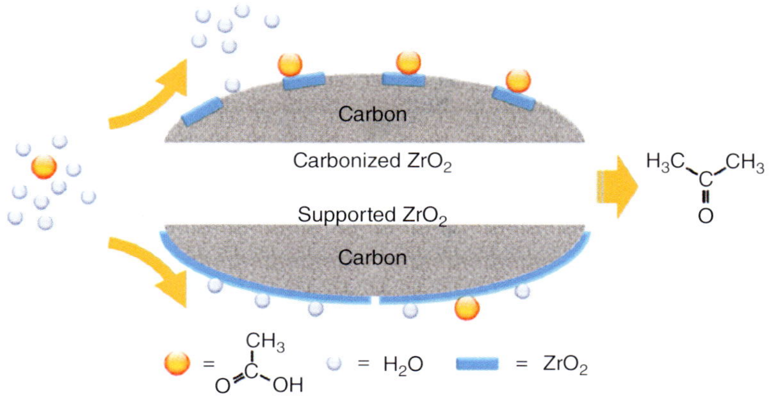

Scheme 6.12 Different catalytic behaviors of carbonized ZrO_2 and ZrO_2 supported on CNT during aqueous-phase acetic acid ketonization. Source: Reproduced with permission from Wu et al. 2017 [412].

6.3.2 Zeolites

6.3.2.1 ZSM-5

The MFI structure is constructed as a combination of two interconnected channel systems, i.e. sinusoidal 10-membered rings (MR) channels running along the direction of the *a*-axis interconnected with the 10-MR straight channels that run down the *b*-axis [413]. ZSM-5 is well known as a shape-selective catalyst in many industrial processes due to its well-defined microporous structure. Even so, for those reactions that involve the formation of bulky compounds such as C–C coupling reactions, this microporosity becomes a disadvantage, hindering mass transfer of reactants to active sites located within micropores and of the products from the micropores back to the bulk phase [414–418]. Moreover, the high density of strong acid sites on HZSM-5 can facilitate the generation of side products and coke formation, which leads to catalyst deactivation and low selectivity toward desired products [419–421]. Several strategies have been reported in the literature to improve the catalytic performance of ZSM-5 zeolites via modifying the solid structure, diffusion pathway, and acid density/strength [422], including the following:

- Introducing mesoporosity into ZSM-5 to enhance the diffusion rate of reactants or bulky products as well as the accessibility to acid sites [216, 416, 418, 423–436].
- Synthesizing nanoparticles of ZSM-5 crystals to shorten the diffusion path [428, 437–439].
- Controlling and modifying the type (Brønsted vs. Lewis), strength, and density of acid sites as well as the pore opening size via incorporating additive compounds, heteroatoms, or oxides into ZSM-5 [440–453].
- Synthesizing catalysts with special morphologies to shorten diffusion pathways via increasing the meso-surface [417, 454].

6.3.2.1.1 Mesoporous ZSM-5 Diffusion limitations associated with the microporosity of ZSM-5 zeolites have been the major problem leading to low conversion and selectivity toward bulky condensation products resulting from alkylation [414, 416, 431, 432] or acylation reactions [417]. Indeed, the inactivity of H-ZSM5 toward the formation of large molecules can be attributed to its small pore openings [418]. In order to enhance the diffusion of reactants and products, as well as the accessibility to acid sites located inside the porous structure of ZSM-5, mesoporosity can be introduced into ZSM-5 catalysts. Several methods to accomplish this have been proposed in the literature, including either direct synthetic [416, 455–457] or post-synthetic – such as alkali-treatment and acid wash [434] – approaches. Methods for introducing mesoporosity into zeolites have been summarized in a review authored by Wei et al. [433].

Milina et al. [434] have investigated the effect of the mesoporosity of ZSM-5 synthesized by alkaline treatment on the activity of this material in the alkylation of benzyl alcohol (BA) with toluene and cyclohexylbenzene. It was shown that as mesoporosity was introduced into the parent ZSM-5, the BA conversion significantly increased (i.e. from 12% to 83%). Interestingly, a higher BA conversion was obtained by increasing the mesopore surface, which demonstrates a direct involvement of the mesoporosity in the rate enhancement. The authors have suggested that the value of mesoporous surface area could be potentially used as a more precise proxy for controlling the activity of hierarchical zeolites than mesopore volume. Notably, the application of mesoporous zeolite single crystals has also been studied for the alkylation of benzene and ethylene to ethylbenzene (EB) [435, 436].

6.3.2.1.2 Nanoparticle ZSM-5 In general, nanoparticles lead to an increase in the active sites on the external surface, which shortens the diffusion path of reactants and products, so high activity and less deactivation can be achieved [202, 437, 438, 458]. For example, while commercial ZSM-5 catalysts showed poor activity for the acylation of anisole and acetic anhydride [202], nanoparticulate ZSM-5 (with a particle size of 18.3 nm) yielded 90% conversion under the following reaction conditions: 100 °C, 0.15 g catalyst, anisole/acetic anhydride = 8 : 1. Moreover, the catalytic activity of nanoparticulate ZSM-5 has been reported to increase with decreasing particle size, which has been ascribed to the increase in external surface [458].

6.3.2.1.3 Modifying Acidity of ZSM-5 via Metal or Oxide Additives The main purpose of this strategy is to modify and control the acidic properties (acid density, Brønsted/Lewis acid sites ratio) of the parent ZSM-5 zeolite to improve yield and selectivity toward a particularly desired product as well as to impede a number of side reactions that could potentially deactivate the catalyst. The latter includes highly condensed coupling products caused by the excess acidity of strong acid sites. For example, in the alkylation of olefins with activated aromatic substrates such as phenol, toluene, or xylene, the unmodified ZSM-5 can facilitate the formation of multi-alkylated species even at low temperature, leading to low selectivity of desired products and quick deactivation [421, 459]. Therefore, in this case a low acid density is generally desired. Another side reaction to be avoided when methanol is used as the alkylation agent is the conversion of methanol to olefins or aromatics [460]. The formed olefins can either create coke or react with benzene to form alkylated products, which would decrease the selectivity toward desired products such as *p*-xylene [415, 416, 432, 461, 462]. Incidentally, the isomerization of *p*-xylene to *o*- and *m*-xylene – which occurs on the external Brønsted acid sites [449, 451, 463, 464] – should also be avoided, as this process also reduces the selectivity of the desired *p*-xylene product. Finally, another side reaction to avoid when toluene is used as an alkylating substrate with an alcohol (e.g. methanol in *p*-xylene production) is the disproportionation of toluene, which can occur on strong acid sites to form benzene, resulting in a decrease in xylene selectivity [465].

A wide range of metals and oxides have been incorporated as additives into parent ZSM-5 zeolites. Depending on the target reactions and products, a lower density of Brønsted acidity either on the external surface or inside the porous structure would be desired [460, 465–469]. Additives used to date include metals such as Mg [443, 449, 451, 452, 467], Sr [451, 470], Ba, Co, Ni [451, 471], Al, Fe, Ga [18, 469, 471–473], B [18, 474, 475], P [475, 476], Zn [459, 469, 477], Ag [471], Cu, Rh, Co, Zr, V, Pd, Cr, La, and Ce [478] and oxides such as SiO_2 [421, 452, 466, 479], P_2O_5 [466, 468], MgO [450, 466], ZnO [431, 477, 479], Fe_2O_3, Bi_2O_3, La_2O_3, ZrO_2 [468], B_2O_3 [450, 468], Co_3O_4–La_2O_3 [480], and P_2O_5–ZnO/ZSM-5 [468], as well as organic substances [481, 482]. It should be noted that the metals added can exist in different oxidation states depending on the synthetic conditions employed. Moreover, metals can be present as nano-metal clusters, cations, oxides, or mixed oxides with the framework Al [471]. The presence of such species inside the zeolite channels can modify the topology of the parent materials, influencing the transport to and from – and thus the accessibility of – the active sites.

Mg and Mg-based Oxides: Ding et al. [467] have investigated the alkali treatment of HZMS-5 followed by doping with Mg via impregnation. The alkali treatment generates mesopores together with additional extra framework Al (EFAL). It has been proposed that the interaction of $Mg(OH)^+$ species with the EFAL on the surface gives rise to extra framework Lewis acid sites (Scheme 6.13). Higher Lewis/Brønsted (L/B) ratios led to higher catalytic activity in the ethylation of benzene, as well as high selectivity toward EB. For example, a 1 wt% Mg/alkali-treated ZSM-5 sample, which possessed the highest L/B ratio (3.79) among the samples tested, showed the highest productivity (30%) and selectivity to EB (92%). However, higher loadings of Mg (>5%) decreased both Brønsted and Lewis acidity due to the condensation of $Mg(OH)^+$ species into MgO, which can block the active sites [423].

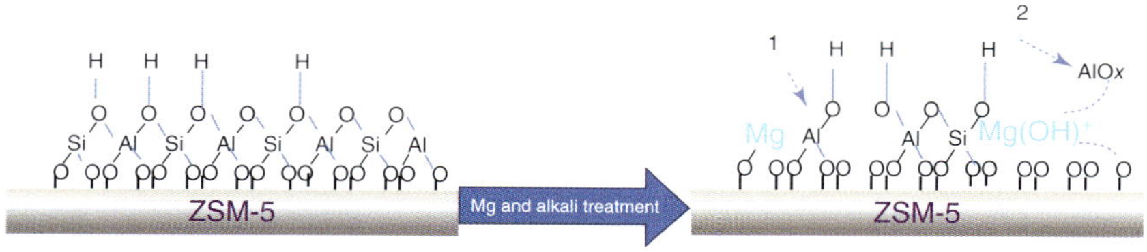

Scheme 6.13 The formation of Lewis acid sites by alkali-treatment and Mg-promotion: (1) framework Al Lewis site, (2) extra framework Al Lewis site [467].

Li et al. [443] have proposed solid-state ion exchange as an efficient method for incorporating metal cations into zeolites, a technique that has also been reported by other authors [483–489]. The addition of Mg is reported to (i) decrease the concentration of strong Brønsted acid sites, (ii) increase the Lewis acidity due to the replacement of Mg^{2+} for protons, and (iii) create diffusional resistance within the zeolite channels, which can improve the selectivity toward *p*-xylene due to its smaller size compared with the other isomers. The Mg–ZSM-5 catalyst exhibited a lower toluene conversion, but an excellent selectivity toward *p*-xylene, reaching 92% selectivity at Mg loadings above 3.98 wt%. Similar effects of Mg-modification have also been observed for the selective production of *p*-xylene via toluene alkylation with methanol [450] or toluene disproportionation [452].

B and P-based Oxides: Chen and Feng [474] have modified ZSM-5 with boron by adding boric acid during catalyst synthesis for toluene alkylation. Their report states that the addition of B reduces the amount of total acid sites as well as strong acid sites, resulting in a decrease of toluene conversion but an increase of *para*-selectivity (82%). Similarly, ZSM-5 modified with P and B via impregnation has also shown a decrease not only in the density of strong acid sites on the external and the internal surfaces but also in the effective pore size of the zeolite channels [475]. As a result, side products of the alkylation reaction such as multi-alkylated adducts could be minimized. The addition of P to ZSM-5 also enhances the aldol condensation rate of methyl acetate and formaldehyde [453] as well as the yield of – and selectivity to – xylene in the alkylation of benzene with methanol [476]. The effects of phosphorus modification on the porosity, accessibility, acidity, and hydrothermal stability of ZSM-5 and other zeolites have been explicitly discussed by van der Bij and Weckhuysen in a recent review [490].

Al, Ga, Fe, Zn, and Their Oxides: Choudhary et al. [472] and Coq et al. [473] have reported that HZMS-5 shows no activity in the alkylation of benzene and benzyl chloride. However, the partial or complete substitution of the framework Al by Fe or Ga has been shown capable of enhancing the alkylation activity of the zeolite. The effect has been attributed to synergistic effects between non-framework Ga or Fe oxide species and adjacent Brønsted acid sites [491, 492]. Specifically, the redox functions of non-framework Ga and Fe oxides assist in activating the benzene nucleus, rendering it more active for the nucleophilic attack of an activated benzyl chloride on the adjacent Brønsted acid site. The acid properties of MFI zeolites have also been modified by the substitution of Al, Ga, Fe, and B for the aldol condensation of acetaldehyde and formaldehyde [18]. The addition of M^{3+} cations generates Lewis acid sites, which have been shown to participate in the cooperative mechanism with Brønsted acid sites of the aldol condensation reaction.

Phatanasri et al. [469] have incorporated Fe or Zn into H-MFI zeolites through a rapid crystallization method, finding that the modified catalysts with high density of moderately acidic sites exhibited the highest *p*-xylene selectivity. It has been reported that the modification of hierarchically porous ZSM-5 zeolites with Zn could simultaneously improve their catalytic activity in the alkylation of benzene with alcohols and minimize side reactions – such as methanol to olefins or polymerization – by decreasing their Brønsted/Lewis acid sites (B/L) ratio and crystalline particle size [202, 431, 458, 459, 477].

Si and Si-based Oxides: The disproportionation of toluene to form benzene is another side reaction that usually occurs when toluene is used as the alkylating substrate, leading to the decrease of *p*-xylene selectivity [465]. The required strength of the acid sites for the alkylation and the associated side reactions follows the order: toluene disproportionation > alkylation > methanol to olefins. As the density of strong Brønsted acid sites decreases, the selectivity to benzene (the product of toluene disproportionation) diminishes [465]. It has been shown that only medium-strength sites are required for alkylation, while strong sites solely promote side reactions. The composite material of ZSM-5 and MCM-41 reported by Alabi et al. [493] shows that the incorporation of silica reduces the density of strong Brønsted acid sites, which curbs toluene disproportionation and improves the yield of (and selectivity to) alkylated products.

The role of the external surface of ZSM-5 in determining the selectivity of alkylation reactions has been widely reported in the literature [448, 449]. Several methods have been proposed to reduce the external active surface of the material, such as coating ZSM-5 with polycrystalline silicalite crystals [442], silicalite-1 [494], or SiO_2 [461, 479, 495]. For example, H-ZSM-5 crystals coated with oriented poly-crystallites via hydrothermal synthesis exhibited an excellent (>99.9%) selectivity toward *p*-xylene and good resistance against coke formation in the alkylation of toluene with methanol [442]. Other materials such as kaolinite [421], silicon polymer [452], P_2O_5, and MgO [466] have also been used to selectively deactivate the external acid sites of zeolite crystals.

Noble Metals: The addition of noble metals with high hydrogenation activity – such as Pt and Pd – leads to significant improvements in the catalytic stability of ZSM-5 zeolites [496–498]. For instance, Pt has been used to impede the deactivation of modified ZSM-5 catalysts for alkylation of toluene with methanol. Indeed, Pt-modified ZSM-5 maintained stable activity for over 180 hours on stream, while in the absence of Pt, a quick loss of catalytic activity was observed after 60 hours on stream. This stabilization effect of Pt has been attributed to its ability to promote the hydrogenation of olefins or unsaturated hydrocarbons, the side products of methanol conversion [414, 496].

Other Metals and Oxides: Other elements such as Mg, Sr, Ba, Co, and Ni have been used to enhance *p*-selectivity for methanol–toluene alkylation [451], while additives such as Fe_2O_3, Bi_2O_3, B_2O_3, La_2O_3, and ZrO_2 have been employed for the alkylation of benzene with CH_3Br [468]. The effect of different metals such as Cu, Co, Ce, Zr, V, Pd, Cr, La, and Rh on the activity of ZSM-5 in the acylation of phenol with acetic anhydride has also been reported [478], along with that of Ag, Ni, Fe on the activity of ZSM-5 in the acylation of anisole and propanoic acid [471]. Moreover, the modification of ZSM-5 used for the selective production of *p*-xylene with other additives such as Co_3O_4 and La_2O_3 [419, 480], CeO_2 [419], and P–Zn [468, 495] has also been investigated.

Organic Substances: Haw and coworkers [481, 482] have examined the hydrosilation of Brønsted acid sites on HZSM-5 with phenylsilane and phenyltrimethyl silane (Scheme 6.14). The functionalization is reversible, which means that the Brønsted acid sites could be fully regenerated by heating the catalyst at high temperature (773 K) under flowing air without any detectable change in catalytic activity. Moreover, the weak bond strength of Si–H compared with Si–O enables the surface to react with other organic compounds such as acetone as shown in Scheme 6.14. This technique might be useful in controlling the density of Brønsted acid sites, which could minimize side reactions and enhance the selectivity toward desired products as discussed earlier. Moreover, the Brønsted acid

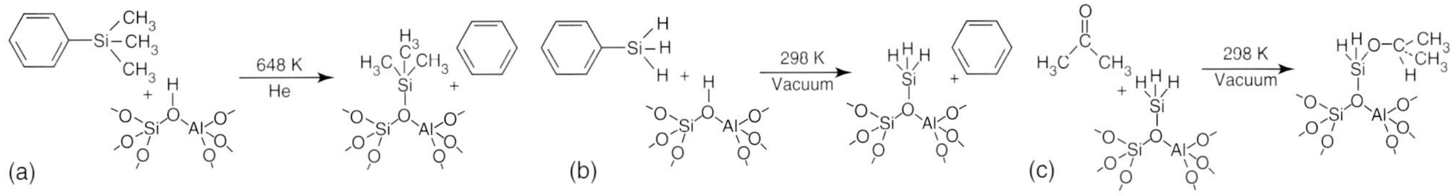

Scheme 6.14 The hydrosilylation of Brønsted acid sites with (a) phenyltrimethyl silane, (b) phenylsilane, and (c) the coupling of a framework hydrosilyl group with acetone. Source: Reproduced with permission from Abubakar et al. 2006 [481].

sites could be easily regenerated, which renders the catalyst adaptable for different applications. The acidity of ZSM-5 could also be enhanced via modification with sulfuric acid groups, as has been reported elsewhere [499–501].

6.3.2.1.4 Modifying Parent ZSM-5 Through Changes in Topology The generation of external active sites by controlling catalyst morphology has been reported as an effective way to increase catalytic activity in acylation reactions [417, 454]. Kim et al. [417] have investigated the activity of nanosponge MFI zeolites with a narrow distribution of mesopores for the liquid phase acylation of aromatic compounds with anhydride-type acylation agents. The catalyst is constructed of crystalline layers (2.5 nm thick) of MFI zeolite, which are self-connected through 4 nm pillars. Large mesopore volume was detected (0.5 cm^3/g) with a high density of external Brønsted acid sites (53 μmol/g) compared with bulk MFI (7 μmol/g) and beta zeolite (6 μmol/g). As a result, the MFI nanosponge exhibited activity three times higher relative to that of the beta zeolite (Table 6.1). Moreover, the high density of Brønsted acid sites located on the external surface allows the nanosponge catalyst to be relatively active in reactions involving bulky substrates (such as naphthalene) or bulky acylating agents (such as hexanoic anhydride). Less deactivation was also observed on the nanosponge catalyst, which maintained 97% of its initial activity after five reaction–regeneration cycles. This has been attributed to a large fraction of active sites being deposited on the mesopore walls, since that facilitates regeneration via solvent washing.

6.3.2.2 HY

Zeolite Y belongs to the family of zeolites with Faujasite (FAU) structure with the general formula $|(Ca, Mg, Na_2)_{29}(H_2O)_{240}|[Al_{58}Si_{134}O_{384}]$ [502]. The high acidity and high void fraction (0.48) of HY zeolites combined with their three-dimensional pore structures and large super cages have made them valuable catalysts for petrochemical applications [503, 504] as well as for biomass conversion [505]. Nevertheless, two important challenges arise from the different nature of biomass compared with petrochemical feedstocks [505–508]: the low stability of zeolite HY in hot liquid water [509] and the high degree of microporosity of zeolite HY, which could become an obstacle as mass transport limitations become relevant in the presence of bulky organic molecules derived from biomass depolymerization [505–508]. Recently proposed strategies to overcome these difficulties are discussed hereinafter. Briefly, while the hot water tolerance of zeolite HY can be improved dramatically by hydrophobization with organosilanes [510–513], mass transport properties can be enhanced via introduction of a secondary level of porosity (hierarchical zeolites) [514–517].

6.3.2.2.1 Modification of HY via Organosilane Hydrophobization The poor hydrothermal stability of zeolite Y limits its applications in biomass conversion and bio-oil upgrading [509]. Zhang et al. have proposed that the hydrophilic silanol groups on the surface of the zeolites facilitate the condensation of water, which initiates the deconstruction of the crystalline structure. Therefore, decreasing the number of defects on the zeolite by adjusting the

Table 6.1 Conversion of substrate in the FC acylation over the MFI nanosponge and bulk beta catalysts at 20 hours of reaction time[a] [417].

Entry	Substrate	Acylating agent	Temperature (K)	Substrate conversion (g_{cat}^{-1}) MFI nano-sponge	Bulk beta
1	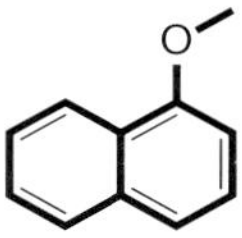	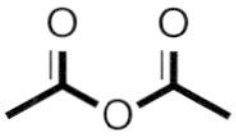	393	14	n.d.[b]
2[c]	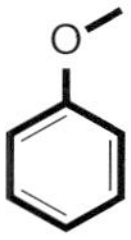	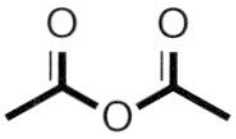	393	62	20
3	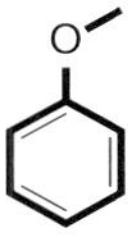	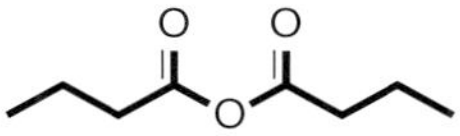	413	10	n.d.
4	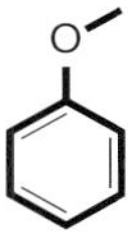	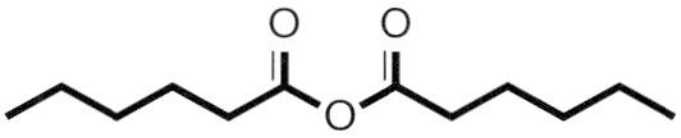	413	6	n.d.
5	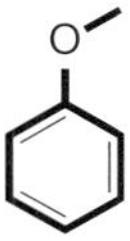	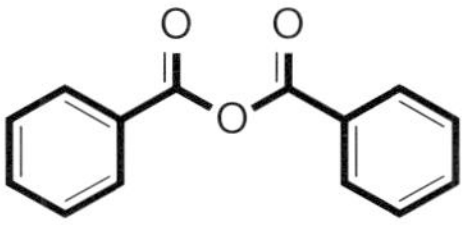	423	17	n.d.
6	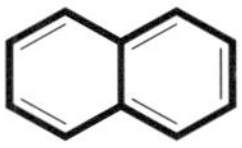	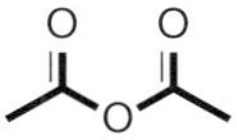	433	2	n.d.
7	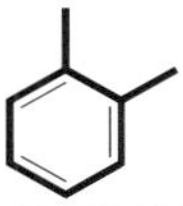	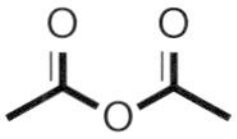	403	11	n.d.

a) Reaction conditions: 1 mmol of aromatic compound, 2 mmol of acylating agent, 4 ml of nitrobenzene, 50 mg of catalyst.
b) Not detected.
c) Reaction conditions: 4 mmol of anisole, 8 mmol of acetic anhydride, 4 ml of nitrobenzene, 50 mg of catalyst.

synthesis protocols or via post-treatments such as hydrophobization could effectively help improve zeolite stability [512].

Zeolite HY functionalized with octadecyl-trichlorosilane (HY-OTS) has been used in the alkylation of *m*-cresol with 2-propanol in biphasic decalin-water [510, 511]. The hydrophilic "untreated" zeolite lost its catalytic activity after three hours of reaction at 200 °C, whereas the hydrophobic HY remained highly

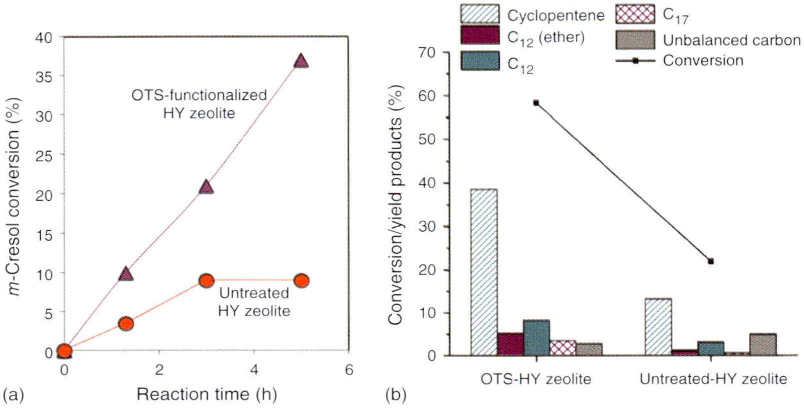

Figure 6.16 (a) Conversion of *m*-cresol as a function of reaction time during alkylation with 2-propanol at 200 °C and 700 psi in He over two HY zeolites (untreated and OTS-functionalized). Source: Zapata et al. 2012 [510]. Reproduced with permission of American Chemical Society. (b) Conversion and product distribution from alkylation of *m*-cresol with cyclopentanol over functionalized H-USY and untreated H-USY zeolites (Si/Al = 30). The reactions were carried out in biphasic system (decalin-water) at 160 °C, 20 bar N_2, reaction time three hours and *m*-cresol/cyclopentanol = 3/1. Source: Pino et al. 2018 [513]. Reproduced with permission of Elsevier.

active even after five hours of reaction [510] (Figure 6.16a). Tellingly, X-ray diffraction patterns demonstrated that the untreated zeolite HY lost almost all its crystallinity, while the hydrophobized HY-OTS remained largely intact. Hydrophobized zeolites have also been employed in the alkylation of *m*-cresol with cyclopentanol [513]. The hydrophobic zeolite showed higher activity and stability than its untreated counterpart at similar reaction conditions (Figure 6.16b). N_2 physisorption and XRD measurements showed that the significant losses in activity experienced by the untreated zeolite Y were due to the collapse of the crystalline structure.

6.3.2.2.2 Modification of HY By Introduction of Mesoporosity (Hierarchical Zeolite HY)

Hierarchical (mesoporous) zeolites feature an additional level of porosity in addition to the intrinsic microporosity of these materials, which can provide enhanced access to active sites located in inner pores [514–519] (Figure 6.17). Multiple synthesis methods have been developed to prepare hierarchical zeolites including bottom-up and top-down strategies [514, 515]. The former refers to modifications of hydrothermal synthesis protocols to obtain the desired mesoporous structures via direct synthesis, while the latter refers to various post-synthetic treatments that include, most commonly, dealumination by acid treatment and desilication by base treatment [516, 517, 520, 521]. In the case of zeolite HY, post-synthesis modification strategies to create mesopores have proved to be more efficient, flexible, and scalable [516, 517, 522].

Numerous publications in the literature have focused on the effect of zeolite HY mesoporosity on the activity and selectivity of C–C coupling reactions such as alkylation [523–527], hydroxyalkylation [159, 173, 528], aldol condensation [529, 530], acylation [226, 261, 531–534], and ketonization [535]. One of the

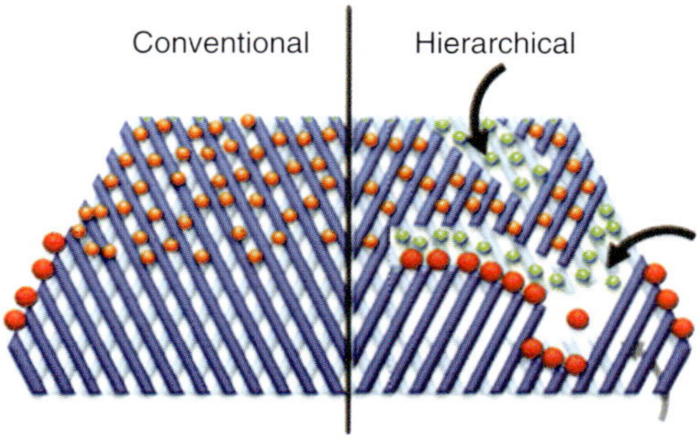

Figure 6.17 Schematic representation of introduction of mesopores in a zeolite material. The large red spheres represent bulky molecules which can only adsorb on the pore mouth. The green spheres represent smaller molecules with enhanced mass transport rates. The orange spheres represent molecules that suffer from single-file diffusion. Source: Ennaert et al. 2016 [506]. Reproduced with permission of Royal Society of Chemistry.

most common benefits of hierarchical zeolites described in the literature is the improvement in mass transport properties [226, 536]. For example, Mériaudeau et al. studied the alkylation of benzene with long-chain alkenes on HY zeolites with different frameworks and mesoporous structures [536]. They observed that the dealuminated HY zeolite showed a higher rate of alkylation and a lower rate of deactivation. Wagholikar et al. also observed an increase in the rate of acylation of anisole with long-chain carboxylic acids with increasing mesoporosity in HY zeolites [226]. Zeolite mesoporosity could also alleviate catalyst deactivation. For example, Colón et al. studied the alkylation of naphthalene with *iso*-propanol in various HY-based catalysts [537]. The authors observed lower coke formation in the dealuminated mesoporous HY compared with the untreated zeolite. It was proposed that dealumination has a twofold effect: lowering the density of acid sites – which consequentially decreases the yield of poly-alkylated naphthalene products – and creating the mesoporosity that enhances the transport of poly-alkylated products that otherwise would be trapped in the zeolite pores and generate coke. Finally, improvements in product selectivity can also be obtained through the introduction of mesoporosity. Indeed, Anand et al. observed that as the mesoporosity of HY zeolites increased, the selectivity to para-isomers increased and the selectivity to ortho-isomers decreased in the alkylation of phenol with cyclohexanol and cyclohexene [524]. Similar effects have been observed in the aldol condensation of acetaldehyde in ethanol [529] and in the alkylation of aromatics with 5-(hydroxymethyl)furfural (HMF) on hierarchical USY with large 3D pores [538].

6.3.2.3 HBEA

The structure of zeolite beta comprises 3D channels interconnecting large (12-membered ring) pore openings with an average size of 0.55–7 nm [539–542]. This structure is an intergrowth of two distinct but closely related polymorphs, A and B, as demonstrated in Figure 6.18 [539, 540].

6.3.2.3.1 Synthesis and Application of Zeolite Beta with Controlled Polymorphs Based on the polymorphs A and B, four other polymorphs were proposed [544],

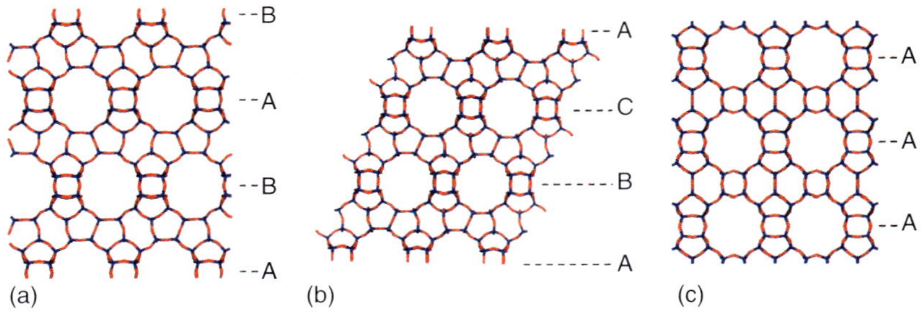

Figure 6.18 Framework structures of (a) polymorph A, (b) polymorph B, and (c) polymorph C of zeolite beta, showing the different stackings of the 12-ring pores as (a) ABAB…, (b) ABCABC…, and (c) AA…. Source: Corma et al. 2008 [543]. Reproduced with permission of American Chemical Society.

including C [539] (as shown in Figure 6.18), C_H [541], D [545, 546], and E [545, 546]. Typical zeolite beta has 45% A, 55% B, and trace amounts of C. The polymorphs define the channel system, which in turn controls the properties of the zeolite. Polymorph A has one chiral and two straight channels, polymorph B has one achiral sinusoidal and two straight channels, and polymorph C has three straight channels [546–548]. Polymorphs D and E contain double four-membered ring cages, which are not found in polymorphs A, B, or C_H [544]. The synthesis and characterization of other polymorphs of zeolite beta have also been reported [546, 549, 550].

The differences in channel sizes and structures between different polymorphs can be leveraged to achieve a desired outcome. For example, in the acylation of 2-methoxynaphthalene (2-MN) with acetic anhydride over zeolite beta, polymorph C limits the diffusion of 2-MN due to the fact that it has smaller pores than polymorphs A and B and leads to a smaller total acylation yield. However, due to its smaller pores, polymorph C displays higher selectivity toward 2-acetylmethoxynaphthalene (2-AMN) than toward 1-acetylmethoxynaphthalene (1-AMN) [551, 552]. Therefore, depending on the desired outcome – total acylation yield or selectivity to a specific product – specific polymorphs can be selected.

6.3.2.3.2 Other Structure Modification Methods Due to its large pore structure, zeolite beta has been utilized in many different reactions – such as acylation, alkylation, and aldol condensation – especially of bulky molecules. In many cases, reducing the diffusion path in zeolites has shown great advantages given that it enhances the accessibility of reactants to active sites, reduces deactivation, minimizes secondary reactions, and increases product yields [216, 253, 435, 553–558]. Therefore, zeolite beta variants with mesoporous–microporous or hierarchical structures have been extensively explored because of their shortened diffusion paths. Hierarchical zeolite beta can be obtained by direct synthesis [250, 559–568] or by a post-synthetic demetallation approach [253, 256, 569–573].

6.3.2.3.3 Sn-Beta and Other Isomorphous Substituted Beta Zeolites In aluminosilicate zeolite beta, the charge associated with an AlO_4^- tetrahedron is compensated by a proton, resulting in Brønsted acid sites. When Al is substituted by Sn^{IV}, the resulting stannosilicate is a Lewis-acidic zeolite. Sn-beta zeolite has exhibited high catalytic performance in different chemistries for biomass conversion such as catalytic oxidation [574–576], the Meerwein–Ponndorf–Verley reduction reaction [577–579], glucose isomerization to fructose [580–585] and epimerization to mannose [584, 586], conversion of sugars to lactates [587], and Diels–Alder reactions [588]. Many studies have combined Sn-beta with other Brønsted acidic catalysts for desired consecutive reactions such as one-pot conversion of glucose to HMF [581], xylose to furfural [589], furfural to valerolactone [590], etc.

Two types of Lewis acid sites have been proposed to exist in Sn-beta zeolite, namely, fully coordinated framework $Sn(-Si-O-)_4$ (closed sites) and the partially hydrolyzed framework Sn center that generates $(-Si-O-)_3Sn-OH$ (open sites) [591]. Depending on the nature of the chemistry, different types of sites are required. In Baeyer–Villiger oxidation (BV) and MPV reduction reactions, the hydrolyzed Sn open sites are more active than the framework Sn closed sites [579, 591]. For reactions that are favorable on Brønsted acid sites such as acylation, the silanol group on the hydrolyzed open site was found to be more active than the Lewis-acidic Sn centers [263]. Other isomorphous substituted zeolite beta samples have also been investigated – such as Zr-beta [579, 592], Ti-beta [593–595], Hf-beta [263], etc. – however, Sn-beta exhibits the strongest Lewis acidity among them.

References

1 Iglesia, E., Barton, D.G., Biscardi, J.A. et al. (1997). Bifunctional pathways in catalysis by solid acids and bases. *Catal. Today* 38 (3): 339–360.

2 Wang, S. and Iglesia, E. (2016). Substituent effects and molecular descriptors of reactivity in condensation and esterification reactions of oxygenates on acid–base pairs at TiO_2 and ZrO_2 surfaces. *J. Phys. Chem.* 120 (38): 21589–21616.

3 Kikhtyanin, O., Čapek, L., Smoláková, L. et al. (2017). Influence of Mg–Al mixed oxide compositions on their properties and performance in aldol condensation. *Ind. Eng. Chem. Res.* 56 (45): 13411–13422.

4 Pérez, C.N., Monteiro, J.L.F., López Nieto, J.M., and Henriques, C.A. (2009). Influence of basic properties of Mg, Al-mixed oxides on their catalytic activity in knoevenagel condensation between benzaldehyde and phenylsulfonylacetonitrile. *Quím. Nova* 32: 2341–2346.

5 Prescott, H.A., Li, Z.-J., Kemnitz, E. et al. (2005). Application of calcined Mg–Al hydrotalcites for Michael additions: an investigation of catalytic activity and acid–base properties. *J. Catal.* 234 (1): 119–130.

6 Di Cosimo, J.I., Díez, V.K., Xu, M. et al. (1998). Structure and surface and catalytic properties of Mg–Al basic oxides. *J. Catal.* 178 (2): 499–510.

7 Philipp, R. and Fujimoto, K. (1992). FTIR spectroscopic study of carbon dioxide adsorption/desorption on magnesia/calcium oxide catalysts. *J. Phys. Chem.* 96 (22): 9035–9038.

8 Morterra, C., Ghiotti, G., Boccuzi, F., and Coluccia, S. (1978). An infrared spectroscopic investigation of the surface properties of magnesium aluminate spinel. *J. Catal.* 51 (3): 299–313.
9 Ordóñez, S., Díaz, E., León, M., and Faba, L. (2011). Hydrotalcite-derived mixed oxides as catalysts for different C–C bond formation reactions from bioorganic materials. *Catal. Today* 167 (1): 71–76.
10 Faba, L., Díaz, E., and Ordóñez, S. (2012). Aqueous-phase furfural-acetone aldol condensation over basic mixed oxides. *Appl. Catal., B* 113–114: 201–211.
11 Smoláková, L., Frolich, K., Kocík, J. et al. (2017). Surface properties of hydrotalcite-based Zn(Mg)Al oxides and their catalytic activity in aldol condensation of furfural with acetone. *Ind. Eng. Chem. Res.* 56 (16): 4638–4648.
12 Tichit, D., Lhouty, M.H., Guida, A. et al. (1995). Textural properties and catalytic activity of hydrotalcites. *J. Catal.* 151 (1): 50–59.
13 Wang, S., Goulas, K., and Iglesia, E. (2016). Condensation and esterification reactions of alkanals, alkanones, and alkanols on TiO_2: elementary steps, site requirements, and synergistic effects of bifunctional strategies. *J. Catal.* 340: 302–320.
14 Shen, W., Tompsett, G.A., Xing, R. et al. (2012). Vapor phase butanal self-condensation over unsupported and supported alkaline earth metal oxides. *J. Catal.* 286: 248–259.
15 Liu, H., Bayat, N., and Iglesia, E. (2003). Site titration with organic bases during catalysis: selectivity modifier and structural probe in methanol oxidation on Keggin clusters. *Angew. Chem. Int. Ed.* 42 (41): 5072–5075.
16 Di Cosimo, J.I., Díez, V.K., and Apesteguía, C.R. (1996). Base catalysis for the synthesis of α,β-unsaturated ketones from the vapor-phase aldol condensation of acetone. *Appl. Catal., A* 137 (1): 149–166.
17 Fan, D., Dong, X., Yu, Y., and Zhang, M. (2017). A DFT study on the aldol condensation reaction on MgO in the process of ethanol to 1,3-butadiene: understanding the structure-activity relationship. *Phys. Chem. Chem. Phys.* 19 (37): 25671–25682.
18 Dumitriu, E., Hulea, V., Fechete, I. et al. (2001). The aldol condensation of lower aldehydes over MFI zeolites with different acidic properties. *Microporous Mesoporous Mater.* 43 (3): 341–359.
19 Dumitriu, E., Azzouz, A., Hulea, V. et al. (1997). Synthesis, characterization and catalytic activity of SAPO-34 obtained with piperidine as templating agent. *Microporous Mater.* 10 (1): 1–12.
20 Dumitriu, E., Hulea, V., Bilba, N. et al. (1993). Synthesis of acrolein by vapor phase condensation of formaldehyde and acetaldehyde over oxides loaded zeolites. *J. Mol. Catal.* 79 (1): 175–185.
21 Kikhtyanin, O., Bulánek, R., Frolich, K. et al. (2016). Aldol condensation of furfural with acetone over ion-exchanged and impregnated potassium BEA zeolites. *J. Mol. Catal. A: Chem.* 424: 358–368.
22 Tago, T., Konno, H., Ikeda, S. et al. (2011). Selective production of isobutylene from acetone over alkali metal ion-exchanged BEA zeolites. *Catal. Today* 164 (1): 158–162.

23 Aragon, J.M., Vegas, J.M.R., and Jodra, L.G. (1994). Self-condensation of cyclohexanone catalyzed by Amberlyst-15. Study of diffusional resistances and deactivation of the catalyst. *Ind. Eng. Chem. Res.* 33 (3): 592–599.

24 Bui, T.V., Sooknoi, T., and Resasco, D.E. (2017). Simultaneous upgrading of furanics and phenolics through hydroxyalkylation/aldol condensation reactions. *ChemSusChem* 10 (7): 1631–1639.

25 Tayade, K.N. and Mishra, M. (2013). A study on factors influencing cross and self product selectivity in aldol condensation over propylsulfonic acid functionalized silica. *Catal. Sci. Technol.* 3 (5): 1288–1300.

26 Chen, H., Wang, Y., Wang, Q. et al. (2014). Bifunctional organic polymeric catalysts with a tunable acid–base distance and framework flexibility. *Sci. Rep.* 4: 6475.

27 Jeong, M.S. and Frei, H. (2000). Acetaldehyde as a probe for the chemical properties of aluminophosphate molecular sieves. An in situ FT-IR study. *J. Mol. Catal. A: Chem.* 156 (1): 245–253.

28 Kikhtyanin, O., Kubička, D., and Čejka, J. (2015). Toward understanding of the role of Lewis acidity in aldol condensation of acetone and furfural using MOF and zeolite catalysts. *Catal. Today* 243: 158–162.

29 Herrmann, S. and Iglesia, E. (2017). Elementary steps in acetone condensation reactions catalyzed by aluminosilicates with diverse void structures. *J. Catal.* 346: 134–153.

30 Boekfa, B., Pantu, P., Probst, M., and Limtrakul, J. (2010). Adsorption and tautomerization reaction of acetone on acidic zeolites: the confinement effect in different types of zeolites. *J. Phys. Chem. C* 114 (35): 15061–15067.

31 Solans-Monfort, X., Bertran, J., Branchadell, V., and Sodupe, M. (2002). Keto–enol isomerization of acetaldehyde in HZSM5. A theoretical study using the ONIOM2 method. *J. Phys. Chem. B* 106 (39): 10220–10226.

32 Biaglow, A.I., Sepa, J., Gorte, R.J., and White, D. (1995). A ^{13}C NMR study of the condensation chemistry of acetone and acetaldehyde adsorbed at the Brønsted acid sites in H-ZSM-5. *J. Catal.* 151 (2): 373–384.

33 Xu, T., Munson, E.J., and Haw, J.F. (1994). Toward a systematic chemistry of organic reactions in zeolites: in situ NMR studies of ketones. *J. Am. Chem. Soc.* 116 (5): 1962–1972.

34 Gorte, R.J. and White, D. (1997). Interactions of chemical species with acid sites in zeolites. *Top. Catal.* 4 (1): 57–69.

35 Sierra, L.R., Kassab, E., and Evleth, E.M. (1993). Calculation of hydroxymethylation of a zeolite model. *J. Phys. Chem.* 97 (3): 641–646.

36 Šepa, J., Lee, C., Gorte, R.J. et al. (1996). Carbonyl ^{13}C shielding tensors and heats of adsorption of acetone adsorbed in silicalite and the 1:1 stoichiometric complex in H–ZSM-5. *J. Phys. Chem.* 100 (47): 18515–18523.

37 Kikhtyanin, O., Kelbichová, V., Vitvarová, D. et al. (2014). Aldol condensation of furfural and acetone on zeolites. *Catal. Today* 227: 154–162.

38 Kikhtyanin, O., Chlubna, P., Jindrova, T., and Kubicka, D. (2014). Peculiar behavior of MWW materials in aldol condensation of furfural and acetone. *Dalton Trans.* 43 (27): 10628–10641.

39 Nguyen Thanh, D., Kikhtyanin, O., Ramos, R. et al. (2016). Nanosized TiO_2—a promising catalyst for the aldol condensation of furfural with acetone in biomass upgrading. *Catal. Today* 277: 97–107.
40 Migues, A.N., Vaitheeswaran, S., and Auerbach, S.M. (2014). Density functional theory study of mixed aldol condensation catalyzed by acidic zeolites HZSM-5 and HY. *J. Phys. Chem. C* 118 (35): 20283–20290.
41 Coitino, E.L., Tomasi, J., and Ventura, O.N. (1994). Importance of water in aldol condensation reactions of acetaldehyde. *J. Chem. Soc., Faraday Trans.* 90 (12): 1745–1755.
42 Biaglow, A.I., Gorte, R.J., and White, D. (1994). ^{13}C NMR studies of acetone in dealuminated faujasites: a probe for nonframework alumina. *J. Catal.* 150 (1): 221–224.
43 Panov, A.G. and Fripiat, J.J. (1998). Acetone condensation reaction on acid catalysts. *J. Catal.* 178 (1): 188–197.
44 Venuto, P.B. (1994). Organic catalysis over zeolites: a perspective on reaction paths within micropores. *Microporous Mater.* 2 (5): 297–411.
45 Khlebnikova, E., Bekker, A., Ivashkina, E. et al. (2015). Thermodynamic analysis of benzene alkylation with ethylene. *Procedia Chem.* 15: 42–48.
46 Nurmakanova, A., Salischeva, A., Chudinova, A. et al. (2014). Comparison between alkylation and transalkylation reactions using ab initio approach. *Procedia Chem.* 10: 430–436.
47 Holbrey, J.D. and Seddon, K.R. (1999). Ionic liquids. *Clean Prod. Processes* 1 (4): 223–236.
48 Chauvin, Y., Hirschauer, A., and Olivier, H. (1994). Alkylation of isobutane with 2-butene using 1-butyl-3-methylimidazolium chloride—aluminium chloride molten salts as catalysts. *J. Mol. Catal.* 92 (2): 155–165.
49 Valkenberg, M.H., DeCastro, C., and Hoelderich, W.F. (2001). Special publication. *R. Soc. Chem.* 266: 242.
50 Shinde, A.B., Shrigadi, N.B., and Samant, S.D. (2004). *tert*-Butylation of phenols using *tert*-butyl alcohol in the presence of $FeCl_3$-modified montmorillonite K10. *Appl. Catal., A* 276 (1): 5–8.
51 Plechkova, N.V. and Seddon, K.R. (2008). Applications of ionic liquids in the chemical industry. *Chem. Soc. Rev.* 37 (1): 123–150.
52 Modrogan, E., Valkenberg, M.H., and Hoelderich, W.F. (2009). Phenol alkylation with isobutene – influence of heterogeneous Lewis and/or Brønsted acid sites. *J. Catal.* 261 (2): 177–187.
53 Sarala Devi, G., Giridhar, D., and Reddy, B.M. (2002). Vapour phase *O*-alkylation of phenol over alkali promoted rare earth metal phosphates. *J. Mol. Catal. A: Chem.* 181 (1): 173–178.
54 Reddy, B.M., Devi, G.S., and Sreekanth, P.M. (2002). Alkali promoted rare earth metal phosphates for vapour phase *O*-alkylation of α- and β-naphthols with methanol. *Res. Chem. Intermed.* 28 (6): 595–601.
55 Bautista, F.M., Campelo, J.M., Garcia, A. et al. (1993). Anion treatment (F^- or $SO_4{}^{2-}$) of $AlPO_4$–Al_2O_3 (25 wt.% Al_2O_3) catalysts: IV. Catalytic performance in the alkylation of phenol with methanol. *Appl. Catal., A* 99 (2): 161–173.

56 Durgakumari, V., Narayanan, S., and Guczi, L. (1990). Alkylation of phenol with methanol over AlPO and SAPO molecular sieves. *Catal. Lett.* 5 (4): 377–384.

57 Pierantozzi, R. and Nordquist, A.F. (1986). Selective *O*-alkylation of phenol with methanol. *Appl. Catal.* 21 (2): 263–271.

58 Fumio, N. and Isao, K. (1977). A study of catalysis by metal phosphates. IV. The alkylation of phenol with methanol over metal phosphate catalysts. *Bull. Chem. Soc. Jpn.* 50 (3): 614–619.

59 Campelo, J.M., Garcia, A., Luna, D. et al. (1988). $AlPO_4–TiO_2$ catalysts IV. The alkylation of phenol with methanol. In: *Studies in Surface Science and Catalysis* (eds. M. Guisnet et al.), 249–256. Elsevier.

60 Cavani, F., Monti, T., and Paoli, D. (2000). Characterisation and reactivity of supported and unsupported B/P/O catalysts in the O-alkylation of diphenols with methanol. In: *Studies in Surface Science and Catalysis* (eds. A. Corma et al.), 2633–2638. Elsevier.

61 Cavani, F. and Monti, T. (2000). *Catalysis of Organic Reactions* (ed. M.E. Ford). New York, NY: Marcel Dekker.

62 Santacesaria, E., Grasso, D., Gelosa, D., and Carrá, S. (1990). Catalytic alkylation of phenol with methanol: factors influencing activities and selectivities: I. Effect of different acid sites evaluated by studying the behaviour of the catalysts: γ-alumina, nafion-H, silica-alumina and phosphoric acid. *Appl. Catal.* 64: 83–99.

63 Santacesaria, E., Serio, M.D., Ciambelli, P. et al. (1990). Catalytic alkylation of phenol with methanol: factors influencing activities and selectivities: II. Effect of intracrystalline diffusion and shape selectivity on H-ZSM5 zeolite. *Appl. Catal.* 64: 101–117.

64 Tleimat-Manzalji, R., Bianchi, D., and Pajonk, G.r.M. (1993). Catalytic alkylation of phenol with methanol on a microporous xerogel and a macroporous aerogel of high surface area aluminas. *Appl. Catal., A* 101 (2): 339–350.

65 Fu, Y., Baba, T., and Ono, Y. (1998). Vapor-phase reactions of catechol with dimethyl carbonate. Part I. *O*-methylation of catechol over alumina. *Appl. Catal., A* 166 (2): 419–424.

66 Fu, Y., Baba, T., and Ono, Y. (1998). Vapor-phase reactions of catechol with dimethyl carbonate. Part II. Selective synthesis of guaiacol over alumina loaded with alkali hydroxide. *Appl. Catal., A* 166 (2): 425–430.

67 Fu, Y., Baba, T., and Ono, Y. (1999). Vapor-phase reactions of catechol with dimethyl carbonate. Part IV: synthesis of catechol carbonate over alumina loaded with cesium hydroxide. *Appl. Catal., A* 178 (2): 219–223.

68 Kiwi-Minsker, L., Porchet, S., Doepper, R., and Renken, A. (1997). Catalyst acid/base properties regulation to control the selectivity in gas-phase methylation of catechol. In: *Studies in Surface Science and Catalysis* (eds. H.U. Blaser, A. Baiker and R. Prins), 149–156. Elsevier.

69 Porchet, S., Kiwi-Minsker, L., Doepper, R., and Renken, A. (1996). Catalyst development for the selective methylation of catechol. *Chem. Eng. Sci.* 51 (11): 2933–2938.

70 Kiwi-Minsker, L., Porchet, S., Moeckli, P. et al. (1996). Selective methylation of catechol: catalyst development and characterisation. In: *Studies in Surface Science and Catalysis* (eds. J.W. Hightower et al.), 171–180. Elsevier.

71 Samolada, M.C., Grigoriadou, E., Kiparissides, Z., and Vasalos, I.A. (1995). Selective O-alkylation of phenol with methanol over sulfates supported on γ-Al_2O_3. *J. Catal.* 152 (1): 52–62.

72 Chantal, P.D., Kaliaguine, S., and Grandmaison, J.L. (1985). Reactions of phenolic compounds over HZSM-5. *Appl. Catal.* 18 (1): 133–145.

73 Lee, S.C., Lee, S.W., Kim, K.S. et al. (1998). *O*-alkylation of phenol derivatives over basic zeolites. *Catal. Today* 44 (1): 253–258.

74 Marczewski, M., Perot, G., and Guisnet, M. (1988). Alkylation of aromatics II. Alkylation of phenol with methanol on various zeolites. In: *Studies in Surface Science and Catalysis* (eds. M. Guisnet et al.), 273–282. Elsevier.

75 Marczewski, M., Bodibo, J.P., Perot, G., and Guisnet, M. (1989). Alkylation of aromatics: part I. Reaction network of the alkylation of phenol by methanol on ushy zeolite. *J. Mol. Catal.* 50 (2): 211–218.

76 García, L., Giannetto, G., Goldwasser, M.R. et al. (1996). Phenol alkylation with methanol: effect of sodium content and ammonia selective poisoning of an HY zeolite. *Catal. Lett.* 37 (1): 121–123.

77 Balsama, S., Beltrame, P., Beltrame, P.L. et al. (1984). Alkylation of phenol with methanol over zeolites. *Appl. Catal.* 13 (1): 161–170.

78 Beltrame, P., Beltrame, P.L., Carniti, P. et al. (1987). Alkylation of phenol with anisole over zeolites or γ-alumina. *Appl. Catal.* 29 (2): 327–334.

79 Parton, R.F., Jacobs, J.M., Ooteghem, H.V., and Jacobs, P.A. (1989). Comparison of the alkylation of anisole and phenol with methanol on pentasil and ultrastable zeolites. In: *Studies in Surface Science and Catalysis* (eds. H.G. Karge and J. Weitkamp), 211–221. Elsevier.

80 Namba, S., Yashima, T., Itaba, Y., and Hara, N. (1980). Selective formation of p-cresol by alkylation of phenol with methanol over Y type zeolite. In: *Studies in Surface Science and Catalysis* (eds. B. Imelik et al.), 105–111. Elsevier.

81 Fu, Z.H. and Ono, Y. (1993). Selective O-methylation of phenol with dimethyl carbonate over X-zeolites. *Catal. Lett.* 21 (1): 43–47.

82 Renavd, M., Chantal, P.D., and Kaliaguine, S. (1986). Anisole production by alkylation of phenol over ZSM5. *Can. J. Chem. Eng.* 64 (5): 787–791.

83 Moon, G., Böhringer, W., and O'Connor, C.T. (2004). An investigation into factors which influence the formation of *p*-cresol in the methanol alkylation of phenol over MCM-22 and ZSM-5. *Catal. Today* 97 (4): 291–295.

84 Romero, M.D., Ovejero, G., Rodríguez, A. et al. (2004). O-methylation of phenol in liquid phase over basic zeolites. *Ind. Eng. Chem. Res.* 43 (26): 8194–8199.

85 Sad, M.E., Duarte, H.A., Padró, C.L., and Apesteguía, C.R. (2014). Selective synthesis of *p*-ethylphenol by gas-phase alkylation of phenol with ethanol. *Appl. Catal., A* 486: 77–84.

86 González-Borja, M.Á. and Resasco, D.E. (2015). Reaction pathways in the liquid phase alkylation of biomass-derived phenolic compounds. *AlChE J.* 61 (2): 598–609.

87 Sad, M.E., Padró, C.L., and Apesteguía, C.R. (2008). Synthesis of cresols by alkylation of phenol with methanol on solid acids. *Catal. Today* 133–135: 720–728.

88 Karthik, M., Tripathi, A.K., Gupta, N.M. et al. (2004). Characterization of Co,Al-MCM-41 and its activity in the *t*-butylation of phenol using isobutanol. *Appl. Catal., A* 268 (1): 139–149.

89 Sato, S., Takahashi, R., Sodesawa, T. et al. (1999). Ortho-selective alkylation of phenol with 1-propanol catalyzed by CeO_2–MgO. *J. Catal.* 184 (1): 180–188.

90 Velu, S. and Swamy, C.S. (1996). Selective C-alkylation of phenol with methanol over catalysts derived from copper–aluminium hydrotalcite-like compounds. *Appl. Catal., A* 145 (1): 141–153.

91 Devassy, B.M., Shanbhag, G.V., Lefebvre, F., and Halligudi, S.B. (2004). Alkylation of *p*-cresol with *tert*-butanol catalyzed by heteropoly acid supported on zirconia catalyst. *J. Mol. Catal. A: Chem.* 210 (1): 125–130.

92 Yadav, G.D. and Kumar, P. (2005). Alkylation of phenol with cyclohexene over solid acids: insight in selectivity of O- versus C-alkylation. *Appl. Catal., A* 286 (1): 61–70.

93 Gagea, B.C., Parvulescu, A.N., Parvulescu, V.I. et al. (2003). Alkylation of phenols and naphthols on silica-immobilized triflate derivatives. *Catal. Lett.* 91 (1): 141–144.

94 Grabowska, H., Miśta, W., Trawczyński, J. et al. (2001). A method for obtaining thymol by gas phase catalytic alkylation of *m*-cresol over zinc aluminate spinel. *Appl. Catal., A* 220 (1): 207–213.

95 Velu, S. and Sivasanker, S. Alkylation of m-cresol with methanol and 2-propanol over calcined magnesium-aluminium hydrotalcites. *Res. Chem. Intermed.* 24: 657–666.

96 Chantal, P.D., Kaliaguine, S., and Grandmaison, J.L. (1984). Reactions of phenolic compounds on H-ZSM-5. In: *Studies in Surface Science and Catalysis* (eds. S. Kaliaguine and A. Mahay), 93–100. Elsevier.

97 Parton, R.F., Jacobs, J.M., Huybrechts, D.R., and Jacobs, P.A. (1989). Shape-selective catalysis in zeolites with organic substrates containing oxygen. In: *Studies in Surface Science and Catalysis* (eds. H.G. Karge and J. Weitkamp), 163–192. Elsevier.

98 Moffat, J.B. (1978). Phosphates as catalysts. *Catal. Rev.* 18 (2): 199–258.

99 Svelle, S., Visur, M., Olsbye, U. et al. (2011). Mechanistic aspects of the zeolite catalyzed methylation of alkenes and aromatics with methanol: a review. *Top. Catal.* 54 (13): 897.

100 Ivanova, I.I. and Corma, A. (1997). Surface species formed and their reactivity during the alkylation of toluene by methanol and dimethyl ether on zeolites as determined by in situ ^{13}C MAS NMR. *J. Phys. Chem. B* 101 (4): 547–551.

101 Vos, A.M., Nulens, K.H.L., De Proft, F. et al. (2002). Reactivity descriptors and rate constants for electrophilic aromatic substitution: acid zeolite catalyzed methylation of benzene and toluene. *J. Phys. Chem. B* 106 (8): 2026–2034.

102 Svelle, S., Kolboe, S., Olsbye, U., and Swang, O. (2003). A theoretical investigation of the methylation of methylbenzenes and alkenes by halomethanes over acidic zeolites. *J. Phys. Chem. B* 107 (22): 5251–5260.

103 Štich, I., Gale, J.D., Terakura, K., and Payne, M.C. (1999). Role of the zeolitic environment in catalytic activation of methanol. *J. Am. Chem. Soc.* 121 (14): 3292–3302.

104 Shah, R., Gale, J.D., and Payne, M.C. (1996). Methanol adsorption in zeolites a first-principles study. *J. Phys. Chem.* 100 (28): 11688–11697.

105 Haase, F. and Sauer, J. (2000). Ab initio molecular dynamics simulation of methanol interacting with acidic zeolites of different framework structure. *Microporous Mesoporous Mater.* 35–36: 379–385.

106 Mihaleva, V.V., van Santen, R.A., and Jansen, A.P.J. (2001). A DFT study of methanol adsorption in 8T rings of chabazite. *J. Phys. Chem. B* 105 (29): 6874–6879.

107 Haase, F. and Sauer, J. (1995). Interaction of methanol with Broensted acid sites of zeolite catalysts: an ab initio study. *J. Am. Chem. Soc.* 117 (13): 3780–3789.

108 Blaszkowski, S.R. and van Santen, R.A. (1996). The mechanism of dimethyl ether formation from methanol catalyzed by zeolitic protons. *J. Am. Chem. Soc.* 118 (21): 5152–5153.

109 Blaszkowski, S.R. and van Santen, R.A. (1997). Theoretical study of the mechanism of surface methoxy and dimethyl ether formation from methanol catalyzed by zeolitic protons. *J. Phys. Chem. B* 101 (13): 2292–2305.

110 Boronat, M., Martinez, C., and Corma, A. (2011). Mechanistic differences between methanol and dimethyl ether carbonylation in side pockets and large channels of mordenite. *Phys. Chem. Chem. Phys.* 13 (7): 2603–2612.

111 Wen, Z., Yang, D., He, X. et al. (2016). Methylation of benzene with methanol over HZSM-11 and HZSM-5: a density functional theory study. *J. Mol. Catal. A: Chem.* 424: 351–357.

112 Svelle, S., Kolboe, S., Swang, O., and Olsbye, U. (2005). Methylation of alkenes and methylbenzenes by dimethyl ether or methanol on acidic zeolites. *J. Phys. Chem. B* 109 (26): 12874–12878.

113 Wen, Z., Yang, D., Yang, F. et al. (2016). Methylation of toluene with methanol over HZSM-5: a periodic density functional theory investigation. *Chin. J. Catal.* 37 (11): 1882–1890.

114 Saepurahman, Visur, M., Olsbye, U. et al. (2011). In situ FT-IR mechanistic investigations of the zeolite catalyzed methylation of benzene with methanol: H-ZSM-5 versus H-beta. *Top. Catal.* 54 (16): 1293.

115 Wang, W., Jiao, J., Jiang, Y. et al. (2005). Formation and decomposition of surface ethoxy species on acidic zeolite Y. *ChemPhysChem* 6 (8): 1467–1469.

116 Kondo, J.N., Ito, K., Yoda, E. et al. (2005). An ethoxy intermediate in ethanol dehydration on Brønsted acid sites in zeolite. *J. Phys. Chem. B* 109 (21): 10969–10972.

117 Lévesque, P. and Dao, L.H. (1989). Alkylation of benzene using an aqueous solution of ethanol. *Appl. Catal.* 53 (2): 157–167.

118 Van der Mynsbrugge, J., Visur, M., Olsbye, U. et al. (2012). Methylation of benzene by methanol: single-site kinetics over H-ZSM-5 and H-beta zeolite catalysts. *J. Catal.* 292: 201–212.

119 Ogunbadejo, B.A., Osman, M.S., Arudra, P. et al. (2015). Alkylation of toluene with ethanol to para-ethyltoluene over MFI zeolites: comparative study and kinetic modeling. *Catal. Today* 243: 109–117.

120 Ronchin, L., Quartarone, G., and Vavasori, A. (2012). Kinetics and mechanism of acid catalyzed alkylation of phenol with cyclohexene in the presence of styrene divinylbenzene sulfonic resins. *J. Mol. Catal. A: Chem.* 353–354: 192–203.

121 Siffert, S., Gaillard, L., and Su, B.L. (2000). Alkylation of benzene by propene on a series of beta zeolites: toward a better understanding of the mechanisms. *J. Mol. Catal. A: Chem.* 153 (1): 267–279.

122 Corma, A., Martínez-Soria, V., and Schnoeveld, E. (2000). Alkylation of benzene with short-chain olefins over MCM-22 zeolite: catalytic behaviour and kinetic mechanism. *J. Catal.* 192 (1): 163–173.

123 Eckstein, S., Hintermeier, P.H., Olarte, M.V. et al. (2017). Elementary steps and reaction pathways in the aqueous phase alkylation of phenol with ethanol. *J. Catal.* 352: 329–336.

124 Odedairo, T. and Al-Khattaf, S. (2013). Comparative study of zeolite catalyzed alkylation of benzene with alcohols of different chain length: H-ZSM-5 versus mordenite. *Catal. Today* 204: 73–84.

125 Odedairo, T. and Al-Khattaf, S. (2010). Ethylation of benzene: effect of zeolite acidity and structure. *Appl. Catal., A* 385 (1): 31–45.

126 Liu, Y., Baráth, E., Shi, H. et al. (2018). Solvent-determined mechanistic pathways in zeolite-H-BEA-catalysed phenol alkylation. *Nat. Catal.* 1 (2): 141–147.

127 Le Van Mao, R., Levesque, P., McLaughlin, G., and Dao, L.H. (1987). Ethylene from ethanol over zeolite catalysts. *Appl. Catal.* 34: 163–179.

128 Dao, L.H. and Levesque, P. (1989). Benzene or toluene alkylation in presence of aqueous ethanol. In: *Proceedings of the 24th Intersociety Energy Conversion Engineering Conference* (6–11 Auguest 1989). IEEE.

129 Paparatto, G., Moretti, E., Leofanti, G., and Gatti, F. (1987). Toluene ethylation on ZSM zeolites. *J. Catal.* 105 (1): 227–232.

130 Kaeding, W.W., Young, L.B., and Chu, C.-C. (1984). Shape-selective reactions with zeolite catalysts: IV. Alkylation of toluene with ethylene to produce *p*-ethyltoluene. *J. Catal.* 89 (2): 267–273.

131 Nie, X., Liu, X., Gao, L. et al. (2010). SO_3H-functionalized ionic liquid catalyzed alkylation of catechol with *tert*-butyl alcohol. *Ind. Eng. Chem. Res.* 49 (17): 8157–8163.

132 Bregolato, M., Bolis, V., Busco, C. et al. (2007). Methylation of phenol over high-silica beta zeolite: effect of zeolite acidity and crystal size on catalyst behaviour. *J. Catal.* 245 (2): 285–300.

133 Ma, Q., Chakraborty, D., Faglioni, F. et al. (2006). Alkylation of phenol: a mechanistic view. *J. Phys. Chem. A* 110 (6): 2246–2252.

134 Attina, M., Cacace, F., Ciranni, G., and Giacomello, P. (1976). Predominant *O*-alkylation in the gas-phase attack of *t*-butyl cations on phenol and anisole. *J. Chem. Soc., Chem. Commun.* (12): 466–467.

135 Campbell, C.B., Onopchenko, A., and Young, D.C. (1990). Amberlyst-15-catalyzed alkylation of phenolics with branched alkenes. Rearrangement of *tert*-alkylphenols and catechols to sec-alkyl isomers. *Ind. Eng. Chem. Res.* 29 (4): 642–647.

136 Martinez-Espin, J.S., De Wispelaere, K., Westgård Erichsen, M. et al. (2017). Benzene co-reaction with methanol and dimethyl ether over zeolite and zeotype catalysts: evidence of parallel reaction paths to toluene and diphenylmethane. *J. Catal.* 349: 136–148.

137 Maihom, T., Boekfa, B., Sirijaraensre, J. et al. (2009). Reaction mechanisms of the methylation of ethene with methanol and dimethyl ether over H-ZSM-5: an ONIOM study. *J. Phys. Chem. C* 113 (16): 6654–6662.

138 Brogaard, R.Y., Henry, R., Schuurman, Y. et al. (2014). Methanol-to-hydrocarbons conversion: the alkene methylation pathway. *J. Catal.* 314: 159–169.

139 Sad, M.E., Padró, C.L., and Apesteguía, C.R. (2014). Phenol methylation on acid catalysts: study of the catalyst deactivation kinetics and mechanism. *Appl. Catal., A* 475: 305–313.

140 Bhattacharyya, K.G., Talukdar, A.K., Das, P., and Sivasanker, S. (2003). Al-MCM-41 catalysed alkylation of phenol with methanol. *J. Mol. Catal. A: Chem.* 197 (1): 255–262.

141 Cavani, F., Maselli, L., Passeri, S., and Lercher, J.A. (2010). Catalytic methylation of phenol on MgO – surface chemistry and mechanism. *J. Catal.* 269 (2): 340–350.

142 Tope, B.B., Alabi, W.O., Aitani, A.M. et al. (2012). Side-chain alkylation of toluene with methanol to styrene over cesium ion-exchanged zeolite X modified with metal borates. *Appl. Catal., A* 443–444: 214–220.

143 Hattori, H., Amusa, A.A., Jermy, R.B. et al. (2016). Zinc oxide as efficient additive to cesium ion-exchanged zeolite X catalyst for side-chain alkylation of toluene with methanol. *J. Mol. Catal. A: Chem.* 424: 98–105.

144 King, S.T. and Garces, J.M. (1987). In situ infrared study of alkylation of toluene with methanol on alkali cation-exchanged zeolites. *J. Catal.* 104 (1): 59–70.

145 Palomares, A.E., Eder-Mirth, G., and Lercher, J.A. (1997). Selective alkylation of toluene over basic zeolites: an in situ infrared spectroscopic investigation. *J. Catal.* 168 (2): 442–449.

146 Palomares, A.E., Eder-Mirth, G., Rep, M., and Lercher, J.A. (1998). Alkylation of toluene over basic catalysts – key requirements for side chain alkylation. *J. Catal.* 180 (1): 56–65.

147 Chen, H., Li, X., Zhao, G. et al. (2015). Free radical mechanism investigation of the side-chain alkylation of toluene with methanol on basic zeolites X. *Chin. J. Catal.* 36 (10): 1726–1732.

148 Borgna, A., Sepúlveda, J., Magni, S.I., and Apesteguía, C.R. (2004). Active sites in the alkylation of toluene with methanol: a study by selective acid–base poisoning. *Appl. Catal., A* 276 (1): 207–215.

149 Itoh, H., Miyamoto, A., and Murakami, Y. (1980). Mechanism of the side-chain alkylation of toluene with methanol. *J. Catal.* 64 (2): 284–294.
150 Wang, B., Huang, W., Wen, Y. et al. (2011). Styrene from toluene by side chain alkylation over a novel solid acid-base catalyst. *Catal. Today* 173 (1): 38–43.
151 Wang, T., Li, K., Liu, Q. et al. (2014). Aviation fuel synthesis by catalytic conversion of biomass hydrolysate in aqueous phase. *Appl. Energy* 136: 775–780.
152 Deng, Q., Han, P., Xu, J. et al. (2015). Highly controllable and selective hydroxyalkylation/alkylation of 2-methylfuran with cyclohexanone for synthesis of high-density biofuel. *Chem. Eng. Sci.* 138: 239–243.
153 Zhang, X., Deng, Q., Han, P. et al. (2017). Hydrophobic mesoporous acidic resin for hydroxyalkylation/alkylation of 2-methylfuran and ketone to high-density biofuel. *AlChE J.* 63 (2): 680–688.
154 Olah, G.A. (1993). Superelectrophiles. *Angew. Chem. Int. Ed. Engl.* 32 (6): 767–788.
155 Pratihar, S. (2016). Triggering the approach of an arene or heteroarene towards an aldehyde via Lewis acid-aldehyde communication. *Org. Biomol. Chem.* 14 (10): 2854–2865.
156 El-Khawaga, A.M., Roberts, R.M., and Sweeney, K.M. (1985). Transacylations between sterically hindered aromatic ketones and various arenes. *J. Org. Chem.* 50 (12): 2055–2058.
157 Wen, C., Barrow, E., Hattrick-Simpers, J., and Lauterbach, J. (2014). One-step production of long-chain hydrocarbons from waste-biomass-derived chemicals using bi-functional heterogeneous catalysts. *Phys. Chem. Chem. Phys.* 16 (7): 3047–3054.
158 Garade, A.C., Niphadkar, P.S., Joshi, P.N., and Rode, C.V. (2010). Hydroxyalkylation of p-cresol to 2,2′-methylenebis(4-methylphenol) using Sn/Si-MCM-41 catalysts. *Chem. Lett.* 39 (2): 126–127.
159 Alvaro, M., García, H., Sanjuán, A., and Esplá, M. (1998). Hydroxyalkylation of benzene derivatives by benzaldehyde in the presence of acid zeolites. *Appl. Catal., A* 175 (1–2): 105–112.
160 Álvaro, M., Das, D., Cano, M., and Garcia, H. (2003). Friedel–Crafts hydroxyalkylation: reaction of anisole with paraformaldehyde catalyzed by zeolites in supercritical CO_2. *J. Catal.* 219 (2): 464–468.
161 Wu, X., Liu, Y., Liu, R. et al. (2016). Hydroxyalkylation of phenol to bisphenol F over heteropolyacid catalysts: the effect of catalyst acid strength on isomer distribution and kinetics. *J. Colloid Interface Sci.* 481: 75–81.
162 Nowińska, K. and Kaleta, W. (2000). Synthesis of bisphenol-A over heteropoly compounds encapsulated into mesoporous molecular sieves. *Appl. Catal., A* 203 (1): 91–100.
163 Jha, A., Garade, A.C., Mirajkar, S.P., and Rode, C.V. (2012). MCM-41 supported phosphotungstic acid for the hydroxyalkylation of phenol to phenolphthalein. *Ind. Eng. Chem. Res.* 51 (10): 3916–3922.

164 Xia, Q., Xia, Y., Xi, J. et al. (2017). Selective one-pot production of high-grade diesel-range alkanes from furfural and 2-methylfuran over Pd/$NbOPO_4$. *ChemSusChem* 10 (4): 747–753.

165 Chen, M., Yan, J., Tan, Y. et al. (2015). Hydroxyalkylation of phenol with formaldehyde to bisphenol F catalyzed by keggin phosphotungstic acid encapsulated in metal–organic frameworks MIL-100(Fe or Cr) and MIL-101(Fe or Cr). *Ind. Eng. Chem. Res.* 54 (47): 11804–11813.

166 Iovel, I.G. and Lukevics, E. (1998). Hydroxymethylation and alkylation of compounds of the furan, thiophene, and pyrrole series in the presence of H^+ cations (review). *Chem. Heterocycl. Compd.* 34 (1): 1–12.

167 Brown, W.H. and Sawatzky, H. (1956). The condensation of furan and sylvan with some carbonyl compounds. *Can. J. Chem.* 34 (9): 1147–1153.

168 Seppo Pennanen, G.N. (1972). Stydies on the furan series, part I. The acidic condensation of aldeydes with methyl 2-furoate. *Acta Chem. Scand.* 26: 1018–1022.

169 Wei, Y., Li, Y., Tan, Y. et al. (2016). Short-channeled mesoporous Zr-Al-SBA-15 as a highly efficient catalyst for hydroxyalkylation of phenol with formaldehyde to bisphenol F. *Chem. Eng. J.* 298: 271–280.

170 Wang, Q., Wu, Z.M., Li, Y. et al. (2014). The efficient hydroxyalkylation of phenol with formaldehyde to bisphenol F over a thermoregulated phase-separable reaction system containing a water-soluble Bronsted acidic ionic liquid. *RSC Adv.* 4 (63): 33466–33473.

171 Garade, A.C., Kshirsagar, V.S., and Rode, C.V. (2009). Selective hydroxyalkylation of phenol to bisphenol F over dodecatungstophosphoric acid (DTP) impregnated on fumed silica. *Appl. Catal., A* 354 (1): 176–182.

172 Udayakumar, S., Ajaikumar, S., and Pandurangan, A. (2007). Electrophilic substitution reaction of phenols with aldehydes: enhance the yield of bisphenols by HPA and supported HPA. *Catal. Commun.* 8 (3): 366–374.

173 Barthel, N., Finiels, A., Moreau, C. et al. (2000). Hydroxyalkylation of aromatic compounds over protonic zeolites. *Top. Catal.* 13 (3): 269–274.

174 Mayr, H., Kempf, B., and Ofial, A.R. (2003). π-Nucleophilicity in carbon–carbon bond-forming reactions. *Acc. Chem. Res.* 36 (1): 66–77.

175 Appel, R. and Mayr, H. (2011). Quantification of the electrophilic reactivities of aldehydes, imines, and enones. *J. Am. Chem. Soc.* 133 (21): 8240–8251.

176 Appel, R., Chelli, S., Tokuyasu, T. et al. (2013). Electrophilicities of benzaldehyde-derived iminium ions: quantification of the electrophilic activation of aldehydes by iminium formation. *J. Am. Chem. Soc.* 135 (17): 6579–6587.

177 Pratihar, S. and Roy, S. (2011). Reactivity and selectivity of organotin reagents in allylation and arylation: nucleophilicity parameter as a guide. *Organometallics* 30 (12): 3257–3269.

178 Urabe, H. and Sato, F. (2001). Titanium(IV) lewis acids, Chapter 15. In: *Lewis Acids in Organic Synthesis*, 653–798. Wiley-VCH.

179 Ooi, T. and Maruoka, K. (2000). *Modern Carbonyl Chemistry*, Chapter 1, 1–32. Wiley-VCH https://onlinelibrary.wiley.com/doi/book/10.1002/9783527613267.

180 Harikrishnan, A., Sanjeevi, J., and Ramaraj Ramanathan, C. (2015). The cooperative effect of Lewis pairs in the Friedel–Crafts hydroxyalkylation reaction: a simple and effective route for the synthesis of (+/−)-carbinoxamine. *Org. Biomol. Chem.* 13 (12): 3633–3647.

181 Olah, G.A., Prakash, G.K.S., Donald, P. et al. (1989). Protonated (protosolvated) onium ions (onlum dications). *Res. Chem. Intermed.* 12 (2): 141–159.

182 Olah, G.A., Mathew, T., Marinez, E.R. et al. (2001). Acid-catalyzed isomerization of pivalaldehyde to methyl isopropyl ketone via a reactive protosolvated carboxonium ion intermediate. *J. Am. Chem. Soc.* 123 (47): 11556–11561.

183 Olah, G.A. and Klumpp, D.A. (2004). Superelectrophilic solvation. *Acc. Chem. Res.* 37 (4): 211–220.

184 Prakash, G.K.S., Panja, C., Shakhmin, A. et al. (2009). BF_3-H_2O catalyzed hydroxyalkylation of aromatics with aromatic aldehydes and dicarboxaldehydes: efficient synthesis of triarylmethanes, diarylmethylbenzaldehydes, and anthracene derivatives. *J. Org. Chem.* 74 (22): 8659–8668.

185 Olah, G.A., Rasul, G., York, C., and Prakash, G.K.S. (1995). Superacid-catalyzed condensation of benzaldehyde with benzene. Study of protonated benzaldehydes and the role of superelectrophilic activation. *J. Am. Chem. Soc.* 117 (45): 11211–11214.

186 Saito, S., Ohwada, T., and Shudo, K. (1995). Friedel–Crafts-type reaction of benzaldehyde with benzene. Diprotonated benzaldehyde as the reactive intermediate. *J. Am. Chem. Soc.* 117 (45): 11081–11084.

187 Schaarschmidt, A., Hermann, L., and Szemzö, B. (1925). Aldehyde und Äthylen-oxyd bei der Friedel-Craftsschen Synthese. *Ber. Dtsch. Chem. Ges.* 58 (8): 1914–1916.

188 Prakash, G.K.S., Mathew, T., Hoole, D. et al. (2004). *N*-Halosuccinimide/BF_3-H_2O, efficient electrophilic halogenating systems for aromatics. *J. Am. Chem. Soc.* 126 (48): 15770–15776.

189 Arnett, E.M. and Wu, C.Y. (1960). Stereoelectronic effects on organic bases. II.1 base strengths of the phenolic ethers. *J. Am. Chem. Soc.* 82 (21): 5660–5665.

190 Garade, A.C., Kshirsagar, V.S., Mane, R.B. et al. (2010). Acidity tuning of montmorillonite K10 by impregnation with dodecatungstophosphoric acid and hydroxyalkylation of phenol. *Appl. Clay Sci.* 48 (1): 164–170.

191 Jha, A., Garade, A.C., Shirai, M., and Rode, C.V. (2013). Metal cation-exchanged montmorillonite clay as catalysts for hydroxyalkylation reaction. *Appl. Clay Sci.* 74: 141–146.

192 Liu, R., Xia, X., Niu, X. et al. (2015). 12-Phosphotungstic acid immobilized on activated-bentonite as an efficient heterogeneous catalyst for the hydroxyalkylation of phenol. *Appl. Clay Sci.* 105–106: 71–77.

193 Garade, A.C., Kshirsagar, V.S., Jha, A., and Rode, C.V. (2010). Structure–activity studies of dodecatungstophosphoric acid impregnated bentonite clay catalyst in hydroxyalkylation of *p*-cresol. *Catal. Commun.* 11 (11): 942–945.

194 Bai, G., Dou, H., Qiu, M. et al. (2010). Friedel–Crafts hydroxyalkylation of anisole over oxalic acid modified Hβ zeolite. *Catal. Lett.* 138 (3): 187–192.
195 Wu, X., Xia, X., Chen, Y., and Lu, Y. (2016). Mesoporous Al-incorporated silica-pillared clay interlayer materials for catalytic hydroxyalkylation of phenol to bisphenol F. *RSC Adv.* 6 (78): 74028–74038.
196 Chambon, F., Rataboul, F., Pinel, C. et al. (2011). Cellulose hydrothermal conversion promoted by heterogeneous Brønsted and Lewis acids: remarkable efficiency of solid Lewis acids to produce lactic acid. *Appl. Catal., B* 105 (1): 171–181.
197 Devassy, B.M., Lefebvre, F., and Halligudi, S.B. (2005). Zirconia-supported 12-tungstophosphoric acid as a solid catalyst for the synthesis of linear alkyl benzenes. *J. Catal.* 231 (1): 1–10.
198 Mao, H., Li, B., Yue, L. et al. (2011). Aluminated mesoporous silica-pillared montmorillonite as acidic catalyst for catalytic cracking. *Appl. Clay Sci.* 53 (4): 676–683.
199 Wu, X., Xia, X., Liu, R., and Chen, Y. (2016). Hydroxyalkylation of phenol to bisphenol F over Al-pillared clay. *RSC Adv.* 6 (41): 34625–34632.
200 Garade, A.C., Malwadkar, A.V., Niphadkar, P.S. et al. (2014). Effect of SnO_2/Al_2O_3 ratio of Si-based MFI on its acidity and hydrophobicity: application in selective hydroxyalkylation of *p*-cresol. *Catal. Commun.* 44: 29–34.
201 Tan, Y., Li, Y., Wei, Y. et al. (2015). The hydroxyalkylation of phenol with formaldehyde over mesoporous M(Al, Zr, Al–Zr)-SBA-15 catalysts: the effect on the isomer distribution of bisphenol F. *Catal. Commun.* 67: 21–25.
202 Freese, U., Heinrich, F., and Roessner, F. (1999). Acylation of aromatic compounds on H-Beta zeolites. *Catal. Today* 49 (1): 237–244.
203 Derouane, E.G., Crehan, G., Dillon, C.J. et al. (2000). Zeolite catalysts as solid solvents in fine chemicals synthesis: 2. Competitive adsorption of the reactants and products in the Friedel–Crafts acetylations of anisole and toluene. *J. Catal.* 194 (2): 410–423.
204 Berreghis, A., Ayrault, P., Fromentin, E., and Guisnet, M. (2000). Acetylation of 2-methoxynaphthalene with acetic anhydride over a series of dealuminated HBEA zeolites. *Catal. Lett.* 68 (1–2): 121–127.
205 Sartori, G. and Maggi, R. (2006). Use of solid catalysts in Friedel-Crafts acylation reactions. *Chem. Rev.* 106 (3): 1077–1104.
206 Bezouhanova, C.P. (2002). Synthesis of aromatic ketones in the presence of zeolite catalysts. *Appl. Catal., A* 229 (1): 127–133.
207 Yamazaki, T., Makihara, M., and Komura, K. (2017). Zeolite catalyzed highly selective synthesis of 2-methoxy-6-acetylnaphthalene by Friedel-Crafts acylation of 2-methoxynaphthalene in acetic acid reaction media. *J. Mol. Catal. A: Chem.* 426: 170–176.
208 Arata, K. and Hino, M. (1990). Preparation of superacids by metal-oxides and their catalytic action. *Mater. Chem. Phys.* 26 (3–4): 213–237.
209 Arata, K., Nakamura, H., and Shouji, M. (2000). Friedel–Crafts acylation of toluene catalyzed by solid superacids. *Appl. Catal., A* 197 (2): 213–219.

210 Fang, Z., He, W., Tu, T. et al. (2018). An efficient and green pathway for continuous Friedel–Crafts acylation over α-Fe_2O_3 and $CaCO_3$ nanoparticles prepared in the microreactors. *Chem. Eng. J.* 331: 443–449.

211 Blasco, T., Corma, A., Martinez, A., and Martinez-Escolano, P. (1998). Supported heteropolyacid (HPW) catalysts for the continuous alkylation of isobutane with 2-butene: the benefit of using MCM-41 with larger pore diameters. *J. Catal.* 177 (2): 306–313.

212 Shimizu, K.-I., Niimi, K., and Satsuma, A. (2008). Polyvalent-metal salts of heteropolyacid as efficient heterogeneous catalysts for Friedel–Crafts acylation of arenes with carboxylic acids. *Catal. Commun.* 9 (6): 980–983.

213 Wu, S., Wang, W., Fang, Y. et al. (2017). Efficient Friedel–Crafts acylation of anisole over silicotungstic acid modified ZIF-8. *React. Kinet. Mech. Catal.* 122 (1): 357–367.

214 Sartori, G. and Maggi, R. (2011). Update 1 of: use of solid catalysts in Friedel–Crafts acylation reactions. *Chem. Rev.* 111: Pr181–Pr214.

215 Lezcano-Gonzalez, I., Vidal-Moya, J.A., Boronat, M. et al. (2013). Identification of active surface species for Friedel–Crafts acylation and koch carbonylation reactions by in situ solid-state NMR spectroscopy. *Angew. Chem. Int. Ed.* 52 (19): 5138–5141.

216 Bejblová, M., Procházková, D., and Čejka, J. (2009). Acylation reactions over zeolites and mesoporous catalysts. *ChemSusChem* 2 (6): 486–499.

217 Climent, M.J., Corma, A., and Iborra, S. (2010). Zeolites as catalysts for the synthesis of fine chemicals. In: *Zeolites and Catalysis: Synthesis, Reactions and Applications* (eds. J. Čejka, A. Corma and S. Zones), 775–826. Wiley-VCH.

218 Duong, N.N., Wang, B., Sooknoi, T. et al. (2017). Enhancing the acylation activity of acetic acid by formation of an intermediate aromatic ester. *ChemSusChem* 10 (13): 2823–2832.

219 Botella, P., Corma, A., López-Nieto, J.M. et al. (2000). Acylation of toluene with acetic anhydride over beta zeolites: influence of reaction conditions and physicochemical properties of the catalyst. *J. Catal.* 195 (1): 161–168.

220 Guidotti, M., Canaff, C., Coustard, J.M. et al. (2005). Acetylation of aromatic compounds over H-BEA zeolite: the influence of the substituents on the reactivity and on the catalyst stability. *J. Catal.* 230 (2): 375–383.

221 Meneses, L., Fuentealba, P., and Contreras, R. (2005). Relationship between the electrophilicity of substituting agents and substrate selectivity in Friedel–Crafts reactions. *Tetrahedron* 61 (4): 831–836.

222 Kresnawahjuesa, O., Gorte, R.J., and White, D. (2004). Characterization of acylating intermediates formed on H-ZSM-5. *J. Mol. Catal. A: Chem.* 208 (1–2): 175–185.

223 Yuan, B., Li, Z., Liu, Y., and Zhang, S. (2008). Liquid phase acylation of 2-methylnaphthalene catalyzed by H-beta zeolite. *J. Mol. Catal. A: Chem.* 280 (1–2): 210–218.

224 Kumari, V.D., Saroja, G., Ratnamala, A. et al. (2003). Vapor phase acylation of phenol over H beta, CeH beta and SO_4^{2-}/ZrO_2[a]. *React. Kinet. Catal. Lett.* 79 (1): 43–51.

225 Ma, Y.D., Wang, Q.L., Jiang, W., and Zuo, B.J. (1997). Friedal–Crafts acylation of anisole over zeolite catalysts. *Appl. Catal., A* 165 (1–2): 199–206.
226 Wagholikar, S.G., Niphadkar, P.S., Mayadevi, S., and Sivasanker, S. (2007). Acylation of anisole with long-chain carboxylic acids over wide pore zeolites. *Appl. Catal., A* 317 (2): 250–257.
227 Chiche, B., Finiels, A., Gauthier, C. et al. (1986). Friedel–Crafts acylation of toluene and para-xylene with carboxylic-acids catalyzed by zeolites. *J. Org. Chem.* 51 (11): 2128–2130.
228 Herron, J.A., Vann, T., Duong, N. et al. (2017). A systems-level roadmap for biomass thermal fractionation and catalytic upgrading strategies. *Energy Technol.* 5 (1): 130–150.
229 Devos, A., Remion, J., Frisquehesbain, A.M. et al. (1979). Synthesis of acyl halides under very mild conditions. *J. Chem. Soc. D* (24): 1180–1181.
230 Hardee, D.J., Kovalchuke, L., and Lambert, T.H. (2010). Nucleophilic acyl substitution via aromatic cation activation of carboxylic acids: rapid generation of acid chlorides under mild conditions. *J. Am. Chem. Soc.* 132 (14): 5002–5003.
231 Villeneuve, G.B. and Chan, T.H. (1997). A rapid, mild and acid-free procedure for the preparation of acyl chlorides including formyl chloride. *Tetrahedron Lett.* 38 (37): 6489–6492.
232 Schrod, M. and Luft, G. (1981). Kinetic investigations of the synthesis of acetic anhydride by homogeneous catalysis. *Ind. Eng. Chem. Prod. Res. Dev.* 20 (4): 649–653.
233 Moon, H.K., Sung, G.H., Kim, B.R. et al. (2016). One for many: a universal reagent for acylation processes. *Adv. Synth. Catal.* 358 (11): 1725–1730.
234 Gumidyala, A., Wang, B., and Crossley, S. (2016). Direct carbon–carbon coupling of furanics with acetic acid over Bronsted zeolites. *Sci. Adv.* 2 (9): e1601072.
235 Klisáková, J., Červený, L., and Čejka, J. (2004). On the role of zeolite structure and acidity in toluene acylation with isobutyric acid derivatives. *Appl. Catal., A* 272 (1): 79–86.
236 Yadav, G.D. and Krishnan, M.S. (1999). Solid acid catalysed acylation of 2-methoxy-naphthalene: role of intraparticle diffusional resistance. *Chem. Eng. Sci.* 54 (19): 4189–4197.
237 Moreau, P., Finiels, A., Pelorgeas, S. et al. (1997). Liquid-phase acetylation of tetralin with acetyl chloride over zeolites: inhibition of the reaction by the products. *Catal. Lett.* 47 (2): 161–166.
238 Chen, Z.H., Chen, W.Q., Tong, T.X., and Zeng, A.W. (2015). Continuous liquid phase acylation of toluene over HBEA zeolite: solvent effects and origin of the deactivation. *J. Mol. Catal. A: Chem.* 396: 231–238.
239 Derouane, E.G., Dillon, C.J., Bethell, D., and Derouane-Abd Hamid, S.B. (1999). Zeolite catalysts as solid solvents in fine chemicals synthesis – 1. Catalyst deactivation in the Friedel–Crafts acetylation of anisole. *J. Catal.* 187 (1): 209–218.

240 Moreau, P., Finiels, A., and Meric, P. (2000). Acetylation of dimethoxybenzenes with acetic anhydride in the presence of acidic zeolites. *J. Mol. Catal. A: Chem.* 154 (1–2): 185–192.

241 Rohan, D., Canaff, C., Fromentin, E., and Guisnet, M. (1998). Acetylation of anisole by acetic anhydride over a HBEA zeolite—origin of deactivation of the catalyst. *J. Catal.* 177 (2): 296–305.

242 Xiong, Y.N., Chen, W.Q., Ma, J.J. et al. (2015). Methods to delay deactivation of zeolites on furan acylation: continuous liquid-phase technology and solvent effects. *RSC Adv.* 5 (125): 103695–103702.

243 Heitling, E., Roessner, F., and van Steen, E. (2004). Origin of catalyst deactivation in fries rearrangement of phenyl acetate over zeolite H-Beta. *J. Mol. Catal. A: Chem.* 216 (1): 61–65.

244 Andy, P., Garcia-Martinez, J., Lee, G. et al. (2000). Acylation of 2-methoxynaphthalene and isobutylbenzene over zeolite beta. *J. Catal.* 192 (1): 215–223.

245 Guignard, C., Pedron, V., Richard, F. et al. (2002). Acylation of veratrole by acetic anhydride over H beta and HY zeolites – possible role of di- and triketone by-products in the deactivation process. *Appl. Catal., A* 234 (1–2): 79–90.

246 Padro, C.L. and Apesteguia, C.R. (2005). Acylation of phenol on solid acids: study of the deactivation mechanism. *Catal. Today* 107–108: 258–265.

247 Xiong, Y.N., Chen, W.Q., and Zeng, A.W. (2017). Optimization for catalytic performances of H beta zeolite in the acylation of 2-methylfuran by surface modification and solvents effect. *Res. Chem. Intermed.* 43 (3): 1557–1574.

248 Cerveny, L., Mikulcova, K., and Cejka, J. (2002). Shape-selective synthesis of 2-acetylnaphthalene via naphthalene acylation with acetic anhydride over large pore zeolites. *Appl. Catal., A* 223 (1–2): 65–72.

249 Fromentin, E., Coustard, J.M., and Guisnet, M. (2000). Acetylation of 2-methoxynaphthalene with acetic anhydride over a HBEA zeolite. *J. Mol. Catal. A: Chem.* 159 (2): 377–388.

250 Huang, G., Ji, P., Xu, H. et al. (2017). Fast synthesis of hierarchical beta zeolites with uniform nanocrystals from layered silicate precursor. *Microporous Mesoporous Mater.* 248: 30–39.

251 Wine, G., Tessonnier, J.P., Rigolet, S. et al. (2006). Beta zeolite supported on a beta-SiC foam monolith: a diffusionless catalyst for fixed-bed Friedel–Crafts reactions. *J. Mol. Catal. A: Chem.* 248 (1–2): 113–120.

252 Wine, G., Pham-Huu, C., and Ledoux, M.J. (2006). Acylation of anisole by acetic anhydride catalysed by BETA zeolite supported on pre-shaped silicon carbide. *Catal. Commun.* 7 (10): 768–772.

253 Escola, J.M., Serrano, D.P., Sanz, R. et al. (2018). Synthesis of hierarchical beta zeolite with uniform mesopores: effect on its catalytic activity for veratrole acylation. *Catal. Today* 304: 89–96.

254 Losch, P., Hoff, T.C., Kolb, J.F. et al. (2017). Mesoporous ZSM-5 zeolites in acid catalysis: top-down vs. bottom-up approach. *Catalysts* 7 (8): 19.

255 Bejblova, M., Vlk, J., Prochazkova, D. et al. (2007). The effect of type of acid sites in molecular sieves on activity and selectivity in acylation reactions. *Collect. Czech. Chem. Commun.* 72 (5–6): 728–746.

256 Zhang, B., Zhang, Y., Hu, Y. et al. (2016). Microexplosion under microwave irradiation: a facile approach to create mesopores in zeolites. *Chem. Mater.* 28 (8): 2757–2767.
257 Gaertner, C.A., Serrano-Ruiz, J.C., Braden, D.J., and Dumesic, J.A. (2009). Catalytic coupling of carboxylic acids by ketonization as a processing step in biomass conversion. *J. Catal.* 266 (1): 71–78.
258 Gauthier, C., Chiche, B., Finiels, A., and Geneste, P. (1989). Influence of acidity in Friedel–Crafts acylation catalyzed by zeolites. *J. Mol. Catal.* 50 (2): 219–229.
259 Sheemol, V.N., Tyagi, B., and Jasra, R.V. (2004). Acylation of toluene using rare earth cation exchanged zeolite beta as solid acid catalyst. *J. Mol. Catal. A: Chem.* 215 (1–2): 201–208.
260 Reddy, P.R., Subrahmanyam, M., and Durga Kumari, V. (1999). Vapor-phase synthesis of acetophenone in benzene acylation over CeHZSM-5(30) zeolite. *Catal. Lett.* 61 (3–4): 207–211.
261 Gaare, K. and Akporiaye, D. (1996). Modified zeolites as catalysts in the Friedel–Crafts acylation. *J. Mol. Catal. A: Chem.* 109 (2): 177–187.
262 Wei, H.J., Liu, K.F., Xie, S.J. et al. (2013). Determination of different acid sites in Beta zeolite for anisole acylation with acetic anhydride. *J. Catal.* 307: 103–110.
263 Koehle, M., Zhang, Z., Goulas, K.A. et al. (2018). Acylation of methylfuran with Brønsted and Lewis acid zeolites. *Appl. Catal., A* 564: 90–101.
264 Bigi, F., Carloni, S., Flego, C. et al. (2002). HY zeolite-promoted electrophilic acylation of methoxyarenes with linear acid chlorides. *J. Mol. Catal. A: Chem.* 178 (1–2): 139–146.
265 Pham, T.N., Sooknoi, T., Crossley, S.P., and Resasco, D.E. (2013). Ketonization of carboxylic acids: mechanisms, catalysts, and implications for biomass conversion. *ACS Catal.* 3 (11): 2456–2473.
266 Pestman, R., Koster, R.M., van Duijne, A. et al. (1997). Reactions of carboxylic acids on oxides: 2. Bimolecular reaction of aliphatic acids to ketones. *J. Catal.* 168 (2): 265–272.
267 Pestman, R., Koster, R.M., Boellaard, E. et al. (1998). Identification of the active sites in the selective hydrogenation of acetic acid to acetaldehyde on iron oxide catalysts. *J. Catal.* 174 (2): 142–152.
268 Pestman, R., van Duijne, A., Pieterse, J.A.Z., and Ponec, V. (1995). The formation of ketones and aldehydes from carboxylic acids, structure-activity relationship for two competitive reactions. *J. Mol. Catal. A: Chem.* 103 (3): 175–180.
269 Snell, R.W. and Shanks, B.H. (2013). Insights into the ceria-catalyzed ketonization reaction for biofuels applications. *ACS Catal.* 3 (4): 783–789.
270 González, F., Munuera, G., and Prieto, J.A. (1978). Mechanism of ketonization of acetic acid on anatase TiO_2 surfaces. *J. Chem. Soc., Faraday Trans. 1 F* 74 (0): 1517–1529.
271 Randery, S.D., Warren, J.S., and Dooley, K.M. (2002). Cerium oxide-based catalysts for production of ketones by acid condensation. *Appl. Catal., A* 226 (1): 265–280.

272 Dooley, K.M., Bhat, A.K., Plaisance, C.P., and Roy, A.D. (2007). Ketones from acid condensation using supported CeO_2 catalysts: effect of additives. *Appl. Catal., A* 320: 122–133.

273 Hendren, T.S. and Dooley, K.M. (2003). Kinetics of catalyzed acid/acid and acid/aldehyde condensation reactions to non-symmetric ketones. *Catal. Today* 85 (2): 333–351.

274 Wang, S. and Iglesia, E. (2017). Experimental and theoretical assessment of the mechanism and site requirements for ketonization of carboxylic acids on oxides. *J. Catal.* 345: 183–206.

275 Martinez, R., Huff, M.C., and Barteau, M.A. (2004). Ketonization of acetic acid on titania-functionalized silica monoliths. *J. Catal.* 222 (2): 404–409.

276 Pestman, R., Koster, R.M., Pieterse, J.A.Z., and Ponec, V. (1997). Reactions of carboxylic acids on oxides: 1. Selective hydrogenation of acetic acid to acetaldehyde. *J. Catal.* 168 (2): 255–264.

277 Neunhoeffer, O. and Paschke, P. (1939). Über den Mechanismus der Ketonbildung aus Carbonsäuren. *Berichte der Deutschen Chemischen Gesellschaft* 72 (4): 919–929.

278 Nagashima, O., Sato, S., Takahashi, R., and Sodesawa, T. (2005). Ketonization of carboxylic acids over CeO_2-based composite oxides. *J. Mol. Catal. A: Chem.* 227 (1): 231–239.

279 Pham, T.N., Shi, D., Sooknoi, T., and Resasco, D.E. (2012). Aqueous-phase ketonization of acetic acid over Ru/TiO_2/carbon catalysts. *J. Catal.* 295: 169–178.

280 Pulido, A., Oliver-Tomas, B., Renz, M. et al. (2013). Ketonic decarboxylation reaction mechanism: a combined experimental and DFT study. *ChemSusChem* 6 (1): 141–151.

281 Gumidyala, A., Sooknoi, T., and Crossley, S. (2016). Selective ketonization of acetic acid over HZSM-5: the importance of acyl species and the influence of water. *J. Catal.* 340: 76–84.

282 Pham, T.N., Shi, D., and Resasco, D.E. (2014). Kinetics and mechanism of ketonization of acetic acid on Ru/TiO_2 catalyst. *Top. Catal.* 57 (6): 706–714.

283 Pham, T.N., Shi, D., and Resasco, D.E. (2014). Reaction kinetics and mechanism of ketonization of aliphatic carboxylic acids with different carbon chain lengths over Ru/TiO_2 catalyst. *J. Catal.* 314: 149–158.

284 Hasan, M.A., Zaki, M.I., and Pasupulety, L. (2003). Oxide-catalyzed conversion of acetic acid into acetone: an FTIR spectroscopic investigation. *Appl. Catal., A* 243 (1): 81–92.

285 Ignatchenko, A.V., DeRaddo, J.S., Marino, V.J., and Mercado, A. (2015). Cross-selectivity in the catalytic ketonization of carboxylic acids. *Appl. Catal., A* 498: 10–24.

286 Bayahia, H., Kozhevnikova, E.F., and Kozhevnikov, I.V. (2015). Ketonisation of carboxylic acids over Zn–Cr oxide in the gas phase. *Appl. Catal., B* 165: 253–259.

287 Nagashima, O., Sato, S., Takahashi, R. et al. (2006). Formation of cyclopentanone from dimethyl hexanedioate over CeO_2. *Appl. Catal., A* 312: 175–180.

288 Rand, L., Wagner, W., Warner, P.O., and Kovac, L.R. (1962). Reactions catalyzed by potassium fluoride. II. The conversion of adipic acid to cyclopentanone. *J. Org. Chem.* 27 (3): 1034–1035.

289 Oliver-Tomas, B., Renz, M., and Corma, A. (2017). Ketone formation from carboxylic acids by ketonic decarboxylation: the exceptional case of the tertiary carboxylic acids. *Chem. Eur. J.* 23 (52): 12900–12908.

290 Aranda-Pérez, N., Ruiz, M.P., Echave, J., and Faria, J. (2017). Enhanced activity and stability of Ru–TiO_2 rutile for liquid phase ketonization. *Appl. Catal., A* 531: 106–118.

291 Tosoni, S. and Pacchioni, G. (2016). Acetic acid ketonization on tetragonal zirconia: role of surface reduction. *J. Catal.* 344: 465–473.

292 Lu, F., Jiang, B., Wang, J. et al. (2017). Promotional effect of Ti doping on the ketonization of acetic acid over a CeO_2 catalyst. *RSC Adv.* 7 (36): 22017–22026.

293 Snell, R.W., Hakim, S.H., Dumesic, J.A., and Shanks, B.H. (2013). Catalysis with ceria nanocrystals: bio-oil model compound ketonization. *Appl. Catal., A* 464–465: 288–295.

294 Bayahia, H. (2018). Gas-phase ketonization of acetic acid over Co–Mo and its supported catalysts. *J. Taibah Univ. Sci.* 12 (2): 191–196.

295 Bennett, J.A., Parlett, C.M.A., Isaacs, M.A. et al. (2017). Acetic acid ketonization over Fe_3O_4/SiO_2 for pyrolysis bio-oil upgrading. *ChemCatChem* 9 (9): 1648–1654.

296 Coluccia, S. and Tench. A.J. (1980). *Proceedings of the 7th International Congress on Catalysis,* Kodansha, Tokyo p. 1154.

297 Hattori, H. (2004). Solid Base catalysts: generation, characterization, and catalytic behavior of basic sites. *J. Jpn. Pet. Inst.* 47 (2): 67–81.

298 Hattori, H. (1988). Catalysis by basic metal oxides. *Mater. Chem. Phys.* 18 (5): 533–552.

299 Climent, M.J., Corma, A., Iborra, S., and Mifsud, M. (2007). MgO nanoparticle-based multifunctional catalysts in the cascade reaction allows the green synthesis of anti-inflammatory agents. *J. Catal.* 247 (2): 223–230.

300 Yang, Q., Sha, J., Wang, L. et al. (2006). MgO nanostructures synthesized by thermal evaporation. *Mater. Sci. Eng., C* 26 (5): 1097–1101.

301 Niu, H., Yang, Q., Tang, K., and Xie, Y. (2006). A simple solution calcination route to porous MgO nanoplates. *Microporous Mesoporous Mater.* 96 (1): 428–433.

302 Bartley, J.K., Xu, C., Lloyd, R. et al. (2012). Simple method to synthesize high surface area magnesium oxide and its use as a heterogeneous base catalyst. *Appl. Catal., B* 128: 31–38.

303 Song, G., Ding, Y.-D., Zhu, X., and Liao, Q. (2015). Carbon dioxide adsorption characteristics of synthesized MgO with various porous structures achieved by varying calcination temperature. *Colloids Surf., A* 470: 39–45.

304 Song, G., Zhu, X., Chen, R. et al. (2016). An investigation of CO_2 adsorption kinetics on porous magnesium oxide. *Chem. Eng. J.* 283: 175–183.

305 Elvira, G.-B., Francisco, G.-C., Víctor, S.-M., and Alberto, M.-L.R. (2017). MgO-based adsorbents for CO_2 adsorption: influence of structural and textural properties on the CO_2 adsorption performance. *J. Environ. Sci.* 57: 418–428.

306 Yi, X., Wenzhong, W., Yitai, Q. et al. (1996). Deposition and microstructural characterization of MgO thin films by a spray pyrolysis method. *Surf. Coat. Technol.* 82 (3): 291–293.

307 Pilarska, A., Lukosek, M., Siwińska-Stefańska, K. et al. (2014). Use of MgO to promote the oxyethylation reaction of lauryl alcohol. *Pol. J. Chem.*: 36.

308 Pilarska, A., Paukszta, D., Ciesielczyk, F., and Jesionowski, T. (2010). Physico-chemical and dispersive characterisation of magnesium oxides precipitated from the $Mg(NO_3)_2$ and $MgSO_4$ solutions. *Pol. J. Chem.* 12: 52.

309 Phuoc, T.X., Howard, B.H., Martello, D.V. et al. (2008). Synthesis of $Mg(OH)_2$, MgO, and Mg nanoparticles using laser ablation of magnesium in water and solvents. *Opt. Lasers Eng.* 46 (11): 829–834.

310 Portillo, R., Lopez, T., Gomez, R. et al. (1996). Magnesia synthesis via sol–gel: structure and reactivity. *Langmuir* 12 (1): 40–44.

311 Kumar, A. and Kumar, J. (2008). On the synthesis and optical absorption studies of nano-size magnesium oxide powder. *J. Phys. Chem. Solids* 69 (11): 2764–2772.

312 Aramendía, M.a.A., Borau, V., Jiménez, C. et al. (2003). Influence of the preparation method on the structural and surface properties of various magnesium oxides and their catalytic activity in the Meerwein–Ponndorf–Verley reaction. *Appl. Catal., A* 244 (2): 207–215.

313 Hao, Y., Meng, G., Ye, C. et al. (2005). Kinetics-driven growth of orthogonally branched single-crystalline magnesium oxide nanostructures. *J. Phys. Chem. B* 109 (22): 11204–11208.

314 Kumari, L., Li, W.Z., Vannoy, C.H. et al. (2009). Synthesis, characterization and optical properties of $Mg(OH)_2$ micro–/nanostructure and its conversion to MgO. *Ceram. Int.* 35 (8): 3355–3364.

315 Ouraipryvan, P., Sreethawong, T., and Chavadej, S. (2009). Synthesis of crystalline MgO nanoparticle with mesoporous-assembled structure via a surfactant-modified sol–gel process. *Mater. Lett.* 63 (21): 1862–1865.

316 McKenzie, A.L., Fishel, C.T., and Davis, R.J. (1992). Investigation of the surface structure and basic properties of calcined hydrotalcites. *J. Catal.* 138 (2): 547–561.

317 Matsuhashi, H. (2009). Synthesis of novel solid base of MgO covered with metal oxides. *Top. Catal.* 52 (6): 828–833.

318 Tsuji, H., Yagi, F., Hattori, H., and Kita, H. (1994). Self-condensation of *n*-butyraldehyde over solid base catalysts. *J. Catal.* 148 (2): 759–770.

319 Faba, L., Díaz, E., and Ordóñez, S. (2013). Improvement on the catalytic performance of Mg–Zr mixed oxides for furfural–acetone Aldol condensation by supporting on mesoporous carbons. *ChemSusChem* 6 (3): 463–473.

320 Cueto, J., Faba, L., Díaz, E., and Ordóñez, S. (2017). Performance of basic mixed oxides for aqueous-phase 5-hydroxymethylfurfural-acetone aldol condensation. *Appl. Catal., B* 201: 221–231.

321 Aramendia, M.A., Borau, V., Jimenez, C. et al. (1999). Synthesis and characterization of $MgO–B_2O_3$ mixed oxides prepared by coprecipitation; selective dehydrogenation of propan-2-ol. *J. Mater. Chem.* 9 (3): 819–825.

322 Wallot, J., Reynders, P., and Rödel, J. (2008). Liquid-phase sintering of nanocrystalline titania doped with boron oxide: bulk versus thin film. *J. Am. Ceram. Soc.* 91 (12): 3856–3863.

323 Zeng, Q., Li, W., Shi, J.-L. et al. (2007). Effect of B_2O_3 on the sintering and microwave dielectric properties of M-phase $LiNb_{0.6}Ti_{0.5}O_3$ ceramics. *J. Eur. Ceram. Soc.* 27 (1): 261–265.

324 Kim, D.-W., Sun Hong, K., Yoon, C.S., and Kyung Kim, C. (2003). Low-temperature sintering and microwave dielectric properties of $Ba_5Nb_4O_{15}–BaNb_2O_6$ mixtures for LTCC applications. *J. Eur. Ceram. Soc.* 23 (14): 2597–2601.

325 Pham, T.N., Zhang, L., Shi, D. et al. (2016). Fine-tuning the acid–base properties of boron-doped magnesium oxide catalyst for the selective aldol condensation. *ChemCatChem* 8 (23): 3611–3620.

326 López, T., Hernández, J., Gómez, R. et al. (1999). Synthesis and characterization of $TiO_2–MgO$ mixed oxides prepared by the sol–gel method. *Langmuir* 15 (18): 5689–5693.

327 Aramendía, M.a.A., Boráu, V., Jiménez, C. et al. (2004). Synthesis and textural-structural characterization of magnesia, magnesia–titania and magnesia–zirconia catalysts. *Colloids Surf., A* 234 (1): 17–25.

328 Youssef, A.M., Khalil, L.B., and Girgis, B.S. (1992). Decomposition of isopropanol on magnesium oxide/silica in relation to texture, acidity and chemical composition. *Appl. Catal., A* 81 (1): 1–13.

329 Ordomsky, V.V., Sushkevich, V.L., and Ivanova, I.I. (2010). Study of acetaldehyde condensation chemistry over magnesia and zirconia supported on silica. *J. Mol. Catal. A: Chem.* 333 (1): 85–93.

330 Llanos, M.E., Lopez, T., and Gomez, R. (1997). Determination of the surface heterogeneity of $MgO–SiO_2$ sol–gel mixed oxides by means of CO_2 and ammonia thermodesorption. *Langmuir* 13 (5): 974–978.

331 Ciesielczyk, F., Bartczak, P., Klapiszewski, L. et al. (2014). Influence of calcination parameters on physicochemiacal and structual properties of co-precipitated magnesium silicate. *Physicochem. Probl. Miner. Process.* 50 (1): 119–129.

332 Zhu, Q., Wang, B., and Tan, T. (2017). Conversion of ethanol and acetaldehyde to butadiene over $MgO–SiO_2$ catalysts: effect of reaction parameters and interaction between MgO and SiO_2 on catalytic performance. *ACS Sustainable Chem. Eng.* 5 (1): 722–733.

333 Díez, V.K., Apesteguía, C.R., and Di Cosimo, J.I. (2006). Aldol condensation of citral with acetone on MgO and alkali-promoted MgO catalysts. *J. Catal.* 240 (2): 235–244.

334 Abelló, S., Medina, F., Tichit, D. et al. (2005). Study of alkaline-doping agents on the performance of reconstructed Mg–Al hydrotalcites in aldol condensations. *Appl. Catal., A* 281 (1): 191–198.

335 Jin, J., Zhang, Z., Ma, H. et al. (2009). Surface modification of spherical magnesium oxide with ethylene glycol. *Mater. Lett.* 63 (17): 1514–1516.

336 Quesada, J., Faba, L., Díaz, E., and Ordóñez, S. (2017). Role of the surface intermediates in the stability of basic mixed oxides as catalyst for ethanol condensation. *Appl. Catal., A* 542: 271–281.

337 Pilarska, A.A., Klapiszewski, Ł., and Jesionowski, T. (2017). Recent development in the synthesis, modification and application of $Mg(OH)_2$ and MgO: a review. *Powder Technol.* 319: 373–407.

338 León, M., Faba, L., Díaz, E. et al. (2014). Consequences of MgO activation procedures on its catalytic performance for acetone self-condensation. *Appl. Catal., B* 147: 796–804.

339 Gulková, D., Šolcová, O., and Zdražil, M. (2004). Preparation of MgO catalytic support in shaped mesoporous high surface area form. *Microporous Mesoporous Mater.* 76 (1): 137–149.

340 Wang, F., Ta, N., and Shen, W. (2014). MgO nanosheets, nanodisks, and nanofibers for the Meerwein–Ponndorf–Verley reaction. *Appl. Catal., A* 475: 76–81.

341 Utamapanya, S., Klabunde, K.J., and Schlup, J.R. (1991). Nanoscale metal oxide particles/clusters as chemical reagents. Synthesis and properties of ultrahigh surface area magnesium hydroxide and magnesium oxide. *Chem. Mater.* 3 (1): 175–181.

342 Mastuli, M.S., Ansari, N.S., Nawawi, M.A., and Mahat, A.M. (2012). Effects of cationic surfactant in sol-gel synthesis of nano sized magnesium oxide. *APCBEE Procedia* 3: 93–98.

343 Samodi, A., Rashidi, A., Marjani, K., and Ketabi, S. (2013). Effects of surfactants, solvents and time on the morphology of MgO nanoparticles prepared by the wet chemical method. *Mater. Lett.* 109: 269–274.

344 Meshkani, F. and Rezaei, M. (2009). Facile synthesis of nanocrystalline magnesium oxide with high surface area. *Powder Technol.* 196 (1): 85–88.

345 Sutradhar, N., Sinhamahapatra, A., Pahari, S.K. et al. (2011). Controlled synthesis of different morphologies of MgO and their use as solid base catalysts. *J. Phys. Chem. C* 115 (25): 12308–12316.

346 Roggenbuck, J. and Tiemann, M. (2005). Ordered mesoporous magnesium oxide with high thermal stability synthesized by exotemplating using CMK-3 carbon. *J. Am. Chem. Soc.* 127 (4): 1096–1097.

347 Li, Q., Ding, Y., Yu, G. et al. (2003). Fabrication of light-emitting porous hydromagnesite with rosette-like architecture. *Solid State Commun.* 125 (2): 117–120.

348 Roggenbuck, J., Koch, G., and Tiemann, M. (2006). Synthesis of mesoporous magnesium oxide by CMK-3 carbon structure replication. *Chem. Mater.* 18 (17): 4151–4156.

349 Bhagiyalakshmi, M., Lee, J.Y., and Jang, H.T. (2010). Synthesis of mesoporous magnesium oxide: its application to CO_2 chemisorption. *Int. J. Greenhouse Gas Control* 4 (1): 51–56.

350 Li, J., Dai, W.-L., and Fan, K. (2008). Formation of ordered mesoporous MgO with tunable pore diameter and its application as excellent alkaline catalyst in Baeyer–Villiger oxidation. *J. Phys. Chem. C* 112 (45): 17657–17663.

351 Vu, A.-T., Jiang, S., Ho, K. et al. (2015). Mesoporous magnesium oxide and its composites: preparation, characterization, and removal of 2-chloroethyl ethyl sulfide. *Chem. Eng. J.* 269: 82–93.

352 Ho, K., Jin, S., Zhong, M. et al. (2017). Sorption capacity and stability of mesoporous magnesium oxide in post-combustion CO_2 capture. *Mater. Chem. Phys.* 198: 154–161.

353 Vu, A.-T., Park, Y., Jeon, P.R., and Lee, C.-H. (2014). Mesoporous MgO sorbent promoted with KNO_3 for CO_2 capture at intermediate temperatures. *Chem. Eng. J.* 258: 254–264.

354 Bian, S.-W., Baltrusaitis, J., Galhotra, P., and Grassian, V.H. (2010). A template-free, thermal decomposition method to synthesize mesoporous MgO with a nanocrystalline framework and its application in carbon dioxide adsorption. *J. Mater. Chem.* 20 (39): 8705–8710.

355 Feng, J., Zou, L., Wang, Y. et al. (2015). Synthesis of high surface area, mesoporous MgO nanosheets with excellent adsorption capability for Ni(II) via a distillation treating. *J. Colloid Interface Sci.* 438: 259–267.

356 Ling, Z., Zheng, M., Du, Q. et al. (2011). Synthesis of mesoporous MgO nanoplate by an easy solvothermal–annealing method. *Solid State Sci.* 13 (12): 2073–2079.

357 Veldurthi, S., Shin, C.-H., Joo, O.-S., and Jung, K.-D. (2012). Synthesis of mesoporous MgO single crystals without templates. *Microporous Mesoporous Mater.* 152: 31–36.

358 Sharma, J., Sharma, M., and Basu, S. (2017). Synthesis of mesoporous MgO nanostructures using mixed surfactants template for enhanced adsorption and antimicrobial activity. *J. Environ. Chem. Eng.* 5 (4): 3429–3438.

359 Bhagiyalakshmi, M., Hemalatha, P., Ganesh, M. et al. (2011). A direct synthesis of mesoporous carbon supported MgO sorbent for CO_2 capture. *Fuel* 90 (4): 1662–1667.

360 Wang, G., Zhang, L., Dai, H. et al. (2008). P123-assisted hydrothermal synthesis and characterization of rectangular parallelepiped and hexagonal prism single-crystalline MgO with three-dimensional wormholelike mesopores. *Inorg. Chem.* 47 (10): 4015–4022.

361 Cui, H., Wu, X., Chen, Y., and Boughton, R.I. (2014). Synthesis and characterization of mesoporous MgO by template-free hydrothermal method. *Mater. Res. Bull.* 50: 307–311.

362 Peng, W., Li, J., Chen, B. et al. (2016). Mesoporous MgO synthesized by a homogeneous-hydrothermal method and its catalytic performance on gas-phase acetone condensation at low temperatures. *Catal. Commun.* 74: 39–42.

363 Wang, F., Gunathilake, C., and Jaroniec, M. (2016). Development of mesoporous magnesium oxide–alumina composites for CO_2 capture. *J. CO_2 Util.* 13: 114–118.

364 Dumitriu, E., Hulea, V., Chelaru, C. et al. (1999). Influence of the acid–base properties of solid catalysts derived from hydrotalcite-like compounds on the condensation of formaldehyde and acetaldehyde. *Appl. Catal., A* 178 (2): 145–157.

365 Connell, G. and Dumesic, J.A. (1987). The generation of Brønsted and Lewis acid sites on the surface of silica by addition of dopant cations. *J. Catal.* 105 (2): 285–298.

366 Catlow, C.R.A., Ackermann, L., Bell, R.G. et al. (1997). Modelling of structure, sorption, synthesis and reactivity in catalytic systems. *J. Mol. Catal. A: Chem.* 115 (3): 431–448.

367 Kanno, T. and Kobayashi, M. (1997). The difference in the modification effects of Li on two MgO samples of higher and lower Mn contents. *J. Mater. Sci. Lett.* 16 (2): 126–127.

368 Hattori, H. (1995). Heterogeneous basic catalysis. *Chem. Rev.* 95 (3): 537–558.

369 Ramasamy, K.K., Gray, M., Job, H. et al. (2016). Tunable catalytic properties of bi-functional mixed oxides in ethanol conversion to high value compounds. *Catal. Today* 269: 82–87.

370 Lercher, J.A. (1982). Acid-base properties of Al_2O_3/MgO oxides, II. Infrared study of adsorption of pyridine. *React. Kinet. Catal. Lett.* 20 (3):$_2$ 409–413.

371 Sádaba, I., Ojeda, M., Mariscal, R. et al. (2011). Catalytic and structural properties of co-precipitated Mg–Zr mixed oxides for furfural valorization via aqueous aldol condensation with acetone. *Appl. Catal., B* 101 (3): 638–648.

372 Liang, D., Li, G., Liu, Y. et al. (2016). Controllable self-aldol condensation of cyclopentanone over MgO–ZrO_2 mixed oxides: origin of activity & selectivity. *Catal. Commun.* 81: 33–36.

373 Faba, L., Díaz, E., and Ordóñez, S. (2013). Gas phase acetone self-condensation over unsupported and supported Mg–Zr mixed-oxides catalysts. *Appl. Catal., B* 142–143: 387–395.

374 Pupovac, K. and Palkovits, R. (2013). Cu/$MgAl_2O_4$ as Bifunctional catalyst for aldol condensation of 5-hydroxymethylfurfural and selective transfer hydrogenation. *ChemSusChem* 6 (11): 2103–2110.

375 Zukal, A., Pastva, J., and Čejka, J. (2013). MgO-modified mesoporous silicas impregnated by potassium carbonate for carbon dioxide adsorption. *Microporous Mesoporous Mater.* 167: 44–50.

376 Krivtsov, I., Faba, L., Díaz, E. et al. (2014). A new peroxo-route for the synthesis of Mg–Zr mixed oxides catalysts: application in the gas phase acetone self-condensation. *Appl. Catal., A* 477: 26–33.

377 Wang, S., Zhai, Y., Li, X. et al. (2006). Coprecipitation synthesis of MgO-doped ZrO_2 nano powder. *J. Am. Ceram. Soc.* 89 (11): 3577–3581.

378 Climent, M.J., Corma, A., Iborra, S. et al. (2004). Increasing the basicity and catalytic activity of hydrotalcites by different synthesis procedures. *J. Catal.* 225 (2): 316–326.

379 Corma, A., Iborra, S., and Velty, A. (2007). Chemical routes for the transformation of biomass into chemicals. *Chem. Rev.* 107 (6): 2411–2502.

380 Jiang, B. and Huang, G. (2004). Graft polymerization on magnesium oxide surface. *Radiat. Phys. Chem.* 69 (5): 433–438.
381 Valle-Zermeño, R.d., Chimenos, J.M., Formosa, J., and Fernández, A.I. (2012). Hydration of a low-grade magnesium oxide. Lab-scale study. *J. Chem. Technol. Biotechnol.* 87 (12): 1702–1708.
382 Gao, Z., Wei, L., Yan, T., and Zhou, M. (2011). Modification of surface layer of magnesium oxide via partial dissolution and re-growth of crystallites. *Appl. Surf. Sci.* 257 (8): 3412–3416.
383 Ngo, D.T., Sooknoi, T., and Resasco, D.E. (2018). Improving stability of cyclopentanone aldol condensation MgO-based catalysts by surface hydrophobization with organosilanes. *Appl. Catal., B* 237: 835–843.
384 Gopalan, R., Chang, C.H., and Lin, Y.S. (1995). Thermal stability improvement on pore and phase structure of sol–gel derived zirconia. *J. Mater. Sci.* 30 (12): 3075–3081.
385 Tanabe, K. and Yamaguchi, T. (1994). Acid-base bifunctional catalysis by ZrO_2 and its mixed oxides. *Catal. Today* 20 (2): 185–197.
386 Yamaguchi, T. (1994). Application of ZrO_2 as a catalyst and a catalyst support. *Catal. Today* 20 (2): 199–217.
387 Hannink, R.H.J., Kelly, P.M., and Muddle, B.C. (2000). Transformation toughening in zirconia-containing ceramics. *J. Am. Ceram. Soc.* 83 (3): 461–487.
388 Volpato, C.Â.M., Altoé Garbelotto, L.G.D., Fredel, M.C., and Bondioli, F. (2011). Application of zirconia in dentistry: biological, mechanical and optical considerations. In: *Advances in Ceramics – Electric and Magnetic Ceramics, Bioceramics, Ceramics and Environment* (ed. C. Sikalidis), 397–420. IntechOpen.
389 Li, M., Feng, Z., Xiong, G. et al. (2001). Phase transformation in the surface region of zirconia detected by UV Raman spectroscopy. *J. Phys. Chem. B* 105 (34): 8107–8111.
390 Platt, P., Frankel, P., Gass, M. et al. (2014). Finite element analysis of the tetragonal to monoclinic phase transformation during oxidation of zirconium alloys. *J. Nucl. Mater.* 454 (1): 290–297.
391 Santos, V., Zeni, M., Bergmann, C.P., and Hohemberger, J.M. (2008). Correlation between thermal treatment and tetragonal/monoclinic nanostructured zirconia powder obtained by sol–gel process. *Rev. Adv. Mater. Sci.* 17: 62–70.
392 Tyagi, B., Sidhpuria, K., Shaik, B., and Jasra, R.V. (2006). Synthesis of nanocrystalline zirconia using sol–gel and precipitation techniques. *Ind. Eng. Chem. Res.* 45 (25): 8643–8650.
393 Zhang, F., Chupas, P.J., Lui, S.L.A. et al. (2007). In situ study of the crystallization from amorphous to cubic zirconium oxide: rietveld and reverse Monte Carlo analyses. *Chem. Mater.* 19 (13): 3118–3126.
394 Figueroa, S., Desimoni, J., Rivas, P.C. et al. (2005). Hyperfine study on sol–gel derived-hematite doped zirconia. *Chem. Mater.* 17 (13): 3486–3491.
395 Bokhimi, X., Morales, A., García-Ruiz, A. et al. (1999). Transformation of yttrium-doped hydrated zirconium into tetragonal and cubic nanocrystalline zirconia. *J. Solid State Chem.* 142 (2): 409–418.
396 Feng, X. and Keith Hall, W. (1997). FeZSM-5: a durable SCR catalyst for NO_x removal from combustion streams. *J. Catal.* 166 (2): 368–376.

397 Petunchi, J.O., Sill, G., and Hall, W.K. (1993). Studies of the selective reduction of nitric oxide by hydrocarbons. *Appl. Catal., B* 2 (4): 303–321.

398 Qingjie Ge, A.K., Roger, A.C., Li, W., and Xu, H. (2004). Preparation and characterization of modified-ZrO_2 catalysts for the reaction of CO hydrogenation. *J. Nat. Gas Chem.* 13 (1): 41–44.

399 Collazo, A., Covelo, A., and Pérez, C. (2010). Structural transformation of silane-based zirconium-modified sol–gel coatings. *Surf. Interface Anal.* 42 (6–7): 1201–1204.

400 Deshmane, V.G. and Adewuyi, Y.G. (2012). Synthesis of thermally stable, high surface area, nanocrystalline mesoporous tetragonal zirconium dioxide (ZrO_2): effects of different process parameters. *Microporous Mesoporous Mater.* 148 (1): 88–100.

401 Chin, Y.-H., Alvarez, W.E., and Resasco, D.E. (2000). Sulfated zirconia and tungstated zirconia as effective supports for Pd-based SCR catalysts. *Catal. Today* 62 (2): 159–165.

402 Krajewski, S.R., Kujawski, W., Dijoux, F. et al. (2004). Grafting of ZrO_2 powder and ZrO_2 membrane by fluoroalkylsilanes. *Colloids Surf., A* 243 (1): 43–47.

403 Picard, C., Larbot, A., Tronel-Peyroz, E., and Berjoan, R. (2004). Characterisation of hydrophilic ceramic membranes modified by fluoroalkylsilanes into hydrophobic membranes. *Solid State Sci.* 6 (6): 605–612.

404 Kujawa, J., Cerneaux, S., and Kujawski, W. (2014). Characterization of the surface modification process of Al_2O_3, TiO_2 and ZrO_2 powders by PFAS molecules. *Colloids Surf., A* 447: 14–22.

405 Rudakova, A.V., Maevskaya, M.V., Emeline, A.V., and Bahnemann, D.W. (2016). Light-controlled ZrO_2 surface hydrophilicity. *Sci. Rep.* 6: 34285.

406 Evans, A.G. and Cannon, R.M. (1986). Overview no. 48: toughening of brittle solids by martensitic transformations. *Acta Metall.* 34 (5): 761–800.

407 Sun, J., Zhu, K., Gao, F. et al. (2011). Direct conversion of bio-ethanol to isobutene on nanosized $Zn_xZr_yO_z$ mixed oxides with balanced acid–base sites. *J. Am. Chem. Soc.* 133 (29): 11096–11099.

408 Crisci, A.J., Dou, H., Prasomsri, T., and Román-Leshkov, Y. (2014). Cascade reactions for the continuous and selective production of isobutene from bioderived acetic acid over zinc-zirconia catalysts. *ACS Catal.* 4 (11): 4196–4200.

409 Pengpanich, S., Meeyoo, V., Rirksomboon, T., and Bunyakiat, K. (2002). Catalytic oxidation of methane over CeO_2–ZrO_2 mixed oxide solid solution catalysts prepared via urea hydrolysis. *Appl. Catal., A* 234 (1): 221–233.

410 Reddy, B.M., Sreekanth, P.M., Lakshmanan, P., and Khan, A. (2006). Synthesis, characterization and activity study of $SO_4^{2-}/Ce_xZr_1-xO_2$ solid superacid catalyst. *J. Mol. Catal. A: Chem.* 244 (1): 1–7.

411 Kumar, T.E.M., Shamshuddin, S., Mohamed, Z., and Mubarak, N.M. (2007). Simple and efficient synthesis of jasminaldehyde over modified forms of zirconia: acid-base bifunctional catalysis. *Indian J. Chem. Technol.* 24: 548–554.

412 Wu, K., Yang, M., Pu, W. et al. (2017). Carbon promoted ZrO_2 catalysts for aqueous-phase ketonization of acetic acid. *ACS Sustainable Chem. Eng.* 5 (4): 3509–3516.

413 Díaz, I., Kokkoli, E., Terasaki, O., and Tsapatsis, M. (2004). Surface structure of zeolite (MFI) crystals. *Chem. Mater.* 16 (25): 5226–5232.

414 Hu, H., Zhang, Q., Cen, J., and Li, X. (2014). High suppression of the formation of ethylbenzene in benzene alkylation with methanol over ZSM-5 catalyst modified by platinum. *Catal. Commun.* 57: 129–133.

415 Dong, P., Li, Z., Ji, D. et al. (2018). Catalytic benzene mono-alkylation over three catalysts: improving activity and selectivity with M–Y catalyst. *J. Inclusion Phenom.Macrocyclic Chem.* 90 (1): 149–155.

416 Deng, W., Xuan, H., Zhang, C. et al. (2014). Promoting xylene production in benzene methylation using hierarchically porous ZSM-5 derived from a modified dry-gel route. *Chin. J. Chem. Eng.* 22 (8): 921–929.

417 Kim, J.-C., Cho, K., Lee, S., and Ryoo, R. (2015). Mesopore wall-catalyzed Friedel–Crafts acylation of bulky aromatic compounds in MFI zeolite nanosponge. *Catal. Today* 243: 103–108.

418 Pandey, A.K. and Singh, A.P. (1997). A novel catalytic method for the acylation of aromatics to the corresponding ketones over zeolite catalysts. *Catal. Lett.* 44 (1): 129–133.

419 Sugi, Y., Kubota, Y., Komura, K. et al. (2006). Shape-selective alkylation and related reactions of mononuclear aromatic hydrocarbons over H-ZSM-5 zeolites modified with lanthanum and cerium oxides. *Appl. Catal., A* 299: 157–166.

420 Huang, L., Guo, W., Deng, P. et al. (2000). Investigation of synthesizing MCM-41/ZSM-5 composites. *J. Phys. Chem. B* 104 (13): 2817–2823.

421 Al-Kinany, M.C., Al-Megren, H.A., Al-Ghilan, E.A. et al. (2012). Selective zeolite catalyst for alkylation of benzene with ethylene to produce ethylbenzene. *Appl. Petrochem. Res.* 2 (3): 73–83.

422 Ji, Y., Yang, H., and Yan, W. (2017). Strategies to enhance the catalytic performance of ZSM-5 zeolite in hydrocarbon cracking: a review. *Catalysts* 7 (12): 367.

423 Le Van Mao, R., Xiao, S., Ramsaran, A., and Yao, J. (1994). Selective removal of silicon from zeolite frameworks using sodium carbonate. *J. Mater. Chem.* 4 (4): 605–610.

424 Cundy, C.S., Henty, M.S., and Plaisted, R.J. (1995). Investigation of Na, TPA-ZSM-5 zeolite synthesis by chemical methods. *Zeolites* 15 (4): 342–352.

425 Čižmek, A., Subotić, B., Aiello, R. et al. (1995). Dissolution of high-silica zeolites in alkaline solutions I. Dissolution of silicalite-1 and ZSM-5 with different aluminum content. *Microporous Mater.* 4 (2): 159–168.

426 Ogura, M., Shinomiya, S.-y., Tateno, J. et al. (2000). Formation of uniform mesopores in ZSM-5 zeolite through treatment in alkaline solution. *Chem. Lett.* 29 (8): 882–883.

427 Bjørgen, M., Joensen, F., Spangsberg Holm, M. et al. (2008). Methanol to gasoline over zeolite H-ZSM-5: improved catalyst performance by treatment with NaOH. *Appl. Catal., A* 345 (1): 43–50.

428 Padmanabhan, A., Selvin, R., Hsu, H.-L., and Xiao, L.-W. (2010). Efficient acylation of anisole over hierarchical porous ZSM-5 structure. *Chem. Eng. Technol.* 33 (6): 998–1002.
429 Freyhardt, C.C., Tsapatsis, M., Lobo, R.F. et al. (1996). A high-silica zeolite with a 14-tetrahedral-atom pore opening. *Nature* 381: 295.
430 Kresge, C.T., Leonowicz, M.E., Roth, W.J. et al. (1992). Ordered mesoporous molecular sieves synthesized by a liquid-crystal template mechanism. *Nature* 359: 710.
431 Hu, H., Lyu, J., Wang, Q. et al. (2015). Alkylation of benzene with methanol over hierarchical porous ZSM-5: synergy effects of hydrogen atmosphere and zinc modification. *RSC Adv.* 5 (41): 32679–32684.
432 Hu, H., Zhang, Q., Cen, J., and Li, X. (2015). Catalytic activity of Pt modified hierarchical ZSM-5 catalysts in benzene alkylation with methanol. *Catal. Lett.* 145 (2): 715–722.
433 Wei, Y., Parmentier, T.E., de Jong, K.P., and Zecevic, J. (2015). Tailoring and visualizing the pore architecture of hierarchical zeolites. *Chem. Soc. Rev.* 44 (20): 7234–7261.
434 Milina, M., Mitchell, S., Trinidad, Z.D. et al. (2012). Decoupling porosity and compositional effects on desilicated ZSM-5 zeolites for optimal alkylation performance. *Catal. Sci. Technol.* 2 (4): 759–766.
435 Christensen, C.H., Johannsen, K., Schmidt, I., and Christensen, C.H. (2003). Catalytic benzene alkylation over mesoporous zeolite single crystals: improving activity and selectivity with a new family of porous materials. *J. Am. Chem. Soc.* 125 (44): 13370–13371.
436 Jacobsen, C.J.H., Madsen, C., Houzvicka, J. et al. (2000). Mesoporous zeolite single crystals. *J. Am. Chem. Soc.* 122 (29): 7116–7117.
437 Yamamura, M., Chaki, K., Wakatsuki, T. et al. (1994). Synthesis of ZSM-5 zeolite with small crystal size and its catalytic performance for ethylene oligomerization. *Zeolites* 14 (8): 643–649.
438 Ding, C., Wang, X., Guo, X., and Zhang, S. (2008). Characterization and catalytic alkylation of hydrothermally dealuminated nanoscale ZSM-5 zeolite catalyst. *Catal. Commun.* 9 (4): 487–493.
439 Lovallo, M.C. and Tsapatis, M. (1996). *Advanced Catalysts and Nanostructured Materials*. San Diego, CA: Academic Press.
440 Chen, N.Y., Kaeding, W.W., and Dwyer, F.G. (1979). Para-directed aromatic reactions over shape-selective molecular sieve zeolite catalysts. *J. Am. Chem. Soc.* 101 (22): 6783–6784.
441 Nunan, J., Cronin, J., and Cunningham, J. (1984). Combined catalytic and infrared study of the modification of H-ZSM-5 with selected poisons to give high *p*-xylene selectivity. *J. Catal.* 87 (1): 77–85.
442 Van Vu, D., Miyamoto, M., Nishiyama, N. et al. (2006). Selective formation of para-xylene over H-ZSM-5 coated with polycrystalline silicalite crystals. *J. Catal.* 243 (2): 389–394.
443 Li, Y.-G., Xie, W.-H., and Yong, S. (1997). The acidity and catalytic behavior of Mg-ZSM-5 prepared via a solid-state reaction. *Appl. Catal., A* 150 (2): 231–242.

444 Lersch, P. and Bandermann, F. (1991). Conversion of chloromethane over metal-exchanged ZSM-5 to higher hydrocarbons. *Appl. Catal.* 75 (1): 133–152.

445 Kovacheva, P., Predoeva, A., Arishtirova, K., and Vassilev, S. (2002). Oxidative methylation of toluene with methane using X zeolite catalyst modified with alkali earth oxides. *Appl. Catal., A* 223 (1): 121–128.

446 Mao, D., Yang, W., Xia, J. et al. (2005). Highly effective hybrid catalyst for the direct synthesis of dimethyl ether from syngas with magnesium oxide-modified HZSM-5 as a dehydration component. *J. Catal.* 230 (1): 140–149.

447 Zhao, L., Gao, J., Xu, C., and Shen, B. (2011). Alkali-treatment of ZSM-5 zeolites with different SiO_2/Al_2O_3 ratios and light olefin production by heavy oil cracking. *Fuel Process. Technol.* 92 (3): 414–420.

448 Fraenkel, D. (1990). Role of external surface sites in shape-selective catalysis over zeolites. *Ind. Eng. Chem. Res.* 29 (9): 1814–1821.

449 Sotelo, J.L., Uguina, M.A., Valverde, J.L., and Serrano, D.P. (1993). Kinetics of toluene alkylation with methanol over magnesium-modified ZSM-5. *Ind. Eng. Chem. Res.* 32 (11): 2548–2554.

450 Breen, J., Burch, R., Kulkarni, M. et al. (2005). Enhanced para-xylene selectivity in the toluene alkylation reaction at ultralow contact time. *J. Am. Chem. Soc.* 127 (14): 5020–5021.

451 Faramawy, S. (1999). Selective toluene-methanol alkylation over modified zsm-5 zeolite catalysts. *Pet. Sci. Technol.* 17 (3–4): 249–271.

452 Uguina, M.A., Sotelo, J.L., Serrano, D.P., and Van Grieken, R. (1992). Magnesium and silicon as ZSM-5 modifier agents for selective toluene disproportionation. *Ind. Eng. Chem. Res.* 31 (8): 1875–1880.

453 Zuo, C., Ge, T., Guo, X. et al. (2018). Synthesis and catalytic performance of Cs/P modified ZSM-5 zeolite in aldol condensation of methyl acetate with different sources of formaldehyde. *Microporous Mesoporous Mater.* 256: 58–66.

454 Kim, J.-C., Cho, K., and Ryoo, R. (2014). High catalytic performance of surfactant-directed nanocrystalline zeolites for liquid-phase Friedel–Crafts alkylation of benzene due to external surfaces. *Appl. Catal., A* 470: 420–426.

455 Serrano, D.P., García, R.A., Vicente, G. et al. (2011). Acidic and catalytic properties of hierarchical zeolites and hybrid ordered mesoporous materials assembled from MFI protozeolitic units. *J. Catal.* 279 (2): 366–380.

456 Serrano, D.P., García, R.A., and Otero, D. (2009). Friedel–Crafts acylation of anisole over hybrid zeolitic-mesostructured materials. *Appl. Catal., A* 359 (1): 69–78.

457 Radhika, N.P., Selvin, R., Kakkar, R., and Roselin, L.S. (2018). Nanocrystalline hierarchical ZSM-5: an efficient catalyst for the alkylation of phenol with cyclohexene. *J. Nanosci. Nanotechnol.* 18 (8): 5404–5413.

458 Selvin, R., Hsu, H.-L., and Her, T.-M. (2008). Acylation of anisole with acetic anhydride using ZSM-5 catalysts: effect of ZSM-5 particle size in the nanoscale range. *Catal. Commun.* 10 (2): 169–172.

459 Gao, J., Zhang, L., Hu, J. et al. (2009). Effect of zinc salt on the synthesis of ZSM-5 for alkylation of benzene with ethanol. *Catal. Commun.* 10 (12): 1615–1619.

460 Adebajo, M.O. and Long, M.A. (2003). The contribution of the methanol-to-aromatics reaction to benzene methylation over ZSM-5 catalysts. *Catal. Commun.* 4 (2): 71–76.

461 Tsai, T.-C., Wang, I., Huang, C.-K., and Liu, S.-D. (2007). Study on ethylbenzene and xylene conversion over modified ZSM-5. *Appl. Catal., A* 321 (2): 125–134.

462 Finsy, V., Verelst, H., Alaerts, L. et al. (2008). Pore-filling-dependent selectivity effects in the vapor-phase separation of xylene isomers on the metal–organic framework MIL-47. *J. Am. Chem. Soc.* 130 (22): 7110–7118.

463 Whyte, T.E. Jr., (1983). Catalytic materials: relationship between structure and reactivity. *Chem. Eng. News Arch.* 61 (16): 22.

464 Kaeding, W.W. (1985). Shape-selective reactions with zeolite catalysts: V. Alkylation or disproportionation of ethylbenzene to produce *p*-diethylbenzene. *J. Catal.* 95 (2): 512–519.

465 Zhu, Z., Chen, Q., Xie, Z. et al. (2006). The roles of acidity and structure of zeolite for catalyzing toluene alkylation with methanol to xylene. *Microporous Mesoporous Mater.* 88 (1): 16–21.

466 Tan, W., Liu, M., Zhao, Y. et al. (2014). Para-selective methylation of toluene with methanol over nano-sized ZSM-5 catalysts: synergistic effects of surface modifications with SiO_2, P_2O_5 and MgO. *Microporous Mesoporous Mater.* 196: 18–30.

467 Ding, W., Cui, Y., Li, J. et al. (2014). Promoting effect of dual modification of H-ZSM-5 catalyst by alkali treating and Mg doping on catalytic performances for alkylation of benzene with ethanol to ethylbenzene. *RSC Adv.* 4 (91): 50123–50129.

468 Zhang, Y.-Y., Li, Y.-F., Chen, L. et al. (2014). A new catalytic process for the synthesis of para-xylene through benzene methylation with CH_3Br. *Catal. Commun.* 54: 6–10.

469 Phatanasri, S., Praserthdam, P., and Punsupsawat, T. (2000). Influence of Fe or Zn loading method on toluene methylation over MFI-type zeolite catalysts. *Korean J. Chem. Eng.* 17 (4): 414.

470 El Morsi, A., Omar, A.M.A., and Almehbad, N.Y. (2014). Development of alkylation toluene with methanol for fuel on modified ZSM-5 zeolites by amphoteric surfactant. *J. Surf. Eng. Mater. Adv. Technol.* 4 (01): 41.

471 Bernardon, C., Ben Osman, M., Laugel, G. et al. (2017). Acidity versus metal-induced Lewis acidity in zeolites for Friedel–Crafts acylation. *C.R. Chim.* 20 (1): 20–29.

472 Choudhary, V.R., Jana, S.K., and Kiran, B.P. (1999). Alkylation of benzene by benzyl chloride over H-ZSM-5 zeolite with its framework Al completely or partially substituted by Fe or Ga. *Catal. Lett.* 59 (2): 217–219.

473 Coq, B., Gourves, V., and Figuéras, F. (1993). Benzylation of toluene by benzyl chloride over protonic zeolites. *Appl. Catal., A* 100 (1): 69–75.

474 Chen, L.Z. and Feng, Y.Q. (1992). Toluene alkylation over ZSM-5 zeolite catalysts with skeletal and nonskeletal boron. *Zeolites* 12 (4): 347–350.

475 Chandawar, K.H., Kulkarni, S.B., and Ratnasamy, P. (1982). Alkylation of benzene with ethanol over ZSM5 zeolites. *Appl. Catal.* 4 (3): 287–295.
476 Lyu, J., Hu, H., Tait, C. et al. (2017). Benzene alkylation with methanol over phosphate modified hierarchical porous ZSM-5 with tailored acidity. *Chin. J. Chem. Eng.* 25 (9): 1187–1194.
477 Niu, X., Gao, J., Miao, Q. et al. (2014). Influence of preparation method on the performance of Zn-containing HZSM-5 catalysts in methanol-to-aromatics. *Microporous Mesoporous Mater.* 197: 252–261.
478 Subba Rao, Y.V., Kulkarni, S.J., Subrahmanyam, M., and Rama Rao, A.V. (1995). An improved acylation of phenol over modified ZSM-5 catalysts. *Appl. Catal., A* 133 (1): L1–L6.
479 Hibino, T., Niwa, M., and Murakami, Y. (1991). Shape-selectivity over HZSM-5 modified by chemical vapor deposition of silicon alkoxide. *J. Catal.* 128 (2): 551–558.
480 Gao, K., Li, S., Wang, L., and Wang, W. (2015). Study of the alkylation of benzene with methanol for the selective formation of toluene and xylene over Co_3O_4-La_2O_3/ZSM-5. *RSC Adv.* 5 (56): 45098–45105.
481 Abubakar, S.M., Lee, J.C., Marcus, D.M. et al. (2006). Functionalization of zeolite ZSM-5 by hydrosilylation and acetone coupling: synthesis, MAS NMR, and theory. *J. Phys. Chem. B* 110 (30): 14598–14603.
482 Song, W., Marcus, D.M., Abubakar, S.M. et al. (2003). Trimethylsilylation of framework Brønsted acid sites in microporous zeolites and silico-aluminophosphates. *J. Am. Chem. Soc.* 125 (46): 13964–13965.
483 Fyfe, C.A., Kokotailo, G.T., Graham, J.D. et al. (1986). Demonstration of contact induced ion exchange in zeolites. *J. Am. Chem. Soc.* 108 (3): 522–523.
484 Kucherov, A.V. and Slinkin, A.A. (1987). Introduction of Cr(V), Mo(V) and V(IV) ions in cationic positions of high-silica zeolites by a solid-state reaction. *Zeolites* 7 (1): 38–42.
485 Kucherov, A.V. and Slinkin, A.A. (1987). Co-introduction of transition metal ions into cationic positions of H-ZSM-5 by a solid-state reaction. *Zeolites* 7 (1): 43–46.
486 Beyer, H.K., Karge, H.G., and Borbély, G. (1988). Solid-state ion exchange in zeolites: part I. Alkaline chlorides/ZSM-5. *Zeolites* 8 (1): 79–82.
487 Kucherov, A.V. and Slinkin, A.A. (1986). Introduction of transition metal ions in cationic positions of high-silica zeolites by a solid state reaction. Interaction of copper compounds with H-mordenite or H-ZSM-5. *Zeolites* 6 (3): 175–180.
488 Borbély, G., Beyer, H.K., Radics, L. et al. (1989). Solid-state ion exchange in zeolites: part IV. Evidence for contact-induced ion exchange between hydrated NaY zeolite and metal chlorides. *Zeolites* 9 (5): 428–431.
489 Karge, H.G., Mavrodinova, V., Zheng, Z., and Beyer, H.K. (1991). Cracking of *n*-decane over Lanthanum Y catalysts: comparison of Lanthanum Y catalysts obtained by solid-state ion exchange and ion exchange in solution. *Applied Catalysis.* 75 (1): 343–357.
490 van der Bij, H.E. and Weckhuysen, B.M. (2015). Phosphorus promotion and poisoning in zeolite-based materials: synthesis, characterisation and catalysis. *Chem. Soc. Rev.* 44 (20): 7406–7428.

491 Choudhary, V.R., Kinage, A.K., Sivadinarayana, C., and Guisnet, M. (1996). Pulse reaction studies on variations of initial activity/selectivity of O_2 and H_2 pretreated Ga-modified ZSM-5 type zeolite catalysts in propane aromatization. *J. Catal.* 158 (1): 23–33.

492 Choudhary, V.R., Kinage, A.K., Sivadinarayana, C. et al. (1996). H-gallosilicate (MFI) propane aromatization catalyst: influence of Si/Ga ratio on acidity, activity and deactivation due to coking. *J. Catal.* 158 (1): 34–50.

493 Alabi, W., Atanda, L., Jermy, R., and Al-Khattaf, S. (2012). Kinetics of toluene alkylation with methanol catalyzed by pure and hybridized HZSM-5 catalysts. *Chem. Eng. J.* 195–196: 276–288.

494 Nishiyama, N., Miyamoto, M., Egashira, Y., and Ueyama, K. (2001). Zeolite membrane on catalyst particles for selective formation of *p*-xylene in the disproportionation of toluene. *Chem. Commun.* (18): 1746–1747.

495 Zhang, J., Qian, W., Kong, C., and Wei, F. (2015). Increasing para-xylene selectivity in making aromatics from methanol with a surface-modified Zn/P/ZSM-5 catalyst. *ACS Catal.* 5 (5): 2982–2988.

496 Zhao, Y., Tan, W., Wu, H. et al. (2011). Effect of Pt on stability of nano-scale ZSM-5 catalyst for toluene alkylation with methanol into *p*-xylene. *Catal. Today* 160 (1): 179–183.

497 Zhao, Y., Wu, H., Tan, W. et al. (2010). Effect of metal modification of HZSM-5 on catalyst stability in the shape-selective methylation of toluene. *Catal. Today* 156 (1): 69–73.

498 Wang, Y., Liu, M., Zhang, A. et al. (2017). Methanol usage in toluene methylation over Pt modified ZSM-5 catalyst: effects of total pressure and carrier gas. *Ind. Eng. Chem. Res.* 56 (16): 4709–4717.

499 Ghiaci, M., Seyedeyn-Azad, F., and Kia, R. (2004). Fast and efficient synthesis of ZSM-5 in a broad range of SiO_2/Al_2O_3 without using seeding gel. *Mater. Res. Bull.* 39 (9): 1257–1264.

500 Massah, A.R., Kalbasi, R.J., and Shafiei, A. (2012). ZSM-5-SO_3H as a novel, efficient, and reusable catalyst for the chemoselective synthesis and deprotection of 1,1-diacetates under eco-friendly conditions. *Monatsh. Chem.* 143 (4): 643–652.

501 Massah, A.R., Kalbasi, R.J., Khalifesoltani, M., and Moshtagh Kordesofla, F. (2013). ZSM-5-SO(3)H: an efficient catalyst for acylation of sulfonamides amines, alcohols, and phenols under solvent-free conditions. *ISRN Org. Chem.* 2013: 951749.

502 Baerlocher, C., McCusker, L.B., and Olson, D.H. (2007). *Atlas of Zeolite Framework Types*. Elsevier Science.

503 Corma, A. (2003). State of the art and future challenges of zeolites as catalysts. *J. Catal.* 216 (1): 298–312.

504 Ribeiro, F.R. (ed.) (1984). *Zeolites: Science and Technology*. Netherlands: Springer https://www.springer.com/gp/book/9789024729357.

505 Kubička, D. and Kikhtyanin, O. (2015). Opportunities for zeolites in biomass upgrading – lessons from the refining and petrochemical industry. *Catal. Today* 243: 10–22.

506 Ennaert, T., Van Aelst, J., Dijkmans, J. et al. (2016). Potential and challenges of zeolite chemistry in the catalytic conversion of biomass. *Chem. Soc. Rev.* 45 (3): 584–611.
507 Resasco, D.E., Wang, B., and Crossley, S. (2016). Zeolite-catalysed C–C bond forming reactions for biomass conversion to fuels and chemicals. *Catal. Sci. Technol.* 6 (8): 2543–2559.
508 Serrano, D.P., Melero, J.A., Morales, G. et al. (2018). Progress in the design of zeolite catalysts for biomass conversion into biofuels and bio-based chemicals. *Catal. Rev.* 60 (1): 1–70.
509 Ravenelle, R.M., Schüβler, F., D'Amico, A. et al. (2010). Stability of zeolites in hot liquid water. *J. Phys. Chem.* 114 (46): 19582–19595.
510 Zapata, P.A., Faria, J., Ruiz, M.P. et al. (2012). Hydrophobic zeolites for biofuel upgrading reactions at the liquid–liquid interface in water/oil emulsions. *J. Am. Chem. Soc.* 134 (20): 8570–8578.
511 Zapata, P.A., Huang, Y., Gonzalez-Borja, M.A., and Resasco, D.E. (2013). Silylated hydrophobic zeolites with enhanced tolerance to hot liquid water. *J. Catal.* 308: 82–97.
512 Zhang, L., Chen, K., Chen, B. et al. (2015). Factors that determine zeolite stability in hot liquid water. *J. Am. Chem. Soc.* 137 (36): 11810–11819.
513 Pino, N., Bui, T., Hincapié, G. et al. (2018). Hydrophobic zeolites for the upgrading of biomass-derived short oxygenated compounds in water/oil emulsions. *Appl. Catal., A* 559: 94–101.
514 Pérez-Ramírez, J., Christensen, C.H., Egeblad, K. et al. (2008). Hierarchical zeolites: enhanced utilisation of microporous crystals in catalysis by advances in materials design. *Chem. Soc. Rev.* 37 (11): 2530–2542.
515 Hua, Z.L., Zhou, J., and Shi, J.L. (2011). Recent advances in hierarchically structured zeolites: synthesis and material performances. *Chem. Commun.* 47 (38): 10536–10547.
516 Verboekend, D. and Pérez-Ramírez, J. (2011). Design of hierarchical zeolite catalysts by desilication. *Catal. Sci. Technol.* 1 (6): 879–890.
517 Verboekend, D., Nuttens, N., Locus, R. et al. (2016). Synthesis, characterisation, and catalytic evaluation of hierarchical faujasite zeolites: milestones, challenges, and future directions. *Chem. Soc. Rev.* 45 (12): 3331–3352.
518 Hartmann, M. (2004). Hierarchical zeolites: a proven strategy to combine shape selectivity with efficient mass transport. *Angew. Chem. Int. Ed.* 43 (44): 5880–5882.
519 Khitev, Y.P., Kolyagin, Y.G., Ivanova, I.I. et al. (2011). Synthesis and catalytic properties of hierarchical micro/mesoporous materials based on FER zeolite. *Microporous Mesoporous Mater.* 146 (1): 201–207.
520 Verboekend, D., Vilé, G., and Pérez-Ramírez, J. (2012). Hierarchical Y and USY zeolites designed by post-synthetic strategies. *Adv. Funct. Mater.* 22 (5): 916–928.
521 Pérez-Ramírez, J. (2012). Imagination has no limits. *Nat. Chem.* 4: 250.
522 Verboekend, D., Keller, T.C., Milina, M. et al. (2013). Hierarchy brings function: mesoporous clinoptilolite and L zeolite catalysts synthesized by tandem acid–base treatments. *Chem. Mater.* 25 (9): 1947–1959.

523 Kocal, J.A., Vora, B.V., and Imai, T. (2001). Production of linear alkylbenzenes. *Appl. Catal., A* 221 (1): 295–301.
524 Anand, R., Gore, K.U., and Rao, B.S. (2002). Alkylation of phenol with cyclohexanol and cyclohexene using HY and modified HY zeolites. *Catal. Lett.* 81 (1): 33–41.
525 Anand, R., Maheswari, R., Gore, K.U., and Chumbhale, V.R. (2002). Selective alkylation of catechol with *t*-butyl alcohol over HY and modified HY zeolites. *Catal. Commun.* 3 (8): 321–326.
526 Craciun, I., Reyniers, M.-F., and Marin, G.B. (2007). Effects of acid properties of Y zeolites on the liquid-phase alkylation of benzene with 1-octene: a reaction path analysis. *J. Mol. Catal. A: Chem.* 277 (1): 1–14.
527 Craciun, I., Reyniers, M.-F., and Marin, G.B. (2012). Liquid-phase alkylation of benzene with octenes over Y zeolites: kinetic modeling including acidity descriptors. *J. Catal.* 294: 136–150.
528 Algarra, F., Corma, A., Garcia, H., and Primo, J. (1995). Acid zeolites as catalysts in organic reactions. Highly selective condensation of 2-alkylfurans with carbonylic compounds. *Appl. Catal., A* 128 (1): 119–126.
529 Zhang, L., Pham, T.N., Faria, J., and Resasco, D.E. (2015). Improving the selectivity to C_4 products in the aldol condensation of acetaldehyde in ethanol over faujasite zeolites. *Appl. Catal., A* 504: 119–129.
530 Komatsu, T., Mitsuhashi, M., and Yashima, T. (2002). Aldol condensation catalyzed by acidic zeolites. In: *Studies in Surface Science and Catalysis* (eds. R. Aiello, G. Giordano and F. Testa), 667–674. Elsevier.
531 Corma, A., JoséCliment, M., García, H., and Primo, J. (1989). Design of synthetic zeolites as catalysts in organic reactions: acylation of anisole by acyl chlorides or carboxylic acids over acid zeolites. *Appl. Catal.* 49 (1): 109–123.
532 Finiels, A., Calmettes, A., Geneste, P., and Moreau, P. (1993). Kinetic study of the acylation of thiophene with acyl chlorides in liquid phase over HY zeolites. In: *Studies in Surface Science and Catalysis* (eds. M. Guisnet et al.), 595–600. Elsevier.
533 Richard, F., Carreyre, H., and Pérot, G. (1996). Zeolite-catalyzed acylation of heterocyclic compounds: acylation of benzofuran and 2-methylbenzofuran in a fixed bed reactor. *J. Catal.* 159 (2): 427–434.
534 Kouwenhoven, H.W. and van Bekkum, H. (2008, 2008). Acylation of aromatics. In: *Handbook of Heterogeneous Catalysis*, 3578–3591. Wiley Online Library.
535 Mori, H., Mizuno, N., Tajima, M. et al. (1991). One-step formation of ethyl methyl ketone from 1-butene and water on ultrastable Y-type zeolite catalyst. *Catal. Lett.* 10 (1): 35–39.
536 Mériaudeau, P., Ben Taarit, Y., Thangaraj, A. et al. (1997). Zeolite based catalysts for linear alkylbenzene production: dehydrogenation of long chain alkanes and benzene alkylation. *Catal. Today* 38 (2): 243–247.
537 Colón, G., Ferino, I., Rombi, E. et al. (1998). Liquid-phase alkylation of naphthalene by isopropanol over zeolites. Part 1: HY zeolites. *Appl. Catal., A* 168 (1): 81–92.

538 Arias, K.S., Climent, M.J., Corma, A., and Iborra, S. (2015). Synthesis of high quality alkyl naphthenic kerosene by reacting an oil refinery with a biomass refinery stream. *Energy Environ. Sci.* 8 (1): 317–331.

539 Newsam, J.M., Treacy, M.M.J., Koetsier, W.T., and De Gruyter, C.B. (1988). Structural characterization of zeolite beta. *Proc. R. Soc. London, Ser. A* 420 (1859): 375–405.

540 Treacy, M.M.J. and Newsam, J.M. (1988). Two new three-dimensional twelve-ring zeolite frameworks of which zeolite beta is a disordered intergrowth. *Nature* 332: 249.

541 Higgins, J.B., LaPierre, R.B., Schlenker, J.L., Rohrman, A.C., Wood, J.D., Kerr, G.T., and Rohrbaugh, W.J. *The Framework Topology of Zeolite Beta*. 1988. United States, https://www.sciencedirect.com/science/article/pii/S0144244988802197.

542 Willhammar, T. and Zou, X. (2013). Stacking disorders in zeolites and open-frameworks – structure elucidation and analysis by electron crystallography and X-ray diffraction. *Z. Kristallogr. – Crys. Mater.* 228 (1): 11.

543 Corma, A., Moliner, M., Cantin, A. et al. (2008). Synthesis and structure of polymorph B of zeolite beta. *Chem. Mater.* 20 (9): 3218–3223.

544 Tong, M., Zhang, D., Fan, W. et al. (2015). Synthesis of chiral polymorph A-enriched zeolite beta with an extremely concentrated fluoride route. *Sci. Rep.* 5: 11521.

545 Camblor, M.A., Barrett, P.A., Diaz-Cabanas, M.J. et al. (2001). High silica zeolites with three-dimensional systems of large pore channels. *Microporous Mesoporous Mater.* 48 (1–3): 11–22.

546 Burton, A.W., Elomari, S., Chan, I. et al. (2005). Structure and synthesis of SSZ-63: toward an ordered form of zeolite beta. *J. Phys. Chem. B* 109 (43): 20266–20275.

547 Zhang, G.Q., Wang, B.C., Zhang, W.P. et al. (2016). Synthesis of polymorph A-enriched beta zeolites in a HF-concentrated system. *Dalton Trans.* 45 (15): 6634–6640.

548 Camblor, M.A., Corma, A., and Valencia, S. (1996). Spontaneous nucleation and growth of pure silica zeolite-beta free of connectivity defects. *Chem. Commun.* (20): 2365–2366.

549 Yu, Z.-B., Han, Y., Zhao, L. et al. (2012). Intergrown new zeolite beta polymorphs with interconnected 12-ring channels solved by combining electron crystallography and single-crystal X-ray diffraction. *Chem. Mater.* 24 (19): 3701–3706.

550 Leiva, S., Sabater, M.J., Valencia, S. et al. (2005). A new synthesis method for the preparation of ITQ-7 zeolites and the characterisation of the resulting materials. *C.R. Chim.* 8 (3): 369–378.

551 Botella, P., Corma, A., and Sastre, G. (2001). Al-ITQ-7, a shape-selective zeolite for acylation of 2-methoxynaphthalene. *J. Catal.* 197 (1): 81–90.

552 Botella, P., Corma, A., Navarro, M.T. et al. (2003). On the shape selective acylation of 2-methoxynaphthalene over polymorph C of beta (ITQ-17). *J. Catal.* 217 (2): 406–416.

553 Schmidt, I., Christensen, C.H., Hasselriis, P. et al. (2005). Mesoporous zeolite single crystals for catalytic hydrocarbon conversion. In: *Molecular Sieves: From Basic Research to Industrial Applications, Pts a and b* (eds. J. Cejka, N. Zilkova and P. Nachtigall), 1247–1254. https://www.sciencedirect.com/science/article/abs/pii/S0167299105804719.

554 Corma, A. (1997). From microporous to mesoporous molecular sieve materials and their use in catalysis. *Chem. Rev.* 97 (6): 2373–2419.

555 Christensen, C.H., Schmidt, I., Carlsson, A. et al. (2005). Crystals in crystals-nanocrystals within mesoporous zeolite single crystals. *J. Am. Chem. Soc.* 127 (22): 8098–8102.

556 Kustova, M.Y., Rasmussen, S.B., Kustov, A.L., and Christensen, C.H. (2006). Direct NO decomposition over conventional and mesoporous Cu-ZSM-5 and Cu-ZSM-11 catalysts: improved performance with hierarchical zeolites. *Appl. Catal., B* 67 (1–2): 60–67.

557 Christensen, C.H., Schmidt, I., and Christensen, C.H. (2004). Improved performance of mesoporous zeolite single crystals in catalytic cracking and isomerization of *n*-hexadecane. *Catal. Commun.* 5 (9): 543–546.

558 Na, K. and Somorjai, G.A. (2015). Hierarchically nanoporous zeolites and their heterogeneous catalysis: current status and future perspectives. *Catal. Lett.* 145 (1): 193–213.

559 Xiao, F.S., Wang, L.F., Yin, C.Y. et al. (2006). Catalytic properties of hierarchical mesoporous zeolites templated with a mixture of small organic ammonium salts and mesoscale cationic polymers. *Angew. Chem. Int. Ed.* 45 (19): 3090–3093.

560 Serrano, D.P., Aguado, J., Escola, J.M. et al. (2006). Hierarchical zeolites with enhanced textural and catalytic properties synthesized from organofunctionalized seeds. *Chem. Mater.* 18 (10): 2462–2464.

561 Egeblad, K., Kustova, M., Klitgaard, S.K. et al. (2007). Mesoporous zeolite and zeotype single crystals synthesized in fluoride media. *Microporous Mesoporous Mater.* 101 (1): 214–223.

562 Möller, K., Yilmaz, B., Jacubinas, R.M. et al. (2011). One-step synthesis of hierarchical zeolite beta via network formation of uniform nanocrystals. *J. Am. Chem. Soc.* 133 (14): 5284–5295.

563 Cho, K., Na, K., Kim, J. et al. (2012). Zeolite synthesis using hierarchical structure-directing surfactants: retaining porous structure of initial synthesis gel and precursors. *Chem. Mater.* 24 (14): 2733–2738.

564 Liu, B., Tan, Y., Ren, Y. et al. (2012). Fabrication of a hierarchically structured beta zeolite by a dual-porogenic surfactant. *J. Mater. Chem.* 22 (35): 18631–18638.

565 Tosheva, L. and Valtchev, V.P. (2005). Nanozeolites: synthesis, crystallization mechanism, and applications. *Chem. Mater.* 17 (10): 2494–2513.

566 Kuechl, D.E., Benin, A.I., Knight, L.M. et al. (2010). Multiple paths to nanocrystalline high silica beta zeolite. *Microporous Mesoporous Mater.* 127 (1): 104–118.

567 Larlus, O., Mintova, S., Wilson, S.T. et al. (2011). A powerful structure-directing agent for the synthesis of nanosized Al- and high-silica zeolite beta in alkaline medium. *Microporous Mesoporous Mater.* 142 (1): 17–25.

568 Martínez-Franco, R., Paris, C., Martínez-Armero, M.E. et al. (2016). High-silica nanocrystalline beta zeolites: efficient synthesis and catalytic application. *Chem. Sci.* 7 (1): 102–108.
569 Bernasconi, S., van Bokhoven, J.A., Krumeich, F. et al. (2003). Formation of mesopores in zeolite beta by steaming: a secondary pore channel system in the (001) plane. *Microporous Mesoporous Mater.* 66 (1): 21–26.
570 Verboekend, D. and Pérez-Ramírez, J. (2011). Desilication mechanism revisited: highly mesoporous all-silica zeolites enabled through pore-directing agents. *Chem. Eur. J.* 17 (4): 1137–1147.
571 Groen, J.C., Peffer, L.A.A., Moulijn, J.A., and Pérez-Ramírez, J. (2004). On the introduction of intracrystalline mesoporosity in zeolites upon desilication in alkaline medium. *Microporous Mesoporous Mater.* 69 (1): 29–34.
572 Groen, J.C., Abelló, S., Villaescusa, L.A., and Pérez-Ramírez, J. (2008). Mesoporous beta zeolite obtained by desilication. *Microporous Mesoporous Mater.* 114 (1): 93–102.
573 Groen, J.C., Peffer, L.A.A., Moulijn, J.A., and Pérez-Ramírez, J. (2005). Mechanism of hierarchical porosity development in MFI zeolites by desilication: the role of aluminium as a pore-directing agent. *Chem. Eur. J.* 11 (17): 4983–4994.
574 Renz, M., Blasco, T., Corma, A. et al. (2002). Selective and shape-selective Baeyer–Villiger oxidations of aromatic aldehydes and cyclic ketones with Sn-beta zeolites and H_2O_2. *Chem. Eur. J.* 8 (20): 4708–4717.
575 Corma, A., Navarro, M.T., and Renz, M. (2003). Lewis acidic Sn(IV) centers – grafted onto MCM-41 – as catalytic sites for the Baeyer–Villiger oxidation with hydrogen peroxide. *J. Catal.* 219 (1): 242–246.
576 Li, P., Liu, G., Wu, H. et al. (2011). Postsynthesis and selective oxidation properties of nanosized Sn-beta zeolite. *J. Phys. Chem. C* 115 (9): 3663–3670.
577 Corma, A., Domine, M.E., and Valencia, S. (2003). Water-resistant solid Lewis acid catalysts: Meerwein–Ponndorf–Verley and Oppenauer reactions catalyzed by tin-beta zeolite. *J. Catal.* 215 (2): 294–304.
578 Corma, A., Domine, M.E., Nemeth, L., and Valencia, S. (2002). Al-free Sn-beta zeolite as a catalyst for the selective reduction of carbonyl compounds (Meerwein–Ponndorf–Verley reaction). *J. Am. Chem. Soc.* 124 (13): 3194–3195.
579 Boronat, M., Corma, A., and Renz, M. (2006). Mechanism of the Meerwein–Ponndorf–Verley–Oppenauer (MPVO) redox equilibrium on Sn- and Zr-beta zeolite catalysts. *J. Phys. Chem. B* 110 (42): 21168–21174.
580 Moliner, M., Roman-Leshkov, Y., and Davis, M.E. (2010). Tin-containing zeolites are highly active catalysts for the isomerization of glucose in water. *Proc. Natl. Acad. Sci. U.S.A.* 107 (14): 6164–6168.
581 Nikolla, E., Roman-Leshkov, Y., Moliner, M., and Davis, M.E. (2011). "One-pot" synthesis of 5-(hydroxymethyl)furfural from carbohydrates using tin-beta zeolite. *ACS Catal.* 1 (4): 408–410.
582 Roman-Leshkov, Y., Moliner, M., Labinger, J.A., and Davis, M.E. (2010). Mechanism of glucose isomerization using a solid Lewis acid catalyst in water. *Angew. Chem. Int. Ed.* 49 (47): 8954–8957.

583 Dijkmans, J., Gabriels, D., Dusselier, M. et al. (2013). Productive sugar isomerization with highly active Sn in dealuminated beta zeolites. *Green Chem.* 15 (10): 2777–2785.
584 Bermejo-Deval, R., Orazov, M., Gounder, R. et al. (2014). Active sites in Sn-beta for glucose isomerization to fructose and epimerization to mannose. *ACS Catal.* 4 (7): 2288–2297.
585 Lew, C.M., Rajabbeigi, N., and Tsapatsis, M. (2012). One-pot synthesis of 5-(ethoxymethyl)furfural from glucose using Sn-BEA and amberlyst catalysts. *Ind. Eng. Chem. Res.* 51 (14): 5364–5366.
586 Gunther, W.R., Wang, Y.R., Ji, Y.W. et al. (2012). Sn-beta zeolites with borate salts catalyse the epimerization of carbohydrates via an intramolecular carbon shift. *Nat. Commun.* 3: 1109.
587 Holm, M.S., Saravanamurugan, S., and Taarning, E. (2010). Conversion of sugars to lactic acid derivatives using heterogeneous zeotype catalysts. *Science* 328 (5978): 602–605.
588 Pacheco, J.J. and Davis, M.E. (2014). Synthesis of terephthalic acid via Diels–Alder reactions with ethylene and oxidized variants of 5-hydroxymethylfurfural. *Proc. Natl. Acad. Sci. U.S.A.* 111 (23): 8363–8367.
589 Choudhary, V., Pinar, A.B., Sandler, S.I. et al. (2011). Xylose isomerization to xylulose and its dehydration to furfural in aqueous media. *ACS Catal.* 1 (12): 1724–1728.
590 Bui, L., Luo, H., Gunther, W.R., and Roman-Leshkov, Y. (2013). Domino reaction catalyzed by zeolites with Bronsted and Lewis acid sites for the production of valerolactone from furfural. *Angew. Chem. Int. Ed.* 52 (31): 8022–8025.
591 Boronat, M., Concepcion, P., Corma, A. et al. (2005). Determination of the catalytically active oxidation Lewis acid sites in Sn-beta zeolites, and their optimisation by the combination of theoretical and experimental studies. *J. Catal.* 234 (1): 111–118.
592 Nie, Y.T., Jaenicke, S., and Chuah, G.K. (2009). Zr-zeolite beta: a new heterogeneous catalyst system for the highly selective cascade transformation of citral to (+/−)-menthol. *Chem. Eur. J.* 15 (8): 1991–1999.
593 Blasco, T., Camblor, M.A., Corma, A. et al. (1996). Unseeded synthesis of Al-free Ti-beta zeolite in fluoride medium: a hydrophobic selective oxidation catalyst. *Chem. Commun.* (20): 2367–2368.
594 Blasco, T., Camblor, M.A., Corma, A. et al. (1998). Direct synthesis and characterization of hydrophobic aluminum-free Ti-beta zeolite. *J. Phys. Chem. B* 102 (1): 75–88.
595 Brand, S.K., Labinger, J.A., and Davis, M.E. (2016). Tin silsesquioxanes as models for the "open" site in tin-containing zeolite beta. *ChemCatChem* 8 (1): 121–124.

7

Downstream Conversion of Biomass-Derived Oxygenates to Fine Chemicals

Michèle Besson, Stéphane Loridant, Noémie Perret, and Catherine Pinel

Université Lyon, Université Claude Bernard, CNRS, IRCELYON, Institut de recherches sur la catalyse et l'environnement de Lyon, UMR5256, 2 Avenue Albert Einstein, 69626 Villeurbanne Cedex, France

7.1 Introduction

Many biomass-derived oxygenated feedstocks can be obtained in high yield after deconstruction and depolymerization of lignocellulose constituents, either via biological or chemical routes. They can then be upgraded to existing or new high value-added chemicals by catalytic conversion. The focus of this chapter is to provide a general overview of catalytic strategies that have been developed for conversion of some of these platform chemicals. Most of them belong to the list of the top 12 biomass-derived building blocks identified by the US Department of Energy (DOE) in 2004 [1] and revisited in 2010 [2]. For instance, glycerol (GLY) is a stoichiometric product in the transesterification process yielding biodiesel [3, 4], while lactic acid (LAC) is mostly produced by fermentation of aqueous carbohydrate solutions under anaerobic conditions [5]. Large-scale production of levulinic acid from cellulosic biomass by reaction with dilute mineral acid has already been achieved [6] and the anaerobic fermentation-based production of succinic acid is now a mature technology [7]. Erythritol (ERY) is produced by microbial fermentation of glucose [8], and xylitol (XYL) or sorbitol (SOR) via hydrogenation of the corresponding monosaccharide or directly from cellulose [9–11]. Furfural (FAL) is produced commercially by the acid dehydration of pentoses (such as xylan) [12], whereas 5-hydroxymethylfurfural (HMF) is available via dehydration of C6 carbohydrates [13].

Although numerous homogeneous catalytic systems can be applied, processes involving heterogeneous catalysts seem to be more widespread. The key reactions involved in the processing of biorenewable molecules are dehydration, decarboxylation, hydrogenation/hydrogenolysis, and selective oxidation, among others. Transformation of these polyoxygenated molecules has to address several key challenges. Biosourced molecules contain two or more functional groups, at least one of which is an oxygen-containing function. Therefore, the design of catalysts that transform very selectively biosourced molecules to targeted products is much more complex than for reactions involving monofunctional molecules [14].

Chemical Catalysts for Biomass Upgrading, First Edition. Edited by Mark Crocker and Eduardo Santillan-Jimenez.

Catalysts previously developed for the processing of hydrocarbon feedstocks are usually unsuitable because of this multifunctionality. Furthermore, most catalysts have been developed for the gas-phase or liquid-phase transformation of hydrocarbons in which the reaction medium is typically nonpolar, while the conversion of biosourced molecules occurs in the presence of water generated during fermentation processes or during their dehydration. Poor volatility of the biomass derivatives requires carrying out the process in the condensed phase, often in water. The presence of an aqueous phase introduces supplementary constrains on catalytic materials relative to their hydrothermal stability in liquid water at temperatures as high as 250 °C. The stability of acid catalysts used for gas phase dehydration reactions operating in the presence of water also represents a major challenge. Further, processing of feed streams derived from raw biomass may be of special concern, as acid, basic, and inorganic components of biomass may act as poisons and are potential causes of deactivation.

7.2 Selective Catalytic Oxidation

7.2.1 Introduction

Selective oxidation of GLY and 5-HMF has been intensively explored over the past two decades. A wide range of high-value chemicals can be produced that find broad application in polymers, fine chemicals, and pharmaceutical industries. The direct use of molecular oxygen as the oxidizing agent and water as solvent are preferred to make these reactions interesting from an economic and environmental standpoint. Methods have been developed to control the selectivity to various products with high yields.

Suitable solid catalysts are mainly mono- and bimetallic systems based on Pt, Pd, and Au on various supports, as reported in recent reviews [15, 16]. The mechanism of oxidation over Pt and Au was proposed by Zope et al. [17, 18]. Based on HPLC–MS analysis, ^{18}O isotope labeling, and DFT calculations, it was shown that hydroxide species adsorbed on gold favor the activation of the O—H and C—H bonds within the substrate via proton transfer. The oxidation mechanism involves activation of associatively adsorbed oxygen through the formation and subsequent dissociation of peroxide (OOH*) and hydrogen peroxide (H_2O_2*) intermediates. Thus, O_2 plays an indirect role, namely, the removal of the electrons added to the surface during the adsorption of hydroxide ions and the regeneration of these species (Scheme 7.1).

The addition of external bases (NaOH, KOH, Na_2CO_3) helps the initial deprotonation of the alcohol to form a metal alkoxide and prevents the deactivation of the catalyst by adsorption of the carboxylic (di)acids, but the formation of salts after neutralization is a serious problem. Therefore, some efforts have been dedicated to finding materials and conditions to perform the reactions under base-free conditions. Replacing expensive noble metals with cost-effective base metals without loss of performance has also been recently explored.

$$O_2^* + H_2O^* \rightarrow OOH^* + OH^*$$
$$OOH^* + {}^* \rightarrow O^* + OH^*$$
$$OOH^* + H_2O^* \rightarrow H_2O_2^* + HO^*$$
$$H_2O_2^* + {}^* \rightarrow OH^* + OH^*$$
$$OH^* + e^- \rightarrow HO^- + {}^*$$

Scheme 7.1 The role of oxygen in polyol oxidation over metal active sites (*). Source: Davis et al. 2013 [18]. Reproduced with permission of Royal Society of Chemistry.

7.2.2 Catalytic Oxidation of Glycerol

The extensive functionalization of GLY makes its selective oxidation particularly difficult. The reaction network is quite complex, with multiple C3, C2, and C1 products (Scheme 7.2). Primary alcohol oxidation of GLY affords glyceric acid (GLYAC) that may undergo a further oxidation of the second primary alcohol to form tartronic acid (TARAC). Oxidation of the secondary alcohol yields dihydroxyacetone (DHA). LAC can also be formed by dehydration of glyceraldehyde (GLYAL). The oxidation and C–C cleavage of the latter affords glycolic acid (GLYCAC) and formic acid. The reaction may also yield oxalic acid (OXAC). All these products are valuable and find many applications [3, 15, 19–21]. Depending on the reaction conditions – particularly the pH of the solution and the catalyst employed – the reaction can be directed to the oxidation of the primary or the secondary hydroxyl groups, or dehydration/rearrangement and C–C cleavage reactions may occur.

7.2.2.1 Glycerol to Glyceric Acid (GLYAC)

7.2.2.1.1 Noble Metal-Catalyzed Oxidation in the Presence of Base In the early 1990s, the glycerate salt was formed with fairly high selectivity (60–70%) over Pt/C and Pd/C in basic media under flowing air at atmospheric pressure [22, 23]. Gold catalysts are more resistant to deactivation by oxygen. However, addition of a strong base and high oxygen partial pressures (1–10 bar) are mandatory for meaningful conversion. While there was no GLY conversion in water, in NaOH aqueous solution selectivity to glycerate was 80%, 66%, and 50% (at ca. 50% conversion) for Au, Pt, and Pd supported on graphite, respectively [24, 25]. A series of Au catalysts were supported on various carbon materials (i.e. active carbon, graphite, multiwall carbon nanotubes [MWCNTs], carbon nanofibers [CNFs]) and oxides according to various preparation procedures, namely, sol immobilization, incipient-wetness, and deposition–precipitation. The synthesis conditions [26–28], the gold particle size [29, 30], the surface chemistry of the support [31–36], and the shape of Au nanoparticles (NPs) [37] all have a strong influence on both the activity and selectivity to glycerate. The high selectivity to the C—C bond cleavage glycolate by-product shown by the smallest particles was attributed to their higher activity for hydrogen peroxide formation, which promotes C–C cleavage [29]. By optimizing the conditions (30 °C, 3 bar O_2,

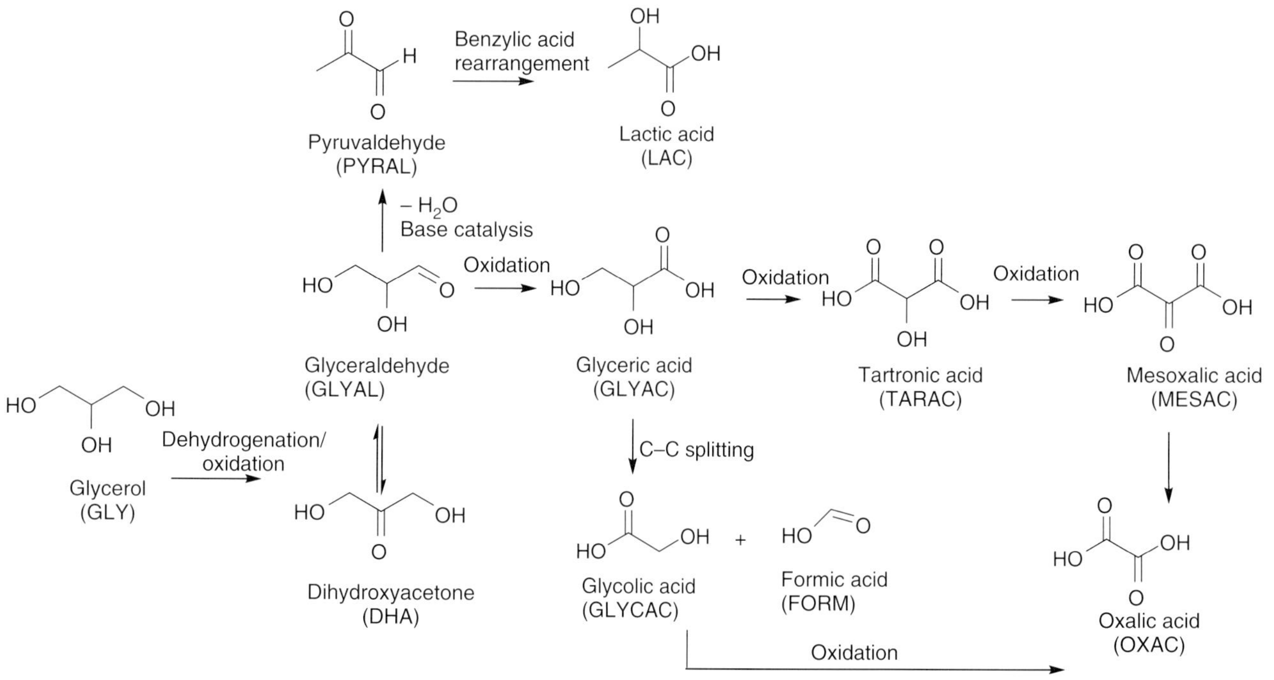

Scheme 7.2 Chemicals produced by oxidation of glycerol.

NaOH/GLY/Au = 16/4/1) and the catalyst preparation, Porta and Prati obtained 92% selectivity to GLYAC at full conversion [26].

Alloyed bimetallic Au–Pd systems significantly improve the activity and selectivity with respect to the monometallic catalysts under alkaline conditions [20, 38–41]. It was suggested by Davies and coworkers that Pd could catalyze the decomposition of hydrogen peroxide, which is formed over Au and is detrimental to the selectivity [41]. The molar ratio of Au/Pd has a strong impact depending on the structure of the bimetallic NPs (alloy or core–shell) [40, 42, 43]. Bimetallic Au–Cu/$CeZrO_x$ catalysts were significantly more active and selective to glycerate than the respective monometallic samples and could be reused [44].

7.2.2.1.2 Base-Free Aqueous Oxidation A significant advance was made when the oxidation of GLY to free GLYAC was successfully carried out under base-free conditions while achieving satisfactory and stable performance. The Pt/C catalysts with smallest Pt particle size (<6 nm) gave the maximum yield of GLYAC and the catalyst could be recycled six times [45]. Pt supported on MWCNTs [46] or CNFs [47] was more active than on C, with similar selectivity to GLYAC (60–70% at 70% conversion at 60 °C under O_2 flow), due to easier accessibility of the metal on the outer walls of the nanostructured supports. The high performance of hydrotalcite (HT)-supported Pt [48] or Pd [49] and of Pt/N-doped CNT [50] was attributed to the basicity of the support, which promoted alkoxide formation. The decrease in conversion observed upon recycling a Pt/mesoporous carbon nitride catalyst was due to the growth and oxidation of Pt particles, leaching of N species, and collapse of the ordered structure of the mesoporous support [51].

Alloying Au with Pt or Pd enhanced activity and catalyst lifetime under base-free conditions [52, 53]. In these studies, considerable attention was also devoted to the effect of the acid/base properties of the support. Among AuPt immobilized on various oxides, acidic supports – such as MCM-41, SiO_2, H-mordenite, and sulfated zirconia – gave a higher selectivity to C3 oxidation products at 80 °C, and the acid site density was found to be key, whereas basic supports (namely, MgO and NiO) promoted activity as well as C—C bond cleavage [54]. The reason behind the improved selectivity to GLYAC for bimetallic Pt–Au(6-4)/HT [55] and Au–Pt (1 : 3)/$Mg(OH)_2$ [56] was reported to be the coexistence of $Mg–OH^-$ and $Mg–CO_3^{2-}$ [57]. However, further study on the role of $Mg(OH)_2$ demonstrated that the outstanding performance of these Mg-based catalysts is simply an effect of support dissolution, of pH control during reaction and, therefore, of an activation of the primary alcohol oxidation [58].

The formation of the $PtCu_3$ alloyed phase was speculated to contribute to the improved performance of a Pt–Cu/C catalyst compared with Pt/C [59]. Enhanced interaction between Pt and Co and better dispersion of Pt in Pt–Co/reduced graphene oxide (RGO) also explains the better performance of this catalyst (82% selectivity at 88% conversion and reusability for five runs without obvious deactivation) compared with Pt/RGO [60].

7.2.2.1.3 Oxidative Esterification Since the GLY stream stemming from biodiesel production contains methanol, oxidative esterification of GLY in this solvent may be attractive. Under base-free conditions, GLY was converted into C3 methyl (di)esters (mainly glycerate and mesoxalate) with over 95% yield using Au supported on several oxide supports, namely, TiO_2, Al_2O_3, and ZnO. In the absence of base, the highest selectivity to methyl glycerate (82% at 72% conversion) was achieved over Au/TiO_2, although this formulation was also the least active [61]. The initial reaction rates were higher in the presence of sodium methoxide; however, deactivation of the catalysts was observed.

Notably, switching from pure GLY to crude GLY from a biodiesel plant highlighted the detrimental effect of impurities. The activity of Au/carbonaceous materials was significantly decreased, although after neutralization of the previously evaporated crude GLY using HCl, results similar to those obtained with commercial GLY were obtained [62]. Comparison of alumina-supported Pt, Pd, Au, and Ag catalysts in the oxidation of GLY to glycerate showed that these catalysts are very sensitive to impurities in crude GLY, as the activity could be decreased up to a factor of 10 for Pt/Al_2O_3 and its selectivity was modified [63].

7.2.2.2 Glycerol to Tartronic Acid (TARAC)

While the formation of GLY over mono- or bimetallic catalysts is easily achieved, the production of the more valuable tartronate (or TARAC) via oxidation of another primary –OH group is more challenging. This is due to the occurrence of C–C cleavage or decarboxylation reactions forming OXAC, LAC, or GLYCAC.

Under basic conditions, a 58% yield of tartronate at total conversion was claimed using Ce- and Bi-promoted Pd [64] and Pd–Pt [65] catalysts supported on carbon. An 80% yield of TARAC was obtained over Bi–Pt/C at pH values of 11–12, albeit starting from a GLYAC salt instead of GLY [66]. Over Au/C, the amount of base affected the selectivity. Indeed, GLYAC formation was favored for NaOH/GLY molar ratios up to 2, while TARAC was formed with high selectivity when increasing this ratio to 4 [30]. Au/TiO_2 [67], Au nanoclusters stabilized by polyvinylpyrrolidone (PVP) [68], and Au–Pd/MgO–Al_2O_3 [69] exhibited less than 50% yield of tartronate. However, promotion of an alloyed Au–Pd/C catalyst with 0.1% Bi provided complete conversion of GLY with an isolated 78% yield of TARAC, after 14 hours at 60 °C under 3 bar O_2 [70]. Addition of Fe to Pt nanoclusters (Fe:Pt = 1 : 1) on ceria considerably enhanced the activity (TOF ca. 20 100 h^{-1} vs. 2 677 h^{-1}) [71]. Deactivation upon recycling was attributed to Fe leaching. Structure reconstruction during annealing under H_2 of a FePt catalyst (Fe/Pt ratio of 4.5) further improved activity (26 340 h^{-1}), yield (58.6%), and stability [72].

Further, under base-free conditions, 80% yield of TARAC at 98% conversion was obtained at 60 °C under 3 bar O_2 over a Au/HY catalyst with very small (1 nm) Au clusters [73].

Very recently, non-noble Co catalysts on MgO–Al_2O_3 surfaces showed complete conversion of GLY with yields of tartronate and oxalate of 64% and 24%, respectively, under mild conditions (70 °C, 1 bar O_2) [74]. However, gradual deactivation of the catalyst during recycling was observed, due to leaching of Al species.

7.2.2.3 Glycerol to Dihydroxyacetone (DHA)

The synthesis of DHA by selective oxidation of the secondary alcohol is usually performed without addition of a base. Kimura et al. demonstrated (using a batch reactor) that in acidic conditions the addition of Bi to a Pt/C catalyst increased the yield from 4% at 37% conversion to 20% at 30% conversion at 50 °C [75]. On various carbonaceous materials, the association of Bi with Pt (Bi/Pt ratio of 0.2) afforded a DHA yield of 37–48% at ca. 80% conversion [23, 76]. The yield was further improved to 52.5% using a fixed-bed reactor [77]. However, the catalyst deactivated due to Bi leaching [78]. A Pt–Bi/MCM-41 catalyst exhibited a higher maximum yield of DHA (54%) than Pt–Bi/C (45%), and DFT calculations showed that Bi should be located as adatoms on Pt [79]. Similarly, addition of Bi to bimetallic Au–Pt/C also changed the reaction pathway from the oxidation to GLYAC to oxidation of the secondary alcohol, yielding DHA [80]. Alloyed Pt–Sb NPs supported on MWCNT (51.4% selectivity at 90% conversion after two hours at 60 °C under flowing O_2) [81] and alloyed Pd–Ag/C (84% selectivity at 52% conversion after 24 hours at 80 °C under 3 bar O_2) [82] were also found to be efficient catalysts. Adsorption of a terminal OH group on the Ag site and the attack of the neighboring secondary OH group by dissociatively adsorbed oxygen on the Pd sites was assumed. A high yield of 75–80% DHA was obtained in water over Au supported on CuO without using a base at 50 °C under 20 bar O_2 [83]. By varying the molar ratio of Mg/Al in Au/MgO–Al_2O_3, the results at low conversion demonstrated that the most acidic and least basic catalysts exhibited the highest activity and selectivity to DHA (82%), inversely to the selectivity to GLYAC [84].

7.2.2.4 Glycerol to Mesoxalic Acid (MESAC)

Mesoxalic acid (MESAC), a chelating agent and building block, has rarely been reported as a product stemming from the catalytic oxidation of GLY over metallic catalysts in aqueous basic solution. Nevertheless, it may be obtained in two steps, i.e. by first oxidizing GLY to TARAC and then using a PtBi/C catalyst to oxidize the secondary hydroxyl group to produce only mesoxalate in 50% yield [85, 86]. Another approach is the oxidative esterification of GLY in methanol at 100 °C over titania- or iron oxide-supported Au in the presence of 10% NaOMe to obtain dimethyl mesoxalate with selectivity up to 89% [87].

7.2.2.5 Glycerol to Glycolic Acid (GLYCAC)

Only a few papers describe C—C bond cleavage to produce GLYCAC from GLY. A yield of 49% GLYCAC at complete conversion was obtained over a 1% Au/graphite catalyst at 120 °C under 7 bar O_2 under basic conditions (NaOH/GLY = 2) [88]. Under strongly basic conditions, a 3.6% Ag/Al_2O_3 catalyst afforded a selectivity of 57.1% to GLYCAC at 85% conversion at 60 °C under 5 bar O_2 [89]. The catalyst was stable in a continuous process and showed relatively good resistance to the impurities in crude GLY. Over an Ag catalyst promoted with Pt (Ag95–Pt5/CeO_2), a selectivity to GLYCAC of 51% was obtained at 54% conversion [90].

Non-noble metal catalysts such as 5Co/Cu-based materials exhibited up to 34% selectivity to glycolate at 96% conversion [91]. A synergy between CoO(OH) in contact (most likely) with CuO species was proposed.

HT-supported Au NPs were reported to convert GLY to GLYCAC in water under mild conditions (53% selectivity at 78% conversion, 60 °C, O_2 flow), whereas Au/C under the same conditions showed no activity [92].

Finally, using H_2O_2 as the oxidant, GLY was successfully oxidized to glycolate with a yield of 56% over 1% Au/graphite [16].

7.2.2.6 Glycerol to Lactic Acid (LAC)

Another interesting valorization route for GLY is transformation to valuable D,L-LAC and its ester derivatives. This occurs via oxidation to GLYAL and DHA, followed by base-catalyzed dehydration and benzylic acid rearrangement of these intermediates (Scheme 7.2). The competitive reaction is the oxidation to GLYAC, which dictates the final selectivities to LAC and GLYAC. This reaction has been recently well documented [5, 93].

Under strong basic conditions (NaOH/GLY = 4), LAC was obtained over noble metal catalysts including bimetallic Au–Pt (1 : 1)/TiO_2 (26% yield) [94], Au/CeO_2 (81% yield) [95], and AuPt/CeO_2 (79% yield, the catalyst being reusable at least for five times without loss of LAC selectivity) [96]. Promotion of Pt/TiO_2 by Ni [97] and of Pt/C by Cu [98] yielded 62% and 55% LAC, respectively. Upon recycling of $Pt_1Ni_1O_x$/TiO_2 conversion declined, whereas 0.5% Cu–1.0% Pt/C has satisfactory reusability with no Pt and only 1.3% Cu leaching. As regards non-noble metal catalysts, Co_3O_4/CeO_2 was reported to afford a selectivity to LAC of 79.8% at 85.7% conversion in aqueous basic medium at a high temperature of 250 °C [99].

A few recent studies on the base-free oxidation of GLY to LAC suggested that the combination of metallic NPs (to activate the hydroxyl groups to form GLYAL and DHA) and a Lewis acid (to accelerate the key steps of dehydration and benzylic rearrangement) produced LAC very efficiently. At elevated temperatures (180 °C), LAC formation was significant (60% LAC at 80% conversion) using Au–Pt/zeolite USY-600 [100]. Au–Pd/TiO_2 combined with $AlCl_3$ was able to convert completely GLY with a 47.6% yield of LAC at 160 °C under 25 bar O_2 for two hours; however, severe hydrolysis of $AlCl_3$ and poisoning by Cl^- deactivated the catalyst [101]. A bifunctional Pt catalyst on a Sn-MFI zeolite afforded 89.8% conversion and 80.5% selectivity to LAC at 100 °C. Unfortunately, leaching of Sn Lewis acidic sites occurred [102]. Pt NPs stabilized by PVP and deposited on TiO_2 Lewis acid solid achieved 63% yield and 99% conversion at 150 °C after 48 hours [103].

$H_3PMo_{12}O_{40}$ (HPMo) loaded on carbon was highly efficient in water giving 94% selectivity to LA at 98% conversion at 60 °C under 5 bar O_2 for five hours. The catalyst could be recycled five times without leaching and displayed good performance to convert both neat and crude GLY (85% and 58% yield, respectively) [104]. Also, introduction of Al^{3+} and Cr^{3+} Lewis acid sites (LAS) into HPMo could produce LAC with 90.5% selectivity at 93.7% conversion under mild reaction conditions [105]. A silver-exchanged Ag_3PMo catalyst was found to be methanol-tolerant in crude GLY, relatively stable upon 12 reaction cycles, and capable of converting crude GLY with 98% conversion and 93% selectivity after 24 hours at 60 °C under 10 bar O_2 [106].

Methyl lactate was obtained in high yields using Au/zeolite USY-600 (73% at 95% conversion, 160 °C, 47 bar air, 10 hours) [61] or Au/Sn-USY (79% at 88% conversion, 160 °C, 5 bar O_2, 10 hours) in MeOH without base [107]. The promoting effect of Sn acts by impacting Au dispersion and introducing acid sites, a decrease in selectivity to methyl lactate being observed in the third run during recycling.

GLY could also be effectively converted into lactate in alkaline solution over Pt-based [108] and CuO/ZrO_2 [109] catalysts at relatively low temperatures (180–200 °C) under an inert atmosphere, without using an oxidant.

7.2.3 Oxidation of 5-Hydroxymethylfurfural (HMF)

5-HMF oxidation occurs sequentially through several intermediate products, including 2,5-diformylfuran (DFF), 5-hydroxymethyl-2-furancarboxylic acid (HMFCA), 5-formyl-2-furancarboxylic acid (FFCA), and 2,5-furandicarboxylic acid (FDCA) as shown in Scheme 7.3 [1, 110]. The initial conversion occurs in two parallel reactions that oxidize either the aldehyde function to form HMFCA or the alcohol to form DFF. Further oxidation of both intermediates affords FFCA, which is finally oxidized to FDCA. The challenge is to selectively produce a single compound preferentially above all the others. Among the oxidation products, FDCA and DFF have received the most attention. FDCA has a large potential as a replacement for petroleum-derived terephthalic acid, a widely used component in various polyesters – e.g. polyethylene terephthalate (PET) – and may also be important for novel polyamides. DFF and HMFCA have also high potential in the production of polymers of commercial importance and as starting material for synthetic drugs, antifungal agents, etc.

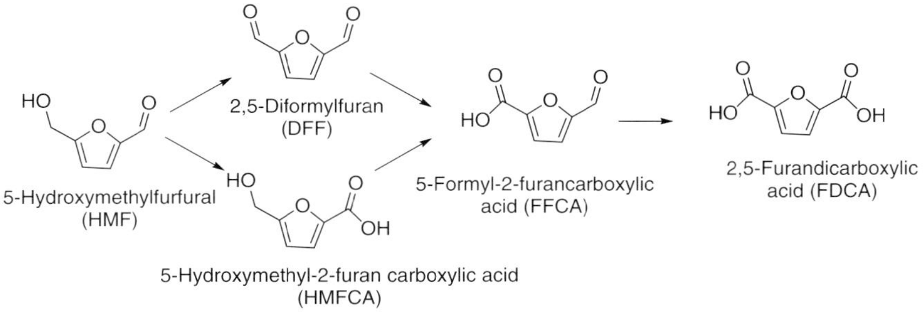

Scheme 7.3 Chemicals produced by oxidation of 5-hydroxymethylfurfural (HMF).

7.2.3.1 HMF to 2,5-Furandicarboxylic Acid (FDCA)

As for GLY, the selective oxidation to FDCA has been mainly performed using supported noble metal catalysts, with and without the addition of a homogeneous base. HMF is highly unstable in water at high pH values, where it may undergo several side reactions leading to the formation of humins; consequently, high HMF/metal molar ratios (in the range 20–100) are used to promote the catalytic reaction. Moreover, FDCA is poorly soluble in base-free aqueous solutions, so a low concentration of HMF should be used and longer reaction times are necessary. Saliently, some recent reviews on the oxidation of HMF to FDCA have been published [111–113].

7.2.3.1.1 Oxidation with Addition of a Base Typically, the conversion of HMF to the salt of FDCA (denoted simply as FDCA) has been achieved using various supported Pt, Pd, and Au catalysts with an excess amount of base additive [114–118]. Modification of Pt/C by Bi (Bi/Pt around 0.2) increased the yield to FDCA [114] and significantly improved the resistance of Pt/C to oxygen poisoning [118]. After four hours, a nearly quantitative FDCA yield was obtained at 100 °C under 40 bar air with Na_2CO_3/HMF = 2. Apart from active carbon supports, metals have been deposited on RGO [119], N- and P-co-doped graphene [120], and magnetic core–shell Fe_3O_4@carbon structures [121, 122], resulting in close to quantitative yield of the desired FDCA under optimized conditions with the catalysts showing good recyclability. The surface functional groups of carbon materials affected the active metal sites. For example, stabilized colloidal Au NPs immobilized on high surface area graphite (HSAG) demonstrated higher activity (TOF = 1195 h^{-1}) and selectivity toward FDCA on the most basic supports than on acidic oxidized HSAG (TOF = 25 h^{-1}), at 90 °C under 10 bar O_2 in the presence of 2 equiv $NaHCO_3$. This was mainly due to the differences in adsorption of hydroxyls and charged reaction intermediates [123]. Different O- and N-containing carbons supports for Ru showed the complex effect of the nature of functional groups [124]. Various oxide supports – such as TiO_2 and ZrO_2 [125], CeO_2 [126], Bi-containing ceria [127], and a cerium coordination polymer [128] – were also used for loading the metal. Au NPs (1 nm size) encapsulated in the supercage of HY zeolite afforded FDCA quantitatively (>99%) in alkaline aqueous medium at 60 °C under 3 bar O_2 after six hours and were reusable with only a small loss of activity [129].

Au-containing bimetallic NPs were reported to be more efficient catalyst systems in alkaline medium and could generally be reused without a decrease in FDCA yield. Higher activity for HMF oxidation resulting in quantitative yield of FDCA was obtained with a Au–Pd (8 : 2)/C catalyst at 60 °C under 30 bar O_2 in the presence of 2 equiv NaOH [130]. Alloyed Au–Pd (1 : 1) NPs on a basic anion-exchange resin generated up to 93% FDCA yield at 100 °C under 10 bar O_2 in an equimolar Na_2CO_3 solution [131]; moreover, when supported on a mesoporous polyionic liquid, the reaction resulted in 99% FDCA yield at 90 °C under atmospheric flowing O_2 in 4 equiv K_2CO_3 solution [132]. The enhanced activity of Au–Cu/TiO_2 (Au/Cu = 1), which afforded a 99% FDCA yield, was attributed to the isolation of Au sites by Cu [133, 134]. The synergistic effect of alloying Ni with Pd (90 : 10) resulted in the total conversion of HMF with 86% selectivity at 80 °C under air bubbling after four hours with 1 equiv of Na_2CO_3 [135].

Recently, the possibility to perform the reaction over less expensive non-noble metal catalysts has been demonstrated, a MnO_x–CeO_2 composite affording a FDCA yield of 91% under basic conditions [136], $Mn_{0.75}Fe_{0.25}$ mixed oxides at

4 equiv NaOH in a two-step procedure (HMF → FFCA → FDCA) resulting in 90% yield [137], and a stable polymeric Fe^{III}-porphyrin-based catalyst yielding 79% FDCA in water [138].

7.2.3.1.2 Base-Free Oxidation Base-free aerobic aqueous oxidation has also been efficiently applied. The reaction can be catalyzed by a suspension of PVP-stabilized Pt [139] or Pd [140] NPs to achieve FDCA yields >95%. Highly basic supports were also tested, such as HT for supporting Au (HMF/Au = 40) [141] or Pd (HMF/Au = 21) [142]; however, extensive leaching of Mg^{2+} was detected [67]. It was reported later that the alkali-free HT-supported Au catalyst produced only HMFCA in the absence of a base and that the oxidation up to FDCA reported in previous work was catalyzed by leached alkali contaminants or coexisting soluble brucite stemming from the catalyst synthesis [143]. Over Pt/MgO–C, a 97% yield was achieved at 110 °C under 10 bar O_2 and a 75% isolated yield was achieved in a scale-up experiment; however, high Mg^{2+} leaching was found upon reuse of the catalyst [144]. Au NPs immobilized on an HT-C (HT-C 2 : 1) composite [145] were found to be stable catalysts compared with the HT or C-supported formulations. Monometallic Pt [146] and bimetallic Au–Pd [147] supported on O-functionalized CNT afforded 94% and 98% FDCA yield, respectively, at 95–100 °C under 5 bar O_2 in pure water; however, the reaction was slow, the carbonyl/quinone and phenol groups on CNT being identified as key promotors. Pt/N-doped carbon catalysts modified by ethylenediamine [148], and Ru/N-doped graphenic materials (HMF/Ru = 10) [149], improved selectivity to FDCA compared with formulations supported on other carbonaceous materials. La-doped CaMgAl-layered double hydroxide (LDH) as support for AuPd was more active than the CaMgAl-LDH- or MgAl-LDH-supported catalysts, also being highly stable, recyclable, and yielding FDCA almost quantitatively at 100 °C under 5 bar O_2 [150].

7.2.3.1.3 Oxidative Esterification Because of the low solubility of FDCA in water, the corresponding diester – namely, 2,5-FDCA dimethyl ester – was synthesized in excellent yields via oxidative esterification in MeOH, over Au/TiO_2 at 130 °C under 4 bar O_2 with addition of 8% MeONa [115], or over Au/CeO_2 at 130 °C under 10 bar O_2 under base-free conditions [151], the nanometric ceria support acting through a Ce^{4+}/Ce^{3+} redox process.

7.2.3.2 HMF to 2,5-Diformylfuran (DFF)

The selective oxidation of HMF to DFF has also been extensively studied using Ru or different V and Mn oxide-based catalysts and has been recently reviewed [113]. These catalysts have been generally used in organic solvents – such as dimethyl sulfoxide (DMSO), *N,N*-dimethylformamide (*N,N*-DMF), methyl isobutyl ketone (MIBK, an extracting solvent for effective production of HMF in biphasic medium), toluene, isopropanol, etc. – despite water being the most environmentally friendly solvent. Nevertheless, performing the reactions under high pressures of air or oxygen may pose some problems in practical applications at larger scale.

Ru catalysts that do not easily perform oxidation of an alcohol or aldehyde moiety to the carboxylic acid (even in water) may be highly selective for the oxidation of HMF to DFF [110, 152].

Interestingly, using water as solvent, the LDH $Mn_{0.70}Cu_{0.05}Al_{0.25}$ gave 87% selectivity to DFF at 90% conversion in the absence of additives and the catalyst displayed high stability after five runs at 90 °C under 8 bar O_2 [153]. Also in water, a hierarchical structure of VO_x nanobelt-arrayed microspheres afforded 95.4% selectivity at 93.7% conversion and 130 °C under 30 bar O_2 after one hour [154]. The remarkable performance of the VO_x microspheres was ascribed to high exposure of the (010) facet, steric advantage (as proved by DFT calculations), and strong hydrogen bonding between the lattice oxygen of vanadyl group (V—O) sites and the OH group of HMF.

7.2.3.3 HMF to 5-Hydroxymethyl-2-furancarboxylic Acid (HMFCA) or 5-Formyl-2-furancarboxylic Acid (FFCA)

Very few studies have been performed on the oxidation of HMF to HMFCA. The reaction over Au catalysts in highly alkaline aqueous solutions starts with a very fast oxidation of HMF to HMFCA; however, the rate of the subsequent oxidation to FFCA is much lower and, thus, a high selectivity of 85% toward HMFCA may be observed over Au/Fe_2O_3 at 130 °C under bubbling O_2 [126]. HMFCA can also be obtained with 87% yield using recyclable Mo acetylacetonate immobilized on montmorillonite K-10 clay in toluene after three hours [155] or with >64% yield over $Ce_{1-x}Bi_xO_2$ mixed oxides in aqueous basic media [127]. An $MgO–CeO_2$ mixed oxide was claimed to reach 94% selectivity toward HMFCA at 74% conversion and 110 °C under 10 bar O_2 in water without any additive [156]. Further, to obtain FFCA selectively, the continuous oxidation of an 0.83 wt% Na_2CO_3 aqueous solutions of HMF was performed over Pt/C, which yielded a 90% selectivity at nearly total conversion of HMF when the appropriate liquid flow rate was used [116]. Finally, a catalyst based on copper/cerium oxides also achieved a 90% selectivity toward FFCA at almost total conversion in water [157].

7.3 Hydrogenation/Hydrogenolysis

7.3.1 Introduction

The catalytic hydrogenation or hydrogenolysis of polyols, carboxylic acids, FAL, and HMF opens an alternative route to the production of useful diols – including 1,3-propanediol (1,3-PDO), propyleneglycol (PG), butanediols (BDO), 1,5-pentanediol (PeDO), 1,6-hexanediol (HDO), etc. and various chemical building blocks.

7.3.2 Hydrogenolysis of Polyols

The hydrogenolysis reactions involve the cleavage of C—O or C—C bonds. The challenge lies in the selective cleavage of C—O vs. C—C bonds, the observed selectivity to products depending on both the nature of the catalyst and on the reaction conditions.

7.3.2.1 Hydrodeoxygenation of Polyols

In order to obtain most target commodity chemicals, it is necessary to decrease the oxygen content of the feedstock material. C–O hydrogenolysis may be useful for that purpose; however, due to the multiple C—O bonds in the feed, selectivity issues have to be addressed.

The C–O hydrogenolysis of GLY can give PG and 1,3-PDO as well as 1- and 2-propanol and propane. Several reviews have focused on the hydrogenolysis of GLY to either PG or 1,3-PDO [15, 158–162]. Most studies have been performed under H_2 pressure, but *in situ* generation of hydrogen via aqueous phase reforming or catalytic transfer hydrogenation (using ethanol, isopropanol, or formic acid as hydrogen donor) are also attractive approaches [162]. Both noble and transition metal catalysts have been successfully used. Regarding non-noble metals, Cu has been largely used (ca. 70% of the articles published to date focus on this metal) since it favors dihydroxylation over C—C bond cleavage leading to high selectivity to PG [163]. Ni catalysts (including Raney® Ni) were also used [164]. Noble metal catalysts (mainly Ru and Pt) are also very efficient for PG production. For the synthesis of the more challenging 1,3-PDO, bimetallic catalysts associating a noble metal (Pt, Ir, Rh) with a reducible metal oxide species (W, Re, Mo) were developed [162, 165, 166]. To date, Pt/WO_x/AlOOH has exhibited the best performance, generating 66% yield of 1,3-PDO at full conversion [167].

Far fewer reports have focused on the hydrogenolysis of C4–C6 polyols. The longer the carbon chain, the less selective the reaction is since many alternative reaction pathways are possible. A larger number of products are formed, not only due to the potential C–O hydrogenolysis of each hydroxyl group but also to possible C–C cleavage by retro aldol reactions or the formation of cyclic products via internal dehydration (see Scheme 7.4) [168]. Furthermore, due to the presence of chiral centers in these substrates, isomerization may also be observed (via dehydrogenation/hydrogenation). The results of some recent hydrogenolysis studies of aqueous solutions of ERY in the presence of mono- or bimetallic catalysts are summarized in Table 7.1.

In the early 1990s, the group of Montassier studied the hydrogenolysis of ERY over a Raney Cu catalyst. The initial selectivity at low conversion toward

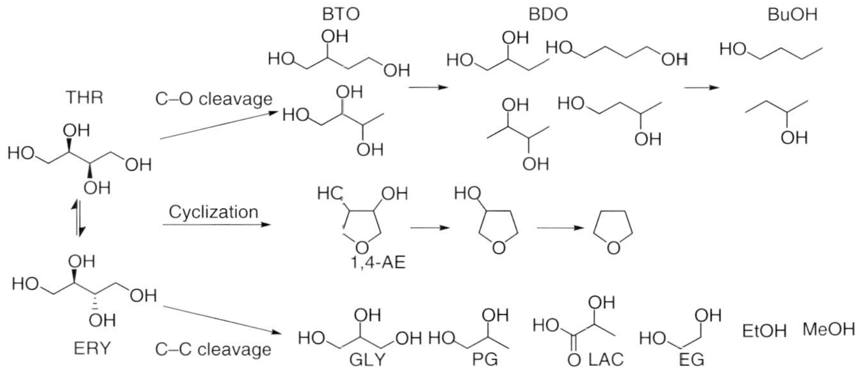

Scheme 7.4 Products formed during hydrodeoxygenation of ERY.

Table 7.1 Hydrodeoxygenation of ERY.

Catalyst (T, °C)	Conversion (%)	Carbon selectivity (%)						References
		BTO	BDO	BuOH	1,4-AE	THR	Others	
Raney® Cu (240)	49	15	3	<1	2	80	0	[163]
Rh/C (100)	6	62	10	2	25	0	<1	[165]
Ru/C (100)	66	26	12	0	7	49	6	[165]
Ir/SiO_2 (140)	3	22	0	0	70	0	9	[165]
MoO_x–Ir/SiO_2 (100)	37	72	19	9	0	0	0	[165]
ReO_x–Ir/SiO_2 (100)	73	21	46	29	0	0	4	[165]
ReO_x–Rh/TiO_2 (200)	80	24	36	5	12	11	8	[169]
ReO_x–Rh/ZrO_2 (200)	82	28	26	4	12	8	11	[169]
ReO_x–Rh/ZrO_2 (150)	80	46	28	6	2	10	8	[169]
ReO_x–Pd/CeO_2 (160)[a)]	98	0	93	<1	0	0	7	[170]

BTO, butanetriols; BDO, butanediols; BuOH, butanols; 1,4-AE, 1,4-anhydroerythritol; THR, threitol; others, C1–C3 and butane.
a) In dioxane.

BDO (1,2-BDO + 2,3-BDO) reached 54% at 240 °C under 40 bar of H_2 [171]. At higher conversion, the main product formed was the isomer threitol (THR). Various C- and SiO_2-supported monometallic (Pd, Pt, Rh, Ir) catalysts were investigated in the reaction of 20 wt% aqueous solutions of ERY at 100–140 °C under 80 bar of H_2 and exhibited low catalytic activity (<6% conversion). The nature of the support exerts a significant influence: Rh/SiO_2 was more active than Rh/C (12% vs. 6% conversion); moreover, 1,4-AE was favored in the presence of Rh/C (selectivity 25%), whereas linear deoxygenated C4 products (BTO + BDO + BuOH, selectivity 88%) were obtained over Rh/SiO_2. Ru/C showed a high activity (66% conversion after 24 hours), but the main product was THR (selectivity 49%) and the selectivity toward linear deoxygenated C4 products was only 38%. Bimetallic catalysts combining a noble metal with an oxophilic promoter (Re, W, and Mo) remarkably improved the conversion in the order: $Ir/SiO_2 < Ir–MoO_x/SiO_2 \approx Ir–WO_x/SiO_2 < Ir–ReO_x/SiO_2$. The addition of a promoter to Ir/SiO_2 also enhanced the selectivity to BTO + BDO + BuOH from 22% to ca. 100% at total conversion. The proposed mechanism involved the reaction of hydride species formed on the noble metal surface with alkoxide intermediates adsorbed on the rhenium oxide clusters in interaction with the noble metal particles. Moreover, the promoters exhibited similar total carbon selectivity toward all deoxygenated products (BTO + BDO + BuOH), whereas some differences were observed between each family of C—O bond cleavage products (BTO vs. BDO vs. BuOH) as illustrated in Table 7.1. The activity of Rh–ReO_x catalysts in ERY hydrogenolysis was dependent on the support (TiO_2 or ZrO_2), and the highest selectivity to BTO and BDO at 80% conversion reached 46% and 28%, respectively, in the presence of 3.7 wt% Rh–3.5 wt% ReO_x/ZrO_2 at 150 °C under 80 bar [169].

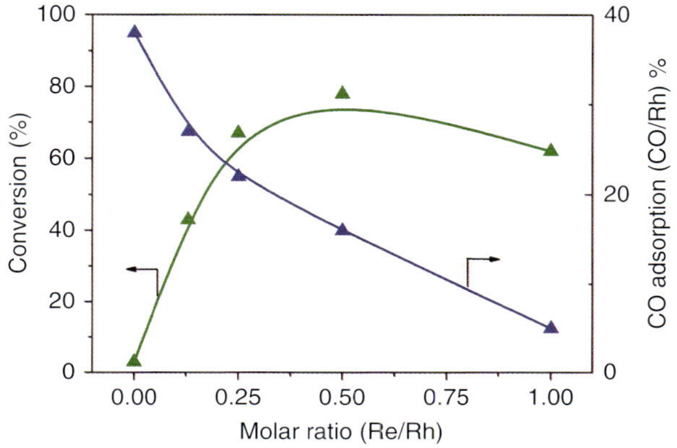

Figure 7.1 Evolution of conversion of glycerol and amount of CO adsorption as a function of the molar ratio (Re/Rh). (▲) CO adsorption (CO/Rh); (▲) glycerol conversion. Source: Adapted from Ref. [172].

Rh–Re bimetallic catalysts have also been optimized for hydrodeoxygenation. For instance, the effect of the Re amount added to Rh/SiO_2 was correlated with CO chemisorption and GLY hydrogenolysis (see Figure 7.1) [172]. The conversion exhibited a volcano-shaped curve and the optimum Re/Rh ratio was 0.5. The CO probe was not chemisorbed over ReO_x/SiO_2 catalyst and a regular decrease in CO chemisorption was observed when Re/Rh increased. Thus, a cooperative effect between the two metals was proposed at low promoter-to-metal ratio, whereas at high amounts of promoter Rh active sites were no longer accessible for hydrogen activation.

The effect of H_2 pressure (in the range 20–120 bar) on the initial reaction rate was investigated for ERY hydrogenolysis at 200 °C over ReO_x–Ir/SiO_2, ReO_x–Rh/ZrO_2, and ReO_x–Rh/TiO_2. The reaction order with respect to H_2 pressure was estimated to be close to 1 for the first two catalysts. On the other hand, it was close to zero over the TiO_2-supported catalyst. Differences in dispersion and in interactions between noble metal, rhenium, and the support may explain the different behaviors [165, 169].

Recently, Jin et al. have also examined a Cu/CaO–Al_2O_3 catalyst [173]. A selectivity of 39% toward 1,2-BDO (after 10 hours of reaction) was observed at 85% conversion. The temperature programmed reduction (TPR) profiles suggest strong interaction between Cu species and the CaO–Al_2O_3 support. The catalyst contained both CaCuAlO and spinel CuAlO phases. By-products were also formed with a global selectivity of 40%, the fact that LAC was formed in the presence of this catalyst being particularly noteworthy. The basic character of the catalyst might favor the production of LAC via the Cannizzaro reaction. It was suggested that dehydrogenation occurred first, followed by dehydration and/or retroaldol reactions.

Using dioxane as solvent, full conversion of ERY was attained over a ceria-supported ReO_x–Pd catalyst and the maximum selectivity toward BDOs reached 93% [170]. It was proposed that Pd and CeO_2 activate hydrogen, while

Table 7.2 Hydrogenolysis of sorbitol or xylitol in the absence of base.

Catalyst	Substrate	*T* (°C)	*P* (bar)	Conversion (%)	Sel. (%)	References
CuO–ZnO	SOR	180	130	92	63	[174]
Raney® Cu	SOR	210	124	99	78	[175]
ReO_x–Ir/SiO_2	XYL	140	80	95	89	[176]
	SOR	140	80	96	93	[176]

Sel.: Selectivity to deoxygenated products.

high valent ReO_x species are the active sites for the deoxydehydration of vicinal hydroxyl groups.

Similar to ERY hydrogenolysis, selective hydrodeoxygenation under neutral conditions of C5–C6 linear polyols – namely, XYL and SOR – was studied, the results being summarized in Table 7.2. Gallezot and coworkers used a commercial CuO–ZnO (33 : 67) mixed oxide for conversion of SOR to deoxyhexitols [174]. Under optimum conditions (180 °C, 130 bar H_2), SOR was converted into a mixture of deoxyhexitols (1,2-hexanediol, 1,6-hexanediol, 1,2,3-hexanetriol, 1,2,6-hexanetriol, 1,2,4,6-hexanetetraol) with a combined selectivity of 63% at 92% conversion. Recently, a sorbitol solution was hydrogenolyzed over a Raney Cu catalyst in a trickle-bed reactor [175]. Using a liquid hourly space velocity (LHSV) of $0.5\,h^{-1}$ and a H_2 flow of 800 ml/min at 210 °C and 124 bar, the selectivity toward C6 deoxygenated products (1,2,6- and 1,2,3-hexanetriol, 1,2,5,6-hexanetetraol) reached 78% at 99% conversion. Supported bimetallic catalysts comprising ReO_x and either Pd or Ir were studied in the hydrogenolysis reaction of SOR and XYL in biphasic conditions (H_2O/*n*-dodecane) with the aim to produce alkanes [176]. After 24 hours at 140 °C under 80 bar, almost full conversion of 11 wt% solutions was achieved. The maximum selectivity toward linear (partially deoxygenated) C5 products (pentanols + pentanediols + pentanetriols) was 89%, the selectivity toward linear C6 deoxygenated products (hexanols + hexanediols + hexanetriols) from sorbitol being 93%. ReO_x–Rh/ZrO_2 has also been observed to be efficient in this reaction (A. Said, University of Lyon, Lyon, unpublished results). In aqueous solution, up to 57% yield of deoxygenated hexitols were obtained, and it was conclusively determined that the reaction occurred through successive deoxygenation.

7.3.2.2 C–C Hydrogenolysis of Polyols

Polyols (XYL or SOR) can undergo both C—C and C—O bond cleavage to produce C2 and C3 polyols (EG, PG, and GLY). Selective hydrogenolysis was reported to proceed efficiently over metal catalysts (Ru, Ni) in association with a soluble alkali hydroxide or a solid basic oxide (see Table 7.3), a base promoter being necessary to catalyze the C–C cleavage of intermediate dehydrogenation products. Indeed, the proposed mechanism involves as a first step the dehydrogenation of a primary alcohol to the aldehyde, which is then converted to shorter polyols via a base-catalyzed retro aldol reaction as shown in Scheme 7.5.

Table 7.3 Hydrogenolysis of xylitol or sorbitol in the presence of a base.

Catalyst	Reaction conditions	Conversion (%)	Selectivity (%)				References
			EG	PG	GLY	LAC	
2% ReO_x–8% Ni/C	10% SOR; 250 °C; 40 bar H_2; $Ba(OH)_2$	92	16	31	7	16	[178]
6% Ni–NaY	15% SOR; 220 °C; 60 bar H_2; $Ca(OH)_2$	75	7	69	4	nd	[179]
3% Cu/ZrO_2	10% SOR; 220 °C; 50 bar H_2; $La(OH)_3$	28	14	26	5	17	[180]
90% $Cu–SiO_2$	10% XYL; 200 °C; 40 bar H_2; $Ca(OH)_2$	90	23	26	4	15	[181]
3% Ru/C	10% XYL; 200 °C; 40 bar H_2; $Ca(OH)_2$	20	32	25	10	17	[182]
51% $Ni–Mg_{1.29}Al_{0.06}O_{1.38}$	5% SOR; 180 °C; 20 bar H_2	81	10	19	31	nd	[184]
50% Ni–MgO	5% SOR; 180 °C; 20 bar H_2	65	10	19	35	nd	[184]
3% Ru/CNF	20% SOR; 220 °C; 80 bar H_2; CaO	86	19	32	10	nd	[185]
3% Ru–19% MnO/C	20% XYL; 200 °C; 60 bar H_2	80	18	17	18	1	[177]

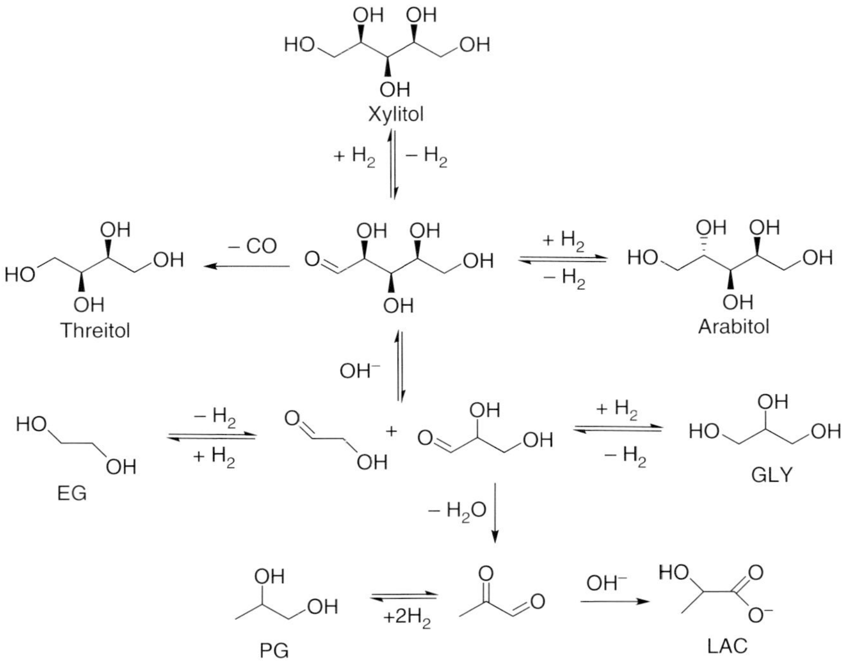

Scheme 7.5 Mechanism of hydrogenolysis of xylitol to shorter polyols in the presence of bifunctional metallic catalyst. Source: Adapted from Ref. [177].

In addition, further Cannizzaro reactions occur in alkaline conditions, through which LAC can be produced in significant amounts. In addition, products from decarbonylation reactions are also observed.

In the presence of homogeneous bases – such as $Ba(OH)_2$, $Ca(OH)_2$, or $La(OH)_3$ – LAC is formed irrespective of the nature of the active metal [178–182]. LAC formation can be limited when only small amounts of soluble base are added (20% LAC selectivity in the presence of 250 mM $Ca(OH)_2$ vs. <5% in the presence of 50 mM $Ca(OH)_2$ during the hydrogenolysis of 0.7 M XYL solutions) [183]. LAC formation is also significantly reduced when solid bases (e.g. CaO) or basic supports (like HT and MnO) are used [173, 177, 184, 185]. Notably, in the case of the HT, the catalytic activity was correlated with the number of basic sites [184].

The use of alcoholic solvents (90% H_2O/10% ROH) had a positive effect on the selectivity to glycols, up to 70% selectivity to (EG + PG + GLY) being attained at 60% conversion at the detriment of decarbonylation products.

7.3.3 Hydrogenation of Carboxylic Acids

Levulinic (LevA) and succinic (SucA) acids are interesting platform molecules for the production of useful lactones or biorenewable diols (see Scheme 7.6).

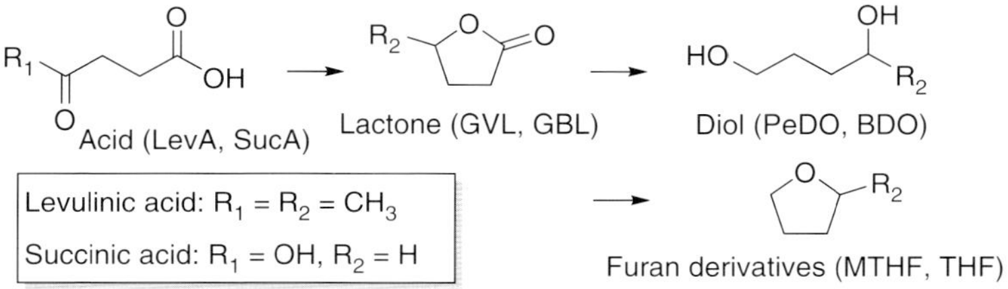

Scheme 7.6 Reaction pathway associated with the hydrogenation of carboxylic acids (levulinic acid, succinic acid) toward lactones (γ-valerolactone, γ-butyrolactone), diols (1,4-pentanediol, 1,4-butanediol), and furan derivatives (2-methyltetrahydrofuran, tetrahydrofuran).

7.3.3.1 Levulinic Acid

γ-Valerolactone (GVL), 1,4-PeDO, and 2-methyltetrahydrofuran (MTHF) can be obtained during the hydrogenation of LevA. These compounds are also a platform to fuels and a variety of value-added chemicals (e.g. polymers, food additives). They are also used as solvents, most notably for biomass conversion. The hydrogenation of LevA toward GVL has been extensively studied over the past decades and many reviews have been published [186–192]. Hence, this section will focus on the significant findings reported since 2012.

7.3.3.1.1 Catalyst Development Over noble metal catalysts (Pd, Pt, Au, and Ru), yields of GVL above 95% are commonly observed, Ru/C exhibiting the best catalytic activity. However, the long-term stability of this catalyst is an issue for possible industrial application, as sintering and leaching of the Ru NPs have been observed, especially in aqueous phase reactions. Therefore, the latest studies have focused on the development of more stable catalysts.

The incorporation of a second metal (Sn, Re, Pd, Ni) or an alkali promoter (e.g. Cs^+) [193] can improve the stability of Ru catalysts. A wide range of bimetallic catalysts have been tested in water or alcohol – such as Ru–Re/C [194], Ru–Sn/C [195], and Au–Ru/TiO_2 [196] – and in dioxane (e.g. Ru–Pd/TiO_2) [197]. Ni–Ru supported on ZrO_2, Al_2O_3, and ordered mesoporous carbon were also tested in solvent-free conditions [198, 199]. The bimetallic catalysts were more stable than the monometallic formulations, although quite frequently they were also less active [194, 195, 197, 200]. Recent studies have shown that Pd-based catalysts such as Pd/SiO_2 [201] and Pd/CeO_2 [202] can be stable in water or propanol. Saliently, the addition of a promoter increases the activity of Pd–Re/hexagonal mesoporous silica (HMS) doped with Ti [200] and of Au–Pd/TiO_2 [197].

Performing this reaction in the absence of noble metals is a step toward the development of a more sustainable process. Hence, recent studies have focused on supported Ni and Cu catalysts, such as Ni/$MgAlO_{2.5}$ [203], Cu/Ni/Mg/Al [204], Ni/boehmite [205], and Cu/ZrO_2 [206]. The latter are usually less active than noble metals, higher temperature being often employed in order to obtain full conversion toward GVL. In water, the deactivation has been usually attributed to metal leaching [205, 206]. It is worth noting that Ni–MoO_x/C exhibited catalytic activity comparable to Ru catalysts under solvent free conditions at 50 bar and 250 °C [207]. Catalysts based on elements from the groups 4–6 have also been tested for activity in this reaction, Mo_2C [208] and ZrO_2/SBA-15 [209] generating yields of GVL of up to 80–90%.

The support can also play a crucial role determining catalyst stability. TiO_2 and ZrO_2 are known to be stable under hydrothermal conditions. In various solvents, catalysts supported on ZrO_2 were more stable than on Al_2O_3, C, or TiO_2 (e.g. Cu, Ni–Ru, Ru) [198, 206, 210]. The lost activity of Ru/TiO_2 and Ru/Al_2O_3 can be partially recovered through reduction, while the deactivation was irreversible on the more electronegative SiO_2 due to the agglomeration of the NPs [211]. In water, Ru/OMS-2 (manganese octahedral molecular sieves) could be reused four times without loss of activity in the aqueous phase conversion of LA in batch reactors [212].

7.3.3.1.2 Toward a Sustainable Process The hydrogenation of LevA has mainly been conducted using alcohols, dioxane, and water as solvents. The use of alcohols is usually associated with lower yields of GVL due to the occurrence of competitive esterification reactions. Dioxane is often used as it has a higher boiling point than GVL, which facilitates its recovery by distillation. Nevertheless, water is a more adequate solvent when working with biomass-based platform chemicals. Notably, GVL can be extracted from aqueous solution using ethyl acetate [205]. Moreover, it has been shown that the activity of Ru catalysts is increased when working in aqueous solution, which might be due to hydrogen-bonded water molecules reducing the activation energy of the reaction [213, 214]. Since LevA hydrogenation is a fast reaction, diffusion limitations can be an issue over Ru/C [215]. Membrane reactors consisting of hydrophobic expanded polytetrafluoroethylene (PTFE) with deposited Ru particles have been developed for this reaction in order to limit the negative impact of these phenomena [216].

Catalytic systems efficient at low temperature using high concentrations of substrates are favored as profitable and sustainable processes for GVL synthesis. Reactions have been conducted with high concentrations of LevA in water (e.g. 50 wt% over Pd/SiO_2) [201] or even in solvent-free conditions over Ru/C [217], MoO_x–Ni/C [207], Ni–Ru/ZrO_2 [198], and Ru–Ni/MSP C [199]. Over Ru/C, temperatures as low as 50 °C in water [211] or 25 °C for solvent-free reactions have been employed [217]. Unfortunately, few studies report the effect of impurities on catalytic performance, albeit it is worth noting that Dumesic and coworkers observed that a ReO_x–Ru/C catalyst was stable in the presence of sulfuric acid, a common catalyst used in the conversion of cellulose [194]. Complete conversion of LevA toward GVL can also be achieved when coupling the reaction with GLY or formic acid as hydrogen donors [198, 218], the later compound being cogenerated with LA during the dehydration of glucose. Finally, it is worth noting that the hydrogenation of LevA has also been studied in the gas phase over Ni and Cu catalysts [219–222].

7.3.3.1.3 Control of the Selectivity Toward GVL, PeDO, and MTHF The formation of GVL occurs via dehydration and hydrogenation steps, which are favored by Lewis acids and metallic sites, respectively. The hydrogenation of LevA toward PeDO or MTHF have been studied to a lesser extent, and they usually require higher temperatures and pressures. In ethanol, up to 96% of PDO was obtained over Cu/ZrO_2 (200 °C, 60 bar) [223]. In water, yields of 82% and 70% were obtained over ReO_x–Ru/C (160 °C, 150 bar) [224] and 4% Rh–MoO_x/SiO_2 (80 °C, 60 bar) [225]. MTHF can be obtained by direct hydrogenation of GVL or cyclization of PDO, Cu/ZrO_2 catalysts generating 91% of MTHF in ethanol (240 °C, 60 bar) [223].

7.3.3.2 Succinic Acid

Succinic acid (SucA or SA) can be hydrogenated to valuable products, such as γ-butyrolactone (GBL), 1,4-BDO, and tetrahydrofuran (THF). These chemicals are mainly used as solvents, as starting materials for pharmaceutical compounds, or for the synthesis of polymers. A wide range of mono- and bimetallic catalysts based on Au, Pd, Re, and Ru have been studied for SucA hydrogenation. Full conversion of SA can easily be achieved, although high selectivity toward a specific product is challenging as GBL is an intermediate to both BDO and THF. Moreover, successive C—O and C—C bond cleavage can occur and form minor by-products such as butyric acid, propionic acid, butanol, and gaseous products. This section focuses on the studies published in the last decade, as information on earlier studies can be found in previously published reviews [15, 226, 227]. Selectivity and activity have been shown to depend on the nature of the metal and support, the catalysts properties (e.g. metal dispersion, acidity, etc.), and the reaction conditions in terms of solvent, temperature, and pressure.

7.3.3.2.1 Catalysts Development Au- and Pd-based catalysts usually favor the formation of GBL, of which high yields (>94%) have been obtained over Au/TiO_2 [228], Pd/AlOOH [229], and Pd/TiO_2 [230] catalysts. GBL is then converted to BDO and up to 30% of BDO can be obtained over Pd/MCM-41 [231]. In water, Re favors the formation of THF at high temperatures (e.g. up to 66% over Re/C at 240 °C) [232, 233], albeit hardly any conversion is observed when working at lower temperatures (160–180 °C) over Re/TiO_2 [234] or Re/C [235]. Over Ru-based catalysts, various products can be observed depending on the reaction conditions. Indeed, over Ru/C, up to 46% of THF can be obtained in dioxane (240 °C, 80 bar) [236], while propanoic acid is favored in water (80 bar, 180 °C) [235]. Under higher pressure (150 bar, 180 °C), GBL is formed first and then there is a dramatic loss of carbon (63%) [237]. Ru-based catalysts are known to be more active than Pd-based catalysts but they can also favor cracking reactions.

The rate of SucA conversion was shown to depend on metal dispersion over supported Pd [230, 231, 238], Re [239], and Ru [236] catalysts. The nature of the precursor as well as the catalyst preparation and activation method can affect the particle size distribution and metal/support interaction, thus also impacting catalytic activity [230]. Moreover, supports with acidic properties favor dehydration and the formation of GBL [240, 241].

The addition of an oxophilic promoter, such as Re, improves the yield toward BDO. Indeed, over supported ReO_x–Pd catalysts, the incorporation of Re facilitates ring opening, while over ReO_x–Ru, it prevents over-reduction [235, 237]. The use of non-noble metal catalysts for this reaction is rather limited due to their lower activity. ReO_xCu/C has been tested in dioxane and 20% BDO was formed [242]. It is worth noting that 95% yield of BDO was obtained during the conversion of the intermediate GBL over Cu–Co/TiO_2 [243]. As was the case for monometallic catalysts, a decrease in particle size was associated with an increase in activity for bimetallic catalysts – including ReO_x–Pd/TiO_2 [234], ReO_x–Ru/C [235], and ReO_x–Cu/MC [242]. The preparation method affects the location of Re particles, Re oxidation states, and dispersion, which in turn impact catalytic activity [234, 244, 245]. In water, the best yield of BDO (82–83%) was obtained over ReO_x–Pd/TiO_2 [234] and Ru–Sn/AC [246], while the highest yields in THF were obtained over ReO_x–Pd/C [232] and Re/C [233].

7.3.3.2.2 Toward a Sustainable Process While most studies have been conducted in organic solvents (e.g. dioxane, alcohols, etc.), the open literature dealing with the hydrogenation of SucA in water is limited. Some of the highest GBL, BDO, and THF yields obtained over the past few years are reported in Table 7.4. The reactivity of SucA is lower compared with LevA due to the nature of the first functional group to be hydrogenated. Therefore, higher temperatures and pressures are usually employed.

The use of water facilitates the direct catalytic upgrading of SucA, as fermentation processes are performed in the aqueous phase. In the latest studies, the recyclability of the catalysts was assessed and a small drop in activity was observed

Table 7.4 Most promising results obtained since 2008 in the aqueous phase hydrogenation of SucA.

Catalysts	Yields	*T* (°C)	*P* (bar)	Concentration (wt%)	References
Pd/TiO_2	95% GBL	160	150	5	[230]
Re/C	63% THF	240	80	10	[233]
ReO_x–Pd/C	65% THF	240	80	10	[232]
ReO_x–Pd/TiO_2	83% BDO	160	150	5	[234]
Ru–Sn/AC	82% BDO 71% BDO[a)]	180 170	100 124	2 5	[246]

a) Continuous reactor.

for Pd/TiO_2 [230] and Re/C [233]. The deactivation was not due to metal leaching but rather to an increase in particle size. It is worth noting that Liang and coworkers also studied the effect of acetic acid and pyruvic acid – which are side products present in the fermentation broth – on catalyst performance [233, 235]. These authors noticed a competitive adsorption of the acids with SA, which was associated with a decrease in activity over ReO_x–Ru/C. Most studies dealing with the hydrogenation of SucA were conducted in batch reactors (except for the work of Beckham and coworkers [246]) and employed concentrations of SucA in the aqueous phase below 10 wt%. This differs from the work conducted with LevA, which is mainly due to the lower reactivity of SucA and to the fact that less attention has been paid to this platform chemical. In the same manner, there are few studies dealing with the hydrogenation of other acids such as itaconic acid, arabinonic acid, glutamic acid, and maleic acid [15].

7.3.4 Selective Hydrogenation of Furanic Compounds

Hydrogenation of FAL provides a route to products such as furfuryl alcohol (FOL) for furanic resins, 2-(hydroxymethyl)tetrahydrofuran (HMTHF) used as a green solvent totally miscible with water, and cyclopentanone (CPN) used for the synthesis of pharmaceuticals, insecticides, and rubber chemicals, as reviewed by Yan et al. [247]. The selective hydrogenation of FAL to FOL is usually carried out over Cu-based catalysts in a vapor phase continuous process [248]. In water at 110 °C under 6 bar of H_2, 5 wt% Pd–5 wt% Cu/MgO or $Mg(OH)_2$ exhibited complete conversion and >98% selectivity, which remained at the same level in five catalytic cycles without reactivation of the catalyst [249].

Further hydrogenation of the furanic ring in FOL results in HMTHF. This compound is usually produced from FAL in two separate steps involving FOL hydrogenation over a noble metal [250]. In a single-pot reaction, complete and direct conversion of FAL was reported with a yield of HMTHF of 95% over a Pd/Si-MFI molecular sieve catalyst in isopropanol [251], or with 78% yield over Pd–Ir–ReO_x/SiO_2 in water [252]. Ring opening hydrogenolysis produces 1,2- or 1,5-PeDO depending on the catalyst employed. Ru/MnO_x achieved a 42% yield of 1,2-PeDO in aqueous solution [253]. Selective ring opening of HMTHF to

1,5-PeDO was achieved by using supported noble metal (i.e. Rh or Ir) catalysts modified with MoO_x or ReO_x, developed by Tomishige et al. [166, 254–257] as well as others [258]. As for GLY hydrogenolysis, the optimum Re/Rh and Mo/Rh ratio was 0.5 and 0.13, respectively. The highest yield of 1,5-PeDO was 94%. Alternatively, supported noble metals [259, 260] as well as Cu and Ni supported on Co_3O_4 and ZrO_2 [261] convert FOL to afford CPN and cyclopentanol (CPL) as the main products in water.

With two substituents on the furan ring, hydrogenation of HMF may also produce several useful diols or compounds for new polymers, fine chemicals, flavors, fragrances, and fuels (see Scheme 7.7) [110]. This transformation in aqueous solution has been recently reviewed by Hu et al. [262] and by Tang et al. [263]. The product selectivity depends on the catalyst (i.e. hydrogenation ability, acid–base properties, etc.) and on the reaction conditions.

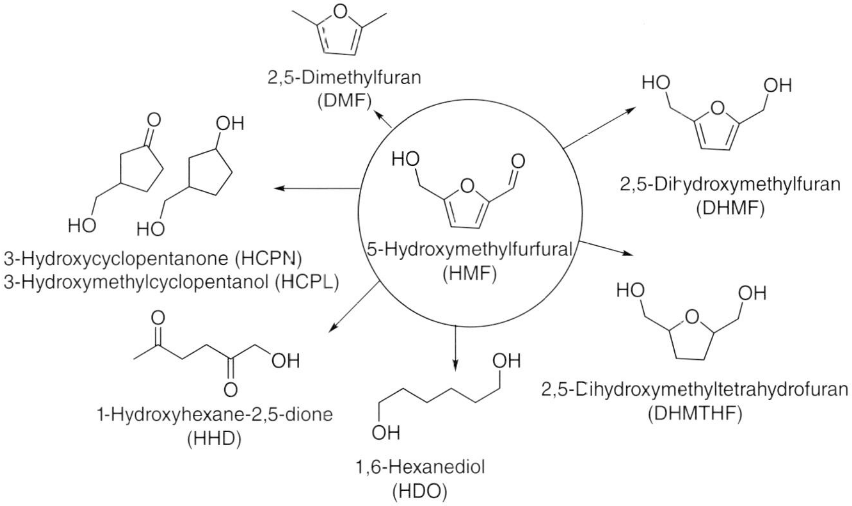

Scheme 7.7 Potential chemicals produced by selective hydrogenation of 5-hydroxymethylfurfural.

Hydrogenation of the aldehyde group produces 2,5-dihydroxymethylfuran (DHMF), an intermediate for the production of bio-based furan polyesters [262–264]. In aqueous solution, a number of supported noble metal catalysts show outstanding performance, including Au/Al_2O_3 at 120 °C and 65 bar [265], $Ru/ZrO_2–SiO_2$ used under milder conditions (25 °C and 5 bar) [266], ReO_x–Ir/SiO_2 at 30 °C and 8 bar [267], Ir/TiO_2 at 50 °C and 60 bar [268], recyclable Pt/MCM-41 at 35 °C and 8 bar [269], and Au/Fe_2O_3 at 80 °C and 30 bar [270]. DHMF yields of >90% and up to 99% could be obtained in a few hours. Base metal catalysts also achieve high yields of DHMF; however, the reaction must be carried out in alcohol or dioxane under more severe conditions (100–120 °C, 15–70 bar H_2) over catalysts such as Cu–ZnO [271], Cu/SiO_2 [272], and NiFe/CNT [273].

The fully hydrogenated compound 2,5-dihydroxymethyltetrahydrofuran (DHMTHF) is used either as a solvent substitute for THF or as a monomer [262, 263]. DHMTHF could be selectively obtained using harsh conditions and catalysts that do not perform reductive deoxygenation to 2,5-dimethylfuran (DMF). Over Raney Ni in alcohol solvent at room temperature under 4 bar H_2, the yield of DHMTHF was >93% [274]. In a 1-butanol/water biphasic system, HMF was hydrogenated over Ru supported on materials with a high isoelectric point, such as ceria, $MgO-ZrO_2$, and γ-Al_2O_3, with 88–91% DHMTHF yield at 130 °C under 27 bar in 12 hours [275]. In pure water solvent, Pd-based bimetallic catalysts were more active than Raney nickel. Alloyed $Ni-Pd/SiO_2$ at 40 °C under 80 bar [276], and $Pd-Ir/SiO_2$ at 2 °C under 80 bar [277] showed a remarkable synergy between both metals and yielded 96% and 95% DHMTHF, respectively. Ir promotes the adsorption of C—O, whereas the Pd surface strongly interacts with the furan ring, the used catalyst showing almost the same performance as the fresh material after regeneration by calcination and reduction. Pd immobilized on basic NH_2-functionalized metal organic framework (Pd/MIL-101(Al)-NH_2) affected the total hydrogenation in water with 96% yield at 30 °C under 10 bar in 12 hours [278]. A non-noble HT-like Ni–Al catalyst could yield as high as 99% DHMTHF at 80 °C under 20 bar after eight hours [279].

1,6-HDO is a valuable monomer for polyesters and polyurethane as well as a precursor to caprolactam (the monomer for nylon-6) [262, 263]. De Vries and coworkers have explored multistep routes to HDO [280]. In the most successful one (see Scheme 7.8), HMF was first hydrogenated with Raney nickel to afford a quantitative yield of DHMTHF in ethanol at 100 °C under 90 bar after 14 hours. Further, DHMTHF was totally hydrogenated to 1,6-HDO (86% yield in water at 120 °C and 80 bar of H_2 for 20 hours) using a bifunctional ReO_x-Rh/SiO_2 catalyst developed by Tomishige and coworkers for the highly selective hydrogenation of FOL to 1,5-PeDO [254, 257]. In the presence of acidic Nafion, the reaction proceeded via the intermediates 1,2,6-hexanetriol (1,2,6-HTO) and tetrahydropyran-2-methanol (2-THPM). In contrast, the direct one-pot reaction in water using ReO_x-Rh/SiO_2 catalyst at 120 °C and 80 bar attained only a 7% HDO yield after 14 hours.

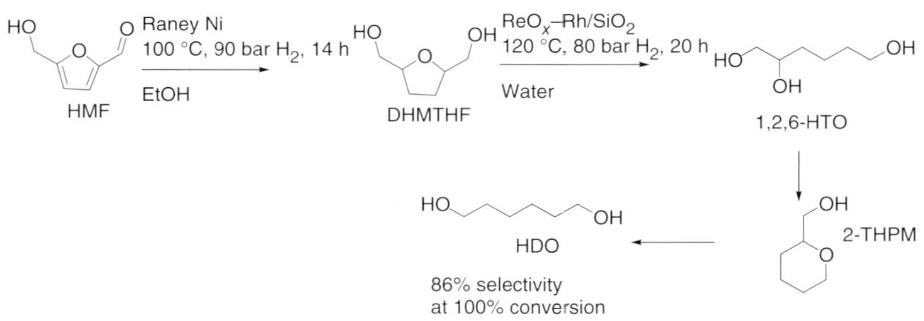

Scheme 7.8 Two-step hydrogenation of HMF via DHMTHF to HDO. Source: Adapted from Ref. [280].

Based on this work, Zhang and coworkers loaded two separate layers of Pd/SiO_2 and $ReO_x–Ir/SiO_2$ in a fixed-bed reactor, in which 57.8% HDO yield was achieved directly from HMF in a THF/water mixture solvent at 100 °C and 70 bar of H_2 [281]. The yield progressively decreased to around 50% and 40% after 15 and 24 hours on stream, respectively, due to leaching of ReO_x during reaction. A single catalyst – namely, $Pd–Ir–ReO_x/SiO_2$ – afforded only 15.2% HDO. The HDO diol can be converted further into caprolactone and caprolactam [230] or oxidized to adipic acid [282].

Hydrolytic ring opening of HMF can result in 1-hydroxyhexane-2,5-dione (HHD) diketone. Under 30 bar CO_2 and 10 bar H_2, HHD was successfully produced in water with a 77% yield over Pd/C at 120 °C for 15 hours [283].

3-Hydroxymethylcyclopentanone (HCPN), which finds applications in the synthesis of fragrances, pesticides, polymers, etc. [262], is synthesized by a ring-rearrangement reaction. Ohayama et al. reported up to 32% HCPN yield in water at 140 °C and 30 bar H_2 over Au immobilized on different oxide supports [284]. Using Pt/SiO_2 in combination with metal oxides, the metallic phase and the acid/base sites are involved in multiple hydrogenation and ring rearrangement pathways, Ta_2O_5 with moderate Lewis acid strength giving the best results [285]. HCPN could also be synthesized in water with a high yield (86%) being obtained at 180 °C and 20 bar of H_2 over a non-noble $Cu–Al_2O_3$ catalyst, while 3-hydroxymethylcyclopentanol (HCPL) could be obtained using $Co–Al_2O_3$ catalyst at 140 °C, the yield being 94% [286]. Both catalysts could be recycled after regeneration. Further, reductive deoxygenation of HMF produces DMF, which is suitable as a liquid transportation fuel. Lastly, an alternative route to the use of H_2 pressure uses transfer hydrogenation of formic acid as a hydrogen source [262].

7.3.5 Reductive Amination of Acids and Furans

The review of Sels and coworkers [287] reflects the growing attention paid to the synthesis of amines from biomass. For example, the reductive amination of FAL or HMF with NH_3 can generate the valuable furfurylamine or 5-hydroxymethyl furfurylamine. Around 90% yield can be observed over Rh/Al_2O_3 under mild conditions (20 bar, 80 °C) [288]. When using aminopropanol and aminoacids, the selectivities varied (between 10% and 80%) over carbon-supported FeNi (10 bar, 150 °C) [289]. The formation of the imine was shown to be the rate determining step. Recently, a few studies have focused on the reductive amination of levulinic acids to N-substituted-5-methyl-2-pyrrolidones. Varied reaction conditions have been employed in terms of temperature (100–150 °C), pressure (3–85 bar) and solvent (GVL, 2-MTHF, ethanol, DMSO, ethyl acetate, and solvent-free) [289–292]. Supported noble metal (Pt [291], Ir [292]), and Ni [289, 290] catalysts have been tested. Yields of pyrrolidones of up to 99% were obtained over $Pt–MoO_x/TiO_2$ using a wide range of benzylamines and aliphatic amines. Reductive amination reactions can even be conducted without catalysts when the reaction is performed under harsher conditions (up to 200 °C) and formic acid is used as hydrogen donor [293, 294]. Ethyl levulinate can also be used as substrate [295, 296].

7.4 Catalyst Design for the Dehydration of Biosourced Molecules

7.4.1 Introduction

Acid- and base-catalyzed dehydration reactions of mono-alcohols to olefins are well-known organic reactions, and numerous studies have been devoted to the correlation between acid–base and catalytic properties [297–302]. The three elimination mechanisms – E1, E2, and E1cb (see Scheme 7.9) – display different kinetics of α carbon deprotonation and of H_2O loss. The E1 and E1cb mechanisms are observed on acidic and basic solids, respectively. The E2 mechanism occurs on the acid–base pairs of an amphoteric solid. The E1 corresponds to a monomolecular elimination following the formation of an intermediate carbocation, while E1cb is a bimolecular elimination with a simultaneous α carbon deprotonation and H_2O elimination through a concerted mechanism (single transition state). In an E1cb mechanism, the α carbon is first deprotonated, leading to the formation of one carbanion. The difference between these mechanisms is mainly kinetic; indeed, in the case of the E1 or the E1cb mechanisms, the reaction rate depends only on the carbocation or carbanion concentrations, while E2 depends both on the starting product and the base concentrations. It should also be kept in mind that although the formation of carbocations is favored on acid sites, they must be stabilized by a base (and vice versa in the case of carbanions).

Water generated by fermentation processes or by dehydration of biosourced molecules can modify the catalytic activity through irreversible textural and/or

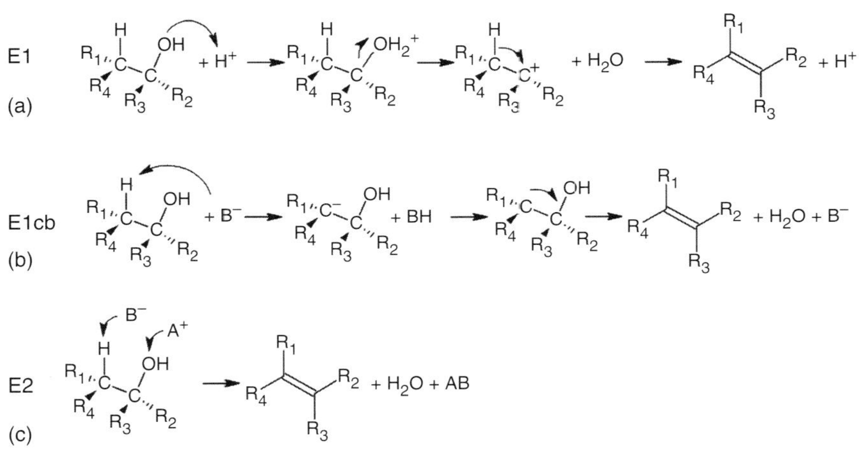

Scheme 7.9 (a) E1, (b) E1cb, and (c) E2 elimination mechanisms for the dehydration of alcohols.

structural evolution, reactant solvation, competitive adsorption and stabilization of intermediates and transition states, product dilution or solvation, or modification of the nature of acidic sites [303]. Hence, the multifunctionality of bioresources and the presence of water often lead to classical acidic catalysts showing poor selectivity and stability [304–307], which highlights the need to develop new water-tolerant solid acids with tailored porosity [308] and acid–base properties. In Sections 7.4.2 to 7.4.4, the recent progress in the design of acidic catalysts for the dehydration of biosourced molecules such as GLY, LAC, and sorbitol is reviewed.

7.4.2 Glycerol to Acrolein

Dehydration of GLY to acrolein has attracted significant interest in recent years due to a projected rise in global biodiesel production offering new valorization opportunities. Improvements in the performance of solid acid catalysts, which have been mostly achieved in the gas phase, have been reviewed in several papers [309–313]. The most efficient catalytic systems, leading to very high acrolein selectivity (>80%), are supported $H_4SiW_{12}O_{40}$ heteropolyacids and supported tungsten oxides. The major challenge to be overcome is that catalysts need to be stable for several months, the issue of catalyst stability being related to the selectivity to hydroxyacetone (or acetol, the main by-product) and to other by-products such as acetone, allylic alcohol, acetaldehyde, and phenol, all of which are precursors of coke [314–317]. Additionally, coke can be obtained by condensation or acetylation of GLY [318], which requires maintaining a very high conversion (low GLY concentration). Acidity, basicity, and porosity are the main physicochemical properties influencing the catalytic performance.

Chai et al. first proposed that the most suitable Hammett acidity to dehydrate GLY to acrolein ranges between −8.2 and −3 [315, 319]. In fact, acrolein formation is favorable on strong Brønsted acid sites (BAS), while strong LAS can lead to the formation of hydroxyacetone and coke [318, 320]. For instance, a catalyst such as $Cs_{2.5}H_{0.5}PW_{12}O_{40}$ containing only BAS gave a selectivity to acrolein close to 97% (and 3% hydroxyacetone) after one hour of reaction, while catalysts containing predominantly LAS such as a Cr–Zn mixed oxide gave 34% acrolein and 42% hydroxyacetone at the same reaction time [321]. The internal oxygen of GLY can be protonated by BAS (E1 mechanism) giving 1,3-dihydroxypropene that is tautomerized to 3-hydroxypropanal in a second step. The latter undergoes further acid-catalyzed dehydration to yield acrolein. On the other hand, concerted transfer of the terminal OH group of GLY to one LAS and migration of the proton from the internal carbon atom to a bridging O atom of LAS gives 2,3-dihydroxypropene, which is then tautomerized to yield acetol [321]. Acetol formation would also be possible over basic sites by dehydrogenation, dehydration, and rehydrogenation of GLY [322]. In fact, it has been shown that GLY conversion increases with acid site density (see Figure 7.2a), while the selectivity to acrolein strongly depends on the acid/base balance at ratios below 6 (being almost constant above that value as shown in Figure 7.2b) [323]. This implies that a catalyst must contain a very low amount (typically <0.5 μmol/m^2) of basic sites in order to display a high selectivity to acrolein (>70%). Interestingly, strong acid

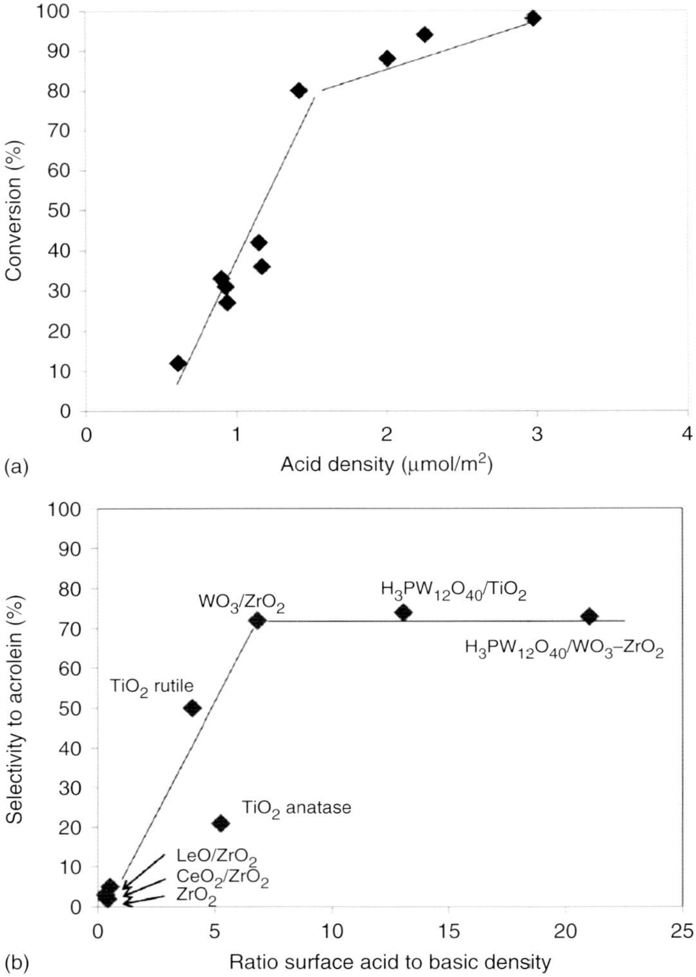

Figure 7.2 (a) Conversion of glycerol as a function of acid site density and (b) correlation between selectivity to acrolein and surface acid to basic density ratio as determined by microcalorimetry using NH_3 and SO_2 as probe molecules. Source: Reprinted with permission from Katryniok et al. 2013 [311]. Copyright 2010, American Chemical Society.

sites are progressively deactivated, but the remaining weak or very weak acid sites are able to convert GLY, albeit with a lower intrinsic activity [324].

Catalyst deactivation is caused by the surface formation of coke, which blocks pores, prevents GLY from reacting on the active sites, and hinders the diffusion of the products formed [325–328]. Furthermore, GLY condensation followed by polymerization can occur in micropores and small mesopores [329, 330]. Different groups have pointed out the importance of using mesoporous catalysts to improve stability over time [311, 314, 325, 328, 331–334]. For instance, heteropolyacids supported on silica (CARiACT Q10) containing large pores (10 nm) have shown higher yields due to a slower decrease in conversion [314]. Atia et al.

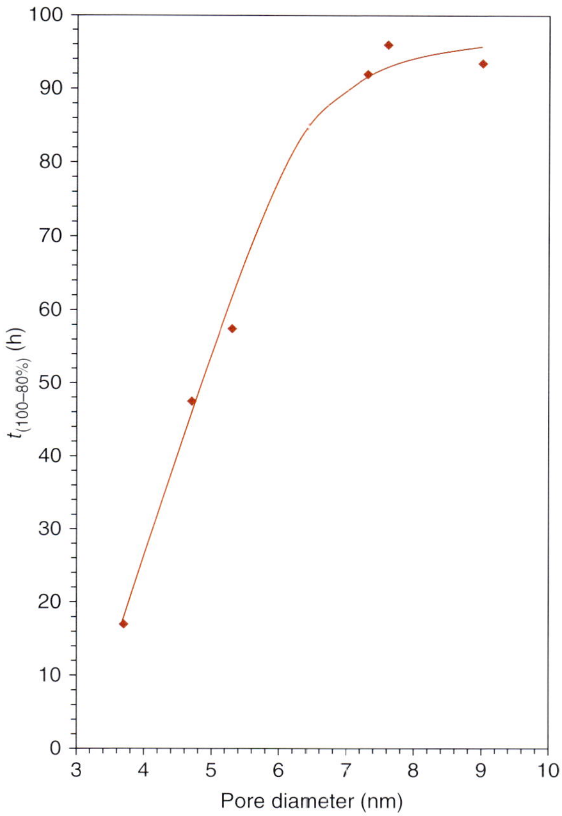

Figure 7.3 Correlation between the $t_{(100-80\%)}$ parameter and the pore diameter established for tungstated zirconias. Reaction temperature: 300 °C, gas feed composition: glycerol/H_2O/N_2: 2.3/46.3/51.4, gas hourly space velocity (GHSV): 2900 h^{-1}.

investigated the catalytic performance of $H_4SiW_{12}O_{40}$ supported on aluminas with different pore sizes (5 and 12 nm) using identical conditions to Tsukuda et al. [331]. Al_2O_3 with a pore diameter of 12 nm led to higher acrolein selectivity and improved catalyst lifetime. Katryniok et al. reported improved stability of $H_4SiW_{12}O_{40}/ZrO_2$-modified SBA-15 containing 8 nm pores.

A clear correlation between stability and pore diameter has also been established for tungstated zirconias [335]. As shown in Figure 7.3, the time necessary to decrease GLY conversion from 100% to 80% ($t_{100-80\%}$) increased five times when increasing the pore diameter from 4 to 9 nm.

Improved catalytic activity was also obtained using zeolites containing some degree of mesoporosity [336, 337]. In this case, coke was formed in the whole porous network while it was strictly limited to the surface for purely microporous zeolites, thus only blocking the pore entrance [317]. More recently, zeolites capable of displaying a longer life were obtained by synthetizing hierarchical porous structures comprising macropores [338–341]. In a similar fashion, hierarchically mesoporous acid nanospheres exhibited better catalytic performance toward the dehydration of GLY to acrolein than HZSM-5 and AlMCM-41, a correlation between coke formation and the average pore diameter being established [342].

Finally, further improvements can be achieved through addition of oxygen or hydrogen to the reaction mixture. O_2 addition was shown capable not only of eliminating some by-products – such as hydroxyacetone, propanal, and phenol – but also of limiting coke formation [317, 343–346]. Coking can also be reduced by doping the acid catalysts with small amounts of noble metals such as Pd and Pt, the addition of hydrogen to the reaction mixture then leading to the hydrogenation of coke precursors [321, 347].

7.4.3 Lactic Acid to Acrylic Acid

LAC and acrylic acid are key platform molecules for a large range of applications, polymer synthesis being particularly noteworthy. Given that the production of LAC (which amounted to 800 000 tons in 2013) is not able to satisfy the demand of acrylic acid (5.75 Mt in 2014), dehydration of LAC to acrylic acid offers an alternative bio-based route. This reaction is rendered challenging because the acid and alcohol functions of LAC can each react, leading to by-products including acetaldehyde, pyruvic acid, propionic acid, 2,3-pentanedione, lactide, and polylactides [348]. The performance of different catalytic systems has been compared in several review papers [21, 349, 350]. Notably, important discrepancies can be found in the literature, which can arise from different analytical methods as well as from the LAC concentration and the vaporization/condensation process leading to the formation of lactide and polylactides in different amounts [350, 351]. The two best catalytic systems are zeolites modified with salts and phosphates (including hydroxyapatites).

The acid–base properties of the catalysts interfere with the adsorption/desorption steps of the catalytic cycle since both the reactant and the product are acids. Furthermore, high acidity favors the formation of acetaldehyde by decarbonylation (the activation energy of which is 115 kJ/mol versus 137 kJ/mol for acrylic acid formation). Nevertheless, an acetaldehyde yield as high as 84% was reached under supercritical conditions (385 °C, 345 bar) after addition of 100 mM of sulfuric acid. Classical dehydration catalysts such as zeolites [352–354], aluminum phosphate [355], aluminum sulfate [356], magnesium aluminate [357], carbon compounds [358], and supported heteropolyacid (HPAs) [309, 356] – all of which contain strong acid sites – are selective to acetaldehyde [359].

NaY zeolites modified by potassium halogen salts (KX) were investigated by Sun et al. [360, 361]. Since Lewis acidity decreased when the cation size was increased [362], substitution of Na^+ by K^+ cations decreased acidity and the acrylic acid yield was improved. Furthermore, X^- anions increase the basicity of adjacent O^{2-} anions, which was proposed to prevent the formation of acetaldehyde. Modification of H-ZSM-5 catalysts with NaOH and Na_2HPO_4 was shown to strongly improve the selectivity to acrylic acid (from 2.5% to 78%) and the catalyst stability by converting strong acid sites into medium strength acid sites as well as by generating medium basic sites, mesopores, and macropores [353]. It was also shown that NaY zeolite modified by sequential dealumination and alkaline treatment affords a solid featuring a hierarchical distribution of micro- and mesopores of reduced Lewis acidity and increased basicity due to the presence of well-dispersed sodium ions interacting with external siloxy groups. These properties were crucial for achieving higher selectivity (75%) and minimizing the activity loss observed in a six hour run [352]. The deactivation of catalysts with a high uptake capacity of polar compounds (such as NaY zeolites) is due to the accumulation of acidic compounds – such as LAC, acrylic acid, and open-chain LAC polymers – the acidic character of the deposits leading to a decline in the selectivity to acrylic acid over time by favoring acid-catalyzed side reactions. Ion exchange of Na^+ by K^+ or Cs^+ reduces the tendency toward cluster formation due to steric reasons [363].

It is now well-established that the formation of acrylic acid requires both moderately strong (or weak) acidic and basic sites, as strong acid sites favor condensation reactions and strong basic sites favor the decarbonylation of LAC [352, 364]. For instance, a selectivity to acrylic acid of only 2.5% was reported for MgO, which is a purely basic catalyst [365]. Furthermore, a clear correlation between the selectivity to acrylic acid achieved in the presence of alkaline earth phosphates and the acid–base balance has been established by Blanco et al. [348]. As shown in Figure 7.4, when the acid/base balance increases from 1.3 to 1.9, the selectivity to acrylic acid decreases from 49% to 25% while the one to acetaldehyde increases from 27% to 44%.

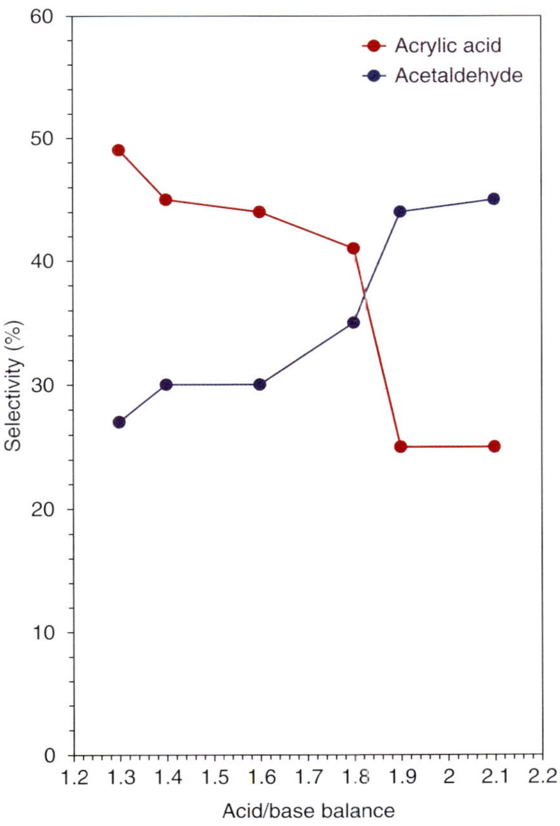

Figure 7.4 Evolution of the selectivity values to acrylic acid and acetaldehyde with the acid/base balance (determined from NH_3 and CO_2–TPD measurements) for various alkaline earth phosphates.

An adequate balance between moderately acidic and basic sites is also necessary for hydroxyapatites, suggesting that acid–base pairs are active in such compounds [365–367]. Interestingly, the acid–base properties of hydroxyapatites can be modulated varying the calcination temperature and the M/P ratio [366]. Finally, DFT calculations have shown that the dehydration of LAC is more favorable through a concerted than a stepwise mechanism over sodium triphosphate [368]. Acid–base pairs might also be involved in an E2 mechanism with interactions of P–O$^-$ and P–OH species with the protons of the methyl and the OH groups of LAC, respectively. In short, acrylic acid formation is favored by acid–base pairs in which acidic and basic sites are of weak and/or moderate strength while acetaldehyde formation needs stronger acidity or basicity.

7.4.4 Sorbitol to Isosorbide

Isosorbide (1,4:3,6-dianhydrohexitol) is a heterocyclic compound that can be produced by double dehydration of D-sorbitol (SOR) in the liquid phase. Isosorbide finds applications in a large number of technical specialty fields as plasticizer, monomer, solvent, or reagent in the synthesis of pharmaceuticals. Industrial units with capacities of several thousand tons are now producing isosorbide.

The synthesis of isosorbide involved the initial formation of 1,4-sorbitan via S_N2 substitution reactions with a protonated hydroxyl as the leaving group (C1 position) followed by a cyclization with the carbon atom in the C4 position. Racemic 3,6-sorbitan, which can also yield isosorbide, is generally observed in very low amounts in spite of a very similar activation barrier [369–373]. 1,5- and 2,5-sorbitans yield mainly humins. Without catalyst, the maximum yield of 1,4-sorbitan in hot liquid water increased with decreasing reaction temperature (for instance, 80% at 227 °C and 57% at 317 °C) [372, 374]. The dehydration of 1,4-sorbitan to isosorbide requires higher reaction temperatures and longer reaction times.

The main achievements in isosorbide synthesis from D-sorbitol have been reviewed recently [11].

The reaction is usually conducted at atmospheric or autogenous pressure and moderate temperature (<200 °C). Solvent-free conditions are preferable in order to limit the formation of humins [375]. High yields (>60%) were obtained using various catalytic systems such as zeolites [375–377], sulfonic acid resins [375, 378, 379], sulfonated carbon [380], sulfonated acid-modified mesoporous-silicas [373, 381], sulfated metal oxides [382–385], phosphates [372, 386, 387], supported heteropolyacids [388], and other acidic catalysts [371, 389]. Such variety indicates that the acidic properties (nature, number, and strength) of the catalysts are not critical even if the reaction is favored by moderate and strong BAS [369, 389]. The activity of Lewis acidic catalysts depends on which metal is used and on the generation of BAS on the metal center [369]. Very strong Brønsted acidity leads to catalyst deactivation [372].

Finally, the key role of the hydrophilic/hydrophobic balance on catalytic activity has been underlined in several papers [375–377, 390]. It was highlighted for both superhydrophobic mesoporous polymer-based acid catalysts and for hydrophilic sulfonic acid-functionalized microbead silicas loaded with 60.5% of (3-mercaptopropyl)trimethoxysilane, which afforded isosorbide yields of 88% and 84%, respectively [373, 379].

7.5 Conclusions and Outlook

Numerous catalytic processes have been developed for the preparation of chemicals from a set of biomass-derived oxygenates. A wide range of important intermediates and end-products can be obtained, which are used as green solvents, precursors for chemical, pharmaceutical, food and cosmetic industries, commodity chemicals, and current or new building blocks for polymers.

The catalysts developed use a variety of catalytic functions, including acid sites, base sites, metal hydrogenation/dehydrogenation/hydrogenolysis sites, etc., to effect selective transformations. The porosity controlling the access of reactants to the active sites, the interactive functions at the surface, and the hydrophobicity provided by the support material may all exert great influence on catalyst performance. The reaction parameters of the processes also have an important effect. Intensive research is still needed for heterogeneous catalysis to be able to compete – in terms of product yields and cost competitiveness – with long-standing petrochemical processes. For the aforementioned transformations employing noble metal catalysts, efforts should be directed to developing and employing catalysts based on earth-abundant and inexpensive materials. However, much attention should still be focused on the stability of the catalysts under the harsh reaction conditions involved and the prospect of catalyst recycling. The adsorption behavior of oxygenated functional groups and water on the catalyst surface play a crucial role and must be taken into account. Particular emphasis must also be given to the effect of contaminants in the feed.

Much work has been done on the efficient transformation of different platform molecules. A major target for the foreseeable future should be two-step or multi-step transformations of biomass derivatives without isolation of the intermediate platform molecules in a single reactor, e.g. formation and further conversion of FAL, HMF, LevA, or SOR. The challenge is then to carefully design efficient and robust catalysts possessing bi- or multi-functionality that allow multiple reactions to be combined, including isomerization, dehydration, selective hydrogenation, or selective oxidation in a one-pot process. This is a real challenge, since these reactions usually take place under different reaction conditions.

References

1 Werpy, T. and Petersen, G. (2004). *Top Value Added Chemicals From Biomass*, Results of Screening for Potential Candidates from Sugars and Synthesis Gas, vol. 1. Washington, DC: US Department of Energy 2004.

2 Bozell, J.J. and Petersen, G.R. (2010). Technology development for the production of biobased products from biorefinery carbohydrates – the US Department of Energy's "Top 10" revisited. *Green Chem.* 12: 539–554.

3 Zhou, C.H., Beltramini, J.N., Fan, Y.X., and Lu, G.Q. (2008). Chemoselective catalytic conversion of glycerol as a biorenewable source to valuable commodity chemicals. *Chem. Soc. Rev.* 37: 527–549.
4 Ciriminna, R., Pina, C.D., Rossi, M., and Pagliaro, M. (2014). Understanding the glycerol market. *Eur. J. Lipid Sci. Technol.* 116: 1432–1439.
5 Blanco, E., Loridant, S., and Pinel, C. (2016). Valorization of lactic acid and derivatives to acrylic acid derivatives: review of mechanistic studies. In: *Reaction Pathways and Mechanisms in Thermocatalytic Biomass Conversion II* (eds. M. Schlaf and Z.C. Zhang), 47–62. Springer.
6 Hayes, D.J., Fitzpatrick, S., Hayes, M.H.B., and Ross, J.R.H. (2006). The biofine process – production of levulinic acid, furfural, and formic acid from lignocellulosic feedstocks. In: *Biorefineries-Industrial Processes and Products: Status Quo and Future Directions*, vol. 1 (eds. B. Kamm, P.R. Gruber and M. Kamm), 139–164. Weinheim: Wiley-VCH.
7 Nghiem, N.P., Kleff, S., and Schwegmann, S. (2017). Succinic acid: technology development and commercialization. *Fermentation* 3: 26–40.
8 Jeya, M., Lee, K.M., Tiwari, M.K. et al. (2009). Isolation of a novel high erythritol-producing *Pseudozyma* tsukubaensis and scale-up of erythritol fermentation to industrial level. *Appl. Microbiol. Biotechnol.* 83: 225–231.
9 Corma, A., Iborra, S., and Velty, A. (2007). Chemical routes for the transformation of biomass into chemicals. *Chem. Rev.* 107 (6): 2411–2502.
10 Kobayashi, H., Ito, Y., Komanoya, T. et al. (2011). Synthesis of sugar alcohols by hydrolytic hydrogenation of cellulose over supported metal catalysts. *Green Chem.* 13: 326–333.
11 Dussenne, C., Delaunay, T., Wiatz, V. et al. (2017). Synthesis of isosorbide: an overview of challenging reactions. *Green Chem.* 19: 5332–5344.
12 Li, X., Jia, P., and Wang, T. (2016). Furfural: a promising platform compound for sustainable production of C4 and C5 chemicals. *ACS Catal.* 6: 7621–7640.
13 Xue, Z., Ma, M.-G., Li, Z., and Mu, T. (2016). Advances in the conversion of glucose and cellulose to 5-hydroxymethylfurfural over heterogeneous catalysts. *RSC Adv.* 6: 98874–98892.
14 Medlin, J.W. (2011). Understanding and controlling reactivity of unsaturated oxygenates and polyols on metal catalysts. *ACS Catal.* 1: 1284–1297.
15 Besson, M., Gallezot, P., and Pinel, C. (2014). Conversion of biomass into chemicals over metal catalysts. *Chem. Rev.* 114: 1827–1870.
16 Sankar, M., Dimitratos, N., Miedziak, P.J. et al. (2012). Designing bimetallic catalysts for a green and sustainable future. *Chem. Soc. Rev.* 41: 8099–8139.
17 Zope, B.N., Hibbits, D.D., Neurock, M., and Davis, R.J. (2010). Reactivity of the gold/water interface during selective oxidation catalysis. *Science* 330: 74–78.
18 Davis, S.E., Ide, M.S., and Davis, R.J. (2013). Selective oxidation of alcohols and aldehydes over supported metal nanoparticles. *Green Chem.* 15: 17–45.
19 Katryniok, B., Kimura, H., Skrzynska, E. et al. (2011). Selective catalytic oxidation of glycerol: perspectives for high value chemicals. *Green Chem.* 13: 1960–1979.

20 Villa, A., Dimitratos, N., Chan-Thaw, C.E. et al. (2015). Glycerol oxidation using gold-containing catalysts. *Acc. Chem. Res.* 48: 1403–1412.
21 Mäki-Arvela, P., Simakova, I.L., Salmi, T., and Murzin, D.Y. (2014). Production of lactic acid/lactates from biomass and their catalytic transformations to commodities. *Chem. Rev.* 114: 1909–1971.
22 Kimura, H., Kimura, A., Kokubo, I. et al. (1993). Palladium based multi-component catalytic systems for the alcohol to carboxylate oxidation reaction. *Appl. Catal., A* 95: 143–169.
23 Garcia, R., Besson, M., and Gallezot, P. (1995). Chemoselective catalytic oxidation of glycerol with air on platinum metals. *Appl. Catal., A* 127: 165–176.
24 Carrettin, S., McMorn, P., Johnston, P. et al. (2002). Selective oxidation of glycerol to glyceric acid using a gold catalyst in aqueous sodium hydroxide. *Chem. Commun.* 7: 696–697.
25 Carrettin, S., McMorn, P., Johnston, P. et al. (2003). Oxidation of glycerol using supported Pt, Pd and Au catalysts. *Phys. Chem. Chem. Phys.* 5: 1329–1336.
26 Porta, F. and Prati, L. (2004). Selective oxidation of glycerol to sodium glycerate with gold-on-carbon catalyst: an insight into reaction selectivity. *J. Catal.* 224: 397–403.
27 Villa, A., Wang, D., Su, D., and Prati, L. (2009). Gold sols as catalysts for glycerol oxidation: the role of the stabilizer. *ChemCatChem* 1: 510–514.
28 Villa, A., Wang, D., Veith, G.M. et al. (2013). Sol immobilization technique: a delicate balance between activity, selectivity and stability for gold catalysts. *Catal. Sci. Technol.* 3: 3036–3041.
29 Ketchie, W.C., Fang, Y.-L., Wong, M.S. et al. (2007). Influence of gold particle size on the aqueous-phase oxidation of carbon monoxide and glycerol. *J. Catal.* 250: 94–101.
30 Demirel-Gülen, S., Lucas, M., and Claus, P. (2005). Liquid phase oxidation of glycerol over carbon supported gold catalysts. *Catal. Today* 102–103: 166–172.
31 Rodrigues, E.G., Carabineiro, S.A.C., Delgado, J.J. et al. (2012). Gold supported on carbon nanotubes for the selective oxidation of glycerol. *J. Catal.* 285: 83–91.
32 Gil, S., Marchena, M., Sanchez-Silva, L. et al. (2011). Effect of operation conditions on the selective oxidation of glycerol with catalysts based on Au supported on carbonaceous materials. *Chem. Eur. J.* 178: 423–435.
33 Prati, L., Villa, A., Chan-Thaw, C.E. et al. (2011). Gold catalyzed liquid phase oxidation of alcohol: the issue of selectivity. *Faraday Discuss.* 152: 353–365.
34 Rodrigues, E.G., Delgado, J.J., Chen, X. et al. (2012). Selective oxidation of glycerol catalyzed by gold supported on multiwalled carbon nanotubes with different surface chemistries. *Ind. Eng. Chem. Res.* 51: 15884–15894.
35 Demirel, S., Kern, P., Lucas, M., and Claus, P. (2007). Oxidation of mono- and polyalcohols with gold: comparison of carbon and ceria supported catalyst. *Catal. Today* 122: 292–300.
36 Prati, L., Villa, A., Lupini, A.R., and Veith, G.M. (2012). Gold on carbon: one billion catalysts under a single label. *Phys. Chem. Chem. Phys.* 14: 2969–2978.

37 Wang, D., Villa, A., Su, D. et al. (2013). Carbon-supported gold nanocatalysts: shape effect in the selective glycerol oxidation. *ChemCatChem* 5: 2717–2723.
38 Bianchi, C., Canton, P., Dimitratos, N. et al. (2005). Selective oxidation of glycerol with oxygen unsing mono and bimetallic catalysts based on Au, Pd and Pt metals. *Catal. Today* 102: 203–212.
39 Dimitratos, N., Porta, F., and Prati, L. (2005). Au, Pd (mono and bimetallic) catalysts supported on graphite using the immobilisation method – synthesis and catalytic testing for liquid phase oxidation of glycerol. *Appl. Catal., A* 291: 210–214.
40 Villa, A., Campione, C., and Prati, L. (2007). Bimetallic gold/palladium catalysts for the selective liquid phase oxidation of glycerol. *Catal. Lett.* 115: 133–136.
41 Ketchie, W., Murayama, M., and Davies, R.J. (2007). Selective oxidation of glycerol over carbon-supported AuPd catalysts. *J. Catal.* 250: 264–273.
42 Wang, D., Villa, A., Porta, F., and Su, D.S. (2008). Bimetallic gold/palladium catalysts: correlation between nanostructure and synergistic effects. *J. Phys. Chem. C* 112: 8617–8622.
43 Dimitratos, N., Lopez-Sanchez, J.A., Anthonykutty, J.M. et al. (2009). Oxidation of glycerol using gold–palladium alloy-supported nanocrystals. *Phys. Chem. Chem. Phys.* 11: 4952–4961.
44 Kaminski, P., Ziolek, M., and van Bokhoven, J.A. (2017). Mesoporous cerium-zirconium oxides modified with gold and copper – synthesis, characterization and performances in selective oxidation of glycerol. *RSC Adv.* 7: 7801–7819.
45 Liang, D., Gao, J., Wang, J. et al. (2009). Selective oxidation of glycerol in a base-free aqueous solution over different sized Pt catalysts. *Catal. Commun.* 10: 1586–1590.
46 Gao, J., Liang, D., Chen, P. et al. (2009). Oxidation of glycerol with oxygen in a base-free aqueous solution over Pt/AC and Pt/MWNTs catalysts. *Catal. Lett.* 130: 185–191.
47 Zhang, M.Y., Nie, R.F., Wang, L. et al. (2015). Selective oxidation of glycerol over carbon nanofibers supported Pt catalysts in a base-free aqueous solution. *Catal. Commun.* 59: 5–9.
48 Tsuji, A., Rao, K.T.V., Nishimura, S. et al. (2011). Selective oxidation of glycerol by using a hydrotalcite-supported platinum catalysts under atmospheric oxygen pressure in water. *ChemSusChem* 4: 542–548.
49 Hamid, S.B.A., Basiron, N., Yehye, W.A. et al. (2016). Nanoscale Pd-based catalysts for selective oxidation of glycerol with molecular oxygen: structure-activity correlations. *Polyhedron* 120: 124–133.
50 Chen, S., Qi, P., Chen, J., and Yuan, Y. (2015). Platinum nanoparticles supported on N-doped carbon nanotubes for the selective oxidation of glycerol to glyceric acid in a case-free aqueous solution. *RSC Adv.* 5: 31566–31574.
51 Wang, F.F., Shao, S., Liu, C.L. et al. (2015). Selective oxidation of glycerol over Pt supported on mesoporous carbon nitride in base-free aqueous solution. *Chem. Eng. J.* 264: 336–343.

52 Prati, L., Spontoni, P., and Gaiassi, A. (2009). From renewable to fine chemicals through selective oxidation: the case of glycerol. *Top. Catal.* 52: 288–296.
53 Villa, A., Veith, G.M., and Prati, L. (2010). Selective oxidation of glycerol under acidic conditions using gold catalysts. *Angew. Chem. Int. Ed.* 49: 4499–4502.
54 Villa, A., Campisi, S., Mohammed, K.M.H. et al. (2015). Tailoring the selectivity of glycerol oxidation by tuning the acid–base properties of Au catalysts. *Catal. Sci. Technol.* 5: 1126–1132.
55 Tongsakul, D., Nishimura, S., and Ebitani, K. (2013). Platinum/gold alloy nanoparticles-supported hydrotalcite catalyst for selective aerobic oxidation of polyols in base-free aqueous solution at room temperature. *ACS Catal.* 3: 2199–2207.
56 Brett, G.L., He, Q., Hammond, C. et al. (2011). Selective oxidation of glycerol by highly active bimetallic catalysts at ambient temperature under base-free conditions. *Angew. Chem. Int. Ed.* 50: 10136–10139.
57 Xu, C., Du, Y., Li, C. et al. (2015). Insight into effect of acid/base nature of supports on selectivity of glycerol oxidation over supported Au–Pt bimetallic catalysts. *Appl. Catal., B* 164: 334–343.
58 Fu, J., He, Q., Miedziak, P.J. et al. (2018). The role of $Mg(OH)_2$ in the so-called "base-free" oxidation of glycerol with AuPd catalysts. *Chem. Eur. J.* 24 https://doi.org/10.1002/chem.201704151.
59 Liang, D., Gao, J., Wang, J. et al. (2011). Bimetallic Pt–Cu catalysts for glycerol oxidation with oxygen in a base-free aqueous solution. *Catal. Commun.* 12: 1059–1062.
60 Zhang, M., Shi, J., Ning, W., and Hou, Z. (2017). Reduced graphene oxide decorated with PtCo bimetallic nanoparticles: facile fabrication and application for base-free oxidation of glycerol. *Catal. Today* 298: 234–240.
61 Purushothaman, R.K.P., van Haveren, J., Melian-Cabrera, I. et al. (2014). Base-free one-pot chemocatalytic conversion of glycerol to methyl lactate using supported gold catalysts. *ChemSusChem* 7: 1140–1147.
62 Gil, S., Marchena, M., Fernandez, C.M. et al. (2013). Catalytic oxidation of crude glycerol using catalysts based on Au supported on carbonaceous materials. *Appl. Catal., A* 450: 189–203.
63 Skrzynska, E., Zaid, S., Girardon, J.S. et al. (2015). Catalytic behavior of four different supported noble metals in the crude glycerol oxidation. *Appl. Catal., A* 499: 89–100.
64 Kimura, H., (1996). Production of ketomalonic acid. JP Patent 1,994,315,623, to Kao Corp Japan, filed 25 November 1994 and issued 11 June 1996.
65 Fordham, P., Besson, M., and Gallezot, P. (1995). Selective catalytic oxidation of glyceric acid to tartronic and hydroxypyruvic acid. *Appl. Catal., A* 133: L179–L184.
66 Kimura, H., Imanaka, T., and Yokota, Y. (1994). Production of tartronic acid. JP Patent 199,395,253, to Kao Corp Japan, filed 23 March 1993 and issued 4 October 1994.
67 Zope, B.N., Davis, S.E., and Davies, R.J. (2012). Influence of reaction conditions on diacid formation during Au-catalyzed oxidation of glycerol and hydroxymethylfurfural. *Top. Catal.* 55: 24–32.

68 Yudha, S., Dhital, S.R., and Sakurai, H. (2011). Gold- and gold-palladium/poly(1-vinylpyrrolidin-2-one) nanoclusters as quasi-homogeneous catalyst for aerobic oxidation of glycerol. *Tetrahedron Lett.* 52: 2633–2637.

69 Liang, W., Wei, Z., Shangjing, Z. et al. (2012). Mg–Al oxides supported bimetallic Au–Pd nanoparticles with superior catalytic properties in aerobic oxidation of benzyl alcohol and glycerol. *Chin. J. Chem.* 30: 2189–2197.

70 Villa, A., Wang, D., Veith, G.M., and Prati, L. (2012). Bismuth as a modifier of Au–Pd catalyst: enhancing selectivity in alcohol oxidation by suppressing parallel reaction. *J. Catal.* 292: 73–80.

71 Jin, X., Zhao, M., Yan, W. et al. (2016). Anisotropic growth of PtFe nanoclusters induced by lattice-mismatch: efficient catalysts for oxidation of biopolyols to carboxylic acid. *J. Catal.* 337: 272–283.

72 Jin, X., Yan, H., Zeng, C. et al. (2017). Phase transformed PtFe nanocomposites show enhanced catalytic performances in oxidation of glycerol to tartronate. *Ind. Eng. Chem. Res.* 56: 13157–13164.

73 Cai, J.H., Ma, H., Zhang, J.J. et al. (2014). Catalytic oxidation to tartronic acid over Au/HY catalyst under mild conditions. *Chin. J. Catal.* 35: 1653–1660.

74 Jin, X., Zhao, M., Zeng, C. et al. (2016). Oxidation of glycerol to dicarboxylic acids using cobalt catalysts. *ACS Catal.* 6: 4576–4583.

75 Kimura, H., Tsuto, K., Wakisaka, T. et al. (1993). Selective oxidation of glycerol on a platinum–bismuth catalyst. *Appl. Catal., A* 96: 217–228.

76 Hu, W., Knight, D., Lowry, B., and Varma, A. (2010). Selective oxidation of glycerol to dihydroxyacetone over Pt–Bi/C catalyst: optimization of catalyst and reaction conditions. *Ind. Eng. Chem. Res.* 49: 10876–10882.

77 Kimura, H. (1993). Selective oxidation of glycerol on a platinum–bismuth catalyst using a fixed bed reactor. *Appl. Catal., A* 105: 147–158.

78 Wörz, N., Brandner, A., and Claus, P. (2010). Platinum bismuth-catalyzed of glycerol: kinetics and the origin of selective deactivation. *J. Phys. Chem.* 114: 1164–1172.

79 Xiao, Y., Greeley, J., Varma, A. et al. (2017). An experimental and theoretical study of glycerol oxidation to 1,3-dihydroxyacetone over bimetallic PtBi catalysts. *AIChE J.* 63: 705–715.

80 Villa, A., Campisi, S., Chan-Thaw, C.E. et al. (2015). Bismuth modified bimetallic catalysts for dihydroxyacetone production. *Catal. Today* 249: 103–108.

81 Nie, R., Liang, D., Shen, L. et al. (2012). Selective oxidation of glycerol with oxygen in base-free solution over MWCNTs supported PtSb alloy nanoparticles. *Appl. Catal., B* 127: 212–220.

82 Hirasawa, S., Watanabe, H., Kizuka, T. et al. (2013). Performance, structure, and mechanism of Pd–Ag alloy catalyst for selective oxidation of glycerol. *J. Catal.* 300: 205–216.

83 Liu, S.S., Sun, K.Q., and Xu, B.Q. (2014). Specific selectivity of Au-catalyzed oxidation of glycerol and other C_3-polyols in water without the presence of a base. *ACS Catal.* 4: 2226–2230.

84 Yuan, Z., Gao, Z., and Xu, B.Q. (2015). Acid–base property of the supporting materials controls the selectivity of Au catalyst for glycerol oxidation in base-free water. *Chin. J. Catal.* 36: 1543–1551.
85 Kimura, H. (1996). Polyketomalonates by catalytic oxidation of glycerol over a CeBiPt catalyst. II. *J. Polym. Sci., Part A: Polym. Chem.* 34: 3607–3614.
86 Fordham, P., Besson, M., and Gallezot, P. (1997). Catalytic oxidation with air of tartronic acid to mesoxalic acid on bismuth-promoted platinum. *Catal. Lett.* 46: 195–199.
87 Taarning, E., Madsen, A.T., Marchetti, J.M. et al. (2008). Oxidation of glycerol and propanediols in methanol over heterogeneous gold catalysts. *Green Chem.* 10: 408–414.
88 Sankar, M., Dimitratos, N., Knight, D.W. et al. (2009). Oxidation of glycerol to glycolate by using supported gold and palladium nanoparticles. *ChemSusChem* 2: 1145–1151.
89 Skrzynska, E., Zaid, S., Addad, A. et al. (2016). Performance of Ag/Al_2O_3 catalysts in the liquid phase oxidation of glycerol – effect of preparation method and reaction conditions. *Catal. Sci. Technol.* 6: 3182–3196.
90 Zaid, S., Skrynska, E., Addad, A. et al. (2017). Development of silver based catalysts promoted by noble metal M (M = Au, Pd or Pt) for glycerol oxidation in liquid phase. *Top. Catal.* 60: 1072–1081.
91 Dodekatos, G. and Harun, T. (2017). Effect of post-treatment on structure and catalytic activity of CuCo-based materials for glycerol oxidation. *ChemCatChem* 9: 610–619.
92 Takagaki, A., Tsuji, A., Nishimura, S., and Ebitani, K. (2011). Genesis of catalytically active gold nanoparticles supported on hydrotalcite for base-free selective oxidation of glycerol in water with molecular oxygen. *Chem. Lett.* 42: 150–152.
93 Razali, N. and Abdullah, A.Z. (2017). Production of lactic acid from glycerol via chemical conversion using solid catalyst: a review. *Appl. Catal., A* 543: 234–246.
94 Shen, Y., Zhang, S., Li, H. et al. (2010). Efficient synthesis of lactic acid by aerobic oxidation of glycerol on $Au–Pt/TiO_2$ catalysts. *Chem. Eur. J.* 16: 7368–7371.
95 Lakshmanan, P., Upare, P.P., Le, N.T. et al. (2013). Facile synthesis of CeO_2-supported gold nanoparticle catalysts for selective oxidation of glycerol into lactic acid. *Appl. Catal., A* 468: 260–268.
96 Purushothaman, R.K.P., van Haveren, J., van Es, D.S. et al. (2014). An efficient one pot conversion of glycerol to lactic acid using bimetallic gold–platinum catalysts on a nanocrystalline CeO_2 support. *Appl. Catal., B* 147: 92–100.
97 Li, Y., Chen, S., Xu, J. et al. (2014). Ni promoted Pt and Pd catalysts for glycerol oxidation to lactic acid. *CLEAN – Soil, Air, Water.* 42: 1140–1144.
98 Zhang, C., Wang, T., Liu, X., and Ding, Y. (2016). Cu-promoted Pt/activated carbon catalyst for glycerol oxidation to lactic acid. *J. Mol. Catal. A: Chem.* 424: 91–97.

99 Palacio, R., Torres, S., Lopez, D., and Hernandez, D. (2018). Selective glycerol conversion to lactic acid on Co_3O_4/CeO_2 catalysts. *Catal. Today* 302: 196–202.

100 Purushothaman, R.K.P., van Haveren, J., Mayoral, A. et al. (2014). Exploratory catalyst screening studies on the base free conversion of glycerol to lactic acid and glyceric acid in water using bimetallic Au–Pt nanoparticles on acidic zeolites. *Top. Catal.* 57: 1445–1453.

101 Xu, J., Zhang, H., Zhao, Y. et al. (2013). Selective oxidation of glycerol to lactic acid under acidic conditions using $AuPd/TiO_2$ catalyst. *Green Chem.* 15: 1520–1525.

102 Cho, H.J., Chang, C., and Fan, W. (2014). Base free, one-pot synthesis of lactic acid from glycerol using a bifunctional Pt/Sn–MFI catalyst. *Green Chem.* 16: 3428–3433.

103 Komanoya, T., Suzuki, A., Nakajima, K. et al. (2016). A combined catalyst of Pt nanoparticles and TiO_2 with water-tolerant Lewis acid sites for one-pot conversion of glycerol to lactic acid. *ChemCatChem* 8: 1094–1099.

104 Tao, M., Yi, X., Delidovich, I. et al. (2015). Heteropolyacid-catalyzed oxidation of glycerol into lactic acid under mild base-free conditions. *ChemSusChem* 8: 4195–4201.

105 Tao, M., Zhang, D., Xeng, X. et al. (2016). Lewis-acid-promoted catalytic cascade conversion of glycerol to lactic acid by polyoxometalates. *Chem. Commun.* 52: 3332–3335.

106 Tao, M. et al. (2016). Designation of highly efficient catalysts for one pot conversion of glycerol to lactic acid. *Sci. Rep.* 6: 29840.

107 Lu, T., Fu, X., Zhou, L. et al. (2017). Promotion effect of Sn on Au/Sn-USY catalysts for one-pot conversion of glycerol to methyl lactate. *ACS Catal.* 7: 7274–7284.

108 Ftouni, J., Villandier, N., Auneau, F. et al. (2015). From glycerol to lactic acid under inert conditions in the presence of platinum-based catalysts: the influence of support. *Catal. Today* 257: 267–273.

109 Yang, G.-Y., Ke, Y.-H., Ren, H.-F. et al. (2016). The conversion of glycerol to lactic acid catalyzed by ZrO_2-supported CuO catalysts. *Chem. Eng. J.* 283: 759–767.

110 van Putten, R.J., van der Waal, J.C., de Jong, E. et al. (2013). Hydroxymethylfurfural, a versatile platform chemical made from renewable resources. *Chem. Rev.* 113: 1499–1597.

111 Zhang, J., Li, J., Tang, Y. et al. (2015). Advances in catalytic production of bio-based polyester monomer 2,5-furandicarboxylic acid derived from lignocellulosic biomass. *Carbohydr. Polym.* 130: 420–428.

112 Zhang, Z. and Deng, K. (2015). Recent advances in the catalytic synthesis of 2,5-furandicarboxylic acid and its derivatives. *ACS Catal.* 5: 6529–6544.

113 Zhang, Z. and Huber, G.W. (2018). Catalytic oxidation of carbohydrates into organic acids and furan chemicals. *Chem. Soc. Rev.* 47: 1351–1390.

114 Verdeguer, P., Merat, N., and Gaset, A. (1993). Oxydation catalytique du HMF en acide 2,5-furane dicarboxylique. *J. Mol. Catal.* 85: 327–344.

115 Taarning, E., Nielsen, I.S., Egeblad, K. et al. (2008). Chemicals from renewables: aerobic oxidation of furfural and hydroxymethylfurfural over gold catalysts. *ChemSusChem* 1: 75–78.
116 Lilga, M.A., Hallen, R.T., and Gray, M. (2010). Production of oxidized derivatives of 5-hydroxymethylfurfural (HMF). *Top. Catal.* 53: 1264–1269.
117 Davis, S.E., Houk, L.R., Tamargo, E.C. et al. (2011). Oxidation of 5-hydroxymethylfurfural over supported Pt, Pd and Au catalysts. *Catal. Today* 160: 55–60.
118 Ait Rass, H., Essayem, N., and Besson, M. (2013). Selective aqueous phase oxidation of 5-hydroxymethylfurfural to 2,5-furandicarboxylic acid over Pt/C catalysts: influence of the base and effect of bismuth promotion. *Green Chem.* 15: 2240–2251.
119 Niu, W.Q., Wang, D., Yang, G.H. et al. (2014). Pt nanoparticles loaded on reduced graphene oxide as an effective catalyst for the direct oxidation of 5-hydroxymethylfurfural (HMF) to produce 2,5-furandicarboxylic acid (HMFCA) under mild conditions. *Bull. Chem. Soc. Jpn.* 87: 1124–1129.
120 Chen, C., Li, X., Wang, L. et al. (2017). Highly porous nitrogen- and phosphorus-codoped graphene: an outstanding support for Pd catalysts to oxidize 5-hydroxymethylfurfural into 2,5-furandicarboxylic acid. *ACS Sustainable Chem. Eng.* 5: 11300–11306.
121 Zhang, Y., Xue, Z., Wang, J. et al. (2016). Controlled deposition of Pt nanoparticles on Fe_3O_4@carbon microspheres for efficient oxidation of 5-hydroxymethylfurfural. *RSC Adv.* 6: 51229–51237.
122 Mei, N., Liu, B., Zheng, J.D. et al. (2015). A novel magnetic palladium catalyst for the mild aerobic oxidation of 5-hydroxymethylfurfural into 2,5-furandicarboxylic acid in water. *Catal. Sci. Technol.* 5: 3194–3202.
123 Donoeva, B., Masoud, N., and de Jongh, P.E. (2017). Carbon support surface effects in the gold-catalyzed oxidation of 5-hydroxymethylfurfural. *ACS Catal.* 7: 4581–4591.
124 Kerdi, F., Ait Rass, H., Pinel, C. et al. (2015). Evaluation of surface properties and pore structure of carbon on the activity of supported Ru catalysts in the aqueous-phase aerobic oxidation of HMF to FDCA. *Appl. Catal., A* 506: 206–219.
125 Ait Rass, H., Essayem, N., and Besson, M. (2015). Selective aerobic oxidation of 5-HMF into 2,5-furan dicarboxylic acid with Pt catalysts supported on TiO_2- and ZrO_2-based supports. *ChemSusChem* 8: 1206–1217.
126 Casanova, O., Iborra, S., and Corma, A. (2009). Biomass into chemicals: aerobic oxidation of 5-hydroxymethyl-2-furfural into 2,5-furandicarboxylic acid with gold nanoparticle catalysts. *ChemSusChem* 2: 1138–1144.
127 Miao, Z.Z., Zhang, Y.B., Pan, X.Q. et al. (2015). Superior catalytic performance of $Ce_{1-x}Bi_xO_{2-\delta}$ solid solution and Au/$Ce_{1-x}Bi_xO_{2-\delta}$ for 5-hydroxymethylfurfural conversion in alkaline aqueous solution. *Catal. Sci. Technol.* 5: 1314–1322.
128 Gong, W., Zheng, H., and Ji, P. (2017). Platinum deposited on cerium coordination polymer for catalytic oxidation of hydroxymethylfurfural producing 2,5-furandicarboxylic acid. *RSC Adv.* 7: 34776–34782.

129 Cai, J.H., Ma, H., Zhang, J.J. et al. (2013). Gold nanoclusters confined in a supercage of Y zeolite for aerobic oxidation of HMF under mild conditions. *Chem. Eur. J.* 19: 14215–14223.
130 Villa, A., Schiavoni, M., Campisi, S. et al. (2013). Pd-modified Au on carbon as an effective and durable catalyst for the direct oxidation of HMF to 2,5-furandicarboxylic acid. *ChemSusChem* 6: 609–612.
131 Antonyraj, C.A., Huyn, N.T.T., Park, S.-K. et al. (2017). Basic-anion-exchange resin (AER)-supported AuPd alloy nanoparticles for the oxidation of 5-hydroxymethyl-2-furfural (HMF) into 2,5-furan dicarboxylic acid (FDCA). *Appl. Catal., A* 547: 230–236.
132 Wang, Q., Hou, W., Li, S. et al. (2017). Hydrophilic mesoporous poly(ionic liquid)-supported Au–Pd alloy nanoparticles towards aerobic oxidation of 5-hydroxymethylfurfural to 2,5-furandicarboxylic acid under mild conditions. *Green Chem.* 19: 3820–3830.
133 Pasini, T., Piccinini, M., Blosi, M. et al. (2011). Selective oxidation of 5-hydroxymethyl-2-furfural using supported gold–copper nanoparticles. *Green Chem.* 13: 2091–2099.
134 Albonetti, S., Pasini, T., Lolli, A. et al. (2012). Selective oxidation of 5-hydroxymethyl-2-furfural over TiO_2-supported gold-copper catalysts prepared from preformed nanoparticles: effect of Au/Cu ratio. *Catal. Today* 195: 120–126.
135 Gupta, K., Rai, R.K., Dwivedi, A.D., and Singh, S.K. (2017). Catalytic aerial oxidation of biomass-derived furans to furan carboxylic acids in water over bimetallic nickel–palladium alloy nanoparticles. *ChemCatChem* 9: 2760–2767.
136 Han, X., Li, C., Liu, X. et al. (2017). Selective oxidation of 5-hydroxymethylfurfural to 2,5-furandicarboxylic acid over MnO_x–CeO_2 composite catalysts. *Green Chem.* 19: 996–1004.
137 Neatu, F., Marin, R.S., Florea, M. et al. (2015). Selective oxidation of 5-hydroxymethylfurfural over non-precious metal heterogeneous catalysts. *Appl. Catal., B* 180: 751–757.
138 Saha, B., Gupta, D., Abu-Omar, M.M. et al. (2013). Porphyrin-based porous organic polymer-supported iron(III) catalyst for efficient aerobic oxidation of 5-hydroxymethylfurfural into 2,5-furandicarboxylic acid. *J. Catal.* 299: 316–320.
139 Siankevich, S., Savoglidis, G., Fei, Z. et al. (2014). A novel platinum nanocatalyst for the oxidation of 5-hydroxymethylfurfural into 2,5-furandicarboxylic acid under mild conditions. *J. Catal.* 315: 67–74.
140 Sito, B., Schneider, M., Pohl, M.-M. et al. (2014). Synthesis, characterization, and application of PVP-Pd NP in the aerobic oxidation of 5-hydroxymethylfurfural (HMF). *Catal. Lett.* 144: 498–506.
141 Gupta, N.K., Nishimura, S., Takagaki, A., and Ebitani, K. (2011). Hydrotalcite-supported gold-nanoparticle-catalyzed highly efficient base-free aqueous oxidation of 5-hydroxymethylfurfural into 2,5-furandicarboxylic acid under atmospheric oxygen pressure. *Green Chem.* 13: 824–827.
142 Wang, Y.B., Yu, K., Lei, D. et al. (2016). Basicity-tuned hydrotalcite-supported Pd catalysts for aerobic oxidation of

5-hydroxymethyl-2-furfural under mild conditions. *ACS Sustainable Chem. Eng.* 4: 4752–4761.

143 Ardemani, L., Cibin, G., Dent, A.J. et al. (2015). Solid base catalysed 5-HMF oxidation to 2,5-FDCA over Au/hydrotalcites: fact or fiction. *Chem. Sci.* 6: 4940–4945.

144 Han, X., Geng, L., Guo, Y. et al. (2016). Base-free aerobic oxidation of 5-hydroxymethylfurfural to 2,5-furandicarboxylic acid over Pt/C–O–Mg catalyst. *Green Chem.* 18: 1597–1604.

145 Gao, T., Gao, T., and Cao, Q. (2017). Base-free aerobic oxidation of 5-hydroxymethylfurfural to 2,5-furandicarboxylic acid in water by hydrotalcite-activated carbon composite supported gold catalysts. *Mol. Catal.* 439: 171–179.

146 Zhou, C.M., Deng, W.P., Wan, X.Y. et al. (2015). Functionalized carbon nanotubes for biomass conversion: the base-free aerobic oxidation of 5-hydroxymethylfurfural to 2,5-furandicarboxylic acid over platinum supported on a carbon nanotube catalyst. *ChemCatChem* 7: 2853–2863.

147 Wan, X., Zhou, C., Chen, J. et al. (2014). Base-free aerobic oxidation of 5-hydroxymethyl-furfural to 2,5-furandicarboxylic acid in water catalyzed by functionalized carbon nanotube-supported Au–Pd alloy nanoparticles. *ACS Catal.* 4: 2175–2185.

148 Han, X., Li, C., Guo, Y. et al. (2016). N-doped carbon supported Pt catalysts for base-free oxidation of 5-hydroxymethylfurfural to 2,5-furandicarboxylic acid. *Appl. Catal., A* 526: 1–8.

149 Ramirez-Barria, C., Lopez-Olmo, C., Guerrero-Ruiz, A., and Rodriguez-Ramos, I. (2017). Direct catalytic effect of nitrogen functional groups exposed on graphenic materials when acting cooperatively with Ru nanoparticles. *RSC Adv.* 7: 44568–44577.

150 Gao, Z., Xie, R., Fan, G. et al. (2017). Highly efficient and stable bimetallic AuPd over La-doped Ca–Mg–Al layered double hydroxide for base-free aerobic oxidation of 5-hydroxymethylfurfural in water. *ACS Sustainable Chem. Eng.* 5: 5852–5861.

151 Casanova, O., Iborra, S., and Corma, A. (2009). Biomass into chemicals: one-pot-base free oxidative esterification of 5-hydroxymethyl-2-furfural into 2,5-dimethylfuroate with gold on nanoparticulated ceria. *J. Catal.* 265: 109–116.

152 Gandini, A. and Belgacem, M.N. (1997). Furans in polymer chemistry. *Prog. Polym. Sci.* 22: 1203–1379.

153 Neatu, F., Petrea, N., Petre, R. et al. (2016). Oxidation of 5-hydroxymethyl furfural to 2,5-diformylfuran in aqueous media over heterogeneous manganese based catalysts. *Catal. Today* 278: 66–73.

154 Yan, Y., Li, K., Zhao, J. et al. (2017). Nanobelt-arrayed vanadium oxide hierarchical microspheres as catalysts for selective oxidation of 5-hydroxymethylfurfural toward 2,5-diformylfuran. *Appl. Catal., B* 207: 358–365.

155 Zhang, Z., Liu, B., Lv, K. et al. (2014). Aerobic oxidation of biomass derived 5-hydroxymethylfurfural into 5-hydroxymethyl-2-furancarboxylic acid

catalyzed by a montmorillonite K-10 clay immobilized molybdenum acetylacetonate complex. *Green Chem.* 16: 2762–2770.
156 Ventura, M., Dibenedetto, A., and Aresta, M. (2018). Heterogeneous catalysts for the selective aerobic oxidation of 5-hydroxymethylfurfural to added value products in water. *Inorg. Chim. Acta* 470: 11–21.
157 Ventura, M., Aresta, M., and Dibenedetto, A. (2016). Selective aerobic oxidation of 5-(hydroxymethyl)furfural to 5-formyl-2-furancarboxylic acid in water. *ChemSusChem* 9: 1096–1100.
158 Sun, D., Yamada, Y., Sato, S., and Ueda, W. (2016). Glycerol hydrogenolysis into useful C3 chemicals. *Appl. Catal., B* 193: 75–92.
159 Wang, Y., Zhou, J., and Guo, X. (2015). Catalytic hydrogenolysis of glycerol to propanediols: a review. *RSC Adv.* 5 (91): 74611–74628.
160 Marinas, A., Bruijnincx, P., Ftouni, J. et al. (2015). Sustainability metrics for a fossil- and renewable-based route for 1,2-propanediol production: a comparison. *Catal. Today* 239: 31–37.
161 Martin, A., Armbruster, U., Gandarias, I., and Arias, P.L. (2013). Glycerol hydrogenolysis into propanediols using in situ generated hydrogen – a critical review. *Eur. J. Lipid Sci. Technol.* 115 (1): 9–27.
162 Nakagawa, Y. and Tomishige, K. (2011). Heterogeneous catalysis of the glycerol hydrogenolysis. *Catal. Sci. Technol.* 1 (2): 179–190.
163 Montassier, C., Giraud, D., Barbier, J., and Boitiaux, J.P. (1989). Polyol transformation by liquid phase heterogeneous catalysis over metals. *Bull. Soc. Chim. Fr.* 2: 148–155.
164 van Ryneveld, E., Mahomed, A.S., van Heerden, P.S. et al. (2011). A catalytic route to lower alcohols from glycerol using Ni-supported catalysts. *Green Chem.* 13 (7): 1819–1827.
165 Amada, Y., Shinmi, Y., Koso, S. et al. (2011). Reaction mechanism of the glycerol hydrogenolysis to 1,3-propanediol over $Ir–ReO_x/SiO_2$ catalyst. *Appl. Catal., B* 105 (1–2): 117–127.
166 Tomishige, K., Nakagawa, Y., and Tamura, M. (2017). Selective hydrogenolysis and hydrogenation using metal catalysts directly modified with metal oxide species. *Green Chem.* 19: 2876–2924.
167 Arundhathi, R., Mizugaki, T., Mitsudome, T. et al. (2013). Highly selective hydrogenolysis of glycerol to 1,3-propanediol over a boehmite-supported platinum/tungsten catalyst. *ChemSusChem* 6 (8): 1345–1347.
168 Ten Dam, J. and Hanefeld, U. (2011). Renewable chemicals: dehydroxylation of glycerol and polyols. *ChemSusChem* 4 (8): 1017–1034.
169 Said, A., Da Silva Perez, D., Perret, N. et al. (2017). Selective C–O hydrogenolysis of erythritol over supported $Rh–ReO_x$ catalysts in the aqueous phase. *ChemCatChem* 9 (14): 2768–2783.
170 Ota, N., Tamura, M., Nakagawa, Y. et al. (2015). Hydrodeoxygenation of vicinal OH groups over heterogeneous rhenium catalyst promoted by palladium and ceria support. *Angew. Chem. Int. Ed.* 54 (6): 1897–1900.
171 Montassier, C., Ménézo, J.C., Hoang, L.C. et al. (1991). Aqueous polyol conversions on ruthenium and on sulfur-modified ruthenium. *J. Mol. Catal.* 70 (1): 99–101.

172 Shinmi, Y., Koso, S., Kubota, T. et al. (2010). Modification of Rh/SiO_2 catalyst for the hydrogenolysis of glycerol in water. *Appl. Catal., B* 94 (3–4): 318–326.

173 Jin, X., Shen, J., Yan, W. et al. (2015). Sorbitol hydrogenolysis over hybrid Cu/CaO–Al_2O_3 catalysts: tunable activity and selectivity with solid base incorporation. *ACS Catal.* 5 (11): 6545–6558.

174 Blanc, B., Bourrel, A., Gallezot, P. et al. (2000). Starch-derived polyols for polymer technologies: preparation by hydrogenolysis on metal catalysts. *Green Chem.* 2 (2): 89–91.

175 Sanborn, A. and Binder, T. (2015). Synthesis of R-glucosides sugar alcohols, reduced sugar alcohols and furan derivatives of reduced sugar alcohols. US Patent US2017/0,121,258 A1, filed 10 April 2014 and issued 4 May 2017.

176 Chen, K., Tamura, M., Yuan, Z. et al. (2013). One-pot conversion of sugar and sugar polyols to *n*-alkanes without C–C dissociation over the Ir–ReO_x/SiO_2. Catalyst combined with H-ZSM-5. *ChemSusChem* 6 (4): 613–621.

177 Riviere, M., Perret, N., Cabiac, A. et al. (2017). Xylitol hydrogenolysis over ruthenium-based catalysts: effect of alkaline promoters and basic oxide-modified catalysts. *ChemCatChem* 9 (12): 2145–2159.

178 Zhang, J., Lu, F., Yu, W. et al. (2014). Selective hydrogenative cleavage of C–C bonds in sorbitol using Ni–Re/C catalyst under nitrogen atmosphere. *Catal. Today* 234: 107–112.

179 Banu, M., Sivasanker, S., Sankaranarayanan, T.M., and Venuvanalingam, P. (2011). Hydrogenolysis of sorbitol over Ni and Pt loaded on NaY. *Catal. Commun.* 12 (7): 673–677.

180 Jia, Y. and Liu, H. (2015). Selective hydrogenolysis of sorbitol to ethylene glycol and propylene glycol on ZrO_2-supported bimetallic Pd–Cu catalysts. *Chin. J. Catal.* 36: 1552–1559.

181 Huang, Z., Chen, J., Jia, Y. et al. (2014). Selective hydrogenolysis of xylitol to ethylene glycol and propylene glycol over copper catalysts. *Appl. Catal., B* 147: 377–386.

182 Sun, J. and Liu, H. (2011). Selective hydrogenolysis of biomass-derived xylitol to ethylene glycol and propylene glycol on supported Ru catalysts. *Green Chem.* 13 (1): 135–142.

183 Riviere, M., Perret, N., Delcroix, D. et al. (2018). Solvent effect in hydrogenolysis of xylitol over bifunctional Ru/MnO/C catalysts under alkaline-free conditions. *ACS Sustainable Chem. Eng.* 6: 4076–4085.

184 Du, W.C., Zheng, L.P., Shi, J.J. et al. (2015). Production of C2 and C3 polyols from D-sorbitol over hydrotalcite-like compounds mediated bi-functional Ni–Mg–Al–O_x catalysts. *Fuel Process. Technol.* 139: 86–90.

185 Zhao, L., Zhou, J.H., Sui, Z.J., and Zhou, X.G. (2010). Hydrogenolysis of sorbitol to glycols over carbon nanofiber supported ruthenium catalyst. *Chem. Eng. Sci.* 65 (1): 30–35.

186 Yan, K., Yang, Y., Chai, J., and Lu, Y. (2015). Catalytic reactions of gamma-valerolactone: a platform to fuels and value-added chemicals. *Appl. Catal., B* 179: 292–304.

187 Yan, L., Yao, Q., and Fu, Y. (2017). Conversion of levulinic acid and alkyl levulinates into biofuels and high-value chemicals. *Green Chem.* 19: 5527–5547.

188 Wright, W.R.H. and Palkovits, R. (2012). Development of heterogeneous catalysts for the conversion of levulinic acid to γ-valerolactone. *ChemSusChem* 5: 1657–1667.

189 Liguori, F., Moreno-Marrodan, C., and Barbaro, P. (2015). Environmentally friendly synthesis of γ-valerolactone by direct catalytic conversion of renewable sources. *ACS Catal.* 5: 1882–1894.

190 Climent, M.J., Corma, A., and Iborra, S. (2014). Conversion of biomass platform molecules into fuels additives and liquid hydrocarbon fuels. *Green Chem.* 16: 516–547.

191 Pileidis, F.D. and Titirici, M.-M. (2016). Levulinic acid biorefineries: new challenges for efficient utilization of biomass. *ChemSusChem* 9: 562–582.

192 Mariscal, R., Maireles-Torres, P., Ojeda, M. et al. (2016). Furfural: a renewable and versatile platform molecule for the synthesis of chemicals and fuels. *Energy Environ. Sci.* 9: 1144–1189.

193 Cao, S., Monnier, J.R., and Regalbuto, J.R. (2017). Alkali promotion of alumina-supported ruthenium catalysts for hydrogenation of levulinic acid to γ-valerolactone. *J. Catal.* 347: 72–78.

194 Braden, D.J., Henao, C.A., Heltzel, J. et al. (2011). Production of liquid hydrocarbon fuels by catalytic conversion of biomass-derived levulinic acid. *Green Chem.* 13: 1755–1765.

195 Wettstein, S.G., Bond, J.Q., Alonso, D.M. et al. (2012). RuSn bimetallic catalysts for selective hydrogenation of levulinic acid to γ-valerolactone. *Appl. Catal., B* 117–118: 321–329.

196 Landenna, L., Villa, A., Zanella, R. et al. (2016). Gold-iridium catalysts for the hydrogenation of biomass derived products. *Chin. J. Catal.* 37: 1771–1775.

197 Luo, W., Sankar, M., Beale, A.M. et al. (2015). High performing and stable supported nano-alloys for the catalytic hydrogenation of levulinic acid to γ-valerolactone. *Nat. Commun.* 6: 6540.

198 Al-Naji, M., Yepez, A., Balu, A.M. et al. (2016). Insights into the selective hydrogenation of levulinic acid to γ-valerolactone using supported mono- and bimetallic catalysts. *J. Mol. Catal. A: Chem.* 417: 145–152.

199 Yang, Y., Gao, G., Zhang, X., and Li, F. (2014). Facile fabrication of composition-tuned Ru–Ni bimetallics in ordered mesoporous carbon for levulinic acid hydrogenation. *ACS Catal.* 4: 1419–1425.

200 Testa, M.L., Corbel-Demailly, L., La Parola, V. et al. (2015). Effect of Au and Pd supported over HMS and Ti doped HMS as catalysts for the hydrogenation of levulinic acid to γ-valerolactone. *Catal. Today* 257: 291–296.

201 Yan, K., Lafleur, T., Wu, G. et al. (2013). Highly selective production of value-added γ-valerolactone from biomass-derived levulinic acid using the robust Pd nanoparticles. *Appl. Catal., A* 468: 52–58.

202 Zhang, Y., Chen, C., Gong, W. et al. (2017). Self-assembled Pd/CeO_2 catalysts by a facile redox approach for high-efficiency hydrogenation of levulinic acid into gamma-valerolactone. *Catal. Commun.* 93: 10–14.

203 Jiang, K., Sheng, D., Zhang, Z. et al. (2016). Hydrogenation of levulinic acid to γ-valerolactone in dioxane over mixed MgO–Al_2O_3 supported Ni catalyst. *Catal. Today* 274: 55–59.

204 Gupta, S.S.R. and Kantam, M.L. (2017). Selective hydrogenation of levulinic acid into γ-valerolactone over Cu/Ni hydrotalcite-derived catalyst. *Catal. Today* https://doi.org/10.1016/j.cattod.2017.08.007.

205 Gundekari, S. and Srinivasan, K. (2017). In situ generated Ni(0)@boehmite from NiAl-LDH: an efficient catalyst for selective hydrogenation of biomass derived levulinic acid to γ-valerolactone. *Catal. Commun.* 102: 40–43.

206 Hengne, A.M. and Rode, C.V. (2012). Cu–ZrO_2 nanocomposite catalyst for selective hydrogenation of levulinic acid and its ester to γ-valerolactone. *Green Chem.* 14: 1064–1072.

207 Shimizu, K.-I., Kanno, S., and Kon, K. (2014). Hydrogenation of levulinic acid to γ-valerolactone by Ni and MoO_x Co-loaded carbon catalysts. *Green Chem.* 16: 3899–3903.

208 Quiroz, J., Mai, E.F., and da Silva, V.T. (2016). Synthesis of nanostructured molybdenum carbide as catalyst for the hydrogenation of levulinic acid to γ-valerolactone. *Top. Catal.* 59: 148–158.

209 Kuwahara, Y., Kaburagi, W., Osada, Y. et al. (2017). Catalytic transfer hydrogenation of biomass-derived levulinic acid and its esters to γ-valerolactone over ZrO_2 catalyst supported on SBA-15 silica. *Catal. Today* 281: 418–428.

210 Ftouni, J., Muñoz-Murillo, A., Goryachev, A. et al. (2016). ZrO_2 is preferred over TiO_2 as support for the Ru-catalyzed hydrogenation of levulinic acid to γ-valerolactone. *ACS Catal.* 6: 5462–5472.

211 Abdelrahman, O.A., Luo, H.Y., Heyden, A. et al. (2015). Toward rational design of stable, supported metal catalysts for aqueous-phase processing: insights from the hydrogenation of levulinic acid. *J. Catal.* 329: 10–21.

212 Molleti, J., Tiwari, M.S., and Yadav, G.D. (2018). Novel synthesis of Ru/OMS catalyst by solvent-free method: selective hydrogenation of levulinic acid to γ-valerolactone in aqueous medium and kinetic modelling. *Chem. Eng. J.* 334: 2488–2499.

213 Michel, C., Zaffran, J., Ruppert, A.M. et al. (2014). Role of water in metal catalyst performance for ketone hydrogenation: a joint experimental and theoretical study on levulinic acid conversion into γ-valerolactone. *Chem. Commun.* 50: 12450–12453.

214 Michel, C. and Gallezot, P. (2015). Why is ruthenium an efficient catalyst for the aqueous-phase hydrogenation of biosourced carbonyl compounds? *ACS Catal.* 5: 4130–4132.

215 Piskun, A.S., van de Bovenkamp, H.H., Rasrendra, C.B. et al. (2016). Kinetic modeling of levulinic acid hydrogenation to γ-valerolactone in water using a carbon supported Ru catalyst. *Appl. Catal., A* 525: 158–167.

216 Stanford, J.P., Soto, M.C., Pfromm, P.H., and Rezac, M.E. (2016). Aqueous phase hydrogenation of levulinic acid using a porous catalytic membrane reactor. *Catal. Today* 268: 19–28.

217 Al-Shaal, M.G., Wright, W.R.H., and Palkovits, R. (2012). Exploring the ruthenium catalysed synthesis of γ-valerolactone in alcohols and utilisation of mild solvent-free reaction conditions. *Green Chem.* 14: 1260–1263.

218 Wu, Z., Zhang, M., Yao, Y. et al. (2018). One-pot catalytic production of 1,3-propanediol and γ-valerolactone from glycerol and levulinic acid. *Catal. Today* 302: 217–226.

219 Kumar, V.V., Naresh, G., Sudhakar, M. et al. (2015). Role of Brønsted and Lewis acid sites on Ni/TiO_2 catalyst for vapour phase hydrogenation of levulinic acid: kinetic and mechanistic study. *Appl. Catal., A* 505: 217–223.

220 Kumar, V.V., Naresh, G., Sudhakar, M. et al. (2016). An investigation on the influence of support type for Ni catalysed vapour phase hydrogenation of aqueous levulinic acid to γ-valerolactone. *RSC Adv.* 6: 9872–9879.

221 Sun, D., Ohkubo, A., Asami, K. et al. (2017). Vapor-phase hydrogenation of levulinic acid and methyl levulinate to γ-valerolactone over non-noble metal based catalysts. *Mol. Catal.* 437: 105–113.

222 Varkolu, M., Velpula, V., Burri, D.R., and Kamaraju, S.R.R. (2016). Gas phase hydrogenation of levulinic acid to γ-valerolactone over supported Ni catalysts with formic acid as hydrogen source. *New J. Chem.* 40: 3261–3267.

223 Du, X.-L., Bi, Q.-Y., Liu, Y.-M. et al. (2012). Tunable copper-catalyzed chemoselective hydrogenolysis of biomass-derived γ-valerolactone into 1,4-pentanediol or 2-methyltetrahydrofuran. *Green Chem.* 14: 935–939.

224 Corbel-Demailly, L., Ly, B.-K., Minh, D.-P. et al. (2013). Heterogeneous catalytic hydrogenation of biobased levulinic and succinic acids in aqueous solutions. *ChemSusChem* 6: 2388–2395.

225 Li, M., Li, G., Wang, A. et al. (2014). Aqueous phase hydrogenation of levulinic acid to 1,4-pentanediol. *Chem. Commun.* 50: 1414–1416.

226 Delhomme, C., Weuster-Botz, D., and Kühn, F.E. (2009). Succinic acid from renewable resources as a C_4 building-block chemical – a review of the catalytic possibilities in aqueous media. *Green Chem.* 11: 13–26.

227 Mazière, A., Prinsen, P., Garcia, A. et al. (2017). A review of progress in (bio)catalytic routes from/to renewable succinic acid. *Biofuels, Bioprod. Biorefin.* https://doi.org/10.1002/bbb.1785.

228 Budroni, G. and Corma, A. (2008). Gold and gold–platinum as active and selective catalyst for biomass conversion: synthesis of γ-butyrolactone and one-pot synthesis of pyrrolidone. *J. Catal.* 257: 403–408.

229 Zhang, C., Chen, L., Cheng, H. et al. (2016). Atomically dispersed Pd catalysts for the selective hydrogenation of succinic acid to γ-butyrolactone. *Catal. Today* 276: 55–61.

230 Tapin, B., Epron, F., Especel, C. et al. (2013). Study of monometallic Pd/TiO_2 catalysts for the hydrogenation of succinic acid in aqueous phase. *ACS Catal.* 3: 2327–2335.

231 Chung, S.-H., Park, Y.-M., Kim, M.-S., and Lee, K.-Y. (2012). The effect of textural properties on the hydrogenation of succinic acid using palladium incorporated mesoporous supports. *Catal. Today* 185: 205–210.

232 Shao, Z., Li, C., Di, X. et al. (2014). Aqueous-phase hydrogenation of succinic acid to γ-butyrolactone and tetrahydrofuran over Pd/C, Re/C, and Pd–Re/C catalysts. *Ind. Eng. Chem. Res.* 53: 9638–9645.

233 Di, X., Shao, Z., Li, C. et al. (2015). Hydrogenation of succinic acid over supported rhenium catalysts prepared by the microwave-assisted thermolytic method. *Catal. Sci. Technol.* 5: 2441–2448.

234 Ly, B.K., Minh, D.P., Pinel, C. et al. (2012). Effect of addition mode of Re in bimetallic Pd–Re/TiO_2 catalysts upon the selective aqueous-phase hydrogenation of succinic acid to 1,4-butanediol. *Top. Catal.* 55: 466–473.

235 Di, X., Li, C., Zhang, B. et al. (2017). Role of Re and Ru in Re–Ru/C bimetallic catalysts for the aqueous hydrogenation of succinic acid. *Ind. Eng. Chem. Res.* 56: 4672–4683.

236 Hong, U.G., Kim, J.K., Lee, J. et al. (2014). Hydrogenation of succinic acid to tetrahydrofuran over ruthenium–carbon composite catalysts: effect of HCl concentration in the preparation of the catalysts. *J. Ind. Eng. Chem.* 20: 3834–3840.

237 Minh, D.P., Besson, M., Pinel, C. et al. (2010). Aqueous-phase hydrogenation of biomass-based succinic acid to 1,4-butanediol over supported bimetallic catalysts. *Top. Catal.* 53: 1270–1273.

238 Hong, U.G., Lee, J., Hwang, S., and Song, I.K. (2011). Hydrogenation of succinic acid to γ-butyrolactone (GBL) over palladium–alumina composite catalyst prepared by a single-step sol–gel method. *Catal. Lett.* 141: 332–338.

239 Hong, U.G., Park, H.W., Lee, J. et al. (2012). Hydrogenation of succinic acid to tetrahydrofuran (THF) over rhenium catalyst supported on H_2SO_4-treated mesoporous carbon. *Appl. Catal., A* 415–416: 141–148.

240 Hong, U.G., Hwang, S., Seo, J.G. et al. (2011). Hydrogenation of succinic acid to γ-butyrolactone (GBL) over palladium catalyst supported on alumina xerogel: effect of acid density of the catalyst. *J. Ind. Eng. Chem.* 17: 316–320.

241 Kuwahara, Y., Magatani, Y., and Yamashita, H. (2015). Ru nanoparticles confined in Zr-containing spherical mesoporous silica containers for hydrogenation of levulinic acid and its esters in γ-valerolactone at ambient conditions. *Catal. Today* 258: 262–269.

242 Kang, K.H., Hong, U.G., Jun, J.O. et al. (2014). Hydrogenation of succinic acid to γ-butyrolactone and 1,4-butanediol over mesoporous rhenium–copper–carbon composite catalyst. *J. Mol. Catal. A: Chem.* 395: 234–242.

243 Huang, Z., Barnett, K.J., Chada, J.P. et al. (2017). Hydrogenation of γ-butyrolactone to 1,4-butanediol over CuCo/TiO_2 bimetallic catalysts. *ACS Catal.* 7: 8429–8440.

244 Ly, B.K., Tapin, B., Aouine, M. et al. (2015). Insights into the oxidation state and location of rhenium in Re–Pd/TiO_2 catalysts for aqueous-phase selective hydrogenation of succinic acid to 1,4-butanediol as function of palladium and rhenium deposition methods. *ChemCatChem* 7: 2161–2178.

245 Tapin, B., Epron, F., Especel, C. et al. (2014). Influence of the Re introduction method onto Pd/TiO_2 catalysts for the selective hydrogenation of succinic acid in aqueous-phase. *Catal. Today* 235: 127–133.

246 Vardon, D.R., Settle, A.E., Vorotnikov, V. et al. (2017). Ru–Sn/AC for the aqueous-phase reduction of succinic acid to 1,4-butanediol under continuous process conditions. *ACS Catal.* 7: 6207–6219.

247 Yan, K., Wu, G., Lafleur, T., and Jarvis, C. (2014). Production, properties and catalytic hydrogenation of furfural to fuel additives and value-added chemicals. *Renewable Sustainable Energy Rev.* 38: 663–676.

248 Manikandan, M., Venugopal, A.K., Nagpure, A.S. et al. (2016). Promotional effect of Fe on the performance of supported Cu catalyst for ambient pressure hydrogenation of furfural. *RSC Adv.* 6: 3888–3898.

249 Fulajtarova, K., Sotak, T., Hronec, M. et al. (2015). Aqueous phase hydrogenation of furfural to furfuryl alcohol over Pd–Cu catalysts. *Appl. Catal., A* 502: 78–85.

250 Stevens, J.G., Bourne, R.A., Twigg, M.V., and Poliakoff, M. (2010). Real-time product switching using a twin catalyst system for the hydrogenation of furfural in supercritical CO_2. *Angew. Chem. Int. Ed.* 49: 8856–8859.

251 Biradar, N.S., Hengne, A.M., Birajdar, S.N. et al. (2014). Single-pot formation of THFAL via catalytic hydrogenation of FFR over Pd/MFI catalyst. *ACS Sustainable Chem. Eng.* 2: 272–281.

252 Liu, S., Amada, Y., Tamura, M. et al. (2014). One-pot selective conversion of furfural into 1,5-pentanediol over a Pd-added Ir–ReO_x/SiO_2 bifunctional catalyst. *Green Chem.* 16: 617–626.

253 Zhang, B., Zhu, Y.L., Ding, G.Q. et al. (2012). Selective conversion of furfuryl alcohol to 1,2-pentanediol over a Ru/MnO_x catalyst. *Green Chem.* 14: 3402–3409.

254 Koso, S., Furikado, I., Shimao, A. et al. (2009). Chemoselective hydrogenolysis of tetrahydrofurfuryl alcohol to 1,5-pentanediol. *Chem. Commun.*: 2035–2037.

255 Koso, S., Ueda, N., Shinmi, Y. et al. (2009). Promoting effect of Mo on the hydrogenolysis of tetrahydrofurfuryl alcohol to 1,5-pentanediol. *J. Catal.* 267: 89–92.

256 Nakagawa, Y. and Tomishige, K. (2012). Production of 1,5-pentanediol from biomass via furfural and tetrahydrofurfuryl alcohol. *Catal. Today* 195: 136–143.

257 Chen, K., Mori, K., Watanabe, H. et al. (2012). C–O bond hydrogenolysis of cyclic ethers with OH groups over rhenium-modified supported iridium catalysts. *J. Catal.* 294: 171–183.

258 Chia, M., Pagan-Torres, Y.J., Hibbits, D. et al. (2011). Selective hydrogenolysis of polyols and cyclic ethers over bifunctional surface sites on rhodium–rhenium catalysts. *J. Am. Chem. Soc.* 133: 12675–12689.

259 Hronec, M., Fulajtarova, K., and Micusik, M. (2013). Influence of furanic polymers on selectivity of furfural rearrangement to cyclopentanone. *Appl. Catal., A* 468: 426–431.

260 Hronec, M. and Fulajtarova, K. (2012). Selective transformation of furfural to cyclopentanone. *Catal. Commun.* 24: 100–104.

261 Li, X.L., Deng, J., Shi, J. et al. (2015). Selective conversion of furfural to cyclopentanone or cyclopentanol using different preparation methods of Cu–Co catalysts. *Green Chem.* 17: 1038–1046.

262 Hu, L., Lin, L., Wu, Z. et al. (2017, 2017). Recent advances in catalytic transformation of biomass-derived 5-hydroxymethylfurfural into the innovative fuels and chemicals. *Renewable Sustainable Energy Rev.* 74: 230–257.

263 Tang, X., Wei, J., Ding, N. et al. (2017). Chemoselective hydrogenation of biomass derived 5-hydroxymethylfurfural to diols: key intermediates for

sustainable chemicals, materials and fuels. *Renewable Sustainable Energy Rev.* 77: 287–296.

264 Jiang, Y., Woortman, A.J.J., van Ekenstein, G.O.R.A. et al. (2014). Enzymatic synthesis of biobased polyesters using 2,5-bis(hydroxymethyl)furan as the building block. *Biomacromolecules* 5: 2482–2493.

265 Ohyama, J., Esaki, A., Yamamoto, Y. et al. (2013). Selective hydrogenation of 2-hydroxymethyl-5-furfural to 2,5-bis(hydroxymethyl)furan over gold sub-nanoclusters. *RSC Adv.* 3: 1033–1036.

266 Chen, J.Z., Lu, F., Zhang, J.J. et al. (2013). Immobilized Ru clusters in nanosized mesoporous zirconium silica for the aqueous hydrogenation of furan derivatives at room temperature. *ChemCatChem* 5: 2822–2826.

267 Tamura, M., Tokonami, K., Nakagawa, Y., and Tomishige, K. (2013). Rapid synthesis of unsaturated alcohols under mild conditions by highly selective hydrogenation. *Chem. Commun.* 49: 7034–7036.

268 Cai, H.I., Li, C.Z., Wang, A.Q., and Zhang, T. (2014). Biomass into chemicals: one-pot production of furan-based diols from carbohydrates via tandem reactions. *Catal. Today* 234: 59–65.

269 Chatterjee, M., Ishizaka, M., and Kawanami, H. (2014). Selective hydrogenation of 5-hydroxymethylfurfural to 2,5-bis-(hydroxymethyl)furan using Pt/MCM-41 in an aqueous medium: a simple approach. *Green Chem.* 16: 4734–4739.

270 Ohyama, J., Hayashi, J., Ueda, K. et al. (2016). Effect of FeO_x modification of Al_2O_3 on its supported Au catalyst for hydrogenation of 5-hydroxymethylfurfural. *J. Phys. Chem. C* 120: 15129–15136.

271 Zhu, Y., Kong, X., Zheng, H. et al. (2015). Efficient synthesis of 2,5-dihydroxymethylfuran and 2,5-dimethylfuran from 5-hydroxymethylfurfural using mineral-derived Cu catalysts as versatile catalysts. *Catal. Sci. Technol.* 5: 4208–4217.

272 Cao, Q., Liang, W., Guan, J. et al. (2014). Catalytic synthesis of 2,5-bis-methoxymethylfuran: a promising cetane number improver for diesel. *Appl. Catal., A* 481: 49–53.

273 Yu, L., He, I., Chen, J. et al. (2015). Robust and recyclable nonprecious bimetallic nanoparticles on carbon nanotubes for the hydrogenation and hydrogenolysis of 5-hydroxymethylfurfural. *ChemCatChem* 7: 1701–1707.

274 Connolly, T.J., Considine, J.L., Ding, Z.X. et al. (2010). Efficient synthesis of 8-oxa-3-aza-bicyclo[3.2.1]octane hydrochloride. *Org. Process Res. Dev.* 14: 459–465.

275 Alamillo, R., Tucker, M., Chia, M. et al. (2012). The selective hydrogenation of biomass-derived 5-hydroxymethylfurfural using heterogeneous catalysts. *Green Chem.* 14: 1413–1419.

276 Nakagawa, Y. and Tomishige, K. (2010). Total hydrogenation of furan derivatives over silica-supported Ni–Pd alloy catalyst. *Catal. Commun.* 12: 154–156.

277 Nakagawa, Y., Takada, K., Tamura, M., and Tomishige, K. (2014). Total hydrogenation of furfural and 5-hydroxymethylfurfural over supported Pd–Ir alloy catalyst. *ACS Catal.* 4: 2718–2726.

278 Chen, J.Z., Liu, R.L., Guo, Y.Y. et al. (2015). Selective hydrogenation of biomass based 5-hydroxymethylfurfural over catalysts of palladium immobilized on amine-functionalized metal organic framework. *ACS Catal.* 5: 722–733.

279 Perret, N., Grigoropoulos, A., Zanella, M. et al. (2016). Catalytic response and stability of nickel/alumina for the hydrogenation of 5-hydroxymethylfurfural in water. *ChemSusChem* 9: 521–531.

280 Buntara, T., Noël, S., Phua, P.H. et al. (2011). Caprolactam from renewable resources: catalytic conversion of 5-hydroxymethylfurfural into caprolactone. *Angew. Chem. Int. Ed.* 50: 7083–7087.

281 Xiao, B., Zheng, M.Y., Li, X.S. et al. (2016). Synthesis of 1,6-hexanediol from HMF over double-layered catalysts of $Pd/SiO_2 + Ir{-}ReO_x/SiO_2$ in a fixed-bed reactor. *Green Chem.* 18: 2175–2184.

282 Mounguengui-Diallo, M., Vermersch, F., Perret, N. et al. (2018). Base-free oxidation of 1,6-hexanediol to adipic acid over supported noble metal mono and bimetallic catalysts. *Appl. Catal., A* 551: 88–97.

283 Liu, F., Audemar, M., De Oliveira-Vigier, K. et al. (2014). Palladium/carbon dioxide cooperative catalysis for the production of diketone derivatives from carbohydrates. *ChemSusChem* 7: 2089–2093.

284 Ohyama, J., Kanao, R., Esaki, A., and Satsuma, A. (2014). Conversion of 5-hydroxymethylfurfural to a cyclopentanone derivative by ring rearrangement over supported Au nanoparticles. *Chem. Commun.* 50: 5633–5636.

285 Ohyama, J., Kanao, R., Ohira, Y., and Satsuma, A. (2016). The effect of heterogeneous acid–base catalysis on conversion of 5-hydroxymethylfurfural into a cyclopentanone derivative. *Green Chem.* 18: 676–680.

286 Ramos, R., Grigoropoulos, A., Perret, N. et al. (2017). Selective conversion of 5-hydroxymethylfurfural to cyclopentanone derivatives over $Cu{-}Al_2O_3$ and $Co{-}Al_2O_3$ catalysts in water. *Green Chem.* 19: 1701–1713.

287 Pleckmans, M., Renders, T., Van de Vyver, S., and Sels, B.F. (2017). Bio-based amines through sustainable heterogeneous catalysis. *Green Chem.* 19: 5303–5331.

288 Chatterjee, M., Ishizaka, T., and Kawanami, H. (2016). Reductive amination of furfural to furfurylamine using aqueous ammonia solution and molecular hydrogen: an environmentally friendly approach. *Green Chem.* 18: 487–496.

289 Chieffi, G., Braun, M., and Esposito, D. (2015). Continuous reductive amination of biomass-derived molecules over carbonized filter paper-supported FeNi alloy. *ChemSusChem* 8: 3590–3594.

290 Gao, G., Sun, P., Li, Y. et al. (2017). Highly stable porous-carbon-coated Ni catalysts for the reductive amination of levulinic acid via an unconventional pathway. *ACS Catal.* 7: 4927–4935.

291 Touchy, A.S., Siddiki, S.M.A.H., Kon, K., and Shimizu, K.-I. (2014). Heterogeneous Pt catalysts for reductive amination of levulinic acid to pyrrolidones. *ACS Catal.* 4: 3045–3050.

292 Martínez, J.J., Silva, L., Rojas, H.A. et al. (2017). Reductive amination of levulinic acid to different pyrrolidones on $Ir/SiO_2{-}SO_3H$: Elucidation of reaction mechanism. *Catal. Today* 296: 118–126.

293 Ledoux, A., Kuigwa, L.S., Framery, E., and Andrioletti, B. (2015). A highly sustainable route to pyrrolidone derivatives – direct access to biosourced solvents. *Green Chem.* 17: 3251–3254.

294 Wei, Y., Wang, C., Jiang, X. et al. (2014). Catalyst-free transformation of levulinic acid into pyrrolidinones with formic acid. *Green Chem.* 16: 1093–1096.

295 Vidal, J.D., Climent, M.J., Concepcion, P. et al. (2015). Chemicals from biomass: chemoselective reductive amination of ethyl levulinate with amines. *ACS Catal.* 5: 5812–5821.

296 Vidal, J.D., Climent, M.J., Corma, A. et al. (2017). One-pot selective catalytic synthesis of pyrrolidone derivates from ethyl levulinate and nitro compounds. *ChemSusChem* 10: 119–228.

297 Carey, F.A. and Sundberg, R.J. (eds.) (2007). *Advanced Organic Chemistry Part A: Structure and Mechanism*, 5e. New York, NY: Springer.

298 Winterbottom, J.M. (1981). Hydration and dehydration by heterogeneous catalysts. *Catalysis* 4: 141–174.

299 Molnar, A. and Bartok, M. (2001). Dehydration of alcohols. In: *Fine Chemicals Through Heterogeneous Catalysis* (eds. R.A. Sheldon and H. Van Bekkum), 295–307. Wiley-VCH Verlag GmbH.

300 John, C.S. and Scurrell, M.S. (1977). Catalytic properties of aluminas for reactions of hydrocarbons and alcohols. *Catalysis* 1: 136–167.

301 Lauron-Pernot, H. (2006). Evaluation of surface acido-basic properties of inorganic-based solids by model catalytic alcohol reaction networks. *Catal. Rev. Sci. Eng.* 48: 315–361.

302 Larmier, K., Nicolle, A., Chizallet, C. et al. (2016). Influence of coadsorbed water and alcohol molecules on isopropyl alcohol dehydration on γ-alumina: multiscale modeling of experimental kinetic profiles. *ACS Catal.* 6: 1905–1920.

303 Buniazet, Z., Couble, J., Bianchi, D. et al. (2017). Unravelling water effects on solid acid catalysts: case study of TiO_2/SiO_2 as a catalyst for the dehydration of isobutanol. *J. Catal.* 348: 125–134.

304 Baertsch, C.D., Komala, K.T., Chua, Y.H., and Iglesia, E. (2002). Genesis of Brønsted acid sites during dehydration of 2-butanol on tungsten oxide catalysts. *J. Catal.* 205: 44–57.

305 Macho, V., Kralik, M., Jurecekova, E. et al. (2001). Dehydration of C4 alkanols conjugated with a positional and skeletal isomerisation of the formed C4 alkenes. *Appl. Catal., A* 214: 251–257.

306 Corma, A. and Renz, M. (2004). Sn-Beta zeolite as diastereoselective water-resistant heterogeneous Lewis-acid catalyst for carbon–carbon bond formation in the intramolecular carbonyl–ene reaction. *Chem. Commun.*: 550–551.

307 Hahn, M.W., Copeland, J.R., van Pelt, A.H., and Sievers, C. (2013). Stability of amorphous silica–alumina in hot liquid water. *ChemSusChem* 6: 2304–2315.

308 Rinaldi, R. and Schüth, F. (2009). Design of solid catalysts for the conversion of biomass. *Energy Environ. Sci.* 2: 610–626.

309 Katryniok, B., Paul, S., Bellière-Baca, V. et al. (2010). Glycerol dehydration to acrolein in the context of new uses of glycerol. *Green Chem.* 12: 2079–2098.
310 Martin, A., Armbruster, U., and Atia, H. (2012). Recent developments in dehydration of glycerol toward acrolein over heteropolyacids. *Eur. J. Lipid Sci. Technol.* 114: 10–23.
311 Katryniok, B., Paul, S., and Dumeignil, F. (2013). Recent developments in the field of catalytic dehydration of glycerol to acrolein. *ACS Catal.* 3: 1819–1834.
312 Talebian-Kiakalaieh, A., Aishah Saidina Amin, N., and Hezaveh, H. (2014). Glycerol for renewable acrolein production by catalytic dehydration. *Renewable Sustainable Energy Rev.* 40: 28–59.
313 Galadima, A. and Muraza, O. (2016). A review on glycerol valorization to acrolein over solid acid catalysts. *J. Taiwan Inst. Chem. Eng.* 67: 29–44.
314 Tsukuda, E., Sato, S., Takahashi, R., and Sodesawa, T. (2007). Production of acrolein from glycerol over silica-supported heteropoly acids. *Catal. Commun.* 8: 1349–1353.
315 Chai, S., Wang, H., Liang, Y., and Xu, B. (2007). Sustainable production of acrolein: investigation of solid acid–base catalysts for gas-phase dehydration of glycerol. *Green Chem.* 9: 1130–1136.
316 Graça, I., Comparot, J.-D., Laforge, S. et al. (2009). Effect of phenol addition on the performances of H–Y zeolite during methylcyclohexane transformation. *Appl. Catal., A* 353: 123–129.
317 Gayubo, A.G., Aguayo, A.T., Atutxa, A. et al. (2004). Transformation of oxygenate components of biomass pyrolysis oil on a HZSM-5 zeolite. I. Alcohols and phenols. *Ind. Eng. Chem. Res.* 43: 2610–2618.
318 Kim, Y.T., Jung, K.-D., and Park, E.D. (2010). Gas-phase dehydration of glycerol over ZSM-5 catalysts. *Microporous Mesoporous Mater.* 131: 28–36.
319 Chai, S., Wang, H., Liang, Y., and Xu, B. (2007). Sustainable production of acrolein: gas-phase dehydration of glycerol over Nb_2O_5 catalyst. *J. Catal.* 250: 342–349.
320 Ning, L., Ding, Y., Chen, W. et al. (2008). Glycerol dehydration to acrolein over activated carbon-supported silicotungstic acids. *Chin. J. Catal.* 29: 212–214.
321 Alhanash, A., Kozhevnikova, E.F., and Kozhevnikov, I.V. (2010). Gas-phase dehydration of glycerol to acrolein catalysed by caesium heteropoly salt. *Appl. Catal., A* 378: 11–18.
322 Kinage, A.K., Upare, P.P., Kasinathan, P. et al. (2010). Selective conversion of glycerol to acetol over sodium-doped metal oxide catalysts. *Catal. Commun.* 11: 620–623.
323 Stošić, D., Bennici, S., Couturier, J.-L. et al. (2012). Influence of surface acid–base properties of zirconia and titania based catalysts on the product selectivity in gas phase dehydration of glycerol. *Catal. Commun.* 17: 23–28.
324 Lauriol-Garbey, P., Postole, G., Loridant, S. et al. (2011). Acid–base properties of niobium–zirconium mixed oxide catalysts for glycerol dehydration by calorimetric and catalytic investigation. *Appl. Catal., B* 106: 94–102.

325 Suprun, W., Lutecki, M., Haber, T., and Papp, H. (2009). Acidic catalysts for the dehydration of glycerol: activity and deactivation. *J. Mol. Catal. A: Chem.* 309: 71–78.

326 Wang, F., Dubois, J.L., and Ueda, W. (2009). Catalytic dehydration of glycerol over vanadium phosphate oxides in the presence of molecular oxygen. *J. Catal.* 268: 260–267.

327 Barrault, J., Clacens, J.-M., and Pouilloux, Y. (2004). Selective oligomerization of glycerol over mesoporous catalysts. *Top. Catal.* 27: 137–142.

328 Lauriol-Garbey, P., Loridant, S., Bellière-Baca, V. et al. (2011). Gas phase dehydration of glycerol to acrolein over WO_3/ZrO_2 catalysts: improvement of selectivity and stability by doping with SiO_2. *Catal. Commun.* 16: 170–174.

329 Hulteberg, C., Leveau, A., and Brandin, J.G.M. (2013). Pore condensation in glycerol dehydration. *Top. Catal.* 56: 813–821.

330 Hulteberg, C., Leveau, A., and Brandin, J.G.M. (2017). Pore condensation in glycerol dehydration: modification of a mixed oxide catalyst. *Top. Catal.* 60: 1462–1472.

331 Atia, H., Armbruster, U., and Martin, A. (2008). Dehydration of glycerol in gas phase using heteropolyacid catalysts as active compounds. *J. Catal.* 258: 71–82.

332 Jia, C.-J., Liu, Y., Schmidt, W. et al. (2010). Small-sized HZSM-5 zeolite as highly active catalyst for gas phase dehydration of glycerol to acrolein. *J. Catal.* 269: 71–79.

333 Bozga, E.R., Plesu, V., Bozga, G. et al. (2011). Conversion of glycerol to propanediol and acrolein by heterogeneous catalysis. *Rev. Chim.* 62: 646–654.

334 Kim, Y.T., Jung, K.-D., and Park, E.D. (2011). Gas-phase dehydration of glycerol over silica–alumina catalysts. *Appl. Catal. B* 107: 177–187.

335 Znaiguia, R., Brandhorst, L., Christin, N. et al. (2014). Toward longer life catalysts for dehydration of glycerol to acrolein. *Microporous Mesoporous Mater.* 196: 97–103.

336 Zhou, C.-J., Huang, C.-J., Zhang, W.-G. et al. (2007). Synthesis of micro- and mesoporous ZSM-5 composites and their catalytic application in glycerol dehydration to acrolein. *Stud. Surf. Sci. Catal.* 165: 527–530.

337 Possato, L.G., Diniz, R.N., Garetto, T. et al. (2013). A comparative study of glycerol dehydration catalyzed by micro/mesoporous MFI zeolites. *J. Catal.* 300: 102–112.

338 Yun, D., Kim, T.Y., Park, D.S. et al. (2014). A tailored catalyst for the sustainable conversion of glycerol to acrolein: mechanistic aspect of sequential dehydration. *ChemSusChem* 7: 2193–2201.

339 Zhang, H., Hu, Z., Huang, L. et al. (2015). Dehydration of glycerol to acrolein over hierarchical ZSM-5 zeolites: effects of mesoporosity and acidity. *ACS Catal.* 5: 2548–2558.

340 Huang, L., Qin, F., Huang, Z. et al. (2016). Hierarchical ZSM-5 zeolite synthesized by an ultrasound-assisted method as a long-life catalyst for dehydration of glycerol to acrolein. *Ind. Eng. Chem. Res.* 55: 7318–7327.

341 Beerthuis, R., Huang, L., Raveendran Shiju, N. et al. (2018). Facile synthesis of a novel hierarchical ZSM-5 zeolite: a stable acid catalyst for dehydrating glycerol to acrolein. *ChemCatChem* 10: 211–221.
342 Choi, Y., Park, H., Yun, Y.S., and Yi, J. (2015). Effects of catalyst pore structure and acid properties on the dehydration of glycerol. *ChemSusChem* 8: 974–979.
343 Dubois, J.L. Duquenne, C., and Hoelderich, W. (2006). Process for dehydrating glycerol to acrolein. WO Patent 2,006,087,083, filed 6 January 2006 and issued 24 August 2006.
344 Dubois, J.L., Duquenne, C., Hoelderich, W., and Kervennal, J. (2006). Process for dehydrating glycerol to acrolein. WO Patent 2,006,087,084, filed 6 January 2005 and issued 24 August 2006.
345 Tao, L.-Z., Yan, B., Liang, Y., and Xu, B.-Q. (2013). Sustainable production of acrolein: catalytic performance of hydrated tantalum oxides for gas-phase dehydration of glycerol. *Green Chem.* 15: 696–705.
346 Cavani, F., Guidetti, S., Marinelli, L. et al. (2010). The control of selectivity in gas-phase glycerol dehydration to acrolein catalysed by sulfated zirconia. *Appl. Catal., B* 100: 197–204.
347 Trakarnpruk, W. (2014). Platinum/phosphotungstic acid/(Zr)MCM-41 catalysts in glycerol dehydration. *Mendeleev Commun.* 24: 167–169.
348 Blanco, E., Delichere, P., Millet, J.M.M., and Loridant, S. (2014). Gas phase dehydration of lactic acid to acrylic acid over alkaline-earth phosphates catalysts. *Catal. Today* 226: 185–191.
349 Fan, Y., Zhou, C., and Zhu, X. (2009). Selective catalysis of lactic acid to produce commodity chemicals. *Catal. Rev.* 51: 293–324.
350 Bonnotte, T., Paul, S., Araque, M. et al. (2018). Dehydration of lactic acid: the state of the art. *ChemBioEng Rev.* 5: 34–56.
351 Vu, D.T., Kolah, A.K., Asthana, N.S. et al. (2005). Oligomer distribution in concentrated lactic acid solutions. *Fluid Phase Equilib.* 236: 125–135.
352 Lari, G.M., Puertolas, B., Frei, M.S. et al. (2016). Hierarchical NaY zeolites for lactic acid dehydration to acrylic acid. *ChemCatChem* 8: 1507–1514.
353 Zhang, X., Lin, L., Zhang, T. et al. (2016). Catalytic dehydration of lactic acid to acrylic acid over modified ZSM-5 catalysts. *Chem. Eng. J.* 284: 934–941.
354 Sad, M.E., González Pena, L.F., Padró, C.L., and Apesteguía, C.R. (2018). Selective synthesis of acetaldehyde from lactic acid on acid zeolites. *Catal. Today* 302: 203–209.
355 Tang, C., Peng, J., Li, X. et al. (2015). Efficient and selective conversion of lactic acid into acetaldehyde using a mesoporous aluminum phosphate catalyst. *Green Chem.* 17: 1159–1166.
356 Zhai, Z., Li, X., Tang, C. et al. (2014). Decarbonylation of lactic acid to acetaldehyde over aluminum sulfate catalyst. *Ind. Eng. Chem. Res.* 53: 10318–10327.
357 Tang, C., Zhai, Z., Li, X. et al. (2015). Highly efficient and robust $Mg_{0.388}Al_{2.408}O_4$ catalyst for gas-phase decarbonylation of lactic acid to acetaldehyde. *J. Catal.* 329: 206–217.
358 Tang, C., Peng, J., Li, X. et al. (2016). Sustainable production of acetaldehyde from lactic acid over the carbon catalysts. *Korean J. Chem. Eng.* 33: 99–106.

359 Luo, W., Cai, J., Zhu, L. et al. (2012). Toxic effects of acrylic acid on *Clostridium propionicum* and isolation of acrylic acid-tolerant mutants for production of acrylic acid. *Eng. Life Sci.* 12: 567–573.
360 Sun, P., Yu, D., Fu, K. et al. (2009). Potassium modified NaY: a selective and durable catalyst for dehydration of lactic acid to acrylic acid. *Catal. Commun.* 10: 1345–1349.
361 Sun, P., Yu, D., Tang, Z. et al. (2010). A Y zeolites catalyze dehydration of lactic acid to acrylic acid: studies on the effects of anions in potassium salts. *Ind. Eng. Chem. Res.* 49: 9082–9087.
362 Deka, R.C. and Hirao, K. (2002). Lewis acidity and basicity of cation-exchanged zeolites: QM/MM and density functional studies. *J. Mol. Catal. A: Chem.* 181: 275–282.
363 Näfe, G., López-Martínez, M.-A., Dyballa, M. et al. (2015). Deactivation behavior of alkali-metal zeolites in the dehydration of lactic acid to acrylic acid. *J. Catal.* 329: 413–424.
364 Tam, M.S., Gunter, G.C., Craciun, R. et al. (1997). Reaction and spectroscopic studies of sodium salt catalysts for lactic acid conversion. *Ind. Eng. Chem. Res.* 36: 3505–3512.
365 Matsuura, Y., Onda, A., Ogob, S., and Yanagisawa, K. (2014). Acrylic acid synthesis from lactic acid over hydroxyapatite catalysts with various cations and anions. *Catal. Today* 226: 192–197.
366 Yan, B., Tao, L.-Z., Liang, Y., and Xu, B.-Q. (2014). Sustainable production of acrylic acid: catalytic performance of hydroxyapatites for gas-phase dehydration of lactic acid. *ACS Catal.* 4: 1931–1943.
367 Matsuura, Y., Onda, A., and Yanagisawa, K. (2014). Selective conversion of lactic acid into acrylic acid over hydroxyapatite catalysts. *Catal. Commun.* 48: 5–10.
368 Zhang, Z., Qu, Y., Wang, S., and Wang, J. (2010). Theoretical study on the mechanisms of the conversion of methyl lactate over sodium polyphosphate catalyst. *J. Mol. Catal. A: Chem.* 323: 91–100.
369 Dabbawala, A.A., Mishra, D.K., Huber, G.W., and Hwang, J.-S. (2015). Role of acid sites and selectivity correlation in solvent free liquid phase dehydration of sorbitol to isosorbide. *Appl. Catal., A* 492: 252–261.
370 Li, J., Buijs, W., Berger, R.J. et al. (2014). Sorbitol dehydration in a $ZnCl_2$ molten salt hydrate medium: molecular modeling. *Catal. Sci. Technol.* 4: 152–163.
371 Morita, Y., Furusato, S., Takagaki, A. et al. (2014). Intercalation-controlled cyclodehydration of sorbitol in water over layered-niobium-molybdate solid acid. *ChemSusChem* 7: 748–752.
372 Rusu, O.A., Hoelderich, W.F., Wyart, H., and Ibert, M. (2015). Metal phosphate catalyzed dehydration of sorbitol under hydrothermal conditions. *Appl. Catal., B* 176–177: 139–149.
373 Shi, J., Shan, Y., Tian, Y. et al. (2016). Hydrophilic sulfonic acid-functionalized microbead silica for dehydration of sorbitol to isosorbide. *RSC Adv.* 6: 13514–13521.
374 Yamaguchi, A., Hiyoshi, N., Sato, O., and Shirai, M. (2011). Sorbitol dehydration in high temperature liquid water. *Green Chem.* 13: 873–881.

375 Morales, G., Iglesias, J., Melero, J.A. et al. (2017). Isosorbide production from sorbitol over heterogeneous acid catalysts: screening and kinetic study. *Top. Catal.* 60: 1027–1039.
376 Kobayashi, H., Yokoyama, H., Feng, B., and Fukuoka, A. (2015). Dehydration of sorbitol to isosorbide over H-beta zeolites with high Si/Al ratios. *Green Chem.* 17: 2732–2735.
377 Otomo, R., Yokoi, T., and Tatsumi, T. (2015). Synthesis of isosorbide from sorbitol in water over high-silica aluminosilicate zeolites. *Appl. Catal., A* 505: 28–35.
378 Polaert, I., Cunha Felix, M., Fornasero, M. et al. (2013). A greener process for isosorbide production: kinetic study of the catalytic dehydration of pure sorbitol under microwave. *Chem. Eng. J.* 222: 228–239.
379 Zhang, J., Wang, L., Liu, F. et al. (2015). Enhanced catalytic performance in dehydration of sorbitol to isosorbide over a superhydrophobic mesoporous acid catalyst. *Catal. Today* 242: 249–254.
380 Zou, J., Cao, D., Tao, W. et al. (2016). Sorbitol dehydration into isosorbide over a cellulose-derived solid acid catalyst. *RSC Adv.* 6: 49528–49536.
381 Dabbawala, A.A., Park, J.J., Valekar, A.H. et al. (2015). Arenesulfonic acid functionalized ordered mesoporous silica as solid acid catalyst for solvent free dehydration of sorbitol to isosorbide. *Catal. Commun.* 69: 207–211.
382 Khan, N.A., Mishra, D.K., Ahmed, I. et al. (2013). Liquid-phase dehydration of sorbitol to isosorbide using sulfated zirconia as a solid acid catalyst. *Appl. Catal., A* 452: 34–38.
383 Ahmed, I., Khan, N.A., Mishra, D.K. et al. (2013). Liquid-phase dehydration of sorbitol to isosorbide using sulfated titania as a solid acid catalyst. *Chem. Eng. Sci.* 93: 91–95.
384 Dabbawala, A.A., Mishra, D.K., and Hwang, J.-S. (2013). Sulfated tin oxide as an efficient solid acid catalyst for liquid phase selective dehydration of sorbitol to isosorbide. *Catal. Commun.* 42: 1–5.
385 Xia, J., Yu, D., Hu, Y. et al. (2011). Sulfated copper oxide: an efficient catalyst for dehydration of sorbitol to isosorbide. *Catal. Commun.* 12: 544–547.
386 Cao, D., Yu, B., Zhang, S. et al. (2016). Isosorbide production from sorbitol over porous zirconium phosphate catalyst. *Appl. Catal., A* 528: 59–66.
387 Gu, M., Yu, D., Zhang, H. et al. (2009). Metal (IV) phosphates as solid catalysts for selective dehydration of sorbitol to isosorbide. *Catal. Lett.* 133: 214–220.
388 Sun, P., Yu, D.H., Hu, Y. et al. (2011). $H_3PW_{12}O_{40}/SiO_2$ for sorbitol dehydration to isosorbide: high efficient and reusable solid acid catalyst. *Korean J. Chem. Eng.* 28: 99–105.
389 Tang, Z.-C., Yu, D.-H., Sun, P. et al. (2010). Phosphoric acid modified Nb_2O_5: a selective and reusable catalyst for dehydration of sorbitol to isosorbide. *Bull. Korean Chem. Soc.* 31: 3679–3683.
390 Cubo, A., Iglesias, J., Morales, G. et al. (2017). Dehydration of sorbitol to isosorbide in melted phase with propyl-sulfonic functionalized SBA-15: influence of catalyst hydrophobization. *Appl. Catal., A* 531: 151–160.

8

Conversion of Lignin to Value-added Chemicals via Oxidative Depolymerization

Justin K. Mobley

University of Kentucky, Department of Chemistry, 505 Rose Street, Lexington, KY 40506, USA

8.1 Introduction

The term lignin, derived from the Latin *lignum* (meaning "wood"), was first described by Candolle in 1813 [1]. Lignin is the second most abundant biopolymer on Earth, surpassed only by cellulose, and has long been the ire of those seeking to find significant uses of biomass. Situated between the cellulose and hemicellulose in the plant secondary cell wall, lignin acts as structural glue that gives terrestrial biomass structural rigidity, defense against chemical and biological attack, and aids in water transport (water proofing) in the trachea of the xylem tissue. Unlike most biologically derived compounds, lignin is not highly controlled with regard to its structure. Indeed, lignin is a heterogeneous and amorphous biopolymer that is composed of a series of linkages formed by radical coupling reactions [2]. More specifically, the formation of lignin is the direct result of radical polymerization of three main monolignols: sinapyl (S), coniferyl (G), and *p*-coumaryl (H) alcohols. Although there are other phenylpropanoid subunits that may participate in lignification – including the increasingly well-known ferulate subunit [3, 4] or the lesser-known caffeyl alcohol subunit (so-called C-lignin produced in vanilla seed coats) [5] – the vast majority of lignin is composed of S, G, and H units (see Figure 8.1) [2]. Due to the random nature of lignification, a series of lignin bonding motifs are known (see Figure 8.2). This results in a biopolymer that is not only incredibly challenging to characterize but also difficult to depolymerize.

The depolymerization of lignin has been a topic of significant research for the greater part of a century. Indeed, much of the early work in lignin depolymerization was focused around the utilization of the biopolymer by the pulp and paper industry due to the large amount of lignin formed as a by-product. While this effort was fruitful (leading to the production of emulsifiers, vanillin, and DMSO), the harsh conditions and large degree of lignin condensation severely limit its use. Increasingly, researchers have begun to use more native lignins, the term "native lignin" referring to lignin that is similar to its *in planta* form. Moreover, due to the complexity and heterogeneity of lignin, researchers typically focus on a single bonding motif and apply model compounds to assess reactivity. The most

Chemical Catalysts for Biomass Upgrading, First Edition. Edited by Mark Crocker and Eduardo Santillan-Jimenez.

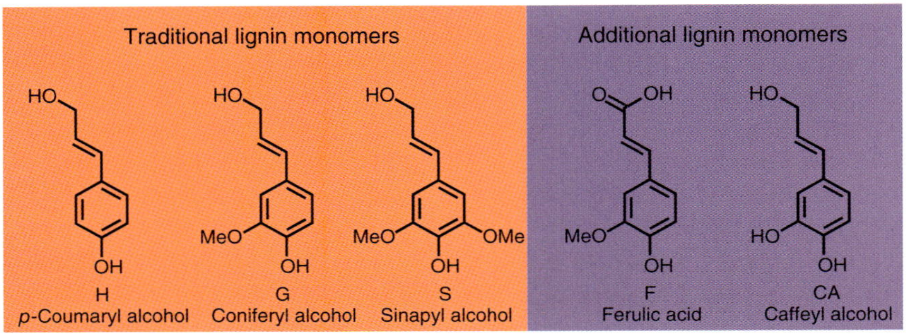

Figure 8.1 Lignin monomers.

common bonding motif is the β-O-4 linkage, which composes 40–60% of the linkages in native lignins, depending on the species [6]. Other common bonding motifs are the β-5 (phenyl coumaran), β-β (resinol), β-1, 5-5, dibenzodioxocin, and spirodieneone. However, by far the most commonly explored linkage is the β-O-4. This is primarily due to the presence of a benzylic alcohol and β-aryl ether connectivity that are easy targets for depolymerization.

In spite of being something that is commonly overlooked by researchers working on lignins, the presence of conjugate esters on native biomass is of great importance. Indeed, while lignins derived from the kraft process are generally lacking in these naturally occurring esters (sometimes referred to as "clip-offs"), their presence in native biomass and in lignins derived from more gentle isolation techniques make for a potentially higher monomer yield from certain lignins, which is generally not discussed in the literature. These subunits are of particular interest when the acylating conjugate is of aromatic nature (e.g. *p*-coumaric acid or *p*-hydroxybenzoic acid) as is seen in some grass (maize) [7, 8] and hardwood (palm and poplar) lignins [9, 10].

The heterogeneity of lignins and the interactions of subunits (monolignols) is highlighted by a recent report by Harman-Ware et al. [11]. Using dehydrogenase polymer (DHP)-lignin model polymers, it was found that H units have a tendency to cluster during synthesis. This can translate to lower molecular weight in G-based polymers and further verifies that H-containing substrates act as capping units. While the translation of this study into the cell wall is uncertain, the implications on transgenic lignins could be considerable. This is especially true if the same were to hold true for Zip™ [3, 4] or high-S lignins [12]. Indeed, the potential for clusters of highly labile or recalcitrant native lignins could translate into significant economic effects with regards to its depolymerization in the biorefinery. Thus, in order to take advantage of this, depolymerization techniques should be either robust enough to depolymerize the lignin regardless of linkage or selected based on the linkage distribution of the lignin.

While somewhat beyond the scope of this chapter, it is nevertheless important to address the extraction of lignin, as it plays an integral role in lignin structure and molecular weight. However, because lignin extraction has been discussed in detail in past reviews [6, 13], it will be discussed only briefly here. Importantly, the extraction strategy employed is directly related to the solubility of lignin [14]. Indeed, lignins extracted via sulfite pulping tend to have increased water

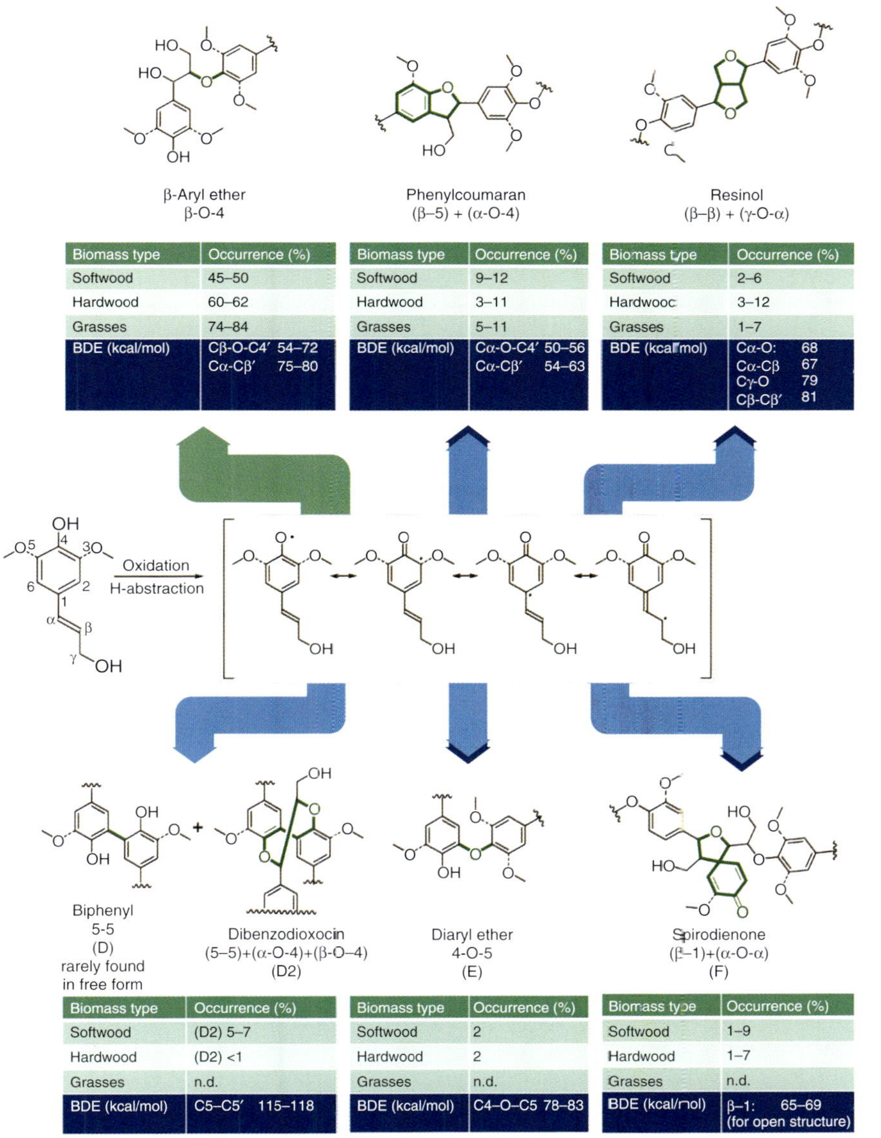

Biomass type	Occurrence (%)
Softwood	45–50
Hardwood	60–62
Grasses	74–84
BDE (kcal/mol)	Cβ-O-C4′ 54–72 Cα-Cβ′ 75–80

Biomass type	Occurrence (%)
Softwood	9–12
Hardwood	3–11
Grasses	5–11
BDE (kcal/mol)	Cα-O-C4′ 50–56 Cα-Cβ′ 54–63

Biomass type	Occurrence (%)
Softwood	2–6
Hardwood	3–12
Grasses	1–7
BDE (kcal/mol)	Cα-O: 68 Cα-Cβ 67 Cγ-O 79 Cβ-Cβ′ 81

Biomass type	Occurrence (%)
Softwood	(D2) 5–7
Hardwood	(D2) <1
Grasses	n.d.
BDE (kcal/mol)	C5–C5′ 115–118

Biomass type	Occurrence (%)
Softwood	2
Hardwood	2
Grasses	n.d.
BDE (kcal/mol)	C4–O–C5 78–83

Biomass type	Occurrence (%)
Softwood	1–9
Hardwood	1–7
Grasses	n.d.
BDE (kcal/mol)	β-1: 65–69 (for open structure)

Figure 8.2 Diagram highlighting the bonding motifs and potential linkages as targets for depolymerization (in green), % occurrence values from the literature, and bond dissociation energies for a range of commonly encountered linkages/bonding motifs in native lignins. It is important to point out that many of these % occurrence values, although reported in the literature, are unlikely or even untenable: β-1 moieties likely do not exceed 1–2% of all structure (substantially lower than the 9% reported); it is impossible to encounter 19–27% of 5–5 units in any lignin (it is probably restricted to a maximum of approximately 9% in softwoods); and the abundance of 4-O-5 linkages in softwoods is almost certainly much lower than the 4–7% claimed. Source: Rinaldi et al. 2016 [26]. Reproduced with permission of Wiley-VCH Verlag GmbH & Co. KGaA.

solubility, while those extracted via the kraft process are typically only soluble in alkaline water and polar organic solvents (such as DMSO, DMF, and pyridine). Other organosolv lignins have a wider range of solubility in organic solvents (alcohols and acetone in addition to DMSO, DMF, and pyridine); however, they tend to lack water solubility. Some other lignin extraction techniques include γ-valerolactone (GVL) [15, 16], mild acidolysis [17], copper alkaline hydrogen peroxide (Cu-AHP) [18], ammonia fiber expansion (AFEX) [19], extractive ammonia (EA) [20], and enzyme extraction [21]. All of these extraction processes imprint their own molecular weight and solubility profile on lignin. In some cases, the extraction technique can lead to oxidation or even wildly different lignin structures. Indeed, acidic conditions tend to dehydrate C_α–OH leading to enol ethers and Hibbert's ketones or condensation of bonds [22, 23]. Additionally, Cu-AHP tends to lead to an increased amount of oxidized bonds [24].

Due to the heterogeneous nature of lignin and its possible degradation during extraction, the results of lignin depolymerization techniques can be difficult to compare. There is a general need for a comparative study of lignin depolymerization techniques using the same lignin and extraction technique. A similar study – discussed in the succeeding text – was performed by Stahl and coworkers [25], who applied their two-step oxidation/hydrolysis method of lignin depolymerization to several different lignins obtained using a variety of extraction techniques. While this makes for an incredibly useful source of comparison, the evaluation of this depolymerization strategy vs. different methods is still absent from the literature.

8.1.1 Cautionary Statements

One of the major issues with the lignin literature is the clear lack of understanding of the lignin polymer. Indeed, many authors take the description of the biopolymer as "random" too literally, which results in some wildly inaccurate representative lignin structures. This is increasingly dangerous from the perspective of lignin model systems, as many authors use representative lignin models that are either absent or impossible in the biopolymer. In view of the latter, one should be careful to make sure that the lignin structure or model used is representative of lignin itself.

Similar issues arise in studies that ignore the presence of phenols in lignin. Unfortunately, under oxidative depolymerization conditions, phenols have the propensity to undergo radical reactions to form more recalcitrant condensed linkages [27]. While the use of methoxylated β-ethers – i.e. bearing this functionality in the 4-position of the A ring (see Figure 8.3) – make for good models of interior lignins, ignoring the phenolic B-ring product is ill-advised given that depolymerized β-ethers will likely yield a copious amount of phenolic products.

Lastly, the use of appropriate solvents for lignin oxidation is paramount. Indeed, while lignin model dimers – such as the commonly used β-aryl ether linkage – are soluble in a range of solvents, most lignins lack this dissolution property. However, most lignins are at least partly soluble in solvents such as DMSO, DMF, and pyridine. While other less common solvents may also dissolve lignin, it is advised to stay away from nonpolar solvents such as toluene. In short, results obtained with solvents that do not dissolve lignin should be considered with skepticism.

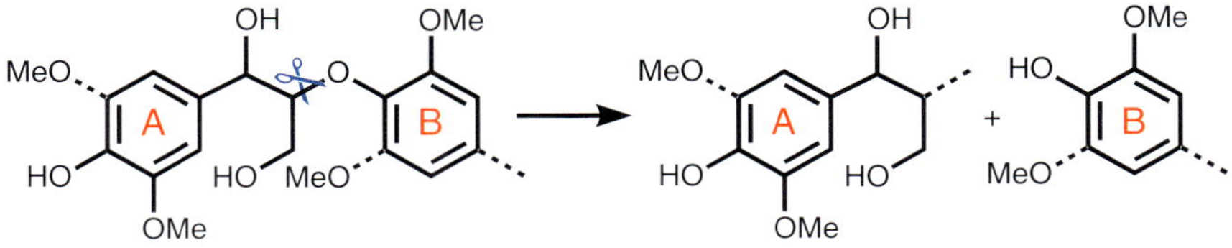

Figure 8.3 Cleavage of the C_{β}—O_4 bond leads to phenolics.

8.2 Catalytic Systems for the Oxidative Depolymerization of Lignin

8.2.1 Enzymes and Bio-mimetic Catalysts

It is well known that lignin is degraded in nature by fungi (namely, the white-rot variety) [28]. These lignin-degrading fungi evolved around 250–300 million years ago during the Permian period and helped to relieve the tremendous carbon sink (which would eventually lead to coal), resulting from the inability to degrade lignin during the Carboniferous period (ca. 300–360 million years ago) [29, 30]. Indeed, white-rot fungi, the most commonly studied being the basidiomycete *Phanerochaete chrysosporium* [31], degrade lignin via the secretion of enzymes such as lignin peroxidase (LiP) and manganese-dependent peroxidase (MnP) with *in situ* generation of hydrogen peroxide. Thus, it follows that many researchers have studied the use of peroxidase enzymes and their active cofactors for lignin degradation. Both LiP and MnP operate via a metalloporphyrin prosthetic group – iron protoporphyrin IX (Fe^{3+}-PPIX) – for lignin depolymerization. During the course of oxidation, the Fe^{3+}-PPIX cofactor undergoes a 2-electron oxidation with hydrogen peroxide to give the oxo-Fe^{4+}-PPIX π-cation radical active species. The reactive species then performs two 1-electron oxidations on the substrate (lignin, radical mediator, or Mn^{2+}) to complete the catalytic cycle [32]. Unfortunately, LiP is prone to over-oxidation, thus limiting its industrial application [32]. While both LiP and MnP are characterized as PPIX-containing glycosylated heme proteins, their catalytic cycle is slightly different. Indeed, LiP abstracts single electrons from the aromatic rings of lignin (or of lignin oxidation mediator) resulting in radical cations and subsequent cleavage of the polymer. MnP, on the other hand, operates through the oxidation of chelated Mn(II) cations to form chelate-stabilized Mn(III) ions, which diffuse from the enzyme to oxidize lignin (Scheme 8.1) [33].

The activity and isolation of enzymes, such as those from the white-rot fungi *P. chrysosporium*, are the biological inspiration for the use of metalloporphyrins for lignin oxidation [34]. The bulky nature of the protein protects the iron porphyrin cofactor from coupling to form the catalytically inactive μ-oxo dimer species and from other oxidative degradation processes. However, metalloporphyrins can also be substituted with phenyl substituents at the meso-sites, which prevent oxidative degradation of the catalysts. Moreover, the porphyrin ring can also be functionalized with solubilizing groups, making these biomimetic catalysts highly tunable [35].

Shimada et al. [36] described the use of an iron porphyrin catalyst for the oxidation of a β-1 model compound. Like the biological analogue, the catalyst was able

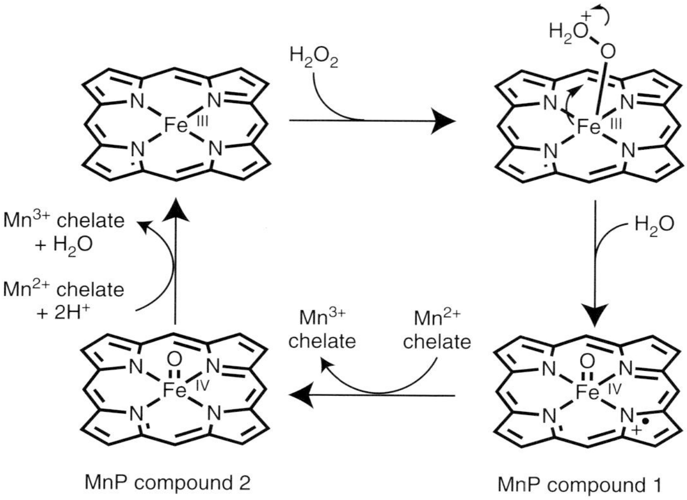

Scheme 8.1 Catalytic cycle of MnP.

to cleave the C_α—C_β bond, leading to monomeric products. Meier and coworkers also published a series of articles describing the effect of metalloporphyrin oxidation on lignin model compounds [37–40]. Similar to Shimada et al., when the iron tetraphenylporphyrin chloride (TPP(Fe)Cl)/tert-butylhydroperoxide (tBHP) catalyst system was applied to β-1 and β-5 model compounds, the oxidation led to cleavage of C—C bonds and the production of monomeric aldehydes and/or acids [37]. This is in contrast with the oxidation of a β-O-4 model, which underwent selective oxidation of the benzylic position to a ketone (Table 8.1) [37, 40]. Other examples of biomimetic metalloporphyrin catalysts include a report by Crestini et al. [41] who found that a Mn-porphyrin/H_2O_2 system was more active for the oxidation of non-phenolic 5–5′ linkages than the analogous Fe system. The main product of this reaction is quinone, which is proposed to form via a hemiacetal intermediate. Application of a cationic Mn-porphyrin to the oxidation of kraft lignin showed a small decrease in the aliphatic and guaiacyl –OH content. Additionally, Cui et al. [42] performed oxidations on a number of model compounds that mimic the major linkages in lignin (β-O-4, β-1, β-5, 5-5) as well as veratryl alcohol using a biomimetic water-soluble iron porphyrin catalyst, which achieved a similar product mixture to lignolytic cultures from *P. chrysosporium*. When applied to kraft lignin (Indulin AT), degradation was observed by gel permeation chromatography (GPC), although the authors were not specific as to the extent.

Metallophthalocyanins have also been used as an alternative to metalloporphyrins due to their relatively inexpensive synthesis. Indeed, Cui and Dolphin [34] utilized Fe and Mn metallophthalocyanins for the oxidation of veratryl alcohol with a variety of oxidants. Similar to Crestini et al.'s findings [41], the best results were obtained with the Mn catalyst when *m*-chloroperbenzoic acid (*m*-CPBA) was used as a terminal oxidant. Unfortunately, yields were low in all cases, which is likely due to the poor stability of the catalyst in the oxidizing medium.

Table 8.1 Oxidation of lignin model compounds using TPP(Fe)Cl/tBHP according to Shimada (gray background) [36] and Meier (white background) [37].

Lignin model compound	Products
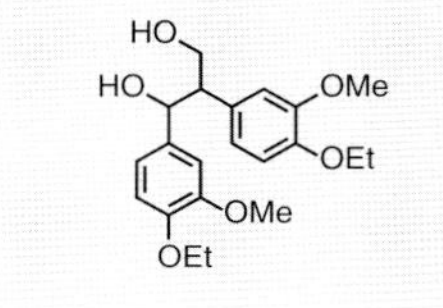	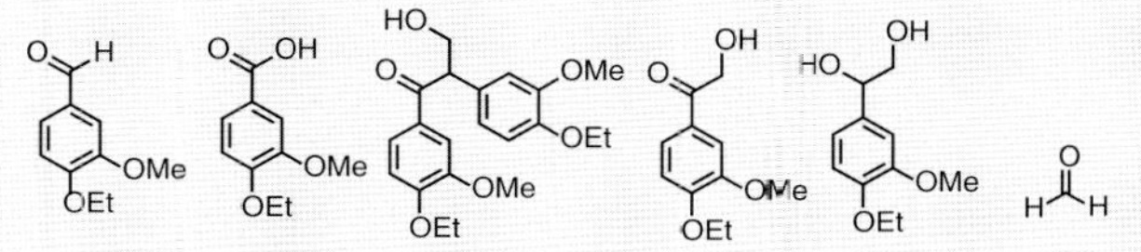 54%, 5%, 3%, 25%, 3%, 12%
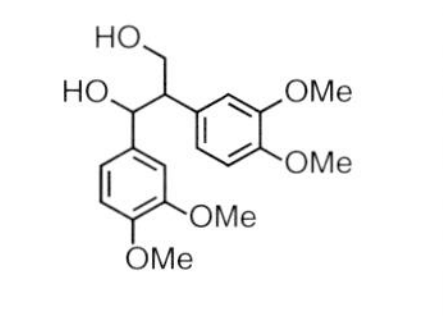	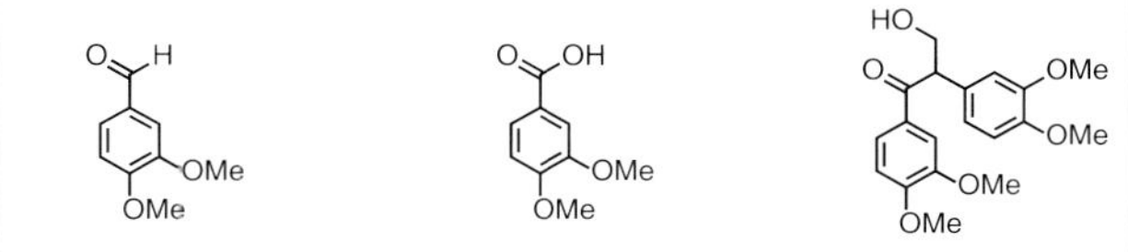 50%, 15%, 8%
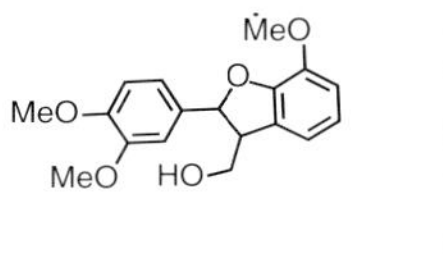	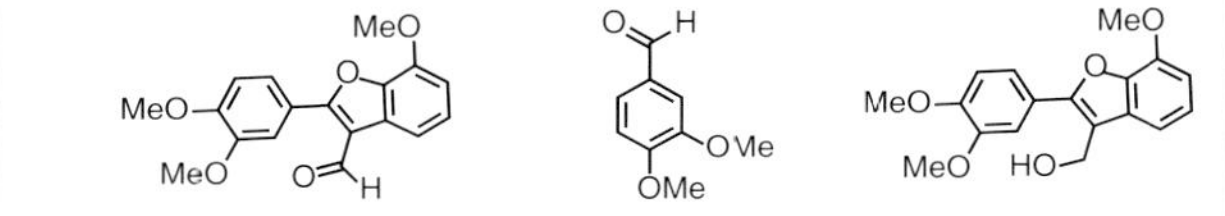 22%, 16%, 3%
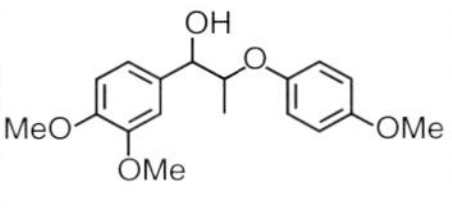	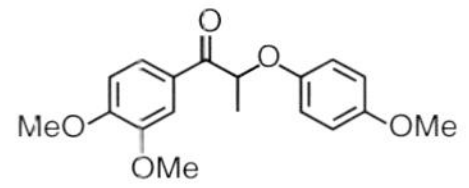 22%

8.2.2 Cobalt Schiff Base Catalysts

Schiff base catalysts – such as cobalt(salen) (salen = [*N*,*N*′-bis(salicylidene) ethane-1,2-diaminato]) complexes – are another class of catalysts that readily perform oxidation on lignin and lignin model compounds. Upon exposure to molecular oxygen, these Co(salen) catalysts form cobalt-superoxo or peroxo dimer complexes, which are the active species for oxidation reactions (Figure 8.4) [6]. Compared with metalloporphryins, Co(salens) are relatively cheap, more stable, and easier to synthesize while offering similar tenability [6].

Perhaps one of the most auspicious examples of Co(salen) oxidation of lignin model compounds can be found in a report by Bozell et al. [43], who found that Co(salen) as well as Co(N-Me salpr) and Co(salen-pyridine) complexes were effective catalysts for the formation of benzoquinones from monomeric lignin phenolic compounds. Interestingly, Bozell et al. found that 5-coordinate Co(N-Me salpr) and Co(salen-pyridine) complexes were more effective catalysts for the oxidation of S-type *para*-substituted phenolic benzylic alcohols than the 4-coordinate Co(salen) complex analogue. Conversely, no conversion was seen when the aldehydic analogue was oxidized using the 5-coordinate catalyst, while the 4-coordinate catalyst effectively converted syringaldehyde to the corresponding benzoquinone. Moreover, when a methoxy-substituted 4-coordinate Co(salen) catalyst was used, activity was slightly inhibited. This suggests the aldehydic substrates undergo coordination prior to the formation of the

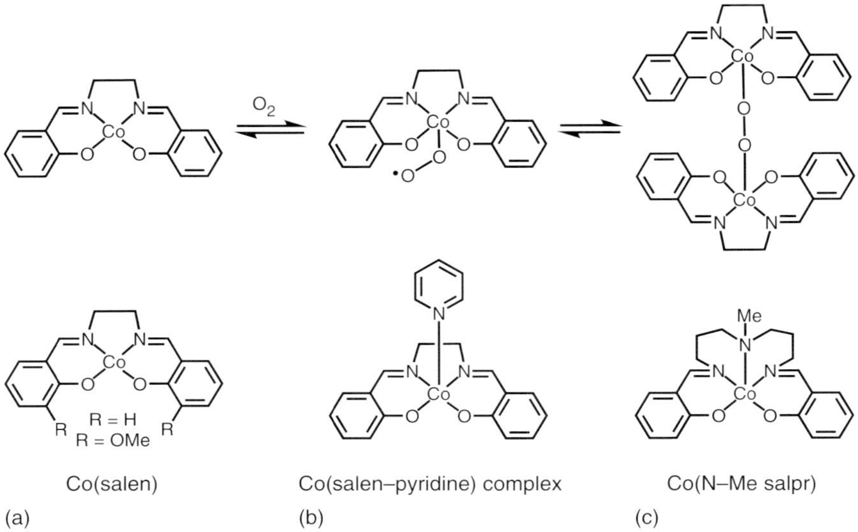

Figure 8.4 (a) Co(salen) complex, (b) cobalt superoxo complex, and (c) cobalt peroxo complex according to Bozell el al. [43].

active super-oxo-Co(salen) species. This is due to the fact that electron-donating groups added to the salen ligand are expected to lower the Lewis acidity of the Co via inductive effects, therefore causing the aldehydic substrate to bind more poorly. On the other hand, S-type *para*-substituted benzylic alcohols likely undergo proton abstraction from the phenolic O–H by the Co-superoxo species, which is stabilized by the aromatic ring (Figure 8.5). Subsequently, a second Co-superoxo species or O_2 adds to the *para*-position of the phenol, which then rearranges to give the benzoquinone as well as L_nCoOOH or L_nCoOH. Finally, the cobalt species either breaks down to regenerate the starting catalyst (L_nCoOOH) or can act as catalyst in its own right (L_nCoOH). Interestingly, when G-type phenols were used, very little activity was seen with any catalyst. Indeed, 5-coordinate catalysts were completely ineffective, while the 4-coordinate catalysts exhibited only slight activity. The authors suggest that this may be due to the inability of the Co(salen) catalysts to generate the phenoxy radical required for the formation of the quinone. Alternatively, the authors suggest that the transition state during hydrogen removal is insufficiently polar when the second electron-donating methoxy group is not present, or that the G-type phenol does not have significant enough electron donation potential to lower the bond energy of the phenolic O–H; however, these suggestions are purely speculative. Conversely, Drago et al. [44] found the Co(salen)-like catalyst Co(N-Me-salpr) to be an effective catalyst for the oxidation of isoeugenol at 60 °C under 75 psi of O_2 in toluene for ca. 3.5 hours. The products formed from this reaction were vanillin (50%) and a β-5 dimer product (20%). The difference between Drago et al.'s [44] and Bozell et al.'s [43] findings suggest that the reactivity of the catalyst is highly dependent on the *para*-substituent. In a recent contribution

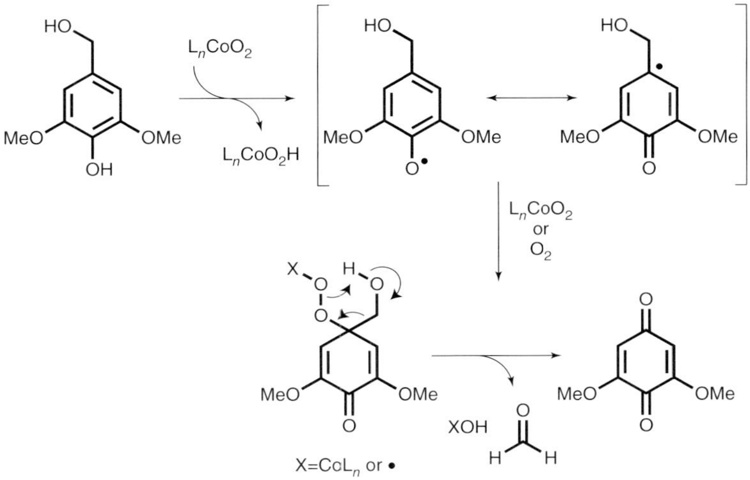

Figure 8.5 Proposed mechanism for the formation of quinone from the oxidation of p-substituted phenolics according to Bozell et al. [43].

from Biannic and Bozell [45], a 4-coordinate Co(salen) catalyst with a bulky nitrogen base substituent was found to be an effective catalyst for the oxidation of both S- and G-type model compounds. More importantly, upon oxidation of a β-O-4 model compound containing both S- and G-type rings, both S-type and G-type benzoquinones were formed in good and modest yields, respectively (Table 8.2). Application of this reaction system to a tulip poplar lignin obtained via organosolv fractionation led to ca. 3.5% isolated small molecule products. This low yield is likely due to the low abundance of phenolic moieties in the lignin isolated via organosolv fractionation.

In order to produce a more industrially relevant catalyst, Badamali et al. [46] immobilized/entrapped Co(salen) on SBA-15 (mesoporous silica). The Co(salen) complex is believed to be immobilized via H-bonding with free silanol groups on the surface of the SBA-15 support. This is substantiated by a blue shift (ca. 25 nm) in the DRUV–Vis (diffuse reflectance ultraviolet–visible) spectrum – corresponding to support interactions in the charge-transfer region of the spectrum (280–460 nm) – with an intense peak at 375 nm. Apocynol (4-(1-hydroxyethyl)-2-methoxyphenol) was oxidized with hydrogen peroxide and the Co(salen)/SBA-15 catalyst under microwave irradiation, achieving full conversion to phenolic dimers and quinones in 40 minutes. The catalyst proved to be nearly twice as active as the homogeneous analogue, converting 88% of the apocynol with 50% selectivity (of identifiable products) to acetovanillone in 30 minutes. In a follow-up contribution, Badamali et al. [47] utilized the same catalyst to oxidize a lignin β-O-4 model, obtaining 2-methoxyphenol as the major product. Other products from the reaction were not discussed. Optimal catalyst activity was achieved in 20 minutes with a turnover number (TON) of 67, which is ca. 5.5 times more active than the corresponding homogeneous

Table 8.2 Oxidation of S- and G-type model compounds with base-substituted Co-Schiff base catalyst.

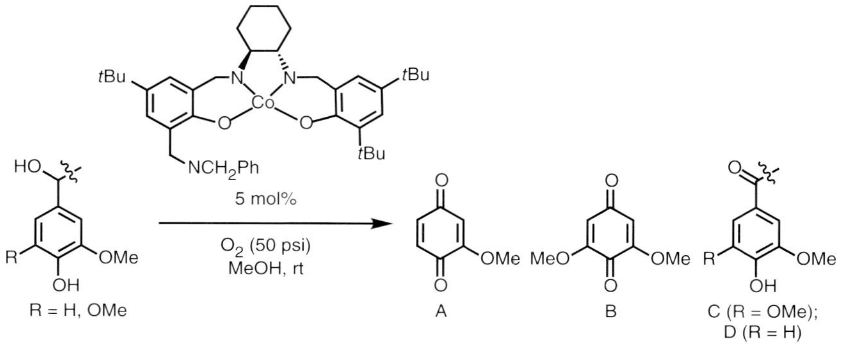

Entry	Substrate	Time (h)	A (%)[a]	B (%)	C or D (%)
1	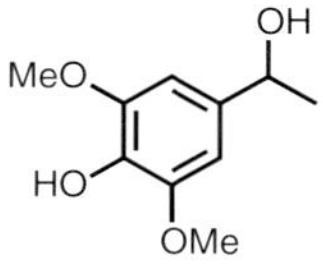	16	–	72	11
2	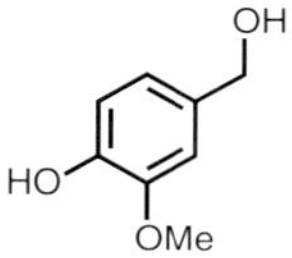	16	83	–	–
3[b]	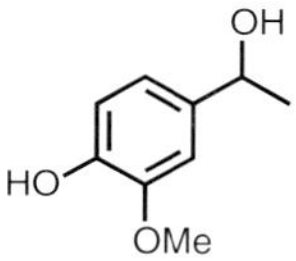	24	51	–	22
4[c]	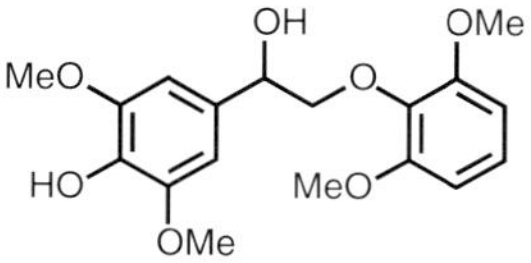	16	–	81	Traces
5	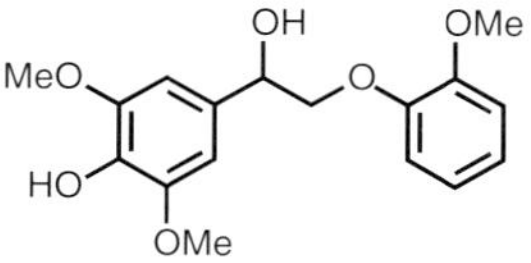	16	17	86	Traces
6[b]	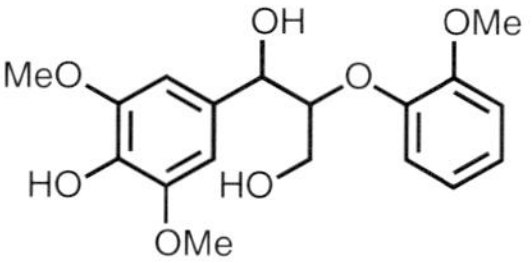	48	21	64	Traces

a) Yields are given for isolated material.
b) 10 mol% of catalyst used.
c) Yields relative to 2 equiv of product formed.
Source: Adapted with permission from [45]. Copyright 2013, American Chemical Society.

catalyst. Under conventional heating only high molecular weight products were obtained. Moreover, conversions calculated from the TON reported by the authors were low (ca. 21%).

8.2.3 Vanadium Catalysts

Organometallic vanadium catalysts have also been heavily studied for the oxidation and cleavage of lignin model compounds. Indeed, Son and Toste [48] utilized a vanadium Schiff base catalyst (Figure 8.6) for the oxidation of a lignin β-O-4 model compound to form a ketone as well as monomeric products produced through C—O bond cleavage. Hanson et al. [49] also utilized this catalyst, showing that product selectivity was not dependent on the presence of phenolic moieties in the starting material. This suggests that catalysis likely occurs via a different reaction pathway than for the other Schiff base catalysts mentioned previously (vide supra). The mechanism proposed by Son and Toste [48] involves hydrogen abstraction from the benzylic O–H, followed by the consecutive elimination of aryloxyl and hydroxyl radicals to form an enone (Figure 8.6, purple). Conversely, when Hanson et al. [49] used an 8-quinolinate-ligated vanadium complex, a strong dependence on the presence of phenolic moieties was observed (Figure 8.6, green). Indeed, while the non-phenolic model underwent benzylic oxidation as well as water elimination (to afford product D in Figure 8.6), the phenolic model produced a benzoquinone product resulting from C(alkyl)—C(phenyl) bond cleavage.

Hanson et al. [50] also obtained carbon–carbon bond cleavage on several lignin model compounds utilizing (dipic)V^V(O)O^{i}Pr as a catalyst at 100 °C in air. Good conversions were obtained (≥94%); however, in most cases reaction times were quite long (6–7 days). Recently, Xu and coworkers [51] performed oxidation on the simple lignin β-O-4 model dimer 2-phenoxy-1-phenylethanol with vanadyl

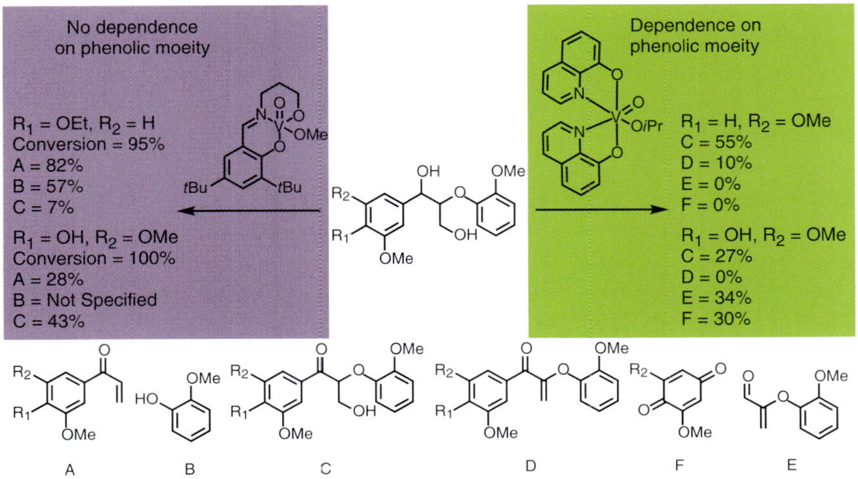

Figure 8.6 Oxidation of lignin model compounds with vanadium catalysts according to Son and Toste [48] and Hanson et al. [49] (purple) and Hanson et al. (green) [49].

acetylacetonate using O_2 as the terminal oxidant. Reactivity of the substrate was found to be heavily dependent on the solvent. When Et_3N or DMSO was used as solvent, the ketone product of benzylic alcohol oxidation was found to be the main product, although conversions were low (ca. 20%). Acetonitrile and ethyl acetate yielded mostly C_α—C_β bond cleavage products along with significantly improved conversion. Organic acids (acetic acid, propanoic acid, and butyric acid) were found to be the most active solvents, yielding products derived from C_α—C_β bond cleavage. The authors performed ^{51}V NMR on the reaction systems in the absence of substrate using the solvents acetic acid, acetonitrile/acetic acid (90/10), and acetonitrile/Et_3N (90/10) to test for the presence of V^V species (V^{IV} is NMR inactive due to its paramagnetism). Interestingly, when acetic acid was present, a signal at ca. −540 ppm was detected, while in the presence of Et_3N, the signal was found at −589 ppm. This suggests that coordinated Et_3N likely acts as a stronger electron-donating species, which more effectively shields the V^V. This likely also affects the redox activity of the catalysts, causing the difference in activity when Et_3N is used compared to acetic acid, although the actual role of acetic acid is yet to be elucidated.

Similar results were seen when $(HQ)_2V(O)(O^iPr)$ was used as a catalyst for β-1 compounds [52]. Indeed, a marked difference was observed between oxidation of phenolic and non-phenolic models. Whereas non-phenolic models produced mainly benzylic oxidation products, phenolic models produced a large amount of quinone. Interestingly, when the aforementioned (dipic)$V^V(O)O^iPr$ catalyst was used on the phenolic β-1 model compound, the main product was the vinyl aldehyde derivative. This is contrasted with the use of (dipic)$V^V(O)O^iPr$ on non-phenolic β-O-4 compounds, which produces mainly the carbonyl products, and further to the non-oxidative C—O bond-cleavage enone product [53].

8.2.4 Methyltrioxorhenium (MTO) Catalysts

Perhaps one of the most active catalysts is methyltrioxorhenium (MTO), which has been shown by many groups to oxidize lignin and lignin model compounds effectively [32, 54]. Crestini et al. [55] studied the oxidation of lignin and lignin model compounds with MTO and hydrogen peroxide. This system was effective at cleaving the alkyl side chain as well as the aromatic ring in β-O-4 lignin model compounds. While complete degradation of the model was seen, low mass balances, believed to be due to over oxidation, were characteristic of the reaction. The main identifiable product of the reaction was the phenolic fragment derived from the B-ring of the dimer; however the aldehyde product of C_α—C_β bond cleavage and vinylic products were also obtained. The authors also carried out the oxidation of technical lignins (hydrolytic sugar cane lignin, red spruce kraft lignin, and hardwood organosolv lignin). The reactivity of the lignins was evaluated by phosphitylation followed by ^{31}P NMR spectroscopic analysis. In all cases significant side chain and aromatic ring oxidation was seen. In a follow-up report, Crestini et al. [56] immobilized MTO on poly(4-vinylpyridine) and polystyrene. The heterogenized catalysts showed similar conversions to homogeneous MTO, and mass balances were moderately improved. The authors suggested that the improved performance of the catalyst may be due to the tunability of the catalyst via support effects (nitrogen ligation, reticulation grade, support oxidation state).

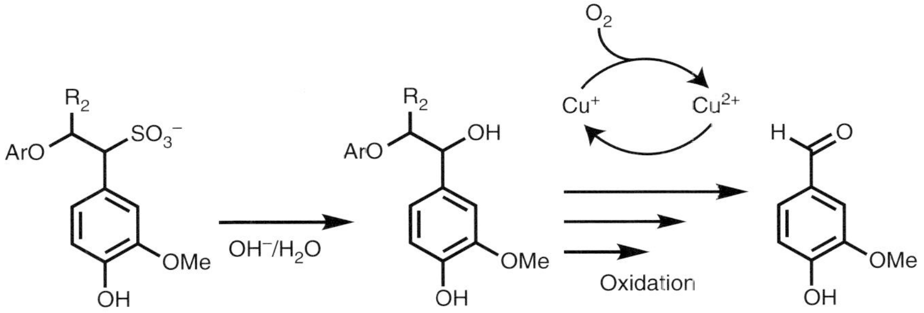

Figure 8.7 Borregaard vanillin process. Source: Reproduced with permission from Calvo-Flores and Dobado 2010 [58]. Copyright 2010, Wiley-VCH Verlag GmbH & Co KGaA, Weinheim.

8.3 Commercial Products from Lignin

Much of the early work in lignin utilization was performed by the pulp and paper industry. Indeed, the pulp and paper industry alone produced more than 50 million tons of lignin in 2004 [57]. Of the 50 million tons produced annually, only ca. 2% is used commercially, the remainder being used to generate process heat. However, niche markets have been found for lignin streams produced by pulping, such as the production of vanillin. Borregaard LignoTech produces vanillin commercially from lignosulfonate using a copper catalyst (Figure 8.7) [58, 59]. While the process is proprietary, it is known to operate via Cu oxidation of sulfonated lignin. Other commercial products include agricultural binders, dispersants, oil field chemicals, etc.

8.4 Stepwise Depolymerization of β-O-4 Linkages

8.4.1 Benzylic Oxidation

Recently, several researchers have shown stepwise oxidative depolymerization to be very effective when applied to both model β-O-4 linkages and lignin itself. Stepwise oxidative depolymerization typically involves the oxidation of the benzylic alcohol position in lignin, which weakens the bond strength of the C_{β}—O_4 bond by ca. 87 kJ/mol [60]. Indeed, several groups have reported the use of selective organo-catalysts for the oxidation of this linkage. TEMPO (2,2,6,6-tetramethylpiperidine-*N*-oxyl) and DDQ (2,3-dichloro-5,6-dicyano-1,4-benzoquinone) have been found to be ideal candidates for this reaction, especially in combination with a nitrite species as a cocatalyst and oxygen as the terminal oxidant [40, 61–65]. Stepwise oxidative lignin depolymerization was pioneered by Stahl and coworkers who studied a range of nitroxyl radical and metal catalysts for the oxidation of a methoxylated β-O-4 lignin model compound (Table 8.3), finding acetamido-TEMPO/HNO_3/HCl/O_2 to be an ideal reaction system for the selective oxidation of the benzylic alcohol [64]. When applied to the aforementioned model, the reaction selectively yielded the benzylically oxidized product in a 95% yield. Moreover, when applied to lignin

Table 8.3 Oxidation of β-O-4 lignin model compounds with nitroxyl radical catalysts as reported by Stahl et al. [64].

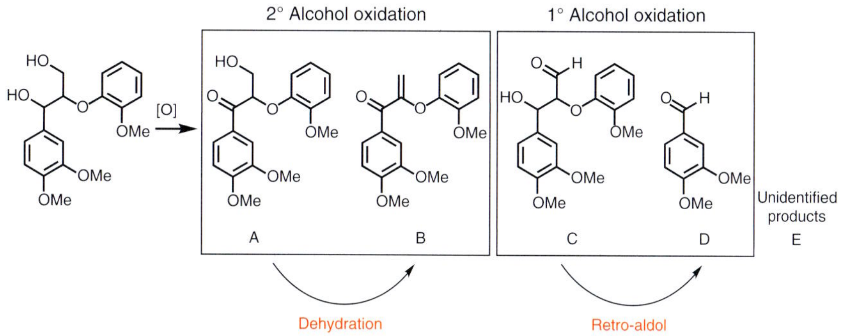

Entry	N-nitroxyl radical or metal salt (mol%)	Ligand (mol%)	Additives (mol%)			Oxidation	Solvent	T (°C)	Time (h)	Conversion (%)[a]	Yield (%)[a]				
			1	2	3						A	B	C	D	E
1[b]	TEMPO (4)	–	$NaNO_2$ (6.5)	HCl (13)	–	1 atm O_2	MeCN	60	16	66	64	0	0	0	–
2[b]	AcNH-TEMPO (3.3)	–	$NaNO_3$ (3.3)	–	–	1 atm O_2	$AcOH:H_2O$ (20:1)	80	16	>85	–	–	–	–	–
3[b]	AcNH-TEMPO (3.3)	–	HNO_3 (2)	–	–	1 atm O_2	$AcOH:H_2O$ (20 : 1)	80	16	>85	–	–	–	–	–
4[b]	TEMPO (5)	–	HNO_3 (20)	–	–	Air	$AcOH:H_2O$ (20:1)	rt	24	41	31	0	0	2	–
5[b]	AcNH-TEMPO (6.6)	–	HNO_3 (30)	–	–	Air	$AcOH:H_2O$ (20:1)	rt	24	50	42	0	5	1	–
6[b]	AcNH-TEMPO (6.6)	–	$NaNO_2$ (100)	HNO_3 (10)	–	Air	$AcOH:H_2O$ (20:1)	rt	12	46	24	10	6	1	–
7[b]	AcNH-TEMPO (5)	–	$NaNO_3$ (10)	HCl (10)	–	Air	$MeCN:H_2O$ (19 : 1)	rt	24	9	8	0	0	0	–
8[b]	4-MeO-TEMPO (5)	–	HNO_3 (10)	HCl (10)	–	1 atm O_2	$MeCN:H_2O$ (19 : 1)	45	24	86	79	0	0	2	–
9[b]	AcNH-TEMPO (5)	–	HNO_3 (10)	HCl (10)	–	1 atm O_2	$AcOH:H_2O$ (19 : 1)	45	24	54	30	4	0	2	–
10[b]	AcNH-TEMPO (5)	–	HNO_3 (10)	HCl (10)	–	1 atm O_2	$EtOAc:H_2O$ (19 : 1)	45	24	44	13	2	0	3	–

11[b)]	AcNH-TEMPO (5)	–	HNO_3 (10)	HCl (10)	–	Air	MeCN:H_2O (19 : 1)	rt	24	74	68	0	0	0	–
12[b)]	AcNH-TEMPO (5)	–	HNO_3 (10)	HBr (10)	–	Air	MeCN:H_2O (19 : 1)	rt	24	91	72	11	0	4	–
13[b)]	AcNH-TEMPO (5)	–	HNO_3 (10)	HCl (10)	–	1 atm O_2	MeCN	45	24	89	83	0	0	0	–
14[b)]	TEMPO (5)	–	HNO_3 (10)	HCl (10)	–	1 atm O_2	MeCN:H_2O (19 : 1)	45	20	98	90	0	0	0	–
15[b)]	AcNH-TEMPO (5)	–	HNO_3 (10)	HCl (10)	–	1 atm O_2	MeCN:H_2O (19 : 1)	45	20	100	95	0	0	0	–
16[b)]	AZADO (5)	–	HNO_3 (10)	HCl (10)	–	1 atm O_2	MeCN:H_2O (19 : 1)	45	20	98	92	0	0	2	–
17[c)]	$Cu(OTf)(MeCN)_4$ (10)	bpy (10)	TEMPO (10)			1 atm O_2	MeCN	60	24	77	9	0	2	25	13
18[c)]	$Cu(OTf)(MeCN)_4$ (10)	bpy (10)	TEMPO (10)	NMI (20)		1 atm O_2	MeCN	60	24	56	0	0	1	30	9
19[c)]	CuBr (10)	bpy (10)	4-MeO-TEMPO (10)			1 atm O_2	MeCN	60	24	78	2	26	2	30	13
20[c)]	CuBr (10)	bpy (10)	TEMPO (10)	DABCO (10)		1 atm O_2	MeCN	60	24	86	6	23	3	31	11
21[c)]	CuBr (10)	bpy (10)	TEMPO (10)	–	–	1 atm O_2	MeCN	60	24	89	10	0	0	37	14
22[c)]	CuI (10)	bpy (10)	–	–	–	1 atm O_2	MeCN	60	20	32	23	0	0	3	0
23[c)]	$Cu(OTf)_2$ (5)	4,4′-$t$$Bu_2$-bpy (5)	TEMPO (10)	DABCO (10)	NMI (5)	1 atm O_2	MeCN	60	24	70	0	21	2	22	0
24[c)]	$Cu(TFA)_2$ (5)	4,4′-$t$$Bu_2$-bpy (5)	TEMPO (10)	DABCO (10)	NMI (5)	1 atm O_2	MeCN	60	24	71	1	21	3	24	3
25[c)]	$Pd(OAc)_2$ (5)	–	–	–	–	1 atm O_2	DMSO	60	15	44	42	0	0	0	0
26[c)]	$Pd(OAc)_2$ (5)	2,9-Dimethyl-1,10-phenathroline (10)	–	–	–	1 atm O_2	DMSO:H_2O (40 : 60)	60	15	87	56	0	0	0	0
27[c)]	$Fe(NO_3)_3$ (5)	–	TEMPO (10)	NaCl (10)	–	1 atm O_2	DCE	60	20	100	79	0	0	2	0
28[c)]	$Fe(NO_3)_3$ (10)	Salen (10)	–	–	–	1 atm O_2	MeCN	60	20	0	0	0	0	0	0
29[c)]	$Fe(acac)_3$ (10)	Phen (10)	K_2CO_3 (100)	NaOH (50)		1 atm O_2	Toluene	60	20	43	19	0	0	8	0
30[c)]	$RuCl_2(PPh_3)_3$ (4)	–	TEMPO (12)			1 atm O_2	Cl–C_6H_5	60	20	30	17	0	3	5	0

a) Conversion and yields were determined by ^{1}H NMR spectroscopy vs. mesitylene as internal standard and relaxation time 25 seconds.
b) Reactions performed in 1 ml of solvent.
c) Reactions performed in 1.1 ml of solvent.
Source: Stahl et al. [64]. Reprinted with permission from the American Chemical Society.

(enzyme lignin extracted from aspen), ca. 91% of the benzylic alcohols in the β-O-4 linkages were found to be oxidized (according to volume integration of the HSQC spectrum). Alcohol oxidation with TEMPO involves a complex mechanism in which the TEMPO radical is typically first oxidized to the nitronium cation by a nitrite cocatalyst in the presence of oxygen. The nitronium cation then oxidizes the benzylic alcohol to the corresponding carbonyl compound and reduces the nitronium ion to TEMPO-H. TEMPO-H is finally reoxidized to the active nitronium species by the nitrite cocatalyst in the presence of oxygen (Scheme 8.2). As mentioned previously, other researchers have also found TEMPO or other nitroxyl radical-based catalysts to be effective for the oxidation of benzylic alcohols in lignin [40, 60–62]. Indeed, Meier et al. utilized a variety of TEMPO-based reaction systems finding that TEMPO/$NaNO_2$/NaCl/O_2 was the most effective for β-O-4 oxidation (obtaining an 84% yield of the benzylically oxidized β-O-4 model).

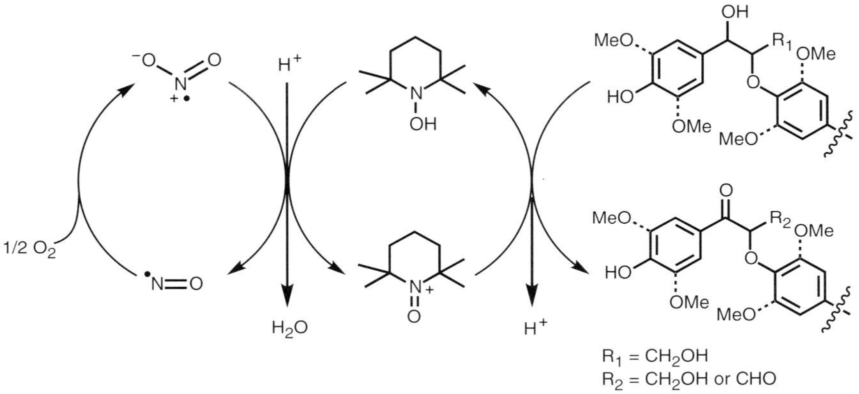

Scheme 8.2 Mechanism of TEMPO oxidation of a β-O-4 model compound in the presence of a nitrite cocatalyst.

In a follow-up report, Meier et al. compared the use of TEMPO in conventional solvents vs. ionic liquids, finding that no oxidation took place in [C1C4im]Cl or [P4444]Cl, whereas good to excellent yields were obtained in dichloromethane (DCM) (53–96% yield of the benzylic oxidation product on β-O-4 models) [38]. Patil and Yan [62] screened a variety of nitroxyl radical catalysts for the oxidation of lignin β-O-4 model compounds finding that the less sterically hindered ABNO (9-azabicyclo[3.3.1]nonane *N*-oxyl) and Me-AZADO (1-methyl-2-azaadamantane-*N*-oxyl) resulted in benzylic oxidation as well as in 1,2-keto esters and alkoxyamines. Additional screening of β-5 and β-1 model compounds (in DCM) revealed that TEMPO/$NaNO_2$/NaCl/O_2 is also effective in the oxidation of β-1 models, producing the benzylic ketone as well as the 1,2-diketone in a 46% and 12% yield, respectively. When a TEMPO/Cu(I)/O_2 system was applied to the same β-1 model, the majority of products were the result of C–C cleavage (benzylic aldehydes and acids), suggesting that the Cu species plays a larger role than simple reoxidation of the TEMPO [37]. As expected, the β-5 model produced primarily benzofuran carboxaldehyde.

Further work by Patil and Yan found that TEMPO/CuCl/bpy/NMI/O_2 could also be used for the oxidation of lignin model compounds. Indeed, under these conditions the C_α—C_β bonds of the β-O-4 models used were oxidatively cleaved to the corresponding acid and phenol through an alkoxyl amine intermediate (Figure 8.8) [61]. Additionally, Wang et al. utilized a TEMPO/$VOSO_4$/O_2 system for the oxidation of lignin model compounds. While this catalyst worked well for the oxidation of two-carbon β-O-4 models (i.e. those missing the γ carbon), when a three-carbon β-O-4 model was used, only a 53% yield of the benzylically oxidized product was obtained. Additionally, HSQC analysis of the oxidized organosolv Alcell lignin showed only a 36% yield of the oxidized product [60].

Sedai et al. [52] studied the oxidation of phenolic and non-phenolic β-1 model compounds using a CuOTf/TEMPO/2,6-lutidine/O_2 system, finding that C_α—C_β bond cleavage products were the dominant products obtained from the non-phenolic model. This contrasts with results obtained with CuCl/TEMPO/O_2, which produced similar products to those obtained over one-electron oxidants from a non-phenolic β-O-4 model compound [53]. Two possible mechanisms were proposed to explain these results. Either primary alcohol oxidation followed by retro-aldol reaction breaks the C_α—C_β bond or one-electron oxidation is followed by C—C bond fission due to the formation of a radical cation. In either case, the dominant products are the corresponding A-ring aldehyde and phenol rather than the benzylic ketone and/or γ-aldehyde. Interestingly, when a phenolic β-1 model was used, the primary product was the benzylic ketone (78%). Contrastingly, when a stoichiometric amount of CuOTf/TEMPO was employed, the major products were the A-ring quinone (82%) as well as the B-ring aldehyde (25%). Similar results were obtained using CuOTf/TEMPO on a phenolic β-O-4 model compound [66]. Indeed, when using a stoichiometric amount of reagent, the major product was the A-ring quinone (46%), while over a catalytic amount of CuOTf/TEMPO, the major product was the benzylic ketone (44%). Moreover, under catalytic conditions, the non-phenolic β-O-4 model compound produced mainly C_α—C_β bond cleavage products. While the reason for the difference in product selectivity under stoichiometric vs. catalytic conditions is not clear, the difference between phenolic and non-phenolic models likely involves a quinone methide intermediate during the oxidation of phenols, which cannot be formed from the non-phenolic models. Moreover, C_α—C_β bond cleavage products present in non-phenolic models may be the results of either retro-aldol reaction [67] or single-electron transfer [68], as is described in Scheme 8.3 [66]. In any case, it is abundantly

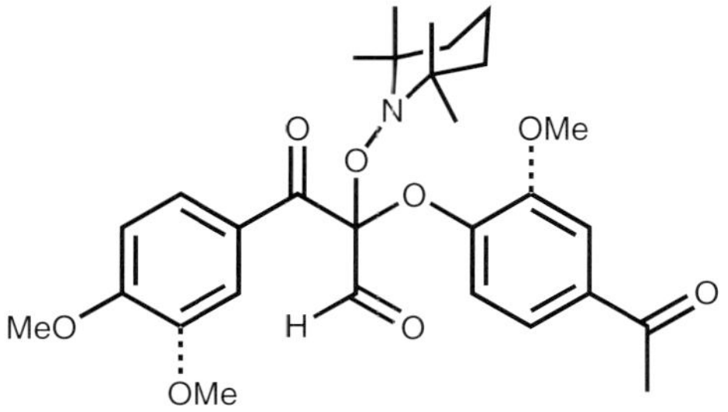

Figure 8.8 Alkoxyl amine intermediate as described by Patil and Yan [61].

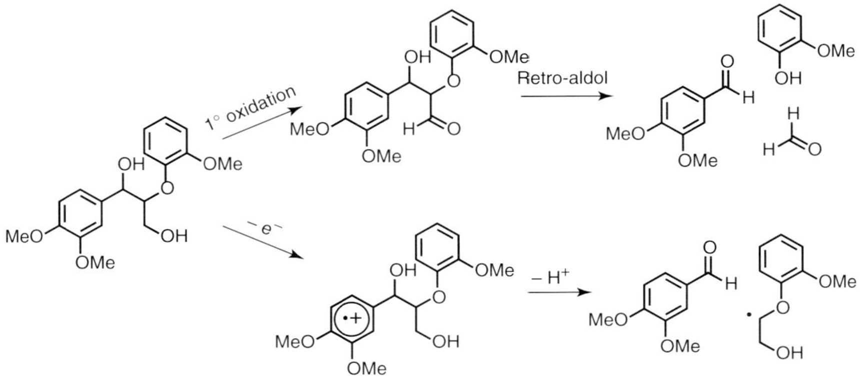

Scheme 8.3 Possible mechanisms of oxidative C_α—C_β bond cleavage of a β-O-4 model compound as proposed by Sedai and Baker [66].

clear that TEMPO oxidations in the presence of Cu cocatalyst operate under a significantly different mechanism than TEMPO/nitrite oxidations. This also suggests that the use of TEMPO-based catalytic systems could be tuned to fit the needs of the substrate employed.

Mechanochemical techniques have also been used in combination with TEMPO-based catalysts. Indeed, Bolm and coworkers [69] explored the use of a ball mill as well as a vibrating disk mill in the oxidative depolymerization of lignin. This method, which is scalable, was highly active for the oxidation of model β-O-4 dimers. Interestingly, when the A-ring was methoxylated in the 4-position, selective benzylic oxidation was obtained in a >94% yield. However, 4-hydroxy substituted A-rings yielded the depolymerized product. Indeed, quinone and phenol were obtained in good yield under these conditions. When applied to organosolv beechwood lignin, the authors reported a yield of monomer products as high as 16% using an OH-TEMPO/KBr/Oxone system.

Another attractive organo-catalyst for benzylic oxidation of lignin is DDQ. Indeed, in a technique pioneered by Westwood and coworkers [63], a catalytic amount of DDQ – used in the presence of *tert*-butylnitrite (co-catalyst) and O_2 – was found to selectively oxidize lignin and lignin models to their corresponding benzylic ketone in near-quantitative yields. In a follow-up publication, Westwood and coworkers describe the effects of a stoichiometric amount of DDQ and the ability to tailor the oxidation system [70]. While this is not a catalytic route to lignin oxidation, it does lend itself to further study, especially with regard to oxidation of other less abundant linkages such as the resinol (β-β) linkage [71]. Indeed, the resinol linkage appears to undergo conversion to an unusual pyranone linkage (see Scheme 8.4) in the presence of excess DDQ, as demonstrated by the characteristic NMR shifts of the pyranone structure located at δ_C/δ_H 52.9/4.65 for the methylene group and δ_C/δ_H 147.3/8.21 for the pyranone in DMSO-d6 [71]. While the use of stoichiometric reagents is beyond the scope of this chapter, it is important to note that this commonly used oxidation reagent may have an unintended effect on lignin. Fang and Meier [37] studied DDQ oxidation of less abundant lignin linkages (i.e. the β-5

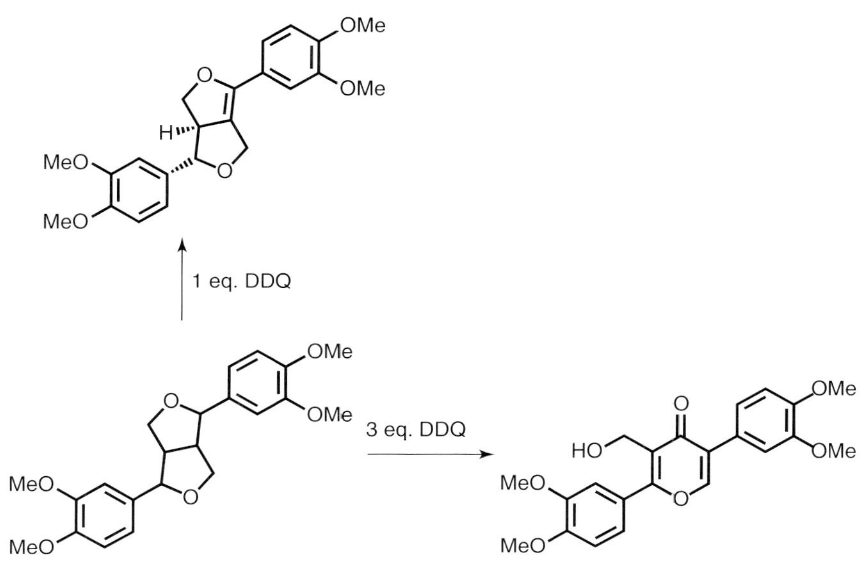

Scheme 8.4 DDQ oxidation of resinol (β-β) linkage [71].

linkage and the β-1 linkage), finding that the β-1 benzylic –OH was selectively oxidized to the corresponding ketone in 84–93% yield. In the case of β-5 model compounds, dehydrogenation of the dihydrobenzofuran was observed, along with a moderate amount of oxidation at the γ-OH to form the corresponding aldehyde. Importantly, no evidence of repolymerization due phenolic radical coupling was observed in either linkage model. Wang et al. also studied the use of a DDQ-based oxidation system for the benzylic oxidation of lignin and lignin model compounds followed by hydrotreating with NiMo-sulfide, finding it to be highly effective for lignin depolymerization (Section 8.4.2) [72].

Activated DMSO has also been investigated for the oxidation of lignin model compounds and native lignin. Recent work by Gao et al. [73] demonstrated the oxidation activity of a highly unusual Swern catalyst operating at elevated temperature under microwave irradiation. This reaction is truly atypical given that Swern-type (activated DMSO) oxidations typically occur at low temperature using stoichiometric reagents, such as oxalyl chloride or SO_3. Interestingly, Gao et al. found $MoO_2Cl_2(DMSO)_2$ to be highly active for the selective benzylic oxidation of *para*-methoxylated lignin β-O-4 model compounds (93% yield in 10 minutes). When phenolic model compounds were evaluated, a very different set of products were obtained, the authors observing the dehydration of the C_α–OH as well as the oxidation of the C_γ–OH to form an enal (81% on a GG model in 10 minutes). The proposed mechanism involves the sequential formation of a quinone methide intermediate, the rearomatization of the aromatic ring, and the oxidation of the C_γ–OH to ultimately afford the enal. Crocker and coworkers [74] observed a similar difference in reactivity between phenolic and non-phenolic model compounds under Swern conditions, albeit using stoichiometric reagents. Moreover, Gao et al. also observed that S-type phenolic β-O-4 model compounds underwent cleavage of the $C_β$—O bond leading to cinnamaldehyde derivatives and the corresponding phenols in good to excellent yields (32–87%) [73]. Preliminary results of this reaction system on Aspen lignin resulted in a 25% yield of well-defined monomers (15% combined yield of sinapaldehyde and coniferaldehyde), while high-S lignin derived from F5H-upregulated transgenic poplar resulted in a 31% yield of monomers (22% yield of syringaldehye). While this yield is not as high as is expected from high-S lignin, the authors suggest this may be due to internal oxidation, which prevents the end-wise pealing reaction from continuing. Thus, if a secondary technique (such as those discussed hereafter) were to be applied, it is conceivable that the yields may be much higher.

8.4.2 Secondary Depolymerization

Several interesting techniques exist for the depolymerization of post-benzylically oxidized lignin. Enol ether formation/hydrolysis [65], zinc reduction [63], Baeyer–Villiger (BV) oxidation/hydrolysis [40, 75, 76], Dakin oxidation [64], Beckmann rearrangement/hydrolysis [77], and copper–ligand oxidation [60] are all promising candidates for this reaction (Scheme 8.5). The unfortunate down side of many of these approaches, especially those comprising oxidative steps, is the production of phenols, which tend to polymerize under oxidative conditions producing intractable mixtures of repolymerized products [37, 40, 78]. However, the system devised by Stahl is particularly promising given its ability to depolymerize lignin under redox neutral conditions (vide infra). Furthermore, if oxidation of the lignin could be achieved prior to lignin extraction, it is conceivable that conjugate esters already present in the lignin could act as acid catalyst for enol ether hydrolysis under simple thermolytic conditions. In any case, secondary depolymerization techniques represent an excellent approach to obtain high yields of monomers from lignin.

The use of secondary processes for lignin depolymerization has only recently begun to be studied. Indeed, after Stahl and coworkers found that facile lignin oxidation of benzylic alcohols in lignin could be achieved with AcNH-TEMPO, several groups have endeavored to find strategies to take advantage of the newly weakened $C_β$—O bond. Wang and coworkers have described several methods to do so, including hydrogenolysis (vide infra) [72, 79] and Cu-based systems [60, 80]. An initial report by Wang et al. describes the use of a homogeneous copper catalyst coordinated by nitrogen-containing ligands. They found that 1,10-phenanthroline and 2,2′-bipyridine were the best ligands yielding 93% and 92% conversion of the lignin model compound, respectively [60]. The products of the reaction were benzoic acid, benzoylformic acid, and phenol. In a follow-up contribution, the authors describe the use of $Cu(AcO)_2$ in the presence of an acid promotor for the oxidative cleavage of $C_α$–$C_β$ [80]. Unlike their previous Cu system, the acid-promoted $Cu(AcO)_2$ catalyst demonstrated remarkable efficiency for the oxidative depolymerization of both β-O-4 and β-1 lignin model compounds. This approach afforded phenols and acid esters (formed by acid esterification with the solvent) in the case of β-O-4 compounds, and aldehydes as well as acid esters in the case of β-1 models. Moreover, a kinetic isotope effect on the $C_β$–H was observed, suggesting that coordination of the acid produces

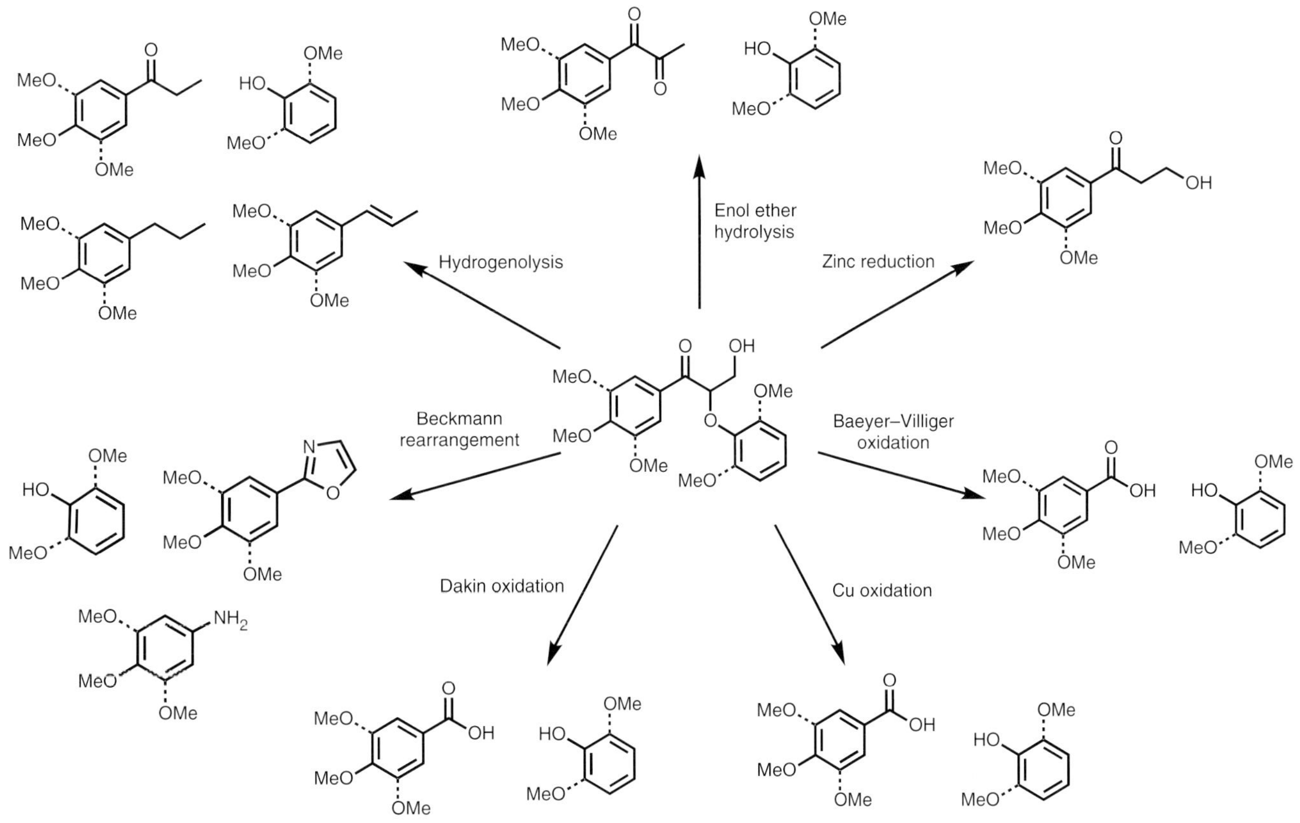

Scheme 8.5 Depolymerization post-benzylic oxidation

a significantly weaker C_β—H bond (by 124 and 29 kJ/mol on simple β-1 and β-O-4 models, respectively). In the first step of the proposed mechanism, a hydroperoxide is formed via coordination of $BF_3 \cdot OEt_2$ to the carbonyl oxygen with concurrent weakening of the C_β—H bond. While it is not explicitly mentioned by the authors, it is assumed that the Cu catalyst plays a significant role in hydroperoxide formation. Then, homolysis of the hydroperoxide O—O bond followed by a single electron transfer to the C_α generates a benzoyl radical (and a formyl radical in the case of β-O-4). Lastly, further reaction of the radical with the solvent forms the ester as the final product. While the authors mention an acid intermediate, they do not describe where the OH for the formation of said intermediate is obtained [80].

One of the most recent secondary depolymerization approaches proposed is the use of the Beckmann rearrangement. This was first achieved by Zhang and coworkers [77], who formed an oxime from the benzylic ketone via nucleophilic addition to hydroxylamine hydrochloride. The latter step was then followed by aprotic acid-catalyzed rearrangement of the oxime with $SOCl_2$ to form the corresponding amide. Interestingly, depending on the geometric isomer, the selectivity of the products varied. Indeed, the Z-type isomer favored the formation of the typical Beckmann amide, while the E-type isomer favored the formation of the cleavage products (benzonitrile and phenol). The authors attributed the fact that amide formation was favored over depolymerization to the stabilization of the carbenium ion by the strongly electron-donating *trans*-OH group. Also interesting is the formation of the oxazole product from the E-oxime 3-carbon model compound. This was suggested to be the result of a nucleophilic attack of the carbenium ion by the γ-OH group to form a five-membered stable intermediate, which is further converted to the oxazole after loss of the aryloxy group and aromatization. Overall, the Beckmann rearrangement was highly effective (75–96% yield of N-substituted products), and the authors demonstrated the successful hydrolysis of the Beckmann amide with NaOH in a one-pot style reaction from the oxime with a 61–96% yield of cleavage products.

Several groups have also studied the BV oxidation of lignin and/or lignin models for the oxidative depolymerization of lignin. This technique is particularly attractive given that any linking carbonyl could be susceptible to BV oxidation, which could provide a platform for oxidative cleavage of a variety of linkages including the β-O-4, β-1, or even other linkages with newly formed carbonyls. A preliminary report by Meier and coworkers found that BV oxidation with formic peracid formed *in situ* produced the corresponding A-ring acid (produced via hydrolysis of the newly generated ester) from a β-O-4 model compound in a 78% yield [40]. Unfortunately, the corresponding B-ring phenol product likely undergoes oxidative polymerization, forming unidentified products. Follow-up studies were performed on kraft lignin resulting in a 10% combined yield of two products (methyl vanillate and methyl 5-carbomethoxyvanillate, produced by derivatization) [39]. In a recent work by Zhang and coworkers [76], a homogeneous selenium catalyst $(PhCH_2Se)_2$ in combination with H_2O_2 was used for the BV oxidation of lignin model compounds, which was compared with traditional stoichiometric oxidation in the presence of *m*CPBA. As expected, the migratory aptitude depended heavily on the substitution on the A-ring as well as on the presence or absence of the C_γ–OH. Furthermore, near-quantitative conversions were obtained on both β-O-4 and β-1 model compounds using quantitative *m*CPBA, resulting in one or both of the ester products in excellent yield (92% of BV products). Analogous results were obtained with the catalytic Se/H_2O_2 reaction system (81% yield of BV products); however, the β-1 model had significantly decreased conversion and yields. Similar findings were also obtained by Andrade et al. [81] who utilized the same $(PhCH_2Se)_2$ catalyst in combination with H_2O_2. Indeed, at room temperature, with 4 eqiv of H_2O_2, the authors obtained up to a 94% yield of the BV product of a simple lignin monomer model compound. Additionally, the selectivity of the catalytic system was much lower than that of the *m*CPBA system [76]. The difference in the selectivity of the catalytic and the stoichiometric systems suggests that the migratory aptitude of the substrate R groups – i.e. which group can accept a partial positive charge [82, 83] – is not the sole feature that determines migration. Thus, it follows that the strength of the oxidant or steric hindrance of the substrate/oxidant combination may also play a role. Importantly, neither the stoichiometric nor the catalytic method caused the *in situ* hydrolysis of the ester, so the authors used a mild base-in-alcohol hydrolysis method to cleave the ester. This, unlike other BV oxidation methods, lead to near-quantitative yields of phenol (B-ring) as well as acid (A-ring). The significance of this cannot be understated, as other oxidative depolymerization techniques have led to large amounts of oxidative polymerization of the phenol. Given that most of the products obtained from this method will inherently be phenolic (and not methoxy substituted like the conveniently used models), finding a reaction system that avoids phenol repolymerization is paramount. Unfortunately, while this catalyst performed well, separation issues as well as the use of a toxic homogeneous selenium catalyst may be a limitation to its industrial use. Likewise, Crocker and coworkers [75] performed BV oxidation on lignin model compounds using a Sn-β-zeolite catalyst and hydrogen peroxide. Their system gave moderate to good yields on simple acyclic ketones; however, when more complex lignin models were trialed, polymerization due to direct ring hydroxylation was observed. Interestingly, it was found that regioselective aryl migration was preferred over alkyl migration, as confirmed via a crystal structure of the product. Unfortunately, the yields are quite low (21–32% for β-O-4 models and 9–22% for β-1 models), and more work is needed to optimize the reaction system for lignin (e.g. using a lignin-appropriate solvent, obtaining a higher yield of products, and preventing phenol polymerization). Nevertheless, the use of a heterogeneous catalyst for this oxidation is of interest for commercial applications. Interestingly, Andrade and coworkers [81] found that an immobilized benzene selenic acid is also effective for the BV oxidation of 2-carbon lignin model compounds (86% yield of BV products). This system, which employs the catalyst in a fixed bed system using H_2O_2 as the terminal oxidant at room temperature, is ideal for industrial use. However, the catalyst will need to be optimized for use on 3-carbon lignin models, given that the dehydration product (C_β–C_γ) was dominant (ca. 90%).

Similar to BV oxidation, Stahl and coworkers [64] performed Dakin oxidation (AHP) on a benzylically oxidized lignin model compound. This reaction, which was chosen as a proof of concept, proved remarkably effective for the conversion of a methoxylated, benzylically oxidized lignin model compound, yielding the veratric acid and guaiacol in 88% and 42% yield, respectively. However, the reaction was not reported on lignin.

Perhaps the most highly cited work on depolymerization of post-benzylically oxidized lignin is that by Stahl and coworkers [65], who utilized sodium formate (3 equiv) in aqueous formic acid at 110 °C. Indeed, in their seminal work Stahl et al. reported a remarkably high production of monomers (61.2 wt% ethyl acetate soluble and 52% well-defined aromatic compounds) from aspen lignin isolated by enzymatic hydrolysis. Their suggested mechanism involves formylation of the γ-OH, followed by elimination to form the enol ether (rate-limiting step), and acid-catalyzed hydrolysis to form the 1,2-diketone and phenol. As discussed previously, this method suggests that it may be possible to benzylically oxidize the lignin prior to extraction and use already present conjugate esters as catalysts for elimination. Recently, Stahl and coworkers [25] reported an extraordinarily useful comparative study of a variety of lignins from different species and extraction techniques subjected to the aforementioned method. This is noteworthy, since lignin isolated via enzymatic hydrolysis is impractical on a commercial scale. Not surprisingly, lignins isolated with gentle extraction techniques such as mild acidolysis yielded higher amounts of monomer products (42 wt%), whereas other lignin isolation techniques (GVL, EA, Cu-AHP) yielded lower amounts of monomers (3–31 wt%). Moreover, poplar lignin appears to be more amenable to oxidation–hydrolysis than maize and maple lignin, regardless of the extraction technique employed. While the reason for activity differences between extraction techniques is likely due – at least in part – to the degree of loss of β-O-4 linkages during lignin isolation, the activity differences between species is yet to be elucidated. However, this is an increasingly important area of study, given that attention has shifted toward standardizing techniques based on monomer yields from lignin. Moreover, as lignin varies not only by source but also by extraction technique, it is becoming increasingly necessary to provide monomer yields from a variety of lignins. This gives the community a standard by which the effectiveness of the described depolymerization technique can be judged.

Stephenson and coworkers [84–86] published a series of papers on the use of photocatalytic redox catalysis to effectively break the $C_β$—O bond of oxidized β-O-4 linkages. This reaction, in which $[Ir(ppy)_2(dtbbpy)]PF_6$ was used in combination with *N*,*N*-Diisopropylethylamine (DIPEA)/formic acid and visible light at room temperature, effectively depolymerized both phenolic and non-phenolic β-O-4 model compounds in near-quantitative yields without significant polymerization of the phenol [86]. This process was further adapted to a continuous flow reaction setup [86] and used in conjunction with an electrocatalyic oxidation system for the oxidation of benzylic alcohols with *N*-hydroxyphthalimide (NHPI)/2,6-lutidine [84]. While this approach was highly effective for β-O-4 model compounds, only a 2–3% yield of GC–MS identified monomers was obtained when applied to dioxosolv pine lignin. However, GPC analysis showed that the lignin was highly depolymerized, resulting in a yield of ca. 55% monomers and ca. 43% oligomers. The authors suggested that the low GC–MS monomer yield is likely due to the need for consecutive β-O-4 linkages to be broken in order to release a single monomer unit. Thus, it is reasonable to hypothesize that the use of lignin derived from hybrid poplar with overexpression of the ferulate 5-hydroxylase (so-called "high-S" lignin) [12] would result in a significantly higher yield of GC–MS identifiable monomers.

Wang and coworkers [87–89] also explored a series of photocatalysts for the depolymerization of lignin and lignin model compounds. Indeed, $ZnIn_2S_4$ was used to oxidatively cleave lignin models under blue light in a 71–91% yield [89]. When applied to organosolv lignin, a 10% yield of *p*-hydroxyacetophenone derivatives was formed. Interestingly, this catalyst is proposed to work under self-hydrogenation transfer hydrogenolysis. Thus, lignin alcohols are dehydrogenated to supply hydrogen for reductive depolymerization. Additional studies by Wang and coworkers have shown that CuO_x clusters supported on ceria/anatase nanotubes selectively cleave C—C bonds oxidatively in β-1 models under visible light irradiation [88]. DFT calculations along with Raman and UV–Vis diffuse reflectance spectroscopy suggest that photocatalytic activity is improved by an increase in the surface defect concentrations on the ceria caused by CuO_x clusters present on the support. More recently, Wang and coworkers have applied mesoporous graphitic carbon nitride to the photocatalytic oxidation of lignin β-1 and β-O-4 models under oxidative conditions [87]. This led to the cleavage of these models in good to excellent yields. Unfortunately, yields of phenolic products were low due to overoxidation and/or polymerization. Interestingly, the mechanism of the reaction was shown to begin with hydrogen abstraction of the C_β–H followed by formation of a hydroperoxide that rearranges to give the benzylic aldehyde (or acid) and formate ester as the major products.

While somewhat beyond the scope of this chapter, it is still worth noting the surprising effectiveness of formaldehyde for the generation of a quasi-stable acetal under reducing conditions, which led to the complete depolymerization of the transgenically modified high-S poplar lignin [23]. This technique is particularly interesting in that it effectively stabilized the hydrogenolytic intermediate preventing the presence of the carbo-cation. Building on this strategy, Luterbacher and coworkers utilized DDQ in order to oxidatively deprotect lignin, which had been acetal-protected with propionaldehyde prior to extraction, leading to a benzylic ketone [90]. This was then followed by use of the formic acid/sodium formate hydrolysis strategy pioneered by Stahl and coworkers [65]; a 47% and 52% yield of monomers from catalytic (20 wt%) and super-stoichiometric (150 wt%) loadings of DDQ was obtained, respectively, from high-S poplar. When the technique was applied to birch, the monomer yield was 36% and 31% for super-stoichiometric (150 wt%) and catalytic (20 wt%) DDQ loadings, respectively. Wang and coworkers investigated a similar strategy in which the benzylic C_α position was stabilized via selective oxidation of the alcohol to the carbonyl, which has a similar protecting effect [72]. This should prevent condensation reactions caused by the formation of the type of benzylic carbocations typically observed during acid-catalyzed lignin extraction [22, 23, 91, 92]. While this method is not suitable for all hydrodeoxygenation catalysts (e.g. those which reduce the carbonyl back to an alcohol), the authors found that NiMo-sulfide was an effective catalyst for this transformation, giving isolated products from model compounds in good yields. It should be noted that the DDQ method used by Wang and coworkers is somewhat different than that used by other researchers. Indeed, Wang and coworkers used a DDQ/NHPI/$NaNO_2$/O_2 reaction system that is proposed to oxidize lignin via a radical mechanism involving hydrogen abstraction from the benzylic C–H, and DDQ is used as a cocatalyst to reoxidize NHPI to phthalimide-*N*-oxyl (PINO). Benzylic oxidation was found to be highly effective for the oxidation of two-carbon β-O-4 models (compounds without a γ-carbon); however, three-carbon β-O-4 models were found to be significantly less active, resulting in benzylic ketones in a 35–61% yield. Interestingly, the aforementioned DDQ-based oxidation system was also active in forming an aldehyde from an α-O-4 model, albeit the yield was quite low (12%). Hydrotreating benzylically oxidized β-O-4 models over sulfided NiMo at 180 °C for six hours resulted in excellent conversions (<97%), the main products being A-ring acetophenone and propiophenone derivatives as well as the corresponding B-ring phenols. Also obtained in a significant yield were A-ring phenyl alkanes. Application of this approach to a preoxidized β-O-4 model polymer resulted in a 92% conversion to a mixture of phenolic acetophenone and ethyl benzene. Moreover, a 32% yield of phenolic monomers was obtained from oxidized birch powder. This was a 10-fold increase in the yield of monomer products obtained over the unoxidized birch powder. Importantly, the improved results obtained with birch powder highlight the potential of performing oxidation on pre-extracted biomass prior to the isolation or depolymerization of the lignin, even under highly acidic conditions. However, it is at present unclear whether any valuable sugars are also affected by the oxidation. Additionally, Westwood and coworkers found that benzylic oxidation with DDQ followed by hydrotreating with a variety of base metal and noble metal catalysts (W_2C/AC, 20% Ni/AC, 5% Pd/AC, 5% Pt/AC, 5% Rh/AC) gave reduced monomer yields compared with the unoxidized lignin [93]. The authors attribute the reduced yields to condensation of the lignin from the oxidation procedure. However, more work needs to be done in order to determine whether oxidation followed by hydrotreating is a viable strategy, especially if oxidation procedures that do not lead to lignin condensation reactions are applied.

Westwood and coworkers also described a reductive route to post-benzylically oxidized lignin and lignin model compounds [63]. This route is particularly attractive given that the newly generated phenols should be stable in solution. The method, which involves DDQ oxidation (vide supra) followed by zinc reduction, is performed in a one-pot fashion. The process was proven to work effectively on a variety of non-phenolic lignin model compounds, producing γ-hydroxypropiophenone derivatives in an 81–92% yield. Unfortunately, when applied to β-O-4 model polymers, only a 23–27% yield was obtained. Moreover, application of the process to organosolv Birch lignin resulted in a low (5%) yield of monomeric products.

8.5 Heterogeneous Catalysts for Lignin Depolymerization

While very promising results have been obtained using homogeneous catalysts for lignin oxidation, the industrial use of these catalysts is hampered by the difficulty of separating the catalysts from an inherently complex product mixture. While this is not necessarily an insurmountable challenge, heterogeneous catalysts are inherently more desirable due to their ease of separation. To this end, several research groups have explored the use of heterogeneous catalysts for the oxidation and depolymerization of lignin. Indeed, Wang and coworkers [94] explored the use of Pd/CeO_2, finding it to be an effective catalyst both for the oxidation of the benzylic alcohol group present in β-O-4 linkages and for $C_\alpha—C_\beta$ bond cleavage. Indeed, at 458 K under an O_2 atmosphere, they obtained ca. 90% conversion of simple lignin model dimers. Promisingly, when this method was extended to organosolv lignin, the authors obtained a yield of small molecules approaching 8.5%.

In a series of publications, Samec and coworkers [95, 96] used a Pd/C catalyst for the selective dehydrogenation of alcohols in the β-O-4 linkage (Scheme 8.6). Although Pd/C was found to be relatively ineffective (18% yield) for the selective dehydrogenative oxidation of benzylic alcohols present in the β-O-4 linkage under mild conditions (80 °C), yields were significantly enhanced (to 68%) through the use of 0.4 equiv of glucose. Upon analysis of the reaction mixture, no apparent oxidation of the glucose was seen. Further probing of the role of glucose was performed with other polyols (i.e. sorbitol and pinacol), which had a similar effect to glucose. The use of pinacol ruled out the possibility of the polyol acting as a mild hydrogen donor and the authors hypothesized that the polyol may have a ligand-like effect that helps prevent the aggregation of palladium particles. Importantly, the activity of the catalyst was unaffected by the use of phenolic model compounds, which, as mentioned previously, have the tendency to polymerize in oxidative systems. In addition, the authors proposed a one-pot-type system for the depolymerization of lignin in which the benzylic alcohol oxidation precedes the base-initiated dehydration of the γ-alcohol, in turn followed by the reductive cleavage of the enol ether intermediate utilizing regenerated Pd/C and formic acid/ammonium formate. The aforementioned method also bears striking resemblance to the method by Stahl and coworkers in which a formic acid/sodium formate system was used to generate and hydrolyze the aforementioned oxidized enol ether [65].

Andrade and coworkers used a Au catalyst for the oxidation of benzylic alcohols in lignin model compounds. Their catalysts, Au/SiO_2, utilized a supplemental base (KOH) and H_2O_2 (40 equiv) as the oxidant at room temperature [81]. Under continuous flow conditions, the authors reported a 74% yield of the benzylic ketone product with no observation of the γ-aldehyde byproduct, higher yields being seen on monomeric lignin model compounds. Interestingly, Andrade et al. reported a gold nanoparticle size of ca. 20 nm, which is relatively large for active gold nanoparticles. Indeed, it is well known that gold nanoparticles lose activity above 5 nm [97]. Thus, a more active catalyst is possible with optimization of the synthesis procedure. Unfortunately, most catalytic systems for benzylic oxidation utilizing gold need the input of a supplementary base (NaOH, KOH, etc.), homogeneous bases being unnecessary in ideal reaction systems (vide infra).

Beckham and coworkers investigated the use of Ni supported on layered double hydroxides (LDHs) for lignin depolymerization [98, 99]. Indeed, LDHs are a logical extension of other studies involving base-catalyzed lignin depolymerization.

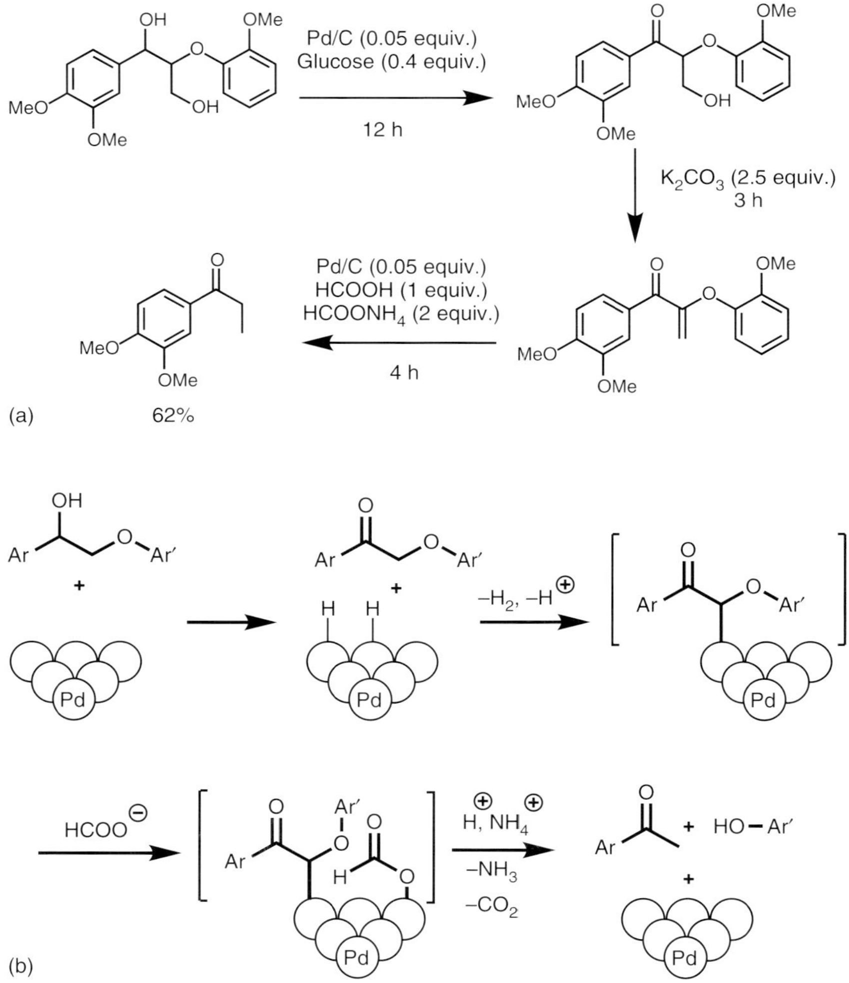

Scheme 8.6 (a) Aerobic oxidation followed by dehydration and reductive depolymerization of a β-O-4 model compound [95]. The reaction was performed at 80 °C in a mixture of EtOAc/H_2O (4:1 v/v). Note that reaction was performed in a one-pot fashion and that the Pd/C was removed and regenerated with H_2 prior to reuse in the final step. (b) Proposed mechanism for lignin model compound depolymerization. Source: Galkin et al. 2014 [96]. Reproduced with permission of Wiley-VCH Verlag GmbH & Co. KGaA.

In those studies, a strong base such as NaOH is used to depolymerize lignin [100, 101]. Unfortunately, such reaction systems are very corrosive and would require the use of specialized reactors. On the other hand, heterogeneous LDHs provide a strongly basic environment without the need of specialized equipment. While the aforementioned reports all use catalyst systems that are expected to operate using a base-catalyzed depolymerization mechanism, it is important to note that LDHs are very effective oxidative catalysts. Indeed, Beckham and coworkers noted that their catalyst system, which involves the use of Ni/Mg-Al LDH at 270 °C, relies heavily on interlayer nitrates, which are typically seen in low concentrations in LDHs. Tellingly, NO_3-incorporated Mg-Al LDH was noted to have comparable or superior activity to $Ni(NO_3)_2$-loaded Mg-Al LDH with model compound systems. Moreover, when applied to deacetylated-disk-refined-enzymatically hydrolyzed lignin (DDE) and dilute acid-pretreated enzymatically hydrolyzed lignin (DAP) from corn stover, a 5–9 wt% yield of monomers was obtained.

Given the possible use of LDHs as a solid base, several research groups have examined LDHs as catalyst supports for lignin model compound oxidation with varying degrees of success. Crocker and coworkers recently described the use of LDH catalysts for the oxidation of benzylic alcohols in lignin model compounds, finding them to be very effective for simple monomeric models under aerobic conditions [102]. While relatively high conversions were obtained on dimeric models, the selectivity and yields were low.

In addition to the homogeneous vanadium catalysts discussed previously, Corma and coworkers [103] also developed a Cu-containing LDH catalyst, which used vanadate (VO_3^-) as the interlayer species. This catalyst was used for the oxidation of a lignin model β-O-4 dimer with molecular oxygen as the oxidant, resulting in $C_\alpha—C_\beta$ bond cleavage. The main products of the reaction were the corresponding aldehyde and carboxylic acid. When applied to organosolv beech lignin approximately 70% (calculated from the reported values) was converted to small molecular weight oligomers (ca. 300 Da). Unfortunately, the authors found that significant metal leaching occurred during the course of the reaction and the catalytically active species was likely homogeneous in nature. Potentially more interesting is the use of Ru on nanohydrotalcite (Ru/n-HT) for the oxidation of veratryl alcohol. Indeed, near-quantitative conversion to veratryl aldehyde was obtained in one hour at 80 °C using O_2 as the terminal oxidant [104]. Unfortunately, the study was only performed on lignin monomers, and it is unclear how the catalyst would perform on more realistic lignin models or on lignin itself.

Crocker and coworkers have also investigated the oxidation of lignin and lignin model compounds over Au/LDHs, finding these catalysts to be very effective for the oxidative depolymerization of lignin β-O-4 model compounds [105]. Indeed, near-complete conversion (ca. 90%) of a methoxylated β-O-4 model compound was achieved in 24 hours at 120 °C using O_2 as the terminal oxidant. Interestingly, the products of the reaction appear to be the result of benzylic oxidation followed by retro-aldol reaction to form an A-ring acid and a B-ring phenol, which condense to form a readily cleavable ester. When applied to kraft lignin (Indulin AT) derived from pine, a 20% yield of ethyl acetate-soluble products and a 10% yield of monomers were obtained (the two major products being vanillin and vanillic acid, 3.4% each). When applied to GVL lignin derived from maple, a 56% yield of ethyl acetate-soluble products was obtained with a 40% yield of GC–MS identifiable monomers. Notably, this represents the largest amount of monomeric products obtained from lignin with a heterogeneous catalyst under oxidative conditions reported to date. This is particularly remarkable given that Stahl et al. achieved a ca. 3% yield of monomers from GVL maple (the same lignin) using their literature-leading AcNH-TEMPO/formic acid system.

8.6 Outlook

While oxidative methods for the efficient depolymerization of lignin to afford fine chemicals have been explored for the better part of a century, only recently do methods seem to have come far enough to make value-added products from lignin in a way that might be practicable in the biorefinery. Indeed, methods such as TEMPO oxidation followed by hydrolysis as well as oxidative depolymerization over gold catalysts may make the economics of lignin valorization viable. Moreover, the creative use of methods such as the Beckmann rearrangement as well as the incorporation of nitrogen in lignin depolymerization products should provide a new set of platform chemicals for researchers to investigate. It is important to note that the use of lignin depolymerization products as replacements for platform chemicals is key to lignin valorization. Indeed, without it even very efficient lignin depolymerization processes will be of limited use commercially. While the field certainly shows room for improvement, considerable progress has been made and the commercialization of products from lignin now appears possible. This is especially true for products such as vanillin, and perhaps soon the old saying that "you can make anything from lignin, except for money" may finally be proven wrong. In any case, more work still needs to be done in order to make oxidative depolymerization methods more robust, i.e. able to depolymerize more than one linkage. Moreover, an experimental review of the most effective depolymerization methods needs to be applied across a variety of lignins stemming from different sources and extraction techniques. This will allow for the proper strategy to be applied according to the feedstock and pretreatment conditions employed.

Acknowledgments

This work is derived, in part, from a PhD dissertation at the University of Kentucky [106].

References

1 Candolle, A.P.D. (1813). *Théorie élémentaire de la botanique; ou, Exposition des prinicpes de la classification naturelle et de l'art de décrire et d'étudier les végétaux*. Paris: Déterville.
2 Boerjan, W., Ralph, J., and Baucher, M. (2003). Lignin biosynthesis. *Annu. Rev. Plant Biol.* 54: 519–546.
3 Karlen, S.D., Zhang, C., Peck, M.L. et al. (2016). Monolignol ferulate conjugates are naturally incorporated into plant lignins. *Sci. Adv.* 2: e1600393.
4 Wilkerson, C.G., Mansfield, S.D., Lu, F. et al. (2014). Monolignol ferulate transferase introduces chemically labile linkages into the lignin backbone. *Science* 344: 90–93.
5 Chen, F., Tobimatsu, Y., Havkin-Frenkel, D. et al. (2012). A polymer of caffeyl alcohol in plant seeds. *Proc. Natl. Acad. Sci. U.S.A.* 109: 1772–1777.

6 Zakzeski, J., Bruijnincx, P.C.A., Jongerius, A.L., and Weckhuysen, B.M. (2010). The catalytic valorization of lignin for the production of renewable chemicals. *Chem. Rev.* 110: 3552–3599.

7 Ralph, J., Hatfield, R.D., Quideau, S. et al. (1994). Pathway of *p*-coumaric acid incorporation into maize lignin as revealed by NMR. *J. Am. Chem. Soc.* 116: 9448–9456.

8 Ralph, J. (2010). Hydroxycinnamates in lignification. *Phytochem. Rev.* 9: 65–83.

9 Lu, F., Karlen, S.D., Regner, M. et al. (2015). Naturally *p*-hydroxybenzoylated lignins in palms. *Bioenergy Res.* 8: 934–952.

10 Smith, R.A., Gonzales-Vigil, E., Karlen, S.D. et al. (2015). Engineering monolignol *p*-coumarate conjugates into poplar and Arabidopsis lignins. *Plant Physiol.* 169: 2992–3001.

11 Harman-Ware, A.E., Happs, R.M., Davison, B.H., and Davis, M.F. (2017). The effect of coumaryl alcohol incorporation on the structure and composition of lignin dehydrogenation polymers. *Biotechnol. Biofuels* 10: 281.

12 Stewart, J.J., Akiyama, T., Chapple, C.C. et al. (2009). The effects on lignin structure of overexpression of ferulate 5-hydroxylase in hybrid poplar. *Plant Physiol.* 150: 621–635.

13 da Costa Sousa, L., Chundawat, S.P.S., Balan, V., and Dale, B.E. (2009). 'Cradle-to-grave' assessment of existing lignocellulose pretreatment technologies. *Curr. Opin. Biotechnol.* 20: 339–347.

14 Sameni, J., Krigstin, S., and Sain, M. (2017). Solubility of lignin and acetylated lignin in organic solvents. *BioResources* 12: 1548–1565.

15 Luterbacher, J.S., Azarpira, A., Motagamwala, A.H. et al. (2015). Lignin monomer production integrated into the γ-valerolactone sugar platform. *Energy Environ. Sci.* 8: 2657–2663.

16 Luterbacher, J.S., Rand, J.M., Alonso, D.M. et al. (2014). Nonenzymatic sugar production from biomass using biomass-derived γ-valerolactone. *Science* 343: 277–280.

17 Hibino, T., Shibata, D., Ito, T. et al. (1994). Chemical properties of lignin from *Aralia cordata*. *Phytochemistry* 37: 445–448.

18 Bhalla, A., Bansal, N., Stoklosa, R.J. et al. (2016). Effective alkaline metal-catalyzed oxidative delignification of hybrid poplar. *Biotechnol. Biofuels* 9: 1–10.

19 Balan, V., Bals, B., Chundawat, S.P. et al. (eds.) (2009). *Lignocellulosic Biomass Pretreatment Using AFEX*, vol. 581, 61–77. Totowa, NJ: The Humanan Press Inc.

20 da Costa Sousa, L., Foston, M., Bokade, V. et al. (2016). Isolation and characterization of new lignin streams derived from extractive-ammonia (EA) pretreatment. *Green Chem.* 18: 4205–4215.

21 Hu, Z., Yeh, T.-F., Chang, H.-m. et al. (2006). Elucidation of the structure of cellulolytic enzyme lignin. *Holzforschung* 60: 389–397.

22 Sturgeon, M.R., Kim, S., Lawrence, K. et al. (2014). A mechanistic investigation of acid-catalyzed cleavage of aryl-ether linkages: implications for lignin Depolymerization in acidic environments. *ACS Sustainable Chem. Eng.* 2: 472–485.

23 Shuai, L., Amiri, M.T., Questell-Santiago, Y.M. et al. (2016). Formaldehyde stabilization facilitates lignin monomer production during biomass depolymerization. *Science* 354: 329–333.
24 Li, Z., Bansal, N., Azarpira, A. et al. (2015). Chemical and structural changes associated with cu-catalyzed alkaline-oxidative delignification of hybrid poplar. *Biotechnol. Biofuels* 8: 1–12.
25 Das, A., Rahimi, A., Ulbrich, A. et al. (2018). Lignin conversion to low-molecular-weight aromatics via an aerobic oxidation-hydrolysis sequence: comparison of different lignin sources. *ACS Sustainable Chem. Eng.* 6: 3367–3374.
26 Rinaldi, R., Jastrzebshi, R., Clough, M.T. et al. (2016). Paving the way for lignin valorisation: recent advances in bioengineering, biorefining and catalysis. *Angew. Chem. Int. Ed.* 55: 8164–8215.
27 Sohail, A., Kumar Karitkey, Y., and Chauhan, S.M.S. (2012). Role of hydroxyl functionalized ionic liquids in oxygenation and oxidation of phenols with H_2O_2 catalysed by iron(III) Porphyrins. *Curr. Org. Chem.* 16: 2989–3001.
28 Leonowicz, A., Matuszewska, A., Luterek, J. et al. (1999). Biodegradation of lignin by white rot fungi. *Fungal Genet. Biol.* 27: 175–185.
29 Hayatsu, R., Winans, R.E., McBeth, R.L. et al. (1979). Lignin-like polymers in coals. *Nature* 278: 41.
30 Floudas, D., Binder, M., Riley, R. et al. (2012). The Paleozoic origin of enzymatic lignin decomposition reconstructed from 31 fungal genomes. *Science* 336: 1715.
31 Cui, F. and Dolphin, D. (1995). Iron porphyrin catalyzed oxidation of lignin model compounds: oxidation of phenylpropane and phenylpropene model compounds. *Can. J. Chem.* 73: 2153–2157.
32 Crestini, C., Crucianelli, M., Orlandi, M., and Saladino, R. (2010). Oxidative strategies in lignin chemistry: a new environmental friendly approach for the functionalisation of lignin and lignocellulosic fibers. *Catal. Today* 156: 8–22.
33 Bonnarme, P. and Jeffries, T.W. (1990). Mn(II) regulation of lignin peroxidases and manganese-dependent peroxidases from lignin-degrading white rot fungi. *Appl. Environ. Microbiol.* 56: 210–217.
34 Cui, F. and Dolphin, D. (1995). Metallophthalocyanines as possible lignin peroxidase models. *Bioorg. Med. Chem.* 3: 471–477.
35 Ma, R., Xu, Y., and Zhang, X. (2015). Catalytic oxidation of biorefinery lignin to value-added chemicals to support sustainable biofuel production. *ChemSusChem* 8: 24–51.
36 Shimada, M., Habe, T., Umezawa, T. et al. (1984). The C–C bond-cleavage of a lignin model-compound, 1,2-diarylpropane-1,3-diol, with a heme-enzyme model catalyst tetraphenylporphyrinatoiron(iii) chloride in the presence of *tert*-butylhydroperoxide. *Biochem. Biophys. Res. Commun.* 122: 1247–1252.
37 Fang, Z. and Meier, M.S. (2018). Toward the oxidative deconstruction of lignin: oxidation of β-1 and β-5 linkages. *Org. Biomol. Chem.* 16: 2330–2341.
38 Yao, S.G., Meier, M.S., Pace Iii, R.B., and Crocker, M. (2016). A comparison of the oxidation of lignin model compounds in conventional and

ionic liquid solvents and application to the oxidation of lignin. *RSC Adv.* 6: 104742–104753.

39 Yao, S.G., Mobley, J.K., Ralph, J. et al. (2018). Mechanochemical treatment facilitates two-step oxidative depolymerization of kraft lignin. *ACS Sustainable Chem. Eng.* 6 (5): 5990–5998.

40 Patil, N.D., Yao, S.G., Meier, M.S. et al. (2015). Selective cleavage of the C_{α}-C_{β} linkage in lignin model compounds via Baeyer-Villiger oxidation. *Org. Biomol. Chem.* 13: 3243–3254.

41 Crestini, C., Saladino, R., Tagliatesta, P., and Boschi, T. (1999). Biomimetic degradation of lignin and lignin model compounds by synthetic anionic and cationic water soluble manganese and iron porphyrins. *Bioorg. Med. Chem.* 7: 1897–1905.

42 Cui, F., Wijesekera, T., Dolphin, D. et al. (1993). Biomimetic degradation of lignin. *J. Biotechnol.* 30: 15–26.

43 Bozell, J.J., Hames, B.R., and Dimmel, D.R. (1995). Cobalt-Schiff base complex-catalyzed oxidation of parasubstituted phenolics – preparation of benzoquinones. *J. Org. Chem.* 60: 2398–2404.

44 Drago, R.S., Corden, B.B., and Barnes, C.W. (1986). Novel cobalt(II)-catalyzed oxidative cleavage of a carbon–carbon double bond. *J. Am. Chem. Soc.* 108: 2453–2454.

45 Biannic, B. and Bozell, J.J. (2013). Efficient cobalt-catalyzed oxidative conversion of lignin models to benzoquinones. *Org. Lett.* 15: 2730–2733.

46 Badamali, S.K., Luque, R., Clark, J.H., and Breeden, S.W. (2009). Microwave assisted oxidation of a lignin model phenolic monomer using co(salen)/SBA-15. *Catal. Commun.* 10: 1010–1013.

47 Badamali, S.K., Luque, R., Clark, J.H., and Breeden, S.W. (2011). Co(salen)/SBA-15 catalysed oxidation of a β-O-4 phenolic dimer under microwave irradiation. *Catal. Commun.* 12: 993–995.

48 Son, S. and Toste, F.D. (2010). Non-oxidative vanadium-catalyzed C–O bond cleavage: application to degradation of lignin model compounds. *Angew. Chem. Int. Ed.* 49: 3791–3794.

49 Hanson, S.K., Wu, R.L., and Silks, L.A. (2012). C–C or C–O bond cleavage in a phenolic lignin model compound: selectivity depends on vanadium catalyst. *Angew. Chem. Int. Ed.* 51: 3410–3413.

50 Hanson, S.K., Baker, R.T., Gordon, J.C. et al. (2010). Aerobic oxidation of lignin models using a base metal vanadium catalyst. *Inorg. Chem.* 49: 5611–5618.

51 Ma, Y., Du, Z., Liu, J. et al. (2015). Selective oxidative C–C bond cleavage of a lignin model compound in the presence of acetic acid with a vanadium catalyst. *Green Chem.* 17: 4968–4973.

52 Sedai, B., Diaz-Urrutia, C., Baker, R.T. et al. (2013). Aerobic oxidation of *β*-1 lignin model compounds with copper and oxovanadium catalysts. *ACS Catal.* 3: 3111–3122.

53 Sedai, B., Diaz-Urrutia, C., Baker, R.T. et al. (2011). Comparison of copper and vanadium homogeneous catalysts for aerobic oxidation of lignin models. *ACS Catal.* 1: 794–804.

54 Li, C., Zhao, X., Wang, A. et al. (2015). Catalytic transformation of lignin for the production of chemicals and fuels. *Chem. Rev.* 115: 11559–11624.
55 Crestini, C., Pro, P., Neri, V., and Saladino, R. (2005). Methyltrioxorhenium: a new catalyst for the activation of hydrogen peroxide to the oxidation of lignin and lignin model compounds. *Bioorg. Med. Chem.* 13: 2569–2578.
56 Crestini, C., Caponi, M.C., Argyropoulos, D.S., and Saladino, R. (2006). Immobilized methyltrioxo rhenium (MTO)/H_2O_2 systems for the oxidation of lignin and lignin model compounds. *Bioorg. Med. Chem.* 14: 5292–5302.
57 Gosselink, R.J.A., de Jong, E., Guran, B., and Abacherli, A. (2004). Co-ordination network for lignin – standardisation, production and applications adapted to market requirements (EUROLIGNIN). *Ind. Crops Prod.* 20: 121–129.
58 Calvo-Flores Francisco, G. and Dobado José, A. (2010). Lignin as renewable raw material. *ChemSusChem* 3: 1227–1235.
59 Bjørsvik, H.-R. and Minisci, F. (1999). Fine chemicals from Lignosulfonates. 1. Synthesis of vanillin by oxidation of Lignosulfonates. *Org. Process Res. Dev.* 3: 330–340.
60 Wang, M., Lu, J., Zhang, X. et al. (2016). Two-step, catalytic C–C bond oxidative cleavage process converts lignin models and extracts to aromatic acids. *ACS Catal.* 6: 6086–6090.
61 Patil, N. and Yan, N. (2016). *Investigate cleavage of β-O-4 linkage in lignin model compounds by aerobic oxidation of Cα and Cγ hydroxyl groups. Tetrahedron Lett.* 57: 3024–3028.
62 Patil, N. and Yan, N. (2016). *Study of the nitroxyl radical catalyst in aerobic oxidative cleavage and functionalization of lignin model compounds. Catal. Commun.* 84: 155–158.
63 Lancefield, C.S., Ojo, O.S., Tran, F., and Westwood, N.J. (2015). Isolation of functionalized phenolic monomers through selective oxidation and C–O bond cleavage of the β-O-4 linkages in lignin. *Angew. Chem. Int. Ed.* 54: 258–262.
64 Rahimi, A., Azarpira, A., Kim, H. et al. (2013). Chemoselective metal-free aerobic alcohol oxidation in lignin. *J. Am. Chem. Soc.* 135: 6415–6418.
65 Rahimi, A., Ulbrich, A., Coon, J.J., and Stahl, S.S. (2014). Formic-acid-induced depolymerization of oxidized lignin to aromatics. *Nature* 515: 249–252.
66 Sedai, B. and Baker, R.T. (2014). Copper catalysts for selective C–C bond cleavage of β-O-4 lignin model compounds. *Adv. Synth. Catal.* 356: 3563–3574.
67 Tarabanko, V.E., Fomova, N.A., Kuznetsov, B.N. et al. (1995). On the mechanism of vanillin formation in the catalytic oxidation of lignin with oxygen. *React. Kinet. Catal. Lett.* 55: 161–170.
68 Baciocchi, E., Fabbri, C., and Lanzalunga, O. (2003). Lignin peroxidase-catalyzed oxidation of nonphenolic Trimeric lignin model compounds: fragmentation reactions in the intermediate radical Cations. *J. Org. Chem.* 68: 9061–9069.
69 Dabral, S., Wotruba, H., Hernández, J.G., and Bolm, C. (2018). Mechanochemical oxidation and cleavage of lignin β-O-4 model compounds and lignin. *ACS Sustainable Chem. Eng.* 6: 3242–3254.

70 Guo, H., Miles-Barrett, D., Neal, A.R. et al. (2018). Unravelling the enigma of lignin OX: can the oxidation of lignin be controlled? *Chem. Sci.* 9: 702–711.
71 Tran, F., Lancefield, C.S., Kamer, P.C.J. et al. (2015). Selective modification of the β–β linkage in DDQ-treated Kraft lignin analysed by 2D NMR spectroscopy. *Green Chem.* 17: 244–249.
72 Zhang, C., Li, H., Lu, J. et al. (2017). Promoting lignin depolymerization and restraining the condensation via an oxidation–hydrogenation strategy. *ACS Catal.* 7: 3419–3429.
73 Gao, R., Li, Y., Kim, H. et al. (2018). Selective oxidation of lignin model compounds. *ChemSusChem* 11: 2045–2050.
74 Mobley, J.K., Yao, S.G., Crocker, M., and Meier, M. (2015). Oxidation of lignin and lignin beta-O-4 model compounds via activated dimethyl sulfoxide. *RSC Adv.* 5: 105136–105148.
75 Jennings, J.A., Parkin, S., Munson, E. et al. (2017). Regioselective Baeyer-Villiger oxidation of lignin model compounds with tin beta zeolite catalyst and hydrogen peroxide. *RSC Adv.* 7: 25987–25997.
76 Wang, Y.L., Wang, Q.Y., He, J.H., and Zhang, Y.T. (2017). Highly effective C–C bond cleavage of lignin model compounds. *Green Chem.* 19: 3135–3141.
77 Wang, Y., Du, Y., He, J., and Zhang, Y. (2018). Transformation of lignin model compounds to N-substituted aromatics via Beckmann rearrangement. *Green Chem.* 20 (14): 3318–3326.
78 Heap, L., Green, A., Brown, D. et al. (2014). Role of laccase as an enzymatic pretreatment method to improve lignocellulosic saccharification. *Catal. Sci. Technol.* 4: 2251–2259.
79 Luo, N., Wang, M., Li, H. et al. (2016). Photocatalytic oxidation–Hydrogenolysis of lignin β-O-4 models via a dual light wavelength switching strategy. *ACS Catal.* 6: 7716–7721.
80 Wang, M., Li, L.H., Lu, J.M. et al. (2017). Acid promoted C–C bond oxidative cleavage of [small beta]-O-4 and [small beta]-1 lignin models to esters over a copper catalyst. *Green Chem.* 19: 702–706.
81 Santos, W.C.C., Dias, K.A., Santos, L.P. et al. (2018). Evaluating gold and selenium chemistry for selective transformations of lignin model compounds. *Adv. Synth. Catal.* 360: 1376–1383.
82 Renz, M. and Meunier, B. (1999). 100 years of Baeyer–Villiger oxidations. *Eur. J. Org. Chem.* 1999: 737–750.
83 Doering, W.v.E. and Speers, L. (1950). The Peracetic acid cleavage of unsymmetrical Ketones1. *J. Am. Chem. Soc.* 72: 5515–5518.
84 Bosque, I., Magallanes, G., Rigoulet, M. et al. (2017). Redox catalysis facilitates lignin Depolymerization. *ACS Cent. Sci.* 3: 621–628.
85 Kärkäs, M.D., Bosque, I., Matsuura, B.S., and Stephenson, C.R.J. (2016). Photocatalytic oxidation of lignin model systems by merging visible-light Photoredox and palladium catalysis. *Org. Lett.* 18: 5166–5169.
86 Nguyen, J.D., Matsuura, B.S., and Stephenson, C.R.J. (2014). A photochemical strategy for lignin degradation at room temperature. *J. Am. Chem. Soc.* 136: 1218–1221.

87 Liu, H., Li, H., Lu, J. et al. (2018). Photocatalytic cleavage of C–C bond in lignin models under visible light on Mesoporous graphitic carbon nitride through π–π stacking interaction. *ACS Catal.* 8: 4761–4771.

88 Hou, T., Luo, N., Li, H. et al. (2017). Yin and Yang dual characters of CuOx clusters for C–C bond oxidation driven by visible light. *ACS Catal.* 7: 3850–3859.

89 Luo, N., Wang, M., Li, H. et al. (2017). Visible-light-driven self-hydrogen transfer Hydrogenolysis of lignin models and extracts into phenolic products. *ACS Catal.* 7: 4571–4580.

90 Lan, W., de Bueren, J.B., and Luterbacher, J.S. (2019). Highly selective oxidation and depolymerization of α,γ-diol-protected lignin. *Angew. Chem. Int. Ed.* 58: 2649–2654.

91 Lahive, C.W., Deuss, P.J., Lancefield, C.S. et al. (2016). Advanced model compounds for understanding acid-catalyzed lignin Depolymerization: identification of renewable aromatics and a lignin-derived solvent. *J. Am. Chem. Soc.* 138: 8900–8911.

92 Deuss, P.J., Scott, M., Tran, F. et al. (2015). Aromatic monomers by *in situ* conversion of reactive intermediates in the acid-catalyzed depolymerization of lignin. *J. Am. Chem. Soc.* 137: 7456–7467.

93 Guo, H., Miles-Barrett, D.M., Zhang, B. et al. (2019). Is oxidation–reduction a real robust strategy for lignin conversion? A comparative study on lignin and model compounds. *Green Chem.* 21: 803–811.

94 Deng, W., Zhang, H., Wu, X. et al. (2015). Oxidative conversion of lignin and lignin model compounds catalyzed by CeO_2-supported Pd nanoparticles. *Green Chem.* 17: 5009–5018.

95 Dawange, M., Galkin, M.V., and Samec, J.S.M. (2015). Selective aerobic Benzylic alcohol oxidation of lignin model compounds: route to aryl ketones. *ChemCatChem* 7: 401–404.

96 Galkin, M.V., Sawadjoon, S., Rohde, V. et al. (2014). Mild heterogeneous palladium-catalyzed cleavage of β-O-4′-ether linkages of lignin model compounds and native lignin in air. *ChemCatChem* 6: 179–184.

97 Mobley, J.K. and Crocker, M. (2015). Catalytic oxidation of alcohols to carbonyl compounds over hydrotalcite and hydrotalcite-supported catalysts. *RSC Adv.* 5: 65780–65797.

98 Sturgeon, M.R., O'Brien, M.H., Ciesielski, P.N. et al. (2014). Lignin depolymerisation by nickel supported layered-double hydroxide catalysts. *Green Chem.* 16: 824–835.

99 Kruger, J.S., Cleveland, N.S., Zhang, S. et al. (2016). Lignin Depolymerization with nitrate-intercalated Hydrotalcite catalysts. *ACS Catal.* 6: 1316–1328.

100 Roberts, V.M., Stein, V., Reiner, T. et al. (2011). Towards quantitative catalytic lignin depolymerization. *Chem. Eur. J.* 17: 5939–5948.

101 Katahira, R., Mittal, A., McKinney, K. et al. (2016). Base-catalyzed Depolymerization of biorefinery Lignins. *ACS Sustainable Chem. Eng.* 4: 1474–1486.

102 Mobley, J., Jennings, J., Morgan, T. et al. (2018). Oxidation of Benzylic alcohols and lignin model compounds with layered double hydroxide catalysts. *Inorganics* 6.

103 Mottweiler, J., Puche, M., Rauber, C. et al. (2015). Copper- and vanadium-catalyzed oxidative cleavage of lignin using dioxygen. *ChemSusChem* 8: 2106–2113.

104 Baguc Ismail, B., Celebi, M., Karakas, K. et al. (2017). Nanohydrotalcite supported ruthenium nanoparticles: highly efficient heterogeneous catalyst for the oxidative valorization of lignin model compounds. *ChemistrySelect* 2: 10191–10198.

105 Song, Y., Mobley, J.K., Motagamwala, A.H. et al. (2018). Gold-catalyzed conversion of lignin to low molecular weight aromatics. *Chem. Sci.* 9 (42): 8127–8133.

106 Mobley, J.K. (2016). Towards catalytic oxidative depolymerization of lignin. PhD theses and dissertations – Chemistry. p. 246. University of Kentucky. https://uknowledge.uky.edu/chemistry_etds/59.

9

Lignin Valorization via Reductive Depolymerization

Yang (Vanessa) Song [1,2,3]

[1] *University of Kentucky, Center for Applied Energy Research, 2540 Pesearch Park Drive, Lexington,, Kentucky 40511, USA*
[2] *Centre College, 600 West Walnut St. Danville, KY 40422, USA*
[3] *University of Kentucky Department of Chemistry,, 505 Rose St, Lexington, KY 40506, USA*

9.1 Introduction

Increased concerns surrounding the magnitude of fossil fuel reserves, together with the negative environmental impacts of fossil fuel use, have forced society to search for renewable fuel alternatives [1]. Among the options for renewable energy resources, lignocellulosic biomass is particularly attractive due to its high energy content, as well as the fact that it does not compete with the global food supply [2]. In parallel to cellulose conversion, which selectively produces glucose and cellobiose for fuels and chemicals with good selectivity [3, 4], lignin valorization has long been of interest; however, the inherent structural recalcitrance and heterogeneity of lignin represent technical barriers that to date have not been overcome. As such, the fate of lignin as a usable substrate has been a low priority during most biomass pretreatment strategies, and most lignin recovery processes used within any specific biorefinery have been selected to optimize the carbohydrate stream [4–8]. Nonetheless, lignin valorization remains critical for the global fuel economy as it is directly correlated to the cost of cellulosic ethanol production [5, 9–11]. A particular challenge lies in the development of new techniques for the rational deconstruction of lignins, with the goal of minimizing reactive intermediates that condense to form refractory bonds [12–14]. Although a variety of isolated lignins (technical lignins) is available (kraft lignin, soda lignin, organosolv lignin, etc.), during biomass fractionation lignin is prone to structural degradation that typically results in ether bond depletion. This results in a concurrent increase in the relative concentration of recalcitrant carbon–carbon bonds as compared with native lignin, lowering the potential for lignin depolymerization [15]. In this context, early-stage catalytic conversion of lignin (ECCL) or "lignin-first" approaches, which combine lignin isolation and depolymerization in one step, has shown success by using lignocellulose, instead of isolated lignins, as the starting material, in order to avoid formation of complex lignin mixtures [2–4, 15–19].

Chemical Catalysts for Biomass Upgrading, First Edition. Edited by Mark Crocker and Eduardo Santillan-Jimenez.

With the goal of elucidating the structure of lignin, biomass fractionation under reductive conditions has a long history dating back to the 1930s, beginning with work by Hibbert and coworkers [20–24]. However, considerable advancements took place in the twenty-first century with major improvements in reaction conditions, catalyst design, the application of green solvents, and alternatives to hydrogen gas to allow for improved sustainability [25]. In this chapter, the most prevalent reductive methods for lignin depolymerization are discussed with a focus on late-stage and early stage catalytic conversion of lignin. Monomer yield is commonly used as a measure of conversion efficiency; thus, only reductive depolymerization methods that produce monomers are compared. The chapter is organized in three main parts. In the first, late-stage reductive lignin depolymerization, or depolymerization of isolated lignins, is discussed, the discussion being separated into sections according to the hydrogen source: hydroprocessing (exogenous hydrogen gas), liquid-phase reforming (hydrogen generated *in situ* from the solvent or lignin itself), and others (hydrosilanes, zinc, sodium). Subsequently, reductive catalytic fractionation (RCF), one method under the umbrella term "lignin first" or "ECCL", is defined and discussed in detail by evaluating the effect of reaction conditions, lignocellulose source, and applied catalyst(s). The closing section presents a brief conclusion and outlook in the broader context of the lignocellulosic biorefinery.

9.2 Late-stage Reductive Lignin Depolymerization

Late-stage catalytic conversion of lignin has been defined as "research into catalytic conversion of (technical) lignins beginning with degraded polymers obtained from wood pulping processes or, more recently, the waste stream from cellulosic bioethanol plants" [3]. Similar to oxidative lignin depolymerization, reductive depolymerization methods applied to isolated lignins also primarily target the alkyl aryl ether bonds (β-O-4 and α-O-4) as well as side chain groups. To this end, redox catalysts in combination with a hydrogen source are typically utilized. The net result in all cases involves (i) hydrogenolysis of the ether bond, (ii) removal of C_α–OH groups, and (iii) possible removal of C_γ–OH groups [15]. Substituted methoxyphenols and small oligomeric fragments are typical products of these primary reactions; however, cyclohexanol, cycloalkanes, and even some light paraffins and olefins are possible products if secondary hydrogenation occurs (Figure 9.1). When lignin is hydrocracked in the presence of hydrogen, the latter can either be in the form of gas or can be derived from a H-donating species. When hydrogen gas is used, the term "hydroprocessing" is employed, and the method can be further divided into three categories: (i) mild, (ii) harsh, and (iii) bifunctional. Furthermore, the term "liquid phase reforming" is applied when hydrogen is derived from the solvent or from the lignin itself. Other reagents such as sodium, zinc, and hydrosilanes have also been reported to act as reducing agents for lignin depolymerization. During the process of reductive depolymerization of lignin, the polymer is also deoxygenated to various degrees via hydrodeoxygenation (HDO) [15].

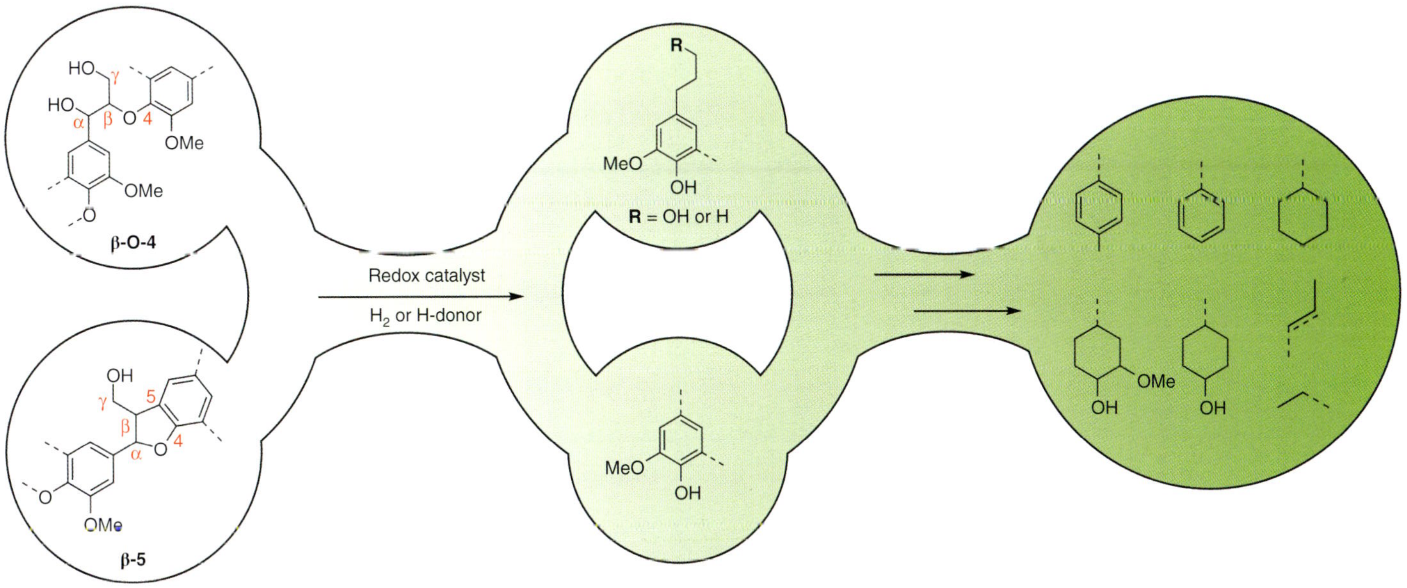

Figure 9.1 Reductive lignin linkage cleavage and select products.

9.2.1 Mild Hydroprocessing

Mild hydroprocessing is typically performed at temperatures ≤300 °C in the liquid phase using water, an organic solvent, or solvent mixtures over a noble or base metal catalyst. The mild conditions, in contrast to harsh hydroprocessing (*vide infra*), preserve the methoxy groups and generate various *p*-substituted methoxyphenols [13, 18, 26–52]. Fast pyrolysis gives a similar monomer yield compared with mild hydroprocessing; however, Faix and Meier [32] demonstrated that the product mixture is much less complex for mild hydroprocessing. Undoubtedly, the monomer product yield and selectivity are affected by catalyst, reaction conditions, and solvent. Generally speaking, the total monomer yield for mild hydroprocessing does not exceed 20 wt%, although occasionally higher monomer yields have been reported when a cocatalyst such as a Lewis acid or Lewis base was employed (Table 9.1) [37–39].

The structure of the isolated lignin has the most pronounced influence on both the monomer structure and yield. Bouxin et al. [27] performed mild hydroprocessing on lignins isolated using various methods – ethanol organosolv, ammonia fiber explosion/expansion (AFEX)-ethanosolv, soda lignin, and *a*mmonia *r*ecycled *p*ercolation (ARP) – using Pt/Al_2O_3 as catalyst and water/methanol (1 : 1 v/v) as solvent. These authors observed that the least condensed lignin (that possessing the highest β-O-4 content), which in this case was poplar ARP lignin, resulted in the highest monomer yield (14 wt%). However, when the same group performed a follow-up study applying the same Pt/Al_2O_3 catalyst and reaction conditions to seven different types of isolated lignins obtained from four biomass species [26], they discovered that a correlation between monomer yield and β-O-4 content was still observed, although the results were not conclusive. Other factors besides the β-O-4 content must play a role since lignins with an overall lower β-O-4 content (miscanthus ionosolv lignin and oak dioxane organsolv lignin) generated higher monomer yields (32 and 55 wt%, respectively) compared with ARP lignin, which gave the highest monomer yield in their previous study (14 wt%).

Van den Bosch et al., as well as Torr et al., investigated the impact of the lignin isolation procedure on the monomer yield [18, 28]. Van den Bosch et al. compared native birch lignin in birch wood sawdust and technical *e*thanol *o*rganosolv *l*ignin (EOL) over a Ru/C catalyst in methanol at 250 °C. Drastic differences in monomer yields were observed as only 3 wt% was recovered from EOL, while native birch

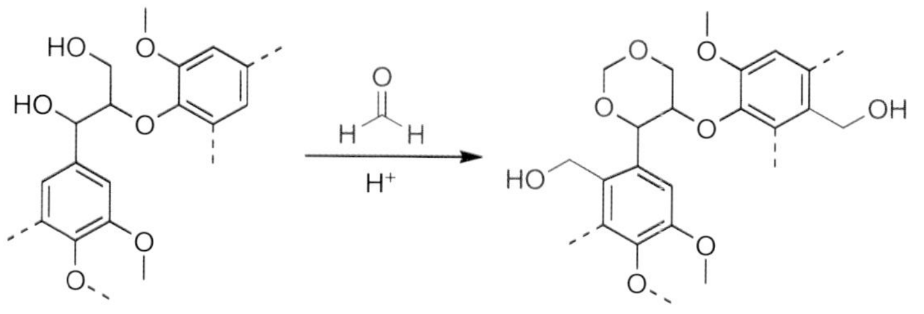

Figure 9.2 Structure of FA-stabilized β-O-4 unit.

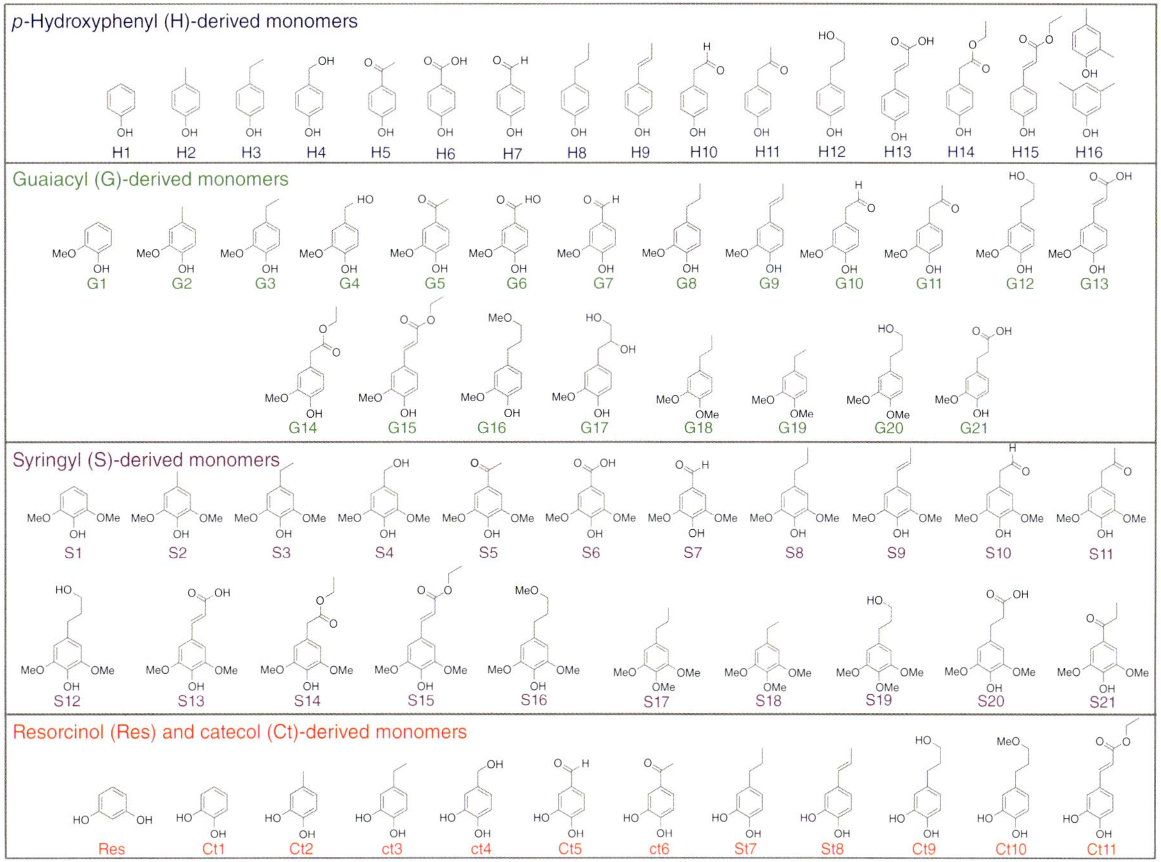

Figure 9.3 Structures and codes correspond to those listed under "Monomer structures" column in Tables 9.1–9.5.

Table 9.1 Results of literature methods utilizing mild hydroprocessing for reductive lignin depolymerization.

Catalyst	Solvent	Temperature (°C)	H_2 (bar)[a)]	Time	Lignin	Monomer yield (wt%)	Monomer structures[b)]	References
Pd/C	EtOH/water (v/v = 9 : 1)	390	100 (280 at reaction temperature)	15 min	Beech *m*illed *w*ood *l*ignin (MWL)	15.47	H/G/S: 1–3,8,12	[32]
					Spruce MWL	11.15	G: 1–3, 8	
					Bamboo MWL	16.36	H/G/S: 1–3, 8, 12	
					Teak MWL	12.63		
$Cu_2Cr_2O_5$	1,4-Dioxane	240–260	100	48 h	Birch MWL	21.2	H/G/S: 2, 3, 8, 12	[41]
					Oak MWL	16.7		
$Cu_2Cr_2O_5$	1,4-Dioxane	225	135	47 h	Spruce MWL	19.9	H/G/S: 2, 3, 8, 12	[42]
					Pine MWL	17.7		
$CoMoO_x$ + NaOH	1,4-Dioxane/3 wt% aq. NaOH (v/v = 1 : 1)	250	69	1 h	Sweetgum HCl hydrolysis extracted lignin	9.3	G/S/Ct: 1–3	[43]
Pd/C	1,4-Dioxane/water (v/v = 1 : 1)	195	35	24 h	Enzyme mild acidolysis extracted pine lignin	21	G: 8, 12	[28]
Ru/C	MeOH	250	30	3 h	Birch EOL	3	G/S: 1–3, 8, 12	[18]
Raney® Ni	Water	400	10	1 h	Spruce enzymatic hydrolysis lignin	2	H/G/Ct: 1	[44]
Ru/C	EtOH/water (v/v = 65 : 35)	275	20	1.5 h	Corn stalk enzymatic hydrolysis/EOL	4.5	H/G: 3	[35]
		250		1.5 h	Corn stalk enzymatic hydrolysis/EOL	3.5 1.5–1.6		

Pt (or Pd)/C					Bamboo lignin	1.3		
Cu-PMO (porous metal oxide)	MeOH	180	40	14 h	Candlenut tree methanol organosolv lignin	54.8	Ct: 7, 9, 10	[40]
		140		20 h		63.7		
Pt/Al_2O_3	MeOH/water (v/v = 1 : 1)	300	20	2 h	Poplar EOL	7	G/S: 1–3, 8, 9, 12, 16	[27]
					Poplar ARP lignin	14		
					Wheat straw AFEX-EOL	10		
					Wheat straw soda lignin	6	G/S: 1–3, 8	
Pt/Al_2O_3	Water	300	20	2 h	Poplar ARP lignin	11	G/S: 1–3, 8, 9, 12, 16	[33]
	MeOH/water (v/v = 1 : 1)		20		Poplar ARP lignin	17		
	MeOH		20 (160 at reaction temperature)		Poplar ARP lignin	44		
Pt/Al_2O_3	MeOH/water (v/v = 1:1)	300	20	2 h	Oak EOL	55	G/S: 1–3, 8, 9, 12, 16	[26]
					Miscanthus ionosolv lignin	32		
WP/C	EtOH/water (v/v = 1 : 1)	280	20	2 h	Softwood kraft lignin	7	G: 1–3, 5, 8	[45]
$Ni_{85}Ru_{15}$ NPs (nanoparticle)	Water	130	10	12 h	Birch dioxane organosolv lignin	7	G/S: 12 + others	[46]

(Continued)

Table 9.1 (Continued)

Catalyst	Solvent	Temperature (°C)	H_2 (bar)[a)]	Time	Lignin	Monomer yield (wt%)	Monomer structures[b)]	References
Ni_7Au_3 NPs	Water	170	10	1 h	Birch dioxane organosolv lignin	14	G/S: 12 + others	[47]
Ni_7Au_3 NPs	Water	160	10	4	Birch organosolv lignin	8	G/S: 8, 12	[39]
Ni_7Au_3 NPs + NaOH						11		
Pt/C	*t*-Butanol	350	30 (150 at reaction temperature	40 min	Protobind™ 1000	4.2	H/G: 3; S: 1, 2, 8	[48]
					Poplar organosolv lignin	4.9	G: 3; S: 1,2,8	
					Wheat straw hydrothermal lignin	1.11	H/G: 3; S:1,8	
Pt/C, Pt/C, Ru/C, Ni/C	MeOH, EtOH, IPA (isopropyl alcohol)	350	30	40 min	Protobind 1000	9–11	H/G/S: 1–3, 5, 8, 9	[34]
Pd/C	MeOH	260	40	5 h	Softwood kraft lignin	7	Phenols/guaiacols	[37]
Pd/C + $CrCl_3$		260				29		
		280				35		
Ru/C	MeOH	260	40	4 h	Pine EOL	6.1	Phenols/guaiacols	[38]
Ru/C + NaOH						12.7		
Ru (or Pd)/C + NaOH, KOH, or Na_2CO_3						8.4–12.7		
Pd/C + $CrCl_3$	MeOH	280	40	5 h	Softwood kraft lignin	16	Phenols/guaiacols	[52]
					lignosulfonate	8		
					Pennisetum EOL	19		

TiN–Ni	MeOH	150	25	1.5–4.5 min	Softwood Indulin AT kraft lignin	3.2	G: 1, 2, 4, 6, 7, 17	[49]
Ni/ZrO, ZrO_2, or TiO_2	*n*-Dodecane	250	40	160 min	Corncob enzymatic hydrolysis lignin	5–11	H/G/S: alkyls, oxygenates	[30]
$Ni/SiO_2–Al_2O_3$	Dioxane, benzene, toluene					9–12		
	Benzene					12		
	Naphthalene, EtOH, THF					5–6		
Pt/Al_2O_3	EtOH/water (w/w = 1 : 1)	200	30	4 h	Indulin AT kraft lignin (pine)	6.2	G: 1–3, 7, 8	[51]
Ru/C	THF	200	40	6 h	Beech dioxane organosolv lignin	7.10	G/S: 3, 8, 12	[29]
					FA-protected beech dioxane organosolv lignin	45.06	G/S:3, 8, 12 G: 18–20; S: 17–19	
		250		15 h	F5H-poplar dioxane organosolv lignin	24.22	G/S: 3, 8, 12	
					FA-protected F5H-poplar dioxane organosolv lignin	77.66	G/S:3, 8, 12 G: 18–20; S: 17–19	
					FA-protected spruce dioxane organosolv lignin	20.56	G: 3, 8, 18, 19	

a) Pressure at room temperature unless otherwise noted.
b) Codes are referenced to Figure 9.3, unless otherwise noted.
Source: Schutyser et al. 2018 [15]. Reproduced with permission of The Royal Society of Chemistry.

lignin yielded 50 wt% monomers. On the other hand, Torr et al. observed comparable yields applying native pine lignin in sawdust form and *e*nzymatic *m*ild *a*cidolysis *l*ignin (EMAL) from pine wood over Pd/C in dioxane/water (1 : 1, v/v) at 195 °C, suggesting mild isolation procedures do not increase lignin reactivity. While EMAL gives lignin with structure and molecular weight very close to those of native lignin [53, 54], the isolation method is industrially impractical [15]. Shuai et al. recently developed a formaldehyde (FA)-protected lignin extraction that preserves most of the β-O-4 linkages, thus preserving the reactivity of lignin (Figure 9.2) [29]. Using beech, high-syringyl transgenic (F5H) poplar, and spruce, Shuai et al. performed hydrogenolysis on FA protected/unprotected dioxane organosolv lignin (tetrahydrofuran (THF) as solvent, 250 °C), and native lignin in whole biomass (MeOH as solvent, 250 °C) using Ru/C as catalyst. The total monomer yield obtained from FA-protected isolated beech wood lignin was similar to that obtained from the native lignin (47.19 and 48.26 wt%, respectively, based on Klason lignin); however, when unprotected isolated lignin was used in the subsequent mild hydrogenolysis, the yield of monomer decreased to 7.10 wt%. Similar trends were observed for F5H-poplar and spruce (Figure 9.3).

9.2.2 Harsh Hydroprocessing

Compared with mild hydroprocessing, harsh hydroprocessing of lignin typically occurs at temperatures ≥320 °C and at hydrogen pressures ≥35 bar [15]. Most of the reactions occur between the solid catalyst and the lignin, and many are performed in a solvent-free environment [55–62]. Examples of the catalysts and reaction environments used are summarized in Table 9.2, many studies employing conventional hydrotreating catalysts such as cobalt/molybdenum oxides ($CoMoO_x$), or nickel/molybdenum oxides ($NiMoO_x$) supported on alumina that have been pre-sulfided using reagents such as dimethyl disulfide (DMDS) [56, 57], carbon disulfide [59], or (di)methylsulfide [62]. On the other hand, noble metal and other base metal catalysts have also been tested [55, 58]. Monomer yields above 20 wt% have been regularly achieved; however, a broad product distribution with low selectivity toward individual products is typically observed.

At such high reaction temperatures, the methoxy groups typically do not survive the harsh hydroprocessing conditions, generating phenols, methylated phenols, and phenols with alkyl chains, and deoxygenated aromatics as major products [15]. The highest monomer yield reported using a conventional hydrotreating catalyst was achieved by Kumar et al. using DMDS-sulfided NiMo/MgO–La_2O_3 for Indulin AT kraft lignin conversion, obtaining 26.4 wt% monomer yield [56]. For noble metal catalysts in solvent-free conditions, Kloekhorst and Heeres obtained ~22 wt% monomer yield employing Pd/C, Ru/C, or Ru/TiO_2 [58]. However, the highest yield for harsh hydroprocessing conditions reported was achieved using DMDS-sulfided NiW/C by Narani et al., which gave 35 wt% monomer yield [57].

Table 9.2 Results of literature methods utilizing harsh hydroprocessing for reductive lignin depolymerization.

Catalyst[a)]	Solvent	Temperature (°C)	H_2 (bar)[b)]	Time	Lignin	Monomer yield[c)] (wt%)	Monomer structures	References
$NiMoO_x/SiO_2–Al_2O_3$	None	420	100 (210 at reaction temperature)	1 h	Spruce organocell lignin	16.3	H: 1–3	[60]
$NiMoO_x/SiO_2–Al_2O_3 + Cr_2O_3/Al_2O_3$	None	400	140	6 h	Spruce organocell lignin	12.3	H: 1–3, 8, 16	[61]
$NiMoO_x/SiO_2–Al_2O_3 + Cr_2O_3/Al_2O_3$	None	395	90	40 min	Indulin AT kraft lignin (pine)	8.3	H: 1–3, 8, 16	[59]
S-$NiMoO_x/SiO_2–Al_2O_3 + Cr_2O_3/Al_2O_3$		430	90	1 h		10.8		
		395	100	30 min		10.4		
		395	100	35 min	Spruce organocell lignin	10.5		
		400	100	30 min	Birch kraft lignin	5.9		
S-$CoMoO_x/Al_2O_3$ batch reactor	1-Methylnaphthalene	404	69	1 h	Aspen methanol organosolv lignin	2.2	H: 1, 2	[62]
S-$CoMoO_x/Al_2O_3$ continuous flow reactor		428				10		
Ru/C, Ru/TiO_2, Pd/C, Pd/Al_2O_3	None	400	100	4 h	Alcell lignin	19.1–22.8	Aromatics, alkylphenols, catechols, cyclic and noncyclic alkanes, alcohols, and ketones	[63]
Cu/ZrO_2						11.8		
Ru/C						21.4		[58]

Table 9.2 (Continued)

Catalyst[a)]	Solvent	Temperature (°C)	H_2 (bar)[b)]	Time	Lignin	Monomer yield[c)] (wt%)	Monomer structures	References
S-NiMo (or CoMo)/MgO–La_2O_3	None	350	100	4 h	Indulin AT kraft lignin (pine)	26.4 (18.1)	Alkylphenols, aromatics, cyclic and noncyclic alkanes	[56]
S-NiMo (or CoMo)/Al_2O_3						14.8 (10.8)		
S-NiMo (or CoMo)/C						20.4 (21.5)		
S-NiMo (or CoMo)/ZSM-5						15.5 (18.2)		
S-NiMo (or CoMo)/C	MeOH	320	35	8 h	Indulin AT kraft lignin (pine)	14.5 (9)	Alkylphenols, alkylguaiacols	[57]
S-WO_x(or Ni)/C						20.5 (19)		
S-NiW/C (or ZSM-5, MgO–La_2O_3, MgO–CeO_2, MgO–ZrO_2)						28.5 (18, 26.5, 22.5, 16.5)	Alkylphenols	

a) "**S**-" indicates catalyst(s) have been pre-sulfided.
b) Pressure at room temperature unless otherwise noted.
c) Yields represented in parentheses correspond to respective catalysts in parentheses.

Source: Schutyser et al. 2018 [15]. Reproduced with permission of The Royal Society of Chemistry.

9.2.3 Bifunctional Hydroprocessing

The term "bifunctional" in this method refers to the bifunctional property of the catalyst system, the catalyst exhibiting both acid and metal active sites. Lignin in this case is converted into cycloalkanes by simultaneous hydrolysis (if water is present) and dehydration catalyzed by the acid sites, while hydrogenation/hydrogenolysis is enabled by the metal sites [30, 64–70]. Some phenolic products obtained from lignin depolymerization via this method also typically undergo subsequent HDO, which can generate a variety of substituted cyclic and noncyclic alkanes (Table 9.3) [15]. High selectivity toward cycloalkanes was observed by Kasakov et al. using supported Ni catalysts [66], as well as by Luterbacher et al. using carbon-supported noble metal catalysts in conjunction with H_3PO_4 [65].

For conversion of beech organosolv lignin, Kasakov et al. [66] found that Ni supported on H-zeolites – specifically HBEA (Si/Al = 75) and HZSM-5 (Si/Al = 45) – yielded completely deoxygenated monocyclic and bicyclic alkanes compared with Ni/SiO_2, which also yielded cyclohexanols. The selectivity toward deoxygenated cyclic alkanes achieved by the use of zeolite supports can be attributed to the presence of strong Brønsted acid sites (which are lacking in silica) [66, 71]. Higher selectivity toward bicyclic alkanes was also observed for Ni/HBEA compared with Ni/HZSM-5, due to the smaller pores of HZSM-5, which can hinder alkylation and condensation of OH substituted aromatic monomers and olefins during the hydrogenation/HDO process [72]. In another study by the same group using enzymatic hydrolysis extracted corncob, Lercher and coworkers found that using Ni supported on amorphous silica alumina (Ni/ASA), which has a lower concentration of Brønsted acid sites compared with Ni/HZSM-5, led to a high conversion toward monocyclic and noncyclic alkanes [30]. Furthermore, when the acidity of the catalyst is decreased by employing less acidic oxides as support (ZnO, ZrO_2, TiO_2, SiO_2, Al_2O_3), the selectivity shifts toward phenols. Based on the foregoing, as well as additional studies performed by Yang and coworkers using acidic Y zeolite (HY)-supported Ru or bimetallic RuM (M = Fe, Ni, Cu, or Zn) [67–69] and Wang and Rinaldi using Al-SBA-15 supported Ni [70], it is evident that the conversion of isolated lignin toward cycloalkanes is highly dependent on the concentration of acid sites on the catalyst.

Unlike the previously discussed bifunctional hydroprocessing methods, Luterbacher et al. introduced a two-step conversion of γ-valerolactone (GVL)-extracted corn stover lignin over carbon-supported Ru (or Pd, Pt, Rh) and H_3PO_4 [65]. The first step involves treatment of the extracted lignin in THF/water mixture at 150 °C under H_2, THF then being removed via evaporation followed by addition of heptane to form a biphasic mixture, which was treated at 250 °C under H_2 over the supported noble metal catalyst. Higher selectivity toward cyclohexane propanoic acid (as well as the methyl ester) was observed overall, with the highest carbon yield (48 C%) achieved when Ru/C was used and methanol was introduced during the second step. Lignin isolated using the GVL extraction method can account for up to 75 wt% of the native lignin (determined by Klason lignin content), but the most attractive feature

Table 9.3 Results of literature methods utilizing bifunctional hydroprocessing for reductive lignin depolymerization.

Catalyst	Solvent	Temperature (°C)	H_2 (bar)[a)]	Time	Lignin	Monomer yield[b)]	Monomer structures	References
Ni/SiO_2	*n*-Hexadecane	250	20	6 h	Beech organosolv lignin	27 wt%	–OH substituted and unsubstituted monocyclic alkanes	[66]
Ni/HZSM-5						34 wt%	Mono- and bicyclic alkanes	
Ni/HBEA					Beech organosolv lignin	42 wt%		
		320				70 wt%	Mono- and bicyclic alkanes, hydrocarbons, methane, and ethane	
Ni/ASA	*n*-Dodecane	250	40	160 min	Enzyme hydrolysis extracted corncob lignin	18%	Mono- and bicyclic alkanes, phenols, alcohols, and ketones	[30]
		300	60			46%		
HY zeolite	Water	250	40	4 h	Alkaline extracted corn stover lignin	10 wt%	Cyclic and noncyclic alkanes, aromatics	[68]
Ru/Al_2O_3						14 wt%		
Ru/Al_2O_3–HY zeolite						22 wt%		
						22 wt%	Cyclic and noncyclic alkanes, oxygenated compounds	[64]

Ru/HY zeolite	Water	250	40	4 h	Dilute acid pretreatment pine lignin	26 wt%	Cyclic and noncyclic alkanes, oxygenated compounds	[69]
Ru–Ni/HY zeolite						32 wt%		
Ru/Al_2O_3–HY zeolite								
Ni/Al-SBA-15	Methylcyclohexane	300	70	8 h	Poplar EOL	45 wt%	Mono- and bicyclic alkanes	[70]
Pd or (Pt, Rh)/C + H_3PO_4	(1) THF/water/H_3PO_4 (2) Heptane/water/H_3PO_4	(1) 150 (2) 250	35	(1) 4 h (2) 4 h	GVL extracted corn stover lignin	23 C% (23 C%, 17 C%)	Monocyclic alkanes,	[65]
Ru/C + H_3PO_4				4 h		38 C%	cyclohexane	
						10 C%	propanoic acid, and cyclohexane	
	(1) THF/water/H_3PO_4 (2) Heptane/water/H_3PO_4/MeOH					17 C%	propanoic acid	
						48 C%	methyl ester	

a) Pressure at room temperature unless otherwise noted.
b) Yields represented in parentheses correspond to respective catalysts in parentheses.
Source: Schutyser et al. 2018 [15]. Reproduced with permission of The Royal Society of Chemistry.

of this method aside from the monomer yield is the fact that this process also maximized the carbohydrate yield from biomass that can be further fermented and upgraded [73]. As mentioned in the introduction, the current strategy for lignin valorization is highly dependent on optimization of the cellulosic component, making this method an integrated upgrade of all three biomass components that can be industrially applicable and potentially profitable.

9.2.4 Liquid Phase Reforming

Lignin conversion via liquid phase reforming is performed in the presence of a hydrogen donating agent or hydrogen donating solvent under an inert atmosphere [50, 55, 74–89]. The majority of the studies reported to date have used ethanol as solvent and hydrogen donor, although H-donors such as formic acid, methanol, tetralin, isopropanol, and glycerol have also been studied. Additionally, supported noble metal catalysts such as Pd/C, Ru/C, and Pt/γ-Al_2O_3, as well as mixed metal oxides ($CuMgAlO_x$, Mo_2C/C-nanofibers, etc.) have been investigated for this process. One advantage of the liquid phase reforming method is the fact that aromaticity is preserved in the majority of monomers produced; however, a great variety of products are observed with low selectivity toward a single product (Table 9.4).

Total monomer yields >20 wt% are frequently reported, the highest yield (86 wt%) being achieved by Hensen and coworkers using softwood kraft lignin as substrate, ethanol as solvent/hydrogen donor, and $CuMgAlO_x$ as catalyst [91]. Compared with noble metal catalysts, a metal oxide-supported non-noble metal catalyst protects the monomers and oligomers from repolymerization by alkylation with the solvent (or capping of the phenolic units). In addition, an important aspect of the use of ethanol is the fact that it scavenges FA formed during lignin decomposition. Generally, the use of ethanol as solvent enables higher monomer yields than other solvents. Li and coworkers [77, 78] used softwood kraft lignin as substrate over carbon-supported α-molybdenum carbide (α-MoC_{1-x}/C) catalyst in a variety of alcohol solvents (ethanol, methanol, isopropanol), and they observed the highest monomer yield in ethanol (28 wt%). In another study performed by the same group testing various Mo-based catalysts [78, 79], the authors discovered that metallic Mo supported on alumina (Mo/Al_2O_3) gave the highest overall aromatic monomer yield (33.3 wt%); additionally, the simplicity of the catalyst preparation method for Mo/Al_2O_3 is another advantage over α-MoC_{1-x}/C (incipient impregnation vs. temperature programmed reaction procedure). Using Alcell® lignin, which consists of a mixture of harwood EOL, Kloekhorst et al. reached a 48 wt% monomer yield using Ru/C catalyst in methanol/formic acid (20 : 1, v/v) under 1 bar H_2 atmosphere [55]. Contrary to previously described scenarios, in this study the use of methanol as solvent exceeded ethanol in terms of monomer production. This was attributed to ethanol dehydration to ethylene, which lowers the H_2 concentration, and thus the HDO rate.

Aiming to study the effect of an acid cocatalyst on liquid phase reforming, Jiang et al. converted enzymatic hydrolysis/alkaline extracted bamboo lignin using methanol/water (5 : 2, v/v) as solvent over Raney® Ni combined with acid zeolites as catalyst [80]. The authors observed an overall increased yield

Table 9.4 Results of literature methods utilizing liquid phase reforming for reductive lignin depolymerization.

Catalyst	Solvent	Temperature (°C)	Pressure (bar)	Time	Lignin	Monomer yield (wt%)	Monomer structures	References
Pd/C + Nafion SAC-13	Water/formic acid (v/v = 24 : 1)	300	95 at reaction temperature	2 h	H_2SO_4 pretreatment/EOL from spruce	4	Res, Ct1, G1,	[90]
					SO_2 pretreatment/EOL from spruce	5		
					Acid hydrolysis lignin from spruce	5 6		
					Concentrated acid hydrolysis lignin from spruce	6		
					Desulfonated kraft lignin from spruce	11		
Pd/C	EtOH/formic acid	350	Not indicated	4 h	Switchgrass EOL	21	G: 1–4, 8	[82]
$Pt/Al_2O_3 + H_2SO_4$	EtOH/water (wt/wt = 1 : 1)	225	95 (He)	1.5 h	Indulin AT kraft lignin	18	H/G: 1, 3, 5, 7, 8 Cat: 1, 3, 5–7	[50, 81]
Pt/Al_2O_3 + heteropoly acid						13		
Pt/Al_2O_3 + NaOH						13		
$Pt/Al_2O_3 + H_2SO_4$					Alcell lignin	9	H/G/S: 1, 3, 4, 9 Cat: 1, 3, 4, 8	
$Pt/Al_2O_3 + H_2SO_4$					Sugarcane bagasse lignin	16	H/G/S: 1, 3, 9, 15 Cat: 1, 3, 8, 11	
$CuMgAlO_x$	EtOH	300	10 (N_2)	4 h 8 h	Protobind 1000	17 23	G: 1–3, alkyl phenols, aromatics, cyclic alkanes and alkenes	[75]

(Continued)

Table 9.4 (Continued)

Catalyst	Solvent	Temperature (°C)	Pressure (bar)	Time	Lignin	Monomer yield (wt%)	Monomer structures	References
		340		1 h		36	alkyl phenols, aromatics, cyclic alkanes and alkenes	[76]
		380		8 h	Protobind 1000	60	Aromatics, alkyl phenols, cyclic alkanes and alkenes	[91]
					Alcell lignin	62		
					Softwood kraft lignin	86		
	MeOH/EtOH (v/v = 1 : 1)	300		4 h	Protobind 1000	9	G1, alkyl phenols, cyclic alkanes and alkenes	
	MeOH					6		[75]
NbN	EtOH	340	10 (N_2)	< 1 h	Protobind 1000	17	Alkyl phenols, aromatics, cyclic alkenes	[83]
TiN					Protobind 1000	19		
					Wheat straw EOL	16		
					Poplar EOL	15		
					Spruce EOL	21		
Ni/Al-SBA-15	Tetralin, isopropyl alcohol, glycerol, formic acid	150	Not indicated	30 min	Olive tree pruning EOL	<1	G/S: 1, 5, 7	[84–86]
MoC/C	EtOH	280	1 N_2	6 h	Softwood kraft lignin	28	Aromatics, benzyl alcohols, and alkyl guaiacols	[77, 78]
	MeOH					2		
	Water					5		
	Isopropyl alcohol					3		

Mo/Al_2O_3	EtOH	280	1 N_2	6 h	Softwood kraft lignin	33	Aromatics, benzyl alcohols, and alkyl guaiacols	[78, 79]
Mo_2N/Al_2O_3						28		
Pd/C	MeOH/formic acid (v/v = 10 : 1)	280	1 N_2	2 h	Wheat alkali lignin	11	H/G/S: alkyls and oxides	[87]
					Sigma-Aldrich lignin	7		
Pd/C + formic acid	Water	264	1 N_2	6 h	Softwood kraft lignin	26	Ct1, H1, G: 1–4	[88]
Ru/C + formic acid	Isopropyl alcohol/formic acid (wt/wt = 1;1)	400	1 H_2	4 h	Alcell lignin	32	Aromatics, alkyl phenols, catechols, cyclic, and noncyclic alkanes	[55]
						48		
Raney Ni/HUSY	MeOH/water (v/v = 5 : 2)	270	1 N_2	30 min	Bamboo enzymatic hydrolysis/alkaline extracted lignin	27.9	H/G/S: 3, 6, 8, 9, 13, 14	[80]
Raney Ni/HZSM-5						25.9		
Raney Ni/HY						24.2		
Raney Ni						12.9		
Cu/Mo-ZSM-5 + NaOH	MeOH/water	220	1 Ar	7 h	Hardwood kraft lignin	21	Phenols, aromatics, cyclic and noncyclic alkanes, oxygenated compounds	[89]
						6		
						8	Phenols	
Ru/C	Isopropyl alcohol	300	20 N_2	3 h	Switchgrass ionic liquid pre-treated enzymatic hydrolysis lignin	15	H/G/S: 1–3, 8, 9, 11	[74]

a) Pressure at room temperature unless otherwise noted.

Source: Schutyser et al. 2018 [15]. Reproduced with permission of The Royal Society of Chemistry.

with the addition of acid zeolites such as HUSY, HZSM-5, and HY (27.9, 25.9, and 24.2 wt%, respectively) compared with Ni only (12.9 wt%). The increase in monomer yield can be attributed to the possibility that Raney Ni acts as the lignin cracking and methanol reforming agent, while the acid zeolite performs ether solvolysis and dehydration. The synergistic effect of the two components prevent coupling reactions between unstable lignin fragments and unreacted lignin, which can produce undesired polymers with higher molecular weight. Weckhuysen and coworkers [50, 81] studied the effect of acid and alkaline additives on Pt/Al_2O_3 catalysts for conversion of Indulin AT kraft lignin in ethanol and water (1 : 1, wt/wt) and found the addition of H_2SO_4 gave the highest monomer yield (18 wt%). In addition, 9 and 16 wt% monomer yields were observed for Alcell lignin and sugarcane bagasse lignin, respectively.

9.2.5 Reductive Lignin Depolymerization Using Hydrosilanes, Zinc, and Sodium

As an alternative to hydrogen as reducing agent, reductive depolymerization of isolated lignins can be achieved using stoichiometric reagents such as hydrosilanes [92], sodium [93], and zinc [94] (Table 9.5). Like many of the methods described previously, these reducing agents also target the β-O-4 and α-O-4 linkages for lignin conversion. Monomer yields for the three methods described are generally lower compared to those utilizing hydrogen; however, lignin conversion using hydrosilanes, zinc, and sodium are generally performed under milder reaction conditions in regards to the reaction temperature and pressure.

Feghali et al. [92] achieved lignin conversion toward the production of propyl- and propanol-substituted catechols and pyrogallols with the use of excess Et_3SiH in the presence of a potent Lewis acid, $B(C_6F_5)_3$, using dichloromethane as solvent. The mechanism for this metal-free reduction method has previously been described by the same group [95] and involves an electrophilic activation of Et_3SiH by the borane cocatalyst that ultimately leads to silylation of all the hydroxy and methoxy groups (Figure 9.4). The silyl groups can be subsequently removed via hydrolysis. Isolated lignins were obtained from a variety of woody feedstocks by formic acid/acetic acid/H_2O treatment (a.k.a. formacell lignin), which the authors used as substrate for lignin depolymerization. Monomer yields ranging from 11 to 41 wt% were achieved for hardwood lignins, these being higher compared with those obtained from softwood lignins, which produced 7–17 wt% monomers.

Prinsen et al. [93] set out to depolymerize a variety of isolated lignins using liquid ammonia at relatively high reaction temperature and pressure (120 °C and 88 bar), observing that most lignins underwent incorporation of nitrogen into the structure. In the case of elephant grass milled wood lignin (MWL), the monomers obtained included a large percentage of amide alkyl phenols. Subsequently, they performed conversion of lignin under reductive conditions by utilizing solvated electrons from the addition of $Na_{(s)}$ in liquid ammonia (Birch reduction). The highest monomer yield was obtained from elephant grass MWL conversion (3.1 wt%) with high selectivity toward alkyl phenols and no nitrogen incorporation in the lignin structure. The reductive cleavage in this

Table 9.5 Results of literature methods utilizing hydrosilanes, zinc, and sodium for reductive lignin depolymerization.

Catalyst	Solvent	Temperature (°C)	Pressure (bar)	Time	Lignin	Monomer yield (wt%)	Monomer structures	References
$Et_3SiH + B(C_6F_5)_3$	Dichloromethane	RT	1 Air	20 h	Formacell lignin from softwood (pine, spruce, and cedar)	7–17[a)]	Propyl and propanol substituted catechols and pyrogallols	[92]
					Formacell lignin from hardwood (oak, birch, beech, poplar, and hybrid plane)	11–41[a)]		
Na(s)	(1) THF (2) Liquid NH_3	20	1 Air	6 h	Elephant grass MWL	3.1	Phenols and alkyl phenols	[93]
					Poplar organosolv lignin	2.4		
					Indulin AT kraft lignin	1.2		
					Protobind 1000	0.6		
(1) $DDQ/{}^tBuONO/O_2$ (2) $Zn + NH_4Cl$	2-Methoxyethanol/1,2-dimethoxyethane (v/v = 2 : 3)	80	1 Air	(1) 14 h (2) 1 h	Birch dioxane organosolv lignin	0.5	G21, S20, S21	[94]

a) Yields have been corrected for $SiEt_3$ groups.

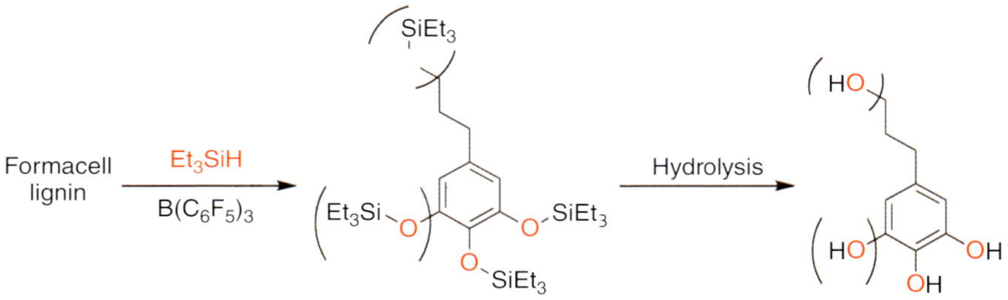

Figure 9.4 Monomer production from reductive lignin depolymerization using hydrosilanes.

case may have been achieved by the addition of the solvated electrons to the carbon in the aryl group, and it has been suggested that either an ammonium ion that complexes the adjacent phenolic oxygen [96] or lignin functional groups can act as the hydrogen donor [97].

The Zn/NH_4Cl method is a two-step lignin depolymerization process, the first step involving oxidation of lignin catalyzed by the 2,3-dichloro-5,6-dicyanobenzoquinone (DDQ)/^{t}BuONO/O_2 system, the subsequent zinc-catalyzed selective degradation of the β-O-4 linkage being inherently reductive. Studies have shown that zinc is a versatile reductant, especially for its application toward the conversion of α-hydroxy ketones [98], which were readily identified in a variety of DDQ oxidized lignins [99]. Lancefield et al. [94] used birch dioxane organosolv lignin as substrate and achieved 6 wt% isolated phenolic monomer yield with **S20** as the major product (5 wt%).

9.3 Reductive Catalytic Fractionation (RCF)

While results presented in Section 9.2 and its subsections show that good monomer yields can be obtained from late-stage lignin catalytic conversion, the tedious isolation process and the involvement of additional chemicals makes the industrial feasibility of this process rather challenging. In recent years, ECCL or "lignin first" approaches have brought monumental advances in the catalytic processing of lignin [100, 101]. ECCL comprises a number of methods that "target conversion of lignin fragments upon their release from the plant cell wall" [3]. RCF, one of the methods that falls under the umbrella term ECCL, will be discussed in detail in this section.

The origin of RCF dates back to the 1940s with studies reported by Hibbert and Pepper [20–24], which were performed with a focus on the analysis of lignin structure. However, it was not until the early 2000s that the focus of RCF research shifted toward the lignocellulose biorefinery. RCF involves solvolytic lignin extraction coupled with simultaneous reductive catalytic depolymerization, which, unlike the methods described in Section 9.2, integrates delignification and lignin depolymerization/stabilization into a one-step process. The process of RCF is closely related to organosolv pulping [15], although instead of producing a high molecular weight lignin precipitate, an uncondensed lignin oil rich in phenolic monomers, dimers, oligomers is obtained [102]. From the perspective of cellulosic ethanol production in the lignocellulosic biorefinery, the RCF process typically preserves the structure and integrity of the cellulosic component [103, 104]; however, it should be noted that the carbohydrate component of lignocellulosic biomass post-RCF is recovered as a (holo)cellulose pulp, together with the spent catalyst. The separation of catalyst and biomass is yet another challenge, in addition to maximizing monomer selectivity and yield, and many studies have opted to tackle the issue with the use of ferromagnetic catalysts [16, 105] or the use of a catalyst basket [19, 106].

The RCF biorefinery involves four essential components: (i) lignocellulosic biomass, (ii) a solvent (either an alcohol or cyclic ether, water often being used as a cosolvent [107]), (iii) a heterogeneous redox catalyst, and (iv) H_2 or a hydrogen donor [15, 102]. These components are typically added to a high-pressure batch reactor and heated (180–250 °C) for 2–24 hours [2, 4, 16, 18, 19, 21, 25, 28, 29, 31, 103, 105–126]. Sels and coworkers [106] recently discovered that delignification, induced by heat, is typically achieved by solvent extraction, which forms reactive fragments that are prone to repolymerization. Redox active catalysts can stabilize these reactive fragments, while also possibly effectuating depolymerization via hydrogenolysis (Figure 9.5) [102]. Results collected from literature that involve the RCF process are presented in Table 9.6, and the following subsections briefly break down the variables and their effect on the delignification process and total monomer yield.

9.3.1 Reaction Conditions

The degree of delignification, carbohydrate retention in the (holo)cellulose pulp, and product yield is highly dependent on the solvent [107, 109, 110, 116], additives [4, 105, 111, 112, 114, 117, 118, 126, 130], and reaction temperature [4, 18, 109, 110]. The total monomer yield and the degree of delignification can generally be enhanced by an increase of reaction temperature, and an increase in the polarity of the solvent used in the reaction can also favor product yield since hydrogen can originate from the solvent [16, 18, 107, 110]. Dilute acid is often employed in enzymatic hydrolysis of cellulose as it catalyzes hemicellulose depolymerization via hydrolysis of the glycosidic bonds [130, 131], and the addition of both Brønsted acid (H_3PO_4), and Lewis acids ($Yb(OTf)_3$ and $Al(OTf)_3$) has also been shown to increase the monomer yield without modification of other reaction conditions.

9.3.2 Lignocellulose Source

As mentioned before, the RCF process produces a "lignin oil" rich in phenolic monomers. The reaction conditions and their modifications are not typically aimed at altering the substituent structures of these phenolic monomers, rather, the lignocellulose source is the major variable that determines the structure of the aromatic monomers. As described in Chapter 8, lignin composition and inter-unit linkage vary among the three main feedstocks (hardwood, softwood, and herbaceous) used in biomass upgrading, including RCF. Studies performed

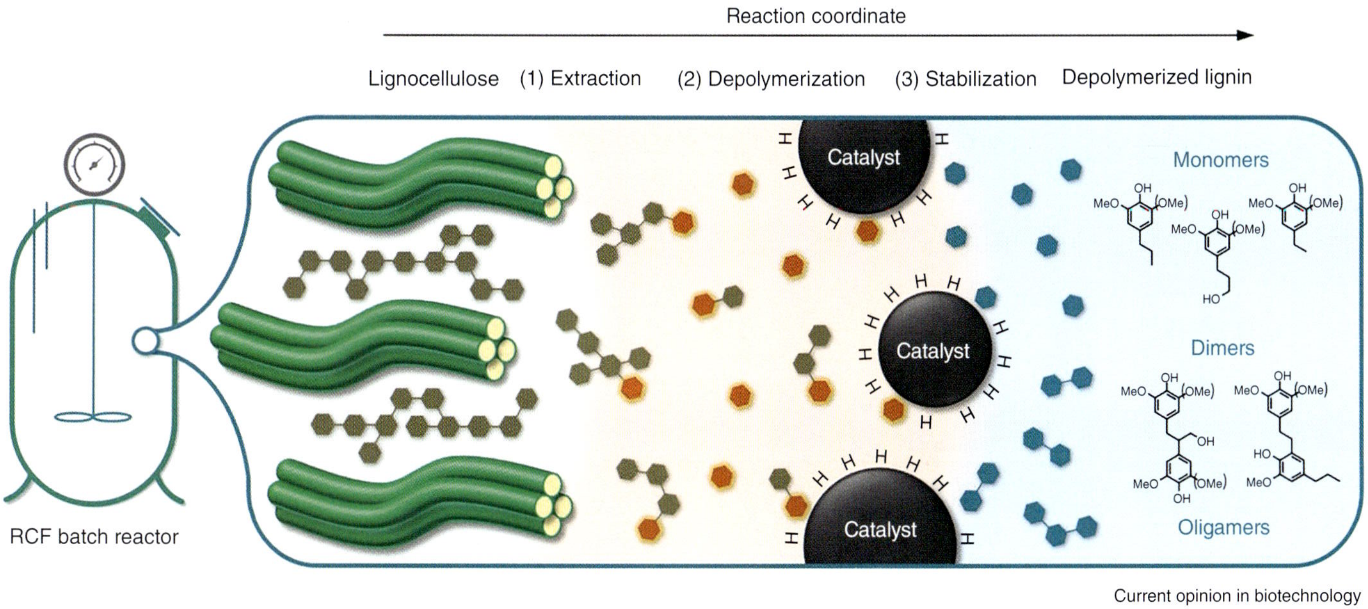

Figure 9.5 Schematic representation of RCF process in high-pressure batch reactor. Source: Reprinted with permission from Renders et al. 2019 [102]. Copyright 2019, Elsevier.

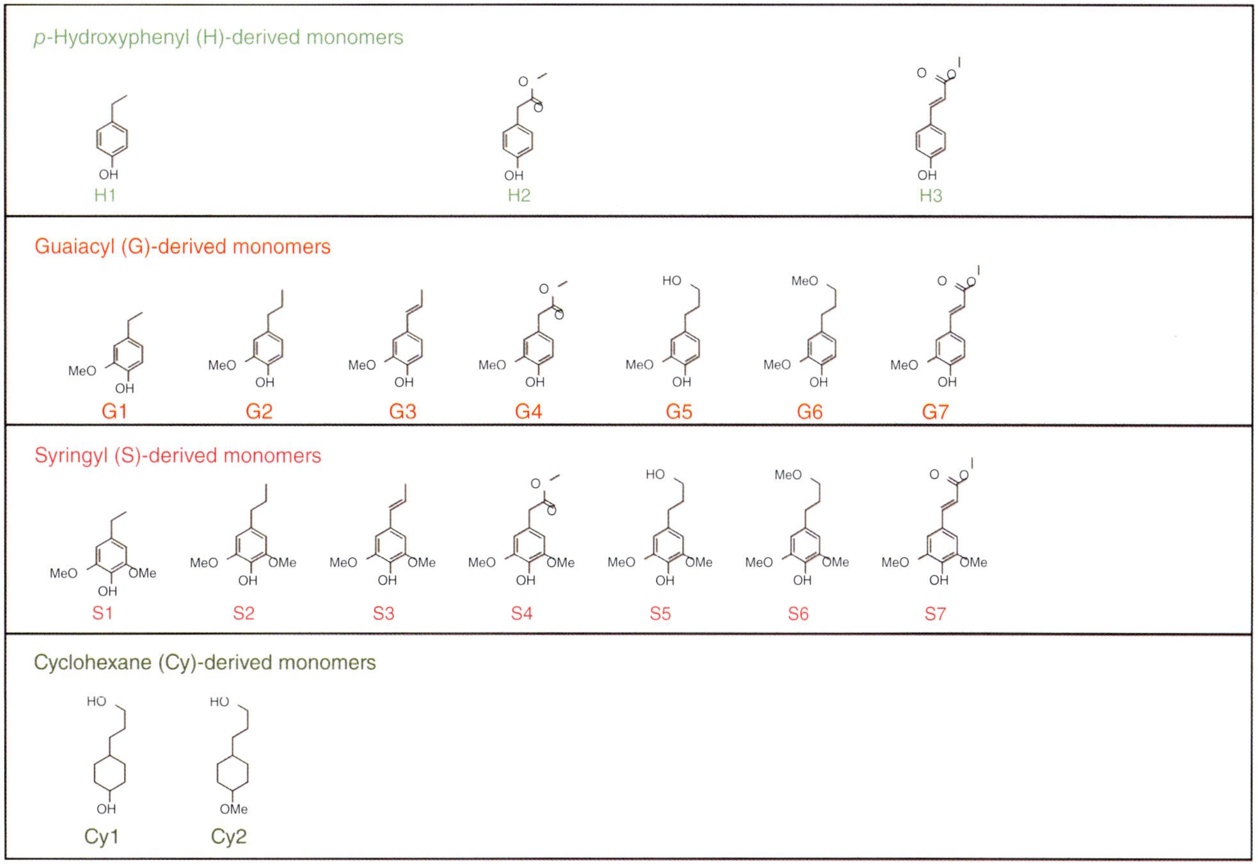

Figure 9.6 Structures and codes correspond to those listed under "Monomer structures" column in Table 9.6.

Table 9.6 Reaction conditions and results of literature methods utilizing reductive catalytic fractionation.

Catalyst	Solvent	Temperature (°C)	H_2 (bar)[a]	Time (h)	Lignocellulose source	Monomer yield	Monomer structures[b]	References
Raney Ni	1,4-Dioxane/water (v/v = 1 : 1)	195	35	5	Aspen	52 wt%	G5; S1, S2, S5	[120]
Pd/C	1,4-Dioxane/water (v/v = 1 : 1)	195	35	10	Spruce	17 wt%	G5	[117]
				5		14.7 wt%	G2, G5	
Pd/C + HCl	1,4-Dioxane/water (v/v = 1 : 1)	195	35	5	Spruce	10.5 wt%	G2, G5	
Pd/C + NaOH						7.2 wt%	G1	
Raney Ni						9.5 wt%	G5	
Raney Ni + HCl						14.7 wt%	G2, G5	
Raney Ni + NaOH						9.4 wt%	G1	
Raney Ni	1,4-Dioxane/water (v/v = 1 : 1)	195	35	5	Spruce	16 wt%	G2, G5	[127]
Rh/C						34 wt%		
Rh/Alumina						13 wt%		
Pd/C						24 wt%	G5	
Raney Ni	1,4-Dioxane/water (v/v = 1 : 1)	195	35	5	Spruce	17 wt%	Cy1, Cy2, G5	[121]
Ru/C						12 wt%		
Ru/Alumina						15 wt%		
Rh/C	1,4-Dioxane/water (v/v = 1 : 1)	195	35	5	Aspen	50 wt%	G/S: 2,5	[122]
Rh/C	1,4-Dioxane/water (v/v = 1 : 1)	195	35	5	Spruce	21 wt%	G1, G2, G5	[118]
Rh/C + HCl						18 wt%	G2, G5	

Rh/C + NaOH						7 wt%	G1, G5	
Rh/C + agitation	1,4-Dioxane/water (v/v = 1 : 1)	195	35	5	Aspen poplar	45 wt%	S2, S5	[128]
Rh/C no agitation						46 wt%		
D-101[c]	1,4-Dioxane	250	50	2	Rice husks	33 wt%	G1, G2, G5; S2; H1	[31]
Ru/C	Water	200	40	4	Birch	4.6 wt%	G/S: 2,5	[114]
Pd/C						25.5 wt%		
Rh/C						19.7 wt%		
Pt/C						33.6 wt%		
Pt/C + H_3PO_4						37.9 wt%		
Pt/C	1,4-Dioxane/water (v/v = 1 : 1)	200	40	4	Birch	41.7 wt%		
Pt/C + H_3PO_4						46.4 wt%		
Rh/C + H_3PO_4						45.5 wt%		
4%Ni–30%W_2C/C	Water	235	60	4	Birch	36.9 wt%	G/S: 2,5	[116]
	Water				Corn stalk	20.6 wt%		
	MeOH				Birch	42.2 wt%		
	Ethylene glycol				Birch	46.5 wt%		
4%Ni–30%W_2C/C	Water	235	60	4	Poplar	32.4 wt%		
					Basswood	37.3 wt%		
					Astree	40.5 wt%		
					Beech	26.1 wt%		
					Xylosma, willow	29.3 wt%		

(Continued)

Table 9.6 (Continued)

Catalyst	Solvent	Temperature (°C)	H_2 (bar)[a]	Time (h)	Lignocellulose source	Monomer yield	Monomer structures[b]	References
					Bagasse	23.4 wt%		
					Pine	10.1 wt%		
Pd–W_2C/C	Water	235	60	4	Birch	28.4 wt%		
Pt–W_2C/C						26.7 wt%		
Ir–W_2C/C						31.5 wt%		
Pd/C						55.1 wt%		
Pt/C						44.1 wt%		
Ir/C						41.1 wt%		
Ru/C						11.6 wt%		
Pd/C	1,4-Dioxane/water (v/v = 1 : 1)	195	35	24	Pine	22 wt%	G2, G5	[28]
					Steam explosion pretreated (SEP) pine	7 wt%		
					SO_2 added SEP pine	2 wt%		
Ni/C	MeOH	200	50	6	Birch	59 wt%	G/S: 2,3	[109]
	MeOH	240	1 (Ar)	6	Birch	49 wt%	G/S: 2	
		200				41 wt%	G/S: 2	
	EtOH	200	1 (Ar)	6	Birch	47 wt%	G/S: 2	
	Ethylene glycol					49 wt%	G/S: 2	
	Isopropyl alcohol					20 wt%	G/S: 2, 3	
	1,4-Dioxane					15 wt%		
	1% MeOH/water					8 wt%	G/S: 2	
	25% MeOH/water					20 wt%	G/S: 2	
Ni/SBA-15	MeOH	200	1 (Ar)	6	Birch	27 wt%	G/S: 2, 3	

Ni/Al_2O_3						19 wt%	G/S: 2, 3	
Cu/C						9 wt%	G/S: 3	
Cu–Cr oxide						2 wt%	G/S: 3	
Pd/C	EtOH/water (v/v = 1 : 1)	195	4 (Ar)	1	Pine	23 mol%	G3	[108]
					Birch	52 mol%	G/S: 3	
Pd/C	EtOH/water (v/v = 1 : 1)	210	1 (Ar)	15	Finnish birch	35 mol%	G/S: 2,3	[103]
					Swedish birch	36 mol%		
					Poplar	22 mol%		
					Spruce	12 mol%		
					Pine	7 mol%		
ZnPd/C	MeOH	225	35	12	Poplar (WT-717)	40 wt%	G/S: 2	[19]
					Poplar (717-F5H)	36 wt%		
					Poplar (WT-NM-6)	44 wt%		
					Poplar (WT-LORRE)	54 wt%		
					White birch	52 wt%		
					Eucalyptus	49 wt%		
					Lodgepole pine	19 wt%	G2	
Pd/C	MeOH	225	35	12	Poplar (WT-717)	59 wt%	G/S: 2,5	[119]
ZnPd/C						44 wt%	G/S: 2	
Pd/C					Poplar (717-F5H)	38 wt%	G/S: 2, 5	
ZnPd/C						36 wt%	G/S: 2	

(Continued)

Table 9.6 (Continued)

Catalyst	Solvent	Temperature (°C)	H_2 (bar)[a]	Time (h)	Lignocellulose source	Monomer yield	Monomer structures[b]	References
Ni/C	MeOH	200	2 (N_2)	6	Birch	32 wt%	G/S: 2	[129]
					Poplar	26 wt%	G/S: 3	
					Eucalyptus	28 wt%	G/S: 2,3	
Ni/C (in cage)	MeOH	225	35	12	Miscanthus	68 wt%	G2, G4, S2, H2	[2]
Ru/C	MeOH	250	30	6	Birch	52 wt%	G/S: 2,5	[18]
	Water					25 wt%	G2, S2	
	MeOH	250	30	3	Birch	50 wt%	G2, S2	
					Poplar	44 wt%	G2, S2	
					Spruce and pine softwood mixture	21 wt%	G2	
					Miscanthus	27 wt%	G2, G4, G5, S2, S5, H2	
Ru/C	MeOH	250	30	3	Birch	48 wt%	G/S: 2,5	[113]
Pd/C						49 wt%	G/S: 5	
Pd/C	MeOH	250	30	3	Poplar	44 wt%	G/S: 2,5	[130]
Pd/C						26 wt%		
Pd/C + H_3PO_4						42 wt%		
Pd/C + NaOH						23 wt%	G/S: 1,2,5	
Pd/C	Water	200	30	3	Birch	44 wt%	G/S: 5	[110]
	MeOH					38 wt%		
	Ethylene glycol					27 wt%		
	EtOH					17 wt%		
	Isopropyl alcohol					12 wt%		
	1-butanol					11 wt%		
	Tetrahydrofuran					6 wt%		

	1,4-Dioxane					5 wt%		
	n-Hexane					2 wt%		
	MeOH					49 wt%		
	Ethylene glycol					50 wt%		
Pd/C	Water	200	20	3	Poplar	34.2 wt%	G/S: 2,5	[107]
	MeOH					28.2 wt%		
	MeOH/water (v/v = 7 : 3)					43.5 wt%		
	EtOH					19.4 wt%		
	EtOH/water (v/v = 5 : 5)					43.3 wt%		
Ru/C	MeOH	250	40	15	Beech	48 wt%	G/S: 2,5	[29]
	THF	250		15	Beech	36 wt%		
	MeOH	250		15	F5H-poplar	77 wt%		
	MeOH	150		20	F5H-poplar	0.2 wt%	S1, S2	
	THF	150		15	Spruce	21 wt%	S2	
Ru/C + HCl	MeOH	250		20	F5H-poplar	47 wt%	G2, G5	
Ni/C	MeOH	250	30	3	Corn stover	29 wt%	H1, H2, G1, G2, G4, S2	[4]
		200		24		29 wt%		
Ru/C		250		3		32 wt%		
		200		24		28 wt%		
Ni/C (HNO_3 oxidized)		200		6		32 wt%	H: 1–3; G:1–4, 7; S2, S3	
Ni/C + H_3PO_4		200		6		38 wt%	H1, H3, G: 1–4, 7; S2, S3, S7	

(Continued)

Table 9.6 (Continued)

Catalyst	Solvent	Temperature (°C)	H_2 (bar)[a]	Time (h)	Lignocellulose source	Monomer yield	Monomer structures[b]	References
Ni/C	MeOH/water (v/v = 6 : 4)	200	20	5	Beech	51.4 wt%	G/S: 2,3	[123]
	MeOH					39.3 wt%		
	1,4-Dioxane					14.3 wt%		
Pd/C	MeOH	180	30	2	Birch	14 wt%	G/S: 2,5	[111]
Pd/C + $Al(OTf)_3$						45 wt%	Not indicated	
Pd/C + $Yb(OTf)_3$						43 wt%	G/S: 1,2,5,6	
Pd/C	MeOH	200	30	1	Birch	24 wt%	G/S: 2,5	
Pd/C + $Yb(OTf)_3$				1	Birch	43 wt%	G/S: 1,2,5,6	
Pd/C + $Yb(OTf)_3$				2	Birch	46 wt%	G/S: 1,2,5,6	
Pd/C				1	Oak	33 wt%	Not indicated	
Pd/C + $Yb(OTf)_3$				1	Oak	47 wt%		
Pd/C				1	Poplar	17 wt%		
Pd/C + $Yb(OTf)_3$				1	Poplar	29 wt%		
Pd/C + $Yb(OTf)_3$				2	Poplar	36 wt%		
Pd/C				1	Scotch pine	10 wt%		
Pd/C + $Yb(OTf)_3$				1	Scotch pine	24 wt%		
Pd/C				1	Wheat chaff	6 wt%		
Pd/C + $Yb(OTf)_3$				1	Wheat chaff	6 wt%		
Pd/C + $Al(OTf)_3$	MeOH	180	30	2	Birch	55 wt%	G/S: 2,5,6	[112]
				4	Oak (300–1000 μm)	46 wt%	Not indicated	
				4	Oak (0–300 μm)	40 wt%		
				4	Douglas fir (300–1000 μm)	17 wt%		

a) Pressure at room temperature unless otherwise noted.
b) Codes are referenced to Figure 9.6.
c) D-101 catalyst consists of a mixture of polyvalent metals modified by addition of alkaline earth metals.
Source: Schutyser et al. 2018 [15]. Reproduced with permission of The Royal Society of Chemistry.

by Van den Bosch et al. [18] and Galkin et al. [103] revealed that a higher β-O-4 ether linkage concentration in native lignin results in higher monomer yield. In addition, Van den Bosch et al. [18] also found a direct correlation between increasing lignin syringyl (S)-content and monomer yield. Mansfield and coworkers have also observed an interdependent relationship between the S-unit concentration and the β-O-4 ether linkage content in the lignin structure [132, 133]. According to studies by Van den Bosch et al. and Galkin et al., the monomer yield generally increases (with a few outliers) from softwood species (up to 30 wt% reported), to herbaceous species (up to 40 wt% reported), to hardwood species (over 50 wt% reported) [15].

The RCF method is generally selective and preserves the aromaticity of the monomeric compounds. Not only does the source of biomass directly affect the yield of *p*-hydroxyl-, guaiacyl-, or syringyl-substitution in the products, the structure of the substituents can also be dependent on process parameters. Most RCF performed on hardwood or softwood species can selectively produce propyl-, propenyl-, propanol-, methoxypropyl-, and ethyl-substituted methoxy phenols, while RCF performed on herbaceous crops is able to generate propionic- and acrylic ester-substituted methoxy phenols in addition to those obtained from woody species (Figure 9.6).

9.3.3 Applied Catalyst

In addition to the biomass feedstock, the nature of the substituent groups on the aromatic structure is highly dependent on the catalyst used, catalyst additives (if used), reaction time, and whether the reaction atmosphere utilizes hydrogen or inert gas [15]. In general, selective propyl-substitution (**G2** and **S2** in Figure 9.6) is a result of hydrogenolysis and dehydration of the γ-OH group, which can be achieved by the following catalyst systems: Ru/C [18, 113], Pt/C [114, 116], ZnPd/C [19, 119], and Rh/C [114] coupled with a hydrogen atmosphere, or Ni/C under inert atmosphere using the solvent as a hydrogen source [109]. On the other hand, preservation of the γ-OH group results in the production of propanol-substituted compounds (**G5** and **S5** in Figure 9.6) and can be enabled by the use of Pd/C [28, 107, 110, 112–114, 116, 117, 119, 127, 130] or Raney Ni [117, 127] in hydrogen atmosphere. Propyl- and propenyl-substituted compounds (**G3** and **S3** in Figure 9.6) were observed as products by Samec and coworkers when Pd/C was used under inert atmosphere coupled with a short reaction time [103, 108]; when the reaction time was extended and temperature was increased, monomer selectivity shifted toward the propenyl-substituted compounds. Based on these observations, Sels and coworkers concluded that the use of pressurized H_2 as well as reaction temperature and time are essential variables in the preservation of γ-OH groups [15]. Addition of strong acids such as HCl and H_2SO_4 [126], or metal triflate catalysts such as $Yb(OTf)_3$ or $Al(OTf)_3$ [112] to the $Pd/C + H_2$ system in methanol modify the propanol substituents, yielding methoxypropenyl-substituted compounds (**G6** and **S6** in Figure 9.6). This is likely due to the etherification of γ-OH groups catalyzed by the metal triflate.

In addition to previously mentioned propionic- and acrylic ester-substituted methoxy phenols (**H2**, **H3**, **G4**, **G7**, **S4**, **S7** in Figure 9.6), ethyl-substituted aromatic products (**H1**, **G1**, and **S2** in Figure 9.6) are also generally unique to RCF that utilize herbaceous crops as the feedstock. However, addition of NaOH to RCF performed on woody biomass can also result in the production of ethyl-substituted aromatic products [117, 118], which are likely to originate from a hydrogenation of the β-O-4 ether linkage, or an enol-ether intermediate [15]. It should also be noted that ethyl-substituted products can also result from decarboxylation of the other products as intermediates [4]. While some ring-saturated products (**Cy1** and **Cy2** in Figure 9.6) have been observed by Pepper and Lee using spruce as the biomass source [121], generally, the aromaticity is effectively preserved in most RCF studies.

9.4 Outlook

Lignin's functionalized phenolic nature renders it an attractive renewable feedstock for future chemical production. Most fractionation methods that target lignin depolymerization can be categorized into either "late-stage catalytic conversion" or "early-stage catalytic conversion," and in either case, cleavage of the β-O-4 ether linkage is a common focal point. For catalytic lignin conversion at the late stage, the holocellulose is first removed and the lignin is isolated as a solid residue or precipitate, the catalyst/reaction conditions being applied with the aim of cleaving the residual C–O ether linkages that have survived the lignin isolation process. A main advantage of late-stage lignin conversion is the fact that modern day analytical methods such as 2D heteronuclear single quantum coherence (HSQC) NMR and gel-permeation chromatography can provide details such as the estimated concentration of the main types of lignin linkages and the molecular weight distribution of these isolated lignins. Not only can these analyses provide useful insights for the development of lignin depolymerization strategies but they can also be used to evaluate the effectiveness of the depolymerization method by revealing differences such as the appearance/disappearance of certain functional groups, or the molecular weight distribution pre- and post-reaction. However, the effectiveness of isolated lignin conversion is limited by the lignin isolation process, which can generate refractory C–C bonds. As a result, many catalyst systems require high reaction temperature and pressure, and reported monomer yields to date are generally lower compared with early-stage catalytic lignin conversion.

RCF of lignin, an example of "ECCL," utilizes whole biomass as substrate. The main advantage in doing so lies in the fact that linkages of interest within the lignin polymer are preserved, and reactive intermediates generated during the solvolytic lignin extraction process can be chemically quenched, drastically decreasing the formation of recalcitrant C–C bonds. Reactions are generally performed at milder temperature and pressure, with overall higher monomer yields obtained compared with the late-stage approach. Moreover, in RCF the carbohydrate components are preserved as a (holo)cellulose pulp and can be further upgraded to cellulosic ethanol, rendering it a viable strategy for an integrated biorefinery.

Unambiguous assessment of the merits of any lignin depolymerization approach is hard to realize given that biomass sources can show significant structural variation even within the same species. Consequently, we would like to encourage the development of "standard" lignocellulose biomass sources. While this is no doubt a difficult task, it is pivotal for the accurate assessment of lignin and lignocellulosic biomass fractionation methods. Nonetheless, with continuous improvements made to existing lignin valorization methods, as well as the development of novel approaches, the commercial implementation of depolymerization processes for lignin valorization is surely not far away.

Acknowledgment

The author would like to acknowledge Dr. Tom Renders (KU Leuven) for providing essential information and advice.

References

1 Beretta, G.P. (2007). World energy consumption and resources: an outlook for the rest of the century. *Int. J. Environ. Technol. Manage.* 7: 99.

2 Luo, H., Klein, I.M., Jiang, Y. et al. (2016). Total utilization of miscanthus biomass, lignin and carbohydrates, using earth abundant nickel catalyst. *ACS Sustainable Chem. Eng.* 4 (4): 2316–2322.

3 Rinaldi, R. (2018). Early-stage conversion of lignin over hydrogenation catalysts. In: *Lignin Valorization: Emerging Approaches* (ed. G.T. Beckham), 108–127. Croydon, London: Royal Society of Chemistry.

4 Anderson, E.M., Katahira, R., Reed, M. et al. (2016). Reductive catalytic fractionation of corn stover lignin. *ACS Sustainable Chem. Eng.* 4 (12): 6940–6950.

5 Holladay, J.E., White, J.F., Bozell, J.J., and Johnson, D. (2007). Top Value-Added Chemicals from Biomass Volume II – Results of Screening for Potential Candidates from Biorefinery Lignin. PNNL-16983, 2.

6 Bozell, J.J., Black, S.K., Myers, M. et al. (2011). Solvent fractionation of renewable woody feedstocks: organosolv generation of biorefinery process streams for the production of biobased chemicals. *Biomass Bioenergy* 35 (10): 4197–4208.

7 Kumar, P., Barrett, D.M., Delwiche, M.J. et al. (2009). Methods for pretreatment of lignocellulosic biomass for efficient hydrolysis and biofuel production. *Ind. Eng. Chem. Res.* 48: 3713–3729.

8 Wyman, C.E., Dale, B.E., Elander, R.T. et al. (2005). Coordinated development of leading biomass pretreatment technologies. *Bioresour. Technol.* 96 (18 Special Issue): 1959–1966.

9 Kautto, J., Realff, M.J., Ragauskas, A.J., and Kässi, T. (2014). Economic analysis of an organosolv process for bioethanol production. *BioResources* 9 (4): 6041–6072.

10 Davis, R., Tao, L., Tan, E.C.D. et al. (2013). Process Design and Economics for the Conversion of Lignocellulosic Biomass to Hydrocarbons: Dilute-Acid

and Enzymatic Deconstruction of Biomass to Sugars and Biological Conversion of Sugars to Hydrocarbons. *NREL/TP-5100-60223*.
11 Ragauskas, A.J., Williams, C.K., Davison, B.H. et al. (2006). The path forward for biofuels and biomaterials. *Science* 311 (5760): 484–489.
12 Sturgeon, M.R., Kim, S., Lawrence, K. et al. (2014). A mechanistic investigation of acid-catalyzed cleavage of aryl-ether linkages: implications for lignin depolymerization in acidic environments. *ACS Sustainable Chem. Eng.* 2 (3): 472–485.
13 Deuss, P.J., Scott, M., Tran, F. et al. (2015). Aromatic monomers by in situ conversion of reactive intermediates in the acid-catalyzed depolymerization of lignin. *J. Am. Chem. Soc.* 137 (23): 7456–7467.
14 Constant, S., Wienk, H.L.J., Frissen, A.E. et al. (2016). New insights into the structure and composition of technical lignins: a comparative characterisation study. *Green Chem.* 18 (9): 2651–2665.
15 Schutyser, W., Renders, T., Van den Bosch, S. et al. (2018). Chemicals from lignin: an interplay of lignocellulose fractionation, depolymerisation, and upgrading. *Chem. Soc. Rev.* 47 (3): 852–908.
16 Ferrini, P. and Rinaldi, R. (2014). Catalytic biorefining of plant biomass to non-pyrolytic lignin bio-oil and carbohydrates through hydrogen transfer reactions. *Angew. Chem. Int. Ed.* 53 (33): 8634–8639.
17 Galkin, M.V., Dahlstrand, C., and Samec, J.S.M. (2015). Mild and robust redox-neutral Pd/C-catalyzed lignol β-O-4′ bond cleavage through a low-energy-barrier pathway. *ChemSusChem* 8 (13): 2187–2192.
18 Van den Bosch, S., Schutyser, W., Vanholme, R. et al. (2015). Reductive lignocellulose fractionation into soluble lignin-derived phenolic monomers and dimers and processable carbohydrate pulps. *Energy Environ. Sci.* 8 (6): 1748–1763.
19 Parsell, T., Yohe, S., Degenstein, J. et al. (2015). A synergistic biorefinery based on catalytic conversion of lignin prior to cellulose starting from lignocellulosic biomass. *Green Chem.* 17 (3): 1492–1499.
20 Godard, H.P., McCarthy, J.L., and Hibbert, H. (1940). Hydrogenation of wood. *J. Am. Chem. Soc.* 62: 988.
21 Godard, H.P., McCarthy, J.L., and Hibbert, H. (1941). Studies on lignin and related compounds. LXII. High pressure hydrogenation of wood using copper chromite catalyst (Part 1). *J. Am. Chem. Soc.* 63: 3061–3066.
22 Bower, J.R., McCarthy, J.L., and Hibbert, H. (1937). Studies on lignin and related compounds. LXIII. Hydrogenation of wood (Part 2). *J. Am. Chem. Soc.* 63: 3066–3068.
23 Bower, J.R., Cooke, L.M., and Hibbert, H. (1943). Studies on lignin and related compounds. LXX. Hydrogenolysis and hydrogenation of maple wood. *J. Am. Chem. Soc.* 65 (6): 1192–1195.
24 Pepper, J.M. and Hibbert, H. (1948). Studies on lignin and related compounds. LXXXVII. High pressure hydrogenation of maple wood. *J. Am. Chem. Soc.* 70 (1): 67–71.
25 Galkin, M.V. and Samec, J.S.M. (2016). Lignin valorization through catalytic lignocellulose fractionation: a fundamental platform for the future biorefinery. *ChemSusChem* 9 (13): 1544–1558.

26 Lancefield, C.S., Rashid, G.M.M., Bouxin, F. et al. (2016, 12). Investigation of the chemocatalytic and biocatalytic valorization of a range of different lignin preparations: the importance of β-O-4 content. *ACS Sustainable Chem. Eng.* 4: 6921–6930.

27 Bouxin, F.P., McVeigh, A., Tran, F. et al. (2015). Catalytic depolymerisation of isolated lignins to fine chemicals using a Pt/alumina catalyst: Part 1 - Impact of the lignin structure. *Green Chem.* 17 (2): 1235–1242.

28 Torr, K.M., van de Pas, D.J., Cazeils, E., and Suckling, I.D. (2011). Mild hydrogenolysis of in-situ and isolated *Pinus radiata* lignins. *Bioresour. Technol.* 102 (16): 7608–7611.

29 Shuai, L., Amiri, M.T., Questell-Santiago, Y.M. et al. (2016). Formaldehyde stabilization facilitates lignin monomer production during biomass depolymerization. *Science* 354 (6310): 329–334.

30 Kong, J., He, M., Lercher, J.A., and Zhao, C. (2015). Direct production of naphthenes and paraffins from lignin. *Chem. Commun.* 51 (99): 17580–17583.

31 Kaplunova, T.S., Abduazimov, K.A., Pulatov, B.K. et al. (1994). Hydrogenolysis of rice husk lignin. *Chem. Nat. Compd.* 29 (4): 530–532.

32 Faix, O. and Meier, D. (1989). Pyrolytic and hydrogenolytic degradation studies on lignocellulosics, pulps and lignins. *Holz Roh Werkst.* 47: 67–72.

33 McVeigh, A., Bouxin, F.P., Jarvis, M.C., and Jackson, S.D. (2016). Catalytic depolymerisation of isolated lignin to fine chemicals: part 2-process optimisation. *Catal. Sci. Technol.* 6 (12): 4142–4150.

34 Kim, J.Y., Park, J., Kim, U.J., and Choi, J.W. (2015). Conversion of lignin to phenol-rich oil fraction under supercritical alcohols in the presence of metal catalysts. *Energy Fuels* 29 (8): 5154–5163.

35 Ye, Y., Zhang, Y., Fan, J., and Chang, J. (2012). Selective production of 4-ethylphenolics from lignin via mild hydrogenolysis. *Bioresour. Technol.* 118: 648–651.

36 Song, Q., Wang, F., and Xu, J. (2012). Hydrogenolysis of lignosulfonate into phenols over heterogeneous nickel catalysts. *Chem. Commun.* 48 (56): 7019–7021.

37 Shu, R., Long, J., Yuan, Z. et al. (2015). Efficient and product-controlled depolymerization of lignin oriented by metal chloride cooperated with Pd/C. *Bioresour. Technol.* 179: 84–90.

38 Long, J., Xu, Y., Wang, T. et al. (2015). Efficient base-catalyzed decomposition and in situ hydrogenolysis process for lignin depolymerization and char elimination. *Appl. Energy* 141: 70–79.

39 Konnerth, H., Ma, D., Yan, N. et al. (2014). Base promoted hydrogenolysis of lignin model compounds and organosolv lignin over metal catalysts in water. *Chem. Eng. Sci.* 123: 155–163.

40 Barta, K., Warner, G.R. Beach, E.S., and Anastas, P.T. (2014). Depolymerization of organosolv lignin to aromatic compounds over Cu-doped porous metal oxides. *Green Chem.* 16 (1): 191–196.

41 Coscia, C.J., Schubert, W.J., and Nord, F.F. (1961). Investigations on lignins and lignification. XXIV. The application of hydrogenation, hydrogenolysis,

and vapor phase chromatography in the study of lignin structure. *J. Org. Chem.* 26 (12): 5085–5091.

42 Olcay, A. (1962). Investigations on lignin and lignification. XXV. Hydrogenation of milled-wood lignins from white pine and blue spruce. *J. Org. Chem.* 27 (5): 1783–1786.

43 Schultz, T.P., Preto, R.J., Pittman, J.L., and Goldstein, I.S. (1982). Hydrotreating of hydrochloric acid lignin in a hydrogen-donor solvent. *J. Wood Chem. Technol.* 2 (1): 17–31.

44 Forchheim, D., Hornung, U., Kempe, P. et al. (2012). Influence of RANEY Nickel on the formation of intermediates in the degradation of lignin. *Int. J. Chem. Eng.* 2012: 1–8.

45 Ma, X., Tian, Y., Hao, W. et al. (2014). Production of phenols from catalytic conversion of lignin over a tungsten phosphide catalyst. *Appl. Catal., A* 481: 64–70.

46 Zhang, J., Teo, J., Chen, X. et al. (2014). A series of NiM (M = Ru, Rh, and Pd) bimetallic catalysts for effective lignin hydrogenolysis in water. *ACS Catal.* 4 (5): 1574–1583.

47 Zhang, J., Asakura, H., Van Rijn, J. et al. (2014). Highly efficient, NiAu-catalyzed hydrogenolysis of lignin into phenolic chemicals. *Green Chem.* 16 (5): 2432–2437.

48 Kim, J.Y., Park, J., Hwang, H. et al. (2015). Catalytic depolymerization of lignin macromolecule to alkylated phenols over various metal catalysts in supercritical tert-butanol. *J. Anal. Appl. Pyrolysis* 113: 99–106.

49 Molinari, V., Clavel, G., Graglia, M. et al. (2016). Mild continuous hydrogenolysis of kraft lignin over titanium nitride–nickel catalyst. *ACS Catal.* 6: 1663–1670.

50 Zakzeski, J., Jongerius, A.L., Bruijnincx, P.C.A., and Weckhuysen, B.M. (2012). Catalytic lignin valorization process for the production of aromatic chemicals and hydrogen. *ChemSusChem* 5 (8): 1602–1609.

51 Wang, X. and Rinaldi, R. (2012). Solvent effects on the hydrogenolysis of diphenyl ether with Raney nickel and their implications for the conversion of lignin. *ChemSusChem* 5 (8): 1455–1466.

52 Shu, R., Long, J., Xu, Y. et al. (2015). Investigation on the structural effect of lignin during the hydrogenolysis process. *Bioresour. Technol.* 200: 14–22.

53 Guerra, A., Filpponen, I., Lucia, L.A., and Argyropoulos, D.S. (2006). Comparative evaluation of three lignin isolation protocols for various wood species. *J. Agric. Food. Chem.* 54: 9696–9705.

54 Tolbert, A., Akinosho, H., Hunsupat, R. et al. (2014). Production pathways for renewable jet fuel: a review of commercialization. *Biofuels, Bioprod. Biorefin.* 8: 836–856.

55 Kloekhorst, A., Shen, Y., Yie, Y. et al. (2015). Catalytic hydrodeoxygenation and hydrocracking of Alcell® lignin in alcohol/formic acid mixtures using a Ru/C catalyst. *Biomass Bioenergy* 80: 147–161.

56 Kumar, C.R., Anand, N., Kloekhorst, A. et al. (2015). Solvent free depolymerization of kraft lignin to alkyl-phenolics using supported NiMo and CoMo catalysts. *Green Chem.* 17 (11): 4921–4930.

57 Narani, A., Chowdari, R.K., Cannilla, C. et al. (2015). Efficient catalytic hydrotreatment of kraft lignin to alkylphenolics using supported NiW and NiMo catalysts in supercritical methanol. *Green Chem.* 17 (11): 5046–5057.
58 Kloekhorst, A. and Heeres, H.J. (2016). Catalytic hydrotreatment of Alcell lignin fractions using a Ru/C catalyst. *Catal. Sci. Technol.* 6 (19): 7053–7067.
59 Oasmaa, A., Alén, R., and Meier, D. (1993). Catalytic hydrotreatment of some technical lignins. *Bioresour. Technol.* 45 (3): 189–194.
60 Meier, D., Ante, R., and Faix, O. (1992). Catalytic hydropyrolysis of lignin: influence of reaction conditions on the formation and composition of liquid products. *Bioresour. Technol.* 40 (2): 171–177.
61 Meier, D., Berns, J., Grünwald, C., and Faix, O. (1993). Analytical pyrolysis and semicontinuous catalytic hydropyrolysis of organocell lignin. *J. Anal. Appl. Pyrolysis* 25 (C): 335–347.
62 Ratcliff, M.A., Johnson, D.K., Posey, F.L. et al. (1988). Hydrodeoxygenation of a lignin model compound. *Res. Thermochem. Biomass Convers.* (3): 941–955.
63 Kloekhorst, A. and Heeres, H.J. (2015). Catalytic hydrotreatment of Alcell lignin using supported Ru, Pd, and Cu catalysts. *ACS Sustainable Chem. Eng.* 3 (9): 1905–1914.
64 Wang, H., Ben, H., Ruan, H. et al. (2017). Effects of lignin structure on hydrodeoxygenation reactivity of pine wood lignin to valuable chemicals. *ACS Sustainable Chem. Eng.* 5 (2): 1824–1830.
65 Luterbacher, J.S., Azarpira, A., Motagamwala, A.H. et al. (2015). Lignin monomer production integrated into the γ-valerolactone sugar platform. *Energy Environ. Sci.* 8 (9): 2657–2663.
66 Kasakov, S., Shi, H., Camaioni, D.M. et al. (2015). Reductive deconstruction of organosolv lignin catalyzed by zeolite supported nickel nanoparticles. *Green Chem.* 17 (11): 5079–5090.
67 Laskar, D.D., Tucker, M.P., Chen, X. et al. (2013). Noble-metal catalyzed hydrodeoxygenation of biomass-derived lignin to aromatic hydrocarbons. *Green Chem.* 16: 897.
68 Wang, H., Ruan, H., Pei, H. et al. (2015). Biomass-derived lignin to jet fuel range hydrocarbons via aqueous phase hydrodeoxygenation. *Green Chem.* 17 (12): 5131–5135.
69 Wang, H., Ruan, H., Feng, M. et al. (2017). One-pot process for hydrodeoxygenation of lignin to alkanes using Ru-based bimetallic and bifunctional catalysts supported on zeolite Y. *ChemSusChem* 10 (8): 1846–1856.
70 Wang, X. and Rinaldi, R. (2016). Bifunctional Ni catalysts for the one-pot conversion of organosolv lignin into cycloalkanes. *Catal. Today* 269: 48–55.
71 Paolo, B., Che, M., and Louis, C. (1998). Molecular approach to the mechanism of deposition–precipitation of the Ni(II) phase on silica. *J. Phys. Chem. B* 102 (15): 2722–2732.
72 Zhao, C., Camaioni, D.M., and Lercher, J.A. (2012). Selective catalytic hydroalkylation and deoxygenation of substituted phenols to bicycloalkanes. *J. Catal.* 288: 92–103.
73 Luterbacher, J.S., Rand, J.M., Alonso, D.M. et al. (2014). Nonenzymatic sugar production from biomass using biomass-derived γ-valerolactone. *Science* 343 (6168): 277–280.

74 Kim, K.H., Simmons, B.A., and Singh, S. (2017). Catalytic transfer hydrogenolysis of ionic liquid processed biorefinery lignin to phenolic compounds. *Green Chem.* 19 (1): 215–224.

75 Huang, X., Korányi, T.I., Boot, M.D., and Hensen, E.J.M. (2014). Catalytic depolymerization of lignin in supercritical ethanol. *ChemSusChem* 7 (8): 2276–2288.

76 Huang, X., Atay, C., Korányi, T.I. et al. (2015). Role of Cu–Mg–Al mixed oxide catalysts in lignin depolymerization in supercritical ethanol. *ACS Catal.* 5 (12): 7359–7370.

77 Ma, R., Hao, W., Ma, X. et al. (2014). Catalytic ethanolysis of kraft lignin into high-value small-molecular chemicals over a nanostructured α-molybdenum carbide catalyst. *Angew. Chem. Int. Ed.* 53 (28): 7310–7315.

78 Ma, X., Ma, R., Hao, W. et al. (2015). Common pathways in ethanolysis of kraft lignin to platform chemicals over molybdenum-based catalysts. *ACS Catal.* 5 (8): 4803–4813.

79 Ma, X., Cui, K., Hao, W. et al. (2015). Alumina supported molybdenum catalyst for lignin valorization: effect of reduction temperature. *Bioresour. Technol.* 192: 17–22.

80 Jiang, Y., Li, Z., Tang, X. et al. (2015). Depolymerization of cellulolytic enzyme lignin for the production of monomeric phenols over Raney Ni and acidic zeolite catalysts. *Energy Fuels* 29 (3): 1662–1668.

81 Jongerius, A.L., Bruijnincx, P.C.A., and Weckhuysen, B.M. (2013). Liquid-phase reforming and hydrodeoxygenation as a two-step route to aromatics from lignin. *Green Chem.* 15 (11): 3049–3056.

82 Xu, W., Miller, S.J., Agrawal, P.K., and Jones, C.W. (2012). Depolymerization and hydrodeoxygenation of switchgrass lignin with formic acid. *ChemSusChem* 5 (4): 667–675.

83 Chen, L., Korányi, T.I., and Hensen, E.J.M. (2016). Transition metal (Ti, Mo, Nb, W) nitride catalysts for lignin depolymerisation. *Chem. Commun.* 52 (60): 9375–9378.

84 Luque, R., Serrano, L., Pineda, A. et al. (2012). Heterogeneously catalysed mild hydrogenolytic depolymerisation of lignin under microwave irradiation with hydrogen-donating solvents. *ChemCatChem* 5 (4): 977–985.

85 Toledano, A., Serrano, L., Balu, A.M. et al. (2013). Fractionation of organosolv lignin from olive tree clippings and its valorization to simple phenolic compounds. *ChemSusChem* 6 (3): 529–536.

86 Toledano, A., Serrano, L., Pineda, A. et al. (2014). Microwave-assisted depolymerisation of organosolv lignin via mild hydrogen-free hydrogenolysis: catalyst screening. *Appl. Catal., B* 145: 43–55.

87 Zhu, G., Qiu, X., Zhao, Y. et al. (2016). Depolymerization of lignin by microwave-assisted methylation of benzylic alcohols. *Bioresour. Technol.* 218: 718–722.

88 Onwudili, J.A. and Williams, P.T. (2014). Catalytic depolymerization of alkali lignin in subcritical water: influence of formic acid and Pd/C catalyst on the yields of liquid monomeric aromatic products. *Green Chem.* 16 (11): 4740–4748.

89 Singh, S.K. and Ekhe, J.D. (2015). Cu-Mo doped zeolite ZSM-5 catalyzed conversion of lignin to alkyl phenols with high selectivity. *Catal. Sci. Technol.* 5 (4): 2117–2124.

90 Liguori, L. and Barth, T. (2011). Palladium-Nafion SAC-13 catalysed depolymerisation of lignin to phenols in formic acid and water. *J. Anal. Appl. Pyrolysis* 92 (2): 477–484.

91 Huang, X., Korányi, T.I., Boot, M.D., and Hensen, E.J.M. (2015). Ethanol as capping agent and formaldehyde scavenger for efficient depolymerization of lignin to aromatics. *Green Chem.* 17 (11): 4941–4950.

92 Feghali, E., Carrot, G., Thuéry, P. et al. (2015). Convergent reductive depolymerization of wood lignin to isolated phenol derivatives by metal-free catalytic hydrosilylation. *Energy Environ. Sci.* 8 (9): 2734–2743.

93 Prinsen, P., Narani, A., and Rothenberg, G. (2017). Lignin depolymerisation and lignocellulose fractionation by solvated electrons in liquid ammonia. *ChemSusChem* 10: 1022–1032.

94 Lancefield, C.S., Ojo, O.S., Tran, F., and Westwood, N.J. (2015). Isolation of functionalized phenolic monomers through selective oxidation and CO bond cleavage of the β-O-4 linkages in Lignin. *Angew. Chem. Int. Ed.* 54 (1): 258–262.

95 Feghali, E. and Cantat, T. (2014). Unprecedented organocatalytic reduction of lignin model compounds to phenols and primary alcohols using hydrosilanes. *Chem. Commun.* 50 (7): 862–865.

96 Éfendieva, N.F., Shorygina, N.N., and Krasovskaya, N.P. (1973). Reductive cleavage of lignocarbohydrate complex by sodium metal in liquid ammonia. *Bull. Acad. Sci. USSR Div. Chem. Sci. (Engl. Transl.)* 22: 2748–2275.

97 Lindgren, B.O. and Nilsson, T. (1974). Oxidation of lignin model compounds with chlorine dioxide and chlorite. Reactions with stilbenes. *Acta Chem. Scand. Ser. B* 28B: 847–852.

98 Yao, Z., Ye, D., Liu, H. et al. (2007). Transformation of aryl acyloin O-alkyl and O-phenyl derivatives to ketones. *Synth. Commun.* 37 (1): 149–156.

99 Song, Y., Motagamwala, A.H., Karlen, S.D. et al. (2019). A comparative study of secondary depolymerization methods on oxidized lignins. *Green Chem.* https://doi.org/10.1039/c9gc01663e.

100 Rinaldi, R., Jastrzebski, R., Clough, M.T. et al. (2016). Paving the way for lignin valorisation: recent advances in bioengineering, biorefining and catalysis. *Angew. Chem. Int. Ed.*: 8164–8215.

101 Li, C., Zhao, X., Wang, A. et al. (2015). Catalytic transformation of lignin for the production of chemicals and fuels. *Chem. Rev.* 115 (21): 11559–11624.

102 Renders, T., Van den Bossche, G., Vangeel, T. et al. (2019). Reductive catalytic fractionation: state of the art of the lignin-first biorefinery. *Curr. Opin. Biotechnol.* 56: 193–201.

103 Galkin, M.V., Smit, A.T., Subbotina, E. et al. (2016). Hydrogen-free catalytic fractionation of woody biomass. *ChemSusChem* 9 (23): 3280–3287.

104 Ferrini, P., Rezende, C.A., and Rinaldi, R. (2016). Catalytic upstream biorefining through hydrogen transfer reactions: understanding the process from the pulp perspective. *ChemSusChem* 9 (22): 3171–3180.

105 Chesi, C., de Castro, I.B.D., Clough, M.T. et al. (2016). The influence of hemicellulose sugars on product distribution of early-stage conversion of lignin oligomers catalysed by Raney Nickel. *ChemCatChem* 8 (12): 2079–2088.

106 Van den Bosch, S., Renders, T., Kennis, S. et al. (2017). Integrating lignin valorization and bio-ethanol production: on the role of Ni-Al_2O_3 catalyst pellets during lignin-first fractionation. *Green Chem.* 19 (14): 3313–3326.

107 Renders, T., Van den Bosch, S., Vangeel, T. et al. (2016). Synergetic effects of alcohol/water mixing on the catalytic reductive fractionation of poplar wood. *ACS Sustainable Chem. Eng.* 4 (12): 6894–6904.

108 Galkin, M.V. and Samec, J.S.M. (2014). Selective route to 2-propenyl aryls directly from wood by a tandem organosolv and palladium-catalysed transfer hydrogenolysis. *ChemSusChem* 7 (8): 2154–2158.

109 Song, Q., Wang, F., Cai, J. et al. (2013). Lignin depolymerization (LDP) in alcohol over nickel-based catalysts via a fragmentation-hydrogenolysis process. *Energy Environ. Sci.* 6 (3): 994–1007.

110 Schutyser, W., Van den Bosch, S., Renders, T. et al. (2015). Influence of bio-based solvents on the catalytic reductive fractionation of birch wood. *Green Chem.* 17 (11): 5035–5045.

111 Huang, X., Zhu, J., Korányi, T.I. et al. (2016). Effective release of lignin fragments from lignocellulose by Lewis acid metal triflates in the lignin-first approach. *ChemSusChem* 9: 3262–3267.

112 Huang, X., Morales Gonzalez, O.M., Zhu, J. et al. (2017). Reductive fractionation of woody biomass into lignin monomers and cellulose by tandem metal triflate and Pd/C catalysis. *Green Chem.* 19 (1): 175–187.

113 Van den Bosch, S., Schutyser, W., Koelewijn, S.F. et al. (2015). Tuning the lignin oil OH-content with Ru and Pd catalysts during lignin hydrogenolysis on birch wood. *Chem. Commun.* 51 (67): 13158–13161.

114 Yan, N., Zhao, C., Dyson, P.J. et al. (2008). Selective degradation of wood lignin over noble-metal catalysts in a two-step process. *ChemSusChem* 1 (7): 626–629.

115 Renders, T., Van den Bosch, S., Koelewijn, S.-F. et al. (2017). Lignin-first biomass fractionation: the advent of active stabilisation strategies. *Energy Environ. Sci.* 10 (7): 1551–1557.

116 Li, C., Zheng, M., Wang, A., and Zhang, T. (2012). One-pot catalytic hydrocracking of raw woody biomass into chemicals over supported carbide catalysts: simultaneous conversion of cellulose, hemicellulose and lignin. *Energy Environ. Sci.* 5: 6383–6390.

117 Pepper, J.M., Steck, W.F., Swoboda, R., and Karapally, J.C. (1966). Hydrogenation of lignin using nickel and palladium catalysts. In: *Lignin Structure and Reactions* (ed. J. Marton), 238–248. American Chemical Society.

118 Pepper, J.M. and Supathna, P. (2006). Lignin and related compounds. VI. A study of variables affecting the hydrogenolysis of spruce wood lignin using a rhodium-on-charcoal catalyst. *Can. J. Chem.* 56 (7): 899–902.

119 Klein, I., Marcum, C., Kenttämaa, H., and Abu-Omar, M.M. (2016). Mechanistic investigation of the Zn/Pd/C catalyzed cleavage and hydrodeoxygenation of lignin. *Green Chem.* 18 (8): 2399–2405.

120 Pepper, J.M. and Steck, W. (1963). The effect of time and temperature on the hydrogenation of aspen lignin. *Can. J. Chem.* 41: 2867–2875.

121 Pepper, J.M. and Lee, Y.W. (1970). Lignin and related compounds. II. Studies using ruthenium and Raney nickel as catalysts for lignin hydrogenolysis. *Can. J. Chem.* 48: 477–479.

122 Pepper, J.M. and Fleming, R.W. (1978). Lignin and related compounds. V. The hydrogenolysis of aspen wood lignin using rhodium-on-charcoal as catalyst. *Can. J. Chem.* 56 (7): 896–898.

123 Chen, J., Lu, F., Si, X. et al. (2016). High yield production of natural phenolic alcohols from woody biomass using a nickel-based catalyst. *ChemSusChem* 9 (23): 3353–3360.

124 Liu, Y., Chen, L., Wang, T. et al. (2015). One-pot catalytic conversion of raw lignocellulosic biomass into gasoline alkanes and chemicals over $LiTaMoO_6$ and Ru/C in aqueous phosphoric acid. *ACS Sustainable Chem. Eng.* 3 (8): 1745–1755.

125 Han, X., Gong, X., Chen, Z. et al. (2016). Direct hydrodeoxygenation of raw woody biomass into liquid alkanes. *Nat. Commun.* 7 (1): 1–10.

126 Huang, X., Ouyang, X., Hendriks, B.M.S. et al. (2017). Selective production of mono-aromatics from lignocellulose over Pd/C catalyst: the influence of acid co-catalysts. *Faraday Discuss.* 202: 141–156.

127 Pepper, J.M. and Lee, Y.W. (1969). Lignin and related compounds. I. A comparative study of catalysts for lignin hydrogenolysis. *Can. J. Chem.* 47: 723–727.

128 Pepper, J.M. and Rahman, M.D. (1987). Lignin and related compounds. XI. Selective degradation of aspen poplar lignin by catalytic hydrogenolysis. *Cellul. Chem. Technol.* 21 (3): 233–239.

129 Klein, I., Saha, B., and Abu-Omar, M.M. (2015). Lignin depolymerization over Ni/C catalyst in methanol, a continuation: effect of substrate and catalyst loading. *Catal. Sci. Technol.* 5 (6): 3242–3245.

130 Renders, T., Schutyser, W., Van den Bosch, S. et al. (2016). Influence of acidic (H_3PO_4) and alkaline (NaOH) additives on the catalytic reductive fractionation of lignocellulose. *ACS Catal.* 6 (3): 2055–2066.

131 Sannigrahi, P., Kim, D.H., Jung, S., and Ragauskas, A. (2011). Pseudo-lignin and pretreatment chemistry. *Energy Environ. Sci.* 4 (4): 1306–1310.

132 Mottiar, Y., Vanholme, R., Boerjan, W. et al. (2016). Designer lignins: harnessing the plasticity of lignification. *Curr. Opin. Biotechnol.* 37: 190–200.

133 Laskar, D.D., Yang, B., Wang, H., and Lee, J. (2013). Pathways for biomass-derived lignin to hydrocarbon fuels. *Biofuels, Bioprod. Biorefin.* 7: 602–626.

10

Conversion of Lipids to Biodiesel via Esterification and Transesterification

Amin Talebian-Kiakalaieh and Amin Nor Aishah Saidina

Universiti Teknologi Malaysia (UTM), Chemical Reaction Engineering Group (CREG), Faculty of Chemical and Energy Engineering, Jalan Hikmah, 81310, Skudai, Johor, Malaysia

10.1 Introduction

Energy is an undeniable requirement for the existence of human life on the Earth. Fluctuations in the crude oil price, environmental concerns, and limited resources around the globe are the main motivations driving the search for alternative energy sources [1, 2]. Rapid industrial development at the beginning of the twentieth century led to a sudden increase in energy demand globally, transportation being reported to rank as the second highest energy consuming sector worldwide [3]. Notably, about 97.6% of current transportation fuels are produced from crude oil, which indicates the importance of fossil fuel resources in the production of energy. Although different types of fossil fuels are the main source of energy for now, it is predicted that the application of renewable energy sources could reduce fossil fuel consumption by 30% in the next two decades [4].

Biodiesel is one of the renewable fuels that has shown acceptable results in the reduction of greenhouse gas (GHG) emissions [5]. Consequently, large numbers of biodiesel plants have been established around the globe to reduce countries' dependence on foreign crude oil resources [6]. The physicochemical characteristics of biodiesel are quite similar to diesel while possessing higher lubricating efficiency and flash point. Biodiesel is considered as an environmentally friendly fuel as it exhibits lower contribution to the release of CO_2 and other particulate emissions compared with fossil fuels [7]. Biodiesel is comprised of fatty acid methyl esters (FAMEs), which can be derived from either edible or nonedible oils. Minimizing the production cost is the greatest challenge in the biodiesel industry. In particular, the application of different feedstocks and catalysts has a significant impact on the production cost as these require about 75% of the total investment. As a result, biodiesel is increasingly produced from nonedible or waste oils instead of edible oils. This eliminates the threat of a food crisis due to the utilization of edible oils, which was one of the major drawbacks of large-scale biodiesel production in the past. Moreover, the use of different sources such as waste cooking oil (WCO) and jatropha oil have significantly reduced (60–90%)

Chemical Catalysts for Biomass Upgrading, First Edition. Edited by Mark Crocker and Eduardo Santillan-Jimenez.

Table 10.1 Common biodiesel feedstocks along with their oil contents.

Oil	Feedstock	Oil content (%)
Edible	Palm	30–60
	Sunflower	25–35
	Rapeseed	38–46
	Soybean	15–20
	Coconut	63–65
	Peanut	45–55
Nonedible	Jatropha seed	35–40
	Pongamia pinnata	27–39
	Neem	20–30
	Castor	53
Other sources	Rubber seed	40–50
	Sea mango	54
	Cotton seed	18–25
	Microalgae	20–70

biodiesel production costs [8]. New biodiesel production facilities tend to use renewable feedstocks to ensure sustainability and maintain biodiesel availability far into the future [9]. Potential oil sources for biodiesel production, along with their oil content, are summarized in Table 10.1 [3].

Along with the importance of feedstock type, the presence of an effective catalyst is necessary for reaction completion [10]. Although there are different methods for the production of bio-derived "diesel," including pyrolysis, dilution, and microemulsification, the transesterification of triglycerides (TGs) is the conventional method, carried out using acid or base catalysts [11, 12]. The transesterification reaction can be made simpler and cheaper by the application of homogeneous catalysts, although they have drawbacks that can only be eliminated by the utilization of heterogeneous catalysts. Heterogeneous catalysts have several merits compared with conventional homogeneous catalysts; for example, they can be easily recycled and reused multiple times [13]. Recent breakthroughs in the field of heterogeneous nanocatalysts are undeniable and have resulted in significant improvements in catalyst activity and stability [14]. Biocatalysts represent another group of catalysts used in the production of biodiesel. Enzymatic catalysts have good thermal stability, but their high cost makes them unattractive for large-scale applications [15].

This chapter summarizes the different raw materials and catalysts used in biodiesel production processes globally. In particular, the importance of algal biomass and heterogeneous catalysts is emphasized. The major aim of the chapter is to consolidate the trends in biodiesel production via transesterification using different catalysts and processes.

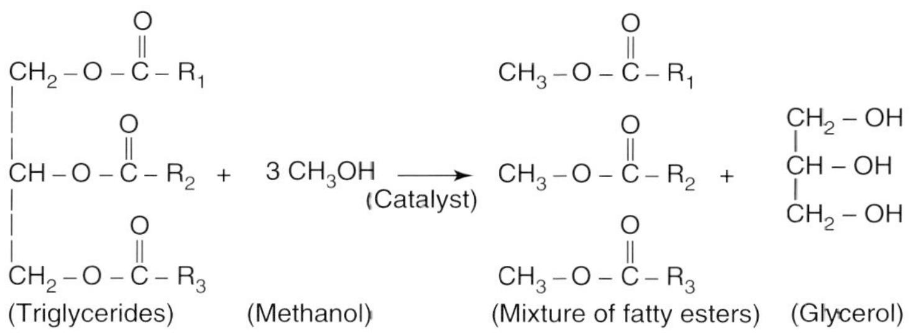

Figure 10.1 Transesterification reaction.

Transesterification is the main commercial method for biodiesel production and requires the use of an alcohol with one to eight carbon atoms (e.g. methanol) and a catalyst [16]. During the transesterification reaction, three reversible reactions take place consecutively in which diglycerides (DGs) and monoglycerides (MGs) are intermediate products [17]. Figure 10.1 shows the conventional transesterification reaction in the presence of methanol and catalyst.

10.2 Different Feedstocks for Biodiesel Production

The feedstock cost is the most important consideration in the large-scale production of biodiesel. Although the application of edible oils may be considered as a potential threat to the food supply, nonedible oils typically require costly pretreatment steps. Nevertheless, utilization of waste or used oils has gained considerable attention due to their lower cost and the potential for waste elimination [18]. Consequently, many studies have investigated the use of WCO, grease, animal fat, and tallow in biodiesel production. The high free fatty acid (FFA) and moisture contents are the major drawbacks of using WCOs and animal fat, since they reduce the yield of biodiesel [19]. Conventionally, feedstocks with FFA and moisture contents higher than 2.5% and 0.5%, respectively, require physical or chemical treatments [20]. Esterification prior to the transesterification process is the most common method for reducing the FFA content of the feedstock. A list of various biodiesel feedstocks along with their properties is reported in Table 10.2.

10.3 Biodiesel Production

Biodiesel production has been transformed due to recent scientific and technological advancements. Biodiesel produced from edible oils is categorized as a first-generation biofuel, while biodiesel obtained from nonedible oils (e.g. jatropha) is categorized as a second-generation biofuel. Biodiesel production from algae (algae oil or microalgae), which is a recent trend in this field, is classified as a third-generation fuel. Biodiesel production from algae will be discussed further in the following section given its potential to solve major issues and drawbacks in the field. Generally, vegetable oils cannot be directly used in

Table 10.2 The properties of different vegetable oils.

Type of oil	Species	V[a)]	D[b)]	F[c)]	H[d)]	A[e)]	C[f)]	CI[g)]	P[h)]
Edible oil	Soybean	32.9	0.91	254	39.6	0.2	37.9	−3.9	−12.2
	Rapeseed	35.1	0.91	246	39.7	2.92	37.6	−3.9	−31.7
	Sunflower	32.6	0.92	274	39.6	—	41.3	18.3	−6.7
	Palm	39.6	0.92	267	—	0.1	42.0	31.0	—
	Peanut	22.72	0.90	271	39.8	3	41.8	12.8	−6.7
	Corn	34.9	0.91	277	39.5	—	37.6	−1.1	−40.0
	Camelina	—	0.91	—	42.2	0.76	—	—	—
	Canola	38.2				0.4	—	—	—
	Cotton	18.2	0.91	234	39.5		41.8	1.7	−15.0
	Pumpkin	35.6	0.92	>230	39	0.55	—	—	—
Nonedible oil	*Jatropha curcas*	29.4	0.92	225	38.5	28	—	—	—
	Pongamia pinnata	27.8	0.91	205	34	5.06	—	—	—
	Sea mango	29.6	0.92	—	40.86	0.24	—	—	—
	Palanga	72.0	0.90	221	39.25	44	—	—	—
	Tallow	—	0.92	—	40.05	—	—	—	—
	Nile tilapia	32.1	0.91	—	—	2.81	—	——	—
	Poultry	—	0.90	—	39.4	—	—	—	—
	WCO	44.7	0.90	—	—	2.5	—	—	—

a) Viscosity at 40 °C.
b) Density (g/cm^3).
c) Flash point (°C).
d) Heating value (MJ/kg).
e) Acid value (mg KOH/g).
f) Cetane number (°C).
g) Cloud point (°C).
h) Pour point (°C).

engines due to their high viscosity, low volatility, and polyunsaturated nature that causes oxidation and limits shelf-life. Thus, different production methodologies (e.g. pyrolysis, micro-emulsion, hydrocarbon blending, and transesterification) have been tested for the production of bio-derived diesel [21, 22]. However, transesterification represents the best approach as it achieves higher conversion efficiency compared with these other methods [23].

10.3.1 Algal Biodiesel Production

The relatively high photosynthetic efficiency of algae compared with terrestrial plants, coupled with their potential for oil production, renders algal biomass a promising raw material for the production of biodiesel. Consequently, interest in biodiesel production from algae has surged in recent years [24]. While the most productive terrestrial plants provide less than 1000 gal of oil per acre, microalgae have the potential – in principle – to yield up to 5000 gal of oil per acre [7], albeit this target remains unobtainable at present. Toward this goal, recent efforts in biological science and technology are aimed at increasing algae growth rates and increasing lipid content [25, 26]. In reality, however, there are significant obstacles (e.g. incomplete knowledge of algae biology, high capital costs for algae cultivation systems) currently impeding the industrialization of algae for the production of low value products such as fuels [27].

Microalgae constitute a group of microorganisms having a simple cellular structure [28, 29] and are among the oldest life-forms on Earth [28, 30]. Microalgae can produce and store TGs within their cells, even under stressful environments [31]. These stored lipids are particularly suited for the production of biodiesel [32]. Compared with other crops, microalgae exhibit high productivity and their cultivation does not compete with food production [28]. There are more than 40 000 different types of microalgae that have been identified to date, some of which contain as much as 20–50% lipid content [30]. Microalgae can be classified into autotrophic microalgae, which only require inorganic nutrients (CO_2, salts and sunlight) for cell growth and development, and heterotrophic microalgae, which require organic nutrients as an energy source [28]. Autotrophic microalgae use photosynthesis to convert CO_2 into biomass. Equation (10.1) indicates the overall photosynthetic reaction:

$$6CO_2 + 12H_2O + \text{photons} \rightarrow C_6H_{12}O_6 + 6O_2 + 6H_2O \tag{10.1}$$

Recently, microalgae have been utilized for a variety of purposes, such as controlling organic waste, wastewater treatment, and storage of solar energy [33]. Industrial exhaust gas that contains CO_2 could be a convenient carbon source for microalgae biomass production. In fact, microalgae represent a promising means of reducing CO_2 gas emissions from industry [26]. In the majority of cases, microalgae productivity is enhanced as the temperature increases up to the optimal range of 20–30 °C. However, there are some types of thermophilic algae that reach their maximum growth rate at much higher temperatures, e.g. 40–60 °C [30]. Nutrients are also essential for microalgae growth [34, 35]. In the process of microalgae cultivation, the appropriate pH range for microalgae growth is typically in the six to eight range. The resulting microalgae biomass can be converted to different types of renewable biofuels [36], such as bio-methane [37], bio-hydrogen, or biodiesel.

10.3.1.1 Nutrients for Microalgae Growth

Microalgae growth and cultivation is highly dependent on the quality and quantity of essential nutrients like nitrogen (N) and phosphorus (P) [33]. To enhance the microalgae growth rate, inorganic fertilizers can be used as the major nutrient source [38]. Interestingly, wastewater can also be used as a nutrient source to cultivate microalgae when N and P are present in significant concentration. Additionally, microalgae can eliminate or reduce the heavy and trace metals in wastewater [30, 39]. Thus, wastewater utilization as a nutrient source for microalgae can not only improve the economics of microalgae biodiesel production but also enhance water quality. Municipal wastewater is another potential source of water and nutrients for microalgae production [40]. However, a concern when using municipal wastewater in the cultivation of microalgae is the presence of toxic compounds.

Industrial wastewater is another nutrient source that can be used for microalgae production. However, the presence of toxic compounds in industrial wastewater may have an adverse impact on microalgae cultivation [41, 42]. Despite this, microalgae currently show significant promise for wastewater treatment. Indeed, it was shown recently that wastewater produced from a carpet mill could be utilized as a nutrient source to cultivate microalgae due to the low concentration of toxic components and the presence of enough N and P to enhance the growth of *Botryococcus braunii*, *Chlorella saccharophila*, and *Pleurochrysis carterae* [43]. Moreover, Origin Oil Inc. has explored the use of microalgae to treat wastewater (e.g. from water flooding and hydraulic cracking) from oil wells [41]. The US onshore oil and gas drilling industry generates about 56 million barrels of wastewater per day, which in principle could be used as a nutrient source for microalgae cultivation, with the potential to produce about 0.7 million gallons of microalgae oil daily [41].

10.3.1.2 Microalgae Cultivation System

Open raceway ponds and closed photobioreactors are the two main types of microalgae cultivation systems. The open raceway pond is economical and is suitable for the treatment of domestic wastewater [42]. Currently, there are three different raceway pond designs, namely, rotating arm, endless loop, and inclined system [26]. Various countries (Israel, United States, China, and Czech Republic) report utilization of these types of ponds.

Photobioreactors, although expensive to build, can reduce the risk of microalgae contamination compared to open raceway ponds [28, 42]. In the last half century, different designs of closed photobioreactors (e.g. tubular, flat-plate, or bubble column) have been utilized [36]. In addition, the combination of a closed photobioreactor and an open raceway pond system, known as a hybrid system, has been explored with the aim of improving microalgae productivity and reducing the overall cost of biomass production [29, 36]. Interestingly, it was confirmed that by applying a hybrid system in microalgae cultivation, a 42% reduction in global warming potential (GWP) and a 38% reduction in fossil energy requirements (FERs) could be achieved in the production of 1 ton of microalgae-derived biodiesel compared with petroleum diesel [29]. Table 10.3 reports a comparison of various types of microalgae cultivation systems [44].

10.3.1.3 Harvesting

In general, microalgae biomass possesses a high water content that should be eliminated as far as possible during downstream processing. The main concern in this regard is the required energy and associated cost to process bulk quantities of wet microalgae. Filtration is commonly used in laboratory-scale studies; however, it cannot be used in large-scale processes because of membrane clogging [36]. Therefore, other technologies for microalgae biomass harvesting have been developed, including gravity sedimentation, centrifugation, flocculation, flotation, and electro coagulation [28, 45]. The main objective of bulk harvesting is the separation of microalgae biomass from the culture medium by flotation, gravity

Table 10.3 Comparison of different microalgae cultivation systems.

Production system		Advantages	Disadvantages
Raceway pond	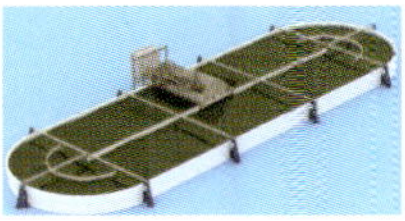	• Low cost • Easy to wash • Low energy required • Easy maintenance	• Low biomass yield • Large land area required • Poor mixing • Culture medium is easily polluted
Tubular photo-bioreactor		• High light surface area • High biomass yield • Suitable for outdoor culture	• Fouling • High cost • Gradient of pH, dissolved O_2 and CO_2 along the tubes
Flat plate photo-bioreactor		• High biomass yield • Low oxygen accumulation • Suitable for outdoor cultivation • Good light penetration • High light surface area	• Fouling • Difficult to scale up • Difficult temperature control
Column photo-bioreactor		• Compact • Low energy requirement • High mass transfer • Well mixed with low shear stress • Low cost, compact, and easy • to operate • Decreases photo-inhibition and photo-oxidation	• Small surface area for illumination • Costly
Hybrid system		• Economical • High productivity • Eliminates contamination • Favors continuous cell division	• Requires skilled manpower to operate the system

sedimentation, or flocculation. Enrichment (concentration) of the slurry is performed in a subsequent thickening stage by means of filtration or centrifugation.

10.3.1.4 Drying

The drying of wet microalgae biomass is an important step before the production of biodiesel. Generally, it is reported that high water content can adversely affect the biodiesel yield. However, recent studies have revealed that even wet algae can provide high lipid yields [46]. Application of natural sunlight (solar drying) is the lowest cost method for drying microalgae biomass, even though it has certain limitations such as the need for a large surface, and unfavorable weather conditions in some regions. Additional heat sources (e.g. natural gas) can be used for the drying of wet microalgae biomass. In addition, spray drying and freeze drying can be used for the drying process, although both of these methods are costly, may destroy valuable microalgae pigments, and/or may reduce the lipid extraction efficiency.

10.3.1.5 Lipid Extraction

The final step before biodiesel production is the extraction of lipids from biomass [47]. Prior to extraction of the intracellular lipids from microalgae biomass, the microalgae cell wall should be disrupted [48]. The cell rupture can be performed through a variety of chemical, mechanical, or biological means, including bead beating, sonification, freezing, osmatic shock, acid lysing, etc. [28, 42]. Finally, the extraction process can be performed by conventional solvent extraction, enzymatic extraction, chemical cool press technology, or supercritical fluid extraction [49].

10.4 Catalytic Transesterification

Both homogeneous and heterogeneous catalysts can be used in the transesterification reaction. In addition, noncatalytic transesterification has been reported to occur under supercritical conditions. Table 10.4 summarizes the merits and limitations of different types of catalysts.

10.4.1 Homogeneous Catalysts

Homogeneous catalysts such as conventional acids and bases (e.g. H_2SO_4, NaOH, KOH, and $KOCH_3$) are the most commonly used catalysts in biodiesel production. The main advantages of this type of catalyst are their simplicity, high activity, and ability to function under mild operating conditions (low temperature and pressure) [10, 22]. However, the generation of large amounts of wastewater is a disadvantage of homogeneously catalyzed processes. In particular, large-scale biodiesel production requires a number of washing and purification steps to produce a high quality product [5].

10.4.1.1 Alkali Catalysts

Alkali catalysts (NaOH and KOH) are often used for the commercial production of biodiesel due to their shorter reaction time (about 4000 times faster than acid catalysts), low cost, and availability. They also require mild reaction conditions (e.g. methanol to oil ratio of 6 : 1 with catalyst loading of 1% at 65 °C) to reach biodiesel yields of about 86–92% [50]. It should be noted that when using base catalysts, the FFA content of the raw material should be low to avoid saponification, which can significantly reduce product yield [51].

Table 10.4 Advantages and disadvantages of different types of catalysts used for transesterification.

Catalyst		Advantages	Disadvantages
Homogeneous catalysts		• Fast reaction • Requires mild reaction conditions • Insensitive to fatty acid and water content in the presence of acid catalysts • Simultaneous transesterification and esterification in the presence of acid catalysts	• Soap formation in the presence of base catalyst • Long reaction time in the presence of acid catalyst • Corrosive • Difficult separation of catalyst from the product
Heterogeneous catalysts	Solid catalysts	• Easy separation • Recycling and reusing of catalyst • High stability of catalyst • Less disposal problems • Non corrosive • Higher selectivity in case of alkaline catalyst	• Leaching of catalyst • High reaction temperature and methanol/oil ratio are required • Diffusion limitation
	Enzymes	• Low temperature is required • Fewer purification steps are required	• Slow reaction rate • Deactivation of enzyme on exposure to alcohol • Not economical

Base catalysts such as $KOCH_3$ and $NaOCH_3$ are also widely used and have been reported in various literature studies. Alcantara et al. [52] reported quantitative conversion (100%) over $NaOCH_3$ for three types of feedstocks, namely, soybean oil, used frying oil, and tallow. Jordanov et al. [53] obtained 85.5% biodiesel yield using $NaOCH_3$ as catalyst and WCO as the feedstock. Leung and Guo [54] compared two different feedstocks, canola oil and WCO, in the presence of NaOH. The optimal reaction condition was 7 : 1 molar ratio of methanol to oil and 1 wt% catalyst at 70 °C in 20 minutes to reach 90.4% conversion in the case of canola oil. Decreasing the temperature from 70 to 40–45 °C increased the conversion to 93.5% at an increased reaction time of 60 minutes. A lower FAME yield of 88.8% was obtained from WCO under similar reaction conditions. Dias et al. [55] investigated the optimal amount of various base catalysts (e.g. KOH, NaOH, and $NaOCH_3$) for different feedstocks, namely, WCO, soybean, and sunflower oils. Their results indicated that 0.4–1.2 wt% of these catalysts was sufficient for the WCO and 0.2–1 wt% was required for the other oils. Figure 10.2 depicts the homogeneous base-catalyzed transesterification mechanism that includes four steps [1]. Initially, the alkoxide ion is generated and directly acts as a strong nucleophile. The main difference between acid and base catalyst activity in transesterification reaction is the formation of electrophilic species vs. strong nucleophiles, respectively.

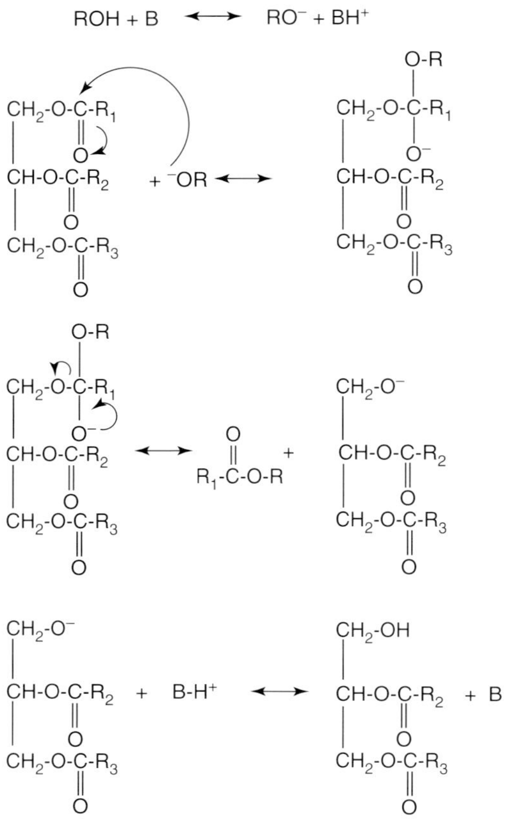

Figure 10.2 Mechanism of homogeneous base-catalyzed transesterification.

10.4.1.2 Acid Catalysts

Acid-catalyzed transesterification typically requires high alcohol to TG ratios and long reaction times to reach high biodiesel yield, the main advantage being that better results are obtained for vegetable oils with FFA >1% [56] than with base catalysts. Acid catalysts are highly sensitive to the water content since high water content can deactivate the catalyst and reduce the product yield [20].

Nye et al. [57] investigated the application of different alcohols (e.g. methanol, ethanol, 1-propanol, 2-propanol, 1-butanol, 2-ethoxy ethanol) for transesterification of WCO in the presence of H_2SO_4 and KOH catalysts. Higher FAME yield was obtained over H_2SO_4 in contrast to the alkali-catalyzed (KOH) reaction. Figure 10.3 exhibits the mechanism of homogeneous acid-catalyzed transesterification of TGs in three steps. First the carbonyl group is protonated, then nucleophilic attack of alcohol occurs to form a tetrahedral intermediate. Finally, the fatty acid ester is eliminated [58].

10.4.1.3 Two-step Esterification–Transesterification Reactions

In the presence of a feedstock with high FFA content of 0.5–3%, a two-step esterification–transesterification process is a suitable solution for the production

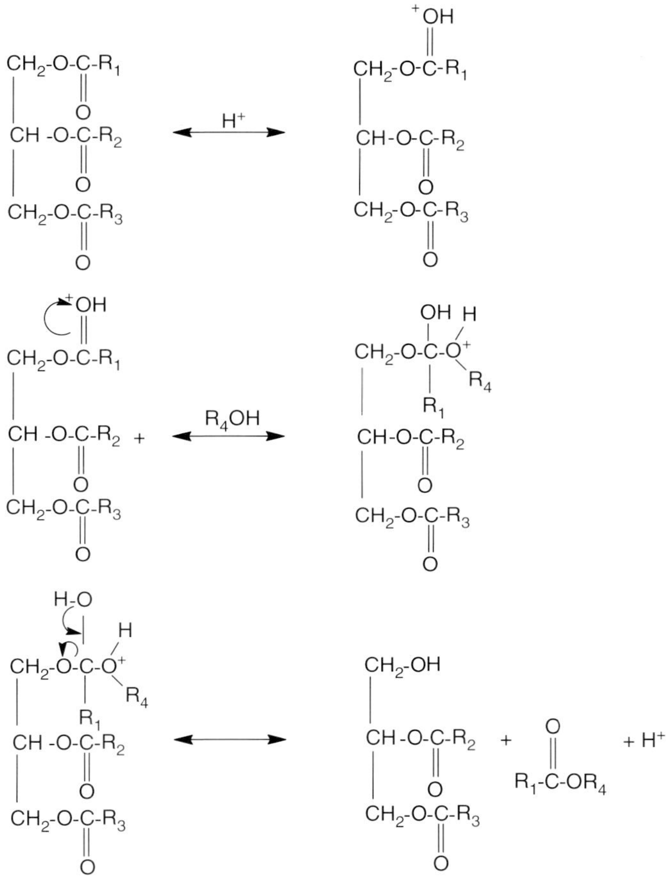

Figure 10.3 Mechanism of homogeneous acid-catalyzed transesterification.

of high quality biodiesel. In short, FFA esterification is performed initially over acid catalysts to reduce the FFA level to <1%, followed by transesterification over base catalysts [59]. This method can eliminate the conventional problems of single-step transesterification such as difficult separation of the product and glycerol and the slow reaction in the presence of homogeneous acid catalysts. Wang et al. [60] compared acid-catalyzed transesterification in the presence of H_2SO_4 and a two-step transesterification reaction using ferric sulfate (0–2%) and NaOH in the first and second steps, respectively. The highest conversion obtained using the single-step method was 90% at a molar ratio of methanol to oil of 20 : 1, at 95 °C in 10 hours. However, in the two-step process the highest conversion was 97.2% at a lower molar ratio of methanol to oil of 10 : 1 and reaction time of four hours at 95 °C. Other advantages of two-step transesterification are that there is no acid waste treatment, the equipment cost is low, and the catalyst can be easily recovered compared with the single-step reaction.

10.4.2 Heterogeneous Catalysts

Homogeneously catalyzed transesterification has certain limitations such as low tolerance toward high FFA and water contents, along with a complex product purification process. However, heterogeneous catalysts have been shown to be effective and convenient for the commercial production of biodiesel at a large scale from different feedstocks. Various types of heterogeneous catalysts such as ion exchange resins [61–63], sulfated oxides [64], transition metal oxides [65], alkaline earth metal oxides [66, 67], mixed metal oxides [68], carbon-based heterogeneous catalysts [69], and enzymes [70] have been used in biodiesel production processes. The main advantages of heterogeneous catalysts are their high tolerance to FFA and water as well as the ease of separation and recycling. Heterogeneous catalysts for biodiesel production can be divided into two classes: acid and base [71]. Solid acid catalysts can simultaneously participate in esterification and transesterification processes, while a solid base catalyst is typically used for transesterification. The major issue reported for heterogeneous catalysts is their mass transfer limitations since they form three phases with oil and alcohol. Application of cosolvents such as ethanol, *n*-hexane, and tetrahydrofuran can solve mass transfer issues by increasing the miscibility of oil and solvent. In addition, the application of structural promoters or catalyst supports to provide increased specific area and pore volume for the reaction of alcohols with large TG molecules is another solution for the elimination of mass transfer limitations [9, 65, 72]. Interestingly, heterogeneous catalysts can be derived from waste, renewable biomass, or natural resources such as egg shell, mollusc shells, ashes and rocks, etc. [18]. Table 10.5 reports the advantages and disadvantages of different types of heterogeneous catalyst.

Table 10.5 Comparison of heterogeneous acid/base catalysts.

Catalyst	Advantages	Disadvantages
• Ion-exchange resins (Nafion, Amberlyst)	• Very high activity, easy regeneration	• Low thermal stability, possible leaching
• TPA ($H_3PW_{12}O_{40}$)	• Very high activity	• Soluble in water
• TPA-Cs ($Cs_{2.5}H_{0.5}PW_{12}O_{40}$)	• Super acid sites	• Low activity per weight
• Zeolites (H-ZSM-5, Y and beta)	• Controllable acidity and hydrophobicity	• Small pore size low activity
• Sulfated metal oxides (zirconia, titania, tin oxide)	• High activity thermally stable	• Deactivates in water, but not in organic phase
• Niobic oxide (Nb_2O_5)	• Water tolerant	• Average activity
• Calcium oxide/Cao	• Low temperatures	• Long reaction times
• Calcium methoxide/$Ca(OMe)_2$	• High yield, reusable	• High ratio of reactants
• Calcium ethoxide/$Ca(OEt)_2$	• High yield, short times	• High ratio of reactants
• Li-doped zinc oxide/ZnO	• Low temperatures	• Long reaction times
• KF loaded on Eu_2O_3	• Short reaction times	• Incomplete yields

10.4.2.1 Solid Acid Catalysts

The main merits of heterogeneous acid catalysts are reported to be their insensitivity to FFA content, their ability to simultaneously catalyze esterification and transesterification reactions, simple catalyst recycling, no product washing, high product yield, low catalyst utilization, environmental friendliness, and elimination or reduction of corrosion [1, 7]. The high effectiveness of heterogeneous catalysts comes from their different strengths of Lewis and Brønsted acidity. Indeed, based on many recent studies, it is evident that the types of acid site (Brønsted or Lewis), as well as the strength of the acid sites (weak or strong), have a significant impact on the reaction conversion, yield of desired products, and also the deactivation of catalyst due to the deposition of coke on the catalyst surface. For example, solid acid catalysts such as sulfated zirconia and tungstated zirconia can achieve higher FAME yield compared with conventional homogeneous catalysts [63]. Cao et al. [73] used high FFA and water content WCO as a feedstock for the production of biodiesel in the presence of tungstophosphoric acid as catalyst. 87% and 97% conversions were achieved in transesterification and esterification reactions, respectively, at 65 °C and 70 : 1 molar ratio of methanol to oil in 14 hours [74]. The applications of select heterogeneous acid catalysts are reported in Table 10.6.

10.4.2.2 Solid Base Catalysts

Heterogeneous solid base catalysts are generally preferred to their solid acid counterparts on account of their activity. There are various single metal

Table 10.6 Application of heterogeneous acid catalysts in biodiesel production.

Feedstock	Catalyst	T[a)] (°C)	MR[b)]	C[c)] (wt%)	t[d)] (h)	Y[e)]	References
Methyl soyate	Ruthenium catalysts	40	—	0.1	2	46	[61]
Waste cooking oil	Zinc stearate immobilized on silica gel (ZS/Si)	200	18 : 1	3	10	98	[75]
Jatropha curcas oil	Sulfated tin oxide (SiO_2 (SO_4^{2-}/SnO_2-SiO_2))	60–80	15 : 1	3	2	97	[76]
Ethyl propanoate and ethylhexanoate	Heteropoly acids (H_2SO_4, Amberlyst-15 and zeolites HY and H-Beta)	60	20 : 1	0.1	1	84	[64]
Sunflower oil/methanol mixtures	Silica-functionalized with 4-ethyl-benzene sulfonic acid groups	200	6 : 1	1.5	—	—	[77]
Palmitic acid	30%HPW–ZrO_2	60	—	0.3[f)]	6	93	[78]

a) Temperature.
b) Molar ratio.
c) Catalyst.
d) Time.
e) Yield.
f) Based on catalyst weight (0.3 g).

oxides such as MgO, CaO, SrO, and BaO that can be used for base-catalyzed transesterification reactions [79]. While the metal ions can act as Lewis acid sites, the basic sites correspond to hydroxyl groups, metal-oxygen ion pairs (e.g. M^{2+}–O^{2-}), and bridging O^{2-} ions. Mixed metal oxides can provide acceptable results in the presence of a feedstock with low FFA content and at mild operating conditions. For example, a ZnO–La_2O_3 catalyst showed high activity and stability with a FAME yield of 92.3% and high reusability (up to 17 times) in a batch reactor [80]. In order to enhance the stability and activity of synthesized catalysts, researchers have focused on the use of doped metal oxides as solid catalysts to improve the surface area, basic site strength, and pore size. Guo et al. [81] obtained more than 95% FAME yield using a 3.0 wt% sodium silicate solid base catalyst, 7.5 : 1 molar ratio of methanol to oil, 60 °C reaction temperature, 60 minutes reaction time, and 250 rpm mixing intensity. Table 10.7 provides a summary of selected applications of solid base catalysts in biodiesel production.

The major issue for heterogeneous catalysis processes is catalyst deactivation. Catalyst deactivation has various impacts on processes such as reduced catalyst life time, loss of activity and selectivity to desired products, interruption of process operation for either replacement or regeneration of catalysts, loss of time

Table 10.7 Application of heterogeneous base catalysts in biodiesel production.

Catalyst	Reaction condition						
	T[b)] (°C)	MR[c)]	t[d)] (min)	C[e)] (wt%)	Y[a)] (%)	Conversion (%)	References
Al/Mg hydrotalcite	100	60 : 1	180	10	<80	—	[82]
KNO_3/Al_2O_3	70	12 : 1	420	6	87	—	[83]
Montmorillonite KSF	190	8 : 1	180	3	79.6	—	[84]
MgO	—	12 : 1	60	0.5	92	—	[85]
	300	39.6 : 1	—	—	91	—	[86]
CaO	100	6 : 1	—	1	—	90	[87]
SnO	65	12 : 1	30	3	90	—	[88]
Waste eggshell	60	18 : 1	60	10	97	—	[89]
Carbon-derived pyrolytic pomelo peel ($25K_2CO_3$/BC)	65	8 : 1	150	6	98	—	[90]
Waste shell-derived carbon	55	6 : 1	120	10	—	95.6	[91]
Waste Turbo jourdani (Turbinidae) shells	—	3 : 1	420	10	—	99	[92]
Ni–ZnO	55	8 : 1	60	11	95.2	—	[93]
Waste egg shell	—	12 : 1	90	5	93.5	—	[94]
Dimethyl carbonate (DMC) with CH_3OK	80	6 : 1	15	2	—	99	[95]

a) Yield.
b) Temperature.
c) Molar ratio.
d) Catalyst.
e) Time.

for production, and loss of investment. There are different types of deactivation: catalyst poisoning by contaminants in the feed; thermal degradation (sintering, evaporation, volatilization); leaching by the reaction mixture; coking (formation of deposits) covering the catalyst surface by carbonaceous species resulting from undesired reactions involving reactants, intermediates, or products; and mechanical damage. Thus, synthesizing a new heterogeneous base catalyst requires comprehensive studies, including extended catalyst testing and analysis to reduce or eliminate any chance of catalyst deactivation during operation.

10.4.3 Enzyme-Catalyzed Transesterification Reactions

Recently, successful applications of enzymatic catalysts (immobilized lipase) for transesterification of various feedstocks have been reported. The main advantages of enzymatic catalysts are easy product removal, no by-product generation, low operating temperature, reusability of catalyst without any separation step, and insensitivity to the water and FFA content of the feedstock [96]. The high cost of enzyme catalysts and the deactivation of enzymes at high operating temperatures are their main drawbacks for large-scale applications [97, 98]. The type of alcohol, molar ratio of alcohol to oil, and the type of lipase all exert a significant impact on biodiesel production with enzyme catalysts. Various types of enzymes can be used as a catalyst in transesterification, examples including *Pseudomonas fluorescens, Candida rugosa, Rhizopus oryzae, Burkholderia, Aspergillus niger, Thermomyces lanuginosus, Pseudomonas cepacia* and *Rhizomucor miehei* [99]. In the case of enzymatic transesterification, lipase is the first choice of enzyme for researchers. The lipase used for transesterification can be intracellular or extracellular in nature [100–103].

The immobilization technique plays a key role in the reduction of production costs such that the enzyme can be reused for several cycles [102]. The common method of immobilization is physical adsorption, as it is nontoxic and inexpensive, and the catalyst can be regenerated [104, 105]. However, lipase immobilized through entrapment is more stable than physically adsorbed lipase [103, 106]. As mentioned earlier, the cost of lipase is the main obstacle for the industrial application of enzymatic catalysts. Thus, immobilized lipase is a suitable catalyst for economic production of biodiesel because it has a higher stability and activity compared with the soluble form of the enzyme. Mittelbach [107] used three types of enzymes (*Candida lipase, Pseudomonas lipase*, and *Mucor lipase*) as catalysts for the transesterification of sunflower oil in the presence of different alcohols (methanol, ethanol, butanol), with and without petroleum ether as solvent. High oil conversions were achieved in the presence of ethanol and butanol, even in the absence of solvent, but only traces of FAME were formed with methanol without solvent. Additionally, *Candida antarctica* was reported [108] as a suitable catalyst for use with secondary alcohols (e.g. *iso*-propanol and 2-butanol) with conversions of about 80%, while *Mucor miehei* was suitable for primary alcohols (e.g. methanol, ethanol, propanol, and butanol) with conversions of around 95% in the presence of hexane as solvent. Low FAME yields were reported for the application of methanol in some enzymatic transesterification reactions, which relates

Table 10.8 Application of heterogeneous enzymatic catalysts in biodiesel production.

Enzyme	Oil	Alcohol	$C^{a)}/Y^{b)}$(%)	Cost	References
Novozyme 435	Vegetable oil	Methanol	$C = 90–93$	Moderate	[110]
	Sunflower oil	Methanol	$C = 97$	High	[111]
	Soybean oil	Methyl acetate	$C = 92$	High	[112]
	Soybean oil	Methanol	$C = 98$	High	[113]
	Rapeseed oil	Methanol	$C = 95$	High	[114]
	Cotton seed	Methanol	$C = 97$	High	[115]
	Jatropha oil	Ethyl acetate	$C = 91.3$	High	[116]
	Canola oil	Methanol	—	—	[117]
	Rice bran oil	Methanol	—	—	[118]
	Olive oil	Methanol	—	—	[119]
Rhizopus oryzae	Vegetable oil	Methanol	$C = 90$	—	[120]
R. oryzae	Soybean oil	Methanol	$C = 90$	Low	[121]
Pseudomonas fluorescens	Jatropha oil	Ethanol	—	—	[122]
Rhizomucor miehei	Palm oil	Methanol	—	—	[123]
Candida rugosa	Jatropha oil	Ethanol	—	—	[122]
P. cepacia	Jatropha oil	Methanol and Ethanol	—	—	[124]
Magnetic immobilized lipase	Soybean oil	Methanol	$Y = 92.8$	Moderate	[125]
Lipase + ionic liquid	Soybean oil	Methanol	$Y = 82$	High	[126]
Immobilized lipase on silica aerogel	Hevea brasi liensis (Oil)	Methanol	$Y = 93$	High	[127]

a) Conversion.
b) Yield.

to the inhibitory effects caused by methanol on the immobilized enzymes. However, stepwise addition of methanol can prevent deactivation of the enzyme and overcome this problem. Recently, the combination of lipase (*C. rugosa*) catalysts with magnetic nanoparticles showed acceptable results (FAME yield of 94%) for biodiesel production from soybean oil and olive oil [109]. Table 10.8 summarizes various applications of enzyme-catalyzed transesterification reactions.

10.5 Supercritical Transesterification Processes

Biodiesel production using conventional catalysts can suffer from drawbacks such as the difficult recovery of by-products, high sensitivity to the feedstock's water and FFA content, and low reaction rates. Utilization of enzyme-catalyzed processes cannot solve these issues completely due to the high cost of the enzyme and the long reaction times required [128]. Thus, researchers have explored supercritical methods in the absence of catalysts to eliminate or decrease these drawbacks in biodiesel production [129]. The idea of using supercritical conditions for biodiesel production was initially proposed by Saka and Kusdiana [130]. No catalyst is required and near 100% conversion can be achieved at very short reaction times. However, this method is not economical for industrial applications since it requires high temperatures and pressures. Use of cosolvents such as CO_2 or hexane is a key factor that can reduce the heat and pressure requirements as well as the capital costs [131]. Jian-Zhong et al. [132] compared a conventional supercritical process with a supercritical process incorporating cosolvents. The conventional method was carried out at various reaction temperatures (260, 300, and 350 °C), a constant molar ratio of methanol to oil of 42 : 1, and a pressure of up to 20 MPa. In the absence of hexane, the FAME yield was no more than 67.7%; however, the yield increased significantly to 85.5% upon the addition of 25 wt% of hexane. Alcohol and oil did not mix together to form a single phase; thus, hexane was added to the mixture to improve the miscibility.

Supercritical transesterification in the presence of KOH as catalyst showed excellent results. The highest yield in supercritical condition without catalyst was <5% but when the mass ratio of KOH to oil was 0.1% at 160 °C and 10 MPa, the yield increased dramatically to 98% FAME. The application of other solvents such as propane [133] and CO_2 [131] has also been reported. Application of a cosolvent that is soluble in methanol and oil in the supercritical condition is another form of this process. This type of reaction does not leave residual catalyst in the glycerol and the ester and can be completed in just 5–10 minutes. For instance, about 99% conversion can be achieved in the presence of tetrahydrofuran as a cosolvent and with high FFA raw material such as animal fat and crude palm oil [134].

Demirbas [135] compared FAME production by catalytic and supercritical methanol methods. Less than 90% yield of FAME was obtained at the optimal condition of 90 : 1 molar ratio, 87 °C reaction temperature, and 6 wt% CaO as catalyst. For the supercritical methanol process, the optimum condition was a reaction temperature of 287 °C and a 41 : 1 molar ratio, achieving 99.6% FAME selectivity at 96.5% conversion. The supercritical method has several advantages, including the fact that FFAs and glycerides react at similar rates, diffusion limitations are eliminated by the homogeneous phase, the process tolerates high water content, and the absence of a catalyst in the separation step, as well as the high molar ratio of methanol to oil that produces high conversions in a short reaction time. Despite having all these merits, the major disadvantages of this method are the high pressure (25–40 MPa) and high reaction temperature (350–400 °C) required, and hence the high operating cost.

10.6 Alternative Processes for Biodiesel Production

10.6.1 Ultrasonic Processes

Ultrasonic processes use sound waves in the range 20–100 kHz, beyond the limit of human hearing (10–12 kHz). Ultrasonic waves have been used in medical research since the early twentieth century [136], but their industrial applications have only significantly increased in recent decades in materials processing, extraction, and atomization [137]. Sound waves can generate high local temperatures (>5000 K) and pressures (>1000 atm) and can significantly enhance reaction rates, mass transfer, and catalytic surface areas [138]. Ultrasound treatment can break up aggregates of a given heterogeneous catalyst into smaller particles, forming new sites for subsequent reactions. Another benefit in transesterification reactions is improved mixing of the oil and alcohol, this being the main factor that improves the FAME yield since the oil and alcohol are not completely miscible in conventional processes.

Ultrasonic power, frequency, type of catalyst, and alcohol are some of the reaction parameters that exert an impact on ultrasonic transesterification processes. The main advantages of ultrasonic irradiation are shorter reaction times, lower molar ratio of alcohol to oil, lower energy consumption (50%), lower catalyst requirement (enzyme), improved reaction rate, enhanced conversion, improved yield, activation of alternate reaction pathways, simple equipment setup, better process economy, simple separation and purification processes, and high quality glycerol production. Ultrasonic processes do have some disadvantages, including the fact that the reaction temperature is typically slightly higher for long reactions and higher catalyst loading may be needed due to the fact that ultrasonic irradiation can facilitate catalyst leaching [139]. Studies have shown that the influence of ultrasound on transesterification reactions is of a purely physical nature. The formation of a fine emulsion between the oil and alcohol due to microturbulence generated by cavitation bubbles generates enormous interfacial area, which accelerates the reaction [140]. In principle, this approach is suitable for large-scale industrial biodiesel production, although it does require more research and study before commercialization. Salamatinia et al. [141] reported a 95% biodiesel yield using an ultrasonic process for the transesterification of palm oil over a heterogeneous catalyst. The use of ultrasound dramatically reduced the required catalyst loading and reaction time to 2.8 wt% and 50 minutes, respectively. The optimum condition was 9 : 1 molar ratio of alcohol to oil, and 70% and 80% ultrasonic amplitude for BaO and SrO catalyst, respectively. In addition, Yu et al. [138] investigated the transesterification of soybean oil in the presence of an immobilized form of Novozym 435 catalyst assisted by ultrasonic irradiation and ultrasonic irradiation with vibration. Enzyme activity and reaction rate were dramatically increased by ultrasonic irradiation with vibration, a 96% yield of FAME being obtained at 0.5 wt% water content, 1 : 1 *tert*-amyl alcohol to oil volume ratio, 6 : 1 methanol to oil molar ratio, and 50% of ultrasonic power (500 W), at a temperature of 40 °C and a reaction time of four hours.

10.6.2 Microwave-Assisted Processes

Microwave ovens were introduced to households in the middle of the twentieth century, and the industrial use of these appliances has dramatically increased in recent years. Microwave irradiation has higher frequency than radio waves and much lower than X-rays with wavelengths of between 1 mm and 1 m and frequencies ranging from 300 MHz to 300 GHz [142]. In principle, the application of microwaves can afford significant energy savings in comparison with conventional reactor systems. Microwaves do not have the ability to directly break bonds, and their interaction with materials is manifested as the generation of heat. This heat can be reflected, transmitted, or absorbed [143].

The main advantages of microwave-based biodiesel production processes are higher product quality and yield, low energy consumption [144], environmental friendliness, shorter reaction time, lower molar ratio of alcohol to oil, fewer by-products, and more effective energy transfer (energy is transferred in the form of electromagnetic energy rather than thermal energy). Motasemi and Ani [145] reported that the production of 1 kg biodiesel using microwave irradiation required 0.47 kWh of electricity, against an average of 2.13 kWh electrical energy that can be produced from 1 kg biodiesel. Thus, a microwave-based process can produce about 1.7 kWh/kg of extra electrical energy, demonstrating the sustainability of this process. Table 10.9 illustrates that the continuous microwave system is more energy efficient than conventional heating and batch microwave systems. Jalilainnosrati et al. [146] investigated the synthesis of fatty acid ethyl esters (FAEE) by a two-step *in situ* esterification/transesterification of *Jatropha curcas* L. seeds using a microwave system. FFA was reduced from 14% to <1% in the first step using H_2SO_4 as catalyst after 35 minutes of microwave irradiation. The second step was initiated by addition of 5 N KOH in ethanol to the organic phase from the first step. The highest conversion to FAEE was 97.29% under the optimum conditions of <0.5 mm seed size, 12.21 minutes irradiation time, 8.15 ml KOH, and 331.52 rpm agitation speed in a 110 W microwave system.

Table 10.9 Comparison of energy consumption in microwave and conventional processes.

Method	Energy requirement (kJ/l)[a]
Conventional heating[b]	94.3
Microwave continuous flow (7.2 l/min feedstock flow)	26.0
Microwave continuous flow (2 l/min feedstock flow)[c]	60.3 (92.3)[d]
Microwave heating (4.6 l batch reaction)[e]	90.1

a) Normalized for energy consumed per liter of biodiesel prepared.
b) On the basis of values from the joint US Department of Agriculture and US Department of Energy 1998 study into life cycle inventory of biodiesel and petroleum diesel for use in an urban bus.
c) Assuming a power consumption of 1700 W and a microwave input of 1045 W.
d) Assuming a power consumption of 2600 W and a microwave input of 1600 W.
e) Assuming a power consumption of 1300 W, a microwave input of 800 W, a time to reach 50 °C of 3.5 minutes, and a hold time at 50 °C of 1 minute.

Table 10.10 FAEE yields at different molar ratios using microwave irradiation.

	FAEE yields (%) at different molar ratios					
Time (s)	3 : 1	6 : 1	9 : 1	12 : 1	15 : 1	18 : 1
10	7.19	23.90	100	100	100	100
20	5.44	34.27	100	100	100	100
30	3.82	27.84	100	100	100	100
40	5.93	31.32	100	100	100	100
50	6.69	39.21	100	100	100	100
60	6.67	32.31	100	100	100	100
80	2.52	41.16	100	100	100	100
100	8.67	24.87	100	100	100	100

Source: Shahid and Jamal [148]. Reproduced with permission of Elsevier.

Lertsathapornsuk et al. [147] applied microwave irradiation to the continuous transesterification of WCO to biodiesel using NaOH as catalyst. A 97% yield of FAEE was achieved at the optimal reaction condition of 12 : 1 molar ratio of ethanol to oil and 3 wt% catalyst in just 30 seconds. In another study, complete conversion of WCO was achieved at 9 : 1 molar ratio of ethanol to oil after 10 seconds. Researchers reported that the reaction time with microwave irradiation was 10 times shorter than when using conventional methods. Table 10.10 shows the FAEE yields obtained at different molar ratios [148]. Also, continuous production of biodiesel using microwave irradiation was examined for different feedstocks (coconut, rice bran, and used cooking palm oil). 100%, 94%, and

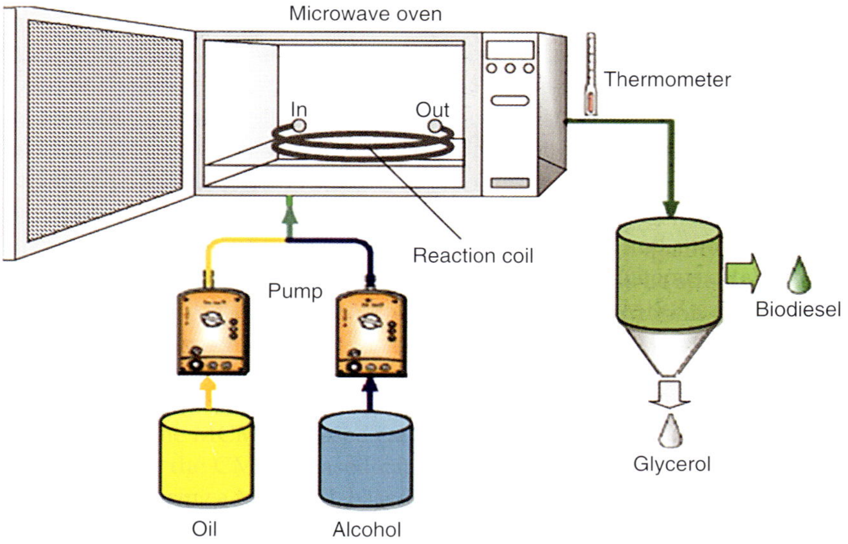

Figure 10.4 Schematic diagram of a continuous microwave process. Source: Lertsathapornsuk et al. 2008 [147]. Reproduced with permission of Elsevier.

83% conversion was obtained for coconut oil, rice bran oil, and used cooking oil, respectively. Figure 10.4 indicates the schematic diagram of the continuous microwave process [147].

10.7 Summary

The depletion of fossil fuel resources, unstable crude oil price, and environmental concerns are the main motivations for finding alternative fuels that should be environmentally friendly, cheap, widely available, and technically acceptable. Biodiesel is one of the best alternatives to petroleum-based fuels. A variety of approaches have been studied for biodiesel production, encompassing the use of homogeneous and heterogeneous acid and base catalysts, enzymatic catalysts, and noncatalytic methods. Various types of vegetable oils and animal fats can be used for biodiesel production, and global energy demand has led to the search for sustainable oil resources such as microalgae for the production of biodiesel. While the transesterification of oils and fats using homogeneous acid and base catalysts represents the current norm in biodiesel production, these methods can lead to serious issues in the downstream purification steps since they are highly sensitive to the FFA and water content of the feedstock. The use of enzymatic catalysts has shown interesting results, however, in general the cost of enzymes makes their use uneconomical, even for immobilized systems. The application of supercritical conditions gives good results but requires high temperatures and pressures, and does not represent an economical approach at the industrial scale. Thus, most research continues to focus on the synthesis of novel heterogeneous acid or base catalysts for biodiesel production since they can be recycled and reused many times, thereby significantly reducing biodiesel production costs. Recent developments such as the application of ultrasonication technology and microwave irradiation, combined with the use of flow reactors, tend to overcome heat and mass transfer limitations and represent another promising area of research.

References

1 Talebian-Kiakalaieh, A., Saidina Amin, N.A., and Mazaheri, H. (2013). A review on novel processes of biodiesel production from waste cooking oil. *Appl. Energy* 104: 683–710.

2 Aransiola, E.F., Ojumu, T.V., Oyekola, O.O. et al. (2014). Review of current technology for biodiesel production: state of the art. *Biomass Bioenergy* 61: 276–297.

3 Baskar, G. and Aiswarya, R. (2016). Trends in catalytic production of biodiesel from various feedstocks. *Renewable Sustainable Energy Rev.* 57: 496–504.

4 Atabani, A.E., Silitonga, A.S., Badruddin, I.A. et al. (2012). A comprehensive review on biodiesel as an alternative energy resource and its characteristics. *Renewable Sustainable Energy Rev.* 16: 2070–2093.

5 Semwal, S., Arora, A.K., Badoni, R.P., and Tuli, D.K. (2011). Biodiesel production using heterogeneous catalysts. *Bioresour. Technol.* 102: 2151–2161.
6 Tariq, M., Ali, S., and Khalid, N. (2012). Activity of homogeneous and heterogeneous catalysts, spectroscopic and chromatographic characterization of biodiesel: a review. *Renewable Sustainable Energy Rev.* 16: 6303–6316.
7 Ahmad, G. and Oki, M. (2014). Biodiesel production from algae by using heterogeneous catalysts: a critical review. *Energy* 78: 72–83.
8 Kiss, A.A. (2009). Novel process for biodiesel by reactive absorption. *Sep. Purif. Technol.* 69 (3): 280–287.
9 Borges, M.E. and Diaz, L. (2012). Recent development on heterogeneous catalysts for biodiesel production by oil esterification and transesterification reactions: a review. *Renewable Sustainable Energy Rev.* 18: 2839–2849.
10 Sharma, Y.C., Singh, B., and Korstad, J. (2011). Latest developments on application of heterogenous basic catalysts for an efficient and ecofriendly synthesis of biodiesel: a review. *Fuel* 90: 1309–1324.
11 Sinha, S., Agarwal, A.K., and Garg, S. (2008). Biodiesel development from rice bran oil: transesterification process optimization and fuel characterization. *Energy Convers. Manage.* 49 (5): 1248–1257.
12 Balat, M. and Balat, H. (2010). Progress in biodiesel processing. *Appl. Energy* 87 (6): 1815–1835.
13 Yu, X., Wen, Z., Li, H. et al. (2011). Transesterification of *Pistacia chinensis* oil for biodiesel catalyzed by CaO–CeO_2 mixed oxides. *Fuel* 80: 1868–1874.
14 Kim, J., Jia, H., and Wang, P. (2006). Challenges in biocatalysis for enzyme-based biofuel cells. *Biotechnol. Adv.* 24: 296–308.
15 Marchetti, J.M., Miguel, V.U., and Errazu, A.F. (2007). Possible methods for biodiesel production. *Renewable Sustainable Energy Rev.* 11: 1300–1311.
16 Demirbas, A. (2005). Biodiesel production from vegetable oils via catalytic and noncatalytic supercritical methanol transesterification methods. *Prog. Energy Combust. Sci.* 31: 466–487.
17 Eckey, E.W. (1956). Esterification and interesterification. *J. Am. Oil Chem. Soc.* 33: 575–579.
18 Nurfitri, I., Maniam, G.P., Hindryawati, N. et al. (2013). Potential of feedstock and catalysts from waste in biodiesel preparation: a review. *Energy Convers. Manage.* 74: 395–402.
19 Jacobson, K., Gopinath, R., Meher, L.C., and Dalai, A.K. (2008). Solid acid catalyzed biodiesel production from waste cooking oil. *Appl. Catal., B* 85: 86–91.
20 Canakci, M. and Van Gerpen, J. (1999). Biodiesel production via acid catalysis. *Trans. ASAE* 42 (5): 1203–1210.
21 Ma, F. and Hanna, M.A. (1999). Biodiesel production: a review. *Bioresour. Technol.* 70 (1): 1–15.
22 Meher, L.C., Sagar, D.V., and Naik, S.N. (2006). Technical aspects of biodiesel production by transesterification-a review. *Renewable Sustainable Energy Rev.* 10: 248–268.
23 Juan, J.C., Kartika, D.A., Wu, T.Y., and Hin, T.Y.Y. (2011). Biodiesel production from jatropha oil by catalytic and non-catalytic approaches: an overview. *Bioresour. Technol.* 102: 452–460.

24 Ahmad, A.L., Mat Yasin, N.H., Derek, C.J.C., and Lim, J.K. (2011). Microalgae as a sustainable energy source for biodiesel production: a review. *Renewable Sustainable Energy Rev.* 15 (1): 584–593.

25 Janaun, J. and Ellis, N. (2010). Perspectives on biodiesel as a sustainable fuel. *Renewable Sustainable Energy Rev.* 14 (4): 1312–1320.

26 Mata, T.M., Martins, A.A., and Caetano, N.S. (2010). Microalgae for biodiesel production and other applications: a review. *Renewable Sustainable Energy Rev.* 14: 217–232.

27 Laurens, L.M.L., Chen-Glasser, M., and McMillan, J.D. (2017). A perspective on renewable bio-energy from photosynthetic algae as feedstock for biofuels and bio-products. *Algal Res.* 24: 261–264.

28 Brennan, L. and Owende, P. (2010). Biofuels from microalgae – a review of technologies for production, processing, and extractions of biofuels and co-products. *Renewable Sustainable Energy Rev.* 14 (2): 557–577.

29 Adesanya, V.O., Cadena, E., Scott, S.A., and Smith, A.G. (2014). Life cycle assessment on microalgal biodiesel production using a hybrid cultivation system. *Bioresour. Technol.* 163: 343–355.

30 Zhu, L. (2015). Microalgal culture strategies for biofuel production: a review. *Biofuels, Bioprod. Biorefin.* 9 (6): 801–814.

31 Narala, R.R., Garg, S., Sharma, K.K. et al. (2016). Comparison of microalgae cultivation in photobioreactor, open raceway pond, and a two-stage hybrid system. *Front. Energy Res.* 4: 29.

32 Barsanti, L. and Gualtieri, P. (2014). *Algae: Anatomy, Biochemistry, and Biotechnology*. Boca Raton, USA: CRC Press.

33 de Oliveira, F.M.F. and Crispim, M.C.B. (2013). Compost extract as a nutrient source for algal cultures. *J. Aquac. Res. Dev.* 4 (5): 100–195.

34 Abdel-Raouf, N., Al-Homaidan, A., and Ibraheem, I. (2012). Microalgae and wastewater treatment. *Saudi J. Biol. Sci.* 19 (3): 257–275.

35 Dalrymple, O.K., Halfhide, T., Udom, I. et al. (2013). Wastewater use in algae production for generation of renewable resources: a review and preliminary results. *Aquat. Biosyst.* 9 (1): 1–11.

36 Schenk, P.M., Thomas-Hall, S.R., Stephens, E. et al. (2008). Second generation biofuels: high-efficiency microalgae for biodiesel production. *Bioenergy Res.* 1 (1): 20–43.

37 Hannon, M., Gimpel, J., Tran, M. et al. (2010). Biofuels from algae: challenges and potential. *Biofuels* 1 (5): 763–784.

38 Lam, M.K. and Lee, K.T. (2012). Potential of using organic fertilizer to cultivate *Chlorella vulgaris* for biodiesel production. *Appl. Energy* 94: 303–308.

39 Larsdotter, K. (2006). Microalgae for phosphorus removal from wastewater in a Nordic climate. PhD thesis. KTH.

40 Li, Y., Chen, Y.F., Chen, P. et al. (2011). Characterization of a microalgae *Chlorella* sp. well adapted to highly concentrated municipal wastewater for nutrient removal and biodiesel production. *Bioresour. Technol.* 102 (8): 5138–5144.

41 Zhou, W., Chen, P., Min, M. et al. (2014). Environment-enhancing algal biofuel production using wastewaters. *Renewable Sustainable Energy Rev.* 36: 256–269.

42 Rawat, I., Kumar, R.R., Mutanda, T., and Bux, F. (2011). Dual role of microalgae: phycoremediation of domestic wastewater and biomass production for sustainable biofuels production. *Appl. Energy* 88 (10): 3411–3424.

43 Pittman, J.K., Dean, A.P., and Osundeko, O. (2011). The potential of sustainable algal biofuel production using wastewater resources. *Bioresour. Technol.* 102 (1): 17–25.

44 Tan, X.B., Lam, M.K., Uemura, Y. et al. (2018). Cultivation of microalgae for biodiesel production: a review on upstream and downstream processing. *Chin. J. Chem. Eng.* 26: 17–30.

45 Chen, C.Y., Yeh, K.L., Aisyah, R. et al. (2011). Cultivation, photobioreactor design and harvesting of microalgae for biodiesel production: a critical review. *Bioresour. Technol.* 102 (1): 71–81.

46 Czartoski, T.J., Perkins, R., Villanueva, J.L., and Richards, G. (2010). Algae biomass fractionation, US Patent. US9296985B2.

47 Pan, Y., Alam, M.A., Wang, Z. et al. (2017). One-step production of biodiesel from wet and unbroken microalgae biomass using deep eutectic solvent. *Bioresour. Technol.* 238: 157–163.

48 Özçimen, D., Gülyurt, M.O., and İnan, B. (2012). *Algal Biorefinery for Biodiesel Production, Biodiesel – Feedstocks, Production and Applications*, 25–57. InTech.

49 Wang, Y. (2013). Microalgae as the third generation biofuel: production, usage, challenges and prospects. Master's thesis. Uppsala University.

50 Vicente, G., Martinez, M., and Aracil, J. (2004). Integrated biodiesel production: a comparison of different homogenous systems. *Bioresour. Technol.* 92: 297–305.

51 Kondamudi, N., Mohapatra, S.K., and Misra, M. (2011). Quintinite as a bifunctional heterogeneous catalyst for biodiesel synthesis. *Appl. Catal., A* 393 (1): 36–43.

52 Alcantara, R., Amores, J., Canoira, L. et al. (2000). Catalytic production of biodiesel from soybean oil. Used frying oil and tallow. *Biomass Bioenergy* 18: 515–527.

53 Jordanov, D.I., Petkov, P.S., Dimitrov, Y.K., and Ivanov, S.K. (2007). Methanol transesterification of different vegetable oils. *Pet. Coal* 49 (2): 21–23.

54 Leung, D.Y.C. and Guo, Y. (2006). Transesterification of neat and used frying oil: optimization for biodiesel production. *Fuel Process. Technol.* 87: 883–890.

55 Dias, J.M., Maria, C.M., Ferraz, A., and Almeida, M.F. (2008). Comparison of the performance of different homogeneous alkali catalysts during transesterification of waste and virgin oils and evaluation of biodiesel quality. *Fuel* 87: 3572–3578.

56 Palligarnai, T.V. and Briggs, M. (2008). Biodiesel production current state of the art and challenges. *J. Ind. Microbio. Biotechnol.* 35: 421–430.

57 Nye, M.J., Williamson, T.W., Deshpande, S. et al. (1983). Conversion of used frying oil to diesel fuel by transesterification: preliminary tests. *J. Am. Oil Chem. Soc.* 60 (8): 1598–1601.

58 Lotero, E., Liu, Y., Lopez, D.E. et al. (2005). Synthesis of biodiesel via acid catalysis. *Ind. Eng. Chem. Res.* 44: 5353–5363.

59 Atadashi, I.M., Aroua, M.K., Abdul, A.A.R., and Sulaiman, N.M.N. (2012). Production of biodiesel using high free fatty acid feedstocks. *Renewable Sustainable Energy Rev.* 16 (5): 3275–3285.

60 Wang, Y., Ou, S., Liu, P. et al. (2006). Comparison of two different processes to synthesize biodiesel by waste cooking oil. *J. Mol. Catal. A: Chem.* 252: 107–112.

61 Marchetti, J.M. and Errazu, A.F. (2008). Comparison of different heterogeneous catalysts and different alcohols for the esterification reaction of oleic acid. *Fuel* 87: 3477–3480.

62 Furuta, S., Matsuhashi, H., and Arata, K. (2006). Biodiesel fuel production with solid amorphous-zirconia catalysis in fixed bed reactor. *Biomass Bioenergy* 30: 870–873.

63 Park, Y.M., Lee, D.W., Kim, D.K. et al. (2008). The heterogeneous catalyst system for the continuous conversion of free fatty acids in used vegetable oils for the production of biodiesel. *Catal. Today* 131: 238–243.

64 Sakai, T., Kawashima, A., and Koshikawa, T. (2009). Economic assessment of batch biodiesel production processes using homogeneous and heterogeneous alkali catalysts. *Bioresour. Technol.* 100: 3268–3276.

65 Sreeprasanth, P.S., Srivastava, R., Srinivas, D., and Ratnasamy, P. (2006). Hydrophobic, solid acid catalysts for production of biofuels and lubricants. *Appl. Catal., A* 314: 148–159.

66 Wang, L. and Yang, J. (2007). Transesterification of soybean oil with nano-MgO or not in supercritical and subcritical methanol. *Fuel* 86: 328–333.

67 Lopez, D.E., Bruce, D.A., and Lotero, J.G.E. (2005). Transesterification of triacetin with methanol on solid acid and base catalysts. *Appl. Catal., A* 295: 97–105.

68 Xu, L., Yang, X., Yu, X. et al. (2008). Preparation of mesoporous polyoxometalate–tantalum pentoxide composite catalyst for efficient esterification of fatty acid. *Catal. Commun.* 9: 1607–1611.

69 Shu, Q., Gao, J., Nawaz, Z. et al. (2010). Synthesis of biodiesel from waste vegetable oil with large amounts of free fatty acids using a carbon based solid acid catalyst. *Appl. Energy* 87: 2589–2596.

70 Hama, S., Yamaji, H., Kaieda, M. et al. (2004). Effect of fatty acid membrane composition on whole-cell biocatalysts for biodiesel-fuel production. *J. Biochem. Eng.* 21: 155–160.

71 Helwani, Z., Othman, M.R., Aziz, N. et al. (2009). Solid catalysis for transesterification of triglycerides with methanol. *Appl. Catal., A* 363: 1–10.

72 Zabeti, M., Daud, W.M.A.W., and Aroua, M.K. (2009). Activity of solid catalysts for biodiesel production: a review. *Fuel Process. Technol.* 90: 770–777.

73 Cao, F., Chen, Y., Zhai, F. et al. (2008). Biodiesel production from high acid value waste frying oil catalyzed by superacid heteropolyacid. *Biotechnol. Bioenergy* 101 (1): 93–100.

74 Talebian-Kiakalaieh, A., Amin, N.A.S., Zarei, A., and Noshadi, I. (2013). Transesterification of waste cooking oil by heteropoly acid (HPA) catalyst: optimization and kinetic model. *Appl. Energy* 102: 283–292.

75 Alsalme, A., Kozhevnikova, E.F., and Kozhevnikov, I.V. (2008). Heteropoly acids as catalysts for liquid-phase esterification and transesterification. *Appl. Catal., A* 349: 170–176.

76 Raita, M., Champreda, V., and Laosiripojana, N. (2010). Biocatalytic ethanolysis of palm oil for biodiesel production using microcrystalline lipase in tert-butanol system. *Process Biochem.* 45: 829–834.

77 Kolaczkowski, S.T., Asli, U.A., and Davidson, M.G. (2009). A new heterogeneous ZnL_2 catalyst on a structured support for biodiesel production. *Catal. Today* 147S: S220–S224.

78 Alcaniz-Monge, J., Bakkali, B.E., Trautwein, G., and Reinoso, S. (2018). Zirconia-supported tungstophosphoric heteropolyacid as heterogeneous acid catalyst for biodiesel production. *Appl. Catal., B* 224: 194–203.

79 Mo, X., Lotero, E., Lu, C. et al. (2008). A novel sulfonated carbon composite solid acid catalyst for biodiesel synthesis. *Catal. Lett.* 123: 1–6.

80 Yan, S., Salley, S.O., and Ng, K.Y.S. (2009). Simultaneous transesterification and esterification of unrefined or waste oils over $ZnO–La_2O_3$ catalyst. *Appl. Catal., A* 353: 203–212.

81 Guo, F., Peng, Z.G., Dai, J.Y., and Xiu, Z.L. (2010). Calcined sodium silicate as solid base catalyst for biodiesel production. *Fuel Process. Technol.* 91: 322–328.

82 Serio, M.D., Ledda, M., Cozzolino, M. et al. (2006). Transesterification of soybean oil to biodiesel by using heterogeneous basic catalysts. *Ind. Eng. Chem. Res.* 45: 3009–3014.

83 Vyas, A.P., Subrahmanyam, N., and Patel, P.A. (2009). Production of biodiesel through transesterification of jatropha oil using KNO_3/Al_2O_3 solid catalyst. *Fuel* 88: 625–628.

84 Kansedo, J., Lee, K.T., and Bhatia, S. (2009). Biodiesel production from palm oil via heterogeneous transesterification. *Biomass Bioenergy* 33: 271–276.

85 Di Serio, M., Tesser, R., Pengmei, L., and Santacesaria, E. (2008). Heterogeneous catalysts for biodiesel production. *Energy Fuels* 22: 207–217.

86 Tateno, T. and Sasaki, T. (2004). Process for producing fatty acid fuels comprising fatty acids esters. US Patent 6,818,026, 16 November 2004.

87 Vujicic, D.J., Comic, D., Zarubica, A. et al. (2010). Kinetics of biodiesel synthesis from sunflower oil over CaO heterogeneous catalyst. *Fuel* 89: 2054–2061.

88 Liu, R., Wang, X., Zhao, X., and Feng, P. (2008). Sulfonated ordered mesoporous carbon for catalytic preparation of biodiesel. *Carbon* 46: 1664–1669.

89 Viriya-empikul, N., Krasae, P., Puttasawat, B. et al. (2010). Waste shells of mollusk and egg as biodiesel production catalysts. *Bioresour. Technol.* 101: 3765–3767.

90 Zhao, C., Lv, P., Yang, L. et al. (2018). Biodiesel synthesis over biochar-based catalyst from biomass waste pomelo peel. *Energy Convers. Manage.* 160: 477–485.

91 Bora, A.P., Dhawane, S.H., Anupam, K., and Halder, G. (2018). Biodiesel synthesis from *Mesua ferrea* oil using waste shell derived carbon catalyst. *Renewable Energy* 121: 195–204.

92 Boonyuen, S., Smith, S.M., Malaithong, M. et al. (2018). Biodiesel production by a renewable catalyst from calcined *Turbo jourdani* (Gastropoda: Turbinidae) shells. *J. Cleaner Prod.* 177: 925–929.
93 Baskar, G., Selvakumar, I.A.E., and Aiswarya, R. (2018). Biodiesel production from castor oil using heterogeneous Ni doped ZnO nanocatalyst. *Bioresour. Technol.* 250: 793–798.
94 Farooq, M., Ramli, A., Naeem, A. et al. (2018). Biodiesel production from date seed oil (*Phoenix dactylifera* L.) via egg shell derived heterogeneous catalyst. *Chem. Eng. Res. Des.* 132: 644–651.
95 Celante, D., Schenkel, J.V.D., and de Castilhos, F. (2018). Biodiesel production from soybean oil and dimethyl carbonate catalyzed by potassium methoxide. *Fuel* 212: 101–107.
96 Hsu, A., Jones, K.C., and Marmer, W.N. (2001). Production of alkyl esters from tallow and grease using lipase immobilized in pyllosilicate sol–gel. *J. Am. Oil Chem. Soc.* 78 (6): 585–588.
97 Shimada, Y., Watanabe, Y., Sugihara, A., and Tominaga, Y. (2002). Enzymatic alcoholysis for biodiesel fuel production and application of the reaction oil processing. *J. Mol. Catal. B: Enzym.* 17: 133–142.
98 Robles-Medina, A., Gonzalez-Moreno, P.A., Esteban-Cerdan, L., and Molina-Grima, E. (2009). Biocatalysis: towards ever greener biodiesel production. *Biotechnol. Adv.* 27: 398–408.
99 Vasudevan, P.T. and Briggs, M. (2008). Biodiesel production-current state of the art and challenges. *J. Ind. Microbiol. Biotechnol.* 35: 421–430.
100 Fukuda, H., Kondo, A., and Noda, H. (2001). Biodiesel fuel production by transesterification of oils. *J. Biosci. Eng.* 92: 405–416.
101 Lam, M.K., Lee, K.T., and Mohamed, A.R. (2010). Homogeneous, heterogeneous and enzymatic catalysis for transesterification of high free fatty acid oil waste cooking oil to biodiesel: a review. *Biotechnol. Adv.* 28: 500–518.
102 Kuan, I.C., Lee, C.C., Tsai, B.H. et al. (2013). Optimizing the production of biodiesel using lipase entrapped in biomimetic silica. *Energies* 6: 2052–2064.
103 Noureddini, H., Gao, X., and Philkana, R.S. (2005). Immobilized *Pseudomonas cepacia* lipase for biodiesel fuel production from soyabean oil. *Bioresour. Technol.* 96: 769–777.
104 Tan, T., Lu, J., Nie, K. et al. (2010). Biodiesel production with immobilized lipase: a review. *Biotechnol. Adv.* 28: 628–634.
105 Kaieda, M., Samukawa, T., Matsumoto, T. et al. (1999). Biodiesel fuel production from plant oil catalyzed by *Rhizopus oryzae* lipase in a water-containing system without an organic solvent. *J. Biosci. Bioeng.* 88: 627–631.
106 Vidinha, P., Augusto, V., Almeida, M. et al. (2006). Sol–gel encapsulation: an efficient and versatile immobilization technique for cutinase in non-aqueous media. *J. Biotechnol.* 121: 23–33.
107 Mittelbach, M. (1990). Lipase catalyzed alcoholysis of sunflower oil. *J. Am. Oil Chem. Soc.* 67: 168–170.
108 Nelson, L.A., Fogila, T.A., and Marmer, W.N. (1996). Lipase-catalyzed production of biodiesel. *J. Am. Oil Chem. Soc.* 73 (8): 1191–1195.

109 Xie, T., Wang, A., Huang, L. et al. (2009). Recent advances in the support and technology used in enzyme immobilization. *Afr. J. Biotechnol.* 8: 4724–4733.

110 Watanabe, Y., Shimada, Y., Sugihara, A. et al. (2000). Continuous production of biodiesel fuel from vegetable oil using immobilized *Candida antarctica* lipase. *J. Am. Oil Chem. Soc.* 77: 355–360.

111 Bako, K.B., Kova, F.C.S., Gubicza, L., and Hansco, J.K. (2002). Enzymatic biodiesel production from sunflower oil by *Candida antarctica* lipase in a solvent free system. *Biocatal. Biotransform.* 20: 437–439.

112 Wei, D., Yuanyuan, X., Dehua, L., and Jing, Z. (2004). Comparative study on lipase catalyzed transformation of soybean oil for biodiesel production with different acyl acceptors. *J. Mol. Catal. B: Enzym.* 30: 125–129.

113 Xu, Y., Du, W., Zeng, J., and Liu, D. (2004). Conversion of soybean oil to biodiesel fuel using lipozyme TL IM in a solvent free medium. *Biocatal. Biotransform.* 22: 45–48.

114 Li, L., Du, W., Liu, D. et al. (2006). Lipase-catalyzed transesterification of rapeseed oils for biodiesel production with a novel organic solvent as the reaction medium. *J. Mol. Catal. B: Enzym.* 43: 58–62.

115 Royon, D., Daz, M., Ellenrieder, G., and Locatelli, S. (2007). Enzymatic production of biodiesel from cotton seed oil using *t*-butanol as a solvent. *Bioresour. Technol.* 98: 648–653.

116 Modi, M.K., Reddy, J.R.C., Rao, B.V.S.K., and Prasad, R.B.N. (2007). Lipase mediated conversion of vegetable oils into biodiesel using ethyl acetate. *Bioresour. Technol.* 98: 1260–1264.

117 Chang, H.M., Liao, H.F., Lee, C.C., and Shieh, C.J. (2005). Optimized synthesis of lipase-catalyzed biodiesel by Novozym 435. *J. Chem. Technol. Biotechnol.* 80: 307–312.

118 Lai, C.C., Zullaikah, S., Vali, S.R., and Ju, Y.H. (2005). Lipase-catalyzed production of biodiesel from rice bran oil. *J. Chem. Technol. Biotechnol.* 80: 331–337.

119 Sanchez, F. and Vasudevan, P.T. (2006). Enzyme catalyzed production of biodiesel from olive oil. *Appl. Biochem. Biotechnol.* 135: 1–14.

120 Ban, K., Kaieda, M., Matsumoto, T. et al. (2001). Whole-cell biocatalyst for biodiesel fuel production utilizing *Rhizopus oryzae* cells immobilized within biomass support particles. *Biochem. Eng.* 8: 39–43.

121 Hama, S., Yamaji, H., Fukumizu, T. et al. (2007). Biodiesel fuel production in a packed-bed reactor using lipase-production *Rhizopus oryzae* cells immobilized within biomass support particles. *Biochem. Eng. J.* 34: 273–278.

122 Shah, S. and Gupta, M.N. (2006). Lipase catalyzed preparation of biodiesel from jatropha oil in a solvent free system. *Process Biochem.* 42: 409–413.

123 Al-Zuhair, S., Jayaraman, K.V., Krishnan, S., and Chan, Y.H. (2006). The effect of fatty acid concentration and water content on the production of biodiesel by lipase. *Biochem. Eng. J.* 30: 212–217.

124 Abigor, R., Uadia, P., Fogila, T. et al. (2000). Lipase-catalyzed production of biodiesel fuel from some *Nigerian lauric* oils. *Biochem. Soc. Trans.* 28: 979–981.

125 Xie, W. and Huang, M. (2018). Immobilization of *Candida rugosa* lipase onto graphene oxide Fe_3O_4 nanocomposite: characterization and application for biodiesel production. *Energy Convers. Manage.* 159: 42–53.

126 Fan, Y., Wang, X., Zhang, L. et al. (2018). Lipase-catalyzed synthesis of biodiesel in a hydroxyl-functionalized ionic liquid. *Chem. Eng. Res. Design.* 132: 199–207.

127 Arumugan, A., Thulasidharan, D., and Jegadeesan, G.B. (2018). Process optimization of biodiesel production from *Hevea brasiliensis* oil using lipase immobilized on spherical silica aerogel. *Renewable Energy* 116: 755–761.

128 Mamoru, I., Chen, B.X., Masashi, E. et al. (2001). Production of biodiesel fuel from triglycerides and alcohol using immobilized lipase. *J. Mol. Catal. B: Enzym.* 16: 53–58.

129 Niza, N.M., Tan, K.T., Lee, K.T., and Ahmad, Z. (2013). Biodiesel production by non-catalytic supercritical methyl acetate: thermal stability study. *Appl. Energy* 101: 198–202.

130 Saka, S. and Kusdiana, D. (2001). Biodiesel fuel from rapeseed oil as prepared in supercritical methanol. *Fuel* 80: 225–231.

131 Han, H.W., Cao, W.L., and Zhang, J.C. (2005). Preparation of biodiesel from soybean oil using supercritical methanol and CO_2 as co-solvent. *Process Biochem.* 40: 3148–3151.

132 Jian-Zhong, Y., Min, X., and Ji-Bin, S. (2008). Biodiesel from soybean oil in supercritical methanol with co-solvent. *Energy Convers. Manage.* 49: 908–912.

133 Cao, W.L., Han, H.W., and Zhang, J.C. (2005). Preparation of biodiesel from soybean oil using supercritical methanol and co-solvent. *Fuel* 84 (4): 347–351.

134 BIOX Process. www.bioxcorp.com (accessed December 2008).

135 Demirbas, A. (2009). Biodiesel from waste cooking oil via base-catalytic and supercritical methanol transesterification. *Energy Convers. Manage.* 50: 923–927.

136 Newman, P.G. and Rozycki, G.S. (1998). The history of ultrasound. *Surg. Clin. North. Am.* 78 (2): 179–195.

137 Cintas, P., Mantegna, S., Gaudino, E.C., and Cravotto, G. (2010). A new pilot flow reactor for high-intensity ultrasound irradiation. Application to the synthesis of biodiesel. *Ultrason. Sonochem.* 17: 985–989.

138 Yu, D., Tian, L., Wu, H. et al. (2010). Ultrasonic irradiation with vibration for biodiesel production from soybean oil by Novozym 435. *Process Biochem.* 45: 519–525.

139 Verziu, M., Florea, M., Simon, S. et al. (2009). Transesterification of vegetable oils on basic large mesoporous alumina supported alkaline fluorides-evidences of the nature of the active site and catalytic performances. *J. Catal.* 263 (1): 56–66.

140 Kalva, A., Sivasankar, T., and Moholkar, V.S. (2009). Physical mechanism of ultrasound assisted synthesis of biodiesel. *Ind. Eng. Chem. Res.* 48 (1): 534–544.

141 Salamatinia, B., Mootabadi, H., Bhatia, S., and Abdullah, A.Z. (2010). Optimization of ultrasonic-assisted heterogeneous biodiesel production from

palm oil: a response surface methodology approach. *Fuel Process. Technol.* 91: 441–448.
142 Orchard, B., Jon, D., and John, C. (2007). Developments in biofuel processing technologies. *World Pumps* 487: 24–28.
143 Varma, R.S. (2001). Solvent-free accelerated organic syntheses using microwaves. *Pure Appl. Chem.* 73 (1): 193–198.
144 Patil, P.D., Gude, V.G., Camacho, L.M., and Deng, S. (2010). Microwave-assisted catalytic transesterification of *Camelina Sativa* oil. *Energy Fuels* 24: 1298–1304.
145 Motasemi, F. and Ani, F.N. (2012). A review on microwave-assisted production of biodiesel. *Renewable Sustainable Energy Rev.* 16: 4719–4733.
146 Jalilianosrati, H., Amin, N.A.S., Talebian-Kiakalaieh, A., and Noshadi, I. (2013). Microwave assisted biodiesel production from *Jatropha curcas* L. seed by two-step in situ process: optimization using response surface methodology. *Bioresour. Technol.* 136: 565–573.
147 Lertsathapornsuk, V., Pairintra, R., Aryusuk, K., and Krisnangkura, K. (2008). Microwave assisted in continuous biodiesel production from waste frying palm oil and its performance in a 100 kW diesel generator. *Fuel Process. Technol.* 89: 1330–1336.
148 Shahid, E.M. and Jamal, Y. (2011). Production of biodiesel: a technical review. *Renewable Sustainable Energy Rev.* 15: 4732–4745.

11

Upgrading of Lipids to Hydrocarbon Fuels via (Hydro)deoxygenation

David Kubička

University of Chemistry and Technology Prague, Technopark Kralupy, Žižkova 7, 278 01 Kralupy nad Vltavou, Czech Republic

In memoriam of Professor Victor Teixeira da Silva, an outstanding scientist and friend.

11.1 Introduction

Production of hydrocarbon fuels relies on the upgrading of various feedstocks with the aim to remove heteroatoms – such as sulfur, nitrogen, and oxygen – and to modify their molecular structure to improve their fuel properties. In traditional (petroleum-based) refineries, hydrodesulfurization (HDS) is the key upgrading operation that ensures environmental acceptability of the fuels due to (i) elimination of SO_x emissions and (ii) protection of catalysts in the exhaust gas converters allowing efficient abatement of pollutants including unburned hydrocarbons, carbon monoxide, and NO_x. While deoxygenation is not a concern when upgrading petroleum distillate fractions, it is of the utmost importance in the upgrading of biomass-derived feedstocks, which typically contain at least 10 wt% of organically bound oxygen. However, deoxygenation is not driven by environmental concerns; rather, it aims at improving the thermochemical stability of fuels and increasing their energy content.

Recent efforts aimed at minimizing the impacts of fossil fuels in the environment – mainly due to the increase in atmospheric CO_2 concentration resulting from fossil fuel combustion – have encouraged worldwide research of biofuels, i.e. renewable (biomass-derived) transportation fuels. Biofuels are generally classified into three generations: the first generation relies on food crops and includes bioethanol from sugar and starch plants as well as biodiesel derived from edible vegetable oils; the second generation is produced from lignocellulosic feedstocks as well as from nonedible oils and waste oils and fats; and the third generation is based on aquatic biomass (e.g. algae). The current shift toward the second- and the third-generation biofuels, as mandated by the proposed RED II directive [1], for instance, enhances the importance of deoxygenation as a robust, versatile technology capable of dealing with a wide variety of feedstocks.

Chemical Catalysts for Biomass Upgrading, First Edition. Edited by Mark Crocker and Eduardo Santillan-Jimenez.

Main deoxygenation reactions of triglycerides

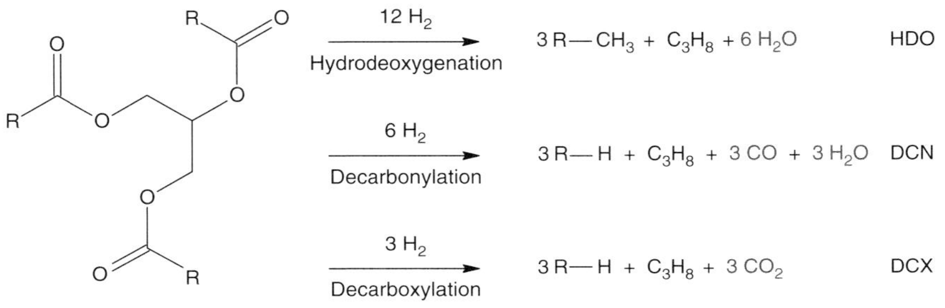

Important side reactions during deoxygenation

$$CO + 3H_2 \rightleftharpoons CH_4 + H_2O \quad \text{Methanation}$$

$$CO_2 + 4H_2 \rightleftharpoons CH_4 + 2\,H_2O \quad \text{Methanation}$$

$$CO + H_2O \rightleftharpoons CO_2 + H_2 \quad \text{Water–gas shift}$$

Figure 11.1 Main deoxygenation reactions and important parallel reactions.

Deoxygenation encompasses four main different types of chemical reactions that take place under reducing conditions – namely, hydrodeoxygenation (HDO), dehydration, decarbonylation (DCN), and decarboxylation (DCX) – ultimately resulting in three main oxygen-containing products, i.e. water, carbon monoxide, and carbon dioxide (Figure 11.1). In addition, other products (such as alcohols) originating from demethoxylation or hydrogenolysis reactions involving ethers and esters, respectively, could be considered as deoxygenation products as well. This makes the deoxygenation chemistry more complex than that of HDS and hydrodenitrification (HDN), which under reducing conditions yield hydrogen sulfide and ammonia, respectively, as the only heteroatom-containing by-products. Obviously, the rich chemistry of deoxygenation has its origins in the complex composition of biomass; however, even relatively simple oxygenates, such as triglycerides (TGs), can undergo a wide range of deoxygenation reactions depending on the catalyst and reaction conditions employed.

Although deoxygenation can be applied to any feedstock containing oxygenates and much attention is currently focused on the deoxygenation of bio-oils – i.e. organic liquids produced by fast pyrolysis of particularly lignocellulosic biomass – the deoxygenation of lipids and lipid-derived feedstocks (mainly comprising TGs and free fatty acids [FFAs]) has already been implemented at industrial scale and is currently the prevailing application of deoxygenation processes. This is a consequence of (i) their well-defined and constant structure and composition and (ii) the possibility to convert the lipids rather easily into hydrocarbon mixtures with superior fuel properties. Therefore, the focus of this chapter will be on the chemistry of deoxygenation and on deoxygenation catalysts using TGs and other lipid-based feedstocks as a model system. Additional details can be found in several reviews published in the last decade [2–10].

11.2 Feedstocks

TGs, which are esters of three fatty acid molecules and a glycerol molecule, are the main constituents of vegetable oils and animal fats together with minor amounts of other components such as phospholipids and tocopherols. The fatty acid moieties present in the most abundantly produced vegetable oils, i.e. palm, soybean, rapeseed (canola), and sunflower oil, are typically between 14 and 22 carbon atoms (C_{16} and C_{18} being the most prevalent ones) in length. Similarly, C_{16} and C_{18} fatty acid chains are the most abundant in the TGs making up animal fats, such as beef tallow, pork lard, and chicken fat [11]. The global annual production of vegetable oils and animal fats has more than doubled in the last 20 years and is currently nearly 200 MMt [12]. The production of animal fats is estimated to be about 14% of the total oil and fat production [11], i.e. ca. 30 Mt. Recently, algal oils have attracted enormous attention due to their remarkable potential, which stems from their superior annual productivity per unit area in comparison with vegetable oils (>50 t/ha for algal oil [13] vs. 3.5–5 t/ha for palm oil [14] and <1 t/ha for other vegetable oils [14]) and the prospect of using nonagricultural land. Other TG feedstocks include used or waste cooking oils and nonedible vegetable oils, such as Jatropha oil. Side-streams from vegetable oil or wood processing (e.g. crude tall oils), which contain significant concentrations of FFAs, can be used as feedstocks to produce fuels as well.

TGs mainly consist of either unsaturated fatty acid moieties (in the case of oils) or saturated fatty acid moieties (in the case of fats), which affects their stability during deoxygenation in the presence of hydrogen as cyclization reactions may occur [15] as well as the overall hydrogen consumption. The overall chain length of the fatty acid moieties (C_{14}–C_{22}) is particularly suitable to produce diesel and/or kerosene fuels that consist of hydrocarbons comprising between 10 and 20 carbon atoms.

The suitable carbon–carbon chain lengths together with their relatively uniform reactivity, absence of water, easy storage and handling, and straightforward production by physical and/or chemical extraction are the salient advantages of TGs over other biomass-derived raw materials, such as bio-oils. Direct conversion of TGs via transesterification has been used to produce fatty acids methyl esters (FAMEs) or biodiesel, mainly because of the ability of the latter to be mixed with (or replace) hydrocarbon fuels [16]. This simple technology allows the processing of highly purified nonacidic feedstocks [16] to afford first-generation biofuels, waste streams being much more difficult to upgrade through biodiesel production. In contrast, deoxygenation technologies can deal not only with waste TG-based feedstocks but also with feeds with high concentrations (or made exclusively) of FFAs, i.e. virtually any lipid-based feedstock can be upgraded by deoxygenation to hydrocarbon fuels. Moreover, sustainability data (in terms of the overall greenhouse gas emissions savings) demonstrate that deoxygenation is a more favorable conversion pathway from an environmental standpoint than transesterification [17].

Finally, the quality of the final biofuel has to be considered as well. Biodiesel can certainly serve as a biofuel, but it has to either be blended with diesel (current EU limit is maximum 7 vol%) to be used as mixed biodiesel–diesel fuel in unmodified vehicles or be used in engines purposely designed or modified to employ neat (100%) biodiesel. Indeed, these fuels usually require adaptations to be made to engine systems. Moreover, due to the oxygen content and unsaturation of most alkyl chains within FAMEs, the chemical, thermal, and oxidative stability as well as the energy content are lower for biodiesel than for diesel fuel [6]. In contrast, hydrocarbons obtained via deoxygenation of TGs are fully compatible with diesel fuel and diesel fuel-based infrastructure and even superior to petroleum-derived diesel as their combustion produces less harmful emissions (due to the absence of aromatics and sulfur).

11.3 Chemistry

Although the structure of TGs is fairly simple, depending on the deoxygenation reaction conditions and the catalyst employed, they are susceptible to a surprisingly wide variety of conversion pathways. The rudimentary differentiation of deoxygenation routes is based on their primary oxygen-containing product lacking an alkyl chain, i.e. inorganic oxygenates. There are three such products: water, carbon monoxide, and carbon dioxide. When carbon oxides are formed, the resulting hydrocarbons have one carbon atom less than the original fatty acid moieties. Since natural oils and fats exclusively contain fatty acid moieties with even carbon number, the resulting hydrocarbons are characterized by an odd carbon number. Depending on the nature of the carbon oxide, the process is called DCN, yielding CO, or DCX, producing CO_2 (Figure 11.1). On the other hand, the formation of water is not accompanied by a carbon atom loss from the original fatty acid moiety, and thus, the hydrocarbons formed have an even number of carbons. HDO and dehydration are the possible reactions affording water, dehydration applying only to the alcohols produced via the stepwise HDO of esters and acids. It should be noted that under hydrotreating conditions, the hydrogenation of carbon oxides may also take place, yielding additional water and methane (Figure 11.1). This intertwined nature of conversion pathways makes the exact evaluation of the underlying chemistry challenging. Also, the formation of water and carbon oxides may inhibit the catalyst performance, e.g. CO is well known as a potent catalyst poison. Finally, the reaction pathway followed has an important effect on the overall hydrogen consumption.

A closer look at the conversion of TGs under hydrotreating conditions reveals that their transformation to hydrocarbons consists of several consecutive and parallel reactions. The most facile reaction in the presence of hydrogen is the saturation of the double bonds. Most studies show that the key intermediate in the deoxygenation of TGs are fatty acids that can be formed either via hydrogenolysis or hydrolysis. This justifies the use of fatty acids as model compounds, albeit the reactivity of the acids has been shown to be lower than that of esters [18]. Considering the absence of water in the feedstocks as well as the mild acidity (if any) of the catalysts employed, hydrogenolysis rather than hydrolysis represents the main reaction through which TGs are converted to fatty acids along the pathway toward hydrocarbons. In principle, the carbon–oxygen bond in esters can be cleaved either in the alpha or beta position with respect to the carboxylic group. Nevertheless, cleavage in the alpha position is not very common, albeit it does take place under rather harsh conditions (>15 MPa) over bimetallic formulations such as CuCr, ZnCr, or CuZn. In fact, the industrial production of a wide range of alcohols, such as fatty acid alcohols or hexanediol, relies on the use of the Adkins catalyst (CuCr) [19]. However, these catalysts are not very active (if at all) in deoxygenation. Thus, hydrogenolysis in the beta position ultimately yielding three fatty acid molecules and propane precedes the actual deoxygenation, as oxygen is not released from the organic molecules during hydrogenolysis. Plausible intermediates of this pathway, including propylstearate and propanediol distearate, have been identified among the reaction products at low conversions [20–22]. Since diglycerides have also been observed [10, 23, 24], hydrolysis may also take place to some extent.

Subsequently, actual deoxygenation proceeds either via DCX, DCN, or HDO, the reaction(s) followed depending both on the catalyst and the reaction conditions applied. DCX/DCN was shown to be favored over group 10 metals (Ni, Pt, Pd), while molybdenum or tungsten-based catalysts (Mo_2C, MoS_2, sulfided NiMo, CoMo, NiW) predominantly promote HDO [25]. Although sulfided catalysts can be generally regarded as selective toward the HDO pathway [25], the selectivity patterns over sulfides are rather complex, NiS being selective to DCX/DCN [20] and sulfided NiW allowing the tuning of the HDO/(DCX + DCN) products ratio by adjusting reaction conditions (particularly hydrogen pressure and, to a lesser extent, reaction temperature) [26]. Indeed, when decreasing the hydrogen pressure from 7 MPa to just 0.7 MPa, the HDO/(DCX + DCN) ratio dropped from ca. 3 to about 0.3 over a sulfided NiW catalyst at 280 °C [26]. As mentioned earlier, FFAs are typical intermediates of both reaction pathways that are often accompanied by alcohols and fatty acid esters [6, 25]. Aldehydes, which constitute other likely intermediates, were observed scarcely [25], plausibly because aldehydes and ketones are the most reactive oxygenates under deoxygenation conditions in general and hydrotreating conditions in particular [27].

Frequently, two or more of the aforementioned pathways may occur simultaneously, which makes the elucidation of the complete transformation scheme complex and challenging (see Figure 11.2). In general, DCX is favored by lower pressure and higher reaction temperature, while HDO is operative at lower reaction temperatures and higher hydrogen pressures. The DCN pathway was shown to be effective when a suitable enol-keto tautomerization equilibrium is established under the reaction conditions. While DCX produces simply carbon dioxide and a hydrocarbon, there are several oxygenated intermediates along the HDO pathway. These include an aldehyde, i.e. the first partial hydrogenation/deoxygenation product that is successively reduced to an alcohol, which in turn represents the second partial hydrogenation/deoxygenation product.

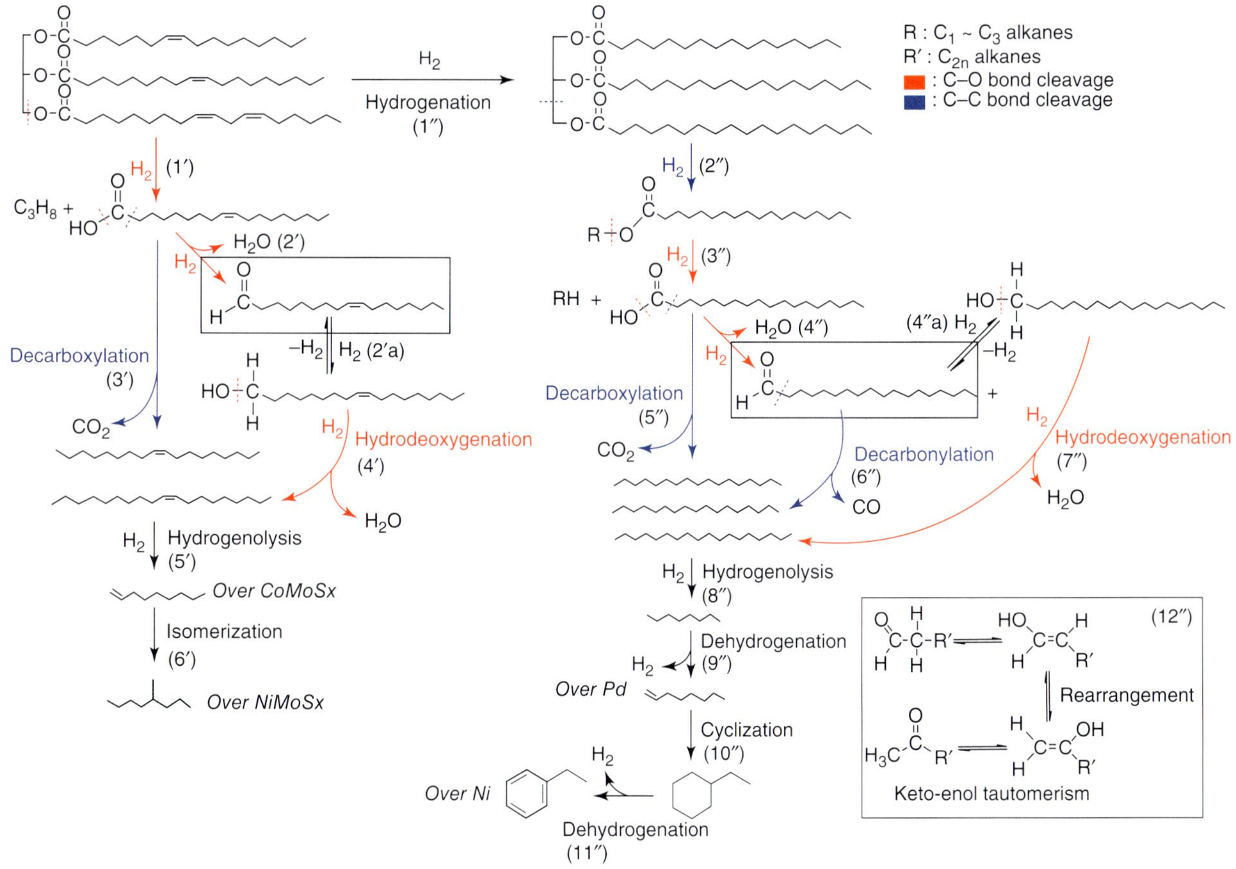

Figure 11.2 Detailed reaction scheme for the deoxygenation pathways of triglycerides. Source: Reprinted with permission from Kim et al. 2014 [28]. Copyright 2014, Elsevier.

The formation of these organic products is accompanied by formation of water. Finally, the alcohol is dehydrated/hydrodeoxygenated to produce a hydrocarbon and an additional water molecule. In addition, the fatty alcohol can also react with a fatty acid to yield a fatty ester and water. Strictly speaking, this esterification can also be considered a deoxygenation reaction. Under specific reaction condition, fatty esters can be produced in yields exceeding 30 wt% [29].

Ultimately, liquid phase hydrocarbons can be obtained from TGs in 81–85 wt% yield, depending on whether the selective DCX (affording the lower yield value) or the HDO pathway is operative. The other final products include propane (5 wt%) and either CO_2 (15 wt%) or water (12 wt%). It should be noted that all yield calculations are based on the complete conversion of tristearin to hydrocarbons, propane, and the primary deoxygenation product while also assuming that CO_2 is not hydrogenated. Consequently, hydrogen consumption ranges from 1 to 3 wt%, which is in the same order of magnitude as hydrogen consumption in hydrocracking and approximately an order of magnitude higher than hydrogen consumption in conventional hydrotreating. This shows that HDO is a hydrogen-intensive process; however, it also implies that by a proper choice of catalyst and/or reaction conditions, significant savings can be achieved. In particular, close attention has to be paid to controlling the successive hydrogenation of the primary CO_2 or CO to methane. Hence, two main selective deoxygenation alternatives have been widely studied: deoxygenation in the presence of H_2 and deoxygenation in the absence of H_2 [25].

In addition, deoxygenation can also be achieved via cracking, often using zeolites as catalysts. The topic has attracted great attention as hydrogen is not needed. However, this route is highly unselective and produces a wide range of hydrocarbons and shorter oxygenates [6, 25], while the catalysts, typically zeolites, have to be frequently reactivated by burning off the coke deposits. Moreover, the wide molecular weight range of the hydrocarbon products, many of which fall outside of the targeted kerosene/diesel fuel range, is undesired. Therefore, this transformation path is outside of the scope of this chapter. Due to the very close relationship between the catalyst nature and the observed product distribution, specific details of deoxygenation reaction pathways will be presented in the succeeding text as part of a discussion of different catalytic systems.

FFAs, fatty acid esters, and specific TGs are generally used as model compounds to represent vegetable oils in deoxygenation studies. Based on an extensive literature review, Gosselink et al. [25] have concluded that while all of them are suitable model compounds for deoxygenation in the presence of hydrogen, FFAs and fatty acid esters are inappropriate model systems for deoxygenation in absence of hydrogen. Furthermore, in order to suppress undesired side reactions (e.g. cracking and coking), the use of hydrogen during deoxygenation has been deemed necessary [25].

11.4 Technologies

The industrial application of a process dedicated to the transformation of TGs into superior quality kerosene and diesel fuel components has been pioneered by the Finnish refining company Neste Oyj. The process is known as NEXBTL, and there are currently four installations, two at the refinery in Porvoo, Finland (210 kt/yr each); one in Rotterdam, The Netherlands (1000 kt/yr); and one in Singapore (1000 kt/yr). The first unit was built in Porvoo in 2007 and the last one (to date) in Rotterdam in 2011. Nowadays, a diesel fuel produced solely by the NEXBTL technology is marketed in Finland as "Neste MY Renewable Diesel." An alternative process, called Ecofining™, was developed by ENI in collaboration with Honeywell-UOP. Since 2013, three commercial units have been installed, one in Norco, USA (500 kt/yr currently being expanded to 900 kt/yr); one in Venice, Italy (375 kt/yr with a second unit in progress); and one in Paramount, USA (125 kt/yr with focus on green jet fuel) [30]. The Finnish company UPM started producing a renewable diesel, called BioVerno, from crude tall oil in its biorefinery in Lapeenranta, Finland (ca. 100 kt/yr) in 2015 [31]. Finally, the refining company Total announced in 2015 the construction of the first industrial version of the Vegan™ process developed by Axens with a renewable diesel production capacity of 500 kt/yr, start-up being expected in 2018 [32]. There are also other potential producers of renewable diesel, such as the Canadian SBI Bioenergy Inc., which has signed an agreement with Shell [33], or the Finnish St1 Nordic Oy, which announced it intends to start renewable diesel production by 2020 [34]. These industrial developments over the past decade demonstrate not only the potential of renewable diesel technologies but also their technical and economic viability.

Apart from the UPM hydrotreatment process that uses exclusively crude tall oil, other technologies rely on the use of a variety of TG-containing feedstocks that allows for more flexibility. Notably, there is a clear strategy on the part of all producers to reduce the carbon footprint of their renewable diesel technologies by relying mostly on waste-based feedstocks, such as waste cooking oils, residual oils, waste fats, etc. For instance, the feeds for the NEXBTL technology consisted of >60% wastes and residues as far back as 2014. Obviously, feed pretreatment and quality control play a crucial role. As the general scheme of industrial deoxygenation technologies is rather similar, the concept will be demonstrated here using the NEXBTL process as an example. After removing the undesired impurities, such as phosphorus as well as alkali and alkaline earth metals, the feedstock enters the hydrotreating reactor where deoxygenation takes place at 300–350 °C and 5 MPa of hydrogen pressure over a sulfided catalyst [35]. The liquid hydrocarbon product stream from the hydrotreating reactor then enters the second-stage reactor where hydroisomerization takes place at 300–350 °C and 4 MPa of hydrogen pressure to convert the *n*-alkanes formed in the deoxygenation stage into *iso*-alkanes with better fuel properties. Specifically, the cold flow properties are considerably improved. Also, by adjusting the reaction conditions in the second stage, mild hydrocracking instead of selective hydroisomerization may take place, thus producing renewable kerosene. Recently, Neste announced that it has started to commercially produce bio-propane, i.e. the by-product of TG deoxygenation, at its facility in Rotterdam, The Netherlands (40 kt/yr) [36].

In addition, aiming at reducing investment costs, coprocessing TG-based feedstocks with petroleum feeds in existing hydrotreating units has gained significant attention. A commercial technology, called H-BIO, was developed by Petrobras over a decade ago. However, despite having been industrially tested at six refineries, it appears not to be currently employed and FAMEs biodiesel is used as the biocomponent in diesel blends [37]. Another coprocessing technology using tall oil as the renewable feed has been installed in the Preem Gothenburg refinery using a catalytic system developed by Haldor-Topsoe [38]. Further plans to produce renewable diesel from crude tall oil have also been announced recently by SunPine in Pitea, Sweden [39].

11.5 Catalysts

The most promising deoxygenation catalysts can be classified into three groups, namely, (i) transition metals, such as Co, Ni, and Pt-group metals; (ii) sulfides of transition metals, mostly nickel and cobalt-promoted molybdenum sulfides; and (iii) transition metal carbides and phosphides, mainly of Ni, Mo, and W. The following sections will discuss the most salient features as well as the respective advantages and disadvantages of these catalysts. First, the sulfided catalysts currently at the heart of commercial technologies will be discussed, after which the two main challengers to these formulations – i.e. metallic catalysts and carbide/phosphide-based catalysts – will be addressed.

The crucial aspects of all catalysts include, besides their activity, their selectivity and the stability of their performance as a function of time on stream (TOS). In deoxygenation, the selectivity to undesired (typically cracking) products, is very low when using the three groups of catalysts mentioned earlier as they typically use nonacidic supports, such as alumina. Hence, the key selectivity issue is the relative extent of HDO and DCX /DCN reactions, which determines both the yield of organic liquids and, at the same time, the extent of hydrogen consumption. Theoretically, the lower the extent of HDO, the lower the hydrogen consumption. As mentioned earlier, this assumes that the subsequent hydrogenation of any CO or CO_2 evolved does not take place. The stability of the catalysts (which dictates the stability of their performance) is a critical parameter for their industrial viability. So far, sulfided (HDS) catalysts are used industrially even though the feedstocks are mostly sulfur-free. Consequently, in order to prevent their fast deactivation, an external source of sulfur has to be added, which is one of the key driving forces for the development of sulfur-free catalysts. Catalyst stability is also closely related to the quality of the feedstock, as some of the natural components of oils and fats have a detrimental impact on the activity of deoxygenation formulations. Thus, all of these issues will be addressed in the following text as part of the discussion of different deoxygenation catalysts.

11.5.1 Sulfided Catalysts

Sulfided (HDS) catalysts are the most widespread hydrogenation (hydrotreating) catalysts in the petroleum upgrading arena. They consist mostly of alumina-supported MoS_2 promoted either with NiS or CoS, albeit catalysts with both promoters are also used in some applications. Alternatively, NiS-promoted

WS_2 is also used, though mostly for hydrocracking applications. Considering the similarity of desulfurization and deoxygenation, it is not surprising that sulfided hydrotreating catalysts have attracted considerable attention as catalysts for the deoxygenation of TGs and have already been used in commercial processes [38].

From a reaction pathway point of view, the HDS catalysts appear to be nonspecific, i.e. they catalyze DCX, DCN, and HDO reactions, although HDO is often the prevailing reaction [18, 40]. However, it has been reported that HDO was practically the only pathway observed when using MoS_2 without any promoters, whereas DCX/DCN was the exclusive reaction pathway when NiS or CoS were employed [20]. This may indicate that the different phases in HDS catalysts favor different reaction pathways. It should also be noted that the activity of both MoS_2 and NiS catalysts was 3.5 and 22 times lower, respectively, than that of a NiMoS catalyst at 260–280 °C and 3.5 MPa [20], which is indicative of a synergy effect analogous to that reported in HDS. The absence of fatty alcohols and aldehydes in the products obtained over NiS indicated that DCX rather than DCN (via keto-enol tautomerization) was the preferred reaction path. Fatty esters were reported to be important intermediates over both NiMoS and MoS_2 catalysts with selectivities of 20–30% being observed at TG conversions of 30–90% [20]. A comparison between promoted (with Ni or Co as promoters) and unpromoted (Mo only) sulfided catalysts was performed by Brillouet et al. [41] using decanoic acid as a model compound. In line with previous reports, both Ni and Co were found to promote DCN and inhibit HDO [41]. It was suggested that an increase in the basicity of sulfur anions neighboring Co or Ni present in the sulfided promoted phase may be responsible for the enhancement of DCN [41]. The work of Brillouet et al. also showed that acids adsorb strongly on the surface, thus inhibiting the deoxygenation of other compounds [41]. The strong adsorption of acids may also explain their lower reactivity relative to esters during deoxygenation [18].

Experimental studies using unsupported sulfides (MoS_2, Ni_3S_2, and Ni-promoted MoS_2) as catalysts for the deoxygenation of ethyl decanoate in the gas phase (250 °C, 1.5 MPa) accompanied by density functional theory (DFT) calculations have confirmed the concurrence of HDO and DCX/DCN over promoted MoS_2, with MoS_2 and Ni_3S_2 favoring HDO and DCX/DCN, respectively [42]. In fact, the presence of the Ni_3S_2 phase in the promoted MoS_2 was deemed to be responsible for the DCX/DCN [42]. DFT calculations revealed the dehydrogenation pathway of aldehyde to alkenoyl and/or ketene intermediate to be favored on the Ni_3S_2 surface relative to MoS_2 surfaces [42]. The latter has been further supported by a recent study of Wagenhofer et al., who suggested that HDO takes place via an aldehyde intermediate on MoS_2 (whose strongly bound electron-poor sulfur did not allow α-hydrogen abstraction) and DCN occurs via the ketene intermediate on Ni-MoS_2 (whose labile electron-rich sulfur favors α-hydrogen abstraction and ketene formation) [43].

In contrast to many studies on supported and unsupported promoted MoS_2 catalysts [20, 42] (and references therein), Zhang et al. reported the conversion of waste cooking oil over an unsupported CoMoS catalyst (300/375 °C, 9 MPa) where DCX/DCN – not HDO – was the main reaction ($C_{18}/C_{17} = 0.6$ vs. 1.5–2.5) [44]. The authors suggested that for supported catalysts, the Lewis acidity of the supports may affect selectivity [44]. This contradicts another study focused on unsupported NiMoS [42], although differences in the reaction setup (batch liquid phase vs. flow gas phase) and conditions (375 °C, 9 MPa vs. 250 °C, 1.5 MPa) employed may account for the discrepancy.

As reported by Kubička [26], the preferred deoxygenation pathway could be completely altered just by varying reaction conditions; indeed, while HDO was the main path over a NiWS catalyst at 310 °C and 7 MPa (C_{18}/C_{17} = ca. 3), DCX prevailed at 280 °C and 0.7 MPa (C_{18}/C_{17} = ca. 0.3) [26]. Furthermore, the presence or absence of H_2S was shown to affect the distribution of deoxygenation products dramatically, DCX reactions being enhanced significantly when dimethyl disulfide (DMDS) was added to the TG feed (to generate H_2S and protect the catalyst from desulfurization and the associated loss of activity) [45]. Tellingly, during the deoxygenation of rapeseed oil over a CoMoS catalyst at 310 °C and 3.5 MPa, the C_{18}/C_{17} ratio decreased from ca. 3.5–4.0 when no DMDS was co-fed to about 1.0–1.5 when DMDS was co-fed, which was attributed to the increased catalyst acidity in presence of H_2S [45]. Finally, the support was also found to influence both the activity and selectivity of supported sulfided catalysts [29]. The use of an inert support (e.g. SiO_2) for the NiMoS active phase resulted in DCX dominating over HDO, whereas HDO was the main reaction pathway when other supports (such as Al_2O_3 or TiO_2) were employed [29]. This was particularly pronounced in the case of NiMoS/TiO_2, with C_{18}/C_{17} exceeding the value of 10 [29]. Apart from the active NiMoS cluster size decreasing based on the support nature in the order $TiO_2 > Al_2O_3 > SiO_2$, differences in the structure of this phase were also ascertained. Specifically, the population of octahedral Ni species in the oxide form of the catalyst (i.e. the precursor before the final activation via sulfidation) was significantly higher in the silica-supported catalyst than in the other two formulations [29]. Chen et al. also found a relationship between the acidity of the support and the preferred deoxygenation pathway on a supported NiMoS active phase. Due to a higher acidity, $NiMoO_4^-$ and MoO_3 sulfided phases dominated on SAPO-11 and DCX/DCN was the preferred deoxygenation route, while Mo^{4+} polymeric octahedral sulfided phases prevailed on Al_2O_3 and HDO was the favored deoxygenation pathway [46].

Kubička and Horáček have clearly demonstrated the loss of activity of a sulfided NiMo catalyst during rapeseed oil deoxygenation when H_2S was absent in the reactor [45], the yield of hydrocarbons dropping 30% over 144 hours TOS without added DMDS (as a source of H_2S) and only 5% over 250 hours when DMDS was constantly added. Moreover, the authors showed catalyst deactivation to be partly reversible, as the deoxygenation performance of the catalyst could be revived after introducing a feed containing DMDS for a short period. However, further cycles with and without DMDS clearly indicated that some of the changes to the catalyst caused in the absence of H_2S (DMDS) are permanent, so a minimum level of H_2S has to be maintained to ensure stable long-term catalyst performance [45].

The inhibition effects of water, hydrogen sulfide, and ammonia, i.e. plausible catalyst poisons present during deoxygenation, were studied for various oxygenates by Laurent and Delmon [47]. Even high concentrations of water were reported to have only a very minor inhibiting effect on the conversion of an ester and a ketone [47]. Hydrogen sulfide was found to promote conversion of carboxylic esters, while it suppressed the activity of NiMoS in the conversion of ketonic groups [47]. On the other hand, ammonia inhibited strongly the conversion of carboxylic esters, but did not affect the hydrogenation of ketonic groups [47]. Although only minor inhibition of deoxygenation reactions by water was observed [47], water was shown to deteriorate significantly the overall catalyst activity, which dropped to one-third of the initial level [48]. This was attributed mainly to partial oxidation of NiS to NiO by water and to a lesser extent to recrystallization of γ-alumina to hydrated boehmite resulting in a decrease in surface area [48]. The adverse effect of water was also observed by Wang et al. [49], who reported the formation of a MoO_3 phase during deoxygenation in the absence of a sulfiding agent due to the desulfidation of the catalyst resulting from the exchange of oxygen (from oxygenates) with the sulfur in the catalyst active phase.

Apart from coke deposition – which is a natural cause of deactivation in all transformations of carbon-containing feedstocks – and the aforementioned catalyst desulfidation and partial oxidation, trace elements inherently present in TG feedstocks (such as P, Na, K, Mg, Ca) can cause severe catalyst deactivation [45]. These elements are associated with phospholipids and deposit on the catalyst surface in the form of phosphates, causing pore blockage and deactivation [45]. The most severe deactivation effect is observed for waste-TG streams with an excess of P (relative to metals), which result in the formation of phosphoric acid (instead of phosphates) and in rapid oligomerization followed by coking [45]. Hence, these trace elements must be removed before the TG feedstocks can be deoxygenated.

Due to their robustness and proven reliability in HDS, sulfided catalysts are currently used commercially to deoxygenate lipid-based feedstocks into hydrocarbon fuels. However, it is necessary to introduce a source of sulfur into the process to keep the catalyst in its active form. Moreover, better selectivity control in terms of HDO vs. DCX/DCN – which necessitates more fundamental knowledge on catalyst structure–activity relationships – and minimized CO_x hydrogenation to improve the overall hydrogen economy are key challenges to be tackled.

11.5.2 Metallic Catalysts

Supported metal catalysts, including transition metals (such as Ni or Co) and noble metals (such as Pt, Pd, or Ru) are effective in the hydrogenation of a wide variety of functionalities including double and triple bonds as well as carbonyl groups. Moreover, the aforementioned metals have also been shown to be active in DCX, which makes them of interest for the upgrading of TGs or fatty acids. Two key lines of research have been followed – deoxygenation under H_2-rich atmospheres [28, 50, 51] and deoxygenation in H_2-poor conditions (or even in absence of H_2) [10, 50, 52, 53]. From the reaction stoichiometry point of view, DCX of saturated fatty acids does not consume any hydrogen to yield saturated hydrocarbons, whereas conversion of a saturated TG would require 3 mol of H_2 to produce saturated hydrocarbons (see Figure 11.1). Thus, many efforts have focused on leveraging this fact. Kubičková et al. [50] upgraded stearic acid and ethyl stearate using 5 wt% Pd/C at 300–360 °C under He, H_2, and 5 vol% H_2 in Ar to obtain *n*-heptadecane along with some heptadecenes. This suggest that both DCX and DCN are operative under the reaction conditions. The highest turnover frequency (TOF) was obtained when using 5 vol% H_2 in Ar [50]. In addition to C_{17} olefins, C_{17} aromatics were also observed in the absence of hydrogen [50], indicating that H_2 was important to prevent dehydrogenation reactions and, hence, protect the catalyst from deactivation. The formation of unsaturated products was higher when ethyl stearate was used [50] as could be expected from stoichiometric considerations, ethylene and ethane also being detected in the gaseous products.

Snåre et al. [53], screened various supported-metal catalysts for activity in the deoxygenation of stearic acid under He, carbon-supported formulations significantly outperforming their alumina-supported analogues [53]. High conversion was also achieved on Ru/MgO, which yielded selectively a C_{35} symmetrical ketone, i.e. a ketonization product of stearic acid, likely due to the basic character of the MgO support although other carriers have also shown high selectivity toward this product at low conversions. On the other hand, the HDO activity of all catalysts was negligible. Due to the inert atmosphere, formation of oligomerization products was pronounced over most catalysts but particularly over Ni-containing formulations (>20% selectivity), which also exhibited substantial cracking activity (ca. 20% selectivity) [53]. The various reaction pathways proposed are summarized in Figure 11.3.

Crocker and coworkers tested carbon-supported Ni, Pt, and Pd catalysts for activity in the conversion of TGs at 350 °C under a N_2 atmosphere [52]. These authors confirmed that oxygen was eliminated as CO and CO_2 [52]. It was

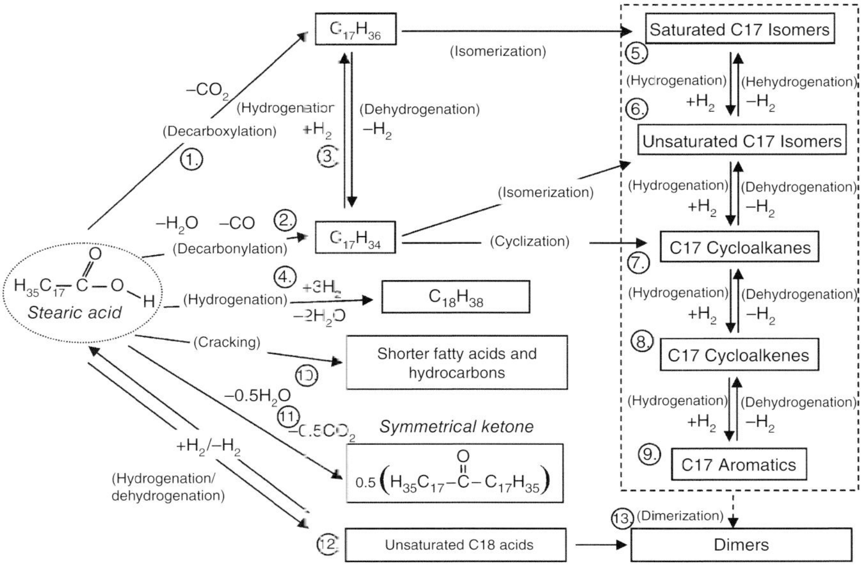

Figure 11.3 Scheme of stearic acid deoxygenation over supported metallic catalysts under inert atmosphere. Source: Reprinted with permission from Snåre et al. 2006 [53]. Copyright 2006, American Chemical Society.

explained that an initial β-elimination step was followed by the DCX of the released fatty acid, C—C bonds scission affording either a C_{17} hydrocarbon (if scission occurred between the ester carbonyl carbon and the α carbon of the hydrocarbon chain) or a C_{15} hydrocarbon and ethylene (if scission took place between the β and γ carbon atoms) [52]. Increasing the degree of unsaturation of the lipids led to increased cracking, especially over the Ni catalyst [52]. Carbon-supported Pt and Pd were less active both in deoxygenation and in cracking than Ni/C [52]. Further investigation of Ni-based catalysts showed that a Ni–Al layered double hydroxide (LDH) catalyst was particularly suited for regeneration as the regenerated catalyst outperformed the fresh formulation, which was attributed to the formation of strong basic sites during regenerative calcination. Also, hydrogen was convincingly shown to curb the deposition of carbonaceous species on the catalyst surface [54]. Moreover, promotion of 20 wt% Ni/Al_2O_3 with 5 wt% Cu resulted in suppressed cracking activity, the formation of diesel-like hydrocarbons being significant especially when upgrading yellow grease [55]. Besides Cu, the use of Sn as promoter was also shown to improve the selectivity of Ni/alumina catalysts to diesel-like hydrocarbons during the conversion of tristearin at 350 °C [56].

Zhao and coworkers studied the transformation of palmitic acid over various Ni-containing catalysts in a batch reactor operated at 260 °C and 12 bar both in the presence and absence of H_2 [57]. They found that reducible supports with oxygen vacancies – such as ZrO_2 – adsorbed palmitic acid via the carboxylic group. This resulted in oxygen elimination and the formation of a ketene, which is hydrogenated to hexadecanal on Ni sites (in the presence of H_2) or reacts with another adsorbed palmitate to yield a symmetrical ketone (in the absence of H_2) [57]. In parallel, palmitic acid is also reduced directly to hexadecanal that is then decarbonylated, both of these reactions occurring on Ni sites [57]. Ni/ZrO_2 exhibited a higher rate of deoxygenation of palmitic acid than Ni/SiO_2 and Ni/Al_2O_3 due to the active contribution of the support to the deoxygenation reaction [57]. Similarly, the rate is further enhanced over zeolite-supported Ni as the dehydration of hexadecanol (which is in equilibrium with hexadecanal) is faster than the other reaction pathways [57]. Consequently, hexadecane is the main reaction product, which is not the case for Ni on supports that do not possess significant Brønsted acidity [57]. This pathway was corroborated by modifying ZrO_2 with SiO_2, which led to the introduction of Brønsted acid sites – accompanied also by a significant surface area increase (from 72 to 193 m^2/g) – and accelerated the deoxygenation of stearic acid by HDO rather than DCN [58]. Likewise, high deoxygenation activity was achieved over Ni supported on hierarchical nanosized MFI exhibiting both excellent Ni dispersion and large number of accessible acid sites [59].

Ni and Co supported on Al- and Si-SBA-15 having large specific surface area (and mild acidity in the case of Al-SBA-15) have also been tested for activity in the deoxygenation of methyl esters in a flow reactor at 300–340 °C and 3 MPa H_2 [51]. In line with the results from Lercher and coworkers [58], Ni/Si-SBA-15 showed lower activity than Ni/Al-SBA-15 and a negligible selectivity to HDO, whereas HDO was the main reaction pathway over Ni/Al-SBA-15 (about 60% of $C_{17} + C_{18}$ hydrocarbons at 300 °C) [51]. Interestingly, Si-SBA-15-supported Co was more active than Ni and showed significant selectivity to HDO (about

20% of C_{17} + C_{18} hydrocarbons at 300 °C), which increased to 70–80% over Co/Al-SBA-15 [51]. Both Al-SBA-15-supported catalysts exhibited low cracking activity, high yields of diesel-range hydrocarbons (>90 wt%) being obtained during the whole six hour TOS period investigated [51]. At 340 °C, the initial yield of desired isomers exceeded 25 wt%, albeit it dropped to about 10 wt% after six hours TOS [51]. Surprisingly, Co/Al-SBA-15 yielded more isomers than Ni/Al-SBA-15 at 300 °C. Nonetheless, the yield decreased rapidly from 15 to 6 wt%, still twice the yield obtained over Ni/Al-SBA-15 that remained constant at about 3 wt% [51].

The isomerization activity of deoxygenation catalysts is of high interest as it would allow converting TGs directly into the most desired diesel-range fuel components, i.e. *iso*-alkanes. Commercially, this is achieved in a two-stage setup, with the *n*-alkanes stemming from the deoxygenation of lipids being upgraded in a second reactor over a commercial isomerization/mild hydrocracking catalyst. In addition to mesoporous supports such as Al-SBA-15, attention has been mainly directed at SAPO materials, such as SAPO-11 and SAPO-31, which exhibit mild acidity thus limiting excessive cracking [60–63]. Kikhtyanin et al. [60] have shown that sunflower oil could be converted exclusively to hydrocarbons in the diesel fuel range over Pd/SAPO-31 at 320–350 °C and 2 MPa H_2 with a high initial selectivity to C_{17} and C_{18} *iso*-alkanes (*iso*–/*n*-alkanes >10). This ratio, however, dropped dramatically within 24 hours TOS to virtually zero indicating severe deactivation, which was attributed to a drop in Pd dispersion (from 50% to 11%) [60]. Full soybean oil conversion and a high isomerization selectivity (*iso*-alkanes representing 60–85% of all produced alkanes) was also observed over Pt/SAPO-11 and Pt/ZSM-22 at 357 °C, 4 MPa H_2 and three hours TOS [63]. While SAPO-11 favored isomerization (<4% selectivity to liquid cracking products), ZSM-22 promoted cracking reactions (50–60% selectivity to liquid cracking products) [63]. Unfortunately, the stability of the catalysts was not reported in this contribution. In the deoxygenation of palm oil, a good liquid product yield – ca. 60% (the theoretical maximum being 80–85%) – and good isomerization selectivity (ca. 50%) were obtained in a 35 hour TOS experiment using 7 wt% Ni/SAPO-11 at 360 °C and 4 MPa H_2 [61].

When metallic catalysts are employed, it is obvious that the product selectivity can be fine-tuned by the choice of (i) the active metal, (ii) the support, and (iii) the reaction conditions in general and the gas atmosphere in particular, all of which determine the preferred reaction pathway. However, the choice of support has uniquely important consequences. Indeed, partially reducible supports may enhance DCN/DCX pathways, while non-reducible supports with Brønsted acid sites (e.g. molecular sieves) may on the one hand shift the preferred pathway toward HDO and on the other hand promote both desirable isomerization and undesirable cracking reactions. The long-term TOS stability of these catalysts remains a challenge to be addressed.

11.5.3 Metal Carbide, Nitride, and Phosphide Catalysts

Metal carbides, nitrides, and phosphides are an emerging class of robust hydrogenation catalysts that offer several advantages over the other types of deoxygenation catalysts. Most importantly, they are much more affordable than noble metal catalysts and do not require the sulfur-containing additives that are essential when sulfided catalysts are employed. As with supported-metal catalysts, deoxygenation over carbide-, nitride-, and phosphide-based formulations has been investigated under both hydrogen-rich and hydrogen-poor conditions.

Molybdenum carbide supported on alumina was investigated as a catalyst in the deoxygenation of sunflower oil (633 K, 5 MPa) by Sousa et al. [64]. Conversion of TGs into alkanes was described as a two-step process consisting of thermal cracking of TGs yielding FFAs followed by hydrogenation of the double bonds and of the carboxylic group of the FFAs [64]. Both CO and CO_2 were absent among the reaction products indicating that DCX and/or DCN reactions did not take place and that HDO was the exclusive deoxygenation reaction pathway [64]. Nonetheless, the occurrence of DCX and/or DCN reactions cannot be completely excluded as methane – which could be formed by CO_x hydrogenation (including that of the CO_x formed through the thermal cracking of the FFAs) – was found in the reaction products [64]. The Mo_2C/Al_2O_3 catalyst was found to provide a more stable performance in terms of both activity and product distribution than sulfided formulations [64]. Moreover, the activity of molybdenum carbide was claimed to be higher than that of molybdenum sulfide [64], which also operates via HDO [20].

Therefore, supported Mo_2C catalyst offer several advantages, including (i) low to negligible formation of CO and CO_2, (ii) stable operation without the necessity to supply a source of sulfur (as is the case for sulfided catalysts), and (iii) direct HDO of FFAs, which obviates the need to convert acids into esters prior to their deoxygenation [64].

Unsupported and active charcoal-supported Mo_2C were some of the first catalysts tested for activity in the deoxygenation of vegetable oils and related model compounds (e.g. stearic acid, methyl stearate) over carbides at 280–300 °C and 1–2 MPa hydrogen pressure [65]. HDO products such as *n*-octadecane were the predominant products obtained over supported Mo_2C with yields around 90% and 75% when esters and stearic acid were used, respectively. At the same time, DCX/DCN products were formed in ca. 4% yield from esters and ca. 13% yield from stearic acid [65]. The performance of the unsupported Mo_2C was inferior to that of the supported Mo_2C [65]. The activated charcoal-supported Mo_2C favored the HDO of esters, aldehydes, and alcohols, whereas Pd nanoparticles supported on $BaSO_4$ favored the DCN pathway producing hydrocarbons with one carbon atom less than in the original lipid [65]. This was also observed for the deoxygenation of hexadecanol [65], which could be explained by invoking the equilibrium between the alcohol and the aldehyde [25]. In the case of stearic acid, the HDO pathway was favored by both catalytic systems. Thus, it was suggested that aldehydes and olefins (rather than carboxylic acids) were the intermediates in the conversion of esters [65]. In contrast to methyl stearate and methyl palmitate, full conversion was not reached when deoxygenating vegetable oils under the same reaction conditions (conversions being 85–95%) [65].

The performance of tungsten and molybdenum carbide supported on carbon nanofibers (CNFs) in the deoxygenation of oleic acid was studied at 350 °C and 5 MPa H_2 pressure by Hollak et al. [66]. These authors found Mo_2C/CNF to be more active and stable than W_2C/CNF. Mo_2C/CNF was highly selective to *n*-octadecane, steric acid being the main reaction intermediate. On the other hand, W_2C/CNF produced olefins in significant yields (up to 50% before being hydrogenated) [66]. This was explained by higher hydrogenation rates being achieved over Mo_2C/CNF, which was also suggested to be the reason behind the superior stability of Mo_2C/CNF in terms of catalytic activity [66]. HDO was found to be the main reaction pathway over both catalysts, DCX products being produced in 7% and 15% molar yield at complete deoxygenation over Mo_2C and W_2C, respectively [66]. This difference was attributed to the higher sensitivity of W_2C/CNF toward oxidation [66].

It was also established that the particle size of the supported carbides (Mo_2C, W_2C) is a decisive factor determining their deoxygenation activity and stability [67]. It was shown that larger carbide particles were more resistant against full oxidation to crystalline metal oxides, thus providing a more stable catalytic phase [67]. At the same time, the TOF over Mo_2C/CNF increased by an order of magnitude when the carbide particle size was increased from 2 to 10 nm [67]. The increase in particle size also resulted in a change in the preferred reaction pathway. Indeed, while the *n*-octadecane produced by HDO was practically the sole hydrocarbon product over 2 nm Mo_2C/CNF, almost equal amounts of C_{17} and C_{18} hydrocarbons were formed over 10 nm Mo_2C/CNF at 300 °C and 3 MPa H_2 [67]. At the same time, the formation of the C_{36} esterification product was greatly diminished. The main oxygenated intermediates included octadecanol and octadecanal and the formation of C_{17} hydrocarbons was assigned to the DCN of the aldehyde [67].

The experiments with octadecanol and octadecanal clearly demonstrated that under a N_2 atmosphere octadecanal was selectively converted to unsaturated C_{17} hydrocarbons over Mo_2C/CNF, whereas stearic acid did not react at all [67]. Octadecanol yielded both C_{17} and C_{18} hydrocarbons (with preference to C_{18} over 2 nm Mo_2C and to C_{17} over 10 nm Mo_2C) indicating that the dehydrogenation of the alcohol to the aldehyde was followed by DCN [67]. In contrast, W_2C/CNF invariably yielded C_{18} hydrocarbons in very large excess likely due to the oxophilic nature of tungsten carbide, the WO_x domains formed on the surface of W_2C/CNF catalyzing alcohol dehydration [67].

Mo_2C supported on different carbon supports – such as reduced graphene oxide (RGO), glassy spherical carbon, activated carbon, and mesoporous carbon was studied by Kim et al. [68]. When using Mo_2C/RGO nanocomposites prepared by the supercritical alcohol route, 30–60% higher hydrocarbon yields were obtained in the HDO of oleic acid at 350 °C, 4.5 MPa H_2 and 2 h^{-1} liquid hourly space velocity (LHSV) [68]. The higher efficiency of the RGO-supported Mo_2C was attributed to a uniform distribution of Mo_2C particles and the large pore size of the support allowing the efficient transport of reactants to the active sites [68]. A decrease in soybean oil conversion of 13% was observed over Mo_2C/RGO during six hours on stream, which was significantly less than the 42% drop in conversion observed over a commercial $CoMoS/Al_2O_3$ catalyst [68]. Nevertheless, it seems that the authors did not add any sulfur to the feed, which would have a detrimental effect on the performance of the $CoMoS/Al_2O_3$ catalyst.

Theoretical DFT studies of the preferred reaction pathways have been performed using short chain (e.g. propionic, butyric) acids as model compounds [69, 70]. The orthorhombic Mo_2C (100) has a surface terminated in either Mo or C. The Mo-terminated surface was shown to favor the HDO path as a result of the low barrier for oxygenates, while the C-terminated surface was preferable for the DCX of propanoic acid due to low barrier for O–H scission [69]. Butyric acid deoxygenation over an hexagonal Mo_2C (101) surface was reported to proceed through the dissociative adsorption of the acid followed by the hydrogenation of the adsorbed acyl to adsorbed butanal [70]. The direct decomposition of acyl to produce propane, CO, and water was found to be less energetically favorable and hence not competitive [70]. Butanal was then hydrogenated to butanol via butoxyl – which was the preferred adsorbed species – followed by butanol dissociation to butyl and hydroxyl species and final hydrogenation of these species to form butane and water [70]. However, it should be noted that neither the structure nor the chain length of the model compounds employed provide a realistic representation of TGs, and steric effects might affect the predicted outcomes.

A NiMo carbide and nitride supported over ZSM-5, i.e. an acidic support, were also used in the deoxygenation of soybean oil at 360–450 °C and 4.5 MPa by Wang et al. [71]. At 450 °C and a LHSV of 1.5 h^{-1}, full conversion of the soybean oil was achieved, but differences in the gas phase products were noticed. The nitride exhibited higher selectivity to CO, CO_2, and CH_4 than the carbide catalyst, indicating not only a higher DCX/DCN activity but also a higher methanation activity [71]. Nonetheless, significant amounts of carboxylic acids were observed in the product mixtures recovered after three and eight days on stream at 360 and 450 °C, respectively, indicating severe deactivation of the NiMo nitride catalyst [71]. The latter was explained invoking strong adsorption of water. In general, the NiMo carbide was found to be more selective to higher hydrocarbons. Based on a comparison between a Mo-carbide/ZSM-5 catalyst (10 wt% loading) and a 3 NiMo-carbide/ZSM-5 formulation (the same metal loading and Ni/Mo ratio = 0.5, 1, and 1.5), the lack of Ni was claimed to be responsible for insufficient hydrogen activation and for the high extent of polymerization reactions eventually leading to reactor plugging [71]. In addition, the yield of organic liquid products decreased as the Ni/Mo ratio increased [71], plausibly due to the hydrogenolytic activity of Ni.

Nguyen et al. [72] reported that – in contrast to 5 wt% Ni/H-Y and 5 wt% Ni/SiO_2 – an unsupported Mo_2N/MoO_2 catalyst (nitride/oxide ratio = 1.2) showed noteworthy resistance to deactivation during the deoxygenation of crude Chlorella algal oil at 300 °C and 3 MPa H_2. However, the conversion over the nitride catalyst was only 14% after six hours of reaction, compared with the 35 and 12% values reached over Ni/H-Y and Ni/SiO_2, respectively [72]. In line with the findings of Wang et al. [71], only low yields (<10%) of hydrocarbons were obtained and fatty acids represented the main reaction products [72].

In addition to carbides and nitrides [25, 64–68, 71–73], transition metal phosphides have recently gained significant attention as deoxygenation catalysts [74–81]. Peroni et al. convincingly showed that HDO was the key reaction pathway during the upgrading of palmitic acid over WP, while DCN of the intermediate aldehyde was a significant reaction route over Ni_2P and MoP catalysts at 4 MPa

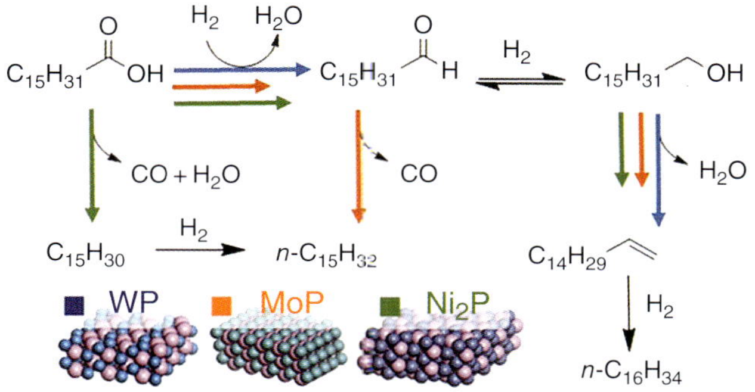

Figure 11.4 Deoxygenation of palmitic acid over different phosphides. Source: Reprinted with permission from Peroni et al. 2017 [77]. Copyright 2017, American Chemical Society.

and 180–300 °C [77] as depicted in Figure 11.4. Furthermore, the dehydration of the intermediate alcohol was correlated with the Lewis acidity of the phosphide catalysts and only Ni_2P catalyzed the dehydration to the ketene followed by the DCN of the latter [77]. However, weak Brønsted acid sites were not strong enough to dehydrate the intermediate alcohol [77]. The activation energies of palmitic acid deoxygenation spanned from 57 kJ/mol for MoP to 129 kJ/mol for WP [77]. The activity could be increased by addition of citric acid, which limited the growth of oxide precursors during the synthesis step and of phosphide crystals during calcination. However, citric acid addition also resulted in partial coverage of the metal surface by residual carbon [77].

Despite the obvious advantage offered by WP in terms of the highest retention of renewable carbon in the liquid deoxygenation products, Ni_2P catalysts have attracted most attention. Indeed, conversion of both fatty acids and their methyl esters has been extensively studied over zeolite- and mesoporous molecular sieve-supported Ni_2P [74, 75, 79, 80], as well as over carbon- and silica-supported Ni_2P [76, 78, 81].

Ni_2P supported on acidic H-ZSM-22 [75] was compared with unsupported $Ni_{12}P_5$ and unmodified H-ZSM-22 (Si/Al = 70) in the solvent-free deoxygenation of palmitic acid at ambient pressure at 310–360 °C under a flow of 5% H_2 in Ar. Full conversion was reached over all catalysts besides the unsupported Ni_xP, while the yield of liquid hydrocarbons increased with the increasing supported Ni_xP phase content from 14% to 46% for 5 wt% Ni_xP/H-ZSM-22 and 60 wt% Ni_xP/H-ZSM-22, respectively [75]. These yields are, however, significantly lower than the theoretical maximum of 83% (assuming only DCX/DCN take place). This is a consequence of gaseous products formation due to cracking reactions [75]. Nonetheless, as the Ni_xP loading increased, less cracking was observed due to the corresponding decrease in catalyst Lewis acidity [75]. The $Ni_{12}P_5$ phosphide phase was suggested to be the most active in DCN and DCX reactions (the optimum crystal size being 20–24 nm), while the acid sites of the support were active in isomerization and cracking reactions [75]. Albeit pentadecane was reported to be the main deoxygenation product over Ni_xP/H-ZSM-22 [75], over Ni_xP/MCM-41 both *n*-hexadecane (a HDO product) and *n*-pentadecane (a DCX/DCN product) were observed during the deoxygenation of methyl palmitate at 3 MPa of H_2 pressure and 290–350 °C [74]. However, this could be attributed to the difference in reaction pressure (0.1 MPa, 5% H_2 in Ar vs. 3 MPa H_2). Indeed, an increase in H_2 pressure has been reported to enhance HDO selectivity at the expense of DCX/DCN selectivity, the $C_{18}/(C_{18}+C_{17})$ ratio increasing from 0.2 to 0.6 upon increasing the pressure from 0.3 to 4 MPa over 15% Ni_xP/SBA-15 at 250 °C [79]. In contrast, the aforementioned ratio did not depend on H_2 pressure, and DCX/DCN was the preferred route over 15% Ni/SBA-15, which afforded deoxygenation product mixtures comprising >95% DCX/DCN products [79].

Interestingly, during the deoxygenation of methyl palmitate, the selectivity to HDO and DCX/DCN was found to be a function of Ni_xP loading on MCM-41, the nC_{16}/nC_{15} ratio decreasing from 60/35 to 35/60 when the Ni_xP loading was increased from 5 to 20 wt% [74]. At the same time, irrespective of the phosphide loading, all catalysts exhibited virtually the same TOF (0.43–0.44 s^{-1}) [74], which suggests that the same sites are active in both HDO and DCX/DCN reactions and/or that the intrinsic activity of different sites is very similar. In addition, at short contact times (3.3 minutes) and low conversions, DCX/DCN prevailed significantly (n-C_{15}/n-C_{16} ca. 6), while at longer contact times (20 minutes), the excess of the DCX/DCN products was only 1.4 [74]. In addition to the hydrogenolysis of the methyl ester yielding an acid and methane, direct formation of aldehyde and methanol was proposed as well [74].

By varying the Ni/P ratio in the range 1.25–3.33, it was established that increasing the concentration of P resulted in a decrease in the conversion of methyl palmitate (from 95% to 60% at 290 °C) and lower selectivity to HDO [74]. Similarly, a significantly lower activity was reported for SBA-15-supported Ni_3P than for Ni_xP/SBA-15 with Ni/P = 0.5–2 during the deoxygenation of methyl oleate at 3 MPa and 250–340 °C [80]. The phosphorus-deficient phase (Ni_3P) also showed higher selectivity to DCX/DCN than to HDO being thus more similar to Ni/SBA-15 than to other Ni_xP/SBA-15 catalysts [80]. The best performance, in terms of conversion and selectivity to HDO products, was provided by catalysts rich in Ni_2P and $Ni_{12}P_5$ phases, whereas an excess of P (Ni/P = 0.5) had a detrimental effect on the catalyst activity due to the formation of large Ni_xP crystallites [80].

No cracking activity was reported over Ni_2P/SBA-15 (all Si-SBA-15) [79] and over Ni_xP/MCM-41 (Al content not reported) [74], which might be due to the absence of acid sites in the supports. A comparison with Ni/SBA-15 demonstrates that the cracking/hydrocracking activity of supported Ni_xP is negligible, which allows for high renewable carbon retention in the liquid products, whereas supported Ni (even on a neutral support) promotes cracking significantly, particularly at higher reaction temperatures [79]. This is corroborated by the long-term testing of Ni_2P (25 wt%) supported on SiO_2 and HY under hydrocracking conditions (370 °C, 3 MPa) [81]. Indeed, the liquid hydrocarbon yields reached 78–82% over both catalysts, but virtually no cracking took place over Ni_2P/SiO_2 (gasoline fraction <1%), while as much as 20 wt% of light hydrocarbon (HC) liquids (boiling below 150 °C) were produced over Ni_2P/HY [81]. However, the extent of the cracking reactions decreased with increasing

TOS from the initial 20 wt% (20 hours TOS) to about 5 wt% (150 hours TOS), while the soybean oil was completely deoxygenated [81]. The decline in the hydrocracking activity was attributed to the adverse effect of the water generated during deoxygenation on catalyst acidity, specifically through the conversion of Lewis acid sites (hydroxylaluminum ions) into neutral $Al(OH)_3$, which blocks the access of reagents to Brønsted acid sites [81].

Silica-supported Ni_2P catalysts having a crystallite size in the range 3–12 nm were prepared using different SiO_2 supports including SBA-15 and MCM-41 and tested for activity in the deoxygenation of methyl laureate at 2 MPa H_2 and 200–340 °C [82]. The TOF was found to be directly proportional to the Ni_2P crystallite size, increasing from 0.064 to 0.151 s^{-1} for a crystallite size of 3 and 12 nm, respectively [82]. The lower TOF values over the smaller Ni_2P crystallites were attributed to a stronger interaction between Ni and P in these crystallites [82]. In contrast to the results reported for Ni_xP/SBA-15 catalysts, DCX/DCN was always preferred over HDO reactions with (DCX + DCN)/HDO in the range 2–4 [82], likely due to the lower H_2 pressure affecting the HDO/(DCX + DCN) ratio [79] and the higher temperature promoting DCX/DCN at the expense of HDO reactions [82]. Based on previous reports [79], an HDO/(DCX + DCN) ratio equal to unity would be expected in the deoxygenation of methyl oleate at 250 °C, so a ratio of about 2 at 300 °C [82] seems to be in line with these results. In contrast with previous work demonstrating that Lewis rather than Brønsted acidity affects the deoxygenation selectivity [77], Brønsted acidity originating from a P surplus has been linked to changes in selectivity [82]. Indeed, the acid Brønsted acid sites were associated with both the DCN and the HDO pathways (involving alcohol dehydration) [82], these acid sites also being deemed responsible for the esterification reaction taking place and resulting in the formation of lauryl laureate [82]. Similar observations were also made for sulfided catalysts, for which the increased acidity due to presence of H_2S (HS^-) promoted DCX/DCN at the expense of HDO reactions and the formation of esters through the reaction between fatty acids and fatty alcohols [45].

As with metallic catalysts, carbides, nitrides, and phosphides of transition metals provide some flexibility to adjust the selectivity to HDO and DCX/DCN pathways, thus allowing some degree of control over hydrogen consumption. They can be operated at a similar range of reaction conditions, which makes them an attractive alternative to noble metal-based catalysts due to their lower cost. Moreover, in contrast to metallic catalysts that preferentially yield DCX/DCN products, HDO can be the main reaction pathway over carbides, nitrides, and phosphides, which results in a higher retention of renewable carbon in the liquid organic products. As with the other catalytic systems discussed, the long-term stability of their performance remains a major challenge to be overcome.

11.6 Conclusions and Outlook

Several advances in the deoxygenation of lipid-based feedstocks has enabled the commercialization of the resulting technology by several companies, and new investments in additional industrial units have been announced. The recent focus on more sustainable feedstocks – such as waste lipids or algal oils – increases the need not only for improving the robustness of the catalysts but also for a meticulous pretreatment of the feedstocks to eliminate catalyst poisons, including alkali and alkaline earth metals and phosphorus. The availability of (sustainable) lipid-based feedstock is one of the factors limiting the widespread use of deoxygenation technology, other factors including the selectivity and long-term stability of deoxygenation catalysts.

Significant progress in understanding the underlying chemistry has been achieved recently. HDO, DCN, and DCX have been identified as the main deoxygenation reactions taking place. The intrinsic selectivity (toward specific reaction paths) of different active phases has been revealed along with the parameters that affect it, the nature and properties of the support, promoters, reaction gas, or reaction conditions having been described. In general, catalysts based on molybdenum sulfides and carbides favor HDO, while DCX and DCN are the preferred reaction pathways over nickel- and noble metal-based catalysts.

Although sulfided catalysts necessitate addition of H_2S (or a source thereof) to maintain their activity, they represent the only deoxygenation catalysts used in commercial deoxygenation technologies. This is due to their robustness and the vast experience of industry with these catalysts. Nonetheless, from both a technical and an environmental point of view, sulfur-free catalysts are highly desirable. Consequently, transition metal catalysts as well as their carbides, nitrides, and phosphides have attracted significant attention as promising alternatives.

The understanding of the intrinsic activity and selectivity of deoxygenation catalysts provides an opportunity to fine-tune their selectivity and to control hydrogen consumption. However, more effort is needed to develop catalysts with a controlled distribution of the desired active sites. Additionally, the stability of these active sites has to be further improved so they can be rendered tolerant to the oxygen-containing inorganic compounds produced, namely, water, CO, and CO_2. The use of promoters and of suitable supports with optimized properties (particularly textural and acid–base properties) is a research area with significant potential for improvement.

Finally, in line with the principles of green chemistry, consolidation of the two stages currently employed to upgrade lipids – i.e. deoxygenation followed by either isomerization or mild hydrocracking to afford high quality diesel fuel and kerosene, respectively – into a single-stage process should be attempted. Despite some promising results, catalysts in general (and isomerization formulations in particular) still suffer from fast deactivation that results in a rapid deterioration of their performance. Thus, the bifunctional catalysts needed to address this issue must be improved further.

References

1 Eur-Lex. Renewable energy directive. http://eur-lex.europa.eu/legal-content/EN/TXT/?uri=CELEX:52016PC0767R%2801%29 (accessed 10 April 2018).

2 Dawes, G.J.S., Scott, E.L., Le Notre, J. et al. (2015). Deoxygenation of biobased molecules by decarboxylation and decarbonylation – a review on the role of heterogeneous, homogeneous and bio-catalysis. *Green Chem.* 17 (6): 3231–3250.

3 Hermida, L., Abdullah, A.Z., and Mohamed, A.R. (2015). Deoxygenation of fatty acid to produce diesel-like hydrocarbons: a review of process conditions, reaction kinetics and mechanism. *Renewable Sustainable Energy Rev.* 42: 1223–1233.

4 James, O.O., Maity, S., Mesubi, M.A. et al. (2012). A review on conversion of triglycerides to on-specification diesel fuels without additional inputs. *Int. J. Energy Res.* 36 (6): 691–702.

5 Kordulis, C., Bourikas, K., Gousi, M. et al. (2016). Development of nickel based catalysts for the transformation of natural triglycerides and related compounds into green diesel: a critical review. *Appl. Catal., B* 181: 156–196.

6 Kubičková, I. and Kubička, D. (2010). Utilization of triglycerides and related feedstocks for production of clean hydrocarbon fuels and petrochemicals: a review. *Waste Biomass Valorization* 1 (3): 293–308.

7 Pattanaik, B.P. and Misra, R.D. (2017). Effect of reaction pathway and operating parameters on the deoxygenation of vegetable oils to produce diesel range hydrocarbon fuels: a review. *Renewable Sustainable Energy Rev.* 73: 545–557.

8 Rogers, K.A. and Zheng, Y. (2016). Selective deoxygenation of biomass-derived bio-oils within hydrogen-modest environments: a review and new insights. *ChemSusChem* 9 (14): 1750–1772.

9 Zhao, C., Bruck, T., and Lercher, J.A. (2013). Catalytic deoxygenation of microalgae oil to green hydrocarbons. *Green Chem.* 15 (7): 1720–1739.

10 Santillan-Jimenez, E. and Crocker, M. (2012). Catalytic deoxygenation of fatty acids and their derivatives to hydrocarbon fuels via decarboxylation/decarbonylation. *J. Chem. Technol. Biotechnol.* 87 (8): 1041–1050.

11 Woodgate, S.L. and van der Veen, J.T. (2014). Fats and oils – animal based. In: *Food Processing. Principles and Applications* (eds. S. Clark, S. Jung and B. Lamsal), 481–500. Wiley.

12 Global production of vegetable oils from 2000/01 to 2017/18. https://www.statista.com/statistics/263978/global-vegetable-oil-production-since-2000-2001/ (accessed 10 April 2018).

13 Sheehan, J., Dunahay, T., Benemann, J., and Roessler, P. (1998). A Look Back at the U.S. Department of Energy's Aquatic Species Program: Biodiesel from Algae. *Contract No.: NREL/TP-580-24190.* National Renewable Energy Laboratory.

14 Johnston, M., Foley, J., Holloway, T. et al. (2009). Resetting global expectations from agricultural biofuels. *Environ. Res. Lett.* 4: 1–9.

15 da Rocha Filho, G.N., Brodzki, D., and Djéga-Mariadassou, G. (1993). Formation of alkanes, alkylcycloalkanes and alkylbenzenes during the catalytic hydrocracking of vegetable oils. *Fuel* 72 (4): 543–549.

16 Ma, F. and Hanna, M.A. (1999). Biodiesel production: a review. *Bioresour. Technol.* 70 (1): 1–15.

17 EU (2009). Directive 2009/28/EC of the European Parliament and of the Council on the promotion of the use of energy from renewable sources.

18 Laurent, E. and Delmon, B. (1994). Study of the hydrodeoxygenation of carbonyl, carboxylic and guaiacyl groups over sulfided CoMo/γ-Al_2O_3 and NiMo/γ-Al_2O_3 catalysts: I. Catalytic reaction schemes. *Appl. Catal., A* 109 (1): 77–96.
19 Turek, T., Trimm, D.L., and Cant, N.W. (1994). The catalytic hydrogenolysis of esters to alcohols. *Catal. Rev. Sci. Eng.* 36 (4): 645–683.
20 Kubička, D. and Kaluža, L. (2010). Deoxygenation of vegetable oils over sulfided Ni, Mo and NiMo catalysts. *Appl. Catal., A* 372 (2): 199–208.
21 Gousi, M., Andriopoulou, C., Bourikas, K. et al. (2017). Green diesel production over nickel-alumina co-precipitated catalysts. *Appl. Catal., A* 536: 45–56.
22 Robota, H.J., Alger, J.C., and Shafer, L. (2013). Converting algal triglycerides to diesel and HEFA jet fuel fractions. *Energy Fuels* 27 (2): 985–996.
23 Liu, Y.Y., Sotelo-Boyas, R., Murata, K. et al. (2011). Hydrotreatment of vegetable oils to produce bio-hydrogenated diesel and liquefied petroleum gas fuel over catalysts containing sulfided Ni–Mo and solid acids. *Energy Fuels* 25 (10): 4675–4685.
24 Choudhary, T.V. and Phillips, C.B. (2011). Renewable fuels via catalytic hydrodeoxygenation. *Appl. Catal., A* 397 (1–2): 1–12.
25 Gosselink, R.W., Hollak, S.A.W., Chang, S.W. et al. (2013). Reaction pathways for the deoxygenation of vegetable oils and related model compounds. *ChemSusChem* 6 (9): 1576–1594.
26 Kubička, D. (2008). Future refining catalysis – introduction of biomass feedstocks. *Collect. Czech. Chem. Commun.* 73 (8–9): 1015–1044.
27 Elliott, D.C. (2007). Historical developments in hydroprocessing bio-oils. *Energy Fuels* 21 (3): 1792–1815.
28 Kim, S.K., Han, J.Y., Lee, H.S. et al. (2014). Production of renewable diesel via catalytic deoxygenation of natural triglycerides: comprehensive understanding of reaction intermediates and hydrocarbons. *Appl. Energy* 116: 199–205.
29 Kubička, D., Horáček, J., Setnička, M. et al. (2014). Effect of support-active phase interactions on the catalyst activity and selectivity in deoxygenation of triglycerides. *Appl. Catal., B* 145: 101–107.
30 Anderson, J. (2017). *Green Transportation Fuels from UOP*. Biomass Research and Development (BR&D) Board.
31 Kotrba, R. (2018). UPM's $206M renewable diesel plant begins commercial production. *Biodiesel Magazine* http://www.biodieselmagazine.com/articles/277480/upms-206m-renewable-diesel-plant-begins-commercial-production (accessed 10 April 2018).
32 Perotta, L. (2016). Hydroconversion of renewable lipids into biojet and biodiesel: a flexible solution becoming a reality. *ERTC 21st Annual Meeting*, Lisbon, Portugal (15 November 2016).
33 Lane, J. (2018). Shell goes carbon negative, signs commercial-scale deal with SBI. *Biofuels Digest*. http://www.biofuelsdigest.com/bdigest/2017/07/02/shell-enters-carbon-negative-signs-commercial-scale-deal-with-sbi/ (accessed 10 April 2018).
34 BioenergyInternational (2018). St1 plans to extend its advanced biofuels production into renewable diesel. *Bioenergy International*. https://

bioenergyinternational.com/biofuels-oils/st1-plans-extend-advanced-biofuels-production-renewable-diesel (accessed 10 April 2018).
35 Hodge, C. (2006). NExBTL renewable diesel. *California Air Resources Board Workshop on Regulatory and Non-Regulatory Fuels-Related Activities*, Sacramento, CA, USA.
36 Guzman, D.D. (2018). Neste starts bio-propane production. *Green Chemicals Blog*. https://greenchemicalsblog.com/2018/03/19/neste-starts-bio-propane-production/ (accessed 10 April 2018).
37 Palombo, F. (2013). Developing technologies for biofuels. *Eco-Mobility 2013, 8th A3PS Conference*, Vienna, Austria (3–4 October 2013).
38 Kubička, D. and Tukač, V. (2013). Hydrotreating of triglyceride-based feedstocks in refineries. In: *Chemical Engineering for Renewables Conversion*, Advances in Chemical Engineering, vol. 42 (ed. D.Y. Murzin), 141–194. San Diego, CA: Elsevier – Academic Press. ISBN: 978-0-12-386505-2.
39 Moran, D. (2018). SunPine to build a new crude tall diesel plant in Sweden. *Biorefineries Blog*. https://biorrefineria.blogspot.cz/2018/04/sunpine-to-build-new-crude-tall-diesel-plant-in-sweden.html (accessed 10 April 2018).
40 Donnis, B., Egeberg, R.G., Blom, P., and Knudsen, K.G. (2009). Hydroprocessing of bio-oils and oxygenates to hydrocarbons. Understanding the reaction routes. *Top. Catal.* 52 (3): 229–240.
41 Brillouet, S., Baltag, E., Brunet, S., and Richard, F. (2014). Deoxygenation of decanoic acid and its main intermediates over unpromoted and promoted sulfided catalysts. *Appl. Catal., B* 148: 201–211.
42 de Brimont, M.R., Dupont, C., Daudin, A. et al. (2012). Deoxygenation mechanisms on Ni-promoted MoS_2 bulk catalysts: a combined experimental and theoretical study. *J. Catal.* 286: 153–164.
43 Wagenhofer, M.F., Barath, E., Gutierrez, O.Y., and Lercher, J.A. (2017). Carbon–carbon bond scission pathways in the deoxygenation of fatty acids on transition-metal sulfides. *ACS Catal.* 7 (2): 1068–1076.
44 Zhang, H.P., Lin, H.F., Wang, W.Z. et al. (2014). Hydroprocessing of waste cooking oil over a dispersed nano catalyst: kinetics study and temperature effect. *Appl. Catal., B* 150: 238–248.
45 Kubička, D. and Horáček, J. (2011). Deactivation of HDS catalysts in deoxygenation of vegetable oils. *Appl. Catal., A* 394 (1–2): 9–17.
46 Chen, H., Wang, Q.F., Zhang, X.W., and Wang, L. (2015). Effect of support on the NiMo phase and its catalytic hydrodeoxygenation of triglycerides. *Fuel* 159: 430–435.
47 Laurent, E. and Delmon, B. (1994). Study of the hydrodeoxygenation of carbonyl, carboxylic and guaiacyl groups over sulfided CoMo/γ-Al_2O_3 and NiMo/γ-Al_2O_3 catalyst: II. Influence of water, ammonia and hydrogen sulfide. *Appl. Catal., A* 109 (1): 97–115.
48 Laurent, E. and Delmon, B. (1994). Influence of water in the deactivation of a sulfided NiMoγ-Al_2O_3 catalyst during hydrodeoxygenation. *J. Catal.* 146 (1): 281–291.
49 Wang, H., Li, G.L., Rogers, K. et al. (2017). Hydrotreating of waste cooking oil over supported CoMoS catalyst catalyst – deactivation mechanism study. *Mol. Catal.* 443: 228–240.

50 Kubičková, I., Snåre, M., Eränen, K. et al. (2005). Hydrocarbons for diesel fuel via decarboxylation of vegetable oils. *Catal. Today* 106 (1): 197–200.
51 Ochoa-Hernandez, C., Yang, Y.X., Pizarro, P. et al. (2013). Hydrocarbons production through hydrotreating of methyl esters over Ni and Co supported on SBA-15 and Al-SBA-15. *Catal. Today* 210: 81–88.
52 Morgan, T., Grubb, D., Santillan-Jimenez, E., and Crocker, M. (2010). Conversion of triglycerides to hydrocarbons over supported metal catalysts. *Top. Catal.* 53 (11–12): 820–829.
53 Snåre, M., Kubičková, I., Mäki-Arvela, P. et al. (2006). Heterogeneous catalytic deoxygenation of stearic acid for production of biodiesel. *Ind. Eng. Chem. Res.* 45 (16): 5708–5715.
54 Santillan-Jimenez, E., Morgan, T., Shoup, J. et al. (2014). Catalytic deoxygenation of triglycerides and fatty acids to hydrocarbons over Ni–Al layered double hydroxide. *Catal. Today* 237: 136–144.
55 Santillan-Jimenez, E., Loe, R., Garrett, M. et al. (2018). Effect of Cu promotion on cracking and methanation during the Ni-catalyzed deoxygenation of waste lipids and hemp seed oil to fuel like hydrocarbons. *Catal. Today* 302: 261–271.
56 Loe, R., Santillan-Jimenez, E., Morgan, T. et al. (2016). Effect of Cu and Sn promotion on the catalytic deoxygenation of model and algal lipids to fuel-like hydrocarbons over supported Ni catalysts. *Appl. Catal., B* 191: 147–156.
57 Peng, B.X., Zhao, C., Kasakov, S. et al. (2013). Manipulating catalytic pathways: deoxygenation of palmitic acid on multifunctional catalysts. *Chem. Eur. J.* 19 (15): 4732–4741.
58 Foraita, S., Liu, Y., Haller, G.L. et al. (2017). Controlling hydrodeoxygenation of stearic acid to *n*-heptadecane and *n*-octadecane by adjusting the chemical properties of $Ni/SiO_2–ZrO_2$ catalyst. *ChemCatChem* 9 (1): 195–203.
59 Schreiber, M.W., Rodriguez-Nino, D., Gutierrez, O.Y., and Lercher, J.A. (2016). Hydrodeoxygenation of fatty acid esters catalyzed by Ni on nano-sized MFI type zeolites. *Catal. Sci. Technol.* 6 (22): 7976–7984.
60 Kikhtyanin, O.V., Rubanov, A.E., Ayupov, A.B., and Echevsky, G.V. (2010). Hydroconversion of sunflower oil on Pd/SAPO-31 catalyst. *Fuel* 89 (10): 3085–3092.
61 Liu, Q.Y., Zuo, H.L., Zhang, Q. et al. (2014). Hydrodeoxygenation of palm oil to hydrocarbon fuels over Ni/SAPO-11 catalysts. *Chin. J. Catal.* 35 (5): 748–756.
62 Ahmadi, M., Nambo, A., Jasinski, J.B. et al. (2015). Decarboxylation of oleic acid over Pt catalysts supported on small-pore zeolites and hydrotalcite. *Catal. Sci. Technol.* 5 (1): 380–388.
63 Wang, C.X., Tian, Z.J., Wang, L. et al. (2012). One-step hydrotreatment of vegetable oil to produce high quality diesel-range alkanes. *ChemSusChem* 5 (10): 1974–1983.
64 Sousa, L.A., Zotin, J.L., and da Silva, V.T. (2012). Hydrotreatment of sunflower oil using supported molybdenum carbide. *Appl. Catal., A* 449: 105–111.

65 Han, J.X., Duan, J.Z., Chen, P. et al. (2011). Molybdenum carbide-catalyzed conversion of renewable oils into diesel-like hydrocarbons. *Adv. Synth. Catal.* 353 (14–15): 2577–2583.
66 Hollak, S.A.W., Gosselink, R.W., van Es, D.S., and Bitter, J.H. (2013). Comparison of tungsten and molybdenum carbide catalysts for the hydrodeoxygenation of oleic acid. *ACS Catal.* 3 (12): 2837–2844.
67 Stellwagen, D.R. and Bitter, J.H. (2015). Structure-performance relations of molybdenum- and tungsten carbide catalysts for deoxygenation. *Green Chem.* 17 (1): 582–593.
68 Kim, S.K., Yoon, D., Lee, S.C., and Kim, J. (2015). Mo_2C/graphene nanocomposite as a hydrodeoxygenation catalyst for the production of diesel range hydrocarbons. *ACS Catal.* 5 (6): 3292–3303. https://doi.org/10.1021/acscatal.5b00335.
69 Kim, S.K., Kim, J., and Lee, S.C. (2017). Surface-termination dependence of propanoic acid deoxygenation on Mo_2C. *Catal. Commun.* 99: 61–65.
70 Shi, Y., Yang, Y., Li, Y.W., and Jiao, H.J. (2016). Theoretical study about $Mo_2C(101)$-catalyzed hydrodeoxygenation of butyric acid to butane for biomass conversion. *Catal. Sci. Technol.* 6 (13): 4923–4936.
71 Wang, H.L., Yan, S.L., Salley, S.O., and Ng, K.Y.S. (2012). Hydrocarbon fuels production from hydrocracking of soybean oil using transition metal carbides and nitrides supported on ZSM-5. *Ind. Eng. Chem. Res.* 51 (30): 10066–10073.
72 Nguyen, H.S.H., Maki-Arvela, P., Akhmetzyanova, U. et al. (2017). Direct hydrodeoxygenation of algal lipids extracted from *Chlorella alga*. *J. Chem. Technol. Biotechnol.* 92 (4): 741–748.
73 Al Alwan, B., Salley, S.O., and Ng, K.Y.S. (2015). Biofuels production from hydrothermal decarboxylation of oleic acid and soybean oil over Ni-based transition metal carbides supported on Al-SBA-15. *Appl. Catal., A* 498: 32–40.
74 Guan, Q., Wan, F., Han, F. et al. (2016). Hydrodeoxygenation of methyl palmitate over MCM-41 supported nickel phosphide catalysts. *Catal. Today* 259: 467–473.
75 Liu, Y.H., Yao, L., Xin, H. et al. (2015). The production of diesel-like hydrocarbons from palmitic acid over HZSM-22 supported nickel phosphide catalysts. *Appl. Catal., B* 174: 504–514.
76 Pan, Z.Y., Wang, R.J., Nie, Z.Y., and Chen, J.X. (2016). Effect of a second metal (Co, Fe, Mo and W) on performance of Ni_2P/SiO_2 for hydrodeoxygenation of methyl laurate. *J. Energy Chem.* 25 (3): 418–426.
77 Peroni, M., Lee, I., Huang, X. et al. (2017). Deoxygenation of palmitic acid on unsupported transition-metal phosphides. *ACS Catal.* 7 (9): 6331–6341.
78 Xin, H., Guo, K., Li, D. et al. (2016). Production of high-grade diesel from palmitic acid over activated carbon-supported nickel phosphide catalysts. *Appl. Catal., B* 187: 375–385.
79 Yang, Y.X., Ochoa-Hernandez, C., O'Shea, V.A.D. et al. (2012). Ni_2P/SBA-15 as a hydrodeoxygenation catalyst with enhanced selectivity for the conversion of methyl oleate into *n*-octadecane. *ACS Catal.* 2 (4): 592–598.

80 Yang, Y.X., Ochoa-Hernandez, C., Pizarro, P. et al. (2015). Influence of the Ni/P ratio and metal loading on the performance of NixPy/SBA-15 catalysts for the hydrodeoxygenation of methyl oleate. *Fuel* 144: 60–70.
81 Zarchin, R., Rabaev, M., Vidruk-Nehemya, R. et al. (2015). Hydroprocessing of soybean oil on nickel-phosphide supported catalysts. *Fuel* 139: 684–691.
82 Yang, Y., Chen, J., and Shi, H. (2013). Deoxygenation of methyl laurate as a model compound to hydrocarbons on Ni_2P/SiO_2, Ni_2P/MCM-41, and Ni_2P/SBA-15 catalysts with different dispersions. *Energy Fuels* 27 (6): 3400–3409.

12

Upgrading of Lipids to Fuel-like Hydrocarbons and Terminal Olefins via Decarbonylation/Decarboxylation

Ryan Loe[1,2], *Eduardo Santillan-Jimenez*[1], *and Mark Crocker*[1,2]

[1] *University of Kentucky, Center for Applied Energy Research, 2540 Research Park Drive, Lexington, KY 40511, USA*
[2] *University of Kentucky, Department of Chemistry, Rose St., Lexington, KY 40506, USA*

12.1 Introduction

The CO_2 emissions stemming from the use of fossil fuels represent one of the main causes of climate change, which puts social and environmental pressure behind the pursuit of alternative fuels. Furthermore, a number of factors – including the non-renewable nature of fossil fuels, the ever-increasing demand for energy resources, and the uneven geographical distribution of oil reserves, as well as geopolitical, technological, and economic developments – can result in an uncertain supply and volatile prices of petroleum-derived fuels. Therefore, many countries have come to consider the development of renewable fuels as a means to attain energy independence and national security. Current petroleum consumption surpasses 96 million barrels of crude oil daily worldwide and that demand is expected to increase by 18% by 2040 [1]. Moreover, while hydrocarbon-based fuels such as jet fuel and diesel are widely used in aviation and trucking, viable alternatives to petroleum-derived fuels for these transportation sectors are currently in short supply. The limited availability of alternative fuels is of great concern, particularly since forecasts indicate that fuel consumption for the aviation sector alone will increase 5% annually until 2030 [2] and that the alternative fuel demands by the transportation sector could reach 900 million tons by 2020, which may result in fuel shortages and further environmental disruption [3]. The most advantageous replacement of petroleum-derived fuels – from both an economic and technological standpoint – would be fuel-like hydrocarbons obtained from a renewable source. Indeed, renewable fuels chemically indistinguishable from (and thus entirely fungible with) petroleum-derived fuels would be fully compatible with the existing engine and fuel distribution infrastructure of nearly every economy around the world [4]. In addition, biofuels have the potential to be carbon neutral, as the amount of CO_2 emitted during biofuel production and use can be equivalent to the amount consumed during the growth of the biomass used as raw material [5]. Additionally, biofuel production could also promote employment and economic

Chemical Catalysts for Biomass Upgrading, First Edition. Edited by Mark Crocker and Eduardo Santillan-Jimenez.

activity in rural regions and help countries reduce their dependence on fuels sourced from other nations [6]. Given the aforementioned environmental, economic, energy independence, and national security benefits, the need to develop renewable “drop-in” hydrocarbon biofuels cannot be overstated.

In view of the projected increase in demand, the conversion of plant oils and animal fats into liquid transportation fuels has attracted much attention, the most mature technology developed to date to effect this transformation being the synthesis of biodiesel. This fully commercialized process involves the transesterification of glycerides and/or the esterification of free fatty acids (FFAs) to form the fatty acid methyl esters (FAMEs) that constitute biodiesel. FAMEs offer several advantages over petroleum-derived fuels as they produce lower amounts of sulfur emissions upon combustion [7], offer increased lubricity [7], display high cetane number (a measure of combustion quality under compression ignition), and the mild reaction conditions involved in their synthesis allow for small-scale production practically everywhere [8]. However, biodiesel also shows a number of important drawbacks. Indeed, the heat content of FAMEs is 9–13% lower than that of petroleum-derived fuels on a mass basis, the fuel can be contaminated with cations stemming from the catalysts used during biodiesel synthesis, and the latter affords considerable amounts of glycerol, which effectively constitutes a waste by-product that needs to be disposed of. Biodiesel also displays poor storage stability due to the fact that some of the FAMEs may contain C=C double bonds [9], which can be hydrated to afford hydroxyl groups that induce polymerization [10]. Further, engine compatibility issues can also arise since water – which has a relatively high solubility in FAMEs – can cause corrosion of metal parts [8, 9], while FAMEs also display poor cold flow properties [9]. Finally, biodiesel production via transesterification is highly sensitive to the feed employed (the presence of FFAs can deactivate the catalysts employed and contaminate the fuel product with soaps [8]), so FFAs must first be removed from glyceride-based feeds before each stream can be converted to FAMEs separately. Altogether, these disadvantages have limited the ability of biodiesel to displace a significant fraction of petroleum-derived fuels.

Notably, many of the drawbacks associated with biodiesel stem from the high oxygen content of FAMEs, which can be higher than 10 wt% [11, 12]. Therefore, research has aimed to develop processes that convert animal fats and plant oils to fuel-like hydrocarbons by removing the oxygen from these feeds via catalytic cracking, hydrodeoxygenation (HDO), or decarbonylation/decarboxylation (deCO$_x$). Catalytic cracking is a simple process that uses acid catalysts such as zeolites and other alumino-silicates or silicoaluminophosphates to yield fuel-like products. However, the reaction is often unselective and creates naphthalene and other aromatic compounds that can lead to coking and the concomitant deactivation of the catalyst [13]. In contrast, HDO removes oxygen in the form of water (see Scheme 12.1) and can be a very selective process resulting in high yields of diesel- and jet fuel-like hydrocarbons. HDO products are chemically identical to petroleum-derived fuels and have been reported to offer improved performance and reduced emissions relative to both biodiesel and petrodiesel [11, 14]. The use of HDO to produce fuel-like hydrocarbons (marketed as green or renewable diesel) also represents a mature technology – Neste Oil’s NExBTL™ and UOP/Eni’s Ecofining™ being

$$\text{(a)}\quad HO{-}\overset{\overset{\displaystyle O}{\|}}{C}{-}(CH_2)_{16}{-}CH_3 + 3H_2 \xrightarrow{\text{Hydrodeoxygenation}} 2H_2O + H_3C{-}(CH_2)_{16}{-}CH_3$$

$$\text{(b)}\quad HO{-}\overset{\overset{\displaystyle O}{\|}}{C}{-}(CH_2)_{16}{-}CH_3 \xrightarrow{\text{Decarbonylation}} CO + H_2O + H_2C{=}CH{-}(CH_2)_{14}{-}CH_3$$

$$\text{(c)}\quad HO{-}\overset{\overset{\displaystyle O}{\|}}{C}{-}(CH_2)_{16}{-}CH_3 \xrightarrow{\text{Decarboxylation}} CO_2 + H_3C{-}(CH_2)_{15}{-}CH_3$$

Scheme 12.1 Equations for (a) hydrodeoxygenation, (b) decarbonylation, and (c) decarboxylation of stearic acid.

noteworthy commercial processes – and the previous Chapter 11 in this book has discussed this particular deoxygenation approach in detail. Although promising, HDO typically involves the use of sulfided catalysts (which may deactivate in the presence of water and contaminate the fuel products with sulfur) and requires high pressures of hydrogen, limiting processing to centralized facilities where this gas can be produced from natural gas [11]. Alternatively, oxygen can be removed from lipids – including glycerides, FFAs, FAMEs, and/or mixtures thereof – in the form of CO and CO_2 via catalytic decarbonylation and decarboxylation, respectively, generating fuel-like hydrocarbon molecules containing one less carbon atom than the original lipid monomer (see Scheme 12.1). Although the carbon loss results in a lower atom efficiency compared with the HDO pathway, the decreased H_2 requirements negate this consequence. Generally, decarbonylation and decarboxylation are considered as $deCO_x$ in the aggregate as opposed to independently, mainly because it is difficult to differentiate the two reaction pathways under most experimental conditions. Indeed, hydrogen is typically present in the reaction atmosphere, resulting in the hydrogenation of any double bonds including the terminal olefins produced via decarbonylation, which makes the liquid products resulting from decarbonylation and decarboxylation indistinguishable. Similarly, the gas products evolved may not constitute a reliable way to differentiate the two reaction pathways, mainly due to the fact that water–gas shift, methanation, and Boudouard reactions can all take place (see Scheme 12.2), even if the occurrence of the reverse water–gas shift (RWGS) reaction can be ruled out by the low value for the RWGS equilibrium constant at typical $deCO_x$ temperatures [15]. The

$$\text{(a)}\quad CO + H_2O \rightleftharpoons CO_2 + H_2$$

$$\text{(b)}\quad 2CO \rightleftharpoons C^* + CO_2$$

$$\text{(c)}\quad CO_2 + 4H_2 \longrightarrow CH_4 + 2H_2O$$

$$\text{(d)}\quad CO + 3H_2 \longrightarrow CH_4 + H_2O$$

Scheme 12.2 (a) Water–gas shift reaction, (b) Boudouard reaction, (c) methanation of CO_2, and (d) methanation of CO.

aforementioned reactions confound CO and CO_2, making it difficult to ascertain the extent to which either decarbonylation or decarboxylation have taken place. Albeit the stoichiometry of the $deCO_x$ reactions shown in Scheme 12.1b,c clearly illustrates that external hydrogen is not required for these reactions to proceed, external hydrogen is generally added to the reaction atmosphere in order to improve catalyst performance. Nevertheless, $deCO_x$ processes display considerably lower hydrogen requirements and afford similar yields of fuel-like hydrocarbons than processes based mainly on HDO, which represent some of the main reasons behind the considerable interest that $deCO_x$ processes have received in the last decade [16, 17].

Against this backdrop, this chapter will focus on the most salient work performed to date on the $deCO_x$ of lipids involving heterogeneous catalysts. Given that a considerable amount of raw feedstock will be needed if renewable hydrocarbons are to become a substantial fraction of liquid transportation fuels and help meet projected demands, the most abundant lipid feed sources will be discussed. In regard to the catalysts, there are numerous reports covering many different formulations active in the $deCO_x$ of lipids. While both homogeneous [18, 19] and heterogeneous [20, 21] catalysts have been used to convert FFAs and esters to hydrocarbons over the last 75 years, interest in this transformation has increased dramatically over the last decade. Moreover, heterogeneous catalysts are more relevant for industrial application due to the ease of catalyst and products separation. Therefore, although several homogeneous catalysts have recently been reported [22–26], the more common work involving the heterogeneously catalyzed $deCO_x$ of lipids to hydrocarbons will be the focal point of this discussion. Furthermore, the literature contains reports of many different types of heterogeneous catalysts – including supported transition metals [27–36] and supported carbides [37–41], as well as phosphided [42–47] and sulfided [48–51] formulations – that exhibit $deCO_x$ activity. However, due to the fact that the vast majority of reports deal with supported Group 10 metals, only these catalysts (including bimetallic formulations containing these metals) will be discussed, along with the effect of different supports and reaction conditions. Finally, the most widely accepted reaction mechanisms will be examined.

12.2 Lipid Feeds

In contrast to biodiesel synthesis – which presupposes the separation of feeds into reasonably pure FFA or glyceride streams prior to their transformation into biodiesel via acid-catalyzed esterification and base-catalyzed transesterification, respectively – the deoxygenation of lipids via $deCO_x$ has the advantage of being feedstock agnostic, as glycerides, FFAs, FAMEs, and/or mixtures thereof can be converted to fuel-like hydrocarbons through a single $deCO_x$ process. As stated previously, the demand for renewable fuels is immense if a significant fraction of the 80 million barrels of oil consumed daily worldwide is to be displaced. Therefore, raw materials that are both abundant and varied are necessary to render renewable fuels a viable alternative. It has already been demonstrated that a number of oils sourced from agricultural crops – including palm [52, 53], sunflower

[54, 55], rapeseed [56], coconut [57], and soybean oil [15, 58] – can be converted to fuel-like hydrocarbons via deCO$_x$. However, the use of edible feedstocks can disrupt the food supply, leading to volatile food prices. Moreover, the clearing of forests and other carbon-rich habitats to create farm land dedicated to these crops can also increase CO_2 emissions and prevent any renewable fuels produced from becoming carbon neutral for decades [59].

Currently, the most widely used renewable fuels are corn ethanol and soybean biodiesel; however, corn and soybeans are both edible feedstocks and the fuels derived from them are net carbon positive, in spite of their lower net greenhouse gas emissions (12% and 41%, respectively) relative to petroleum-derived fuels [60]. Fuels derived from inedible oleaginous plants that can thrive in arid or degraded soils (such as *Jatropha curcas*) represent a more desirable alternative to avoid disrupting the food supply, realize additional greenhouse gas reductions, and minimize agrichemical pollution [60, 61]. Nevertheless, renewable fuels obtained from these sources alone could only replace a limited amount of petroleum-derived fossil fuels, the demand for renewable diesel alone already exceeding the current production levels of terrestrial oleaginous crops. Fortunately, several potential options to supplement the oil supply exist, including microalgae, the latter having been extensively reviewed [62–69]. Unfortunately, both the energy consumption and the cost of cultivating and processing algae must be reduced considerably before algae can be deemed a viable feedstock for the production of renewable fuels [67, 70]. Other inedible oleaginous raw materials include waste streams such as tall oil fatty acid (TOFA), yellow grease (waste cooking oil), and brown grease (grease recovered from grease traps). While the low cost of these feeds is certainly advantageous, their use can also prevent them from reaching landfills or water streams. TOFA can be obtained from vacuum distillation of crude tall oil, which is a side stream from the pulp and paper industry with production approaching 845 000 tons in 2004 in the United States alone [71]. TOFA have been successfully deoxygenated by Murzin and coworkers [71, 72]. Yellow grease is another inedible feedstock – produced at a rate of 100 million gallons per day in the United States [73] – for which deoxygenation to hydrocarbons has been demonstrated [74–76]. Brown grease offers similar advantages, although dewatering must precede its conversion to fuel to prevent the hydrothermal deactivation of the catalyst employed unless formulations showing satisfactory stability and performance in the presence of water are used [77]. Although these challenges with brown grease have hindered large-scale production, successful deoxygenation to fuel-like hydrocarbons has been demonstrated [77].

It should be noted that in contrast with edible oils and fats, which are mostly triglyceride based, waste lipid streams are typically rich in FFAs. Admittedly, deCO$_x$ processes can yield fuel-like hydrocarbons from glycerides, FFAs, or mixtures containing both; however, other characteristics such as the length and the degree of unsaturation of the carbon chains must also be taken into account. In addition to the fact that natural fats typically contain hydrocarbon chains of 8–22 carbons – which makes them ideal for the production of fuels – several authors have reported that chain length does not affect the rate of deoxygenation observed over most catalysts [78–81]. However, Lamb and coworkers [82] and

Martin and coworkers [83] found that over carbon supports, longer chain fatty acids exhibited an increased initial reaction rate that was attributed to increased tendency for adsorption [82]. Martin and coworkers found that as the chain length of the FA increased to 18, significant increases in cracking occurred [83]. Chain unsaturation is a more challenging factor due to the fact that the presence of carbon–carbon double bonds favors cracking reactions resulting in catalyst fouling, which in turn leads to catalyst deactivation, and in hydrocarbon products consisting of shorter hydrocarbon chains. Given that the chain length of jet and diesel fuel range between 9 and 20 carbons, excessive cracking can limit the use of the hydrocarbon products. Notably, the difficulties associated with unsaturated feeds can be partially addressed by the type of catalyst and reaction conditions employed, both of which will be addressed in the following sections.

12.3 deCO$_x$ Catalysts: Active Phases

Over the past 15 years, a number of catalysts have been actively investigated for the deoxygenation of fats and oils. In an initial catalyst screening study performed by Snåre et al. [84], several commercially available catalysts comprising different active phases and supports were tested for activity in the deoxygenation of stearic acid, a model fatty acid compound. Carbon-supported formulations were found to be more active, which was attributed to the amphoteric properties, surface functionalities, and high surface area of the carrier minimizing the effect of coke-induced deactivation. Results obtained over similar supports and, accounting for the metal content, revealed that deoxygenation activity follows the order Pd > Pt > Ni > Rh > Ir > Ru > Os. Therefore, for several years following that report, work on the conversion of lipids to fuel-like hydrocarbons focused on carbon-supported Pd and Pt catalysts. However, Ni-based formulations started garnering attention due to their ability to offer comparable deoxygenation activity to Pd and Pt with certain feeds and reaction conditions [15]. In view of the foregoing, this chapter will focus on supported Pt, Pd, and Ni catalysts – along with bimetallic formulations involving these metals – such as those included in Table 12.1. In instances where the deoxygenation ability of different metals is compared, catalysts with the same support and similar metal loadings are used (whenever possible) in an effort to minimize the effect of these variables.

In the initial catalyst screening study performed by Snåre et al., which included 5 wt% Pt and 5 wt% Pd on a carbon support, the Pd-based catalyst afforded the best results, including quantitative conversion of stearic acid and a selectivity to C_{17} hydrocarbons of 99%. In the case of Ni catalysts, higher loadings are typically employed to counter the lower intrinsic activity of this metal relative to Pd and Pt. Interestingly, Wu et al. recently reported that 5 wt% Pt/C affords a significantly higher conversion of stearic acid than 20 wt% Ni/C at 330 °C in hydrogen- and solvent-free conditions [97]. However, these results contrast with data previously reported by Crocker and coworkers [15], which can be attributed to the fact that these authors reduced the catalyst prior to the deoxygenation reaction

Table 12.1 Selected examples of $deCO_x$ catalysts and their performance in the deoxygenation of lipid feeds.

Catalyst	Feed/catalyst ratio	Feed/solvent	Reactor type	Pressure (bar), gas	Temperature (°C)	Time (h)	Conversion (%)	Selectivity	References
7% Ni/HZSM-5	30 g/3 g	Methyl hexadecanoate	BR, stirred	20, H_2	220	1	>90	83% C_5–C_{16}	[85]
10% Ni/H-β (Si/Al = 75)	1 g/0.2 g	1 g stearic acid/100 ml C_{12}	BR, stirred	40, H_2	260	8	100	~73% C_{18} ~15% C_{17}	[86]
7% Ni/SAPO-11	LHSV = 2 h^{-1}	Palm oil	FB	40, H_2	280	6	100	52.8% *i*-C_{15}–C_{18} 9.6% *n*-C_{15}–C_{18}	[87]
5% Ni/ZrO_2	1 g/0.5 g	1 g stearic acid/100 ml C_{12}	SBR, stirred	40, H_2	260	8	100	2% C_{18} 90% C_{17}	[88]
10% Ni/ZrO_2	1 g/0.5 g	1 g microalgae oil ml C_{12}	SBR, stirred	40, H_2	260	8	100	68% C_{17}	[88]
5% Ni/TiO_2	1 g/0.5 g	1 g stearic acid/100 ml C_{12}	SBR, stirred	40, H_2	260	8	98	5% C_{18} 87% C_{17}	[88]
5% Ni/CeO_2	1 g/0.5 g	1 g stearic acid/100 ml C_{12}	SBR, stirred	40, H_2	260	8	100	0.4% C_{18} 93% C_{17}	[88]
5% Ni/Al_2O_3	1 g/0.5 g	1 g stearic acid/100 ml C_{12}	SBR, stirred	40, H_2	260	8	63	0.7% C_{18} 81% C_{17}	[88]
5%Ni/SiO_2	1 g/0.5 g	1 g stearic acid/100 ml C_{12}	SBR, stirred	40, H_2	260	8	45	1.5% C_{18} 57% C_{17}	[88]

(Continued)

Table 12.1 (Continued)

Catalyst	Feed/catalyst ratio	Feed/solvent	Reactor type	Pressure (bar), gas	Temperature (°C)	Time (h)	Conversion (%)	Selectivity	References
15% Ni/Al_2O_3	~3.1 g/0.5 g	~3.1 g stearic acid/100 ml C_{12}	BR, stirred	8, H_2	290	4	100	80% C_{17}	[89]
20% Ni-5% Cu/Al_2O_3	1.8 g/0.5 g	1.8 g tristearin/22 g C_{12}	SBR, stirred	40, H_2	260	6	97	71% C_{17}	[75]
20% Ni/C	20 g/0.22 g	20 g soybean oil	BR, stirred	6.9, N_2	350	4	92	54% C_8–C_{17}	[15]
5% Pd/C	20 g/0.22 g	20 g Soybean oil	BR, stirred	6.9, N_2	350	4	30	17.8% C_8–C_{17}	[15]
5% Pd/C	4.5 g/1 g	4.5 g stearic acid/86 g C_{12}	SBR, stirred	6, N_2	300	6	100	95% C_{17}	[84]
5% Pd/Al_2O_3	4 g/1 g	4 g stearic acid/12 g C_{12}	BR, stirred	14.2, H_2	350	6	100	90.1% C_{17}	[90]
0.6% Pd-Au/SiO_2	1.49 g/0.5 g	1.49 g/100 ml C_{12}	SBR, stirred	17, 5%H_2/Ar	300	5	96	98% C_{17}	[91]
5% PdAu/SiO_2	4 μl/min/0.5 g	4 μl/min octanoic acid	FB	1, 10%H_2/Ar	260	5	~45	>97% C_7	[92]
1% Pt/C	20 g/0.22 g	20 g soybean oil	BR, stirred	6.9, N_2	350	4	42	34.7% C8–C_{17}	[15]
5% Pt/Al_2O_3	4 g/1 g	4 g stearic acid/12 g C_{12}	BR, stirred	14, H_2	350	6	92.4	45% C_{17}	[93]
5% Pt/SAPO-11	18 g/1 g	Oleic acid[a)]	BR, stirred	20, H_2	325	2	>99	31.8% C_{17} 21% dodecylbenzene	[94]
5%Pt_3-Sn/C[b)]	108 μl/5.4 mg	108 μl/min oleic acid/1 g of H_2O	BR	[a)]	350	2	>95	>90% C_{17}	[95]
5.2% Pt-4.2% Re/C	20 g/0.5 g	20 g stearic acid/80 g of H_2O	BR, stirred	34.5, H_2	300	3	92	~80% C_{17}	[96]

BR, batch reactor; SBR, semi-batch reactor; FB, fixed-bed reactor.

a) Actual amounts used not specified, only a ratio was given.

b) Sn metal loading was 25 mol% of the total metal loading.

and performed their experiments using a stirred reactor. Indeed, when Crocker and coworkers tested 1 wt% Pt/C, 5 wt% Pd/C and 20 wt% Ni/C for activity in the deoxygenation of soybean oil (an unsaturated triglyceride feed) at 350 °C under an inert atmosphere, the Ni catalyst displayed a conversion of 92% and a yield of C_8–C_{17} hydrocarbons of 54%. In contrast, the Pd and Pt catalysts displayed significantly lower conversion values (30% and 23%, respectively). Tellingly, the Ni catalyst also afforded a considerably higher yield of light (C_{1-7}) hydrocarbons than the precious metal catalysts, which is indicative of increased cracking. A follow-up study by Crocker and coworkers evaluated the effect of the reaction atmosphere on the performance of 20 wt% Ni/C and 5 wt% Pd/C [15]. Results indicate that the conversion of FFA and triglyceride model compounds is increased upon addition of H_2 to the reaction atmosphere. However, the Pd catalyst displays a higher selectivity to C_{17} hydrocarbons than Ni irrespective of the atmosphere employed, which parallels the previous report and suggests that Ni exhibits increased cracking activity relative to Pd regardless of the hydrogen partial pressure. In short, in spite of the fact that when normalized on the basis of metal loading, the reactivity of carbon-supported catalysts follows the order Pd > Pt > Ni, reduced Ni catalysts with high (10–30 wt%) metal loadings can outperform noble metal catalysts, particularly when concentrated feeds are employed. Notably, whereas decarboxylation represents the main pathway observed over Pd and Ni catalysts, decarbonylation is favored over Pt formulations [84].

In addition to the work summarized previously on carbon-supported catalysts, several authors have also tested Pt, Pd, and Ni on Al_2O_3 for activity in the conversion of lipids to hydrocarbons, which has led to additional insights. For instance, Berenblyum et al. prepared alumina-supported Pt, Pd, and Ni catalysts with the same 5 wt% metal loading [93], thus enabling the direct comparison of the deoxygenation activity of these metals that is lacking for carbon-supported formulations. When tested for stearic acid deoxygenation at 350 °C in a H_2 atmosphere using a batch reactor, activity followed the order Pd > Pt > Ni, conversion values being 100%, 92%, and 61%, respectively. Berenblyum et al. concluded that decarbonylation constituted the main deoxygenation pathway for all three metals under the conditions employed, the highest olefin yield being observed with Ni. The change in the deoxygenation pathway from decarboxylation to decarbonylation when Ni and Pd are supported on oxidic – as opposed to carbonaceous – supports has also been observed by other authors [82]. Analogous findings vis-à-vis the activity trends discussed previously have also been reported for the upgrading of unsaturated lipids [98]. Indeed, albeit at 325 °C in the presence of hydrogen Madsen et al. found that 5% Pt/Al_2O_3 and 5% Pd/Al_2O_3 displayed similar activity in the conversion of oleic acid, these authors determined that 5% Pd/Al_2O_3, 5% Pt/Al_2O_3, and 5% Ni/Al_2O_3 afforded tripalmitin conversions of 71%, 46%, and 10%, respectively. Interestingly, when Srifa et al. tested the same three catalysts for activity in the deoxygenation of palm oil using an industrially relevant trickle-bed reactor and reaction conditions (300 °C, 50 bar of H_2 and LHSV = 1 h^{-1}), each of the catalysts afforded quantitative conversion [53]. Tellingly, these authors also observed that lower quantities of long-chain hydrocarbons and propane – as well as higher amounts of methane and ethane – were obtained with the Ni catalyst, which is indicative of the increased cracking activity of Ni resulting in the shortening of both the hydrocarbons chains and of the propane stemming from the triglyceride backbone. In short, the order of activity in the catalytic deCO$_x$ of lipids for Group 10 metals on carbonaceous and oxidic supports is Pd > Pt > Ni, a trend that has also been observed in gas phase reactions [99]. Parenthetically, when these three metals supported on a SAPO-11 zeolite were tested in the isomerization of normal to branched alkanes to obtain fuels with improved cold flow properties, the order of reactivity was found to be Pt > Pd > Ni [100], which suggests that Group 10 precious metals are more intrinsically active than Ni in both isomerization and deoxygenation reactions. Moreover, although higher Ni loadings can be used to match the activity of Pt and Pd, exacerbated cracking and the impact of the latter on both selectivity and deactivation must be considered.

In an effort to more thoroughly understand the deCO$_x$ activity of Group 10 metals, several authors have systematically studied the effect of metal loading and dispersion on catalyst performance. Chen et al. tested Pt/SAPO-11 catalysts with loadings ranging from 0.6% to 3% for deoxygenation activity and found that although metal dispersion decreased with increased metal loading, there was an overall increase in the number of active metal sites, which makes the fact that conversion increased with metal loading unsurprising [101]. Murzin and coworkers studied the effect of metal dispersion on the deoxygenation of palmitic and stearic acids over 1 wt% Pd/C and found the optimal Pd dispersion to be 65%. Indeed, while a catalyst with a lower dispersion exhibited lower catalyst activity – which can be explained on the basis that larger particles should result in a reduced number of surface sites – a catalyst with 72% metal dispersion also showed lower activity than the formulation with 65% dispersion in spite of the fact that the average metal particle size of both catalysts was similar. The authors explained this finding invoking a strong metal–support interaction for the 72% dispersion catalyst, resulting in changes in the Pd metal structure required for deoxygenation. Murzin and coworkers also found that 5% Pd/C exhibited better conversion and selectivity than 1 wt% Pd/C in the deoxygenation of stearic acid; however, increasing the Pd loading further did not result in better conversion or selectivity to fuel-like hydrocarbons [84]. Interestingly, Lamb and coworkers found that Pd dispersion values similar to those used by Murzin led to quick catalyst deactivation during the deoxygenation of octanoic acid over silica-supported catalysts [92]. These authors concluded that when silica is used as the carrier, a much lower dispersion of 16% (corresponding to a metal particle size of 7.5 nm) is preferred, which indicates that the optimal metal dispersion is support-dependent for Pd-based catalysts. Similarly, the optimal metal loading has been reported to be support dependent for Ni-based catalysts. Indeed, for the supports on which the effect of metal loading has been assessed, the optimal Ni loadings have been found to be 15, 20, 7, 7, and 15 wt% on Al_2O_3 [89], carbon [97], SAPO-11 [102], HSZM-5 [85], and ZrO_2 [88], respectively. For each of these supports, both lower and higher Ni loadings than the ones listed result in a decrease in the number of active sites. Moreover, increasing the loading to 25 wt% Ni has also been reported to favor the formation of inactive Ni-aluminate on alumina-supported Ni [89]. In short, for a given support, catalyst performance is a function of both metal loading and dispersion.

Bimetallic active phases have attracted a fair amount of interest due to their potential to outperform monometallic formulations in the conversion of lipids to hydrocarbons via deCO$_x$. For instance, Savage and coworkers observed some improvements relative to Pt/C when using a bimetallic Pt–Sn/C catalyst in the deoxygenation of stearic acid under hydrothermal conditions [95]. Indeed, the Pt–Sn/C catalyst exhibited higher selectivity to C_{17} (the main deCO$_x$ product from the stearic acid feed), albeit the addition of Sn also led to a decrease in activity. In contrast, incorporating 5 wt% Re into a Pt–Re bimetallic catalyst doubled the C_{17} obtained from the deoxygenation of stearic acid yield under hydrothermal conditions compared to a monometallic Pt catalyst [96]. Similarly, Sun et al. found Pd/SiO_2 to be selective toward deCO$_x$ products when tested in the in-gas phase deoxygenation of octanoic acid, although significant deactivation rates were observed [92]. However, the authors found that deactivation could be curbed through the addition of Au in a 1 : 1 atomic ratio during catalyst synthesis, which afforded bimetallic particles with Au cores and Pd-rich surfaces. The observed improvement in catalyst stability was attributed to the Au addition resulting in a lower CO adsorption energy on the surface Pd sites. The reduced presence of CO – a known catalyst poison – on the catalyst surface decreases catalyst deactivation, which allowed the Au–Pd catalyst to display a consistent performance in a fixed-bed reactor for more than six hours of time on stream. Noticeable improvements in activity, selectivity, and stability have also been achieved through the addition of Cu to Ni/Al_2O_3 [74, 75]. Indeed, while 20 wt% Ni/Al_2O_3 displayed a conversion and selectivity to C_{17} hydrocarbons of 27% and 63% during the deoxygenation of tristearin at 260 °C in a semi-batch reactor, the corresponding values achieved over 20 wt% Ni-5 wt% Cu/Al_2O_3 were 97% and 71%. In view of its promising performance, the 20 wt% Ni-5 wt% Cu/Al_2O_3 catalyst was also used to upgrade waste and highly unsaturated feeds such as yellow grease and hemp seed oil using a fixed-bed reactor and industrially relevant reaction conditions [74]. Saliently, the Ni–Cu catalyst displayed remarkable stability during eight hours of time on stream – yielding >90% diesel-like hydrocarbons at all reaction times sampled – and significantly less cracking relative to the corresponding Ni-only catalyst. The promotional effects observed were determined to be due (at least in part) to the fact that Cu facilitates the reduction of Ni at lower temperatures, which results in the formation of the metallic surface Ni sites believed to constitute the active site for the deCO$_x$ reaction. It should be noted that whereas the bimetallic catalyst afforded excellent yields of diesel-like hydrocarbons, the monometallic formulation displayed quantitative yields of methane. This marked improvement in selectivity observed over alumina-supported Ni–Cu catalysts has also been observed by Yakovlev et al. in their work with other supports [103], of which CeO_2 and ZrO_2 were found to be most effective, revealing the presence of the support effects that are discussed in the following text in more detail. Nevertheless, the noteworthy improvements achieved over Ni–Cu catalysts are most encouraging in the development of inexpensive catalysts for the deoxygenation of lipid feeds, as they illustrate that the promotion of Ni with other earth-abundant metals can result in catalysts rivaling the performance of precious metal-based formulations.

12.4 deCO$_x$ Catalysts: Support Materials

The 5 wt% Pd/C formulation identified as most promising by Snåre et al. in their 2006 contribution [84] remains a benchmark catalyst in the conversion of lipids to hydrocarbons via deCO$_x$. In general, the fact that good performance can be achieved through the use of carbonaceous supports can be attributed to the large surface area of these carriers, which remains high (>500 m^2/g) even after loading considerable amounts of metal (e.g. 20 wt%). In addition to the fact that a large surface area can help curb the deactivation stemming from the formation of coke deposits and/or the sintering of supported metals, these carriers display amphoteric properties arising from the presence of various surface functional groups [84]. The latter also impact catalyst performance, as confirmed independently by Gosselink et al. and by Chen et al., who observed that improvements in catalytic activity could be realized through the addition of polarized functional groups to carbonaceous supports [104, 105]. However, in the event of coke-induced deactivation carbon-supported catalysts are notoriously difficult to regenerate and recycle, which disfavors their use in industrial settings where coked catalysts are typically regenerated via calcination in hot air (a treatment that would destroy a carbon-supported formulation). In light of this, catalyst development has shifted to more refractory oxidic supports, which can be divided into zeolites and other metal oxides.

Zeolites are acidic and porous aluminosilicate materials with wide applications in catalysis in general and within the petroleum industry in particular. Attempts to use them as deoxygenation catalyst supports have employed various frameworks, including HZSM-5 [85, 86, 89, 94, 106], SAPO-11 [87, 100, 101, 107–109], SBA-15 [100, 106, 110], H-β [86], HY [86, 102, 111], and 5A [81, 112]. Typically, deoxygenation occurs through the HDO pathway and increased cracking and isomerization of the hydrocarbon products are observed, which is unsurprising given the characteristic concentration of Brønsted acid sites on the surface of these materials [85, 89, 108]. Modifications to the latter, such as increasing the Si:Al ratio, can lower their acidity and improve the yield of deCO$_x$ products, albeit at the cost of decreased conversion [86]. Indeed, the high conversion values associated with acid catalysts stem from the fact that increased acidity decreases the electronic charge of support oxygen atoms, which in turn leads to a decrease in electron density of the surface metal. As a result, any adsorbed hydrogen in the form of a H–metal bond should be more readily activated, increasing the hydrogenation rate of the catalyst along with its selectivity toward the HDO pathway [106]. It is worth noting that while some zeolite-supported catalysts yield near equivalent amounts of HDO and deCO$_x$ products – such as 1 wt% Pt/SAPO-11–Al_2O_3 [58] – others can favor deCO$_x$ over HDO, as is the case for Pd supported on an Al-SBA-15 zeolite with a Si:Al ratio of 22 [106]. Murzin and coworkers also developed a 0.6 wt% Pd/SBA-15 catalyst that achieved 96% conversion of stearic acid with a 98% selectivity to deCO$_x$ products, further confirming that tuning acidity can direct the reaction to proceed via the deCO$_x$ pathway [91]. In short, the best performing deCO$_x$ catalysts have low to medium acidity, large pores, and high surface area. However, the HDO product yield is significant for the majority of zeolites tested to date, which renders these supports less than ideal for deCO_x catalysts with the possible exception of modified SBA-15.

Other oxidic carriers can be further divided into two categories, namely, reducible (ZrO_2, TiO_2, and CeO_2) and nonreducible (Al_2O_3 and SiO_2) metal oxides. Through their efforts to identify promising supports for deCO_x catalysts comprising 10 wt% Ni as the active phase, Lercher and coworkers found that the use of the reducible supports mentioned earlier led to complete stearic acid conversion and a C_{17} selectivity between 87% and 96%, thus outperforming 5 wt% Pd/C [88]. In contrast, the nonreducible supports Al_2O_3 and SiO_2 only achieved conversions of 63% and 45%, respectively. The authors attribute the superior activity attained through the use of the reducible supports to the ability of said carriers to become involved in the deoxygenation reaction, which is further discussed in Section 12.6. It should be noted, however, that the Al_2O_3 support employed by Lercher and coworkers had a relatively low surface area (80 m^2/g) compared with Al_2O_3 supports tested by other authors (150–400 m^2/g), the latter leading to better conversions of lipid model compounds to hydrocarbons [89, 113]. Also of note is the fact that although several authors have determined SiO_2 to be a poor support for Ni-based deCO_x catalysts [84, 88, 89, 102], SiO_2 has been proven to be a good support for Pd-based catalysts in gas phase reaction conditions [21], which indicates that support effects are dependent on the nature of the active phase. These caveats for Al_2O_3 and SiO_2 are noteworthy due to the fact that their use as catalyst supports often results in a high selectivity to deCO_x products and decreased cracking, which can be attributed to their moderate acidity. In summary, the best supports for deCO_x catalysts are reducible oxides with moderate acidity and high surface area; thus, efforts to increase the surface area of reducible supports are both clearly indicated and currently underway.

12.5 Reaction Conditions

In addition to the catalyst used, the reaction conditions employed are also a major factor to be considered during the deoxygenation of lipids to hydrocarbons. Therefore, the effect of several variables such as reaction atmosphere, pressure, temperature, and solvents has been studied. In regard to the effect of the reaction atmosphere, it should be noted that H_2 is not stoichiometrically required for the deoxygenation of lipids via deCO_x, as is clearly shown in Scheme 12.1b,c. However, the presence of H_2 in the reaction atmosphere has been found to lead to improvements vis-à-vis both catalyst activity and selectivity to deCO_x products. Hydrogen can be supplied internally from the surface of prereduced catalysts [86, 88, 114–116] and/or through the reforming of the solvent [90, 117], or externally by including H_2 in the reaction atmosphere, which provides the most sizeable and durable benefits [9, 56, 118, 119]. Indeed, deoxygenation – particularly that of unsaturated feeds – occurs more slowly under inert reaction atmospheres [120, 121], and in some cases not at all [56]. However, the main reaction pathway in lipid deoxygenation under inert atmosphere appears to be direct decarboxylation, at least with fatty acid feeds

[122]. As hydrogen partial pressure increases, most authors have reported an increase in conversion along with a shift in the deoxygenation pathway toward decarbonylation [71, 90, 110, 120, 122, 123], with further increases in H_2 pressure tending to favor the HDO pathway [86, 124]. Interestingly, Boda et al. have shown that above certain H_2 pressures the hydrocarbon yield decreases, which may be due to increased competition between H_2 and the feed for adsorption sites on the catalyst surface [125].

Optimizing the hydrogen partial pressure can provide several benefits, such as promoting the hydrogenation of unsaturated feeds. Indeed, a number of authors have shown that hydrogenation of the C=C bonds in the feed takes place prior to its deoxygenation, as saturated species have been found to be the most prominent intermediates even under inert reaction atmospheres [71, 79, 88, 110, 112, 117]. This is of benefit given that the presence of unsaturated species has been shown to form aromatic compounds that risk giving rise to coke (of which they are precursors) and increase the deactivation rate of the catalyst [120, 126, 127]. Given that the CO_x evolved via the deCO$_x$ reaction can remain absorbed on the metal active sites and act as a catalyst poison, any hydrogen present can also assist in the removal of CO_x from the catalyst surface. The latter not only prevents catalyst deactivation by keeping the catalyst surface clean and accessible, but it also shifts the reaction equilibrium to the right, particularly when semi-batch and fixed-bed reactors (in which gaseous products are continuously removed) are employed [117]. In view of the foregoing, the presence of hydrogen can promote the catalytic deoxygenation of lipids via deCO$_x$, albeit the extent to which the latter takes place depends on the catalyst [17], reactor [116], and reaction conditions employed [71]. Nevertheless, care must be taken to ensure that hydrogen is not used in the deoxygenation reaction itself (as in HDO) and/or in side reactions (such as methanation) to avoid high H_2 consumption.

Unsurprisingly, the reaction temperature also influences the activity and selectivity values obtained during the catalytic deoxygenation of lipids, especially in situations where both HDO and deCO$_x$ can occur. Although typical deoxygenation reaction temperatures range between 250 and 400 °C, some authors have observed deoxygenation activity at lower temperatures [102, 111, 128], and others have used temperatures in excess of 400 °C [129]. For those reaction conditions in which deoxygenation can take place via both HDO and deCO$_x$, lower temperatures favor the former, while higher temperatures favor the latter [16, 17, 55, 112, 130]. Elevated reaction temperatures also tend to increase conversion [71, 89, 131, 132], albeit this comes at a cost. Indeed, higher temperatures tend to favor catalytic cracking and dehydration reactions, which are often followed by aromatization or isomerization reactions affording undesirable products [9, 49, 55, 71, 103, 115, 133]. Therefore, the influence of temperature on both conversion and selectivity should not be ignored; however, the catalyst employed as well as other reaction parameters seem to exert more significant effects.

In terms of the reaction solvent, the deoxygenation of lipids has been investigated in both aqueous and organic media, and the reaction pathway (HDO or deCO$_x$) appears to be independent of the solvent employed [95, 133–135]. However, there are conflicting reports regarding the effect of H_2O on catalyst activity. Indeed, whereas Chen et al. observed a dramatic drop in the conversion of oleic acid over a Pt/C catalyst when H_2O was used as the reaction medium [105], Lu and coworkers observed a 14% increase in the conversion of fatty acid esters when using H_2O compared with dodecane [134]. Naturally, water is particularly beneficial for the conversion of triglycerides since H_2O hydrolyses the C—O bond to afford three fatty acid molecules and glycerol, which reduces H_2 consumption [136] and may even provide the reagents necessary to generate hydrogen *in situ* through the dehydration of FFAs [71] and reforming of glycerol [96]. In fact, the prospect of utilizing H_2O both to generate H_2 *in situ* through the water–gas shift reaction and as a solvent during deoxygenation experiments has attracted some interest in recent years [95, 134, 135]. Notably, although it seems that the reaction pathway (HDO or deCO$_x$) is independent of the reaction medium, the relative performance of different catalysts is affected by the solvent employed. For instance, while in aqueous media Pt/C has been reported to be a more active deoxygenation catalyst than Pd/C, the reverse trend has been observed in organic media [79, 133]. Finally, a few studies have shown that among organic solvents, those with lower boiling points tend to lead to better catalytic deoxygenation results [114, 127]. Lamb and coworkers have explained this observation invoking the fact that solvents with high boiling points have lower vapor pressure, which increases the partial pressure of H_2 to a point that may inhibit the deCO$_x$ reaction as discussed earlier [117]. In view of the aforementioned effects of reaction atmosphere, pressure, temperature, and reaction medium – as well as the dependence of these effects on the catalyst employed – the importance of carefully considering the combination of catalyst and reaction conditions in the deoxygenation of lipid to hydrocarbons is clearly indicated.

12.6 Reaction Mechanism

Although the mechanism of lipid deoxygenation to hydrocarbons is dependent on several factors – including the nature of the active phase and the catalyst support as well as the reaction conditions – some general trends have been observed. First, many authors have reported that the saturation of any C=C bonds in the feed occurs prior to deoxygenation, even in the absence of exogenous hydrogen [71, 79, 88, 110, 117, 137]. In fact, Murzin and coworkers have suggested that under inert atmosphere unsaturated lipid chains undergo geometrical and double bond positional isomerizations that, when followed by dehydrogenation reactions ultimately result in the formation of polyunsaturated and aromatic byproducts via intramolecular Diels–Alder reactions [71]. Murzin and coworkers conclude that this reaction chain generates endogenous hydrogen that can saturate the C=C bonds in other unsaturated lipid molecules, which is in overall agreement with reports by other authors [137–140]. Doll and coworkers observed direct decarboxylation of unsaturated feeds leading to olefinic products [28]. Doll and coworkers proposed a tandem isomerization–decarboxylation model in which the C=C bonds undergo positional isomerization. Decarboxylation occurs when the double bonds are in close proximity to the carbonyl carbon, yielding β-alkenes [28]. The unsaturated products are susceptible to cracking, resulting in chain shortening via terminal carbon loss, forming methane. Similar products suggested by Doll's model were observed by Crocker and coworkers with unsaturated feeds [74]. For the resulting saturated lipids, the deoxygenation pathway followed depends on whether the feed is a fatty acid, a triglyceride, or another ester.

During the deoxygenation of saturated triglycerides to hydrocarbons, many authors have observed the formation and subsequent disappearance of fatty acids, which suggest that FFAs represent major intermediate molecules [15, 88, 141]. In aqueous media, the fatty acids are expected to be generated through hydrolysis of the C—O ester bonds to afford glycerol and three fatty acid molecules (Scheme 12.3a) [79, 142]. In organic media or solvent-free conditions, a β-elimination mechanism takes place as proposed by Vonghia et al. for the cracking of triglycerides over activated alumina [143]. Notably, in their study of the latter reaction, Vonghia et al. found two dominate pathways: (i) a β-elimination mechanism that produces a carboxylic acid and an unsaturated difatty ester (see Scheme 12.3b) and (ii) a γ-hydrogen transfer mechanism resulting in a C—C bond cleavage between the α- and β-carbons producing a terminal olefin with two less carbons than the parent fatty acid moiety (Scheme 12.3c). Given that over supported metal catalysts, fatty acids have been observed as intermediates during the conversion of triglycerides to hydrocarbons, while C_{n-2} hydrocarbons have not, it has been concluded that the β-elimination pathway is more plausible under $deCO_x$ reaction conditions [144]. Naturally, the unsaturated glycerol difatty ester produced through this route can subsequently undergo two sequential hydrogenation/β-elimination steps to ultimately yield three fatty acid molecules and a propane molecule stemming from the triglyceride backbone. Nevertheless, although the β-elimination mechanism is the most widely reported, the selective hydrogenolysis of the backbone-ester C—O bond – which yields similar products than β-elimination/hydrogenation as illustrated in Scheme 12.3d – has also been suggested [121].

Fatty acid feeds, as well as fatty acids formed as intermediates during triglyceride conversion, are deoxygenated to hydrocarbons through several different pathways, which are also dependent on the active metal, support, and reaction atmosphere employed. Under inert atmospheres, the dominant reaction pathway is the direct decarboxylation of FFAs to form C_{n-1} hydrocarbons and CO_2 [90, 123]. However, as the H_2 partial pressure increases, decarbonylation becomes more prevalent [120, 122], this route affording H_2O, CO, and a C_{n-1} terminal olefin that can be readily hydrogenated to the corresponding alkane in the presence of hydrogen. Notably, two distinct mechanisms have been proposed for this particular pathway. Indeed, Boda et al. suggested that under conditions favoring decarbonylation, hydrogenolysis of the C—COOH bond occurs to produce formic acid and a C_{n-1} hydrocarbon [125], which is in line with a report by Berenblyum et al. who observed heptadecene and formic acid formation on the surface of Pd clusters in their work on the $deCO_x$ of stearic acid [90]. The formic acid generated is subsequently decarbonylated to CO and H_2O or decarboxylated to CO_2 and H_2, the rate limiting step being the C—C bond cleavage of the fatty acid. The second mechanism was proposed by Lercher and coworkers based on their work on the conversion of lipids to hydrocarbons

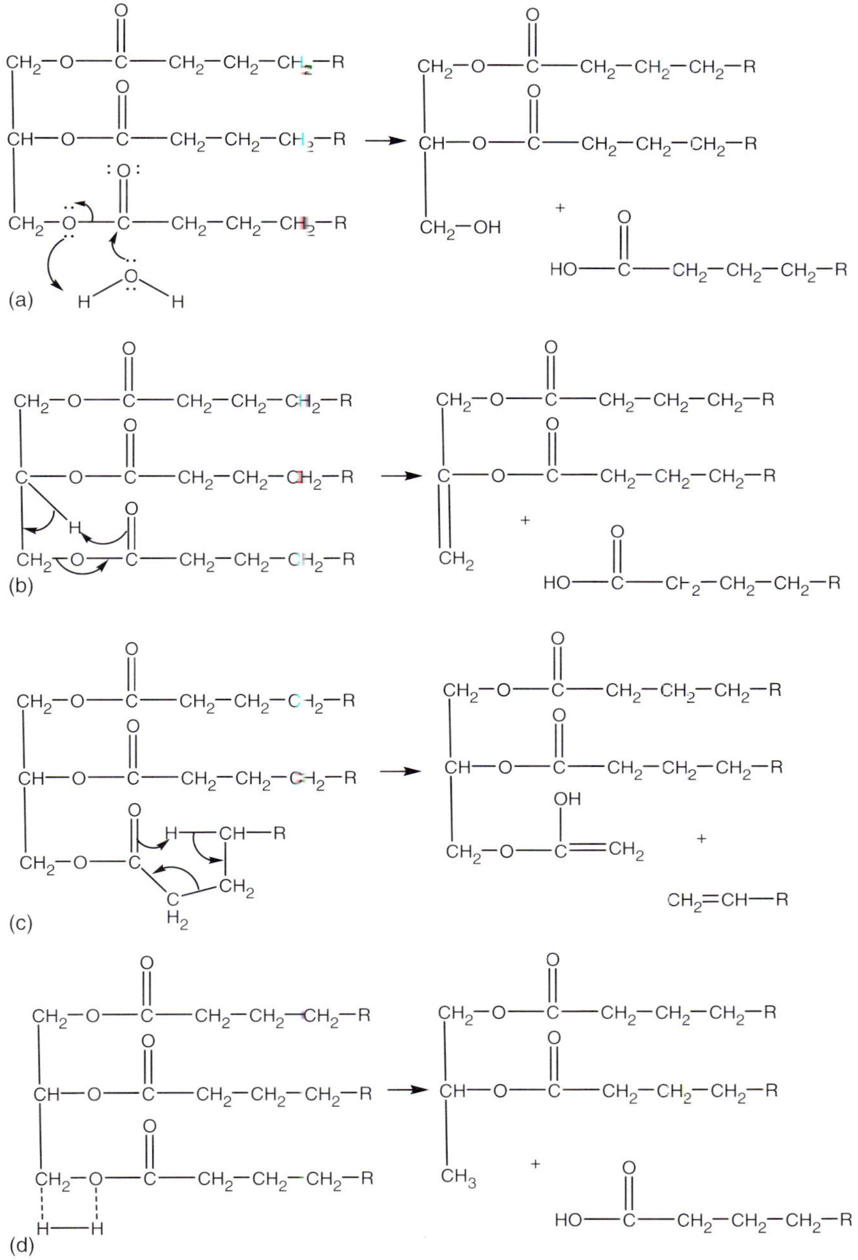

Scheme 12.3 Mechanism for (a) hydrolysis, (b) β-elimination, (c) γ-hydrogen transfer, and (d) hydrogenolysis of triglycerides.

over a Ni/ZrO_2 catalyst [88, 121]. In line with the discussion earlier, Lercher and coworkers proposed that prior to deoxygenation, any C=C double bonds in the feed were hydrogenated, while the selective hydrogenolysis of the backbone-ester C–O bond in triglycerides formed three fatty acids and propane (Scheme 12.3d). The fatty acids formed are then either dissociatively adsorbed on the support at oxygen defect sites to yield a carboxylate species [145–147] or adsorbed on the nickel surface [121]. On the support, abstraction of the α-hydrogen on the carboxylate generates a ketene intermediate, albeit this has only been reported for reducible carriers such as ZrO_2 [88]. The α-hydrogen and the hydrogen stemming from the dissociative adsorption of the fatty acid combine with the oxygen at the defect site, desorption of the water molecule thus formed regenerating the oxygen vacancy. In inert atmospheres, the support-bound ketene intermediate interacts with a neighboring carboxylate species to form an acid anhydride, which decomposes to a symmetrical ketone releasing CO_2 in the process [145]. In H_2 atmospheres, the ketene intermediate migrates from the support to a nickel surface site, where it is hydrogenated to the aldehyde as shown in Scheme 12.4. Miran and coworkers also indicated that a metal oxide support is responsible for the formation of esters from neighboring alcohol products and carboxylic acids. Simultaneously, the fatty acids adsorbed on the nickel surface can undergo hydrogenolysis to form the corresponding aldehyde and H_2O. Lercher and coworkers have determined that while the removal of the first oxygen atom can be effected by the support or the nickel surface, the decarbonylation step resulting in the removal of the second oxygen atom can only occur on nickel [88, 121]. Other authors have also observed the formation of aldehyde intermediates [120, 148] over Pd/Al_2O_3 and Pd/C catalysts, extending the applicability of the pathway proposed by Lercher and coworkers over Ni/ZrO_2 to other metals and supports, albeit (as mentioned earlier) the direct involvement of the carrier in deoxygenation has only been observed with reducible supports.

Tellingly, other oxygenates such as alcohols and esters have also been observed in the conversion of fatty acids to hydrocarbons [81, 121]. The detection of alcohols is not entirely surprising, since in the presence of hydrogen an alcohol is formed via further hydrogenation of an aldehyde and both species exist in equilibrium [121]. Indeed, the hydrogenation process is reversible, meaning that alcohol can be dehydrogenated back to the aldehyde for decarbonylation to ensue. Alternatively, if conditions favoring HDO are employed, the alcohol is hydrogenated/dehydrated to form a hydrocarbon containing the same number of carbon atoms as the fatty acid from which it was derived. However, an alcohol can also undergo esterification with a fatty acid molecule, thus producing the long-chain ester species observed by some authors. In turn, esters can undergo hydrogenolysis to reform the parent alcohol and aldehyde [121], which then undergo the deoxygenation pathway previously described. The multiple reaction pathways involving FFAs discussed previously are illustrated in Scheme 12.5.

Lastly, an alternative pathway for the deoxygenation of non-glyceride esters has been proposed by Murzin and coworkers based on their work on the deCO$_x$ of ethyl stearate over Pd/C [127]. Parenthetically, ethyl and methyl esters are important feeds stemming from lipid extraction methods such as the *in situ* esterification approach employed to recover the oleaginous fraction of algal cells [149].

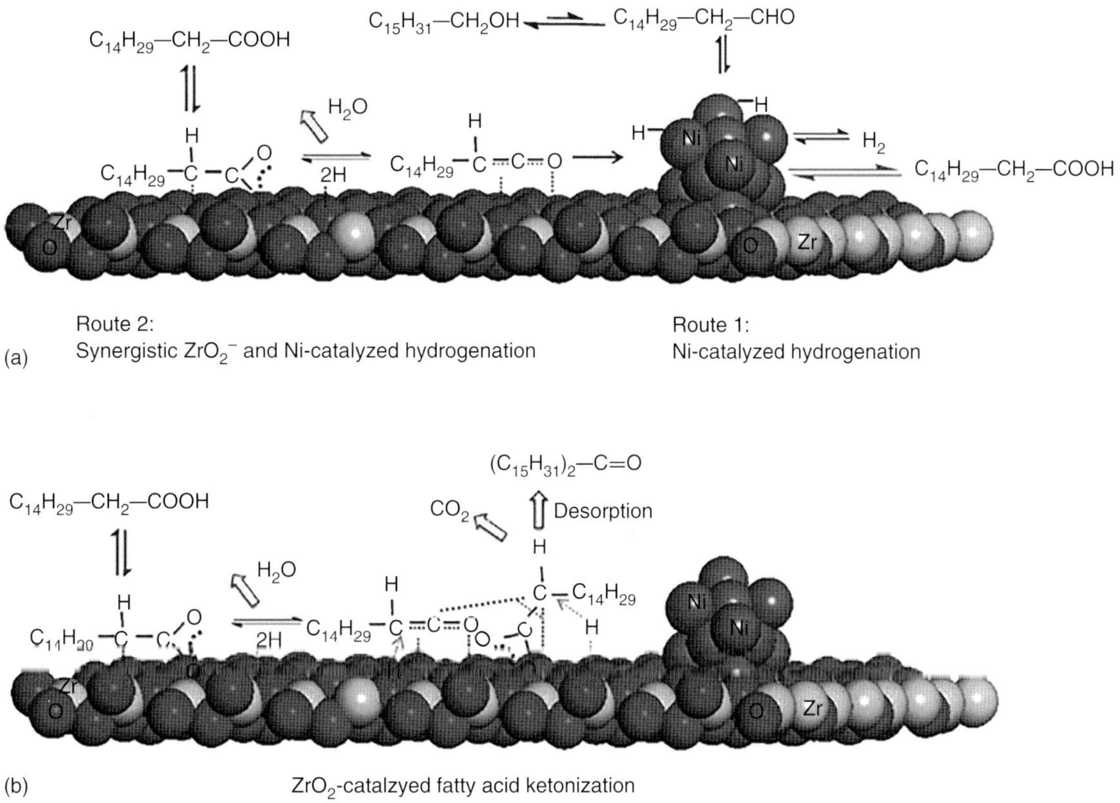

Scheme 12.4 Proposed reaction mechanism for (a) the hydrogenation of palmitic acid to hexadecanal in the presence of H_2, and (b) the ketonization of palmitic acid to palmitone in the absence of H_2. Source: Adapted from Peng et al. (2013) [121].

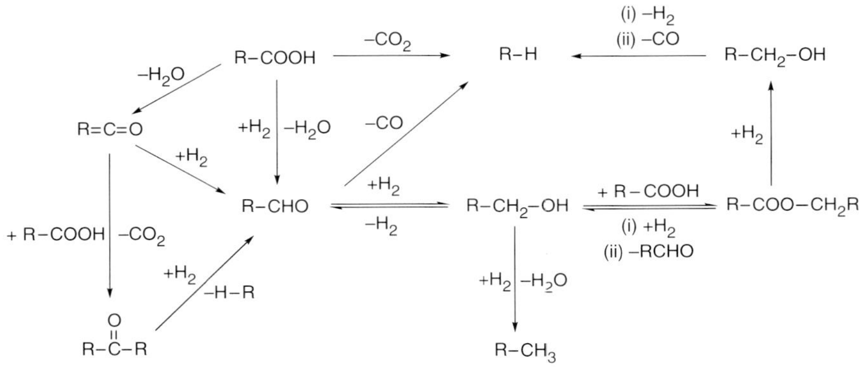

Scheme 12.5 Reaction scheme for stearic acid deoxygenation. Source: Adapted from Loe et al. (2016) [75].

In the presence of H_2, Murzin and coworkers observed the scission of the bond between the carbonyl carbon and the α-carbon of the alkyl chain, resulting in the generation of *n*-heptadecane and an ester moiety that subsequently decomposes to yield ethanol and CO. Murzin and coworkers also observed hydrocarbons with less than 17 carbon atoms in the product mixture, which was explained by suggesting that shorter hydrocarbons can be formed from the alkyl chains via cracking on acid sites associated with the support or by hydrogenolysis over metallic sites on the surface of palladium particles [127].

12.7 Catalyst Deactivation

Catalyst resistance to deactivation remains one of the main challenges associated with deoxygenation reactions. While a number of deactivation pathways often lead to loss of active metal sites – including leaching, oxidation, sintering, or poisoning – decreases in the surface area and/or the porosity of the support also result in lower catalyst activity. However, neither metal oxidation [150, 151] nor metal leaching [55, 84, 116] appear to be very prevalent under most deoxygenation reaction conditions. Sintering of the active metals not only decreases the metal-specific surface area by increasing the particle size, but this process also lessens the metal–support interaction, which can potentially alter the electronic state of the active metal and affect its activity. Moreover, large metal particles can also block the pores of the support, thus hampering the access of reagent molecules to catalytically active sites. Notably, although sintering can be significant for carbon-supported Pd and Pt catalysts used under hydrothermal conditions, this does not appear to result in loss of catalytic activity [133]. In contrast, Ni catalysts are readily deactivated under hydrothermal conditions due to oxidation of the metallic Ni active phase [151]. In organic media, sintering seems to be considerable only for catalysts using silicoaluminophosphates as the support [125], which is fortunate since metal dispersion is known to be important for formulations comprising carbonaceous [152] and oxidic supports [125, 153].

In regard to poisoning, impurities in the feeds, including nitrogen-, phosphorus-, and sulfur-containing compounds, can deactivate the catalyst by irreversibly binding to the active metal and to Lewis acid sites on the catalyst support [78, 114, 154]. Similarly, any CO generated during the deoxygenation reaction can also act as a catalyst poison by strongly binding to the surface of the active metal. Indeed, CO is known to dissociatively adsorb on Ni catalysts [155], forming carbonaceous deposits that can act as precursors to coke.

Notably, the most common pathway for catalyst deactivation during the deoxygenation of lipids to hydrocarbons is the accumulation of organic deposits on the catalyst surface. Indeed, carbonaceous deposits reduce the metal-specific surface area by physically covering metal sites or by blocking the pores of the support, thus impeding the access of reagents to active sites [156]. Naturally, the reaction atmosphere and the feed employed can exacerbate these processes, as higher deactivation rates have been observed under inert reaction atmospheres and when unsaturated feeds are used [157]. Remarkably, van Es and coworkers found that under an inert atmosphere, the addition of 0.1 equiv of mono-unsaturated fatty acids to the reaction medium reduced the deoxygenation activity of Pd catalysts by 60%, irrespective of whether the support was carbonaceous or oxidic [157]. This is consistent with the fact that unsaturated compounds can be dehydrogenated to polyunsaturated molecules that may undergo intramolecular Diels–Alder [71, 105, 140] or oligomerization [126] reactions, forming the aromatic or unsaturated aliphatic products observed by many authors. Polymeric and crystalline types of carbonaceous deposits – commonly referred to as coke – are typically more complex [158] and are more strongly adsorbed than smaller and simpler unsaturated species, as evinced by the catalyst deactivation studies performed by Savage and coworkers [159]. Nevertheless, Savage found that for Pt/C catalysts tested under hydrothermal conditions, unsaturated products represent minor contributors to deactivation and coke formation, the collapse of the pores being mainly responsible for deactivation. Interestingly, Li et al. found in a recent study that the deactivation of zeolite-supported Ni catalysts is associated with a specific mechanism of coke formation in which carbocations generated on Brønsted acid sites on the catalyst surface stack into disordered and filament-like carbon strands similar to carbon nanotubes that evolve into graphitic coke covering the catalyst surface [153]. Unsurprisingly, which of the multiple pathways for coke-induced deactivation takes precedence seems to depend on the catalyst and reaction conditions employed. However, once deactivation has occurred, catalyst regeneration can be difficult. Fortunately, the adsorption of unsaturated products can be curbed to a large degree by the presence of H_2 in the reaction atmosphere [98, 127, 157, 160]. In fact, for catalysts deactivated due to the surface accumulation of unsaturated products, removal of the adsorbed species from the catalyst surface can be achieved by simply exposing the catalyst to H_2 [161], albeit Ping et al. have claimed that a series of solvent washes can also be used to remove most of the organic deposits. Obviously, both unsaturated products and coke can be removed through controlled oxidation [162], which would be the approach favored industrially as it would avoid the use of solvents and excess H_2. As mentioned earlier, this solution would not be viable for carbon-supported catalysts, which makes the use of metal oxides as carriers more advantageous, at least in terms of catalyst regeneration.

12.8 Conclusions and Outlook

Fuel-like hydrocarbons stemming from the deoxygenation of fats and oils offer several advantages compared with both biodiesel- and petroleum-derived fuels [100]. In the deoxygenation of these feeds via the deCO$_x$ pathway, various metallic active phases and support materials have been investigated, Pt-, Pd-, and Ni-based catalysts exhibiting the most promising results. More recently, the study of bimetallic active phases has gained traction as these catalysts have shown improved deoxygenation activity and selectivity to deCO$_x$ products compared with their monometallic counterparts [74, 75, 92, 95, 96, 103]. The reaction conditions employed also affect the product slate in deCO$_x$ reactions, particularly when H_2 is used as the reaction atmosphere. However, in order for the deCO$_x$ pathway to be deemed more advantageous than the HDO pathway, it is important to minimize the H_2 consumption. Indeed, although H_2 is not consumed in the removal of oxygen from the lipid feeds, the CO and CO_2 evolved can in some cases be readily converted to methane by the deCO$_x$ catalyst in the presence of H_2. Admittedly, the hydrocarbon products have also been observed to partake in side reactions such as cracking and isomerization; however, these products may not be problematic as lighter and branched hydrocarbons display superior cold flow properties. If H_2 is not used, several unfavorable side reactions can take place – such as oligomerization, ketonization, cyclization, dehydration, aromatization, and coking – many of which lead to catalyst deactivation and lower yields of desirable products. These negative side reactions are more prevalent with unsaturated lipids, which are common to most feed sources. Therefore, the consumption of some hydrogen is necessary to saturate these lipids. For this reason, simultaneous deoxygenation and reforming of glycerol [163] or methanol [164] to generate H_2 *in situ* has started to attract interest. The issues of catalyst deactivation and regeneration must also be addressed more broadly and thoroughly in future studies. Nevertheless, although improvements to deCO$_x$ catalyst technology are needed, recent developments in this area are promising.

References

1 US Energy Information Administration (2017). Internation Energy Outlook.

2 IEA Bioenergy Task 40 Sustainable Internation Bioenergy Trade (2012). The potential role of biofuels in commercial air transport–biojetfuel.

3 Benazzi, E. and Cameron, C. (2006). Boutique diesel in the on-road market. *Hydrocarbon Process.* 85 (3): DD17–DD18.

4 Chevron (2006). Alternative jet fuels.

5 Pattanaik, B.P. and Misra, R.D. (2017). Effect of reaction pathway and operating parameters on the deoxygenation of vegetable oils to produce diesel range hydrocarbon fuels: a review. *Renew. Sustain. Energy Rev.* 73: 545–557.

6 Naik, S.N., Goud, V.V., Rout, P.K., and Dalai, A.K. (2010). Production of first and second generation biofuels: a comprehensive review. *Renew. Sustain. Energy Rev.* 14 (2): 578–597.

7 Knothe, G. (2010). Biodiesel and renewable diesel: a comparison. *Prog. Energy Combust. Sci.* 36 (3): 364–373.

8 Kordulis, C., Bourikas, K., Gousi, M. et al. (2016). Development of nickel based catalysts for the transformation of natural triglycerides and related compounds into green diesel: a critical review. *Appl. Catal., B* 181: 156–196.

9 Kubičková, I., Snåre, M., Eränen, K. et al. (2005). Hydrocarbons for diesel fuel via decarboxylation of vegetable oils. *Catal. Today* 106 (1–4): 197–200.

10 Tran, N.H., Bartlett, J.R., Kannangara, G.S.K. et al. (2010). Catalytic upgrading of biorefinery oil from micro-algae. *Fuel* 89 (2): 265–274.

11 Santillan-Jimenez, E. and Crocker, M. (2012). Catalytic deoxygenation of fatty acids and their derivatives to hydrocarbon fuels via decarboxylation/decarbonylation. *J. Chem. Technol. Biotechnol.* 87 (8): 1041–1050.

12 Ameen, M., Azizan, M.T., Yusup, S. et al. (2017). Catalytic hydrodeoxygenation of triglycerides: an approach to clean diesel fuel production. *Renew. Sustain. Energy Rev.* 80: 1072–1088.

13 Sotelo-Boyás, R., Trejo-Zárraga, F., and de Jesús Hernández-Loyo, F. (2012). Hydroconversion of triglycerides into green liquid fuels. In: *Hydrogenation* (ed. I. Karamé), 187–216. InTech.

14 Snåre, M., Mäki-Arvela, P., Simakova, I. et al. (2009). Overview of catalytic methods for production of next generation biodiesel from natural oils and fats. *Russ. J. Phys. Chem. B* 3 (7): 1035–1043.

15 Morgan, T., Grubb, D., Santillan-Jimenez, E., and Crocker, M. (2010). Conversion of triglycerides to hydrocarbons over supported metal catalysts. *Top. Catal.* 53 (11–12): 820–829.

16 Huber, G.W., O'Connor, P., and Corma, A. (2007). Processing biomass in conventional oil refineries: production of high quality diesel by hydrotreating vegetable oils in heavy vacuum oil mixtures. *Appl. Catal., A* 329: 120–129.

17 Berenblyum, A.S., Danyushevsky, V.Y., Katsman, E.A. et al. (2010). Production of engine fuels from inedible vegetable oils and fats. *Pet. Chem.* 50 (4): 305–311.

18 Bertram, S.H. (1936). Action of selenium on stearic acid. *ChemWeek* 33: 457–459.

19 Foglia, T. and Barr, P. (1976). Decarbonylation dehydration of fatty acids to alkenes in the presence of transition metal complexes. *J. Am. Oil Chem. Soc.* 53 (12): 737–741.

20 Stern, R. and Hillion, G. (1985). Process for manufacturing a linear olefin from a saturated fatty acid or fatty acid ester. US Patent 4,554,397, filed 24 August 1984 and issued 19 November 1985.

21 Maier, W.F., Roth, W., Thies, I., and Schleyer, P.V.R. (1982). Hydrogenolysis, IV. Gas phase decarboxylation of carboxylic acids. *Eur. J. Inorg. Chem.* 115 (2): 808–812.

22 Knothe, G., Steidley, K.R., Moser, B.R., and Doll, K.M. (2017). Decarboxylation of fatty acids with triruthenium dodecacarbonyl: influence of the compound structure and analysis of the product mixtures. *ACS Omega* 2 (10): 6473–6480.

23 Moser, B.R., Knothe, G., Walter, E.L. et al. (2016). Analysis and properties of the decarboxylation products of oleic acid by catalytic triruthenium dodecacarbonyl. *Energy Fuels* 30 (9): 7443–7451.
24 Li, W., Gao, Y., Yao, S. et al. (2015). Effective deoxygenation of fatty acids over $Ni(OAc)_2$ in the absence of H_2 and solvent. *Green Chem.* 17 (8): 4198–4205.
25 John, A., Dereli, B.S., Ortuño, M.A. et al. (2017). Selective decarbonylation of fatty acid esters to linear α-olefins. *Organometallics* 36 (15): 2956–2964.
26 John, A., Hillmyer, M.A., and Tolman, W.B. (2017). Anhydride-additive-free nickel-catalyzed deoxygenation of carboxylic acids to olefins. *Organometallics* 36 (3): 506–509.
27 Bie, Y., Lehtonen, J., and Kanervo, J. (2016). Hydrodeoxygenation (HDO) of methyl palmitate over bifunctional Rh/ZrO_2 catalyst: Insights into reaction mechanism via kinetic modeling. *Appl. Catal., A* 526: 183–190.
28 Murray, R.E., Walter, E.L., and Doll, K.M. (2014). Tandem isomerization-decarboxylation for converting alkenoic fatty acids into alkenes. *ACS Catal.* 4 (10): 3517–3520.
29 Di, L., Yao, S., Song, S. et al. (2017). Robust ruthenium catalysts for the selective conversion of stearic acid to diesel-range alkanes. *Appl. Catal., B* 201: 137–149.
30 Rodina, V., Ermakov, D.Y., Saraev, A. et al. (2017). Influence of reaction conditions and kinetic analysis of the selective hydrogenation of oleic acid toward fatty alcohols on $Ru\text{-}Sn\text{-}B/Al_2O_3$ in the flow reactor. *Appl. Catal., B* 209: 611–620.
31 He, L., Wu, C., Cheng, H. et al. (2012). Highly selective and efficient catalytic conversion of ethyl stearate into liquid hydrocarbons over a Ru/TiO_2 catalyst under mild conditions. *Catal. Sci. Technol.* 2 (7): 1328–1331.
32 Zhao, X., Wei, L., Cheng, S. et al. (2016). Hydroprocessing of carinata oil for hydrocarbon biofuel over $Mo\text{-}Zn/Al_2O_3$. *Appl. Catal., B* 196: 41–49.
33 Asikin-Mijan, N., Lee, H., Juan, J. et al. (2017). Promoting deoxygenation of triglycerides via Co-Ca loaded $SiO_2\text{-}Al_2O_3$ catalyst. *Appl. Catal., A* 552: 38–48.
34 Cao, Y., Shi, Y., Bi, Y. et al. (2018). Hydrodeoxygenation and hydroisomerization of palmitic acid over bi-functional Co/H-ZSM-22 catalysts. *Fuel Process. Technol.* 172: 29–35.
35 Bian, J., Wang, Y., Zhang, Q. et al. (2017). Fatty acid decarboxylation reaction kinetics and pathway of co-conversion with amino acid on supported iron oxide catalysts. *RSC Adv.* 7 (75): 47279–47287.
36 Kandel, K., Anderegg, J.W., Nelson, N.C. et al. (2014). Supported iron nanoparticles for the hydrodeoxygenation of microalgal oil to green diesel. *J. Catal.* 314: 142–148.
37 Al Alwan, B., Salley, S.O., and Ng, K.Y.S. (2015). Biofuels production from hydrothermal decarboxylation of oleic acid and soybean oil over Ni-based transition metal carbides supported on Al-SBA-15. *Appl. Catal., A* 498: 32–40.
38 Han, J., Duan, J., Chen, P. et al. (2011). Nanostructured molybdenum carbides supported on carbon nanotubes as efficient catalysts for one-step

hydrodeoxygenation and isomerization of vegetable oils. *Green Chem.* 13 (9): 2561.
39 Rocha, A.S., Souza, L.A., Oliveira, R.R. Jr., et al. (2017). Hydrodeoxygenation of acrylic acid using Mo_2C/Al_2O_3. *Appl. Catal., A* 531: 69–78.
40 Stellwagen, D.R. and Bitter, J.H. (2015). Structure–performance relations of molybdenum- and tungsten carbide catalysts for deoxygenation. *Green Chem.* 17 (1): 582–593.
41 Wang, W., Zhang, K., Liu, H. et al. (2013). Hydrodeoxygenation of p-cresol on unsupported Ni–P catalysts prepared by thermal decomposition method. *Catal. Commun.* 41: 41–46.
42 Prins, R. and Bussell, M.E. (2012). Metal phosphides: preparation, characterization and catalytic reactivity. *Catal. Lett.* 142 (12): 1413–1436.
43 Peroni, M., Lee, I., Huang, X. et al. (2017). Deoxygenation of palmitic acid on unsupported transition-metal phosphides. *ACS Catal.* 7 (9): 6331–6341.
44 Pan, Z., Wang, R., Nie, Z., and Chen, J. (2016). Effect of a second metal (Co, Fe, Mo and W) on performance of Ni_2P/SiO_2 for hydrodeoxygenation of methyl laurate. *J. Energy Chem.* 25 (3): 418–426.
45 Liu, C., Hao, Y., Jing, Z. et al. (2016). Hydrodeoxygenation of fatty acid methyl esters and isomerization of products over NiP/SAPO-11 catalysts. *J. Fuel Chem. Technol.* 44 (10): 1211–1216.
46 Alvarez-Galvan, M.C., Blanco-Brieva, G., Capel-Sanchez, M. et al. (2018). Metal phosphide catalysts for the hydrotreatment of non-edible vegetable oils, Catal. *Today* 302: 242–249.
47 Liu, Y., Yao, L., Xin, H. et al. (2015). The production of diesel-like hydrocarbons from palmitic acid over HZSM-22 supported nickel phosphide catalysts. *Appl. Catal., B* 174–175: 504–514.
48 Ferrari, M., Maggi, R., Delmon, B., and Grange, P. (2001). Influences of the hydrogen sulfide partial pressure and of a nitrogen compound on the hydrodeoxygenation activity of a CoMo/carbon catalyst. *J. Catal.* 198 (1): 47–55.
49 Yang, Y., Wang, Q., Zhang, X. et al. (2013). Hydrotreating of C_{18} fatty acids to hydrocarbons on sulphided $NiW/SiO_2–Al_2O_3$. *Fuel Process. Technol.* 116: 165–174.
50 Verma, D., Rana, B.S., Kumar, R. et al. (2015). Diesel and aviation kerosene with desired aromatics from hydroprocessing of jatropha oil over hydrogenation catalysts supported on hierarchical mesoporous SAPO-11. *Appl. Catal., A* 490: 108–116.
51 Priecel, P., Kubička, D., Čapek, L. et al. (2011). The role of Ni species in the deoxygenation of rapeseed oil over NiMo-alumina catalysts. *Appl. Catal., A* 397 (1–2): 127–137.
52 Srifa, A., Viriya-empikul, N., Assabumrungrat, S., and Faungnawakij, K. (2015). Catalytic behaviors of $Ni/\gamma-Al_2O_3$ and $Co/\gamma-Al_2O_3$ during the hydrodeoxygenation of palm oil. *Catal. Sci. Technol.* 5 (7): 3693–3705.
53 Srifa, A., Faungnawakij, K., Itthibenchapong, V., and Assabumrungrat, S. (2015). Roles of monometallic catalysts in hydrodeoxygenation of palm oil to green diesel. *Chem. Eng. J.* 278: 249–258.

54 Onyestyák, G., Harnos, S., Szegedi, Á., and Kalló, D. (2012). Sunflower oil to green diesel over Raney-type Ni-catalyst. *Fuel* 102: 282–288.

55 Kikhtyanin, O.V., Rubanov, A.E., Ayupov, A.B., and Echevsky, G.V. (2010). Hydroconversion of sunflower oil on Pd/SAPO-31 catalyst. *Fuel* 89 (10): 3085–3092.

56 Sugami, Y., Minami, E., and Saka, S. (2016). Renewable diesel production from rapeseed oil with hydrothermal hydrogenation and subsequent decarboxylation. *Fuel* 166: 376–381.

57 Sugami, Y., Minami, E., and Saka, S. (2017). Hydrocarbon production from coconut oil by hydrolysis coupled with hydrogenation and subsequent decarboxylation. *Fuel* 197: 272–276.

58 Herskowitz, M., Landau, M.V., Reizner, Y., and Berger, D. (2013). A commercially-viable, one-step process for production of green diesel from soybean oil on Pt/SAPO-11. *Fuel* 111: 157–164.

59 Fargione, J., Hill, J., Tilman, D. et al. (2008). Land clearing and the biofuel carbon debt. *Science* 319 (5867): 1235–1238.

60 Tilman, D., Hill, J., and Lehman, C. (2006). Carbon-negative biofuels from low-input high-diversity grassland biomass. *Science* 314 (5805): 1598–1600.

61 Vásquez, M.C., Silva, E.E., and Castillo, E.F. (2017). Hydrotreatment of vegetable oils: a review of the technologies and its developments for jet biofuel production. *Biomass Bioenergy* 105: 197–206.

62 Amin, S. (2009). Review on biofuel oil and gas production processes from microalgae. *Energy Convers. Manage.* 50 (7): 1834–1840.

63 Zhao, C., Brück, T., and Lercher, J.A. (2013). Catalytic deoxygenation of microalgae oil to green hydrocarbons. *Green Chem.* 15 (7): 1720–1739.

64 Posten, C. and Schaub, G. (2009). Microalgae and terrestrial biomass as source for fuels—a process view. *J. Biotechnol.* 142 (1): 64–69.

65 Mata, T.M., Martins, A.A., and Caetano, N.S. (2010). Microalgae for biodiesel production and other applications: a review. *Renew. Sustain. Energy Rev.* 14 (1): 217–232.

66 Brennan, L. and Owende, P. (2010). Biofuels from microalgae—a review of technologies for production, processing, and extractions of biofuels and co-products. *Renew. Sustain. Energy Rev.* 14 (2): 557–577.

67 Bwapwa, J.K., Anandraj, A., and Trois, C. (2017). Possibilities for conversion of microalgae oil into aviation fuel: a review. *Renew. Sustain. Energy Rev.* 80: 1345–1354.

68 Shi, F., Wang, P., Duan, Y. et al. (2012). Recent developments in the production of liquid fuels via catalytic conversion of microalgae: experiments and simulations. *RSC Adv.* 2 (26): 9727–9747.

69 Chen, Y., Wu, Y., Hua, D. et al. (2015). Thermochemical conversion of low-lipid microalgae for the production of liquid fuels: challenges and opportunities. *RSC Adv.* 5 (24): 18673–18701.

70 Berenblyum, A.S., Podoplelova, T.A., Shamsiev, R.S. et al. (2012). Catalytic chemistry of preparation of hydrocarbon fuels from vegetable oils and fats. *Catal. Ind.* 4 (3): 209–214.

71 Mäki-Arvela, P.i., Rozmysłowicz, B., Lestari, S. et al. (2011). Catalytic deoxygenation of tall oil fatty acid over palladium supported on mesoporous carbon. *Energy Fuels* 25 (7): 2815–2825.

72 Jeništová, K., Hachemi, I., Mäki-Arvela, P. et al. (2017). Hydrodeoxygenation of stearic acid and tall oil fatty acids over Ni-alumina catalysts: influence of reaction parameters and kinetic modelling. *Chem. Eng. J.* 316: 401–409.

73 Chhetri, A.B., Watts, K.C., and Islam, M.R. (2008). Waste cooking oil as an alternate feedstock for biodiesel production. *Energies* 1 (1): 3–18.

74 Santillan-Jimenez, E., Loe, R., Garrett, M. et al. (2018). Effect of Cu promotion on cracking and methanation during the Ni-catalyzed deoxygenation of waste lipids and hemp seed oil to fuel-like hydrocarbons. *Catal. Today* 302: 261–271.

75 Loe, R., Santillan-Jimenez, E., Morgan, T. et al. (2016). Effect of Cu and Sn promotion on the catalytic deoxygenation of model and algal lipids to fuel-like hydrocarbons over supported Ni catalysts. *Appl. Catal., B* 191: 147–156.

76 Ford, J.P., Thapaliya, N., Kelly, M.J. et al. (2013). Semi-batch deoxygenation of canola- and lard-derived fatty acids to diesel-range hydrocarbons. *Energy Fuels* 27 (12): 7489–7496.

77 Sari, E., DiMaggio, C., Kim, M. et al. (2013). Catalytic conversion of brown grease to green diesel via decarboxylation over activated carbon supported palladium catalyst. *Ind. Eng. Chem. Res.* 52 (33): 11527–11536.

78 Simakova, I., Simakova, O., Mäki-Arvela, P., and Murzin, D.Y. (2010). Decarboxylation of fatty acids over Pd supported on mesoporous carbon. *Catal. Today* 150 (1–2): 28–31.

79 Fu, J., Lu, X., and Savage, P.E. (2011). Hydrothermal decarboxylation and hydrogenation of fatty acids over Pt/C. *ChemSusChem* 4 (4): 481–486.

80 Lestari, S., Mäki-Arvela, P., Simakova, I. et al. (2009). Catalytic deoxygenation of stearic acid and palmitic acid in semibatch mode. *Catal. Lett.* 130 (1–2): 48–51.

81 Yang, L. and Carreon, M.A. (2017). Deoxygenation of palmitic and lauric acids over Pt/ZIF-67 membrane/zeolite 5A bead catalysts. *ACS Appl. Mater. Interfaces* 9 (37): 31993–32000.

82 Ford, J.P., Immer, J.G., and Lamb, H.H. (2012). Palladium catalysts for fatty acid deoxygenation: influence of the support and fatty acid chain length on decarboxylation kinetics. *Top. Catal.* 55 (3–4): 175–184.

83 Mohite, S., Armbruster, U., Richter, M., and Martin, A. (2014). Impact of chain length of saturated fatty acids during their heterogeneously catalyzed deoxygenation. *J. Sustain. Bioenergy Syst.* 4 (3): 183–193.

84 Snåre, M., Kubičková, I., Mäki-Arvela, P. et al. (2006). Heterogeneous catalytic deoxygenation of stearic acid for production of biodiesel. *Ind. Eng. Chem. Res.* 45 (16): 5708–5715.

85 Shi, N., Liu, Q.-Y., Jiang, T. et al. (2012). Hydrodeoxygenation of vegetable oils to liquid alkane fuels over Ni/HZSM-5 catalysts: methyl hexadecanoate as the model compound. *Catal. Commun.* 20: 80–84.

86 Peng, B., Yao, Y., Zhao, C., and Lercher, J.A. (2012). Towards quantitative conversion of microalgae oil to diesel-range alkanes with bifunctional catalysts. *Angew. Chem.* 124 (9): 2114–2117.
87 Liu, Q., Zuo, H., Wang, T. et al. (2013). One-step hydrodeoxygenation of palm oil to isomerized hydrocarbon fuels over Ni supported on nano-sized SAPO-11 catalysts. *Appl. Catal., A* 468: 68–74.
88 Peng, B., Yuan, X., Zhao, C., and Lercher, J.A. (2012). Stabilizing catalytic pathways via redundancy: selective reduction of microalgae oil to alkanes. *J. Am. Chem. Soc.* 134 (22): 9400–9405.
89 Kumar, P., Yenumala, S.R., Maity, S.K., and Shee, D. (2014). Kinetics of hydrodeoxygenation of stearic acid using supported nickel catalysts: effects of supports. *Appl. Catal., A* 471: 28–38.
90 Berenblyum, A.S., Podoplelova, T.A., Shamsiev, R.S. et al. (2011). On the mechanism of catalytic conversion of fatty acids into hydrocarbons in the presence of palladium catalysts on alumina. *Pet. Chem.* 51 (5): 336–341.
91 Lestari, S., Mäki-Arvela, P., Eränen, K. et al. (2010). Diesel-like hydrocarbons from catalytic deoxygenation of stearic acid over supported Pd nanoparticles on SBA-15 catalysts. *Catal. Lett.* 134 (3–4): 250–257.
92 Sun, K., Wilson, A.R., Thompson, S.T., and Lamb, H.H. (2015). Catalytic deoxygenation of octanoic acid over supported palladium: effects of particle size and alloying with gold. *ACS Catal.* 5 (3): 1939–1948.
93 Berenblyum, A.S., Shamsiev, R.S., Podoplelova, T.A., and Danyushevsky, V.Y. (2012). The influence of metal and carrier natures on the effectiveness of catalysts of the deoxygenation of fatty acids into hydrocarbons. *Russ. J. Phys. Chem. A* 86 (8): 1199–1203.
94 Ahmadi, M., Macias, E.E., Jasinski, J.B. et al. (2014). Decarboxylation and further transformation of oleic acid over bifunctional, Pt/SAPO-11 catalyst and Pt/chloride Al_2O_3 catalysts. *J. Mol. Catal. A: Chem.* 386: 14–19.
95 Yeh, T.M., Hockstad, R.L., Linic, S., and Savage, P.E. (2015). Hydrothermal decarboxylation of unsaturated fatty acids over $PtSn_x$/C catalysts. *Fuel* 156: 219–224.
96 Vardon, D.R., Sharma, B.K., Jaramillo, H. et al. (2014). Hydrothermal catalytic processing of saturated and unsaturated fatty acids to hydrocarbons with glycerol for in situ hydrogen production. *Green Chem.* 16 (3): 1507–1520.
97 Wu, J., Shi, J., Fu, J. et al. (2016). Catalytic decarboxylation of fatty acids to aviation fuels over nickel supported on activated carbon. *Sci. Rep.* 6: 27820.
98 Madsen, A.T., Ahmed, E.H., Christensen, C.H. et al. (2011). Hydrodeoxygenation of waste fat for diesel production: study on model feed with Pt/alumina catalyst. *Fuel* 90 (11): 3433–3438.
99 Lugo-José, Y.K., Monnier, J.R., and Williams, C.T. (2014). Gas-phase, catalytic hydrodeoxygenation of propanoic acid, over supported group VIII noble metals: metal and support effects. *Appl. Catal., A* 469: 410–418.
100 Gong, S., Chen, N., Nakayama, S., and Qian, E.W. (2013). Isomerization of n-alkanes derived from jatropha oil over bifunctional catalysts. *J. Mol. Catal. A: Chem.* 370: 14–21.

101 Chen, N., Gong, S., Shirai, H. et al. (2013). Effects of Si/Al ratio and Pt loading on Pt/SAPO-11 catalysts in hydroconversion of jatropha oil. *Appl. Catal., A* 466: 105–115.

102 Zuo, H., Liu, Q., Wang, T. et al. (2012). Catalytic hydrodeoxygenation of vegetable oil over Ni catalysts to produce second-generation biodiesel. *J. Fuel Chem. Technol.* 40 (9): 1067–1073.

103 Yakovlev, V.A., Khromova, S.A., Sherstyuk, O.V. et al. (2009). Development of new catalytic systems for upgraded bio-fuels production from bio-crude-oil and biodiesel. *Catal. Today* 144 (3–4): 362–366.

104 Gosselink, R.W., Xia, W., Muhler, M. et al. (2013). Enhancing the activity of Pd on carbon nanofibers for deoxygenation of amphiphilic fatty acid molecules through support polarity. *ACS Catal.* 3 (10): 2397–2402.

105 Chen, H., Chen, K., Fu, J. et al. (2017). Water-mediated promotion of the catalytic conversion of oleic acid to produce an alternative fuel using Pt/C without a H_2 source. *Catal. Commun.* 98: 26–29.

106 Duan, J., Han, J., Sun, H. et al. (2012). Diesel-like hydrocarbons obtained by direct hydrodeoxygenation of sunflower oil over Pd/Al-SBA-15 catalysts. *Catal. Commun.* 17: 76–80.

107 Smirnova, M.Y., Kikhtyanin, O.V., Smirnov, M.Y. et al. (2015). Effect of calcination temperature on the properties of Pt/SAPO-31 catalyst in one-stage transformation of sunflower oil to green diesel. *Appl. Catal., A* 505: 524–531.

108 Wang, C., Liu, Q., Song, J. et al. (2014). High quality diesel-range alkanes production via a single-step hydrotreatment of vegetable oil over Ni/zeolite catalyst. *Catal. Today* 234: 153–160.

109 Liu, Q., Zuo, H., Zhang, Q. et al. (2014). Hydrodeoxygenation of palm oil to hydrocarbon fuels over Ni/SAPO-11 catalysts. *Chin. J. Catal.* 35 (5): 748–756.

110 Lee, S.-P. and Ramli, A. (2013). Methyl oleate deoxygenation for production of diesel fuel aliphatic hydrocarbons over Pd/SBA-15 catalysts. *Chem. Cent. J.* 7 (1): 149.

111 Zuo, H., Liu, Q., Wang, T. et al. (2012). Hydrodeoxygenation of methyl palmitate over supported Ni catalysts for diesel-like fuel production. *Energy Fuels* 26 (6): 3747–3755.

112 Ahmadi, M., Nambo, A., Jasinski, J.B. et al. (2015). Decarboxylation of oleic acid over Pt catalysts supported on small-pore zeolites and hydrotalcite. *Catal. Sci. Technol.* 5 (1): 380–388.

113 Morgan, T., Santillan-Jimenez, E., Harman-Ware, A.E. et al. (2012). Catalytic deoxygenation of triglycerides to hydrocarbons over supported nickel catalysts. *Chem. Eng. J.* 189–190: 346–355.

114 Mäki-Arvela, P., Snåre, M., Eränen, K. et al. (2008). Continuous decarboxylation of lauric acid over Pd/C catalyst. *Fuel* 87 (17–18): 3543–3549.

115 Snåre, M., Kubičková, I., Mäki-Arvela, P. et al. (2008). Catalytic deoxygenation of unsaturated renewable feedstocks for production of diesel fuel hydrocarbons. *Fuel* 87 (6): 933–945.

116 Lestari, S., Maki-Arvela, P., Bernas, H. et al. (2009). Catalytic deoxygenation of stearic acid in a continuous reactor over a mesoporous carbon-supported Pd catalyst. *Energy Fuels* 23 (8): 3842–3845.
117 Immer, J.G., Kelly, M.J., and Lamb, H.H. (2010). Catalytic reaction pathways in liquid-phase deoxygenation of C_{18} free fatty acids. *Appl. Catal., A* 375 (1): 134–139.
118 Santillan-Jimenez, E., Morgan, T., Shoup, J. et al. (2014). Catalytic deoxygenation of triglycerides and fatty acids to hydrocarbons over Ni–Al layered double hydroxide. *Catal. Today* 237: 136–144.
119 Santillan-Jimenez, E., Morgan, T., Lacny, J. et al. (2013). Catalytic deoxygenation of triglycerides and fatty acids to hydrocarbons over carbon-supported nickel. *Fuel* 103: 1010–1017.
120 Rozmysłowicz, B., Mäki-Arvela, P., Tokarev, A. et al. (2012). Influence of hydrogen in catalytic deoxygenation of fatty acids and their derivatives over Pd/C. *Ind. Eng. Chem. Res.* 51 (26): 8922–8927.
121 Peng, B., Zhao, C., Kasakov, S. et al. (2013). Manipulating catalytic pathways: deoxygenation of palmitic acid on multifunctional catalysts. *Chem. Eur. J.* 19 (15): 4732–4741.
122 Sun, K., Schulz, T.C., Thompson, S.T., and Lamb, H.H. (2016). Catalytic deoxygenation of octanoic acid over silica- and carbon-supported palladium: support effects and reaction pathways. *Catal. Today* 269: 93–102.
123 Immer, J.G. and Lamb, H.H. (2010). Fed-batch catalytic deoxygenation of free fatty acids. *Energy Fuels* 24 (10): 5291–5299.
124 Li, X.-F. and Luo, X.-G. (2015). Preparation of mesoporous activated carbon supported Ni catalyst for deoxygenation of stearic acid into hydrocarbons. *Environ. Prog. Sustain. Energy* 34 (2): 607–612.
125 Boda, L., Onyestyák, G., Solt, H. et al. (2010). Catalytic hydroconversion of tricaprylin and caprylic acid as model reaction for biofuel production from triglycerides. *Appl. Catal., A* 374 (1–2): 158–169.
126 Do, P.T., Chiappero, M., Lobban, L.L., and Resasco, D.E. (2009). Catalytic deoxygenation of methyl-octanoate and methyl-stearate on Pt/Al_2O_3. *Catal. Lett.* 130 (1–2): 9–18.
127 Mäki-Arvela, P., Kubickova, I., Snåre, M. et al. (2007). Catalytic deoxygenation of fatty acids and their derivatives. *Energy Fuels* 21 (1): 30–41.
128 Ding, R., Wu, Y., Chen, Y. et al. (2015). Effective hydrodeoxygenation of palmitic acid to diesel-like hydrocarbons over MoO_2/CNTs catalyst. *Chem. Eng. Sci.* 135: 517–525.
129 Ito, T., Sakurai, Y., Kakuta, Y. et al. (2012). Biodiesel production from waste animal fats using pyrolysis method. *Fuel Process. Technol.* 94 (1): 47–52.
130 Ochoa-Hernández, C., Yang, Y., Pizarro, P. et al. (2013). Hydrocarbons production through hydrotreating of methyl esters over Ni and Co supported on SBA-15 and Al-SBA-15. *Catal. Today* 210: 81–88.
131 Bernas, H., Eränen, K., Simakova, I. et al. (2010). Deoxygenation of dodecanoic acid under inert atmosphere. *Fuel* 89 (8): 2033–2039.
132 Lestari, S., Simakova, I., Tokarev, A. et al. (2008). Synthesis of biodiesel via deoxygenation of stearic acid over supported Pd/C catalyst. *Catal. Lett.* 122 (3–4): 247–251.

133 Fu, J., Lu, X., and Savage, P.E. (2010). Catalytic hydrothermal deoxygenation of palmitic acid. *Energy Environ. Sci.* 3 (3): 311.

134 Fu, J., Yang, C., Wu, J. et al. (2015). Direct production of aviation fuels from microalgae lipids in water. *Fuel* 139: 678–683.

135 Miao, C., Marin-Flores, O., Davidson, S.D. et al. (2016). Hydrothermal catalytic deoxygenation of palmitic acid over nickel catalyst. *Fuel* 166: 302–308.

136 Kaewmeesri, R., Srifa, A., Itthibenchapong, V., and Faungnawakij, K. (2015). Deoxygenation of waste chicken fats to green diesel over Ni/Al_2O_3: effect of water and free fatty acid content. *Energy Fuels* 29: 833–840.

137 Yang, L., Tate, K.L., Jasinski, J.B., and Carreon, M.A. (2015). Decarboxylation of oleic acid to heptadecane over Pt supported on zeolite 5A beads. *ACS Catal.* 5 (11): 6497–6502.

138 Tian, Q., Qiao, K., Zhou, F. et al. (2016). Direct production of aviation fuel range hydrocarbons and aromatics from oleic acid without an added hydrogen donor. *Energy Fuels* 30 (9): 7291–7297.

139 Zhang, J. and Zhao, C. (2016). Development of a bimetallic Pd-Ni/HZSM-5 catalyst for the tandem limonene dehydrogenation and fatty acid deoxygenation to alkanes and arenes for use as biojet fuel. *ACS Catal.* 6 (7): 4512–4525.

140 Rabaev, M., Landau, M.V., Vidruk-Nehemya, R. et al. (2015). Conversion of vegetable oils on Pt/Al_2O_3/SAPO-11 to diesel and jet fuels containing aromatics. *Fuel* 161: 287–294.

141 Chiappero, M., Do, P.T.M., Crossley, S. et al. (2011). Direct conversion of triglycerides to olefins and paraffins over noble metal supported catalysts. *Fuel* 90 (3): 1155–1165.

142 Domínguez-Barroso, M., Herrera, C., Larrubia, M., and Alemany, L. (2016). Diesel oil-like hydrocarbon production from vegetable oil in a single process over $Pt–Ni/Al_2O_3$ and Pd/C combined catalysts. *Fuel Process. Technol.* 148: 110–116.

143 Vonghia, E., Boocock, D.G., Konar, S.K., and Leung, A. (1995). Pathways for the deoxygenation of triglycerides to aliphatic hydrocarbons over activated alumina. *Energy Fuels* 9 (6): 1090–1096.

144 Choudhary, T.V. and Phillips, C.B. (2011). Renewable fuels via catalytic hydrodeoxygenation. *Appl. Catal., A* 397 (1–2): 1–12.

145 Yokoyama, T. and Yamagata, N. (2001). Hydrogenation of carboxylic acids to the corresponding aldehydes. *Appl. Catal., A* 221 (1): 227–239.

146 Yokoyama, T., Setoyama, T., Fujita, N. et al. (1992). Novel direct hydrogenation process of aromatic carboxylic acids to the corresponding aldehydes with zirconia catalyst. *Appl. Catal., A* 88 (2): 149–161.

147 Pestman, R., Koster, R., Boellaard, E. et al. (1998). Identification of the active sites in the selective hydrogenation of acetic acid to acetaldehyde on iron oxide catalysts. *J. Catal.* 174 (2): 142–152.

148 Hollak, S.A., Bitter, J.H., van Haveren, J. et al. (2012). Selective deoxygenation of stearic acid via an anhydride pathway. *RSC Adv.* 2 (25): 9387–9391.

149 Laurens, L., Nagle, N., Davis, R. et al. (2015). Acid-catalyzed algal biomass pretreatment for integrated lipid and carbohydrate-based biofuels production. *Green Chem.* 17 (2): 1145–1158.

150 Ping, E.W., Pierson, J., Wallace, R. et al. (2011). On the nature of the deactivation of supported palladium nanoparticle catalysts in the decarboxylation of fatty acids. *Appl. Catal., A* 396 (1–2): 85–90.
151 Wen, G., Xu, Y., Ma, H. et al. (2008). Production of hydrogen by aqueous-phase reforming of glycerol. *Int. J. Hydrogen Energy* 33 (22): 6657–6666.
152 Simakova, I., Simakova, O., Mäki-Arvela, P. et al. (2009). Deoxygenation of palmitic and stearic acid over supported Pd catalysts: effect of metal dispersion. *Appl. Catal., A* 355 (1–2): 100–108.
153 Song, W., Zhao, C., and Lercher, J.A. (2013). Importance of size and distribution of Ni nanoparticles for the hydrodeoxygenation of microalgae oil. *Chem. Eur. J.* 19 (30): 9833–9842.
154 Nguyen, H.S., Mäki-Arvela, P., Akhmetzyanova, U. et al. (2017). Direct hydrodeoxygenation of algal lipids extracted from *Chlorella alga*. *J. Chem. Technol. Biotechnol.* 92 (4): 741–748.
155 Van Stiphout, P., Stobbe, D., Scheur, F.T.V., and Geus, J. (1988). Activity and stability of nickel-copper/silica catalysts prepared by deposition-precipitation. *Appl. Catal.* 40: 219–246.
156 Chen, L., Li, H., Fu, J. et al. (2016). Catalytic hydroprocessing of fatty acid methyl esters to renewable alkane fuels over Ni/HZSM-5 catalyst. *Catal. Today* 259: 266–276.
157 Hollak, S.A.W., de Jong, K.P., and van Es, D.S. (2014). Catalytic deoxygenation of fatty acids: elucidation of the inhibition process. *ChemCatChem* 6 (9): 2648–2655.
158 Li, Y., Zhang, C., Liu, Y. et al. (2017). Coke formation on the surface of Ni/HZSM-5 and Ni-Cu/HZSM-5 catalysts during bio-oil hydrodeoxygenation. *Fuel* 189: 23–31.
159 Yeh, T., Linic, S., and Savage, P.E. (2014). Deactivation of Pt catalysts during hydrothermal decarboxylation of butyric acid. *ACS Sustain. Chem. Eng.* 2 (10): 2399–2406.
160 Rozmysłowicz, B., Mäki-Arvela, P., Lestari, S. et al. (2010). Catalytic deoxygenation of tall oil fatty acids over a palladium-mesoporous carbon catalyst: a new source of biofuels. *Top. Catal.* 53 (15–18): 1274–1277.
161 Sari, E., Kim, M., Salley, S.O., and Ng, K.Y.S. (2013). A highly active nanocomposite silica-carbon supported palladium catalyst for decarboxylation of free fatty acids for green diesel production: correlation of activity and catalyst properties. *Appl. Catal., A* 467: 261–269.
162 Zhao, X., Wei, L., Cheng, S., and Julson, J. (2017). Review of heterogeneous catalysts for catalytically upgrading vegetable oils into hydrocarbon biofuels. *Catalysts* 7 (3): 83–108.
163 Idesh, S., Kudo, S., Norinaga, K., and Hayashi, J.-I. (2013). Catalytic hydrothermal reforming of jatropha oil in subcritical water for the production of green fuels: characteristics of reactions over Pt and Ni catalysts. *Energy Fuels* 27 (8): 4796–4803.
164 Zhang, Z., Yang, Q., Chen, H. et al. (2018). In situ hydrogenation and decarboxylation of oleic acid into heptadecane over a Cu–Ni alloy catalyst using methanol as a hydrogen carrier. *Green Chem.* 20 (1): 197–205.

13

Conversion of Terpenes to Chemicals and Related Products

Anne E. Harman-Ware

National Renewable Energy Laboratory, 15013 Denver West Parkway, Golden, CO 80401, USA

13.1 Introduction

Terpenes and terpenoids, hydrocarbons present in many types of biomass, have an extensive history of use as fuels, binders, polymers, pharmaceuticals and therapeutics, flavors, and fragrances [1]. The structure of terpenes and terpenoids makes them amenable for use as a feedstock for catalytic conversion to other materials and also useful in their own right in the pharmaceutical, polymer, fuel, and chemical industries. Terpenes occur in pine oleoresin and various essential oils and are generated as by-products from paper pulping processes. Oleoresin and terpene products derived from pine and paper pulping processes are collectively generated in millions of tons every year and are a multibillion dollar industry [2–5]. Given their widespread and historical use, global markets exist and are continually evolving, for products both derived from the direct utilization of terpenes and derived from their conversion to other materials. There are continued efforts worldwide to find efficient and novel processes for the catalytic conversion of terpenes to new, improved, and essential chemicals and materials.

13.2 Terpene Biosynthesis and Structure

Terpenes are oligomers of isoprene units that occur naturally in plants, animals, and insects. Plant-derived terpenes are more abundant and useful than terpenes from animals and will be the focus of the material covered in this text. Terpenoids, which include terpene precursors, are terpenes not only consisting of isoprene units but also containing phosphate, alcohol, ether, carboxylate, aldehyde, hydroxyl, or other functional groups. Biosynthesis of terpenes occurs via mevalonate or non-mevalonate (MEP) pathways (depending on the organism) that generate isopentyl pyrophosphate (IPP) from acetyl-CoA that can be isomerized to dimethylallyl pyrophosphate (DMAPP) by isopentenyl pyrophosphate isomerase. IPP and DMAPP are the activated building blocks that assemble to form isoprenoid oligomers that conform to the "C5" or "biogenetic isoprene rule." The biogenetic isoprene rule, coined by Leopold

Chemical Catalysts for Biomass Upgrading, First Edition. Edited by Mark Crocker and Eduardo Santillan-Jimenez.

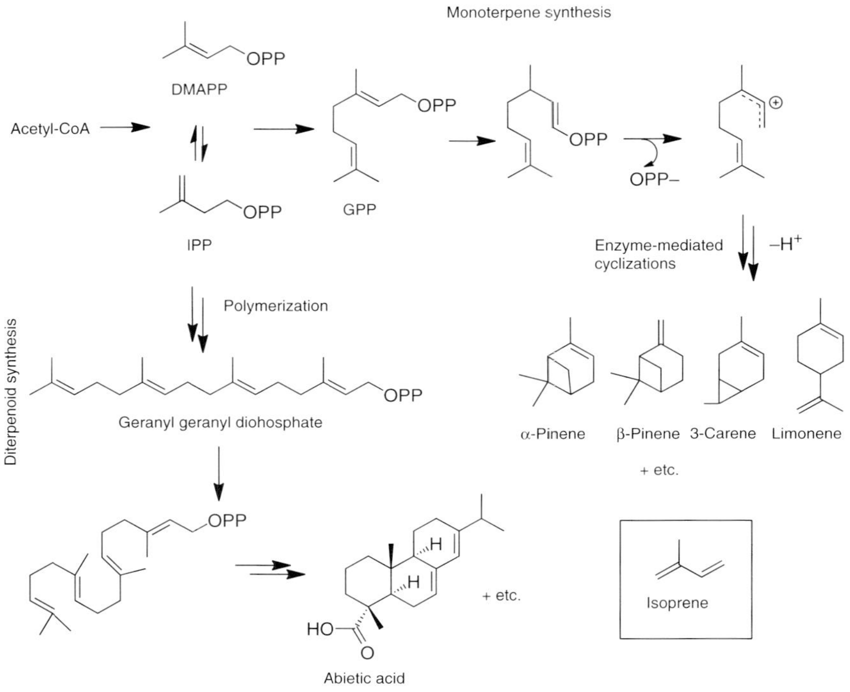

Figure 13.1 Biosynthesis of monoterpenes and diterpenoid resin acids.

Ružička [6], classifies terpenes based on the number of isoprene units, C_5H_8, which are linked together either endwise or in rings. For example, as shown in Figure 13.1, IPP and DMAPP may condense to form geranyl pyrophosphate (GPP) by GPP synthase that ultimately leads to the production of monoterpenes containing two isoprene units such as limonene, α-pinene, and 3-carene. Farnesyl pyrophosphate, geranyl GPP, and other higher isoprenoid building blocks generated by prenyl transferases lead to the formation of higher terpenoids such as sesquiterpenes and diterpenes.

Hemiterpenes are terpenes consisting of a single isoprene unit that include isoprene and IPP/DMAPP, while hemiterpenoids include compounds such as isovaleric acid and tiglic acid. Monoterpenes are formed from two isoprene units, either cyclic or acyclic, and have a molecular formula of $C_{10}H_{16}$. Common monoterpenes and monoterpenoids, shown in Figure 13.2, are typically volatile compounds that can be isolated from essential oils from various types of biomass, including herbs and citrus fruits, and are present in the oleoresins of pine and other conifer trees. Monoterpenoids are monoterpenes containing other functional groups such as ketones, alcohols, or ethers. Sesquiterpenes are made from three isoprene units and have a molecular formula of $C_{15}H_{24}$. Common sesquiterpenes, typically found in essential oils and extracts, include santalol, caryophyllene, bisabolenes, and humulene. Diterpenes are formed from four isoprene units and have a molecular formula of $C_{20}H_{32}$. Typically, diterpenes are oxidized to diterpenoid alcohols, aldehydes, or acids and have

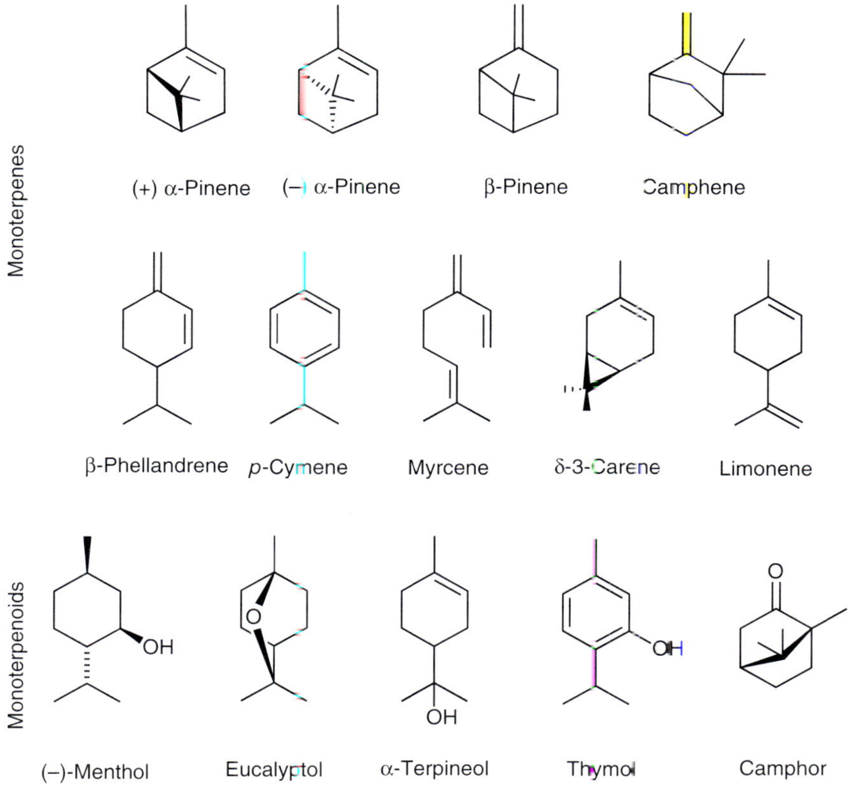

Figure 13.2 Common monoterpenes and monoterpenoids.

cyclic structures [7]. Diterpenoid resin acids, shown in Figure 13.3, are abundant in the oleoresins of coniferous trees.

Triterpenes, consisting of six isoprene units with a molecular formula of $C_{30}H_{48}$, are the molecular class of compounds that include squalene and precursors of sterol compounds. Steroids are classified as triterpenoids as they are triterpenes that contain functional groups such as alcohols and carboxylic acids. Tetraterpenes, formed from eight isoprene units with a molecular formula $C_{40}H_{64}$, as well as tetraterpenoids, are the classes to which carotene and carotenoid compounds belong. Polyisoprene structures form the basis of natural rubber and latex.

Volatile monoterpenes and sesquiterpenes produced in plants serve a wide variety of biological purposes. The types, amount, and spatial distributions of terpenes present in biomass vary depending on the biomass age and type, environmental conditions, defense factors, and biological role of the terpenes. In coniferous trees, mono- and sesquiterpenes are secreted along with diterpenoid resin acids as a defense mechanism upon a wound response (physical or chemical) or herbivore or fungal infestation [7–10]. Volatile terpenes present in and emitted from softwood and hardwood trees, herbaceous leaves, and other

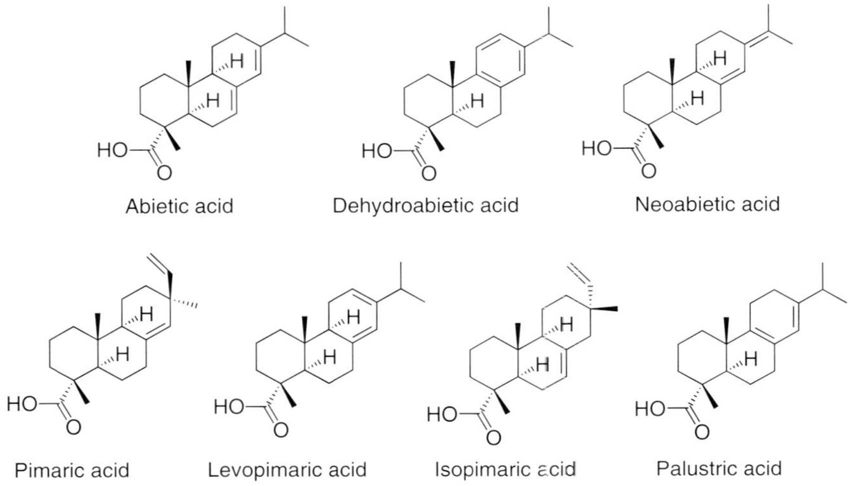

Figure 13.3 Common diterpenoid resin acids.

biomass sources have also shown the capacity to act as antioxidants to protect the cells during periods of oxidative stress induced by environmental factors such as temperature fluctuations and ozone damage [11, 12]. The emission of isoprene and other volatile terpenes as hormone signaling compounds has been associated with accelerated flowering in herbaceous plants such as barley and arabidopsis [13]. The biological roles of larger terpenes in plants are also diverse and play roles in cell defense and signaling as well. Diterpenoid resin acids have been associated with antiherbivore properties in conifers [7]. Phytol is a diterpenoid alcohol that is a constituent of chlorophylls, molecules that are essential for photosynthesis. Triterpenes such as sterols and steroids have roles in the growth and development of biomass and play roles in basic cellular signaling, structure, and functions [14]. Tetraterpenes such as carotenes and carotenoids also play roles in photosynthesis and serve photoprotecting roles [15]. There are many other structural classes that can be used to describe other less abundant terpenes and terpenoids, these being covered in detail elsewhere [16].

13.3 Sources of Terpenes

13.3.1 Conifers and Other Trees

Coniferous trees and shrubs are gymnosperm plants that bear cones containing seeds and whose wood contains lignin composed primarily of coniferyl alcohol monomeric units. Conifers, including pine, spruce, and fir, contain specialized cells that make up resin canals or ducts, particularly within the stem xylem, containing terpenes and other lipids such as fatty acids that are collectively known as oleoresin [9, 17]. Figure 13.4 shows the presence of resin ducts in the phloem of conifer biomass [18]. Insects, fungi, and wounding of the plant elicit the secretion of the oleoresin that acts as an antimicrobial agent and can trap attacking insects.

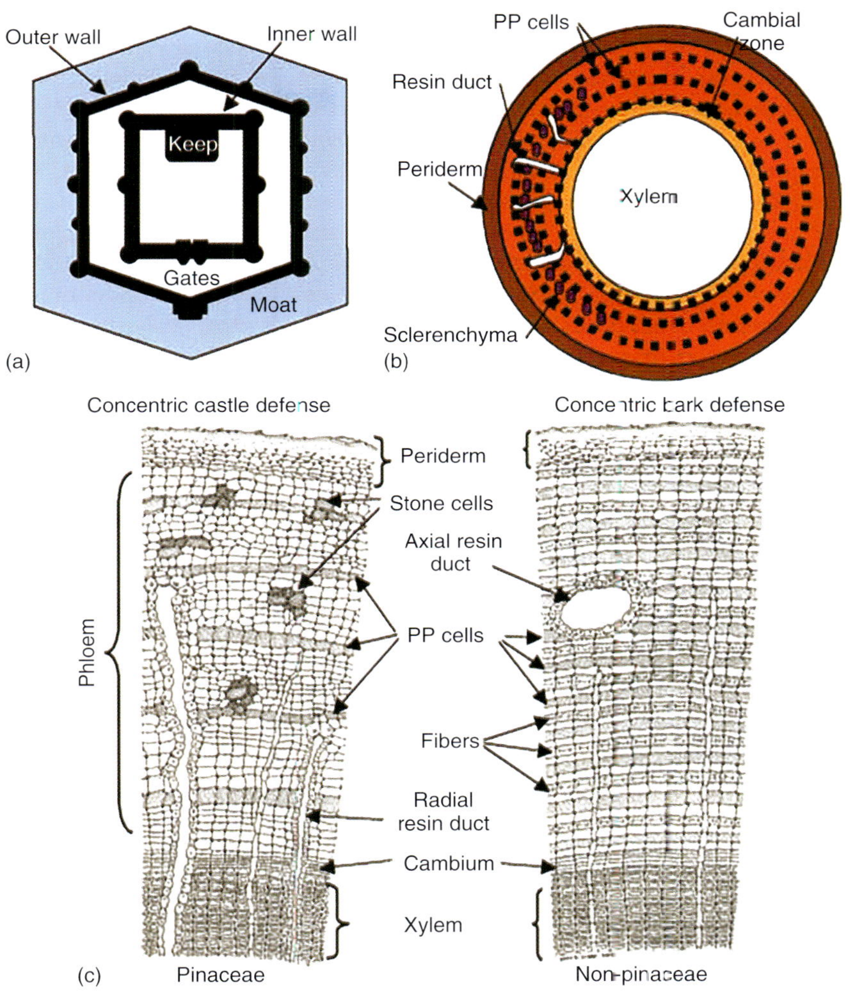

Figure 13.4 Concentric castle analogy for constitutive defenses in conifer bark. (a) Concentric castle; (b) conifer bark concentric defenses; (c) pinaceae and non-pinaceae schematics of phloem cellular structure including the presence of resin ducts. Source: Reprinted with permission from Franceschi et al. [18]. Copyright 2005 John Wiley & Sons.

Terpenes are also present in and emitted from conifer needles for defense and signaling purposes. Pine oleoresin constitutes anywhere between 1 and 4 dry wt% of the biomass, where approximately 20–40% can be monoterpenes, about 60% is diterpenoid resin acids and the remainder is fatty acids and other terpenes and lipids [19–21]. The monoterpenes in pine oleoresin consist primarily of α-pinene (~60%), β-pinene (~20–40%), limonene (1–5%), and small amounts of cymene, Δ^3-carene, terpineol, and other compounds. Fir, spruce, and other coniferous trees also produce oleoresin that contains monoterpenes and diterpenoid resin acids [19]. Caryophyllene, a sesquiterpene, is also produced in small quantities from coniferous trees. Citrus plants and trees, which have evergreen leaves but are not conifers, contain subdermal secretory cavities in their leaves that contain terpenes [22, 23]. The terpenes present in citrus plants and their fruits are typically extracted to produce "essential oils" and other extracts and are described in Section 13.3.2. Subdermal secretory cavities containing terpenes as well as other essential oil components are also present in the leaves of *Eucalyptus* trees [24]. Hardwoods such as poplar also produce terpenes, particularly as emissions from buds and leaves in response to herbivores and oxidative stress [25–27]. Terpene emissions from poplar, particularly isoprene, have also been implicated as serving as thermoprotective compounds for stabilization of chloroplast membranes [28, 29].

13.3.2 Essential Oils and Other Extracts

Terpenes are produced by a wide variety of plants and are commonly isolated or extracted for use as chemicals. Essential oils are the hydrocarbon, mostly nonpolar and often volatile compounds isolated from plants that have a characteristic scent and/or taste. There are many ways to obtain essential oils from their biological sources, including steam distillation and cold pressing [30]. Other extraction methods yield oils that are given different names depending on the extraction technique and the final composition. Depending on the source and extraction method, essential and extracted oils contain a wide variety of terpenes and terpenoids as well as fatty acids, flavonoids, phenolics, tannins, and other compounds. Extractable oils in biomass fulfill physiological roles described previously for terpenes – which include defense as well as growth regulation – and also for signaling purposes such as attracting pollinators [31]. Essential oils and extracts have a large number of pharmacological, culinary, and cosmetic applications, as briefly described in Section 13.5.2.

Citrus-derived essential oils are of particular industrial importance due to their use in flavors, fragrances, and for other nutraceutical and chemical applications. Terpenes are present in the essential oils found in the leaves, flowers, fruit tissues, and peel of citrus plants such as *Citrus limon* (lemon), *Citrus sinensis* (orange), *Citrus aurantifolia* (lime), and *Citrus reticulata* (mandarin) [22, 23, 30, 32, 33]. The primary component of most citrus-derived essential oils is limonene [23], which can comprise up to almost 90% of the oil. Other terpenes present in citrus essential oils include β-pinene, α-terpineol, geranial, and γ-terpinene [23, 30]. Typically, essential oils constitute less than 5 dry wt% of the biomass feedstock, depending on the source [30].

Essential and extracted oils are also obtained from other trees and herbaceous plants. *Eucalyptus* species are a source of essential oils high in 1,8-cineole content, as well as citronellal and various sesquiterpenes [24, 30]. Pine, fir, and other conifers can be distilled to produce essential oils containing α-pinene and other terpenes, although typically pine-derived terpenes are obtained as turpentine distilled from oleoresin sourced by tapping trees or from distilled crude sulfate turpentine (CST), as described in Section 13.4.2 [30, 34]. Mint oils, including cornmint and peppermint oils, contain primarily (−)-menthol and (−)-menthone, whereas spearmint oil is primarily (−)-carvone with some limonene [30]. There are many other plants and animals that produce terpenes that are present in extractable oils, most of which are used raw or as feedstocks for catalytic conversion to other materials in the fine chemical, flavor, and fragrance industry.

13.4 Isolation of Terpenes

13.4.1 Tapping and Extraction

"Tapping" pine or other coniferous trees refers to processes used to recover oleoresin by means of bark removal, cutting, slashing, or (borehole) drilling into living trees to drain and collect oleoresin as it is secreted. Figure 13.5a shows traditional slashing techniques used to collect oleoresin, and Figure 13.5b shows how a borehole tapping process releases pine oleoresin that can be collected in to a bag [35]. Oleoresin that is collected can then be distilled to produce gum turpentine, the volatile fraction containing mostly monoterpenes, and gum rosin containing diterpenoid resin acids, fatty acids, and other nonvolatile components. Tapping processes used to be regularly practiced in the United States to recover gum turpentine and rosin and are still performed in several Asian countries [2, 4, 34]. Since CST and crude tall oil (CTO) from the pulp and paper processing of pine (Section 13.4.2) is more readily and economically available in the United States, traditional tapping methods are not widely practiced. However, recent studies show that borehole drilling of live pine trees in the United States can be economical, given the right conditions, methods, and yield [2].

Distillation, cold pressing, and solvent extraction of conifers, citrus peel waste, herbaceous plants, and other biomass sources has been used to isolate oils containing terpenes, particularly for the fine chemicals industry. Steam or water

(a) (b)

Figure 13.5 (a) Slash tapping methods used to collect oleoresin from conifers. Source: Reprinted with permission from Bohlmann and Keeling [35]. Copyright 2008, John Wiley & Sons. (b) Borehole tapping methods used to collect oleoresin from conifers. Source: Photo courtesy of Gary Peter, University of Florida.

distillation of the biomass leaves behind an oil phase containing terpenes that separates from the aqueous phase after condensation. Further vacuum or fractional distillation can then be used to isolate particular components for higher purity applications. However, some compounds may react or change during heating, so distillation processes must be carefully controlled for the isolation of particular products. Cold pressing processes are mechanical in nature and involve grinding and pressing the biomass to release and recover oils. Although some heat may be generated, cold pressing is a preferable method for the recovery of oils that are heat sensitive, volatile, and reactive or whose flavor and fragrance profiles are otherwise affected by heating or solvent extraction methods. Solvent extraction methods for the recovery of terpenes can incorporate the use of ethanol, nonpolar solvents such as hexane, or other solvents depending on the type of biomass and the recovered products. Supercritical CO_2 has also been studied as a means of isolating and fractionating terpenes from biomass [30, 36].

13.4.2 Terpenes as a By-product of Pulping Processes

The chemical pulping of pine trees during the kraft process generates a cellulose-based pulp used to manufacture paper. Kraft processing of pine and spruce also generates by-products known as CST and CTO that contain terpenes. CST consists of volatile monoterpenes such as α-pinene, β-pinene, and δ^3-carene as well as sulfur-containing compounds, aromatics, and residues. Fractionation by distillation and catalytic desulfurization of CST removes the sulfur, aromatic, and other compounds to yield turpentine containing the monoterpenes. A typical pulping mill produces several thousand tons of CST annually, which constitutes less than 1 wt% of the yield of dry pulp. CTO is generated at less than 5 wt% of pulp yield and contains fatty acids, sterols, and diterpenoid resin acids. CTO undergoes a number of distillation steps to separate the various components into fractions, 30% of which is the "rosin" or tall oil rosin (TOR) that contains the diterpenoid resin acids. The distributions of diterpenoid resin acids present in TOR differ slightly from those present in raw pine oleoresin due to the changes induced during processing. Most of the diterpenoid acids in CTO/TOR are abietane isomers as well as pimaric and isopimaric acids, shown in Figure 13.3 [34].

13.5 Historical Uses of Raw Terpenes

13.5.1 Adhesives and Turpentine

Oleoresin and other oils present in biomass that contain terpenes have been used as adhesives and sealants for millennia. For example, pine resins, tar, and pitch (obtained from the heating of pine oleoresin) containing monoterpenes and diterpenoid resin acids were used as adhesives for swords, ceramics, and other materials according to archaeological evidence dating back thousands of years [37, 38]. Pine resins and pitch were also used for binding materials during soldering, as canoe caulk, and for waterproofing boats and other building materials [1]. The term "naval stores" was eventually coined for the resinous materials derived from pine (typically collected by tapping methods and subsequently heated) that were used to seal ships before the widespread use of petroleum-based chemicals in the twentieth century [39].

Turpentine has primarily been used as a solvent, paint thinner, fuel, and source of monoterpenes for further processing as platform chemicals [34, 40, 41]. The metrics used for classifying turpentine and its potential for use in a particular industry relate to its relative density, flash point, and various other chemical and physical properties. Processes used to obtain turpentine also leave behind the nonvolatile fraction of pine oleoresin, which can be used as an adhesive and as a feedstock for upgrading to other materials as described in Section 13.6 [34].

13.5.2 Flavors, Fragrances, Therapeutics, and Pharmaceutical Applications

Essential oils and extracts containing terpenes have been used for therapeutics, food additives, flavors, and fragrances since antiquity [1, 42, 43]. Citrus essential oils, and limonene in particular, have an extensive history of culinary, chemical, and material applications [1, 44]. Fine fragrances, food and beverages, medicines, cosmetics, and other products have been made with terpene-rich citrus oils [30]. Citrus essential oils have been used as cleaning and preservation products and have been widely shown to have antiseptic, antimicrobial, and antifungal properties [23, 32, 33, 44]. Resins and extracts from coniferous trees have also had fragrance, preservative, medicinal, pharmaceutical, and nutraceutical applications throughout history. Pine or other coniferous resin has been chewed, used to treat wounds, and used as embalming agents for mummification [1, 45]. Coniferous oils and their constituents have also been used in insect repellant and for therapeutic purposes and have been shown to have antimicrobial properties [4, 30, 43, 46]. Individual terpenes isolated from pine, citrus, and other plants have extensive use as raw chemicals in the pharmaceutical and food additives (or nutraceutical) industry. Various clinical studies have been performed to elucidate the potential health benefits and the bioactivity of terpenes and related compounds such as β-sitosterol (prostate health), β-carotene, and lutein (vision) [43, 47]. One of the most important examples of the use of a terpene for medicinal purposes is the chemotherapy medication, paclitaxel, a diterpene derived from the Pacific Yew tree. The World Health Organization has included paclitaxel in its list of essential medicines for the treatment of various types of lung, breast, and ovarian cancers [48]. Terpenes are most commonly used as a feedstock for conversion to other compounds useful in the fragrance, pharmaceutical, and materials industries by means of homogeneous and heterogeneous catalytic processes.

13.6 Catalytic Methods for Conversion of Terpenes to Fine Chemicals and Materials

Terpenes have been widely used as raw feedstocks for the production of flavoring agents, fragrances, fuels, adhesives, and antimicrobials. Since terpenes contain chiral centers and a variety of functional groups, especially olefin groups but also hydroxyl, carboxylic, and carbonyl groups, they are capable of being converted to a wide variety of other useful products with specific properties. Depending on the terpene feedstock, the final product and its preferred properties, various types of catalysts and conditions can be used. Homogeneous (catalyst in the same phase) and heterogeneous (catalyst in a separate phase) catalytic processes have been used to isomerize terpenes into other terpenes or terpenoids as well as to convert them to chemicals and materials. As with most catalytic processes, the design of the catalyst and the operating conditions are carefully considered in an effort to maximize conversion and selectivity to certain products while minimizing the severity of the reaction conditions and the associated processing costs. Since many terpenes have stereocenters and multiple functional groups, selectivity is often the most important attribute in a catalytic process.

13.6.1 Homogeneous Processes

Homogeneous catalysts, including mineral and transition metal acids and bases, organic peroxy acids, organometallics, enzymes, metal halides, and others, have been used in the conversion of terpenes to fine chemicals and other materials. Many homogeneous catalysts are also supported or immobilized on substrates such as polymers or molecular sieves in an effort to utilize them as heterogeneous catalysts while attempting to take advantage of the activity they exhibit in homogeneous reactions. One of the most important metrics of a catalytic process is the selectivity for a particular product, as many terpenes contain olefin, hydroxyl, and carboxylate groups, and frequently more than one type of functionality. Due to their hydrocarbon nature in general, and their olefin functionality in particular, catalysts used for the conversion of terpenes to other products are frequently adopted from traditional processes that use petroleum-derived substrates. While many industrial processes use homogeneous catalysts, there is a significant amount of recent research devoted to the utilization of heterogeneous catalytic processes, which are covered in Section 13.6.2. This section describes a number of homogeneous catalytic processes that are either important as they are utilized industrially or they demonstrate relevant research and principles relating to the conversion of terpenes to other products.

13.6.1.1 Hydration and Oxidation Reactions

Due to its abundance as a by-product of kraft pulping processes, α-pinene is commonly used as a feedstock for the production of other terpenes for the flavor and fragrance industries. For example, aqueous mineral acids such as sulfuric and hydrochloric acid are typically used in industry to catalyze the hydration of α-pinene to produce α-terpineol, an important fragrance compound (Figure 13.6). The industrial hydration of dihydromyrcene is also accomplished using aqueous mineral acid-catalyzed processes to produce dihydromyrcenol for the fragrance industry [49]. Heterogeneous catalysis for the hydration of monoterpenes (Section 13.6.2) has been investigated and utilized in an effort to overcome the disadvantages associated with the homogeneously catalyzed processes (which primarily includes waste generation and management).

The catalytic oxidation of terpenes is also used to produce fine chemicals as well as intermediates used in the synthesis of various types of chemicals and

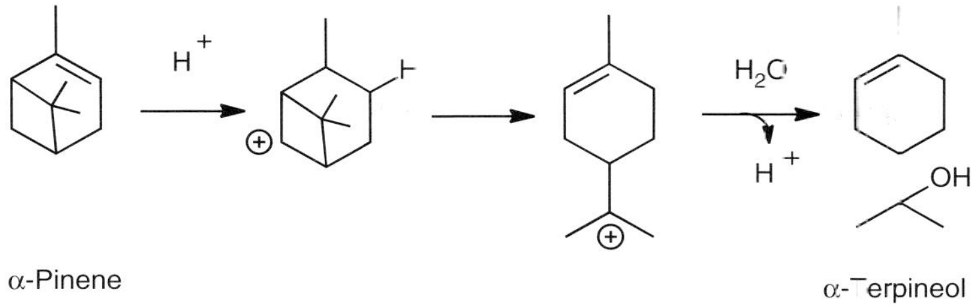

Figure 13.6 Brønsted acid-catalyzed hydration of α-pinene to α-terpineol.

materials. Pd catalysts such as $PdCl_2$ and $Pd(OAc)_2$ can be used for the oxidation (and isomerization) of monoterpenes such as α- and β-pinene and camphene using hydrogen peroxide as an oxidant [49, 50]. Studies by Gusevskaya et al. provide an example of how the activity and selectivity of the catalyst changes depending on the terpene feedstock [50]. While β-pinene (a bicyclic monoterpene with an exocyclic alkene) can be converted to pinocarveol, pinocarveol acetate, and myrtenyl acetate in various ratios depending on the reaction conditions (Figure 13.7a), limonene (a monocyclic monoterpene with both a terminal and endocyclic alkene) was not oxidized in the presence of $Pd(OAc)_2$.

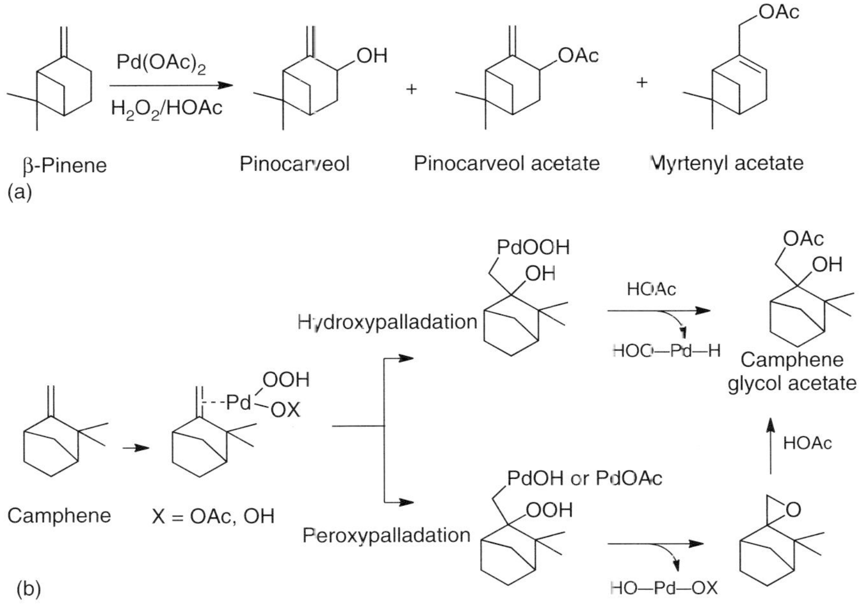

Figure 13.7 (a) Oxidation of β-pinene by $Pd(OAc)_2$ with hydrogen peroxide in acetic acid solution to produce pinocarveol, pinocarveol acetate, and myrtenyl acetate. Reaction conditions varied time, H_2O_2, benzoquinone, and $Pd(OAc)_2$ concentrations, while holding β-pinene concentration constant at 0.5 mol/l and temperature at 60 °C. Source: Gusevskaya et al. 1998 [50]. Reproduced with permission from Elsevier. (b) Camphene oxidation to camphene glycol acetate in acetic acid explained using two possible mechanisms based on hydroxypalladation or peroxypalladation with active Pd catalyst species.

In addition, the conversion of camphene to produce camphene-glycol acetate could be explained by two possible routes as shown in Figure 13.7b. Beneficial effects were observed from the addition of benzoquinone for the conversion of camphene, the former acting as a cocatalyst and stabilizing Pd during the reaction [50]. Notably, the Pd-catalytic system was developed without the use of $CuCl_2$ as a cocatalyst, which sometimes causes unwanted side reactions.

The hydrophobic nature of terpenes can influence the catalytic conversion process, particularly when conversion processes are in aqueous conditions, such as biocatalytic oxidations. Industrial processes are beginning to utilize biocatalysts such as cytochrome P450 monooxygenase enzymes with nonpolar substrates. In-cell biocatalysis offers process advantages based on the reduction of steps and chemicals used in nonpolar substrate pretreatment. However, passive diffusion channels must enable nonpolar substrates to enter cells to allow biocatalysis to occur and to release products. To overcome cell diffusion limitations, Ruff et al. engineered *Escherichia coli* variants with ferric hydroxamate uptake (FhuA) membrane proteins that acted as passive diffusion channels to allow transport of nonpolar substrates such as terpenes inside the cell where they were oxidized by means of P450 BM3 variant enzymes [51]. Figure 13.8 shows a schematic for the

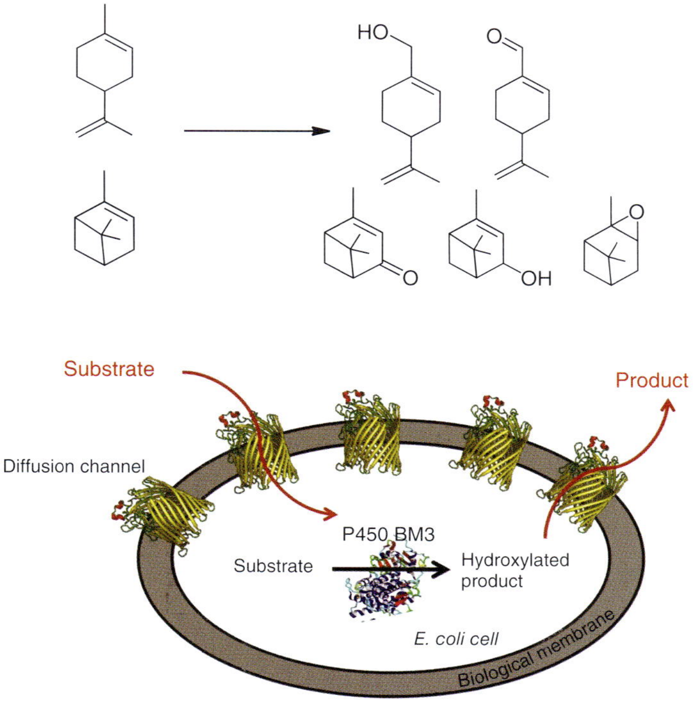

Figure 13.8 Engineered variants of ferric hydroxamate uptake protein (FhuA) in *E. coli* resulting in outer membrane proteins with passive diffusion channels, enabling the uptake of nonpolar substrates such as limonene and α-pinene. Inside the cell, oxidations can occur by cytochrome P450 enzymes, and products can subsequently be released through the diffusion channels. Source: Ruff et al. 2016 [51]. Reprinted with permission from Elsevier.

passage of α-pinene and limonene in to the cell where they can be oxidized to pinene oxide (an epoxide) and perillyl alcohol, respectively, with other minor oxidation products, and subsequently released through the diffusion channels [51].

13.6.1.2 Homogeneous Catalysis for the Epoxidation of Monoterpenes

The epoxidation of α-pinene is important for the conversion of this feedstock to other materials and has been performed industrially using homogeneous *m*-chloroperbenzoic acid or other percarboxylic acids but is also performed over heterogeneous catalysts [52]. The use of both homogeneous and heterogeneous catalysts for the epoxidation of α-pinene in industrial and research applications demonstrates the balance in activity and sustainable practices that are commonly the focus of catalytic conversion processes. For example, Mn(III) porphyrins and Mn(III) acetate have been successfully used as homogeneous catalysts for the oxidation of monoterpenes [49, 53]. However, heterogeneous Mn-containing metal–organic framework (MOF) catalysts based on MIXMIL-53-NH_2(50) have shown similar performance to Mn(III) acetate for the oxidation of α-pinene to its epoxide (31% conversion, 17% yield, 55% selectivity after six hours) with the added advantage of catalyst separation by filtration and potential for recycling for five cycles without loss of activity. Reaction conditions – such as the oxidant and solvent used (in this case, diethyl carbonate with molecular oxygen being favorable) – can also be improved with the use of heterogeneous catalytic processes [53]. Other Mn-based catalysts such as Jacobsen-type Mn–salen complexes have been investigated for the epoxidation of monoterpenes such as limonene to produce the diepoxide [54].

The oxidation of monoterpenes such as α-pinene, β-pinene, and limonene by Re catalysts, particularly using methyltrioxorhenium (MTO) for the epoxidation of olefin groups, has been the focus of a significant amount of research and has been reviewed thoroughly [55]. Reactions are typically performed at low temperatures with H_2O_2 as an oxidant in the presence of a nitrogen base such as a pyridine, the overall reaction being shown in Figure 13.9. The presence of the base increases the reaction rate and prevents ring opening reactions and diol formation and may improve the stability of the catalyst [56]. Dichloromethane is a common solvent used for epoxidation reactions and has also been used in biphasic systems with acetonitrile. The effect of the reaction conditions has been studied extensively for various types of monoterpenes, and it is generally found that the conditions have a huge impact on the conversion and selectivity to epoxides. For example, a thorough study on the effects of the solvent, additives, and conditions of α-pinene epoxidation by Re MTO was conducted by Michel et al. [57]. The kinetics of α-pinene epoxidation under various conditions were also studied as shown in Figure 13.10. The optimized reaction used a ratio of α-pinene:MTO:*t*-butylpyridine:urea hydrogen peroxide of 200 : 1 : 40 : 600 at 0 °C in nitromethane and produced a 95% yield of α-pinene oxide after three hours with a TOF of >600 h^{-1}.

13.6.1.3 Isomerizations

The isomerization of olefins in terpenes is another important process for the production of fine chemicals. Rh catalysts in particular have been used

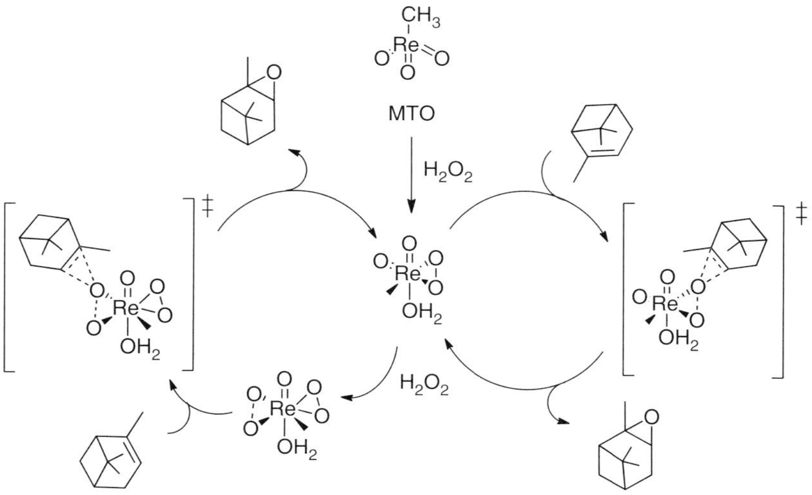

Figure 13.9 Epoxidation of α-pinene by methyltrioxorhenium (MTO) catalyst using H_2O_2 as an oxidant. Both mono-and bisperoxo catalyst species can be active for epoxidation, with the bisperoxo complex likely being more abundant in excess H_2O_2.

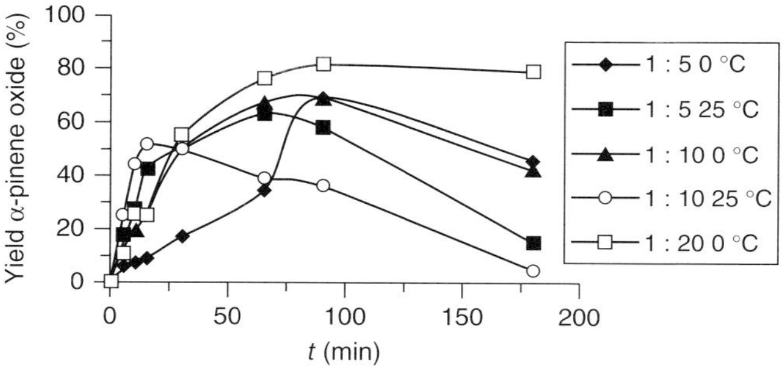

Figure 13.10 Kinetics of α-pinene epoxidation is affected by the molar ratio of MTO:*t*-butylpyridine and temperature. Conditions: α-pinene:MTO: H_2O_2 = 100 : 1 : 300 in dichloromethane. Source: Michel et al. 2011 [57]. Reprinted with permission from Elsevier.

in enantioselective isomerization processes. For example, the pyrolysis of β-pinene can be used to generate myrcene (or myrcene can be obtained from other sources), which is then converted to geranyldiethylamine using lithium diethyl amide catalyst. Next, a Rh(I)–BINAP catalyst (Rh with chelates of *R*- or *S*-2,2′-diphenylphosphino-1,1′-binaphthyl) isomerizes an alkene by converting it from allylamine to the (1*R*,3*R*,4*S*) enamine. The enamine is hydrolyzed (acid) to citronellal, $ZnCl_2$ or $ZnBr_2$ (Lewis acid) subsequently catalyzing ring closure, which is followed by reduction of the remaining alkene to produce (−)-menthol for use in peppermint flavors and fragrances. The final "trio" of catalytic processes, shown in Figure 13.11, was adopted industrially by the Takasago Corporation to produce over 28 000 tons of (−)-menthol over a seven year period, with a turnover number (TON) of greater than 40 000 for the

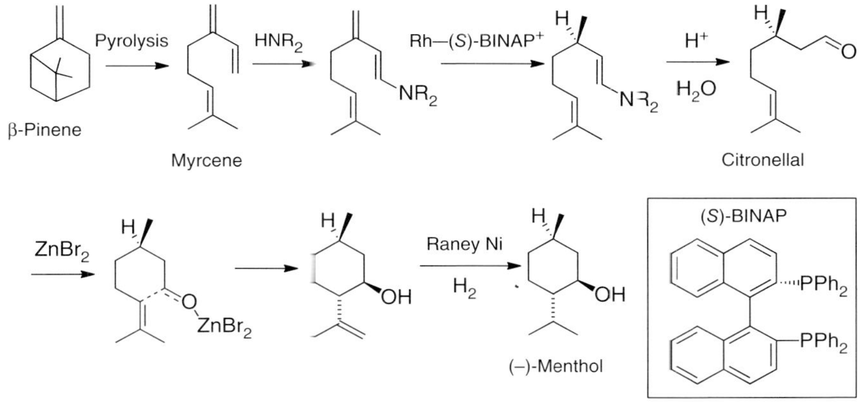

Figure 13.11 Takasago process for the production of (−)-menthol beginning with the pyrolysis of β-pinene to produce myrcene. Rh(I)–*S*-BINAP catalyzes the enantioselective isomerization of the allylamine produced from myrcene, Zn halides catalyze ring closure of the aldehyde after the hydrolysis of the enamine, and Raney Ni reduces the final olefin in the "trio catalytic process."

Rh-catalyzed step [58–61]. Rh catalysts can also be used for the generation of aldehydes used in perfumery [49]. For example, $[Rh_2(\mu\text{-}SR)_2(CO)_2L_2]$ catalysts (where $L = PPh_3$, $P(OPh)_3$, or $P(OMe)_3$) have been used in mild conditions for the hydroformylation of pinenes, isopulegol, and its acetate, as well as limonene [62].

As demonstrated in the Tukasago process, the subsequent conversion of functionalized or oxidized terpenes – including epoxides, alcohols, acetates, aldehydes, etc. – to other products by homogeneous catalysts is also important. Additional uses of homogeneous $ZnCl_2$ and $ZnBr_2$ as Lewis acid catalysts include the industrial isomerization α-pinene oxide to produce campholenic aldehyde (for sandalwood fragrance) and the isomerization of limonene oxide to cyclopentanecarboxaldehyde and dihydrocarvone (fragrance) [49, 63]. The low TON, aqueous catalyst removal step, and waste treatment required for Zn halide-catalyzed processes have led to an extensive amount of research in the heterogeneous catalytic conversion of α-pinene oxide to campholenic aldehyde and other isomerizations, as reviewed thoroughly by Corma et al. and addressed in Section 13.6.2 [63].

13.6.1.4 Production of Terpene Carbonates from CO_2 and Epoxides

Another conversion pathway for terpene oxide feedstocks that has received attention lately involves the cycloaddition of CO_2 to the epoxide to produce cyclic carbonates. CO_2-based carbonates are a potential way to utilize CO_2 as a feedstock (an otherwise wasted carbon source), and the properties of carbonates make them useful in the polymer industry and as solvents [64, 65]. Cyclic terpene carbonates can be generated from a wide variety of terpene scaffolds including limonene oxides and carvone oxide using Lewis acidic catalysts. In a study by Fiorani et al., Al(aminotriphenolate) with a nucleophilic additive was used to generate cyclic terpene carbonates from various terpene oxides, as shown in Figure 13.12 [66]. Their study showed the stereoselective conversion of

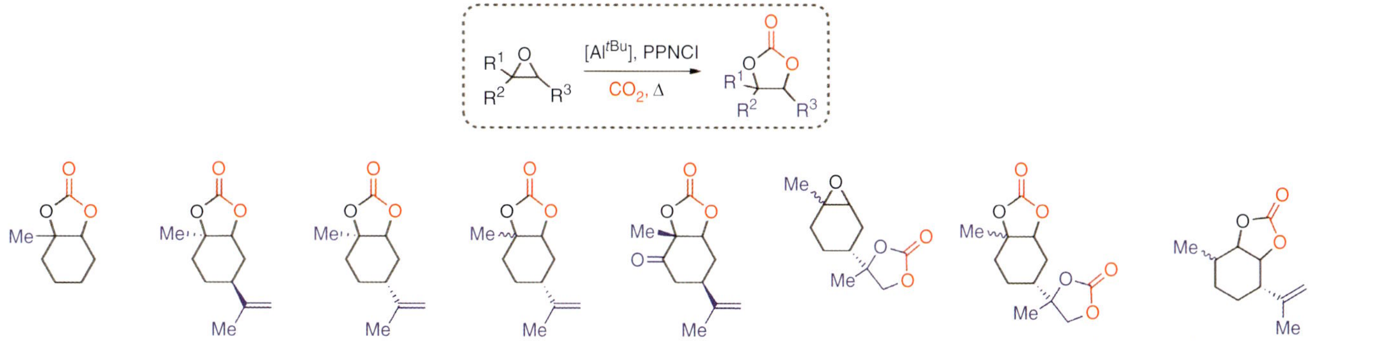

1b: Conv. 75%
Sel. >99%
Yield : 49%

2b: Conv. 73%
Sel. >99%
Yield : 57%
dr >99 : 1 (*trans*)

3b: Conv. 4%
Sel. >99%
dr >99% (*cis*)

4b: Conv. 60%
Sel. >99%
Yield : 43%
dr = 4 : 96 (*trans*)

5b: Conv. 80%
Sel. >99%
Yield = 52%
dr = 20 : 80 (*cis*)

6b: Conv. 90%[a]
Yield : 52%
dr = 48 : 52 (*cis*)

7b: Conv. 92%[b]
Yield : 27% [c]
dr = 80 : 20 (*trans*)

8b: [Al^{Cl}]/TBAB = 0.2/5.0 mol%
Conv. 68%; Yield = 45%
dr = 74 : 26[d]

Figure 13.12 Bicyclic carbonates based on terpene scaffolds produced using Al(aminotriphenolate) (AltBu) complexes. Reaction conditions typically used 1.0 mol% AltBu, 3.0 mol% bis(triphenylphosphine)iminium chloride (PPNCL, nucleophilic additive) in 1.0 ml MEK at 85 °C using 1.0 MPa CO_2 for 66 hours unless otherwise noted. Values obtained by ^{1}H NMR were different for [a] and [b]. [c] Used 120 °C reaction temperature and [d] noted "the major stereoisomer assignment was not possible by NMR owing to a combination of complex patterns." dr, diastereoisomeric ratio. Source: Reprinted with permission from Fiorani et al. [66]. Copyright 2016, John Wiley & Sons.

trans-limonene oxide was favored over the conversion of the *cis* isomer and that the reactions typically took ~66 hours to reach completion.

One application of terpene-based carbonates was demonstrated in an important study by Bähr et al. [67]. Hundred percent conversion of limonene dioxide to limonene dicarbonate was achieved at 140 °C and 30 bar CO_2 pressure after 50 hours with tetrabutylammonium bromide (TBAB) catalyst. Alkylpyridinium iodide supported on SiO_2 did not perform as well as the homogeneous catalyst. The limonene dicarbonate was subsequently used to produce non-isocyanate polyurethanes. Many other catalysts have been investigated for the cycloaddition of CO_2 to epoxides in an effort to reduce the severity of the reaction conditions and increase the turnover frequency (TOF).

A large amount of research has been pioneered by Coates and other researchers at Cornell University – and has been further developed by Hauenstein et al. as well as others – regarding the production of poly(limonene) carbonate from limonene oxide and CO_2 [68–71]. First, several β-diiminate Zn acetate complexes (historically known for copolymerization of CO_2 with epoxides [72, 73]) were investigated by Coates and Moore to produce poly(limonene carbonate) from the copolymerization of different diastereomers of (*R*)-limonene oxide (LO) and CO_2 [69]. While both *cis* and *trans* diastereomers of LO were used, the *trans* starting material more readily underwent nucleophilic attack at the less hindered carbon to generate (1*S*,2*S*,4*R*)-1-methyl-4-(1-methylethenyl)-1,2,-cyclohexanediol (Figure 13.13a), which was subsequently copolymerized with CO_2 using the best complex (#8 in Figure 13.13b). The ligands on the Zn complexes were varied based on steric and electron withdrawing properties and resulted in variations in the activity of the catalyst. Inspired by the work of Coates, Hauenstein et al. optimized the conditions to synthesize poly(limonene carbonate) starting from limonene. They used the same optimized β-diiminate Zn acetate complex to make highly pure *trans*-LO, which generated higher molecular weight (>100 kg/mol) polymers that exhibit favorable properties with respect to hardness, thermal stability, and transparency [70]. Their proposed mechanism involves the alternating insertion of CO_2 and limonene oxide at the catalyst, the latter existing in a dimeric state. Further work by Hauenstein led to the production of poly(limonene carbonate)-derived polymers that have pH-dependent water solubility and improved thermal and antimicrobial properties among other characteristics, making poly(limonene carbonate) a valuable green platform polymer [68].

13.6.1.5 Polymers and Other Materials from Terpenes

Polyesters have also been synthesized using terpenoids. For example, Peña Carrodeguas et al., inspired by Coates' work, synthesized semiaromatic polyesters using terpene-based monomers such as limonene oxide, menthene oxide, and carene oxide via a ring opening copolymerization (ROCOP) of epoxides method with an Fe(III)-based catalyst and (bis(triphenylphosphine)iminium chloride, as shown in Figure 13.14. [74]. The Fe complex catalyst was chosen based on its flexibility for the activation of "sterically more demanding epoxides." The polymers had low polydispersities, molecular weights up to 25 kg/mol, and tunable

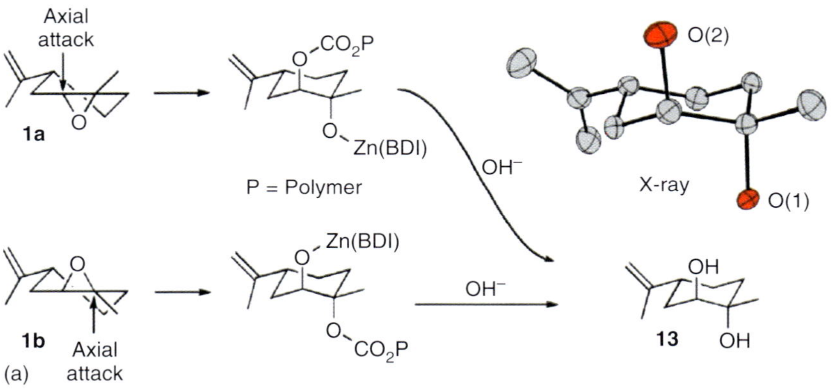

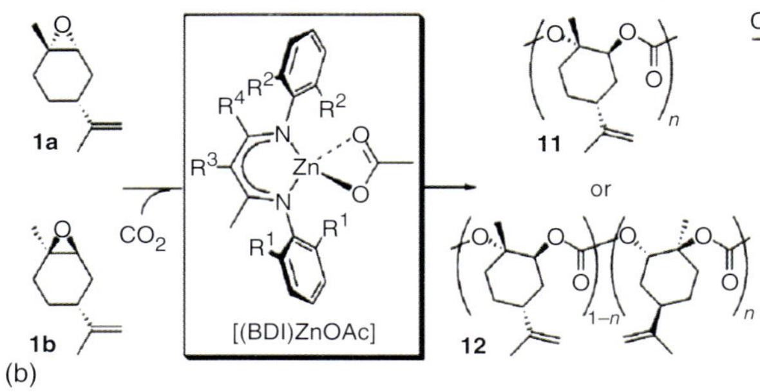

Complex	R^1	R^2	R^3	R^4
2	Et	Et	H	CH_3
3	^{i}Pr	^{i}Pr	H	CH_3
4	Me	Me	H	CH_3
5	Et	^{i}Pr	H	CH_3
6	Et	Et	H	CF_3
7	^{i}Pr	Et	H	CF_3
8	Et	^{i}Pr	H	CF_3
9	Et	Et	CN	CH_3
10	^{i}Pr	Me	CN	CH_3

Figure 13.13 (a) Axial attack on limonene oxide is preferred for diasteroemer **1a** (*trans* isomer) as it is less hindered and hence consumed quicker than **1b** for ring opening during copolymerization. Hydrolytic cleavage produces (1*S*,2*S*,4*R*)-1-methyl-4-(1-methylethenyl)-1,2,-cyclohexanediol (structure 13). (b) Zn complexes tested for the copolymerization of *trans*-limonene oxide (**1a**) and *cis*-limonene oxide (**1b**). Reaction conditions used 0.4 mol% Zn complexes, 25 °C and 100 psi CO_2. Source: Reprinted with permission from Byrne et al. [71]. Copyright 2004, American Chemical Society.

glass transition temperatures (T_g) up to 165 °C and could potentially be used in coating applications.

Many other types of polymers can be synthesized from monoterpenes, particularly by cationic and radical polymerization routes, these having been reviewed extensively by Silvestre and Gandini [75], Gandini and Lacerda [76, 77], Zhu et al. [78], Wilbon et al. [79], Winnacker and Rieger, [80], and Llevot et al. [81]. For example, the cationic polymerization of β-pinene to produce poly(β-pinene) has been performed over a variety of Lewis acid catalysts such as $EtAlCl_2$ or $AlCl_3$ and may also include cocatalysts or additives such as diphenyl ether to prevent chain transfer reactions [82]. Hydrogenated poly(β-pinene) with high molecular weight and good mechanical and thermal properties with potential use as an optical plastic has been successfully synthesized by Satoh et al. [83]. Figure 13.15 shows how β-pinene-based polymers were prepared by means of a $RCl/EtAlCl_2$ catalytic system followed by hydrogenation using Pd/Al_2O_3 (heterogeneous catalyst). The use of a weak Lewis base mediator such as diethyl ether resulted in a more controlled

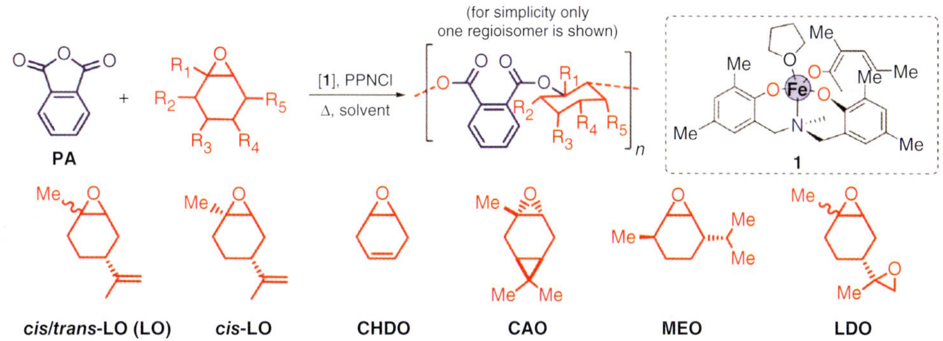

Entry	Sub	[1]/PPNCl (mol %)	Solvent	t (h)	conv[b] (%)	M_n[c] (kg/mol)	$Đ$[c,d]	T_d^{10}[e] (°C)	T_g[f] (°C)
1	LO	0.50, 0.50	THF	24	84	10.5	1.24	255	131
2[g]	LO	0.50, 0.50		24	>99	9.5	1.21		115
3	*cis*-LO	0.50, 0.50	THF	24	>99	16.4	1.33	258	141
4[g]	*cis*-LO	0.50, 0.50		24	>99	9.2	1.44		129
5	CHDO	0.50, 0.50	THF	40	85	24.9	1.54	309	132
6	CHDO	0.50, 0.50		48	>99	19.6	1.42		105
7	CAO	0.50, 0.50	THF	100	79	3.7	1.39	210	130
8[g,h]	CAO	0.50, 0.50		48	89	3.3	1.52		112
9	MEO	0.50, 0.50	THF	24	56	3.2	1.24		155
10[g]	MEO	0.50, 0.50		24	75	5.1	1.28		161
11[g]	MEO	0.30, 0.30		72	75	12.7	1.20	297	165
12[i]	LDO	0.50, 0.50	THF	24	33	8.7	1.94	287	59
13[g,i]	LDO	0.50, 0.50		24	52	6.7	2.41		53

Figure 13.14 Production of semiaromatic polyesters with various terpene oxide scaffolds using Fe-based triaminophenolate catalysts. Reaction conditions: 1.5 mmol PA, solvent (0.50 ml), 65 °C unless stated otherwise, [PA]:[Sub] = 1 : 1.1. [a]Determined by ^{1}H NMR; selectivity for the alternating polymer ≥98%. [b]Determined by GPC in THF (30 °C) using polystyrene standards. [c]$Đ = M_w/M_n$. [d]From thermogravimetric analysis; data refer to Td 10 values at 10 wt% loss. [e]Differential scanning calorimetry (DSC); the data refer to the second heating cycle. [f][PA]:[epox] = 1 : 2. [g]Reaction at 95 °C. [h]Reaction at 45 °C. Source: Reprinted with permission from Peña Carrodeguas et al. [74]. Copyright 2017 American Chemical Society.

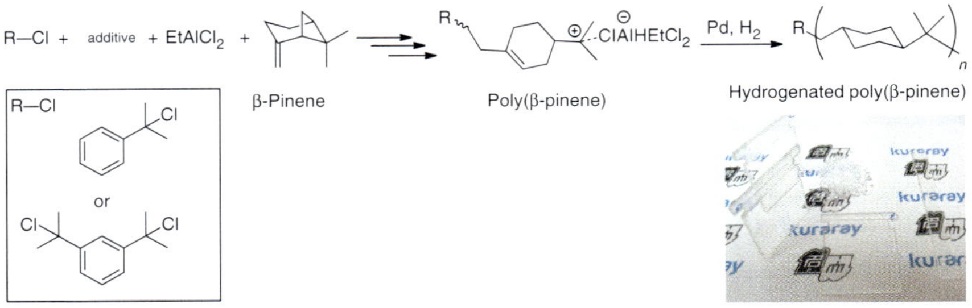

Figure 13.15 Cationic polymerization of β-pinene followed by hydrogenation to produce high molecular weight cycloolefin polymer, shown after extrusion and injection molding at 200 °C. Source: Satoh et al. [83]. Reprinted with permission from Royal Society of Chemistry.

polymerization and lower polydispersity. The presence and type of initiating system (R–Cl in Figure 13.15) and reaction conditions also showed large effects on the polymerization reactions and on the properties of the polymers produced.

Diterpenoid resin acids are also a feedstock for homogeneously catalyzed processes, particularly for the purpose of producing adhesives, resins, and polymeric materials. For example, rosin ester tackifiers can be made from rosin (mostly diterpenoid resin acids) derived from the pulp and paper industry by acid-catalyzed esterification. Additionally, various catalytic methods are used to prevent resin acids from oxidizing and typically result in their oligomerization and/or polymerization [34]. For example, mineral acids or a Lewis acid such as $AlCl_3$ are used to dimerize or oligomerize rosin prior to esterification because the final rosin ester product will have a higher molecular weight and higher softening point [34].

13.6.1.6 "Click Chemistry" Routes for the Production of Materials and Medicinal Compounds from Terpenes

An innovative example of a catalytic process used to make a renewable, degradable resin acid ester was reported by Yao et al. using "click chemistry" catalysis [84]. The combination of ring opening polymerization (ROP) and "click chemistry" catalyzed by CuI/DBU (1,8-diazabicyclo[5.4.0]undec-7-ene) was used to produce a rosin-ester caprolactone grafted polymer, as shown in Figure 13.16. The rosin moiety (feedstock tests used both gum rosin and hydrogenated rosin) contributed to the hydrophobicity of the final product and increased its glass transition temperature [84].

Many compounds with medicinal potential have also been generated using "click chemistry" methods that involve Cu-catalyzed Huisgen 1,3-dipolar cycloadditions (CuAAC) using di- and triterpenoid (particularly steroids, cholesterol, oleanolic acid) feedstocks as reviewed in [85]. In particular, the addition of 1,2,3-triazole moieties has been performed to improve the pharmaceutical properties of terpenoids (i.e. cellular uptake). "Click chemistry" was first described by Sharpless and coworkers [86] and refers to a chemical reaction that must be "modular, wide in scope, gives very high yields, generates only inoffensive by-products that can be removed by nonchromatographic methods and be stereospecific" whereby the process conditions utilize readily

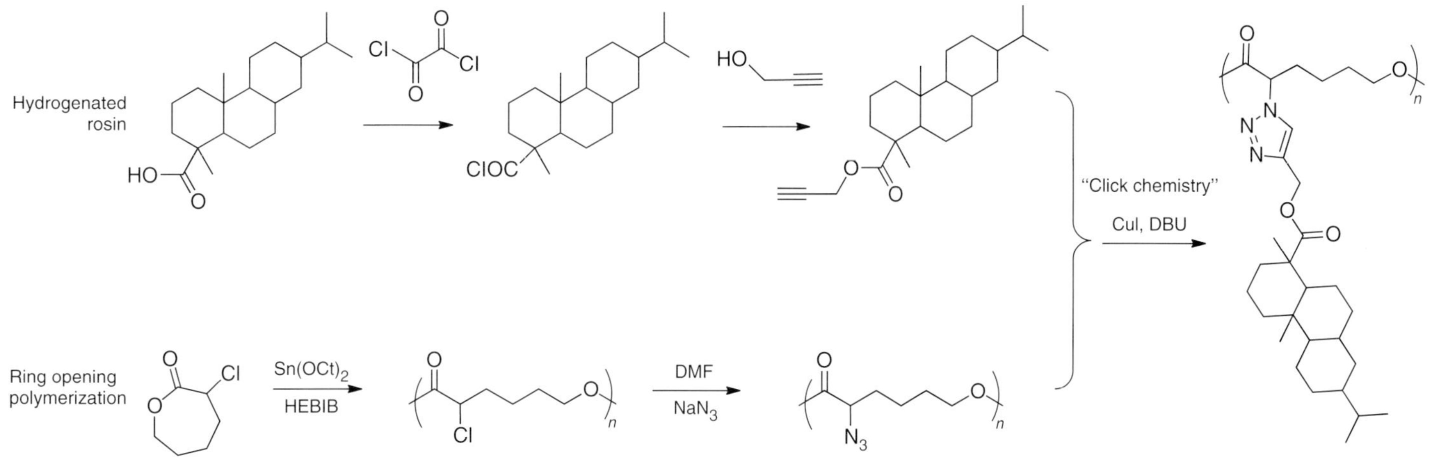

Figure 13.16 Ring opening polymerization (ROP) and "click chemistry" catalyzed by CuI/DBU (1,8-diazabicyclo[5.4.0]undec-7-ene) were combined to produce a rosin ester caprolactone grafted polymer. Hydrogenated rosin is shown as the feedstock

available materials and simple conditions without harmful solvents. The use of CuAAC reactions has become synonymous with "click chemistry" although other catalysts and conditions can qualify for the strict criteria.

Within each class of terpene compounds, there are many types of homogeneous catalytic reactions that the terpenes can undergo, particularly given the diversity of functional groups and stereocenters present in terpenes. Indeed, so many catalysts have been used for the production of terpene-based products that it is impossible to cover them all in one text. However, the diversity of catalysts, reaction conditions, and terpene feedstocks as well as of the uses of the products generated means that there are still many opportunities for the development of novel processes and materials from the homogeneous catalytic conversion of terpenes. As a brief summary and guide, Table 13.1 provides a short list of some

Table 13.1 Tabulated summary of selected homogeneous catalysts used to convert terpenes to chemicals and other products.

Reaction type	Feedstock	Catalyst	Products
Hydration	α-Pinene	Brønsted mineral acids	α-Terpineol
	Dihydromyrcene		Dihydromyrcenol
Oxidation	Camphene	Pd(II)	Camphene-glycol acetate
	Pinenes		Pinocarveol, pinocarveol acetate, myrtenyl acetate
	Pinenes	In-cell P450 enzyme	Pinene oxide
	Limonene		Perillyl alcohol
	α-Pinene, β-pinene and limonene	Percarboxylic acid, Mn(III), Re(MTO)	Pinene oxides, limonene oxide
	β-Pinene and limonene	Re	Pinene oxide, limonene oxide
Cycloaddition of CO_2 to epoxide	Terpene oxide	Al(aminotriphenolate), TBAB	Terpene carbonate or dicarbonate
Isomerization	Terpene derived allylamine	Rh(I)–BINAP	Terpene-derived enamine
	α-Pinene oxide	Zn(II) halides	Campholenic aldehyde
Esterification	Diterpenoid resin acid	Brønsted mineral acids	Resin esters
Polymerization	Terpene oxide	β-Diiminate Zn acetates	Terpene-CO_2-copolymer
	Terpene oxide	Fe(III) complexes	Terpene-PA-polyester
	β-Pinene, limonene	Lewis acids, $AlCl_3$	Poly(β-pinene), poly(limonene)
	Diterpenoid resin acid	Brønsted or Lewis acids	Diterpenoid resin acid oligomers
	Diterpenoid resin acid	Cu(I) "click chemistry"	Rosin-ester caprolactone grafted polymer
Other functional modifications	Diterpenoids, triterpenoids	Cu(I) "click chemistry"	Pharmaceuticals

reaction types, terpene feedstocks, and homogeneous catalysts used for the generation of particular products covered in this section.

13.6.2 Heterogeneous Processes

Heterogeneous catalysts offer several advantages over homogeneous catalysts and the associated processing conditions in the conversion of terpenes to other compounds and materials. For example, heterogeneous catalysts can be more easily separated from the reaction products; they can be frequently regenerated; they can have higher stability than homogeneous catalysts under certain reaction conditions; and they can exhibit high activity and selectivity for certain reactions. However, heterogeneous catalysts can suffer from diffusion limitations, deactivation, and leaching problems. In an effort to overcome the problems associated with the use of homogeneous catalysts, the use and design of heterogeneous catalysts for the conversion of terpenes with high activity and low E-factors has been the focus of extensive research. Metal oxides, layered double hydroxides (LDH), zeolites, molecular sieves, mesoporous materials, activated carbon, and other solid acid and base catalysts have been used to convert terpenes to useful products by means of isomerization, hydration, epoxidation, hydrogenation, dehydrogenation, condensation, and other routes. Table 13.2 summarizes the heterogeneous catalysts and reactions covered in this section. Noteworthy reviews concerning the use of heterogeneous catalysis for the transformation of terpenes have been published by Mäki-Arvela et al. [87] and Corma et al. [63].

13.6.2.1 Isomerization and Hydration of α-Pinene

The isomerization of α-pinene to produce limonene and camphene, shown in Figure 13.17, is an important reaction in the fragrance industry. Camphene is often converted to isoborneol/isobornyl acetate and then camphor [49]. Camphor is used as a fragrance and in various over-the-counter medications to provide symptomatic relief and improvement in sleep for people with upper respiratory infections [88]. Limonene is typically used as a fragrance although many other applications have been reported [44]. The use of acidic catalysts such as TiO_2 and zeolites for the isomerization of α-pinene to limonene and camphene has been utilized in industry and also researched and reviewed extensively. Reaction selectivity depends on the properties of the catalyst and reaction conditions [49, 61, 63]. The design of catalysts and conditions for terpene isomerization typically focuses on improving selectivity to particular products. Typical reaction conditions include temperatures around 100 °C, 1–10 bar pressure (both liquid and gas phase), and the incorporation of various types of acidic supports. The most important attributes of the catalysts that affect the conversion and selectivity of α-pinene to camphene and limonene appear to be the Si/Al ratio in zeolites, the strength and density of Brønsted acid sites, and the catalyst pore structure [63]. At high conversions, camphene is typically produced with a selectivity of 20–60%, and limonene is usually the second most abundant product (around 20% selectivity) followed by trace amounts of terpenines, although relative proportions can change depending on the catalyst and reaction conditions. As an example, Ecormier et al. used the isomerization of α-pinene to study the activity

Table 13.2 Tabulated summary of selected heterogeneous catalysts used to convert terpenes to chemicals and other products.

Reaction type	Feedstock	Catalyst	Products
Isomerization	α-Pinene	TiO_2, zeolites, SiO_2, ZrO_2	Limonene, camphene
	α-Pinene oxide	Supported Ti, Cu, zeolites, molecular sieves, MCM-41, SiO_2	Campholenic aldehyde
Hydration	α-Pinene	Zeolites, activated carbon, heteropolyacids	α-Terpineol
Epoxidation	α-Pinene, limonene	Ti, Co, Mn(II) Schiff on supports, Mg/Al LDH, SiO_2, Al_2O_3, MCM-41, SBA, carbon	α-Pinene oxide, limonene oxide, limonene dioxide
Condensation	Citral	MgO, LDH	Pseudoionone
	Phytol	Zeolites, ion exchange resin, Al_2O_3, SiO_2	Vitamin E
	α-Pinene, camphene and other monoterpenes	Nafion, Montmorillonite	Dimers (terpene-derived fuel)
Dehydrogenation	α-Pinene, limonene	Pd/SiO_2, zeolites, Pd-Ni/HZSM-5, PtO_2	*p*-Cymene
Hydrogenation	Citral	Pt, Ru–Sn, Rh–Sn, Pd, Ni on various supports, SiO_2, TiO_2, Al_2O_3, carbon	Geraniol, nerol, menthol, etc.
	Rosin	Pd, Pt on various supports	Di, tetra-hydrogenated product for rosin ester feed
Hydrodesulfurization	Crude sulfated turpentine	NiMo/C-Al_2O_3	Sulfur-free turpentine (α-pinene, β-pinene, limonene, carene, camphene, etc.)

of hexagonal mesoporous silica (HMS) catalysts grafted with sulfated zirconia (4–15 wt% bulk Zr loading) by wet impregnation followed by sulfuric acid immersion [89]. The authors chose the catalysts on the basis of properties that included high surface area (over 790 m^2/g), stability, tunable number of strongly acidic Lewis/Brønsted sites, control over pore dimensions, and ability to remove the template by solvent extraction. The activity of the catalyst for the isomerization of α-pinene increased with Zr content, although selectivity to limonene or camphene was not addressed. The same authors in another study did note that an

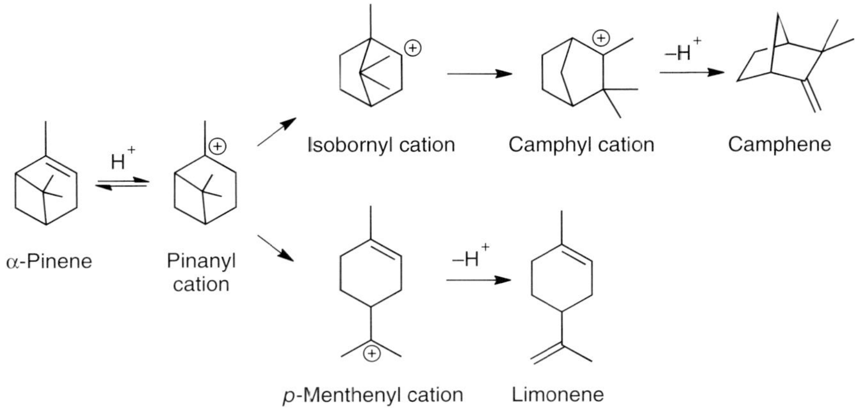

Figure 13.17 Acid-catalyzed isomerization of α-pinene to camphene and limonene.

increase in acid site strength on mesoporous sulfated zirconia led to an increase in selectivity to the monocyclic limonene product and a decrease in selectivity to bicyclic camphene [90]. Heteropolyacid Keggin-type catalysts $H_3PW_{12}O_{40}$ (HPA or HPW) supported on SBA-15 have also been used to isomerize α-pinene in solventless conditions [91]. The pore structure and surface area of the support enabled more uniform dispersion of the HPW crystallites, which afforded uniform acid site densities and minimized clustering. In this case, the selectivity to limonene was inversely proportional to the acid strength of the catalysts tested.

The hydration of α-pinene to produce α-terpineol (lilac fragrance, Figure 13.6) can also be accomplished using solid acidic catalysts such as zeolites, as well as activated carbon and heteropolyacids [41, 49, 61, 63, 92, 93]. For example, Robles-Dutenhefner et al. studied the hydration and acetoxylation of α- and β-pinene and limonene to produce α-terpineol and α-terpenyl acetate using a homogeneous Keggin-type heteropolyacid, $H_3PW_{12}O_{40}$ (HPA), and the heterogeneous version supported on silica [93]. The catalysts showed higher activity (up to 90% conversion, 85% selectivity to the two products depending on conditions and substrate) than the mineral acids that are typically used in industry.

13.6.2.2 Heterogeneous Catalysts for the Epoxidation of Monoterpenes

Another important reaction in the fragrance industry, as discussed in Section 13.6.1.2, is the epoxidation of olefin groups in monoterpenes. Solid catalysts incorporating various metals supported or grafted on LDH, silica, alumina, and mesoporous supports have been used to epoxidize monoterpenes that would typically be converted to other products used in the fragrance industry [49, 63]. As reviewed by Mäki-Arvela et al., the heterogeneous epoxidation of terpenes using Ti supported on various materials, particularly Ti-MCM-41 and Ti-SBA, has been extensively studied [87, 94]. Catalysts incorporating Co have also shown high activity (and recyclability) for the epoxidation of α-pinene, including Co(II) exchanged on zeolites such as ZSM-5 [95]. A Co-containing polyoxymetalate

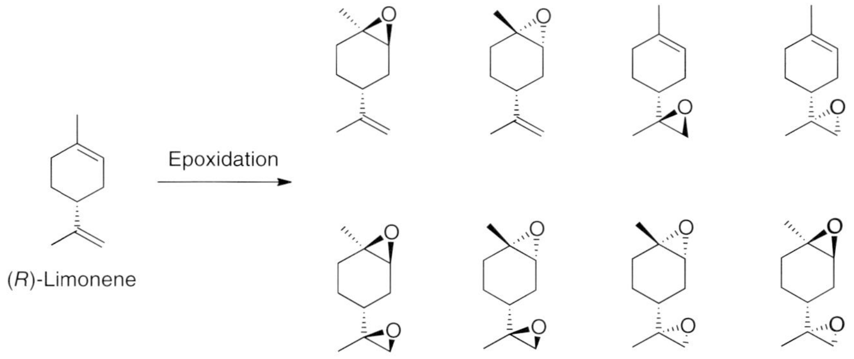

Figure 13.18 The epoxidation of (*R*)-limonene can generate a number of mono- and diepoxide isomers.

supported on mesoporous silica was used for the epoxidation of α-pinene with co-oxidation of isobutyric acid by Maksimchuk et al. [96]. Using functionalized mesoporous cellular foams as catalyst support and optimized conditions, 96% of α-pinene was converted to α-pinene oxide with 90% selectivity (with most of the remaining product being campholenic aldehyde). Without the isobutyric acid, the reaction converted α-pinene to verbenol and verbenone (other useful fragrance compounds), with significantly lower conversion and TOF. The heterogeneous catalysts outperformed their homogeneous counterparts and the MCF was recyclable for two cycles, with a total of about 25% reduction in conversion and selectivity after five cycles.

The epoxidation of limonene can produce a number of products (Figure 13.18) due to the presence of two double bonds, as demonstrated in a study by Bonon et al. using Al_2O_3 as the catalyst [97]. Selective epoxidation of the 1,2 position of limonene has been achieved using LDH in the presence of various additives under mild conditions [98, 99]. Selective epoxidation studies of both α-pinene and limonene were reported by Mavrogiorgou et al. using a Mn(II)-Schiff base catalyst that had been covalently supported on mesoporous MCM-41, SBA-15, mesoporous carbon nanorods, and CMK-3 made using two different synthetic routes [100]. Optimized reactions using H_2O_2 as oxidant and ammonium acetate as an additive proceeded by so-called "single-site heterogeneous catalysis." TOFs for limonene and α-pinene epoxidation on a Mn(II)-CMK-3 catalyst that was synthesized by a single synthetic step were more than twenty times higher (>400 and >700 h^{-1} for α-pinene and limonene production, respectively) than the silicon-based supports. The yields of the 1,2-epoxides from limonene were highest from the MCM-41-based catalyst at 70% and slightly lower for the rapid reaction using the CMK-3-based catalyst (63%). A similar amount of α-pinene was obtained using a MCM-41-based catalyst and a CMK-3 catalyst, yields being 42% and 41%, respectively. The activity of the CMK-3 catalyst was attributed to its carbon-based support and zero porosity, which indicates the presence of diffusion limitations in other catalysts.

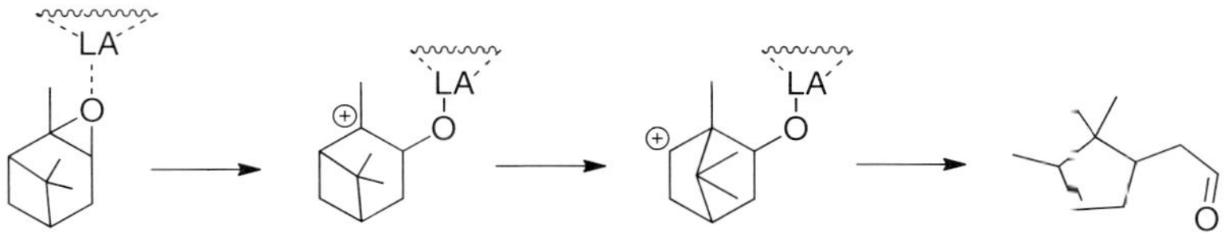

Figure 13.19 The isomerization of α-pinene oxide to produce campholenic aldehyde by heterogeneous Lewis acid (LA) catalyst.

13.6.2.3 Isomerization of α-Pinene Oxide

The isomerization of α-pinene oxide to produce campholenic aldehyde (sandalwood fragrance) has received a great deal of attention and can be performed over Lewis acidic zeolites, molecular sieves, and mesoporous catalysts [49, 87, 101]. Figure 13.19 shows a mechanism used to explain the isomerization based on the presence of Lewis acid active sites (LA) on heterogeneous catalysts [102]. For example, Ti-β zeolite catalysts have successfully been used to convert α-pinene oxide to campholenic aldehyde with 89% selectivity in the liquid phase and 94% selectivity in the gas phase (conversion is 95%), thus representing a more environmentally friendly approach (i.e. no solvent in gas phase) over the typical industrial homogeneous routes [102]. Conversion and selectivity to the aldehyde decreased over time on stream under gas phase conditions, the cause being related to the accumulation of oligomers in the pore system. The catalyst could be completely regenerated up to 100 times. A study by Ravasio et al. covered the combined epoxidation, isomerization, and hydrogenation of a number of terpenes using Ti-MCM-41, Ti/SiO_2, and Cu/SiO_2 catalysts to produce a number of compounds used in the fragrance industry [101]. They reported that simple, amorphous, Lewis acidic alumina–silica catalysts were the most selective for campholenic aldehyde and that Brønsted acidity reduces selectivity for this product. Their research demonstrates how the use of acidic catalysts under different conditions can be used for selective and bifunctional reactions of terpenes for the production of different types of fragrance compounds.

13.6.2.4 Vitamins from Terpenes

The formation of C—C bonds using terpene feedstocks to produce fragrances and other compounds such as vitamins has been the subject of many investigations using heterogeneous catalysts. For example, metal oxides, zeolites, and hydrotalcites have been used to produce pseudoionone and β-ionone (synthetic vitamin precursors) from the aldol condensation of citral and acetone as shown in Figure 13.20a [49, 103]. Climent et al. compared the production of pseudoionones from the condensation of acetone and citral using MgO, hydrotalcite and rehydrated hydrotalcites to demonstrate how the nature of basic sites influences the selectivity and rate of reaction [103]. The Lewis basic sites on MgO and hydrotalcites inhibited selectivity to the pseudoionone product, whereas the rehydrated hydrotalcite, with only moderately basic sites, produced pseudoionone with 99% selectivity at 96% citral conversion. Additionally, the ratio of acetone/citral influenced the initial reaction rate to different degrees for the different basic catalysts. The synthesis of vitamin E acetate (α-tocopherol, Figure 13.20b) from

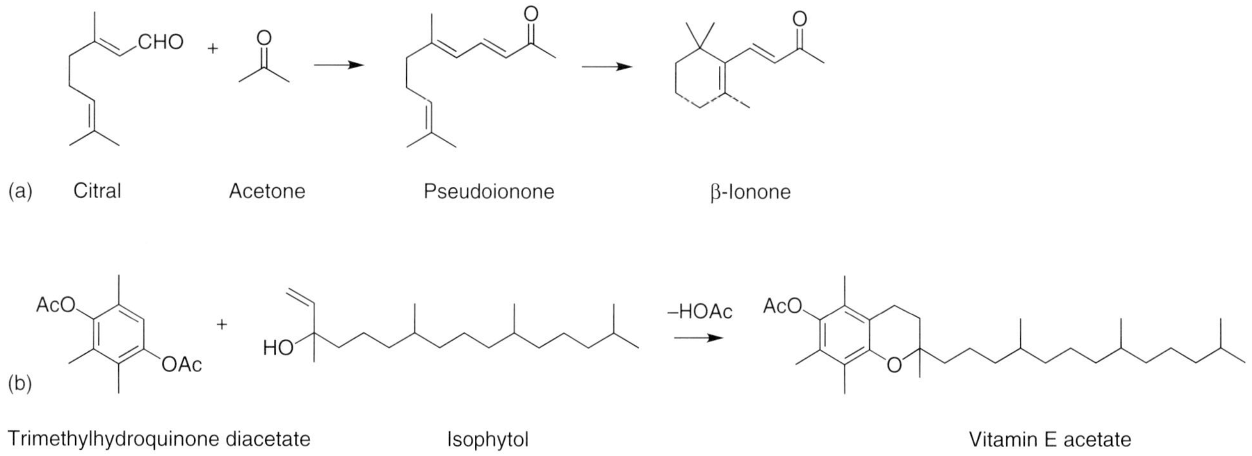

Figure 13.20 (a) Condensation of citral and acetone to produce pseudoionone and β-ionone. (b) Production of vitamin E acetate from isophytol and trimethylhydroquinone acetate.

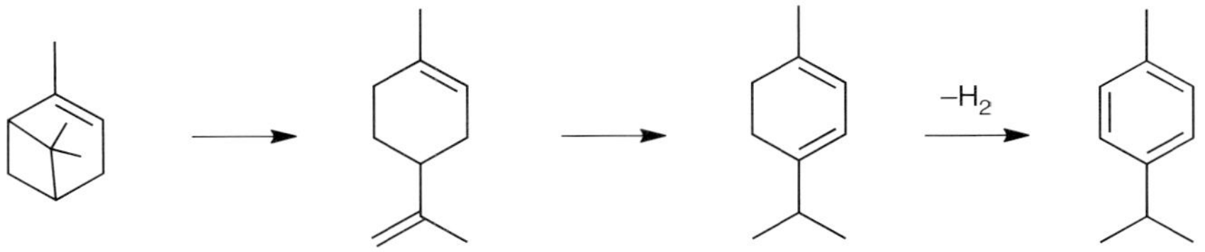

Figure 13.21 Dehydroisomerization of α-pinene to produce *p*-cymene.

isophytol and trimethylhydroquinone diacetate by Friedel–Crafts alkylation can be performed using a variety of acidic homogeneous catalysts such as Zn, Al, and Fe halides [104–106]. However, the use of heterogeneous catalysts such as zeolites, ion exchange resins, alumina, and silica-supported catalysts has also been investigated for the production of vitamin E, as reviewed in Bonrath and Netscher [106]. An important advantage of using heterogeneous catalysts in the synthesis of vitamin E stems from the production of water during the reaction, which can deactivate typical homogeneous catalysts used.

13.6.2.5 Dehydrogenation and Hydrogenation Reactions of Terpenes

Dehydrogenation reactions of terpenes are important for the production of fragrances, chemicals, fuels, and polymers. The dehydrogenation/dehydroisomerization reactions of α-pinene and limonene are used to make *p*-cymene (Figure 13.21), which is used as a fragrance and a solvent and is an intermediate in the production of *p*-cresol. *p*-Cymene has also been used (oxidized) to make bio-based *p*-terephthalate using heterogeneous catalysts [107]. The dehydrogenation of monoterpenes to produce *p*-cymene has been successfully performed using Pd supported on various acidic supports such as SiO_2 and zeolites [49, 63, 87, 108]. An important consideration for this process is the presence of sulfur in the terpene feedstock, which can poison the catalyst, making the use of otherwise cheap and abundant CST limited to reactions using sulfur-resistant catalysts or requiring extensive purification prior to dehydrogenation. Linnekoski et al. used "sulfur-tolerant" Faujasite Y (FAU Y) zeolite catalyst to convert CST to *p*-cymene with 100% conversion of the feedstock and a selectivity of 23% to *p*-cymene [109]. The FAU Y catalyst had high activity and selectivity, which was attributed to a high amount of both Lewis and Brønsted acid sites. Reaction conditions played an important role in the CST conversion and product selectivity as well. The isomerization of *p*-cymene to its *o*- and *m*- isomers reduced the overall selectivity to *p*-cymene, which may be due to diffusional limitations in the zeolite catalysts [87, 109].

Catalytic hydrogenation is used to reduce terpenes such as citral to geraniol and nerol (fragrances), which can also undergo hydrogenation to produce citronellol (see Figure 13.22a) and to reduce diterpenes, particularly to stabilize rosins (Figure 13.22b) [34, 49, 63]. The selective hydrogenation of aldehyde groups or particular C—C bonds is of critical importance for high product yields, especially for terpenes containing multiple functionalities. For example, the conversion of citral to geraniol/nerol or citronellol has been performed over a variety heterogeneous catalysts, including Pt, Ru–Sn, Rh–Sn, Pd, and Ni on various supports

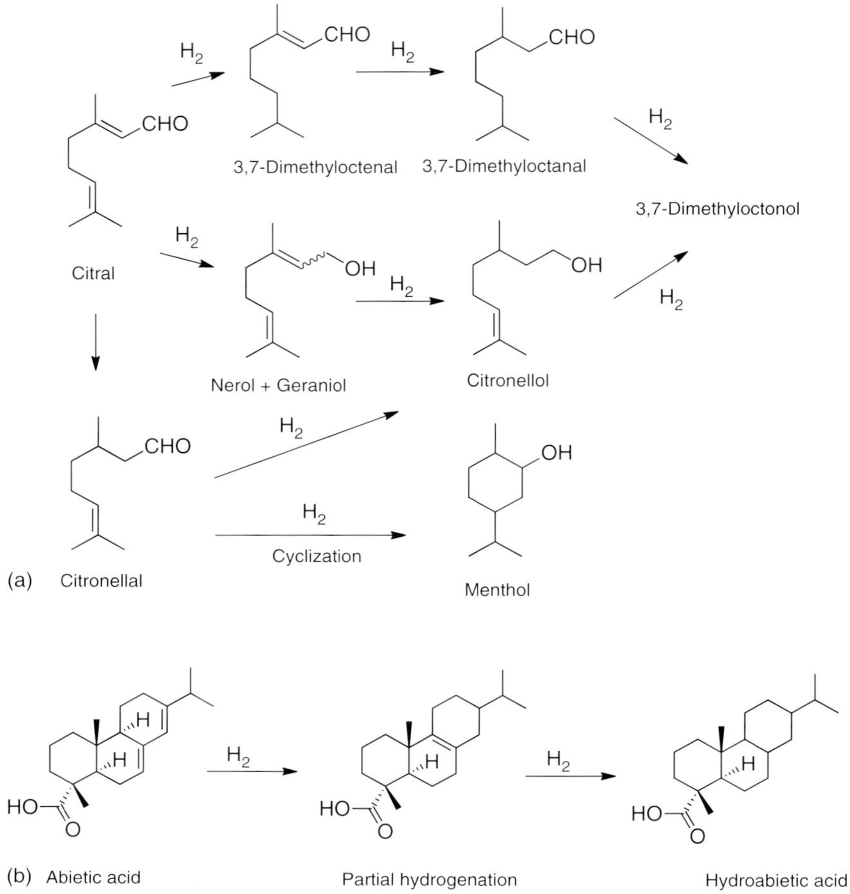

Figure 13.22 (a) Nonselective hydrogenation of citral can produce citronellal, menthol, citronellol, 3,7-dimethyloctenal, 3,7-dimethyloctanal, 3,7-dimethyloctanol, nerol, and geraniol, demonstrating the importance of selective catalysis for the conversion of this feedstock. (b) The hydrogenation of olefins in rosin.

such as activated carbon, SiO_2, TiO_2, and Al_2O_3 [63, 110, 111]. The addition of Sn to Pt- and Rh-based catalysts has proven to be important for the selectivity of citral hydrogenation to the geraniol/nerol product, otherwise citronellal, citronellol, and menthol can be produced. Nonselective industrial hydrogenation of rosin is performed using Pd and Pt catalysts in the presence of H_2 for the reduction of C—C bonds in resin acids such as abietic acid (Figure 13.22b) [34]. Depending on reaction severity, one or multiple C—C bonds may be reduced in diterpenoid resin acids.

13.6.2.6 Conversion of Terpenes to Fuels

Hydrogenation, dehydrogenation, and condensation reactions are also relevant for the conversion of terpenes to fuels. While terpenes have been used as a raw fuel source throughout history, more recent investigations into their various physicochemical properties have shown why particular terpene feedstocks are being identified as "specialty biofuels" [3]. The hydrogenation of myrcene and limonene has been performed over Pd/activated charcoal and Pt/Al_2O_3 to produce 2,6-dimethyloctane and 1-isopropyl-4-methylcyclohexane, respectively, for fuel property studies upon blending with diesel [112]. Crude wood sulfate turpentine has also been upgraded to remove sulfur for various reasons, including gasoline blending studies, using commercial NiMo/C-Al_2O_3 hydrodesulfurization, and NiW–NiO hydrodewaxing catalysts [113]. Moreover, the utilization of terpenes for the production of biofuels is not limited to their use as a single feedstock. Zhang and Zhao have used the dehydroaromatization of limonene to *p*-cymene to produce H_2 *in situ* for the concomitant hydrodeoxygenation of stearic acid to alkanes using heterogeneous Pd-Ni/HZSM-5 [114]. The reaction occurred under relatively mild conditions at 280 °C and only 2 bar H_2. The bimetallic catalyst outperformed catalysts incorporating the individual metals, and the reaction was explained on the basis of dehydroaromatization of limonene occurring on Pd sites and the Ni being the active site for hydrodeoxygenation of stearic acid to produce alkanes, as shown in Figure 13.23. The same researchers also used their novel "carbon-chain filling strategy" that uses limonene to produce aromatics and *in situ* H_2 in the presence of Pd-Ni/HZSM-5 with palm oil and without the use of *ex situ* H_2 [115].

Other catalytic processes used to generate fuels from terpenes are worth mentioning as well. For example, Meylemans et al. used heterogeneous acid catalysts

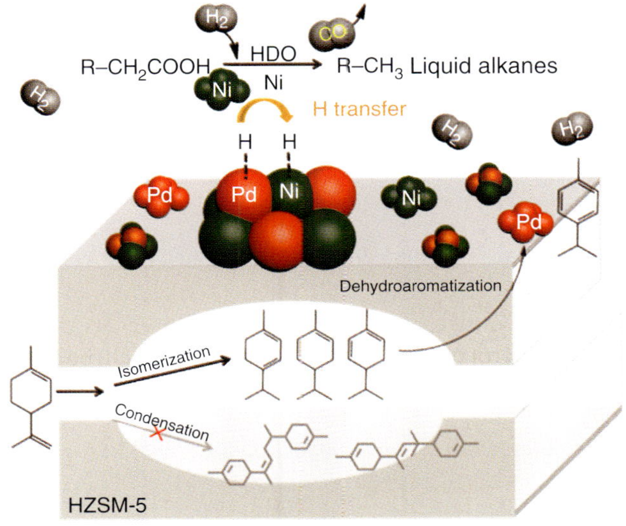

Figure 13.23 Limonene dehydrogenation produces *in situ* H_2 for the hydrodeoxygenation of stearic acid to produce fatty alkanes in the presence of Pd-Ni/HZSM-5 catalysts. Source: Reprinted with permission from Zhang and Zhao [114]. Copyright 2016, American Chemical Society.

Montmorillonite K10 and Nafion SAC-13 to dimerize various monoterpenes and their mixtures to study the fuel properties of the resulting dimers [116]. Their study, coupled with previous results, show how the structures of the monoterpenes influence the reaction mechanisms based on their preference for isomerization and adsorption on the catalysts. In particular, mixtures of monoterpenes (turpentine as well as α-pinene/camphene) showed high yields of dimers from Nafion SAC-13, although the individual α-pinene and camphene did not readily dimerize. The authors speculated that the α-pinene bound to the catalyst first and subsequently cross-coupled with camphene. One important outcome of this study is that it showed that acidic heterogeneous catalysts that have traditionally been used for the isomerization of monoterpenes (described previously) also lead to the production of dimers. Meylemans et al. followed up their study by analyzing other fuel and fuel blend (particularly jet fuel, JP-8) properties of turpentine dimers and pinane that was produced by hydrogenation of α-pinene over heterogeneous PtO_2 [117].

Despite the development of routes for the conversion of terpenes to fuels, their limited global production and high value as fine chemicals and as feedstocks for specialty or high-performing polymers and adhesives, would likely make their use as fuel precursors uneconomical given current fuel prices, although techno-economic analyses would need to be performed. Currently, the conversion of terpenes to resins, polymers, and other chemicals and materials is the focus of a wide variety of catalysis and materials research. Commercially, the vast majority of terpenes are used as – and converted to – fragrances and other fine chemicals such as pharmaceuticals, using a wide variety of catalytic pathways. Given the diversity of functional groups present in terpenes and the variety of catalysts that can be used to convert them, countless permutations exist for their use in the production of other chemicals, and terpenes will undoubtedly continue to be the source of many future investigations across industry and academia.

Acknowledgments

This work was authored (in part) by the Alliance for Sustainable Energy, LLC, the manager and operator of the National Renewable Energy Laboratory for the US Department of Energy (DOE) under Contract No. DE-AC36-08GO28308. Funding provided by the US Department of Energy (DOE) Advanced Research Projects Agency – Energy (ARPA-E) under award No. DE-AR0000209 and by the US Department of Energy Center for Bioenergy Innovation, a US DOE Bioenergy Research Center supported by the Office of Science. The views expressed in the article do not necessarily represent the views of the DOE or the US Government. The US Government retains, and the publisher, by accepting the article for publication, acknowledges that the US Government retains a nonexclusive, paid-up, irrevocable, worldwide license to publish or reproduce the published form of this work or allow others to do so, for US Government purposes.

References

1 Keoke, E.D. and Porterfield, K.M. (2002). *Encyclopedia of American Indian Contributions to the World: 15,000 Years of Inventions and Innovations.* Facts on File.

2 Susaeta, A., Peter, G.F., Hodges, A.W., and Carter, D.R. (2014). Oleoresin tapping of planted slash pine (*Pinus elliottii* Engelm. var *elliottii*) adds value and management flexibility to landowners in the southern United States. *Biomass Bioenergy* 68: 55–61.

3 Mewalal, R., Rai, D.K., Kainer, D. et al. (2017). Plant-derived terpenes: a feedstock for specialty biofuels. *Trends Biotechnol.* 35: 227–240.

4 Rodrigues-Corrêa, K.C.d.S., de Lima, J.C., and Fett-Neto, A.G. (2012). Pine oleoresin: tapping green chemicals, biofuels, food protection, and carbon sequestration from multipurpose trees. *Food Energy Secur.* 1: 81–93.

5 Peter, G. (2015) Sustaining southern pine competitiveness. *Spring Symposium.* University of Florida School of Forest Resources & Conservation: Conference.

6 Ruzicka, L. (1953). The isoprene rule and the biogenesis of terpenic compounds. *Experientia* 9: 357–367.

7 Keeling, C.I. and Bohlmann, J. (2006). Diterpene resin acids in conifers. *Phytochemistry* 67: 2415–2423.

8 Achotegui-Castells, A., Llusià, J., Hódar, J., and Peñuelas, J. (2013). Needle terpene concentrations and emissions of two coexisting subspecies of Scots pine attacked by the pine processionary moth (*Thaumetopoea pityocampa*). *Acta Physiol. Plant.* 35: 3047–3058.

9 Martin, D., Tholl, D., Gershenzon, J., and Bohlmann, J. (2002). Methyl jasmonate induces traumatic resin ducts, terpenoid resin biosynthesis, and terpenoid accumulation in developing xylem of Norway spruce stems. *Plant Physiol.* 129: 1003–1018.

10 Cheng, A.-X., Lou, Y.-G., Mao, Y.-B. et al. (2007). Plant terpenoids: biosynthesis and ecological functions. *J. Integr. Plant Biol.* 49: 179–186.

11 Vickers, C.E., Possell, M., Cojocariu, C.I. et al. (2009). Isoprene synthesis protects transgenic tobacco plants from oxidative stress. *Plant Cell Environ.* 32: 520–531.

12 Loreto, F. and Velikova, V. (2001). Isoprene produced by leaves protects the photosynthetic apparatus against ozone damage, quenches ozone products, and reduces lipid peroxidation of cellular membranes. *Plant Physiol.* 127: 1781–1787.

13 Terry, G.M., Stokes, N.J., Hewitt, C.N., and Mansfield, T.A. (1995). Exposure to isoprene promotes flowering in plants. *J. Exp. Bot.* 46: 1629–1631.

14 Schaller, H. (2003). The role of sterols in plant growth and development. *Prog. Lipid Res.* 42: 163–175.

15 Young, A.J. (1991). The photoprotective role of carotenoids in higher plants. *Physiol. Plant.* 83: 702–708.

16 Breitmaier, E. (2006). *Terpenes: Flavors, Fragrances, Pharmaca, Pheromones.* Wiley.

17 Fahn, A. (1988). Secretory tissues in vascular plants. *New Phytol.* 108: 229–257.

18 Franceschi, V.R., Krokene, P., Christiansen, E., and Krekling, T. (2005). Anatomical and chemical defenses of conifer bark against bark beetles and other pests. *New Phytol.* 167: 353–376.

19 Bertaud, F., Crampon, C., and Badens, E. (2017). Volatile terpene extraction of spruce, fir and maritime pine wood: supercritical CO_2 extraction compared to classical solvent extractions and steam distillation. *Holzforschung* 71: 667. https://www.degruyter.com/view/j/hfsg.2017.71.issue-7-8/hf-2016-0197/hf-2016-0197.xml.

20 Harman-Ware, A.E., Davis, M.F., Peter, G.F. et al. (2017). Estimation of terpene content in loblolly pine biomass using a hybrid fast-GC and pyrolysis-molecular beam mass spectrometry method. *J. Anal. Appl. Pyrolysis*.

21 Harman-Ware, A.E., Sykes, R., Peter, G.F., and Davis, M. (2016). Determination of terpenoid content in pine by organic solvent extraction and fast-GC analysis. *Front. Energy Res.* 4: 2. https://doi.org/10.3389/fenrg.2016.00002.

22 Bennici, A. and Tani, C. (2004). Anatomical and ultrastructural study of the secretory cavity development of *Citrus sinensis* and *Citrus limon*: evaluation of schizolysigenous ontogeny. *Flora-Morphol. Distrib. Funct. Ecol. Plants* 199: 464–475.

23 Espina, L., Somolinos, M., Lorán, S. et al. (2011). Chemical composition of commercial citrus fruit essential oils and evaluation of their antimicrobial activity acting alone or in combined processes. *Food Control* 22: 896–902.

24 Goodger, J.Q.D., Heskes, A.M., Mitchell, M.C. et al. (2010). Isolation of intact sub-dermal secretory cavities from *Eucalyptus*. *Plant Methods* 6: 20.

25 Irmisch, S., Jiang, Y., Chen, F. et al. (2014). Terpene synthases and their contribution to herbivore-induced volatile emission in western balsam poplar (*Populus trichocarpa*). *BMC Plant Biol.* 14: 270.

26 Warnant, P., Mertens, P., and Marche, C. (2004). Screening of poplar biomass for bio-active compounds: a simple method to assess antioxidant activity. *Bioresour. Technol.* 93: 43–48.

27 Arimura, G.-I., Huber, D.P.W., and Bohlmann, J. (2004). Forest tent caterpillars (*Malacosoma disstria*) induce local and systemic diurnal emissions of terpenoid volatiles in hybrid poplar (*Populus trichocarpa* × *deltoides*): cDNA cloning, functional characterization, and patterns of gene expression of (−)-germacrene D synthase, *PtdTPS1*. *Plant J.* 37: 603–616.

28 Schnitzler, J.P., Louis, S., Behnke, K., and Loivamäki, M. (2010). Poplar volatiles – biosynthesis, regulation and (eco)physiology of isoprene and stress-induced isoprenoids. *Plant Biol.* 12: 302–316.

29 Sharkey, T.D. and Yeh, S. (2001). Isoprene emission from plants. *Annu. Rev. Plant Physiol. Plant Mol. Biol.* 52: 407–436.

30 Surburg, H. and Panten, J. (2006). Natural raw materials in the flavor and fragrance industry. In: *Common Fragrance and Flavor Materials*, 177, 177–238, 238. Weinheim: Wiley-VCH.

31 Raguso, R.A., Levin, R.A., Foose, S.E. et al. (2003). Fragrance chemistry, nocturnal rhythms and pollination "syndromes" in *Nicotiana*. *Phytochemistry* 63: 265–284.

32 Viuda-Martos, M., Ruiz-Navajas, Y., Fernández-López, J., and Pérez-Álvarez, J. (2008). Antifungal activity of lemon (*Citrus lemon* L.), mandarin (*Citrus reticulata* L.), grapefruit (*Citrus paradisi* L.) and orange (*Citrus sinensis* L.) essential oils. *Food Control* 19: 1130–1138.

33 Ben Hsouna, A., Ben Halima, N., Smaoui, S., and Hamdi, N. (2017). Citrus lemon essential oil: chemical composition, antioxidant and antimicrobial activities with its preservative effect against Listeria monocytogenes inoculated in minced beef meat. *Lipids Health Dis.* 16: 146.

34 Buisman, G.J.H. and Lange, J.H.M. (2016). Arizona chemical: refining and upgrading of bio-based and renewable feedstocks. In: *Industrial Biorenewables: A Practical Viewpoint* (ed. P.D. de María), 21–62. Hoboken, NJ: Wiley.

35 Bohlmann, J. and Keeling, C.I. (2008). Terpenoid biomaterials. *Plant J.* 54: 656–669.

36 Reverchon, E. (1997). Supercritical fluid extraction and fractionation of essential oils and related products. *J. Supercrit. Fluids* 10: 1–37.

37 Regert, M. and Rolando, C. (2002). Identification of archaeological adhesives using direct inlet electron ionization mass spectrometry. *Anal. Chem.* 74: 965–975.

38 Jerković, I., Marijanović, Z., Gugić, M., and Roje, M. (2011). Chemical profile of the organic residue from ancient amphora found in the Adriatic Sea determined by direct GC and GC-MS analysis. *Molecules* 16: 7936–7948.

39 Outland Iii, R.B. (2004). *Tapping the Pines: The Naval Stores Industry in the American South*. LSU Press.

40 Palmer, R.C. (1943). Solvents from pine. *Ind. Eng. Chem.* 35: 1023–1025.

41 Gscheidmeier, M. and Fleig, H. (2000). Turpentines. In: *Ullmann's Encyclopedia of Industrial Chemistry*, 537–550. Weinheim: Wiley-VCH.

42 Surburg, H. and Panten, J. (2006). Introduction. In: *Common Fragrance and Flavor Materials*, 1–5. Weinheim: Wiley-VCH.

43 Ludwiczuk, A., Skalicka-Woźniak, K., and Georgiev, M.I. (2017). Terpenoids. In: *Pharmacognosy*, Chapter 11 (ed. R. Delgoda), 233–266. Boston, MA: Academic Press.

44 Ciriminna, R., Lomeli-Rodriguez, M., Demma Cara, P. et al. (2014). Limonene: a versatile chemical of the bioeconomy. *Chem. Commun.* 50: 15288–15296.

45 Abdel-Maksoud, G. and El-Amin, A.-R. (2011). A review on the materials used during the mummification processes in ancient egypt. *Mediterr. Archaeol. Archaeom.* 11: 129–150.

46 Zeng, W.C., Zhang, Z., Gao, H. et al. (2012). Chemical composition, antioxidant, and antimicrobial activities of essential oil from pine needle (*Cedrus deodara*). *J. Food Sci.* 77: C824–C829.

47 Singh, B. and Sharma, R.A. (2015). Plant terpenes: defense responses, phylogenetic analysis, regulation and clinical applications. *3 Biotech* 5: 129–151.

48 World Health Organization (2017). *19th WHO Model List of Essential Medicines. 2015*. World Health Organization. Online. Accessed.
49 Monteiro, J. and Veloso, C. (2004). Catalytic conversion of terpenes into fine chemicals. *Top. Catal.* 27: 169–180.
50 Gusevskaya, E., Robles-Dutenhefner, P.A., and Ferreira, V.M.S. (1998). Palladium-catalyzed oxidation of bicyclic monoterpenes by hydrogen peroxide. *Appl. Catal., A* 174: 177–186.
51 Ruff, A.J., Arlt, M., van Ohlen, M. et al. (2016). An engineered outer membrane pore enables an efficient oxygenation of aromatics and terpenes. *J. Mol. Catal. B: Enzym.* 134: 285–294.
52 Sienel, G., Rieth, R., and Rowbottom, K.T. (2000). Epoxides. In: *Ullmann's Encyclopedia of Industrial Chemistry*, 139–154. Weinheim: Wiley-VCH.
53 Raupp, Y.S., Yildiz, C., Kleist, W., and Meier, M.A.R. (2017). Aerobic oxidation of α-pinene catalyzed by homogeneous and MOF-based Mn catalysts. *Appl. Catal., A* 546: 1–6.
54 Cubillos, J., Vásquez, S., and Montes de Correa, C. (2010). Salen manganese (III) complexes as catalysts for R-(+)-limonene oxidation. *Appl. Catal., A* 373: 57–65.
55 Korstanje, T.J. and Gebbink, R.J.M.K. (2012). Catalytic oxidation and deoxygenation of renewables with rhenium complexes. In: *Organometallics and Renewables* (eds. M. Meier, B. Weckhuysen and P. Bruijnincx), 129–174. Berlin, Heidelberg: Springer.
56 Adolfsson, H., Copéret, C., Chiang, J.P., and Yudin, A.K. (2000). Efficient epoxidation of alkenes with aqueous hydrogen peroxide catalyzed by methyltrioxorhenium and 3-cyanopyridine. *J. Org. Chem.* 65: 8651–8658.
57 Michel, T., Betz, D., Cokoja, M. et al. (2011). Epoxidation of α-pinene catalyzed by methyltrioxorhenium(VII): influence of additives, oxidants and solvents. *J. Mol. Catal. A: Chem.* 340: 9–14.
58 Van Leeuwen, P.W.N.M. (2006). *Homogeneous Catalysis: Understanding the Art*. Springer Science & Business Media.
59 Tani, K., Yamagata, T., Tatsuno, Y. et al. (1985). Bis[(*R*)-(+)-binap]rhodium(I) perchlorate, a highly efficient catalyst for the asymmetric isomerization of allylamines. *Angew. Chem. Int. Ed. Engl.* 24: 217–219.
60 Akutagawa, S. (1997). Enantioselective isomerization of allylamine to enamine: practical asymmetric synthesis of (−)-menthol by Rh–BINAP catalysts. *Top. Catal.* 4: 271–274.
61 Surburg, H. and Panten, J. (2006). Individual fragrance and flavor materials. In: *Common Fragrance and Flavor Materials*, 7–175. Weinheim: Wiley-VCH.
62 Ciprés, I., Kalck, P., Park, D.-C., and Serein-Spirau, F. (1991). Carbon monoxide as a building block for organic synthesis. Part III. Selective hydrocarbonylation of monoterpenes to give potentially biologically active aldehydes. *J. Mol. Catal.* 66: 399–407.
63 Corma, A., Iborra, S., and Velty, A. (2007). Chemical routes for the transformation of biomass into chemicals. *Chem. Rev.* 107: 2411–2502.
64 North, M., Pasquale, R., and Young, C. (2010). Synthesis of cyclic carbonates from epoxides and CO_2. *Green Chem.* 12: 1514–1539.

65 Schäffner, B., Schäffner, F., Verevkin, S.P., and Börner, A. (2010). Organic carbonates as solvents in synthesis and catalysis. *Chem. Rev.* 110: 4554–4581.
66 Fiorani, G., Stuck, M., Martín, C. et al. (2016). Catalytic coupling of carbon dioxide with terpene scaffolds: access to challenging bio-based organic carbonates. *ChemSusChem* 9: 1304–1311.
67 Bähr, M., Bitto, A., and Mulhaupt, R. (2012). Cyclic limonene dicarbonate as a new monomer for non-isocyanate oligo- and polyurethanes (NIPU) based upon terpenes. *Green Chem.* 14: 1447–1454.
68 Hauenstein, O., Agarwal, S., and Greiner, A. (2016). Bio-based polycarbonate as synthetic toolbox. *Nat. Commun.* 7: 11862.
69 Coates, G.W. and Moore, D.R. (2004). Discrete metal-based catalysts for the copolymerization of CO_2 and epoxides: discovery, reactivity, optimization, and mechanism. *Angew. Chem. Int. Ed.* 43: 6618–6639.
70 Hauenstein, O., Reiter, M., Agarwal, S. et al. (2016). Bio-based polycarbonate from limonene oxide and CO_2 with high molecular weight, excellent thermal resistance, hardness and transparency. *Green Chem.* 18: 760–770.
71 Byrne, C.M., Allen, S.D., Lobkovsky, E.B., and Coates, G.W. (2004). Alternating copolymerization of limonene oxide and carbon dioxide. *J. Am. Chem. Soc.* 126: 11404–11405.
72 Inoue, S., Koinuma, H., and Tsuruta, T. (2003). Copolymerization of carbon dioxide and epoxide. *J. Polym. Sci., Part B: Polym. Lett.* 7: 287–292.
73 Inoue, S., Koinuma, H., and Tsuruta, T. (2003). Copolymerization of carbon dioxide and epoxide with organometallic compounds. *Die Makromol. Chem.* 130: 210–220.
74 Peña Carrodeguas, L., Martín, C., and Kleij, A.W. (2017). Semiaromatic polyesters derived from renewable terpene oxides with high glass transitions. *Macromolecules* 50: 5337–5345.
75 Silvestre, A.J.D. and Gandini, A. (2008). Terpenes: major sources, properties and applications. *Monomers Polym. Compos. Renew. Resour.* 1: 17–38.
76 Gandini, A. and Lacerda, T.M. (2015). From monomers to polymers from renewable resources: recent advances. *Prog. Polym. Sci.* 48: 1–39.
77 Gandini, A. (2011). The irruption of polymers from renewable resources on the scene of macromolecular science and technology. *Green Chem.* 13: 1061–1083.
78 Zhu, Y., Romain, C., and Williams, C.K. (2016). Sustainable polymers from renewable resources. *Nature* 540: 354–362.
79 Wilbon, P.A., Chu, F., and Tang, C. (2013). Progress in renewable polymers from natural terpenes, terpenoids, and rosin. *Macromol. Rapid Commun.* 34: 8–37.
80 Winnacker, M. and Rieger, B. (2015). Recent progress in sustainable polymers obtained from cyclic terpenes: synthesis, properties, and application potential. *ChemSusChem* 8: 2455–2471.
81 Llevot, A., Dannecker, P.K., von Czapiewski, M. et al. (2016). Renewability is not enough: recent advances in the sustainable synthesis of biomass-derived monomers and polymers. *Chem. Eur. J.* 22: 11510–11521.
82 Kukhta, N.A., Vasilenko, I.V., and Kostjuk, S.V. (2011). Room temperature cationic polymerization of [small beta]-pinene using modified $AlCl_3$ catalyst:

toward sustainable plastics from renewable biomass resources. *Green Chem.* 13: 2362–2364.

83 Satoh, K., Nakahara, A., Mukunoki, K. et al. (2014). Sustainable cycloolefin polymer from pine tree oil for optoelectronics material: living cationic polymerization of [small beta]-pinene and catalytic hydrogenation of high-molecular-weight hydrogenated poly([small beta]-pinene). *Polym. Chem.* 5: 3222–3230.

84 Yao, K., Wang, J., Zhang, W. et al. (2011). Degradable rosin-ester–caprolactone graft copolymers. *Biomacromolecules* 12: 2171–2177.

85 Kacprzak, K., Skiera, I., Piasecka, M., and Paryzek, Z. (2016). Alkaloids and isoprenoids modification by copper(I)-catalyzed huisgen 1,3-dipolar cycloaddition (Click Chemistry): toward new functions and molecular architectures. *Chem. Rev.* 116: 5689–5743.

86 Kolb, H.C., Finn, M.G., and Sharpless, K.B. (2001). Click chemistry: diverse chemical function from a few good reactions. *Angew. Chem. Int. Ed.* 40: 2004–2021.

87 Mäki-Arvela, P., Holmbom, B., Salmi, T., and Murzin, D.Y. (2007). Recent progress in synthesis of fine and specialty chemicals from wood and other biomass by heterogeneous catalytic processes. *Catal. Rev.* 49: 197–340.

88 Paul, I.M., Beiler, J.S., King, T.S. et al. (2010). Vapor rub, petrolatum, and no treatment for children with nocturnal cough and cold symptoms. *Pediatrics* 126: 1092–1099.

89 Ecormier, M.A., Lee, A.F., and Wilson, K. (2005). High activity, templated mesoporous SO_4/ZrO_2/HMS catalysts with controlled acid site density for α-pinene isomerisation. *Microporous Mesoporous Mater.* 80: 301–310.

90 Ecormier, M.A., Wilson, K., and Lee, A.F. (2003). Structure–reactivity correlations in sulphated-zirconia catalysts for the isomerisation of α-pinene. *J. Catal.* 215: 57–65.

91 Frattini, L., Isaacs, M.A., Parlett, C.M.A. et al. (2017). Support enhanced α-pinene isomerization over HPW/SBA-15. *Appl. Catal., B* 200: 10–18.

92 Sanchez, L.M., Thomas, H.J., Climent, M.J. et al. (2016). Heteropolycompounds as catalysts for biomass product transformations. *Catal. Rev. Sci. Eng.* 58: 497–586.

93 Robles-Dutenhefner, P.A., da Silva, K.A., Siddiqui, M.R.H. et al. (2001). Hydration and acetoxylation of monoterpenes catalyzed by heteropoly acid. *J. Mol. Catal. A: Chem.* 175: 33–42.

94 Trong On, D., Kapoor, M.P., Joshi, P.N. et al. (1997). Catalytic epoxidation of α-pinene over bifunctional mesoporous molecular sieves. *Catal. Lett.* 44: 171–176.

95 Tang, B., Lu, X.H., Zhou, D. et al. (2012). Highly efficient epoxidation of styrene and α-pinene with air over Co^{2+}-exchanged ZSM-5 and Beta zeolites. *Catal. Commun.* 21: 68–71.

96 Maksimchuk, N.V., Melgunov, M.S., Chesalov, Y.A. et al. (2007). Aerobic oxidations of α-pinene over cobalt-substituted polyoxometalate supported on amino-modified mesoporous silicates. *J. Catal.* 246: 241–248.

97 Bonon, A.J., Kozlov, Y.N., Bahú, J.O. et al. (2014). Limonene epoxidation with H_2O_2 promoted by Al_2O_3: kinetic study, experimental design. *J. Catal.* 319: 71–86.

98 Aramendía, M.A., Borau, V., Jiménez, C. et al. (2001). Epoxidation of limonene over hydrotalcite-like compounds with hydrogen peroxide in the presence of nitriles. *Appl. Catal., A* 216: 257–265.

99 Bhattacharjee, S., Dines, T.J., and Anderson, J.A. (2004). Synthesis and application of layered double hydroxide-hosted catalysts for stereoselective epoxidation using molecular oxygen or air. *J. Catal.* 225: 398–407.

100 Mavrogiorgou, A., Baikousi, M., Costas, V. et al. (2016). Mn-Schiff base modified MCM-41, SBA-15 and CMK-3 NMs as single-site heterogeneous catalysts: Alkene epoxidation with H2O2 incorporation. *J. Mol. Catal. A: Chem.* 413: 40–55.

101 Ravasio, N., Zaccheria, F., Guidotti, M., and Psaro, R. (2004). Mono- and bifunctional heterogeneous catalytic transformation of terpenes and terpenoids. *Top. Catal.* 27: 157–168.

102 Kunkeler, P.J., van der Waal, J.C., Bremmer, J. et al. (1998). Application of zeolite titanium Beta in the rearrangement of α-pinene oxide to campholenic aldehyde. *Catal. Lett.* 53: 135–138.

103 Climent, M.J., Corma, A., Iborra, S., and Velty, A. (2002). Designing the adequate base solid catalyst with Lewis or Bronsted basic sites or with acid–base pairs. *J. Mol. Catal. A: Chem.* 182–183: 327–342.

104 Netscher, T. (2007). Synthesis of Vitamin E. In: *Vitamins & Hormones*, vol. 76, 155–202. Academic Press.

105 Baldenius, K.-U., von dem Bussche-Hünnefeld, L., Hilgemann, E. et al. (2000). Vitamins, 4. Vitamin E (Tocopherols, Tocotrienols). In: *Ullmann's Encyclopedia of Industrial Chemistry*, 197–210. Weinheim: Wiley-VCH.

106 Bonrath, W. and Netscher, T. (2005). Catalytic processes in vitamins synthesis and production. *Appl. Catal., A* 280: 55–73.

107 Neaţu, F., Culică, G., Florea, M. et al. (2016). Synthesis of terephthalic acid by p-cymene oxidation using oxygen: toward a more sustainable production of bio-polyethylene terephthalate. *ChemSusChem* 9: 3102–3112.

108 Buhl, D., Roberge, D.M., and Hölderich, W.F. (1999). Production of p-cymene from α-limonene over silica supported Pd catalysts. *Appl. Catal., A* 188: 287–299.

109 Linnekoski, J.A., Asikainen, M., Heikkinen, H. et al. (2014). Production of p-cymene from crude sulphate turpentine with commercial zeolite catalyst using a continuous fixed bed reactor. *Org. Process Res. Dev.* 18: 1468–1475.

110 Vilella, I.M.J., de Miguel, S.R., de Lecea, C.S.-M. et al. (2005). Catalytic performance in citral hydrogenation and characterization of PtSn catalysts supported on activated carbon felt and powder. *Appl. Catal., A* 281: 247–258.

111 Coupé, J.N., Jordão, E., Fraga, M.A., and Mendes, M.J. (2000). A comparative study of SiO_2 supported Rh–Sn catalysts prepared by different methods in the hydrogenation of citral. *Appl. Catal., A* 199: 45–51.

112 Tracy, N.I., Chen, D., Crunkleton, D.W., and Price, G.L. (2009). Hydrogenated monoterpenes as diesel fuel additives. *Fuel* 88: 2238–2240.

113 Knuuttila, P. (2013). Wood sulphate turpentine as a gasoline bio-component. *Fuel* 104: 101–108.
114 Zhang, J. and Zhao, C. (2016). Development of a bimetallic Pd-Ni/HZSM-5 catalyst for the tandem limonene dehydrogenation and fatty acid deoxygenation to alkanes and arenes for use as biojet fuel. *ACS Catal.* 6: 4512–4525.
115 Zhang, J. and Zhao, C. (2015). A new approach for bio-jet fuel generation from palm oil and limonene in the absence of hydrogen. *Chem. Commun.* 51: 17249–17252.
116 Meylemans, H.A., Quintana, R.L., and Harvey, B.G. (2012). Efficient conversion of pure and mixed terpene feedstocks to high density fuels. *Fuel* 97: 560–568.
117 Meylemans, H.A., Baldwin, L.C., and Harvey, B.G. (2013). Low-temperature properties of renewable high-density fuel blends. *Energy Fuels* 27: 883–888.

14

Conversion of Chitin to Nitrogen-containing Chemicals

Xi Chen and Ning Yan

National University of Singapore, Department of Chemical and Biomolecular Engineering, 4 Engineering Drive 4, 117585, Singapore

14.1 Waste Shell Biorefinery

Biomass represents the largest carbon resource on Earth and constitutes a promising alternative to fossil oils to furnish the fuels, chemicals, and functional materials necessary for modern society [1–6]. The valorization of land-based lignocellulosic biomass such as cellulose and lignin has been prevalently studied for several decades. On the other hand, there is a trend of shifting the starting material from land-based biomass toward ocean-based biomass [7–9]. With 71% of the planet's surface covered by water, ocean-based biomass is an enormous yet underexplored treasure. For example, around 6–8 million metric tons of crustacean shells are generated annually worldwide, the majority being landfilled or dumped as fishery waste without any utilization [10]. Upon decomposition, the waste shells release CO_2 as well as toxic nitric oxides, posing environmental issues. As a result, it is beneficial to utilize this cheap and sustainable resource from both an economic and ecological standpoint.

Chitin is a biopolymer that accounts for about 15–40% of the weight of crustacean shells depending on the species [11]. Chitin is industrially extracted from crab or shrimp shells as a white powder product after demineralization, deproteination, and decolorization. Existing not only in crustacean shells but also in the skeletons of insects, fungi, etc., chitin is the world's second most abundant biopolymer next to cellulose with an estimated global annual production of 100 billion metric tons. Chemically, chitin is a straight chain polymer consisting of 2-(acetylamino)-2-deoxy-D-glucose (GlcNAc) and glucosamine ($GlcNH_2$) monosaccharides linked by $\beta(1 \rightarrow 4)$ glycosidic bonds [12–14]. Chitosan is a water-soluble derivative of chitin that normally has a degree of deacetylation (DD) above 50%. The structure of chitin closely resembles that of cellulose except for substitution of the –OH group with $-NH_2$ or an acetamido group at the C2 position. As a consequence of its unique structure, chitin contains biologically fixed nitrogen that makes it an ideal resource to renewably produce a series of useful nitrogen-containing (N-containing) compounds that cannot be obtained from other types of biomass. The concept of the shell biorefinery (see Figure 14.1)

Chemical Catalysts for Biomass Upgrading, First Edition. Edited by Mark Crocker and Eduardo Santillan-Jimenez.

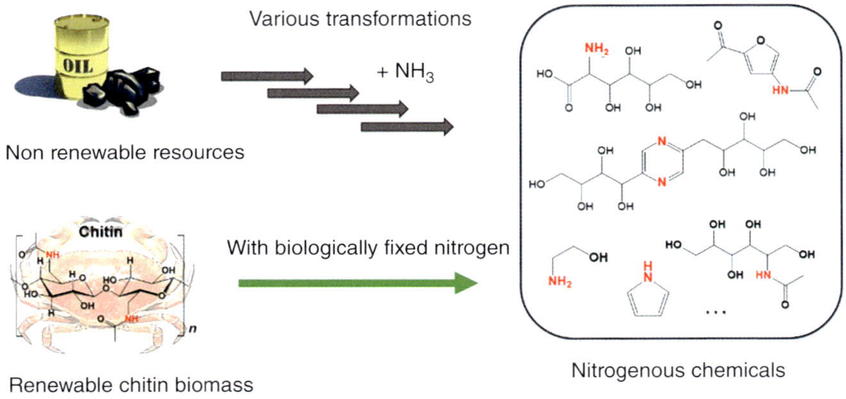

Figure 14.1 Comparison of traditional nonrenewable pathways and shell biorefinery for the synthesis of nitrogenous chemicals.

for valuable chemicals is new, despite the fact that chitin has a long history of application as a functional material in biomedical, cosmetic, and agricultural fields [15–20]. N-containing compounds represent a group of highly valued industrial and fine chemicals with large global markets. However, at present, the manufacture of these N-containing chemicals is nonrenewable, tedious, and energy-consuming. Taking the synthesis of pyrrole as an example, its production requires one isolation step and five chemical transformations from nonrenewable resources such as fossil oils. The nitrogen fixation is through ammonia synthesis, which is a low-energy-efficiency process accounting for about 2–3% of total world energy consumption. The shell biorefinery opens up a new avenue to achieve the sustainable and efficient synthesis of these useful compounds and enlarges the boundaries of traditional biomass refineries [8, 21].

Since the proposal of the concept, the chemical space of bio-based products from chitin biomass (including GlcNAc, $GlcNH_2$, chitosan, and chitin polymer) has grown rapidly [10]. To date, a number of different N-containing chemicals have been successfully obtained in the lab, including amines, amides, heterocycles, etc., along with other carbohydrate products (5-HMF and organic acids), using various transformation strategies such as hydrolysis, hydrogenation/hydrogenolysis, oxidation, etc. Moreover, most of the chitin-derived products are versatile building blocks that can be further upgraded into advanced products that are high-value fine chemicals. The object of this book chapter is to provide a synopsis of the rising field of chitin biorefining for the production of chemicals. First, the detailed transformation routes and methods of chitin biomass into nitrogenous chemicals will be introduced, systematically divided by the structure of the products. Following that, non-N-containing chemicals obtained from chitin biomass will be mentioned to exemplify the wide range of products that can be generated from chitin. Lastly, recent developments in the production of advanced products (fine and specialty chemicals, etc.) via multiple steps from the chitin polymer will be illustrated. Other applications of chitin and chitin derivatives – as functional polymers, catalyst precursors/supports, etc. – will not be covered in this book chapter.

14.2 Production of Amines and Amides from Chitin Biomass

14.2.1 Sugar Amines/Amides

Hydrolysis of chitin and chitosan into oligosaccharides and monomers is a long-standing area of study, predating the proposal and recognition of the chitin biorefinery [22–26]. The hydrolysis products, including $GlcNH_2$, GlcNAc, and chito-oligomers, are all value-added chemicals with broad applications in biomedicine, food, agriculture, materials, nutritional supplements, etc. [27–29]. For example, $GlcNH_2$ is regarded as the best joint supplement whose global market size exceeded US$ 1 billion in 2016, while GlcNAc is biologically active and an ingredient in cosmetics, and chito-oligomers are suitable materials in membrane preparation for drug delivery. Conventionally, their synthesis was achieved by chemical hydrolysis and/or deacetylation of chitin using concentrated acids (37% HCl, 98% H_2SO_4, etc.) with heating for a period of time [30, 31]. Later on, technologies such as microwave irradiation, ultrasonication, etc. were incorporated to improve the conversion efficiency and selectivity [32–34], although these techniques are usually energy intensive. Solvent optimization has proved to be a simple and effective way to enhance the selectivity and reactivity of chitin transformation into glucosamine [35, 36]. In 2017, a novel aprotic solvent/water system was exploited for the highly selective transformation of chitin into glucosamine [37]. About 10 different types of cosolvent were screened, diethylene glycol diethyl ether (DGDE) performed the best (see Figure 14.2a). In a water/DGDE mixture (volume ratio of 1 : 4), $GlcNH_2$ was obtained in an unprecedented yield of 80% at 175 °C within one hour from ball-milled chitin in the presence of 0.1 M H_2SO_4. The presence of an aprotic solvent facilitated hydrolysis as well as deacetylation, resulting in reduced usage of acid and superior selectivity toward $GlcNH_2$. From these studies, the basicity and solvating power of the cosolvent were determined to be the essential factors determining performance (as shown in Figure 14.2b). Based on kinetic studies and control experiments, it was

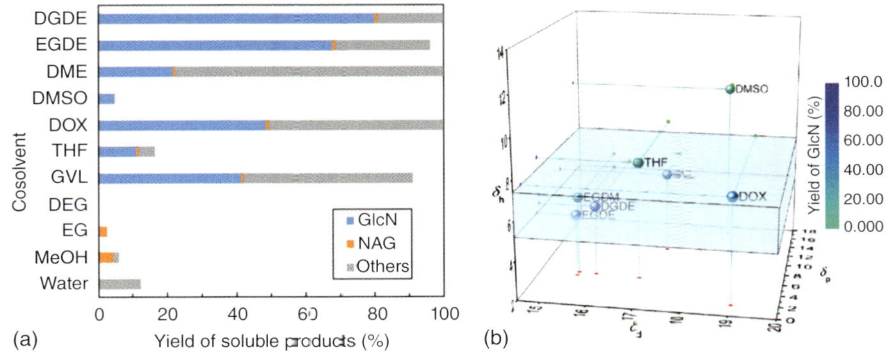

Figure 14.2 (a) The influence of different cosolvent systems; (b) plot of the aprotic polar solvents in Hansen space, with the $GlcNH_2$ yield represented by the color. Source: Adapted with permission from Zhang and Yan [37]. Copyright 2017, Wiley.

proposed that chitin first underwent hydrolysis into chito-oligomers, and then further hydrolysis and deacetylation of chito-oligomers occurred to form GlcNAc and $GlcNH_2$. Among these steps, cleavage of the glycosidic bond was probably the rate-limiting step.

Nowadays, mechanochemical methods have received increasing attention in biomass conversion due to the low amount of catalyst required, synergistic effects, and solvent-free feature [38–42]. Fukuoka's group devised an effective mechanochemical method to hydrolyze chitin into sugars with almost complete retention of the acetamido group [43]. Chitin powder was first impregnated with diethyl ether containing a catalytic amount of H_2SO_4. After solvent evaporation, the acid was evenly attached to the chitin polymer chains. Next, the acid–chitin powder was loaded into a mechanical ball mill and ground for six hours at 500 rpm, leading to chitin depolymerization into 100% soluble *N*-acetyl glucosamine (NAG) and oligomer products in a solvent-free condition. Notably, high yields of soluble products were not realized by using acid impregnation or ball milling alone, which obviously indicates a synergy between chemical and mechanical forces. Afterwards, the soluble products were simply depolymerized in water or methanol without further addition of acid catalyst at 170 °C for one hour, generating GlcNAc and 1-*O*-methyl-*N*-acetylglucosamine in 53% and 70% yields, respectively. The effective retention of the acetyl amide group in the process was ascribed to the imposed tensile stress from the ball mill on the main chains of the chitin polymer rather than the side chains. Very recently, Kerton's group demonstrated mechanochemical depolymerization of chitin using kaolinite, a natural clay [44]. With both acid sites and a layered structure, kaolinite promoted the depolymerization of chitin and led to the formation of 75.8% water-soluble products in six hours in a mixer/mill. The major products were acetylated chito-oligomers with a degree of polymerization (DP) ranging from 1 to 5, the quantifications being conducted by combined use of matrix-assisted laser desorption/ionization time-of-flight mass spectrometry (MALDI-TOF-MS), colorimetric assay, and gel permeation chromatography (GPC). The replacement of a mineral acid with a solid acid facilitates potential recycling and reuse of the catalyst and facile separation of the soluble products. Base-catalyzed, simultaneous depolymerization and deacetylation of chitin under solid-state conditions in a ball mill has also been reported [45]. Using both a base and a ball mill, the amount of base required could be reduced to one-tenth of that needed without a ball mill and the product was more narrowly distributed. The recycling of NaOH was achievable by methanol extraction/evaporation. Moreover, the method has been directly applied to process raw shrimp shells with simple separations to obtain the low molecular weight chitosan (LMWC) products (see Figure 14.3). In the presence of NaOH, simultaneous hydrolysis and deacetylation occurred to completely convert chitin into water-soluble LWMC products at 700 rpm for two hours. With a prolonged reaction time of three hours, the product was highly deacetylated with a DD value of 83.3% and a DP of around 37, which was attractive for materials synthesis and biological uses. Based on solid-state NMR analysis, strong coordinating ability and superior deacetylation ability were crucial factors for the excellent activity of NaOH compared with other base species tested. Based on HPLC and ESI-MS analysis,

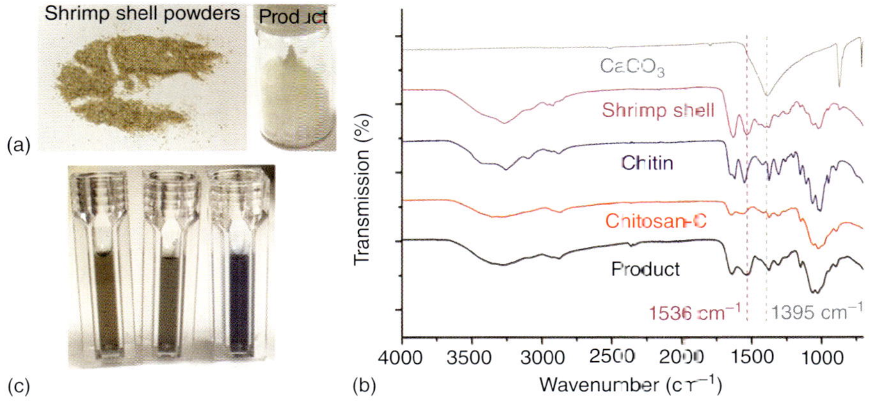

Figure 14.3 (a) Photos of the crude shrimp shell powders and the obtained white LMWC product; (b) FTIR spectra of the shrimp shell, product, and references; (c) images of the color changes of water, the product, and shrimp shell powder hydrolysates with the Bradford reagent (from left to right) to display that negligible proteins were left in the final product. Source: Reproduced with permission from Chen et al. [45]. Copyright 2017, Royal Society of Chemistry.

the reaction followed a base-catalyzed hydrolytic pathway without any formation of oxidation products. In addition, the ESI-MS data revealed that NaOH also played a role in suppressing side reactions, which may be responsible for the uniform product distribution obtained.

In addition, ethylene glycol (EG)-derived amino sugars were obtained from chitin via acid-catalyzed liquefaction [46]. An effective and simple post-separation protocol was applied to extract the main products, namely, hydroxyethyl-2-amino-2-deoxyhexopyranoside (HADP) and hydroxyethyl-2-acetamido-2-deoxyhexopyranoside (HAADP). At 165 °C for 1.5 hours, 75% conversion of chitin was achieved with a total yield (HADP and HAADP) of ca. 30% in the presence of 1.6 mM H_2SO_4 in EG. The EG solvent disrupted the crystalline structure of chitin via interactions with the chitin polymer chains and stabilized the monomer sugars produced from chitin depolymerization. HAADP was initially formed and readily underwent deacetylation to generate HADP as the dominant product. However, due to its exposure to the basic $-NH_2$ groups, the acid catalyst was constantly consumed, leading to the deactivation of the catalytic system after 1.5 hours. Subsequently, an advanced catalytic system for chitin liquefaction was proposed, using formic acid as the solvent, catalyst, and reactant [47]. The liquefaction mechanism involved an initial partial formylation of –OH groups on chitin chains to produce soluble chitin fiber derivatives, thereby reducing the energy barrier for further cleavage of glycosidic bonds. An unconventional nonhydrolytic depolymerization pathway was identified to form formylated oligomers and monomers. The multiple functions of formic acid enabled the liquefaction to proceed under mild conditions with high selectivity and activity. Starting from ball milled chitin, a 32.7% yield of formylated GlcNAc products was obtained at 100 °C after 12 hours. At long reaction times, a single product of 5-(formyloxymethyl)furfural (FMF) could be produced with 34.6% yield.

Figure 14.4 Formic acid-mediated liquefaction of raw shrimp shell under mild conditions. Source: Adapted with permission from Zhang and Yan [47]. Copyright 2016, Royal Society of Chemistry.

Furthermore, the catalytic system was able to convert chitin in raw shrimp shells with simple processing (see Figure 14.4). Moreover, formic acid can be derived from biomass and is highly volatile, which is beneficial for product separation.

14.2.2 Furanic Amines/Amides

In 2012, the formation of a N-containing furan derivative from chitin monomer GlcNAc was first reported by Kerton's group [48]. The product, 3-acetamido-5-acetylfuran (3A5AF), is an analog to 5-HMF in cellulose valorization and possesses the potential to be a versatile platform chemical for downstream upgrading into various useful N-containing compounds. Boric acid was discovered as the most effective promoter for the production of 3A5AF. Meanwhile, several ionic liquids (ILs) were evaluated as the solvent and the Cl^- ion was found to display positive effects. In the presence of boric acid, GlcNAc was fully converted in [Bmim]Cl with a 60% yield of 3A5AF at 180 °C for one hour. Prior to this work, 3A5AF was obtained in only 2% yield as a side product of the thermal pyrolysis of GlcNAc. From kinetic studies the activation energy of the reaction was calculated to be 82.8 kJ/mol, which is comparable with that of glucose dehydration to 5-HMF. In a subsequent study, the effect of chloride salts as a co-additive was investigated in an organic solvent system [49]. NaCl was the best co-additive to be applied together with boric acid, resulting in 62% 3A5AF yield under optimal conditions with microwave heating. A possible reaction pathway for 3A5AF production from GlcNAc was suggested: the pyranose ring of GlcNAc opens to form an open-chain structure and then rearranges to form a furan ring (–OH on C4 position attacks the aldehyde group), after which three dehydrations are required to form the 3A5AF product.

When the dehydration process was not thorough, other N-containing furan compounds were obtained. Osada's group developed a noncatalytic hydrothermal conversion of GlcNAc in a flow reactor, producing Chromogen I and Chromogen III as the dominant products [50]. Minor products consisted of ManNAc, 3,6-anhydro-GNF and 3,6-anhydro-MNF, etc., which were formed via epimerization of GlcNAc and Chromogen I. 3A5AF was not obtained as a product, possibly because dehydration was difficult in aqueous solution. This is a simple and green method that can be accomplished within seconds. However, the reaction time and temperature have to be strictly controlled to achieve high selectivity toward a target product. For example, at 200 °C the yield of Chromogen I was above 20% at a reaction time of 20 seconds, but it dropped to around 5% at about 30 seconds. When the temperature was set at 220 °C, its yield was generally low, regardless of the reaction time. Compared to Chromogen I, Chromogen III was a more highly dehydrated product obtained at relatively high temperature or longer reaction time, probably because of the higher activation energy, as proved by a kinetic modeling study. In the modeling, it was also revealed that the substrate GlcNAc was more susceptible to decomposition in high temperature water than glucose. Thus, a hypothesis was devised that the presence of the acetamido group increased the chemical reactivity. Moreover, these observations provide insights about the sequence of dehydration steps resulting in 3A5AF. As Chromogen I was obtained as the initial product, the first dehydration step should occur at the C2–C3 position. Due to the electron-withdrawing ability of the acetamido group, the hydrogen atom at the C2 position was easily eliminated. The next dehydration step occurred at the C4–C5 position of the furan ring to afford the more stable Chromogen III. The third dehydration step of the side chain of the furan ring led to the formation of 3A5AF. Following this work, the authors attempted to adopt this noncatalytic hydrothermal technique for chitin dimer conversion [51]. However, the chitin dimer was not efficiently transformed because there was insufficient acidity in the reaction system to break the glycosidic linkage. An overall reaction pathway for the production of 3A5AF and Chromogens from chitin biomass is presented in Figure 14.5.

A one-step, direct conversion of chitin polymer into 3A5AF was reported in 2014 [52]. After careful solvent and additive screening, *N*-methyl-2-pyrrolidone (NMP) was selected as the best solvent, while the use of boric acid, LiCl, and HCl in combination gave the highest yield among other additives and/or combinational additives. Under optimal conditions, about 50% conversion of the chitin substrate was realized, whereas the 3A5AF yield only reached 7.5%, possibly due to the structural rigidity of the chitin polymer. The chitin conversion

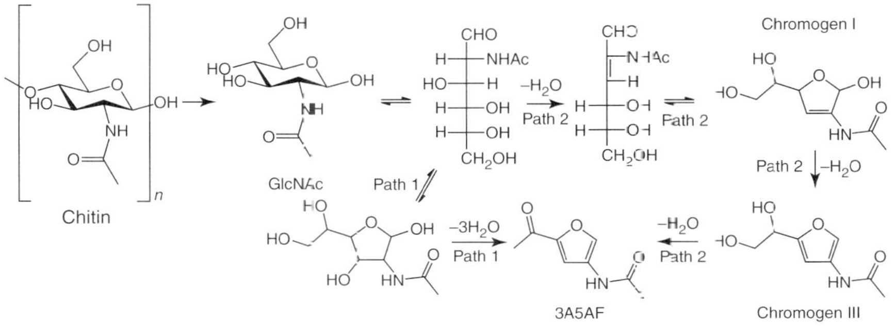

Figure 14.5 Proposed reaction pathway for the production of 3A5AF and chromogens from chitin and chitosan. Source: Adapted with permission from Chen et al. [21]. Copyright 2016, John Wiley & Sons.

and 3A5AF yield were monitored as a function of the reaction time. The product was rather stable for the first 1.5 hours but gradually decomposed after that. Several by-products were observed, indicating the occurrence of side reactions, including levoglucosenone, acetic acid, 4-(acetylamino)-1,3-benzenediol, and humins. Importantly, the work shed light on the reaction pathway and promotional role of boric acid. Based on poisoning tests and comprehensive NMR studies, the formation of a boron-substrate complex was convincingly shown. Boric acid was found to coordinate to two –OH groups on GlcNAc (most probably C4 and C6 positions), facilitating the transition of the substrate from its α-anomer to β-anomer. As a result, the superior performance of boric acid may be ascribed to its ability to generate a boron-substrate intermediate that can undergo subsequent ring-opening and dehydration. Moreover, boric acid interacted to a negligible degree with the acetyl amide group and thus the side chain was maintained in the product. Apart from organic solvents, ILs such as [Bmim]Cl were also capable of promoting this reaction [53]. Indeed, a faster reaction rate was identified in ILs than organic solvents. In an effort to enhance the 3A5AF yield from chitin, the same group utilized pretreatment techniques to lessen the rigidity of the chitin polymer prior to transformation, and specific structure–activity correlations were constructed [54]. Among various structural parameters, crystallinity and hydrogen bonding network were the most vital factors. By using treatment methods – such as ball mill grinding – that could significantly impair these factors, the yield of 3A5AF improved to around 30%, which was significantly higher than the previous value.

As a relatively new chemical, the chemicophysical properties of 3A5AF remain unclear. Therefore, some of its chemical and physical parameters have been investigated by employing experimental and computational technologies, including assessing its acidity, solubility, reactivity, etc. [55]. The pK_a value is around 20, which is comparable to common amides. A solubility test in supercritical CO_2, as well as time-dependent density functional theory (TD-DFT), indicate that 3A5AF is less soluble but more polar than 5-HMF and that it is liable to undergo dimerization. Additionally, reactions using a methyl Grignard agent suggest that the chemical reactivity of 3A5AF is not high, its reactivity being inferior to that of acetophenone. Since 3A5AF is regarded as a potential chemical building block, more fundamental research on its properties are desirable as a useful guide to future applications and synthesis.

14.2.3 Polyol Amines/Amides

A series of polyol amines/amides were generated via hydrogenation/hydrogenolysis of chitin biomass catalyzed by commercial noble metal catalysts such as Pt/C, Pd/C, Ru/C, and Rh/C under pressurized hydrogen gas at elevated temperature in water [56]. Ru/C was superior to other metal catalysts based on both carbon balance and product yields. A recyclability test was conducted for Ru/C, and negligible decrease in performance was observed for four runs. At a relatively low temperature of 80 °C, quantitative conversion of GlcNAc into 2-acetoamido-2-deoxy-D-sorbitol (ADS) was realized. ADS is a potential precursor to synthesize polyamides and polyesteramides. With further increase in temperature, a number of smaller C_2 to C_4 amines/amides were produced via hydrogenolysis, including 2-acetoamido-hexan-1,4,5,6-tetraol (rDNAG), butan-1,2,3,4-tetraol (BTO), and *N*-acetylmonoethanolamine (NMEA), among others. The total yields of these smaller molecules were around 20% at 180 °C for one hour. Although the selectivity to ADS was extremely high, it was challenging to achieve a high selectivity toward a single hydrogenolysis product, due to the existence of multiple parallel pathways at elevated temperatures (see Figure 14.6). In an effort to enhance the yields of C_2–C_4 chemicals, additives such as NaOH and WO_3 were introduced since they are commonly effective for retro-aldol cleavage of sugars. In the presence of additives, the yields of C_6–C_8 products apparently decreased. However, the yields of C_2–C_4 chemicals improved negligibly. Thus, such acid and base additives failed to increase the product selectivity despite promoting decomposition of the longer carbon chains. Lastly, chitin and chitosan polymer were employed as the substrate instead of GlcNAc. A higher reaction temperature of 250 °C was necessary to facilitate the efficient conversion of chitin and chitosan because of their high molecular weight and crystalline structure. Nevertheless, severe side reactions occurred at this high temperature, leading to an 8% total yield of C_2 products along with ca. 10% gaseous products, and no C_6–C_8 products.

Fukuoka's group reported a one-pot, two-step hydrolytic hydrogenation of chitin polymer with mechanochemical pretreatment to produce ADS in high yield [57]. First, H_2SO_4-impregnated chitin powder was ground in a ball mill to lessen the recalcitrance of chitin and convert it into water-soluble GlcNAc and oligomers (total yield ~83%). Different types of acids were screened and H_2SO_4 was selected because of its low cost and high efficiency. Then, a two-step method

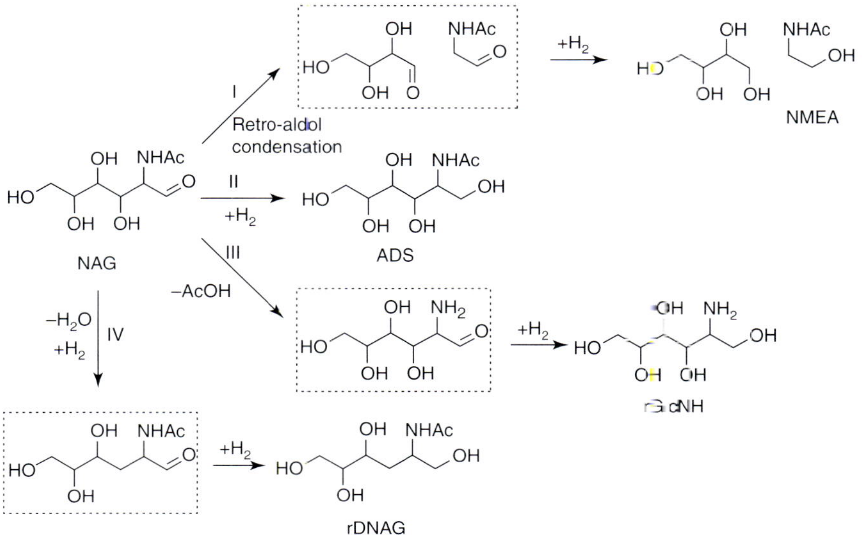

Figure 14.6 Parallel reaction pathways for the conversion of NAG at elevated temperature. Source: Reproduced with permission from Bobbink et al. [56]. Copyright 2015, Royal Society of Chemistry.

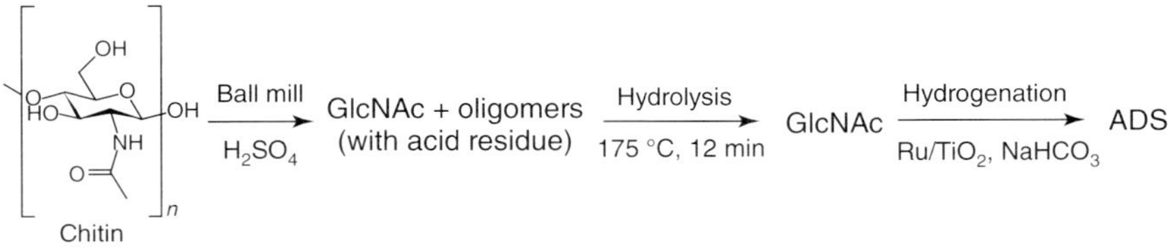

Figure 14.7 A two-step transformation (hydrolysis and hydrogenation) of mechanochemically pretreated chitin into high yield of ADS product.

was adopted to hydrolyze and hydrogenate the pretreated chitin in order to maximize the product yield (see Figure 14.7), because the optimal conditions for each of the reactions were distinct. It was found that hydrolysis required high temperature and acidity (pH = 2), while selective hydrogenation was favored at relatively low temperature and a pH range of 3–4. In the first step, i.e. hydrolysis, 61% GlcNAc monomer was obtained from pretreated chitin upon fast heating in water to 175 °C for 12 minutes. Note that no extra acid catalyst was supplied since the H_2SO_4 residue from the ball mill grinding step provided sufficient acidity. A base (such as $NaHCO_3$) was subsequently added to the system to adjust the pH to 3–4, ADS being formed via hydrogenation in the presence of a Ru/TiO_2 catalyst with high selectivity and an overall yield of around 52% at 120 °C and 40 bar of H_2. Kinetic simulations were undertaken by curve-fitting of the experimental data, which allowed to calculate the pseudo-first-order rate constants for the hydrogenation of GlcNAc to ADS (k_1), of GlcNAc to side products (k_2), and of ADS to side products (k_3). The increase of pH to 3–4 for the hydrogenation step considerably reduced the value of k_2, efficiently suppressing the side reactions of the monomer sugar. It was concluded that a minimal ratio of k_2/k_1 at pH 3–4 was essential to achieve high selectivity toward ADS in the hydrogenation step. In short, by utilizing mechanocatalysis within a two-step strategy, the highest yield of ADS from chitin was achieved.

Amino acid polyols are another type of important polyol amines/amides that could be produced via the oxidation of chitin monomer sugars. Catalysts comprising supported Au nanoparticles (NPs) have been reported to be very effective for the oxidation of GlcN and GlcNAc with O_2 to their corresponding amino acids in aqueous phase at relatively low temperature. Ebitani's group investigated different support materials for Au NPs, including SiO_2, Al_2O_3, MgO, TiO_2, hydrotalcite (HT), and hydroxyapatite (HAP) [58]. In general, basic supports were more adequate for the oxidation catalysts because (i) base sites favored oxidation reactions and (ii) Au NPs were smaller – and thus more active – on basic supports. Moreover, a stronger base support improved the recyclability of the catalyst. For example, Au/MgO maintained its reactivity after three cycles, while Au/HT showed obvious deactivation in the first reuse. Tellingly, X-ray absorption near edge structure (XANES) spectroscopic data showed that the Au NPs stayed at zero valence without oxidation on both MgO and HT supports. Nevertheless, after reuse, the Au NPs on HT aggregated, so the larger average particle size could be the main reason for deactivation. In contrast, the stronger basicity of MgO induced strong interactions between the support and the Au NPs inhibiting the growth of the latter, which explains why Au/MgO displayed better recyclability. Under O_2 flow at 40 °C, GlcN was converted into glucosaminic acid, a 93% yield being achieved after three hours in the presence of a Au/MgO catalyst. Likewise, under O_2 flow at 25 °C, GlcNAc was transformed into *N*-acetyl glucosaminic acid, a 95% yield being obtained after five hours using a Au/HT catalyst. These products are both bioactive and chiral with applications in nutrition, the biomedical field, and asymmetric synthesis. In the future, more research efforts are anticipated to realize the direct synthesis of these useful products from chitin or the chitosan polymer, mainly via the rational design and development of multifunctional catalysts.

14.3 Production of N-heterocyclic Compounds from Chitin Biomass

N-heterocyclic compounds are very valuable chemicals, as they are some of the main constituents of pharmaceuticals and drugs. For example, popular drugs for treating niacin and niacinamide deficiency are pyridine derivatives, while medicines such as Varenicline and Brimonidine contain pyrazine as a core unit. The formation of pyrazine compounds from glucose and proteins has been known for decades as the Maillard reaction, in which proteins function as the nitrogen source. Using chitin or chitosan as the nitrogen sources, a variety of N-heterocyclic chemicals – including pyridine, pyrrole, and pyrazine derivatives – have been obtained via thermal pyrolysis [59–61]. Nevertheless, these methods lacked selectivity as parallel reactions take place leading to very low product yields.

Selective production of pyrazine products – such as deoxyfructosazine (DOF) and fructosazine (FZ) – by acid- or base-catalyzed condensation of chitin monomer sugars have also been reported (an overview of transformation methods being shown in Figure 14.8). Fujii et al. observed the conversion of glucosamine into FZ with a yield of 24.8% at 70 °C in methanol in the presence of metallic sodium [62]. FZ was formed via epimerization and self-condensation of glucosamine. More recently, boron-based catalysts such as phenylboronic acid and boric acid were found to effectively convert glucosamine into DOF in basic water at room temperature [63]. In fact, DOF represented the sole

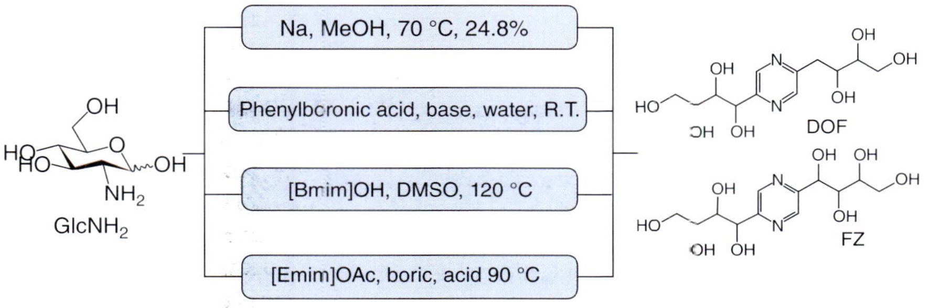

Figure 14.8 Various chemical conversions of chitin monomer into DOF and FZ.

product and could be easily isolated via a simple process with 58% yield. The promotional effect of the boron-based catalyst was illustrated by using NMR (^{11}B and ^{13}C) spectroscopy. The boron atom in the catalyst interacted with the three groups of the glucosamine substrate in an open chain form (the two –OH groups on the C3 and C4 positions and the $-NH_2$ group on the C2 position). Next, self-condensation of the boron-glucosamine intermediate generated a pyrazine ring, subsequent dehydration, and isomerization ultimately affording DOF. In recent years, Jia et al. demonstrated a green and simple approach to transform GlcN into FZ and DOF by using [Bmim]OH – a basic IL – as an environmentally friendly solvent and catalyst [64]. Performing the reaction in the temperature range of 80–120 °C led to higher yields at a reaction time of three hours. Similarly, the product yield was initially improved upon increasing substrate concentration – possibly because the reaction was intermolecular and the chances of condensation were enhanced – but eventually decreased once the concentration surpassed 20% likely due to the excess of acid in the product mixture (glucosamine hydrochloride acid was used as the substrate). In alcoholic [Bmim]OH solution, the total product yield was ca. 35% under optimal reaction conditions. Additional bases such as NaOH, KOH, etc. were introduced in order to enhance product yield; however, these particular efforts were unsuccessful. Lastly, the effect of cosolvents was studied, DMSO as a cosolvent being found to facilitate the conversion of GlcN to DOF and FZ leading to a higher total yield of around 49%.

In a following contribution, [Emim]OAc was used instead of [Bmim]OH, as the former IL could also efficiently catalyze this reaction [65]. Results of mechanistic studies – undertaken via *in situ* NMR – showed [Emim]OAc to boast bifunctional effects. Indeed, the proton exchange between different functional groups was boosted by the acidic hydrogen atom on the imidazole ring of the IL; meanwhile, the anion was a strong hydrogen bonding acceptor capable of activating the reaction and its basic features promoted epimerization and the amount of open-chain GlcN in the solution. The authors of this work suggested that both the hydrogen bonding capability and the lone pair of electrons on the nitrogen were the forces behind the generation of DOF and FZ. More recently, the same authors described the conversion of GlcN to DOF and FZ in [Emim]OAc with boric acid as the additive [66]. An increased DOF yield of 40.2% was achieved in the presence of boric acid after three hours at 90 °C. The boron complex intermediate was probed using coupled 2D NMR, ESI-MS, and DFT calculations, coordination kinetically and thermodynamically favoring the ring-opening and dehydration reactions. The key step was determined to be the dehydration of dihydrofructosazine, the energy barrier associated with this step apparently decreasing upon chelation of the boron additive with the sugars.

Pyrrole is another kind of industrially important N-heterocyclic compound and its production from chitin polymer in a single step under hydrothermal conditions has been reported [67]. Considering the considerable structural differences between pyrrole and chitin, a one-step transformation proved challenging. The highest yield obtained was around 12%, the supply of an external source of nitrogen (such as ammonia) being found to be necessary. The method developed used a tube reactor that could be rapidly heated to >300 °C within seconds, and an aqueous NaOH solution was adopted as the solvent. Optimization work suggested that high temperature and short reaction time favored the formation of pyrrole, albeit a more influential factor was determined to be the addition of ammonia, which provided an excess of nitrogen in the reaction system. Through a number of control experiments, preliminary insights on the formation pathway of pyrrole were gained. The authors concluded that chitin is first hydrolyzed and deacetylated into a GlcN monomer whose rearrangement into a five-membered furan ring is followed by dehydration reactions. Oxygen-to-nitrogen atom exchange then converts the furan ring into a pyrrole ring. Finally, the side chains of the pyrrole ring are detached under high temperature to produce pyrrole as a stable product. In order to increase pyrrole yield, monomer sugars were employed as substrate, but to no apparent improvement. Therefore, the best options to improve yield in future work is to exploit effective catalytic systems and to understand the reaction mechanism in depth.

14.4 Production of Carbohydrates and Acetic Acid from Chitin Biomass

Apart from N-containing compounds, a number of value-added carbohydrates that are typically derived from cellulose can also be synthesized from chitin, which highlights the versatility of this feedstock and its value alongside woody biomass as a biorefinery stream. The most characteristic example is 5-hydroxymethylfurfural (5-HMF), which has been regarded as a top value chemical and an essential intermediate for downstream processes since its first synthesis from cellulosic biomass. Similar furan derivatives were produced from chitin as early as 2009 using a biphasic solvent system comprising 1,2-dichloroethane and a concentrated HCl solution [68]. Acid-catalyzed depolymerization and dehydration of the chitin polymer occurred at elevated temperature, affording 5-chloromethyl furfural (5-CMF) in high yield (45%). 5-CMF can then be easily hydrolyzed to form 5-HMF. Under the reaction conditions employed, a remarkably high 5-CMF yield of 84% can be achieved from cellulose, which indicates that chitin exhibits stronger chemical resistance than cellulose. This is probably due to the fact that the structure of chitin is more rigid and robust.

The straightforward dehydration of chitin and chitosan polymer into 5-HMF was realized by using an effective metal chloride catalyst in aqueous solution under microwave irradiation [69]. Besides 5-HMF, levulinic acid was also identified in the product mixture. Levulnic acid (LevA), a product usually seen in 5-HMF synthesis, was generated through rehydration of 5-HMF. After screening >20 types of metal salts, acids, and bases, $SnCl_4$ was identified as the best formulation. Further investigation suggested that SnO_2 and HCl were the actual active components responsible for the catalytic performance. Notably, both the substrate concentration and the catalyst amount showed considerable influence on product selectivity. Concentrated substrate solutions and high catalyst amounts favored the formation of levulinic acid rather than 5-HMF. Starting from chitosan, 23.9 wt% LevA was obtained at 200 °C after 30 minutes using microwave heating, a concentrated substrate solution and a large amount of catalyst. In contrast, 10 wt% 5-HMF was produced under similar conditions but using a more dilute substrate concentration and a smaller amount of catalyst. Saliently, when chitin was employed as the substrate, the only product observed was LevA and the highest yield achieved was 12.7 wt%. However, by starting from GlcN, the highest yield of LevA was enhanced to 32.0 wt%. A possible reaction pathway was proposed, in which hydrolysis to the GlcN monomer sugar represents a first step that is followed by deamination and dehydration to produce 5-HMF (see Figure 14.9).

More recently, another reaction system has been explored to transform chitin biomass into 5-HMF. Using a concentrated $ZnCl_2$ (67 wt%) aqueous solution at 120 °C, a 21.9% yield of 5-HMF was obtained from GlcN after 1.5 hours [70]. In an effort to improve the yield, several additives were screened, boric acid and $AlCl_3$ proving effective in increasing the yield to 26.5% and 23.0%, respectively. In addition, other substrates such as GlcNAc, LMW chitosan, and chitin were also tested. Surprisingly, the lowest yield of 5-HMF (2.8%) was observed using GlcNAc, the yields of 5-HMF obtained from chitosan and chitin being 10.1% and 9.0%. As suggested by the authors, the reason was probably that the presence of an amino group is critical to induce the promotional effect of Zn cations. Indeed, it was proposed that Zn^{2+} and/or the additive interacted with the functional groups on the C1 and C2 positions to facilitate the isomerization and dehydration reactions necessary to form 5-HMF. Nevertheless, the reason why the 5-HMF yield obtained from chitin was higher than that obtained from GlcNAc remains unclear. In this reaction system, no LevA was observed in the product mixture. Although attempts were made to gain information on the reaction pathway by using *in situ* NMR analysis, no conclusive insights could be reached.

5-HMF has also been obtained from chitin biomass in dilute H_2SO_4 aqueous solution under hydrothermal conditions [71]. The product yield was influenced by reaction parameters including pH, reaction time, and temperature. After optimization, the highest 5-HMF yield achieved was 12.1% with a reaction time of 37 minutes at 174 °C with 2.2 wt% H_2SO_4. In another work, acidic ILs were used as the acid catalyst and solvent for chitin biomass dehydration into

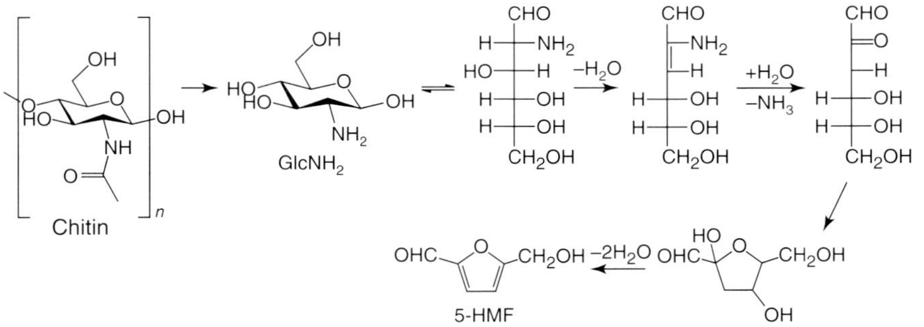

Figure 14.9 Proposed reaction pathway of 5-HMF formation from chitin biomass. Source: Adapted with permission from Chen et al. [21]. Copyright 2016, John Wiley & Sons.

5-HMF [72]. About nine different types of ILs were essayed. The 5-HMF yield was affected by complex factors such as steric hindrance, hydrogen bonding ability, and acidity. Under optimal conditions, 29.5% and 19.3% yields of 5-HMF were obtained from chitosan and chitin, respectively, in N-methylimidazolium hydrogen sulfate ([Hmim]HSO_4) at 180 °C after five hours. In a follow-up contribution, the same group investigated the conversion of GlcNAc into 5-HMF using [Hmim]HSO_4 as catalyst in a water/DMSO solution [73]. A remarkably high yield of 64.6% 5-HMF was realized at 180 °C in six hours. Moreover, the acid catalyst was reused four times without considerable loss of activity. On the other hand, acid-catalyzed dehydration of biomass may produce LevA – as opposed to 5-HMF – as the major product under harsher reaction conditions. Indeed, using a relatively concentrated H_2SO_4 aqueous solution (2 M) and microwave irradiation, LevA was obtained as the dominant product of chitin and chitosan conversion at 190 °C and 0.5 hour reaction time [74]. Likewise, Jeong's group reported the formation of LevA in 39.0% and 29.0% yields from GlcN and chitin, respectively, in a 4 wt% H_2SO_4 aqueous solution at 190 °C [75]. The same group also exploited a Lewis acid zirconium oxychloride ($ZrOCl_2$) catalyst to promote GlcN conversion to LevA with 21.3% yield under hydrothermal conditions [76]. Following these works, the sulfamic acid-catalyzed transformation has also been performed, yielding 33.8% LevA under similar reaction conditions [77].

Acetic acid (HAc) – another important platform chemical – was first produced in high yield from chitin biomass in 2016 [67]. Chitin was found to be remarkably advantageous to generate HAc due to its unique chemical structure relative to other types of biomass [78–80]. A theoretical yield of 25% HAc could be readily achieved via the hydrolysis of the acetamido side chain without any oxidants. High temperature water and strong bases were adopted to depolymerize and cleave the monomer into fragments, while low-cost metal oxides and oxygen gas were employed as the catalyst and/or oxidant to facilitate the oxidation of aldehyde intermediates. Reaction conditions including 300 °C, 2 M NaOH aqueous solution, ball milled CuO, and 5 bar of O_2 led to the highest yields of HAc, i.e. 38.1% and 47.9% from chitin and proto-shrimp shells, respectively, the raw shrimp shells affording an even higher yield due to the additional formation of HAc stemming from the oxidation of proteins. It was found that CuO was converted during the reaction to Cu_2O, which was recovered after the run. In the literature, the oxidation ability of CuO was reported to be stronger than that of the oxygen gas [81]; thus, CuO was likely the main catalyst and oxidant in the reaction system. Besides HAc, a number of other organic acids and N-containing chemicals were noticed as side products. A comprehensive reaction network has been proposed, and the multiple reaction pathways ultimately affording HAc (see Figure 14.10) was the major reason for the high selectivity toward this compound. Results of isotope NMR measurements were in agreement with the theoretical expectation that ~25% of the HAc formed was generated from the acetamido group while the other ~17% of the HAc produced stemmed from the cleavage and oxidation of the C_6 chains.

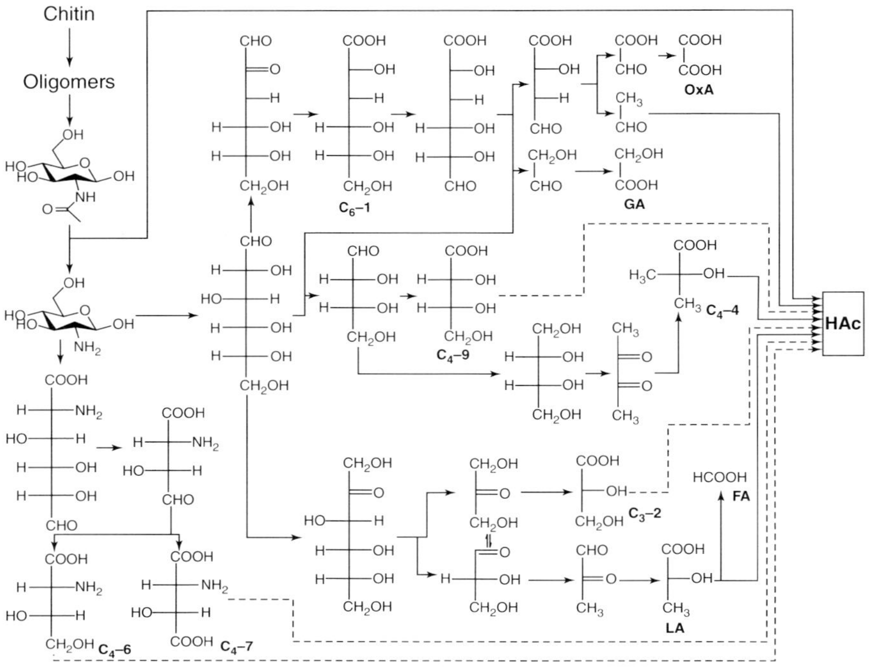

Figure 14.10 The reaction network of chitin conversion into HAc. Source: Reproduced with permission from Gao et al. [67]. Copyright 2016, American Chemical Society.

14.5 Production of Advanced Products from Chitin Biomass

The previous sections have described how a number of valuable platform chemicals can be successfully obtained from chitin biomass. Very recently, research works have extended the potential of N-containing chemical building blocks such as 3A5AF to reach functional fine and specialty compounds. This further validates the chitin-based biorefinery concept and advances the sustainable development of chemicals important to society. For instance, reactions of 3A5AF with a variety of ketones to produce novel heteroaromatic scaffolds have been achieved in 10% HCl aqueous solution at 50–70 °C (see Figure 14.11) [82]. However, the chemical reactivity of 3A5AF was first examined using aldehydes. Indeed, by reacting formaldehyde and acetaldehyde with 3A5AF, a hydroxymethylfuran derivative and difurylmethane were formed in 59% and 64% yield, respectively, with the expected regioselectivity, at 50 °C for three hours. Meanwhile, various dihydrodifuropyridines were obtained in yields around 30–53% at 70 °C when ketones (both chain and cyclic) were used, whereas aromatic ketones were found unsuitable for this process. The authors of this work also proposed a possible mechanism for this reaction in which the ketone and 3A5AF first

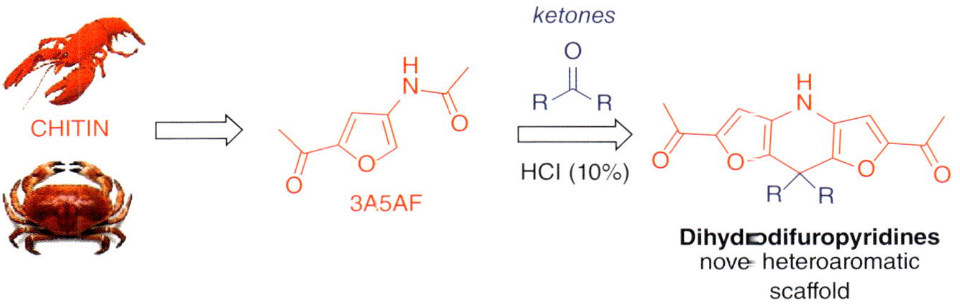

Figure 14.11 The synthesis of heteroaromatic scaffolds from chitin via 3A5AF. Source: Reproduced with permission from Pham et al. [82]. Copyright 2017, Springer Nature.

formed a difurylmethane intermediate, the acetamide side chain of which was hydrolyzed upon heating. Then, tautomerization, intramolecular cyclization and deamination occurred to afford the final heterocyclic products. The cyclization step was probably facilitated by the Thorpe–Ingold effect imparted by the gem-disubstituted carbon connecting the two furans, which explains why no pyridine ring was formed when using aldehydes instead of ketones.

Although there is a pool of chitin biomass-derived small molecules, the incorporation of these compounds in synthetic processes involving multiple steps for the production of fine chemicals has not been widely achieved. 3A5AF, with its unique substitution pattern that is difficult to access using traditional synthetic chemistry, represents an ideal and sustainable motif for the synthesis of bioactive molecules of interest to the medicine and biomedicine fields. In 2018, the sustainable synthesis of the anticancer alkaloid proximicin A was accomplished starting from chitin (see Figure 14.12) [83]. The proximicins have exhibited strong cytostatic effects in various human tumor cell lines and are regarded as outstanding chemotherapeutic leads. As shown in Figure 14.12, the first step is to obtain 3A5AF – the critical building block – from ball-milled chitin in an IL solvent. Then, 3A5AF was transformed into the relevant ester **1** in methanol, where promoters including CuO, I_2, pyridine, and then K_2CO_3 were added. Selective hydrolysis of the amine in **1** was simply catalyzed by HCl in methanol solvent to give the amino-furan **2**. Next, the methyl carbamate was grafted onto the active $-NH_2$ group to form **3** in dimethylcarbonate (DMC) in the presence of KO*t*Bu. Further hydrolysis of **3** led to the generation of **4**, which was then coupled with **2** by using the safe, uronium-based coupling agent COMU in DMC as solvent to afford bisfuran **5**. The final product proximicin A was obtained in a single step from **5** upon stirring with ammonium hydroxide in methanol. Compared with traditional synthetic routes, this total synthesis starting from chitin meets many of the fundamental principles of green chemistry. Toxic and dangerous solvents and reagents are avoided, and renewable nitrogen from chitin biomass is utilized. This contribution accentuates the potentially paramount roles of chitin biomass in the sustainable production of N-containing fine chemicals that are not directly attainable from lignocellulosic biomass.

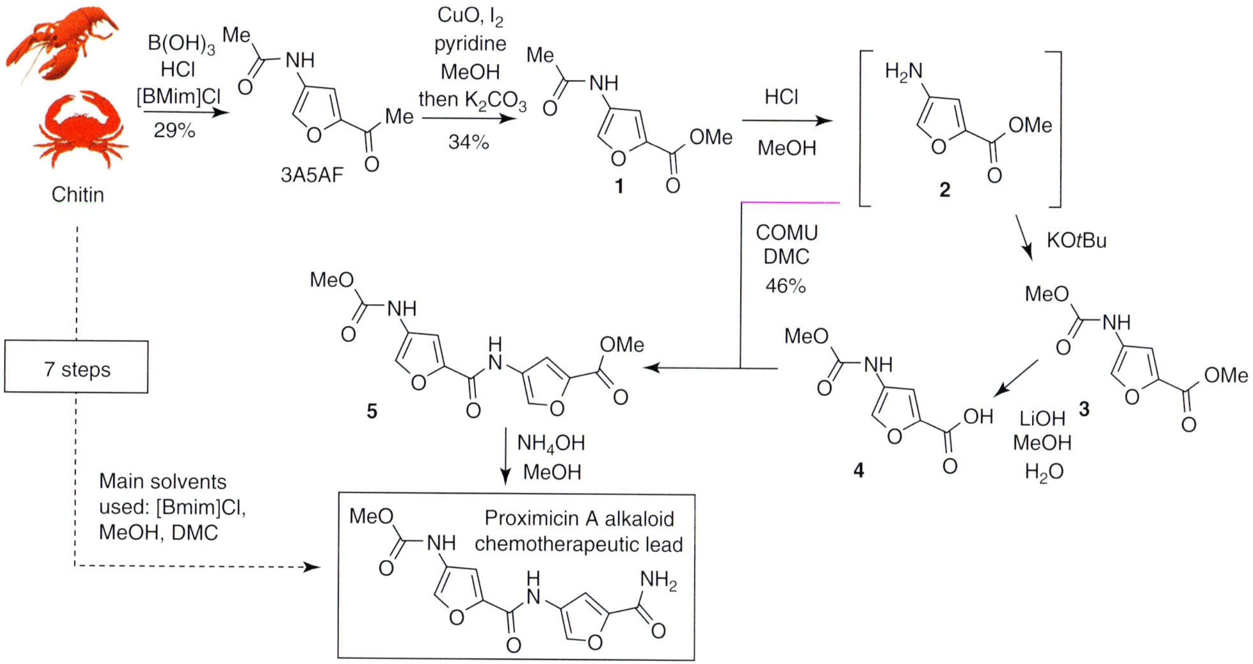

Figure 14.12 Sustainable synthesis of proximicin A starting from chitin. Source: Reproduced with permission from Sadiq et al. [83]. Copyright 2018 John Wiley & Sons.

14.6 Conclusion

The concept and development of the shell biorefinery not merely complements the ability of the woody biomass refinery to produce carbohydrates but extends the chemical space of bio-based products to include valuable nitrogen-containing chemicals, opening new directions that are potentially superior to conventional synthesis methods from nonrenewable resources. Although a number of valuable nitrogenous chemicals can already be obtained from chitin, further efforts are anticipated to exploit improved catalytic systems to generate new types of chemicals with high values, including other platform chemical building blocks. Admittedly, the transformation strategies proposed should be efficient and selective toward a target product to be practicable and economically viable. In addition, multistep conversions of chitin into advanced chemicals are important to establish a chitin-based process chain. Another issue to be addressed in chitin valorization is the high price of this feed at present, which results from the noneconomic extraction method currently employed. Therefore, the development of new crustacean shell fractionation methods to obtain chitin is imperative. An ideal protocol should offer a green, simple, and low-cost way for industrial scale extraction of chitin from waste shell material. More research efforts should be devoted in this direction to make the shell biorefinery concept a reality.

References

1 Sanderson, K. (2011). *Nature* 474: S12–S14.
2 Tuck, C.O., Pérez, E., Horváth, I.T. et al. (2012). *Science* 337: 695–699.
3 Zhu, Y., Romain, C., and Williams, C.K. (2016). *Nature* 540: 354.
4 Besson, M., Gallezot, P., and Pinel, C. (2014). *Chem. Rev.* 114: 1827–1870.
5 Mika, L.T., Cséfalvay, E., and Németh, Á. (2018). *Chem. Rev.* 118: 505–613.
6 Chen, X., Zhang, B., Wang, Y., and Yan, N. (2015). *Chimia* 69: 120–124.
7 Baghel, R.S., Trivedi, N., Gupta, V. et al. (2015). *Green Chem.* 17: 2436–2443.
8 Kerton, F.M., Liu, Y., Omari, K.W., and Hawboldt, K. (2013). *Green Chem.* 15: 860–871.
9 Chen, X. and Yan, N. (2014). *Catal. Surv. Asia*: 1–13. https://doi.org/10.1007/s10563-014-9171-1.
10 Yan, N. and Chen, X. (2015). *Nature* 524: 155–157.
11 Rødde, R.H., Einbu, A., and Vårum, K.M. (2008). *Carbohydr. Polym.* 71: 388–393.
12 Kumar, M.N.R. (2000). *React. Funct. Polym.* 46: 1–27.
13 Dutta, P.K., Dutta, J., and Tripathi, V. (2004). *J. Sci. Ind. Res* 63: 20–31.
14 Thakur, V.K. and Thakur, M.K. (2014). *ACS Sustain. Chem. Eng.* 2: 2637–2652.
15 Kurita, K. (2006). *Mar. Biotechnol.* 8: 203.
16 Jeon, Y.-J., Shahidi, F., and KIM, S.-K. (2000). *Food Rev. Int.* 16: 159–176.
17 Pillai, C., Paul, W., and Sharma, C.P. (2009). *Prog. Polym. Sci.* 34: 641–678.

18 Anitha, A., Sowmya, S., Kumar, P.S. et al. (2014). *Prog. Polym. Sci.* 39: 1644–1667.
19 Jayakumar, R., Prabaharan, M., Kumar, P.S. et al. (2011). *Biotechnol. Adv.* 29: 322–337.
20 Agulló, E., Rodríguez, M.S., Ramos, V., and Albertengo, L. (2003). *Macromol. Biosci.* 3: 521–530.
21 Chen, X., Yang, H., and Yan, N. (2016). *Chem. Eur. J.* 22: 13402–13421.
22 Novikov, V.Y. (2004). *Russ. J. Appl. Chem.* 77: 484–487.
23 Niola, F., Basora, N., Chornet, E., and Vidal, P.F. (1993). *Carbohydr. Res.* 238: 1–9.
24 Xing, R., Liu, S., Yu, H. et al. (2005). *Carbohydr. Res.* 340: 2150–2153.
25 Cabrera, J.C. and Van Cutsem, P. (2005). *Biochem. Eng. J.* 25: 165–172.
26 Li, J., Du, Y., Yang, J. et al. (2005). *Polym. Degrad. Stab.* 87: 441–448.
27 Jung, W.-J. and Park, R.-D. (2014). *Mar. Drugs* 12: 5328–5356.
28 Nakamura, H. (2011). *Carbohydr. Polym.* 84: 835–839.
29 Song, S., Zhou, H., Li, X. et al. (2014). *Chin. J. Org. Chem.* 34: 706–716.
30 Rupley, J. (1964). *Biochim. Biophys. Acta* 83: 245–255.
31 Lee, M.-Y., Var, F., Shin-ya, Y. et al. (1999). *Process Biochem.* 34: 493–500.
32 Ajavakom, A., Supsvetson, S., Somboot, A., and Sukwattanasinitt, M. (2012). *Carbohydr. Polym.* 90: 73–77.
33 Chen, Q., Xiao, W., Zhou, L. et al. (2012). *Polym. Degrad. Stab.* 97: 49–53.
34 Villa-Lerma, G., González-Márquez, H., Gimeno, M. et al. (2013). *Bioresour. Technol.* 146: 794–798.
35 Zhang, Z., Li, C., Wang, Q., and Zhao, Z.K. (2009). *Carbohydr. Polym.* 78: 685–689.
36 Luterbacher, J.S., Rand, J.M., Alonso, D.M. et al. (2014). *Science* 343: 277–280.
37 Zhang, J. and Yan, N. (2017). *ChemCatChem* 9: 2790–2796.
38 Ribeiro, L.S., Orfao, J.J.M., and Pereira, M.F.R. (2015). *Green Chem.* 17: 2973–2980.
39 Tabasso, S., Carnaroglio, D., Calcio Gaudino, E., and Cravotto, G. (2015). *Green Chem.* 17: 684–693.
40 Hilgert, J., Meine, N., Rinaldi, R., and Schüth, F. (2013). *Energy Environ. Sci.* 6: 92–96.
41 Carrasquillo-Flores, R., Käldström, M., Schüth, F. et al. (2013). *ACS Catal.* 3: 993–997.
42 Meine, N., Rinaldi, R., and Schüth, F. (2012). *ChemSusChem* 5: 1449–1454.
43 Yabushita, M., Kobayashi, H., Kuroki, K. et al. (2015). *ChemSusChem* 8: 3760–3763.
44 Margoutidis, G., Parsons, V.H., Bottaro, C.S. et al. (2018). *ACS Sustain. Chem. Eng.* 6: 1662–1669.
45 Chen, X., Yang, H., Zhong, Z., and Yan, N. (2017). *Green Chem.* 19: 2783–2792.
46 Pierson, Y., Chen, X., Bobbink, F.D. et al. (2014). *ACS Sustain. Chem. Eng.* 2: 2081–2089.
47 Zhang, J. and Yan, N. (2016). *Green Chem.* 18: 5050–5058.
48 Drover, M.W., Omari, K.W., Murphy, J.N., and Kerton, F.M. (2012). *RSC Adv.* 2: 4642–4644.

49 Omari, K.W., Dodot, L., and Kerton, F.M. (2012). *ChemSusChem* 5: 1767–1772.
50 Osada, M., Kikuta, K., Yoshida, K. et al. (2013). *Green Chem.* 15: 2960–2966.
51 Osada, M., Kikuta, K., Yoshida, K. et al. (2014). *RSC Adv.* 4: 33651–33657.
52 Chen, X., Chew, S.L., Kerton, F.M., and Yan, N. (2014). *Green Chem.* 16: 2204–2212.
53 Chen, X., Liu, Y., Kerton, F.M., and Yan, N. (2015). *RSC Adv.* 5: 20073–20080.
54 Chen, X., Gao, Y., Wang, L. et al. (2015). *ChemPlusChem* 80: 1565–1572.
55 Liu, Y., Rowley, C.N., and Kerton, F.M. (2014). *ChemPhysChem* 15: 4087–4094.
56 Bobbink, F.D., Zhang, J., Pierson, Y. et al. (2015). *Green Chem.* 17: 1024–1031.
57 Kobayashi, H., Techikawara, K., and Fukuoka, A. (2017). *Green Chem.* 19: 3350–3356.
58 Ohmi, Y., Nishimura, S., and Ebitani, K. (2013). *ChemSusChem* 6: 2259–2262.
59 Knorr, D., Wampler, T.P., and Teutonico, R.A. (1985). *J. Food Sci.* 50: 1762–1763.
60 Metreveli, A.K., Metreveli, P.K., Makarov, I.E., and Ponomarev, A.V. (2013). *High Energy Chem.* 47: 35–40.
61 Franich, R.A., Goodin, S.J., and Wilkins, A.L. (1984). *J. Anal. Appl. Pyrolysis* 7: 91–100.
62 Fujii, S., Kikuchi, R., and Kushida, H. (1966). *J. Org. Chem.* 31: 2239–2241.
63 Rohovec, J., Kotek, J., Peters, J.A., and Maschmeyer, T. (2001). *Eur. J. Org. Chem.* 2001: 3899–3901.
64 Jia, L., Wang, Y., Qiao, Y. et al. (2014). *RSC Adv.* 4: 44253–44260.
65 Jia, L., Pedersen, C.M., Qiao, Y. et al. (2015). *Phys. Chem. Chem. Phys.* 17: 23173–23182.
66 Jia, L., Zhang, Z., Qiao, Y. et al. (2017). *Ind. Eng. Chem. Res.* 56: 2925–2934.
67 Gao, X., Chen, X., Zhang, J. et al. (2016). *ACS Sustain. Chem. Eng.* 4: 3912–3920.
68 Mascal, M. and Nikitin, E.B. (2009). *ChemSusChem* 2: 859–861.
69 Omari, K.W., Besaw, J.E., and Kerton, F.M. (2012). *Green Chem.* 14: 1480–1487.
70 Wang, Y., Pedersen, C.M., Deng, T. et al. (2013). *Bioresour. Technol.* 143: 384–390.
71 Lee, S.-B. and Jeong, G.-T. (2015). *Appl. Biochem. Biotechnol.* 176: 1151–1161.
72 Li, M., Zang, H., Feng, J. et al. (2015). *Polym. Degrad. Stab.* 121: 331–339.
73 Yu, S., Zang, H., Chen, S. et al. (2016). *Polym. Degrad. Stab.* 134: 105–114.
74 Szabolcs, Á., Molnár, M., Dibó, G., and Mika, L.T. (2013). *Green Chem.* 15: 439–445.
75 Jeong, G.-T. (2014). *Ind. Crops Prod.* 62: 77–83.
76 Park, M.-R., Kim, S.-K., and Jeong, G.-T. (2018). *J. Ind. Eng. Chem.* 61: 119–123.
77 Kim, H.S., Kim, S.-K., and Jeong, G.-T. (2018). *RSC Adv.* 8: 3198–3205.
78 Jin, F. and Enomoto, H. (2011). *Energy Environ. Sci.* 4: 382–397.
79 Zhang, J., Sun, M., Liu, X., and Han, Y. (2014). *Catal. Today* 233: 77–82.
80 Deng, W., Zhang, Q., and Wang, Y. (2014). *Catal. Today* 234: 31–41.

81 Amaniampong, P.N., Trinh, Q.T., Wang, B. et al. (2015). *Angew. Chem. Int. Ed.* 54: 8928–8933.
82 Pham, T.T., Chen, X., Yan, N., and Sperry, J. (2017). *Monatsh. Chem. Chem. Mon.* https://doi.org/10.1007/s00706-017-2112-8.
83 Sadiq, A.D., Chen, X., Yan, N., and Sperry, J. (2018). *ChemSusChem* 11: 532–535.

15

Outlook

Eduardo Santillan-Jimenez[1] *and Mark Crocker*[1,2]

[1] *University of Kentucky, Center for Applied Energy Research, Lexington, KY 40511, USA*
[2] *University of Kentucky, Department of Chemistry, Lexington, KY 40506, USA*

The preceding chapters confirm not only that research on the upgrading of biomass to fuels and chemicals has flourished in recent years – driven mainly by environmental and sustainability considerations – but also that chemical catalysts play a major role in this area of study. As mentioned in the preface, this book aims to describe and discuss the catalysts and catalytic processes employed in the upgrading of the most abundant and important types of biomass feedstock. Therefore, the discussion of the emerging trends and current needs provided in this chapter has been organized following a similar scheme. In addition, common and recurrent themes will be emphasized and discussed at the end of this chapter.

Considering the application of catalysts to the pyrolysis of biomass and the upgrading of the bio-oil product, it is evident that additional work is needed to improve not only the yields achieved in the production of bio-oil from lignocellulosic biomass via catalytic fast pyrolysis but also the quality of the bio-oil produced (the oxygen content of which should ideally be <6 wt%). Similarly, advances must be made vis-à-vis the consistency of bio-oil with respect to its composition and improvements must be achieved in terms of its storage and transport stability, as these characteristics are critical for the coprocessing of bio-oil with traditional petroleum feedstock in existing refinery units. To this end, research is also needed involving realistic feeds and commercial catalysts at pilot plant scale, additional progress in the kinetic modeling of the cracking of mixed feeds also being necessary in order to establish optimal reaction conditions in the reaction–regeneration sections. The potential of the coprocessing approach is truly noteworthy, as traditional refineries could progressively contribute to an economy in which fuels and chemicals are simultaneously derived from both petroleum and biomass. Interestingly, besides the technological challenges associated with this approach – including the exacting issue of catalyst deactivation and regeneration – there are economic, social, and ethical implications to be considered. Adopting the "sustainable refinery" model could result in economic stimuli apportioned according to the reduction in CO_2 and other greenhouse gas (GHG) emissions realized and the benefit

Chemical Catalysts for Biomass Upgrading, First Edition. Edited by Mark Crocker and Eduardo Santillan-Jimenez.

to other economic sectors (i.e. agriculture, forestry, etc.). In this regard, it is also important to note that bio-oil produced from diverse biomass sources in a delocalized way, e.g. using mobile units, has the potential to improve the economy of agricultural and sparsely populated regions, which once again highlights the importance of bio-oil storage and transport stability.

Since esterification represents one of the main approaches to stabilize bio-oil, the effect of this treatment on the properties and the subsequent catalytic upgrading of this intermediate should be evaluated more fully. Indeed, the reaction network during bio-oil esterification needs to be more completely elucidated, particularly since little is known at this point about the structures or the detailed reaction pathways of heavy bio-oil components. Indeed, processing these large molecules remains one of the biggest challenges facing bio-oil utilization. Given that steric hindrance represents a problem during the conversion of heavy bio-oil components, the catalysts employed must possess active sites that are accessible to molecules with varied molecular size. Moreover, additional progress on reactor and process engineering will also be required. Although bio-oil esterification is typically performed in batch reactors – which are convenient for achieving good mass balances and for investigating reaction networks – esterification should ideally be performed in continuous reactors in order to increase throughput and lower costs. Consequently, it will be necessary to match the configuration of continuous flow reactors with the reaction kinetics of the major components in bio-oil during esterification. Lastly, since alcohols are generally used in excess in bio-oil esterification, effective, economical, and energy efficient methods to recover the unconsumed alcohols must be devised.

In the catalytic upgrading of C_5–C_6 sugars, efforts are required to standardize reaction conditions, feedstock composition, reactor designs, and performance indicators to permit quantitative comparison of disparate catalytic systems. The variable nature of the feed stream (in terms of component composition and concentration), the solubility of sugars and reaction products, and the dissolution of reactive gases are also challenges to be overcome. Catalytic processes must be adaptable to regular changes in feedstock composition, or a well-defined, mixed output stream must be agreed upon. Biphasic media and cosolvents may overcome some of these issues; however, solvents should be nontoxic, safe to handle, nonflammable, and noncorrosive. The continued development of hydrothermally stable catalysts able to operate in water should be encouraged. A major challenge to selective catalytic sugar transformations is the prevalence of competing side reactions, albeit a number of strategies (e.g. passivation of catalyst supports, tunable acid character and strength, tunable catalyst hydrophobicity, control over surface terminations) are available to improve selectivity. Advanced nanomaterials with hierarchical porous architectures can not only improve mass transport through superior in-pore accessibility but also help stabilize highly dispersed active species. While coke formation and attendant pore-blocking and/or active site-blocking is certainly problematic, catalyst deactivation attributed to metal leaching and/or sintering should also be minimized, which can be achieved by promoting strong metal–support interactions and by encapsulating nanoparticles within porous materials, respectively.

Novel catalyst preparation methods – such as atomic layer deposition (ALD) – should also be investigated. ALD can be leveraged to deposit highly uniform, conformal thin films of single or mixed metals on the surface of porous solids to afford coated materials with tunable properties. Deposition can be achieved with atomic scale precision, giving rise to single atom (site) catalysts that offer exciting opportunities to bridge the gap between homogeneous and heterogeneous catalysis, with the added benefit of precious metal thrifting. Process optimization through flow chemistry should also improve selectivity by allowing better control over reactant/product residence times and by removing primary products from the reaction zone before competitive side reactions initiate. Finally, catalytic C_5–C_6 sugar reforming often involves multistep cascade reactions that are desirably performed in a one-pot fashion. While cascades are advantageous in respect of atom economy, process time, labor and resource management, and waste generation, such one-pot catalytic synthetic processes require multifunctional catalysts if they are to become industrially viable processes. In this context, new concepts for catalyst design through spatially orthogonal chemical functionalization of hierarchically porous frameworks could allow improved control over reaction sequences.

Regarding the use of C—C bond formation reactions during biomass upgrading, most outstanding challenges are particular to specific condensation reactions or catalytic materials. For instance, while the strength of basic and acidic sites must be optimally balanced to achieve the highest aldol condensation rates, the actual role of acid sites in this reaction still needs to be clarified. Indeed, given that Brønsted acid sites could participate in either tautomerization or protonation, studies are needed to compare the activation energy of these steps in an effort to gain fundamental insights on factors determining the reaction pathway and product selectivity. The role of the Lewis acid sites is even less well understood, as it remains unclear whether these sites alone can catalyze aldol condensation and, if so, what the requirements are for an active Lewis acidic catalyst. Additional work is also necessary on both acid-catalyzed alkylation and hydroxyalkylation, particularly to determine the reaction mechanism(s) through which these reactions proceed, as several mechanistic proposals have been put forward to date. The lack of comprehensive studies on the mechanism of Lewis acid-catalyzed hydroxyalkylation on solid surfaces is particularly noteworthy, especially because heterogeneous systems may contain both Brønsted and Lewis acid sites and the role of each type of site – as well as their contribution to the observed activity – remains to be elucidated. Similarly, the role of Brønsted vs. Lewis acid sites in the zeolite-catalyzed acylation reaction needs to be clarified, particularly since studies in this regard have relied on varying the Brønsted/Lewis ratio and observing the effect on catalytic activity. However, this approach can also change the zeolite structure, which might explain the contradicting conclusions reported on the role of Brønsted and Lewis acid sites. Future work should also focus on elucidating the surface ketonization mechanism, which has been the center of controversy for many years.

As is the case with most reactions involving the formation of bulky compounds, the issue of catalyst deactivation during C–C coupling reactions remains a challenge, decrease of the diffusion path length being desirable to enhance the accessibility of reactants to active sites and the desorption of condensation products. The latter is particularly relevant when zeolitic materials are employed, as their microporosity can hinder the mass transfer of reactants to active sites within micropores and of products back to the bulk phase. Therefore, the strategies proposed to date to improve the performance of zeolitic materials in condensation reactions (e.g. structural modifications, zeolite nanoparticle syntheses, leveraging special morphologies, or hierarchical porosities, as well as controlling the type, strength, and density of acid sites) should be developed further, and/or novel approaches should be devised.

In the downstream conversion of biomass-derived oxygenates to fine chemicals, intensive research is still needed for heterogeneous catalysis to be able to compete – in terms of product yields and cost competitiveness – with long-standing petrochemical processes. For those transformations employing noble-metal catalysts, efforts should be directed to developing and employing catalysts based on earth-abundant and inexpensive materials. Moreover, much attention should be focused on the stability of the catalysts under the harsh reaction conditions involved. The adsorption behavior of oxygenated functional groups and water on the catalyst surface play a crucial role and must be taken into account. Particular emphasis must also be given to the effect of contaminants in the feed. A major target for the foreseeable future should be two-step or multi-step transformations of biomass derivatives without isolation of the intermediate platform molecules in a single reactor. The challenge is then to carefully design efficient and robust catalysts possessing bi- or multi-functionality that allow multiple reactions to be combined in a one-pot process. This is a real challenge since these reactions usually take place under different reaction conditions.

Regarding the oxidative depolymerization of lignin to value-added chemicals, additional work is needed to develop catalysts able to depolymerize more than one linkage. Current work tends to focus on the reactive β-O-4 linkage; however, systems that can also cleave β-1, β-5, and β–β linkages should be able to achieve more extensive depolymerization. Moreover, the most effective depolymerization methods need to be assessed across a variety of lignins stemming from different sources and extraction techniques. This will allow for the proper strategy to be applied according to the feedstock and pretreatment conditions employed.

In the case of reductive processing, "lignin-first" approaches that use whole lignocellulosic biomass as starting material and combine lignin isolation and depolymerization into a single step offer great promise, as they avoid the need for a separate isolation step (which may require additional chemicals) as well as the formation of complex lignin mixtures. In addition, reactions are performed under milder conditions that those employed to depolymerize isolated lignins and often afford higher monomer yields. It should be noted that these lignin-first approaches often convert the carbohydrate component of the lignocellulosic biomass employed into a holocellulose pulp containing the spent catalyst. Recovering the catalyst represents a challenge, as is maximizing monomer selectivity and yield. However, the holocellulose pulp can be upgraded to cellulosic ethanol, rendering this a valuable approach for integrated biorefinery systems. As is also the case with the oxidative depolymerization of lignin, the comparison of different reductive depolymerization approaches is rendered difficult by the structural heterogeneity of lignin and lignocellulosic biomass. Consequently, the use of "standardized" versions of these materials is necessary to enable the accurate comparison of different catalytic upgrading methods.

In terms of the conversion of the lipids constituting fats and oils to fuels and chemicals, irrespective of which conversion method is employed (esterification/transesterification, hydrodeoxygenation, or decarboxylation/decarbonylation), the use of more sustainable feeds – such as waste and algal oils – should be favored. To this end, the challenges posed by the presence of catalyst poisons in these feeds will need to be addressed, which should ideally be done by increasing the robustness of the catalysts employed, albeit improvements can also be achieved via feedstock pretreatment focused on the elimination of catalyst poisons. Since these feeds may also include a considerable amount of water, the development of water-tolerant catalytic technology is highly desirable.

In the conversion of lipids to biodiesel, the use of heterogeneous catalysts should be advanced, as the transition from homogeneous to heterogeneous catalysts stands to simplify downstream purification steps as well as facilitate catalyst recovery and reuse, which would reduce costs. As this transition takes place, novel approaches with the potential to overcome heat and mass transfer limitations – including ultrasonication and microwave irradiation combined with the use of flow reactors – may prove advantageous.

In the upgrading of lipids via deoxygenation, replacement of the sulfided catalysts currently employed in commercial HDO technology with sulfur-free catalysts may prove attractive, such catalysts including the transition metal and transition metal carbide, nitride, and phosphide formulations that have attracted significant attention and shown great promise. However, since the ability to fine-tune the selectivity of HDO catalysts offers a means to reduce hydrogen consumption, more effort is needed to develop catalysts with a controlled distribution of the desired active sites stable in the presence of water and CO_x. The latter is also crucial in the upgrading of lipids to fuel-like hydrocarbons and terminal olefins via decarbonylation/decarboxylation, where the minimization of hydrogen consumption must be extended to include that consumed via the methanation of CO_x. In fact, curtailing the occurrence of this and other undesirable side reactions leading to catalyst deactivation and lower yields of desirable products should be prioritized. However, since the use of sustainable feeds – such as waste and algal lipids – is desirable and some of these feeds are highly unsaturated, the consumption of some hydrogen will continue to be required in order to saturate these lipids. Therefore, the generation of hydrogen *in situ* from waste and renewable hydrogen sources (such as glycerol) is also of great interest. Incidentally, the fact that some of the aforementioned feeds may also include a considerable amount of water offers a tantalizing possibility vis-à-vis *in situ* hydrogen generation, namely, that of using any water present for that purpose via reactions such as aqueous phase reforming or water–gas shift. Finally, the deoxygenation and isomerization/mild hydrocracking stages currently employed to upgrade lipids should be integrated into a single-stage process. To this end, the bifunctional catalysts needed must be improved further, particularly in terms of their resistance to deactivation. The use of promoters and supports with optimized textural and acid–base properties is a research area with significant promise in this regard, the goal being to develop bifunctional formulations that are selective, stable, recyclable, and capable of simultaneously catalyzing deoxygenation, isomerization, and perhaps even water–gas shift and aqueous phase reforming for *in situ* H_2 generation.

Albeit the use of terpenes as hydrogen precursors during deoxygenation reactions has been proposed, their main use will likely remain as a feedstock for the production of fine chemicals. In this regard, the development of catalytic processes capable of transforming terpenes not only into compounds used in the flavor and fragrance industry but also into other high value chemicals and materials (such as bioplastics) is necessary. Admittedly, developments in this direction may be hindered by the limited supply of terpenes. Indeed, efforts will need to be made to improve their supply, either by increasing the amount of terpenes obtained from industrial sources (such as the pulp and paper industry), upregulating their production in biomass so terpenes can be produced in larger amounts – which would trigger both environmental and ecological concerns (e.g. increased atmospheric emissions and forest fires) – or by obtaining terpenes from natural sources other than those typically exploited (i.e. pine and eucalyptus). Interestingly, cannabis (including industrial hemp) represents not only a topical and interesting potential source of these compounds but also a perfect example of the ability of terpenes to display synergistic effects with compounds with medicinal properties (terpenes have been shown to increase the therapeutic index of cannabinoids through potentiation and entourage effects) and/or to display these properties in their own right. Although more research is needed to better understand how terpenes can be leveraged for medicinal applications, catalysis will likely play a role in the production of chemicals with pharmaceutical properties from these compounds.

Similar to the case of terpenes, in the conversion of chitin to nitrogenous chemicals further efforts are expected to focus on the development of improved catalytic systems capable of efficiently, selectively, and economically generating new types of chemicals with high value. These will likely include both platform chemicals as well as fine chemicals requiring multistep conversions. However, another issue to be addressed is the cost of this feed, which stems from the high cost of the extraction method currently employed. Hence, new methods to obtain chitin in an easy, inexpensive, and environmentally friendly manner are also necessary.

Unsurprisingly, there are some common and recurrent themes associated with the foregoing trends and needs. For instance, as with most catalytic systems involving carbonaceous reactants, intermediates, and products, the issues of catalyst deactivation and regeneration will need to be addressed and overcome. Although coking is a key issue that should ideally be prevented due to the fact that it leads to lower yields of desirable products both by consuming the organic feedstock and by leading to catalyst deactivation, steps must also be taken to provide for catalyst regeneration if (or rather when) coking takes place. For instance, catalysts must be amenable to regeneration via burn-off of the coke deposits resulting in the total (or close to total) recovery of their activity. In view of this, catalyst deactivation can be considered irreversible in the case of carbon-supported catalysts, and thus, the use of these formulations should be avoided in favor of catalysts amenable to regeneration via the simple processes preferred in industry, such as the combustion of coke. Nevertheless, other catalyst deactivation pathways – including poisoning as well as metal leaching, sintering, and other mechanisms resulting in the loss of the catalytically active phase (e.g. sulfur loss for sulfided catalysts or oxidation for catalysts in which reduced metal sites constitute the active phase) – should also be considered. Albeit in some instances provisions can be made in this regard via feedstock preprocessing, such as cases in which feedstock pretreatment can be leveraged to remove catalyst poisons, these issues would be best addressed through efforts targeting the design and development of more robust formulations. Admittedly, developing catalysts impervious to all of the aforementioned deactivation pathways represents a daunting (if not altogether unrealistic) proposition, so developing catalysts that are both more robust (resistant to deactivation) and amenable to regeneration will be paramount.

Catalyst design and development also offers opportunities to improve the sustainability of the transformations and processes discussed in this book. For instance, formulations capable of functioning not only in the presence of minerals and other catalysts poisons but also in aqueous environments are most desirable. However, although water is often proposed as the ultimate green solvent, water should also be considered as a precious resource that should be used as wisely and sustainably as possible. In this regard, it is important to mention that the use of waste biomass feeds will become critical moving forward, as the use of these feeds avoids competition with food production for water, land, and other resources, thus avoiding the fuel-vs.-food controversy associated with the rise in food prices observed when edible crops are used for fuel production. Admittedly, since the amount of waste biomass available may be insufficient and the composition of the waste biomass available would likely be very variable, it can be argued that a considerable amount of biomass feed will need to be derived from purpose-grown crops produced on marginal land. Nevertheless, minimizing the water requirements associated with the production of any biomass source employed will remain critical and pose its own set of challenges. For instance, the use of marine algae species capable of growing in sea water will require processes and catalysts able to handle and withstand elevated concentrations of minerals. In line with the principles of green chemistry, the use of hazardous or costly materials (including sulfided HDO catalysts and precious metal-based formulations) should be disfavored along with the use of stoichiometric reagents (e.g. alcohols for biodiesel synthesis, supplemental oxidants for the oxidative depolymerization of lignin, etc.), particularly those derived from exogenous and nonrenewable sources – such as the hydrogen used in several processes, which is typically obtained from the reforming of natural gas. In this regard, the generation of hydrogen *in situ* from sources that are both renewable and inexpensive (such as glycerol) during biomass upgrading represents an important target.

Last but certainly not least, issues related to the economic viability and the sustainability of any process proposed will need to be carefully considered. Indeed, whether the technologies espoused are sound in terms of their economics, their energy balance, their carbon efficiency, and their emission profiles represent critical issues that must be addressed. To this end, techno-economic and life cycle analyses are indispensable tools in the assessment of any technology being considered for scale-up and commercialization. Notably, since economic competitiveness can be achieved by targeting higher value products (e.g. renewable aromatics and other platform chemicals) over lower margin (fuel) products, bio-based products may become the economic driver that will allow biomass carbon to find its way into the fuel supply; in other words, bio-products may be economically leveraged to enable biofuels. With this in mind, there is plenty of opportunity for the development not only of new and improved catalysts but also of process improvements related to the separation and isolation of high value products.

Ultimately, the development of new catalytic processes for the conversion of biomass to the next generation of chemicals and fuels will require the collaborative efforts not only of chemists, chemical engineers, and materials scientists but also of agronomists, economists, and experts in logistics, techno-economic analysis, and life cycle assessment.

Index

Chemical Catalysts for Biomass Upgrading, First Edition. Edited by Mark Crocker and Eduardo Santillan-Jimenez.

c

f

g

m

t